The Performance of the ATLAS Detector

ATLAS Collaboration CERN
Editor

The Performance
of the ATLAS Detector

 Springer

Editor
ATLAS Collaboration CERN
Genève
Switzerland
atlas.secretariat@cern.ch

ISBN 978-3-642-22115-6   e-ISBN 978-3-642-22116-3
DOI 10.1007/978-3-642-22116-3
Springer Heidelberg Dordrecht London New York

Library of Congress Control Number: 2011936007

Springer is part of Springer Science+Business Media (www.springer.com)

Table of Contents

Readiness of the ATLAS liquid argon calorimeter for LHC collisions

The ATLAS Collaboration[*,**]

G. Aad[83], B. Abbott[110], J. Abdallah[11], A.A. Abdelalim[49], A. Abdesselam[117], O. Abdinov[10], B. Abi[111], M. Abolins[88], H. Abramowicz[151], H. Abreu[114], B.S. Acharya[162a,162b], D.L. Adams[24], T.N. Addy[56], J. Adelman[173], C. Adorisio[36a,36b], P. Adragna[75], T. Adye[128], S. Aefsky[22], J.A. Aguilar-Saavedra[123a], M. Aharrouche[81], S.P. Ahlen[21], F. Ahles[48], A. Ahmad[146], H. Ahmed[2], M. Ahsan[40], G. Aielli[132a,132b], T. Akdogan[18], T.P.A. Åkesson[79], G. Akimoto[153], A.V. Akimov[94], A. Aktas[48], M.S. Alam[1], M.A. Alam[76], J. Albert[167], S. Albrand[55], M. Aleksa[29], I.N. Aleksandrov[65], F. Alessandria[89a,89b], C. Alexa[25a], G. Alexander[151], G. Alexandre[49], T. Alexopoulos[9], M. Alhroob[20], M. Aliev[15], G. Alimonti[89a], J. Alison[119], M. Aliyev[10], P.P. Allport[73], S.E. Allwood-Spiers[53], J. Almond[82], A. Aloisio[102a,102b], R. Alon[169], A. Alonso[79], M.G. Alviggi[102a,102b], K. Amako[66], C. Amelung[22], V.V. Ammosov[127], A. Amorim[123b], G. Amorós[165], N. Amram[151], C. Anastopoulos[138], T. Andeen[29], C.F. Anders[48], K.J. Anderson[30], A. Andreazza[89a,89b], V. Andrei[58a], X.S. Anduaga[70], A. Angerami[34], F. Anghinolfi[29], N. Anjos[123b], A. Antonaki[8], M. Antonelli[47], S. Antonelli[19a,19b], B. Antunovic[41], F. Anulli[131a], S. Aoun[83], G. Arabidze[8], I. Aracena[142], Y. Arai[66], A.T.H. Arce[14], J.P. Archambault[28], S. Arfaoui[29,a], J.-F. Arguin[14], T. Argyropoulos[9], E. Arik[18,*], M. Arik[18], A.J. Armbruster[87], O. Arnaez[4], C. Arnault[114], A. Artamonov[95], D. Arutinov[20], M. Asai[142], S. Asai[153], R. Asfandiyarov[170], S. Ask[82], B. Åsman[144], D. Asner[28], L. Asquith[77], K. Assamagan[24], A. Astbury[167], A. Astvatsatourov[52], G. Atoian[173], B. Auerbach[173], E. Auge[114], K. Augsten[126], M. Aurousseau[4], N. Austin[73], G. Avolio[161], R. Avramidou[9], D. Axen[166], C. Ay[54], G. Azuelos[93,b], Y. Azuma[153], M.A. Baak[29], G. Baccaglioni[89a,89b], C. Bacci[133a,133b], A. Bach[14], H. Bachacou[135], K. Bachas[29], M. Backes[49], E. Badescu[25a], P. Bagnaia[131a,131b], Y. Bai[32], D.C. Bailey[156], T. Bain[156], J.T. Baines[128], O.K. Baker[173], M.D. Baker[24], F. Baltasar Dos Santos Pedrosa[29], E. Banas[38], P. Banerjee[93], S. Banerjee[167], D. Banfi[89a,89b], A. Bangert[136], V. Bansal[167], S.P. Baranov[94], S. Baranov[65], A. Barashkou[65], T. Barber[27], E.L. Barberio[86], D. Barberis[50a,50b], M. Barbero[20], D.Y. Bardin[65], T. Barillari[99], M. Barisonzi[172], T. Barklow[142], N. Barlow[27], B.M. Barnett[128], R.M. Barnett[14], S. Baron[29], A. Baroncelli[133a], A.J. Barr[117], F. Barreiro[80], J. Barreiro Guimarães da Costa[57], P. Barrillon[114], N. Barros[123b], R. Bartoldus[142], D. Bartsch[20], J. Bastos[123b], R.L. Bates[53], S. Bathe[24], L. Batkova[143], J.R. Batley[27], A. Battaglia[16], M. Battistin[29], F. Bauer[135], H.S. Bawa[142], M. Bazalova[124], B. Beare[156], T. Beau[78], P.H. Beauchemin[117], R. Beccherle[50a], N. Becerici[18], P. Bechtle[41], G.A. Beck[75], H.P. Beck[16], M. Beckingham[48], K.H. Becks[172], I. Bedajanek[126], A.J. Beddall[18,c], A. Beddall[18,c], P. Bednár[143], V.A. Bednyakov[65], C. Bee[83], M. Begel[24], S. Behar Harpaz[150], P.K. Behera[63], M. Beimforde[99], C. Belanger-Champagne[164], P.J. Bell[82], W.H. Bell[49], G. Bella[151], L. Bellagamba[19a], F. Bellina[29], M. Bellomo[118a], A. Belloni[57], K. Belotskiy[96], O. Beltramello[29], S. Ben Ami[150], O. Benary[151], D. Benchekroun[134a], M. Bendel[81], B.H. Benedict[161], N. Benekos[163], Y. Benhammou[151], G.P. Benincasa[123b], D.P. Benjamin[44], M. Benoit[114], J.R. Bensinger[22], K. Benslama[129], S. Bentvelsen[105], M. Beretta[47], D. Berge[29], E. Bergeaas Kuutmann[144], N. Berger[4], F. Berghaus[167], E. Berglund[49], J. Beringer[14], K. Bernardet[83], P. Bernat[114], R. Bernhard[48], C. Bernius[77], T. Berry[76], A. Bertin[19a,19b], N. Besson[135], S. Bethke[99], R.M. Bianchi[48], M. Bianco[72a,72b], O. Biebel[98], J. Biesiada[14], M. Biglietti[131a,131b], H. Bilokon[47], M. Bindi[19a,19b], S. Binet[114], A. Bingul[18,c], C. Bini[131a,131b], C. Biscarat[178], U. Bitenc[48], K.M. Black[57], R.E. Blair[5], J.-B. Blanchard[114], G. Blanchot[29], C. Blocker[22], J. Blocki[38], A. Blondel[49], W. Blum[81], U. Blumenschein[54], G.J. Bobbink[105], A. Bocci[44], M. Boehler[41], J. Boek[172], N. Boelaert[79], S. Böser[77], J.A. Bogaerts[29], A. Bogouch[90,*], C. Bohm[144], J. Bohm[124], V. Boisvert[76], T. Bold[161,d], V. Boldea[25a], A. Boldyrev[97], V.G. Bondarenko[96], M. Bondioli[161], M. Boonekamp[135], J.R.A. Booth[17], S. Bordoni[78], C. Borer[16], A. Borisov[127], G. Borissov[71], I. Borjanovic[72a], S. Borroni[131a,131b], K. Bos[105], D. Boscherini[19a], M. Bosman[11], M. Bosteels[29], H. Boterenbrood[105], J. Bouchami[93], J. Boudreau[122], E.V. Bouhova-Thacker[71], C. Boulahouache[122], C. Bourdarios[114], J. Boyd[29], I.R. Boyko[65], I. Bozovic-Jelisavcic[12b], J. Bracinik[17], A. Braem[29], P. Branchini[133a], G.W. Brandenburg[57], A. Brandt[7], G. Brandt[41], O. Brandt[54], U. Bratzler[154], B. Brau[84], J.E. Brau[113], H.M. Braun[172], B. Brelier[156], J. Bremer[29], R. Brenner[164], S. Bressler[150], D. Breton[114], N.D. Brett[117], D. Britton[53], F.M. Brochu[27], I. Brock[20], R. Brock[88], T.J. Brodbeck[71], E. Brodet[151], F. Broggi[89a,89b], C. Bromberg[88], G. Brooijmans[34],

W.K. Brooks[31b], G. Brown[82], E. Brubaker[30], P.A. Bruckman de Renstrom[38], D. Bruncko[143], R. Bruneliere[48], S. Brunet[41], A. Bruni[19a], G. Bruni[19a], M. Bruschi[19a], T. Buanes[13], F. Bucci[49], J. Buchanan[117], P. Buchholz[140], A.G. Buckley[77,e], I.A. Budagov[65], B. Budick[107], V. Büscher[81], L. Bugge[116], O. Bulekov[96], M. Bunse[42], T. Buran[116], H. Burckhart[29], S. Burdin[73], T. Burgess[13], S. Burke[128], E. Busato[33], P. Bussey[53], C.P. Buszello[164], F. Butin[29], B. Butler[142], J.M. Butler[21], C.M. Buttar[53], J.M. Butterworth[77], T. Byatt[77], J. Caballero[24], S. Cabrera Urbán[165], D. Caforio[19a,19b], O. Cakir[3], P. Calafiura[14], G. Calderini[78], P. Calfayan[98], R. Calkins[5], L.P. Caloba[23a], R. Caloi[131a,131b], D. Calvet[33], P. Camarri[132a,132b], M. Cambiaghi[118a,118b], D. Cameron[116], F. Campabadal Segura[165], S. Campana[29], M. Campanelli[77], V. Canale[102a,102b], F. Canelli[30], A. Canepa[157a], J. Cantero[80], L. Capasso[102a,102b], M.D.M. Capeans Garrido[29], I. Caprini[25a], M. Caprini[25a], M. Capua[36a,36b], R. Caputo[146], D. Caracinha[123b], C. Caramarcu[25a], R. Cardarelli[132a], T. Carli[29], G. Carlino[102a], L. Carminati[89a,89b], B. Caron[2,b], S. Caron[48], G.D. Carrillo Montoya[170], S. Carron Montero[156], A.A. Carter[75], J.R. Carter[27], J. Carvalho[123b], D. Casadei[107], M.P. Casado[11], M. Cascella[121a,121b], C. Caso[50a,50b,*], A.M. Castaneda Hernadez[170], E. Castaneda-Miranda[170], V. Castillo Gimenez[165], N. Castro[123a], G. Cataldi[72a], A. Catinaccio[29], J.R. Catmore[71], A. Cattai[29], G. Cattani[132a,132b], S. Caughron[34], D. Cauz[162a,162c], P. Cavalleri[78], D. Cavalli[89a], M. Cavalli-Sforza[11], V. Cavasinni[121a,121b], F. Ceradini[133a,133b], A.S. Cerqueira[23a], A. Cerri[29], L. Cerrito[75], F. Cerutti[47], S.A. Cetin[18,f], F. Cevenini[102a,102b], A. Chafaq[134a], D. Chakraborty[5], K. Chan[2], J.D. Chapman[27], J.W. Chapman[87], E. Chareyre[78], D.G. Charlton[17], V. Chavda[82], S. Cheatham[71], S. Chekanov[5], S.V. Chekulaev[157a], G.A. Chelkov[65], H. Chen[24], S. Chen[32], T. Chen[32], X. Chen[170], S. Cheng[32], A. Cheplakov[65], V.F. Chepurnov[65], R. Cherkaoui El Moursli[134d], V. Tcherniatine[24], D. Chesneanu[25a], E. Cheu[6], S.L. Cheung[156], L. Chevalier[135], F. Chevallier[135], V. Chiarella[47], G. Chiefari[102a,102b], L. Chikovani[51], J.T. Childers[58a], A. Chilingarov[71], G. Chiodini[72a], M. Chizhov[65], G. Choudalakis[30], S. Chouridou[136], D. Chren[126], I.A. Christidi[152], A. Christov[48], D. Chromek-Burckhart[29], M.L. Chu[149], J. Chudoba[124], G. Ciapetti[131a,131b], A.K. Ciftci[3], R. Ciftci[3], D. Cinca[33], V. Cindro[74], M.D. Ciobotaru[161], C. Ciocca[19a,19b], A. Ciocio[14], M. Cirilli[87], M. Citterio[89a], A. Clark[49], W. Cleland[122], J.C. Clemens[83], B. Clement[55], C. Clement[144], D. Clements[53], Y. Coadou[83], M. Cobal[162a,162c], A. Coccaro[50a,50b], J. Cochran[64], S. Coelli[89a,89b], J. Coggeshall[163], E. Cogneras[16], C.D. Cojocaru[28], J. Colas[4], B. Cole[34], A.P. Colijn[105], C. Collard[114], N.J. Collins[17], C. Collins-Tooth[53], J. Collot[55], G. Colon[84], R. Coluccia[72a,72b], P. Conde Muiño[123b], E. Coniavitis[164], M. Consonni[104], S. Constantinescu[25a], C. Conta[118a,118b], F. Conventi[102a,g], J. Cook[29], M. Cooke[34], B.D. Cooper[75], A.M. Cooper-Sarkar[117], N.J. Cooper-Smith[76], K. Copic[34], T. Cornelissen[50a,50b], M. Corradi[19a], F. Corriveau[85,h], A. Corso-Radu[161], A. Cortes-Gonzalez[163], G. Cortiana[99], G. Costa[89a], M.J. Costa[165], D. Costanzo[138], T. Costin[30], D. Côté[41], R. Coura Torres[23a], L. Courneyea[167], G. Cowan[76], C. Cowden[27], B.E. Cox[82], K. Cranmer[107], J. Cranshaw[5], M. Cristinziani[20], G. Crosetti[36a,36b], R. Crupi[72a,72b], S. Crépé-Renaudin[55], C. Cuenca Almenar[173], T. Cuhadar Donszelmann[138], M. Curatolo[47], C.J. Curtis[17], P. Cwetanski[61], Z. Czyczula[35], S. D'Auria[53], M. D'Onofrio[11], A. D'Orazio[99], P.V.M. Da Silva[23a], C. Da Via[82], W. Dabrowski[37], T. Dai[87], C. Dallapiccola[84], S.J. Dallison[128,*], C.H. Daly[137], M. Dam[35], H.O. Danielsson[29], D. Dannheim[99], V. Dao[49], G. Darbo[50a], G.L. Darlea[25a], W. Davey[86], T. Davidek[125], N. Davidson[86], R. Davidson[71], A.R. Davison[77], I. Dawson[138], J.W. Dawson[5], R.K. Daya[39], K. De[7], R. de Asmundis[102a], S. De Castro[19a,19b], P.E. De Castro Faria Salgado[24], S. De Cecco[78], J. de Graat[98], N. De Groot[104], P. de Jong[105], E. De La Cruz-Burelo[87], C. De La Taille[114], L. De Mora[71], M. De Oliveira Branco[29], D. De Pedis[131a], A. De Salvo[131a], U. De Sanctis[162a,162c], A. De Santo[76], J.B. De Vivie De Regie[114], G. De Zorzi[131a,131b], S. Dean[77], H. Deberg[163], G. Dedes[99], D.V. Dedovich[65], P.O. Defay[33], J. Degenhardt[119], M. Dehchar[117], C. Del Papa[162a,162c], J. Del Peso[80], T. Del Prete[121a,121b], A. Dell'Acqua[29], L. Dell'Asta[89a,89b], M. Della Pietra[102a,g], d. della Volpe[102a,102b], M. Delmastro[29], N. Delruelle[29], P.A. Delsart[55], C. Deluca[146], S. Demers[173], M. Demichev[65], B. Demirkoz[27], J. Deng[161], W. Deng[24], S.P. Denisov[127], C. Dennis[117], J.E. Derkaoui[134c], F. Derue[78], P. Dervan[73], K. Desch[20], P.O. Deviveiros[156], A. Dewhurst[71], B. DeWilde[146], S. Dhaliwal[156], R. Dhullipudi[24,i], A. Di Ciaccio[132a,132b], L. Di Ciaccio[4], A. Di Domenico[131a,131b], A. Di Girolamo[29], B. Di Girolamo[29], S. Di Luise[133a,133b], A. Di Mattia[88], R. Di Nardo[132a,132b], A. Di Simone[132a,132b], R. Di Sipio[19a,19b], M.A. Diaz[31a], F. Diblen[18], E.B. Diehl[87], J. Dietrich[48], S. Diglio[114], K. Dindar Yagci[39], D.J. Dingfelder[48], C. Dionisi[131a,131b], P. Dita[25a], S. Dita[25a], F. Dittus[29], F. Djama[83], R. Djilkibaev[107], T. Djobava[51], M.A.B. do Vale[23a], A. Do Valle Wemans[123b], M. Dobbs[85], D. Dobos[29], E. Dobson[117], M. Dobson[161], J. Dodd[34], O.B. Dogan[18,*], T. Doherty[53], Y. Doi[66], J. Dolejsi[125], I. Dolenc[74], Z. Dolezal[125], B.A. Dolgoshein[96], T. Dohmae[153], M. Donega[119], J. Donini[55], J. Dopke[172], A. Doria[102a], A. Dos Anjos[170], A. Dotti[121a,121b], M.T. Dova[70], A. Doxiadis[105], A.T. Doyle[53], Z. Drasal[125], C. Driouichi[35], M. Dris[9], J. Dubbert[99], E. Duchovni[169], G. Duckeck[98], A. Dudarev[29], F. Dudziak[114], M. Dührssen[48], L. Duflot[114], M.-A. Dufour[85], M. Dunford[30], A. Duperrin[83], H. Duran Yildiz[3,j], A. Dushkin[22],

R. Duxfield[138], M. Dwuznik[37], M. Düren[52], W.L. Ebenstein[44], J. Ebke[98], S. Eckert[48], S. Eckweiler[81], K. Edmonds[81], C.A. Edwards[76], P. Eerola[79,k], K. Egorov[61], W. Ehrenfeld[41], T. Ehrich[99], T. Eifert[29], G. Eigen[13], K. Einsweiler[14], E. Eisenhandler[75], T. Ekelof[164], M. El Kacimi[4], M. Ellert[164], S. Elles[4], F. Ellinghaus[81], K. Ellis[75], N. Ellis[29], J. Elmsheuser[98], M. Elsing[29], R. Ely[14], D. Emeliyanov[128], R. Engelmann[146], A. Engl[98], B. Epp[62], A. Eppig[87], V.S. Epshteyn[95], A. Ereditato[16], D. Eriksson[144], I. Ermoline[88], J. Ernst[1], M. Ernst[24], J. Ernwein[135], D. Errede[163], S. Errede[163], E. Ertel[81], M. Escalier[114], C. Escobar[165], X. Espinal Curull[11], B. Esposito[47], F. Etienne[83], A.I. Etienvre[135], E. Etzion[151], H. Evans[61], L. Fabbri[19a,19b], C. Fabre[29], P. Faccioli[19a,19b], K. Facius[35], R.M. Fakhrutdinov[127], S. Falciano[131a], A.C. Falou[114], Y. Fang[170], M. Fanti[89a,89b], A. Farbin[7], A. Farilla[133a], J. Farley[146], T. Farooque[156], S.M. Farrington[117], P. Farthouat[29], F. Fassi[165], P. Fassnacht[29], D. Fassouliotis[8], B. Fatholahzadeh[156], L. Fayard[114], F. Fayette[54], R. Febbraro[33], P. Federic[143], O.L. Fedin[120], I. Fedorko[29], W. Fedorko[29], L. Feligioni[83], C.U. Felzmann[86], C. Feng[32], E.J. Feng[30], A.B. Fenyuk[127], J. Ferencei[143], J. Ferland[93], B. Fernandes[123b], W. Fernando[108], S. Ferrag[53], J. Ferrando[117], A. Ferrari[164], P. Ferrari[105], R. Ferrari[118a], A. Ferrer[165], M.L. Ferrer[47], D. Ferrere[49], C. Ferretti[87], M. Fiascaris[117], F. Fiedler[81], A. Filipčič[74], A. Filippas[9], F. Filthaut[104], M. Fincke-Keeler[167], M.C.N. Fiolhais[123b], L. Fiorini[11], A. Firan[39], G. Fischer[41], M.J. Fisher[108], M. Flechl[164], I. Fleck[140], J. Fleckner[81], P. Fleischmann[171], S. Fleischmann[20], T. Flick[172], L.R. Flores Castillo[170], M.J. Flowerdew[99], F. Föhlisch[58a], M. Fokitis[9], T. Fonseca Martin[76], D.A. Forbush[137], A. Formica[135], A. Forti[82], D. Fortin[157a], J.M. Foster[82], D. Fournier[114], A. Foussat[29], A.J. Fowler[44], K. Fowler[136], H. Fox[71], P. Francavilla[121a,121b], S. Franchino[118a,118b], D. Francis[29], M. Franklin[57], S. Franz[29], M. Fraternali[118a,118b], S. Fratina[119], J. Freestone[82], S.T. French[27], R. Froeschl[29], D. Froidevaux[29], J.A. Frost[27], C. Fukunaga[154], E. Fullana Torregrosa[5], J. Fuster[165], C. Gabaldon[80], O. Gabizon[169], T. Gadfort[34], S. Gadomski[49], G. Gagliardi[50a,50b], P. Gagnon[61], C. Galea[98], E.J. Gallas[117], M.V. Gallas[29], B.J. Gallop[128], P. Gallus[124], E. Galyaev[40], K.K. Gan[108], Y.S. Gao[142,l], A. Gaponenko[14], M. Garcia-Sciveres[14], C. García[165], J.E. García Navarro[49], R.W. Gardner[30], N. Garelli[29], H. Garitaonandia[105], V. Garonne[29], C. Gatti[47], G. Gaudio[118a], O. Gaumer[49], P. Gauzzi[131a,131b], I.L. Gavrilenko[94], C. Gay[166], G. Gaycken[20], J.-C. Gayde[29], E.N. Gazis[9], P. Ge[32], C.N.P. Gee[128], Ch. Geich-Gimbel[20], K. Gellerstedt[144], C. Gemme[50a], M.H. Genest[98], S. Gentile[131a,131b], F. Georgatos[9], S. George[76], P. Gerlach[172], A. Gershon[151], C. Geweniger[58a], H. Ghazlane[134d], P. Ghez[4], N. Ghodbane[33], B. Giacobbe[19a], S. Giagu[131a,131b], V. Giakoumopoulou[8], V. Giangiobbe[121a,121b], F. Gianotti[29], B. Gibbard[24], A. Gibson[156], S.M. Gibson[117], L.M. Gilbert[117], M. Gilchriese[14], V. Gilewsky[91], D. Gillberg[28], A.R. Gillman[128], D.M. Gingrich[2,m], J. Ginzburg[151], N. Giokaris[8], M.P. Giordani[162a,162c], R. Giordano[102a,102b], P. Giovannini[99], P.F. Giraud[29], P. Girtler[62], D. Giugni[89a], P. Giusti[19a], B.K. Gjelsten[116], L.K. Gladilin[97], C. Glasman[80], A. Glazov[41], K.W. Glitza[172], G.L. Glonti[65], J. Godfrey[141], J. Godlewski[29], M. Goebel[41], T. Göpfert[43], C. Goeringer[81], C. Gössling[42], T. Göttfert[99], V. Goggi[118a,118b,n], S. Goldfarb[87], D. Goldin[39], T. Golling[173], N.P. Gollub[29], A. Gomes[123b], L.S. Gomez Fajardo[160], R. Gonçalo[76], L. Gonella[20], C. Gong[32], S. González de la Hoz[165], M.L. Gonzalez Silva[26], S. Gonzalez-Sevilla[49], J.J. Goodson[146], L. Goossens[29], P.A. Gorbounov[156], H.A. Gordon[24], I. Gorelov[103], G. Gorfine[172], B. Gorini[29], E. Gorini[72a,72b], A. Gorišek[74], E. Gornicki[38], S.V. Goryachev[127], V.N. Goryachev[127], B. Gosdzik[41], M. Gosselink[105], M.I. Gostkin[65], I. Gough Eschrich[161], M. Gouighri[134a], D. Goujdami[134a], M.P. Goulette[49], A.G. Goussiou[137], C. Goy[4], I. Grabowska-Bold[161,d], P. Grafström[29], K.-J. Grahn[145], L. Granado Cardoso[123b], F. Grancagnolo[72a], S. Grancagnolo[15], V. Grassi[89a], V. Gratchev[120], N. Grau[34], H.M. Gray[34,o], J.A. Gray[146], E. Graziani[133a], B. Green[76], T. Greenshaw[73], Z.D. Greenwood[24,i], I.M. Gregor[41], P. Grenier[142], E. Griesmayer[46], J. Griffiths[137], N. Grigalashvili[65], A.A. Grillo[136], K. Grimm[146], S. Grinstein[11], Y.V. Grishkevich[97], L.S. Groer[156], J. Grognuz[29], M. Groh[99], M. Groll[81], E. Gross[169], J. Grosse-Knetter[54], J. Groth-Jensen[79], K. Grybel[140], V.J. Guarino[5], C. Guicheney[33], A. Guida[72a,72b], T. Guillemin[4], H. Guler[85,p], J. Gunther[124], B. Guo[156], A. Gupta[30], Y. Gusakov[65], A. Gutierrez[93], P. Gutierrez[110], N. Guttman[151], O. Gutzwiller[29], C. Guyot[135], C. Gwenlan[117], C.B. Gwilliam[73], A. Haas[142], S. Haas[29], C. Haber[14], R. Hackenburg[24], H.K. Hadavand[39], D.R. Hadley[17], P. Haefner[99], R. Härtel[99], Z. Hajduk[38], H. Hakobyan[174], J. Haller[41,q], K. Hamacher[172], A. Hamilton[49], S. Hamilton[159], H. Han[32], L. Han[32], K. Hanagaki[115], M. Hance[119], C. Handel[81], P. Hanke[58a], J.R. Hansen[35], J.B. Hansen[35], J.D. Hansen[35], P.H. Hansen[35], T. Hansl-Kozanecka[136], P. Hansson[142], K. Hara[158], G.A. Hare[136], T. Harenberg[172], R.D. Harrington[21], O.B. Harris[77], O.M. Harris[137], K. Harrison[17], J. Hartert[48], F. Hartjes[105], T. Haruyama[66], A. Harvey[56], S. Hasegawa[101], Y. Hasegawa[139], K. Hashemi[22], S. Hassani[135], M. Hatch[29], F. Haug[29], S. Haug[16], M. Hauschild[29], R. Hauser[88], M. Havranek[124], C.M. Hawkes[17], R.J. Hawkings[29], D. Hawkins[161], T. Hayakawa[67], H.S. Hayward[73], S.J. Haywood[128], M. He[32], S.J. Head[82], V. Hedberg[79], L. Heelan[28], S. Heim[88], B. Heinemann[14], S. Heisterkamp[35], L. Helary[4], M. Heller[114], S. Hellman[144], C. Helsens[11], T. Hemperek[20], R.C.W. Henderson[71],

M. Henke[58a], A. Henrichs[54], A.M. Henriques Correia[29], S. Henrot-Versille[114], C. Hensel[54], T. Henß[172],
A.D. Hershenhorn[150], G. Herten[48], R. Hertenberger[98], L. Hervas[29], N.P. Hessey[105], A. Hidvegi[144],
E. Higón-Rodriguez[165], D. Hill[5,*], J.C. Hill[27], K.H. Hiller[41], S.J. Hillier[17], I. Hinchliffe[14], M. Hirose[115], F. Hirsch[42],
J. Hobbs[146], N. Hod[151], M.C. Hodgkinson[138], P. Hodgson[138], A. Hoecker[29], M.R. Hoeferkamp[103], J. Hoffman[39],
D. Hoffmann[83], M. Hohlfeld[81], S.O. Holmgren[144], T. Holy[126], J.L. Holzbauer[88], Y. Homma[67], P. Homola[126],
T. Horazdovsky[126], T. Hori[67], C. Horn[142], S. Horner[48], S. Horvat[99], J.-Y. Hostachy[55], S. Hou[149], M.A. Houlden[73],
A. Hoummada[134a], T. Howe[39], J. Hrivnac[114], T. Hryn'ova[4], P.J. Hsu[173], S.-C. Hsu[14], G.S. Huang[110], Z. Hubacek[126],
F. Hubaut[83], F. Huegging[20], E.W. Hughes[34], G. Hughes[71], R.E. Hughes-Jones[82], P. Hurst[57], M. Hurwitz[30],
U. Husemann[41], N. Huseynov[10], J. Huston[88], J. Huth[57], G. Iacobucci[102a], G. Iakovidis[9], I. Ibragimov[140],
L. Iconomidou-Fayard[114], J. Idarraga[157b], P. Iengo[4], O. Igonkina[105], Y. Ikegami[66], M. Ikeno[66], Y. Ilchenko[39],
D. Iliadis[152], Y. Ilyushenka[65], M. Imori[153], T. Ince[167], P. Ioannou[8], M. Iodice[133a], A. Irles Quiles[165], A. Ishikawa[67],
M. Ishino[66], R. Ishmukhametov[39], T. Isobe[153], V. Issakov[173,*], C. Issever[117], S. Istin[18], Y. Itoh[101], A.V. Ivashin[127],
W. Iwanski[38], H. Iwasaki[66], J.M. Izen[40], V. Izzo[102a], J.N. Jackson[73], P. Jackson[142], M. Jaekel[29], M. Jahoda[124],
V. Jain[61], K. Jakobs[48], S. Jakobsen[29], J. Jakubek[126], D. Jana[110], E. Jansen[104], A. Jantsch[99], M. Janus[48],
R.C. Jared[170], G. Jarlskog[79], P. Jarron[29], L. Jeanty[57], K. Jelen[37], I. Jen-La Plante[30], P. Jenni[29], P. Jez[35],
S. Jézéquel[4], W. Ji[79], J. Jia[146], Y. Jiang[32], M. Jimenez Belenguer[29], G. Jin[32], S. Jin[32], O. Jinnouchi[155], D. Joffe[39],
M. Johansen[144], K.E. Johansson[144], P. Johansson[138], S. Johnert[41], K.A. Johns[6], K. Jon-And[144], G. Jones[82],
R.W.L. Jones[71], T.W. Jones[77], T.J. Jones[73], O. Jonsson[29], D. Joos[48], C. Joram[29], P.M. Jorge[123b], V. Juranek[124],
P. Jussel[62], V.V. Kabachenko[127], S. Kabana[16], M. Kaci[165], A. Kaczmarska[38], M. Kado[114], H. Kagan[108], M. Kagan[57],
S. Kaiser[99], E. Kajomovitz[150], L.V. Kalinovskaya[65], A. Kalinowski[129], S. Kama[41], N. Kanaya[153], M. Kaneda[153],
V.A. Kantserov[96], J. Kanzaki[66], B. Kaplan[173], A. Kapliy[30], J. Kaplon[29], M. Karagounis[20], M. Karagoz Unel[117],
V. Kartvelishvili[71], A.N. Karyukhin[127], L. Kashif[57], A. Kasmi[39], R.D. Kass[108], A. Kastanas[13], M. Kastoryano[173],
M. Kataoka[29], Y. Kataoka[153], E. Katsoufis[9], J. Katzy[41], V. Kaushik[6], K. Kawagoe[67], T. Kawamoto[153],
G. Kawamura[81], M.S. Kayl[105], F. Kayumov[94], V.A. Kazanin[106], M.Y. Kazarinov[65], S.I. Kazi[86], J.R. Keates[82],
R. Keeler[167], P.T. Keener[119], R. Kehoe[39], M. Keil[49], G.D. Kekelidze[65], M. Kelly[82], J. Kennedy[98], M. Kenyon[53],
O. Kepka[135], N. Kerschen[29], B.P. Kerševan[74], S. Kersten[172], K. Kessoku[153], M. Khakzad[28], F. Khalil-zada[10],
H. Khandanyan[163], A. Khanov[111], D. Kharchenko[65], A. Khodinov[146], A.G. Kholodenko[127], A. Khomich[58a],
G. Khoriauli[20], N. Khovanskiy[65], V. Khovanskiy[95], E. Khramov[65], J. Khubua[51], G. Kilvington[76], H. Kim[7],
M.S. Kim[2], P.C. Kim[142], S.H. Kim[158], O. Kind[15], P. Kind[172], B.T. King[73], J. Kirk[128], G.P. Kirsch[117], L.E. Kirsch[22],
A.E. Kiryunin[99], D. Kisielewska[37], T. Kittelmann[122], H. Kiyamura[67], E. Kladiva[143], M. Klein[73], U. Klein[73],
K. Kleinknecht[81], M. Klemetti[85], A. Klier[169], A. Klimentov[24], R. Klingenberg[42], E.B. Klinkby[44],
T. Klioutchnikova[29], P.F. Klok[104], S. Klous[105], E.-E. Kluge[58a], T. Kluge[73], P. Kluit[105], M. Klute[54], S. Kluth[99],
N.S. Knecht[156], E. Kneringer[62], B.R. Ko[44], T. Kobayashi[153], M. Kobel[43], B. Koblitz[29], M. Kocian[142], A. Kocnar[112],
P. Kodys[125], K. Köneke[41], A.C. König[104], L. Köpke[81], F. Koetsveld[104], P. Koevesarki[20], T. Koffas[29], E. Koffeman[105],
F. Kohn[54], Z. Kohout[126], T. Kohriki[66], T. Kokott[20], H. Kolanoski[15], V. Kolesnikov[65], I. Koletsou[4], J. Koll[88],
D. Kollar[29], S. Kolos[161,r], S.D. Kolya[82], A.A. Komar[94], J.R. Komaragiri[141], T. Kondo[66], T. Kono[41,q], A.I. Kononov[48],
R. Konoplich[107], S.P. Konovalov[94], N. Konstantinidis[77], S. Koperny[37], K. Korcyl[38], K. Kordas[16], V. Koreshev[127],
A. Korn[14], I. Korolkov[11], E.V. Korolkova[138], V.A. Korotkov[127], O. Kortner[99], P. Kostka[41], V.V. Kostyukhin[20],
M.J. Kotamäki[29], S. Kotov[99], V.M. Kotov[65], K.Y. Kotov[106], Z. Koupilova[125], C. Kourkoumelis[8], A. Koutsman[105],
R. Kowalewski[167], H. Kowalski[41], T.Z. Kowalski[37], W. Kozanecki[135], A.S. Kozhin[127], V. Kral[126],
V.A. Kramarenko[97], G. Kramberger[74], M.W. Krasny[78], A. Krasznahorkay[107], A. Kreisel[151], F. Krejci[126],
A. Krepouri[152], J. Kretzschmar[73], P. Krieger[156], G. Krobath[98], K. Kroeninger[54], H. Kroha[99], J. Kroll[119],
J. Kroseberg[20], J. Krstic[12a], U. Kruchonak[65], H. Krüger[20], Z.V. Krumshteyn[65], T. Kubota[153], S. Kuehn[48],
A. Kugel[58c], T. Kuhl[172], D. Kuhn[62], V. Kukhtin[65], Y. Kulchitsky[90], S. Kuleshov[31b], C. Kummer[98], M. Kuna[83],
A. Kupco[124], H. Kurashige[67], M. Kurata[158], L.L. Kurchaninov[157a], Y.A. Kurochkin[90], V. Kus[124], W. Kuykendall[137],
E. Kuznetsova[131a,131b], O. Kvasnicka[124], R. Kwee[15], M. La Rosa[86], L. La Rotonda[36a,36b], L. Labarga[80], J. Labbe[4],
C. Lacasta[165], F. Lacava[131a,131b], H. Lacker[15], D. Lacour[78], V.R. Lacuesta[165], E. Ladygin[65], R. Lafaye[4], B. Laforge[78],
T. Lagouri[80], S. Lai[48], M. Lamanna[29], C.L. Lampen[6], W. Lampl[6], E. Lancon[135], U. Landgraf[48], M.P.J. Landon[75],
J.L. Lane[82], A.J. Lankford[161], F. Lanni[24], K. Lantzsch[29], A. Lanza[118a], S. Laplace[4], C. Lapoire[83], J.F. Laporte[135],
T. Lari[89a], A.V. Larionov[127], A. Larner[117], C. Lasseur[29], M. Lassnig[29], P. Laurelli[47], W. Lavrijsen[14], P. Laycock[73],
A.B. Lazarev[65], A. Lazzaro[89a,89b], O. Le Dortz[78], E. Le Guirriec[83], C. Le Maner[156], E. Le Menedeu[135], M. Le Vine[24],
M. Leahu[29], A. Lebedev[64], C. Lebel[93], T. LeCompte[5], F. Ledroit-Guillon[55], H. Lee[105], J.S.H. Lee[148], S.C. Lee[149],

M. Lefebvre[167], M. Legendre[135], B.C. LeGeyt[119], F. Legger[98], C. Leggett[14], M. Lehmacher[20], G. Lehmann Miotto[29], X. Lei[6], R. Leitner[125], D. Lelas[167], D. Lellouch[169], J. Lellouch[78], M. Leltchouk[34], V. Lendermann[58a], K.J.C. Leney[73], T. Lenz[172], G. Lenzen[172], B. Lenzi[135], K. Leonhardt[43], C. Leroy[93], J.-R. Lessard[167], C.G. Lester[27], A. Leung Fook Cheong[170], J. Levêque[83], D. Levin[87], L.J. Levinson[169], M.S. Levitski[127], S. Levonian[41], M. Lewandowska[21], M. Leyton[14], H. Li[170], J. Li[7], S. Li[41], X. Li[87], Z. Liang[39], Z. Liang[149,s], B. Liberti[132a], P. Lichard[29], M. Lichtnecker[98], K. Lie[163], W. Liebig[105], D. Liko[29], J.N. Lilley[17], H. Lim[5], A. Limosani[86], M. Limper[63], S.C. Lin[149], S.W. Lindsay[73], V. Linhart[126], J.T. Linnemann[88], A. Liolios[152], E. Lipeles[119], L. Lipinsky[124], A. Lipniacka[13], T.M. Liss[163], D. Lissauer[24], A.M. Litke[136], C. Liu[28], D. Liu[149,t], H. Liu[87], J.B. Liu[87], M. Liu[32], S. Liu[2], T. Liu[39], Y. Liu[32], M. Livan[118a,118b], A. Lleres[55], S.L. Lloyd[75], E. Lobodzinska[41], P. Loch[6], W.S. Lockman[136], S. Lockwitz[173], T. Loddenkoetter[20], F.K. Loebinger[82], A. Loginov[173], C.W. Loh[166], T. Lohse[15], K. Lohwasser[48], M. Lokajicek[124], J. Loken[117], L. Lopes[123b], D. Lopez Mateos[34,o], M. Losada[160], P. Loscutoff[14], M.J. Losty[157a], X. Lou[40], A. Lounis[114], K.F. Loureiro[108], L. Lovas[143], J. Love[21], P. Love[71], A.J. Lowe[61], F. Lu[32], J. Lu[2], H.J. Lubatti[137], C. Luci[131a,131b], A. Lucotte[55], A. Ludwig[43], D. Ludwig[41], I. Ludwig[48], J. Ludwig[48], F. Luehring[61], L. Luisa[162a,162c], D. Lumb[48], L. Luminari[131a], E. Lund[116], B. Lund-Jensen[145], B. Lundberg[79], J. Lundberg[29], J. Lundquist[35], G. Lutz[99], D. Lynn[24], J. Lys[14], E. Lytken[79], H. Ma[24], L.L. Ma[170], G. Maccarrone[47], A. Macchiolo[99], B. Maček[74], J.Machado Miguens[123b], R. Mackeprang[29], R.J. Madaras[14], W.F. Mader[43], R. Maenner[58c], T. Maeno[24], P. Mättig[172], S. Mättig[41], P.J. Magalhaes Martins[123b], E. Magradze[51], C.A. Magrath[104], Y. Mahalalel[151], K. Mahboubi[48], A. Mahmood[1], G. Mahout[17], C. Maiani[131a,131b], C. Maidantchik[23a], A. Maio[123b], S. Majewski[24], Y. Makida[66], M. Makouski[127], N. Makovec[114], Pa. Malecki[38], P. Malecki[38], V.P. Maleev[120], F. Malek[55], U. Mallik[63], D. Malon[5], S. Maltezos[9], V. Malyshev[106], S. Malyukov[65], M. Mambelli[30], R. Mameghani[98], J. Mamuzic[41], A. Manabe[66], L. Mandelli[89a], I. Mandić[74], R. Mandrysch[15], J. Maneira[123b], P.S. Mangeard[88], I.D. Manjavidze[65], A. Manousakis-Katsikakis[8], B. Mansoulie[135], A. Mapelli[29], L. Mapelli[29], L. March[80], J.F. Marchand[4], F. Marchese[132a,132b], M. Marcisovsky[124], C.P. Marino[61], C.N. Marques[123b], F. Marroquim[23a], R. Marshall[82], Z. Marshall[34,o], F.K. Martens[156], S. Marti i Garcia[165], A.J. Martin[75], A.J. Martin[173], B. Martin[29], B. Martin[88], F.F. Martin[119], J.P. Martin[93], T.A. Martin[17], B. Martin dit Latour[49], M. Martinez[11], V. Martinez Outschoorn[57], A. Martini[47], V. Martynenko[157b], A.C. Martyniuk[82], T. Maruyama[158], F. Marzano[131a], A. Marzin[135], L. Masetti[20], T. Mashimo[153], R. Mashinistov[96], J. Masik[82], A.L. Maslennikov[106], G. Massaro[105], N. Massol[4], A. Mastroberardino[36a,36b], T. Masubuchi[153], M. Mathes[20], P. Matricon[114], H. Matsumoto[153], H. Matsunaga[153], T. Matsushita[67], C. Mattravers[117,u], S.J. Maxfield[73], E.N. May[5], A. Mayne[138], R. Mazini[149], M. Mazur[48], M. Mazzanti[89a,89b], P. Mazzanti[19a], J. Mc Donald[85], S.P. Mc Kee[87], A. McCarn[163], R.L. McCarthy[146], N.A. McCubbin[128], K.W. McFarlane[56], H. McGlone[53], G. Mchedlidze[51], R.A. McLaren[29], S.J. McMahon[128], T.R. McMahon[76], R.A. McPherson[167,h], A. Meade[84], J. Mechnich[105], M. Mechtel[172], M. Medinnis[41], R. Meera-Lebbai[110], T.M. Meguro[115], R. Mehdiyev[93], S. Mehlhase[41], A. Mehta[73], K. Meier[58a], B. Meirose[48], A. Melamed-Katz[169], B.R. Mellado Garcia[170], Z. Meng[149,t], S. Menke[99], E. Meoni[11], D. Merkl[98], P. Mermod[117], L. Merola[102a,102b], C. Meroni[89a], F.S. Merritt[30], A.M. Messina[29], I. Messmer[48], J. Metcalfe[103], A.S. Mete[64], J.-P. Meyer[135], J. Meyer[54], T.C. Meyer[29], W.T. Meyer[64], J. Miao[32], L. Micu[25a], R.P. Middleton[128], S. Migas[73], L. Mijović[74], G. Mikenberg[169], M. Mikuž[74], D.W. Miller[142], W.J. Mills[166], C.M. Mills[57], A. Milov[169], D.A. Milstead[144], A.A. Minaenko[127], M. Miñano[165], I.A. Minashvili[65], A.I. Mincer[107], B. Mindur[37], M. Mineev[65], L.M. Mir[11], G. Mirabelli[131a], S. Misawa[24], S. Miscetti[47], A. Misiejuk[76], J. Mitrevski[136], V.A. Mitsou[165], P.S. Miyagawa[82], J.U. Mjörnmark[79], D. Mladenov[22], T. Moa[144], P. Mockett[137], S. Moed[57], V. Moeller[27], K. Mönig[41], N. Möser[20], B. Mohn[13], W. Mohr[48], S. Mohrdieck-Möck[99], R. Moles-Valls[165], J. Molina-Perez[29], G. Moloney[86], J. Monk[77], E. Monnier[83], S. Montesano[89a,89b], F. Monticelli[70], R.W. Moore[2], C.Mora Herrera[49], A. Moraes[53], A. Morais[123b], J. Morel[4], G. Morello[36a,36b], D. Moreno[160], M. Moreno Llácer[165], P. Morettini[50a], M. Morii[57], A.K. Morley[86], G. Mornacchi[29], S.V. Morozov[96], J.D. Morris[75], H.G. Moser[99], M. Mosidze[51], J. Moss[108], R. Mount[142], E. Mountricha[9], S.V. Mouraviev[94], E.J.W. Moyse[84], M. Mudrinic[12b], F. Mueller[58a], J. Mueller[122], K. Mueller[20], T.A. Müller[98], D. Muenstermann[42], A. Muir[166], R. Murillo Garcia[161], W.J. Murray[128], I. Mussche[105], E. Musto[102a,102b], A.G. Myagkov[127], M. Myska[124], J. Nadal[11], K. Nagai[24], K. Nagano[66], Y. Nagasaka[60], A.M. Nairz[29], K. Nakamura[153], I. Nakano[109], H. Nakatsuka[67], G. Nanava[20], A. Napier[159], M. Nash[77,v], N.R. Nation[21], T. Nattermann[20], T. Naumann[41], G. Navarro[160], S.K. Nderitu[20], H.A. Neal[87], E. Nebot[80], P. Nechaeva[94], A. Negri[118a,118b], G. Negri[29], A. Nelson[64], T.K. Nelson[142], S. Nemecek[124], P. Nemethy[107], A.A. Nepomuceno[23a], M. Nessi[29], M.S. Neubauer[163], A. Neusiedl[81], R.N. Neves[123b], P. Nevski[24], F.M. Newcomer[119], C. Nicholson[53], R.B. Nickerson[117], R. Nicolaidou[135], L. Nicolas[138], G. Nicoletti[47], F. Niedercorn[114], J. Nielsen[136],

A. Nikiforov[15], K. Nikolaev[65], I. Nikolic-Audit[78], K. Nikolopoulos[8], H. Nilsen[48], P. Nilsson[7], A. Nisati[131a], T. Nishiyama[67], R. Nisius[99], L. Nodulman[5], M. Nomachi[115], I. Nomidis[152], H. Nomoto[153], M. Nordberg[29], B. Nordkvist[144], D. Notz[41], J. Novakova[125], M. Nozaki[66], M. Nožička[41], I.M. Nugent[157a], A.-E. Nuncio-Quiroz[20], G. Nunes Hanninger[20], T. Nunnemann[98], E. Nurse[77], D.C. O'Neil[141], V. O'Shea[53], F.G. Oakham[28,b], H. Oberlack[99], A. Ochi[67], S. Oda[153], S. Odaka[66], J. Odier[83], G.A. Odino[50a,50b], H. Ogren[61], S.H. Oh[44], C.C. Ohm[144], T. Ohshima[101], H. Ohshita[139], T. Ohsugi[59], S. Okada[67], H. Okawa[153], Y. Okumura[101], M. Olcese[50a], A.G. Olchevski[65], M. Oliveira[123b], D. Oliveira Damazio[24], J. Oliver[57], E. Oliver Garcia[165], D. Olivito[119], A. Olszewski[38], J. Olszowska[38], C. Omachi[67], A. Onofre[123b], P.U.E. Onyisi[30], C.J. Oram[157a], G. Ordonez[104], M.J. Oreglia[30], Y. Oren[151], D. Orestano[133a,133b], I. Orlov[106], C. Oropeza Barrera[53], R.S. Orr[156], E.O. Ortega[129], B. Osculati[50a,50b], C. Osuna[11], R. Otec[126], J. P Ottersbach[105], F. Ould-Saada[116], A. Ouraou[135], Q. Ouyang[32], M. Owen[82], S. Owen[138], V.E. Ozcan[77], K. Ozone[66], N. Ozturk[7], A. Pacheco Pages[11], S. Padhi[170], C. Padilla Aranda[11], E. Paganis[138], C. Pahl[63], F. Paige[24], K. Pajchel[116], A. Pal[7], S. Palestini[29], D. Pallin[33], A. Palma[123b], J.D. Palmer[17], Y.B. Pan[170], E. Panagiotopoulou[9], B. Panes[31a], N. Panikashvili[87], S. Panitkin[24], D. Pantea[25a], M. Panuskova[124], V. Paolone[122], Th.D. Papadopoulou[9], S.J. Park[54], W. Park[24,w], M.A. Parker[27], S.I. Parker[14], F. Parodi[50a,50b], J.A. Parsons[34], U. Parzefall[48], E. Pasqualucci[131a], G. Passardi[29], A. Passeri[133a], F. Pastore[133a,133b], Fr. Pastore[29], G. Pásztor[49,x], S. Pataraia[99], J.R. Pater[82], S. Patricelli[102a,102b], A. Patwa[24], T. Pauly[29], L.S. Peak[148], M. Pecsy[143], M.I. Pedraza Morales[170], S.V. Peleganchuk[106], H. Peng[170], A. Penson[34], J. Penwell[61], M. Perantoni[23a], K. Perez[34,o], E. Perez Codina[11], M.T. Pérez García-Estañ[165], V. Perez Reale[34], L. Perini[89a,89b], H. Pernegger[29], R. Perrino[72a], P. Perrodo[4], S. Persembe[3], P. Perus[114], V.D. Peshekhonov[65], B.A. Petersen[29], J. Petersen[29], T.C. Petersen[35], E. Petit[83], C. Petridou[152], E. Petrolo[131a], F. Petrucci[133a,133b], D. Petschull[41], M. Petteni[141], R. Pezoa[31b], B. Pfeifer[48], A. Phan[86], A.W. Phillips[27], G. Piacquadio[48], M. Piccinini[19a,19b], R. Piegaia[26], J.E. Pilcher[30], A.D. Pilkington[82], J. Pina[123b], M. Pinamonti[162a,162c], J.L. Pinfold[2], J. Ping[32], B. Pinto[123b], O. Pirotte[29], C. Pizio[89a,89b], R. Placakyte[41], M. Plamondon[167], W.G. Plano[82], M.-A. Pleier[24], A. Poblaguev[173], S. Poddar[58a], F. Podlyski[33], P. Poffenberger[167], L. Poggioli[114], M. Pohl[49], F. Polci[55], G. Polesello[118a], A. Policicchio[137], A. Polini[19a], J. Poll[75], V. Polychronakos[24], D.M. Pomarede[135], D. Pomeroy[22], K. Pommès[29], L. Pontecorvo[131a], B.G. Pope[88], D.S. Popovic[12a], A. Poppleton[29], J. Popule[124], X. Portell Bueso[48], R. Porter[161], G.E. Pospelov[99], P. Pospichal[29], S. Pospisil[126], M. Potekhin[24], I.N. Potrap[99], C.J. Potter[147], C.T. Potter[85], K.P. Potter[82], G. Poulard[29], J. Poveda[170], R. Prabhu[20], P. Pralavorio[83], S. Prasad[57], R. Pravahan[7], T. Preda[25a], K. Pretzl[16], L. Pribyl[29], D. Price[61], L.E. Price[5], P.M. Prichard[73], D. Prieur[122], M. Primavera[72a], K. Prokofiev[29], F. Prokoshin[31b], S. Protopopescu[24], J. Proudfoot[5], X. Prudent[43], H. Przysieszniak[4], S. Psoroulas[20], E. Ptacek[113], C. Puigdengoles[11], J. Purdham[87], M. Purohit[24,w], P. Puzo[114], Y. Pylypchenko[116], M. Qi[32], J. Qian[87], W. Qian[128], Z. Qian[83], Z. Qin[41], D. Qing[157a], A. Quadt[54], D.R. Quarrie[14], W.B. Quayle[170], F. Quinonez[31a], M. Raas[104], V. Radeka[24], V. Radescu[58b], B. Radics[20], T. Rador[18], F. Ragusa[89a,89b], G. Rahal[178], A.M. Rahimi[108], D. Rahm[24], S. Rajagopalan[24], M. Rammes[140], P.N. Ratoff[71], F. Rauscher[98], E. Rauter[99], M. Raymond[29], A.L. Read[116], D.M. Rebuzzi[118a,118b], A. Redelbach[171], G. Redlinger[24], R. Reece[119], K. Reeves[172], E. Reinherz-Aronis[151], A. Reinsch[113], I. Reisinger[42], D. Reljic[12a], C. Rembser[29], Z.L. Ren[149], P. Renkel[39], S. Rescia[24], M. Rescigno[131a], S. Resconi[89a], B. Resende[105], P. Reznicek[125], R. Rezvani[156], A. Richards[77], R.A. Richards[88], D. Richter[15], R. Richter[99], E. Richter-Was[38,y], M. Ridel[78], S. Rieke[81], M. Rijpstra[105], M. Rijssenbeek[146], A. Rimoldi[118a,118b], L. Rinaldi[19a], R.R. Rios[39], I. Riu[11], G. Rivoltella[89a,89b], F. Rizatdinova[111], E.R. Rizvi[75], D.A. Roa Romero[160], S.H. Robertson[85,h], A. Robichaud-Veronneau[49], D. Robinson[27], M. Robinson[113], A. Robson[53], J.G. Rocha de Lima[5], C. Roda[121a,121b], D. Rodriguez[160], Y. Rodriguez Garcia[15], S. Roe[29], O. Røhne[116], V. Rojo[1], S. Rolli[159], A. Romaniouk[96], V.M. Romanov[65], G. Romeo[26], D. Romero Maltrana[31a], L. Roos[78], E. Ros[165], S. Rosati[131a,131b], G.A. Rosenbaum[156], E.I. Rosenberg[64], L. Rosselet[49], L.P. Rossi[50a], M. Rotaru[25a], J. Rothberg[137], I. Rottländer[20], D. Rousseau[114], C.R. Royon[135], A. Rozanov[83], Y. Rozen[150], X. Ruan[32], B. Ruckert[98], N. Ruckstuhl[105], V.I. Rud[97], G. Rudolph[62], F. Rühr[58a], F. Ruggieri[133a], A. Ruiz-Martinez[165], L. Rumyantsev[65], N.A. Rusakovich[65], J.P. Rutherfoord[6], C. Ruwiedel[20], P. Ruzicka[124], Y.F. Ryabov[120], V. Ryadovikov[127], P. Ryan[88], G. Rybkin[114], S. Rzaeva[10], A.F. Saavedra[148], H.F.-W. Sadrozinski[136], R. Sadykov[65], H. Sakamoto[153], G. Salamanna[105], A. Salamon[132a], M. Saleem[110], D. Salihagic[99], A. Salnikov[142], J. Salt[165], B.M. Salvachua Ferrando[5], D. Salvatore[36a,36b], F. Salvatore[147], A. Salvucci[47], A. Salzburger[41], D. Sampsonidis[152], B.H. Samset[116], M.A. Sanchis Lozano[165], H. Sandaker[13], H.G. Sander[81], M.P. Sanders[98], M. Sandhoff[172], R. Sandstroem[105], S. Sandvoss[172], D.P.C. Sankey[128], B. Sanny[172], A. Sansoni[47], C. Santamarina Rios[85], L. Santi[162a,162c], C. Santoni[33], R. Santonico[132a,132b], D. Santos[123b], J. Santos[123b], J.G. Saraiva[123b], T. Sarangi[170], E. Sarkisyan-Grinbaum[7], F. Sarri[121a,121b], O. Sasaki[66], T. Sasaki[66], N. Sasao[68], I. Satsounkevitch[90], G. Sauvage[4],

P. Savard[156,b], A.Y. Savine[6], V. Savinov[122], L. Sawyer[24,i], D.H. Saxon[53], L.P. Says[33], C. Sbarra[19a,19b], A. Sbrizzi[19a,19b], D.A. Scannicchio[29], J. Schaarschmidt[43], P. Schacht[99], U. Schäfer[81], S. Schaetzel[58b], A.C. Schaffer[114], D. Schaile[98], R.D. Schamberger[146], A.G. Schamov[106], V.A. Schegelsky[120], D. Scheirich[87], M. Schernau[161], M.I. Scherzer[14], C. Schiavi[50a,50b], J. Schieck[99], M. Schioppa[36a,36b], S. Schlenker[29], J.L. Schlereth[5], P. Schmid[62], M.P. Schmidt[173,*], K. Schmieden[20], C. Schmitt[81], M. Schmitz[20], M. Schott[29], D. Schouten[141], J. Schovancova[124], M. Schram[85], A. Schreiner[63], C. Schroeder[81], N. Schroer[58c], M. Schroers[172], G. Schuler[29], J. Schultes[172], H.-C. Schultz-Coulon[58a], J. Schumacher[43], M. Schumacher[48], B.A. Schumm[136], Ph. Schune[135], C. Schwanenberger[82], A. Schwartzman[142], Ph. Schwemling[78], R. Schwienhorst[88], R. Schwierz[43], J. Schwindling[135], W.G. Scott[128], J. Searcy[113], E. Sedykh[120], E. Segura[11], S.C. Seidel[103], A. Seiden[136], F. Seifert[43], J.M. Seixas[23a], G. Sekhniaidze[102a], D.M. Seliverstov[120], B. Sellden[144], M. Seman[143], N. Semprini-Cesari[19a,19b], C. Serfon[98], L. Serin[114], R. Seuster[99], H. Severini[110], M.E. Sevior[86], A. Sfyrla[163], M. Shamim[113], L.Y. Shan[32], J.T. Shank[21], Q.T. Shao[86], M. Shapiro[14], P.B. Shatalov[95], L. Shaver[6], C. Shaw[53], K. Shaw[138], D. Sherman[29], P. Sherwood[77], A. Shibata[107], M. Shimojima[100], T. Shin[56], A. Shmeleva[94], M.J. Shochet[30], M.A. Shupe[6], P. Sicho[124], A. Sidoti[15], A. Siebel[172], F. Siegert[77], J. Siegrist[14], Dj. Sijacki[12a], O. Silbert[169], J. Silva[123b], Y. Silver[151], D. Silverstein[142], S.B. Silverstein[144], V. Simak[126], Lj. Simic[12a], S. Simion[114], B. Simmons[77], M. Simonyan[4], P. Sinervo[156], N.B. Sinev[113], V. Sipica[140], G. Siragusa[81], A.N. Sisakyan[65], S.Yu. Sivoklokov[97], J. Sjoelin[144], T.B. Sjursen[13], P. Skubic[110], N. Skvorodnev[22], M. Slater[17], T. Slavicek[126], K. Sliwa[159], J. Sloper[29], T. Sluka[124], V. Smakhtin[169], S.Yu. Smirnov[96], Y. Smirnov[24], L.N. Smirnova[97], O. Smirnova[79], B.C. Smith[57], D. Smith[142], K.M. Smith[53], M. Smizanska[71], K. Smolek[126], A.A. Snesarev[94], S.W. Snow[82], J. Snow[110], J. Snuverink[105], S. Snyder[24], M. Soares[123b], R. Sobie[167,h], J. Sodomka[126], A. Soffer[151], C.A. Solans[165], M. Solar[126], E. Solfaroli Camillocci[131a,131b], A.A. Solodkov[127], O.V. Solovyanov[127], R. Soluk[2], J. Sondericker[24], V. Sopko[126], B. Sopko[126], M. Sosebee[7], V.V. Sosnovtsev[96], L. Sospedra Suay[165], A. Soukharev[106], S. Spagnolo[72a,72b], F. Spanò[34], P. Speckmayer[29], E. Spencer[136], R. Spighi[19a], G. Spigo[29], F. Spila[131a,131b], R. Spiwoks[29], M. Spousta[125], T. Spreitzer[141], B. Spurlock[7], R.D.St. Denis[53], T. Stahl[140], R. Stamen[58a], S.N. Stancu[161], E. Stanecka[29], R.W. Stanek[5], C. Stanescu[133a], S. Stapnes[116], E.A. Starchenko[127], J. Stark[55], P. Staroba[124], P. Starovoitov[91], J. Stastny[124], A. Staude[98], P. Stavina[143], G. Stavropoulos[14], P. Steinbach[43], P. Steinberg[24], I. Stekl[126], B. Stelzer[141], H.J. Stelzer[41], O. Stelzer-Chilton[157a], H. Stenzel[52], K. Stevenson[75], G. Stewart[53], M.C. Stockton[17], K. Stoerig[48], G. Stoicea[25a], S. Stonjek[99], P. Strachota[125], A. Stradling[7], A. Straessner[43], J. Strandberg[87], S. Strandberg[14], A. Strandlie[116], M. Strauss[110], P. Strizenec[143], R. Ströhmer[98], D.M. Strom[113], J.A. Strong[76,*], R. Stroynowski[39], J. Strube[128], B. Stugu[13], I. Stumer[24,*], D.A. Soh[149,z], D. Su[142], S.I. Suchkov[96], Y. Sugaya[115], T. Sugimoto[101], C. Suhr[5], M. Suk[125], V.V. Sulin[94], S. Sultansoy[3,aa], T. Sumida[29], X. Sun[32], J.E. Sundermann[48], K. Suruliz[162a,162b], S. Sushkov[11], G. Susinno[36a,36b], M.R. Sutton[138], T. Suzuki[153], Y. Suzuki[66], Yu.M. Sviridov[127], I. Sykora[143], T. Sykora[125], T. Szymocha[38], J. Sánchez[165], D. Ta[20], K. Tackmann[29], A. Taffard[161], R. Tafirout[157a], A. Taga[116], Y. Takahashi[101], H. Takai[24], R. Takashima[69], H. Takeda[67], T. Takeshita[139], M. Talby[83], A. Talyshev[106], M.C. Tamsett[76], J. Tanaka[153], R. Tanaka[114], S. Tanaka[130], S. Tanaka[66], G.P. Tappern[29], S. Tapprogge[81], D. Tardif[156], S. Tarem[150], F. Tarrade[24], G.F. Tartarelli[89a], P. Tas[125], M. Tasevsky[124], E. Tassi[36a,36b], C. Taylor[77], F.E. Taylor[92], G.N. Taylor[86], R.P. Taylor[167], W. Taylor[157b], P. Teixeira-Dias[76], H. Ten Kate[29], P.K. Teng[149], S. Terada[66], K. Terashi[153], J. Terron[80], M. Terwort[41,q], M. Testa[47], R.J. Teuscher[156,h], C.M. Tevlin[82], J. Thadome[172], R. Thananuwong[49], M. Thioye[173], S. Thoma[48], J.P. Thomas[17], T.L. Thomas[103], E.N. Thompson[84], P.D. Thompson[17], P.D. Thompson[156], R.J. Thompson[82], A.S. Thompson[53], E. Thomson[119], R.P. Thun[87], T. Tic[124], V.O. Tikhomirov[94], Y.A. Tikhonov[106], C.J.W.P. Timmermans[104], P. Tipton[173], F.J. Tique Aires Viegas[29], S. Tisserant[83], J. Tobias[48], B. Toczek[37], T. Todorov[4], S. Todorova-Nova[159], B. Toggerson[161], J. Tojo[66], S. Tokár[143], K. Tokushuku[66], K. Tollefson[88], L. Tomasek[124], M. Tomasek[124], F. Tomasz[143], M. Tomoto[101], D. Tompkins[6], L. Tompkins[14], K. Toms[103], G. Tong[32], A. Tonoyan[13], C. Topfel[16], N.D. Topilin[65], E. Torrence[113], E. Torró Pastor[165], J. Toth[83,x], F. Touchard[83], D.R. Tovey[138], S.N. Tovey[86], T. Trefzger[171], L. Tremblet[29], A. Tricoli[29], I.M. Trigger[157a], S. Trincaz-Duvoid[78], T.N. Trinh[78], M.F. Tripiana[70], N. Triplett[64], A. Trivedi[24,w], B. Trocmé[55], C. Troncon[89a], A. Trzupek[38], C. Tsarouchas[9], J.C.-L. Tseng[117], I. Tsiafis[152], M. Tsiakiris[105], P.V. Tsiareshka[90], D. Tsionou[138], G. Tsipolitis[9], V. Tsiskaridze[51], E.G. Tskhadadze[51], I.I. Tsukerman[95], V. Tsulaia[122], J.-W. Tsung[20], S. Tsuno[66], D. Tsybychev[146], M. Turala[38], D. Turecek[126], I. Turk Cakir[3,ab], E. Turlay[114], P.M. Tuts[34], M.S. Twomey[137], M. Tylmad[144], M. Tyndel[128], G. Tzanakos[8], K. Uchida[115], I. Ueda[153], M. Uhlenbrock[20], M. Uhrmacher[54], F. Ukegawa[158], G. Unal[29], D.G. Underwood[5], A. Undrus[24], G. Unel[161], Y. Unno[66], D. Urbaniec[34], E. Urkovsky[151], P. Urquijo[49], P. Urrejola[31a], G. Usai[7], M. Uslenghi[118a,118b], L. Vacavant[83], V. Vacek[126], B. Vachon[85], S. Vahsen[14], J. Valenta[124], P. Valente[131a], S. Valentinetti[19a,19b], S. Valkar[125], E. Valladolid Gallego[165], S. Vallecorsa[150],

J.A. Valls Ferrer[165], R. Van Berg[119], H. van der Graaf[105], E. van der Kraaij[105], E. van der Poel[105], D. Van Der Ster[29], N. van Eldik[84], P. van Gemmeren[5], Z. van Kesteren[105], I. van Vulpen[105], W. Vandelli[29], G. Vandoni[29], A. Vaniachine[5], P. Vankov[73], F. Vannucci[78], F. Varela Rodriguez[29], R. Vari[131a], E.W. Varnes[6], D. Varouchas[14], A. Vartapetian[7], K.E. Varvell[148], L. Vasilyeva[94], V.I. Vassilakopoulos[56], F. Vazeille[33], G. Vegni[89a,89b], J.J. Veillet[114], C. Vellidis[8], F. Veloso[123b], R. Veness[29], S. Veneziano[131a], A. Ventura[72a,72b], D. Ventura[137], M. Venturi[48], N. Venturi[16], V. Vercesi[118a], M. Verducci[171], W. Verkerke[105], J.C. Vermeulen[105], M.C. Vetterli[141,b], I. Vichou[163], T. Vickey[170], G.H.A. Viehhauser[117], M. Villa[19a,19b], E.G. Villani[128], M. Villaplana Perez[165], J. Villate[123b], E. Vilucchi[47], M.G. Vincter[28], E. Vinek[29], V.B. Vinogradov[65], S. Viret[33], J. Virzi[14], A. Vitale[19a,19b], O.V. Vitells[169], I. Vivarelli[48], F. Vives Vaques[11], S. Vlachos[9], M. Vlasak[126], N. Vlasov[20], H. Vogt[41], P. Vokac[126], M. Volpi[11], G. Volpini[89a,89b], H. von der Schmitt[99], J. von Loeben[99], H. von Radziewski[48], E. von Toerne[20], V. Vorobel[125], A.P. Vorobiev[127], V. Vorwerk[11], M. Vos[165], R. Voss[29], T.T. Voss[172], J.H. Vossebeld[73], N. Vranjes[12a], M. Vranjes Milosavljevic[12a], V. Vrba[124], M. Vreeswijk[105], T. Vu Anh[81], D. Vudragovic[12a], R. Vuillermet[29], I. Vukotic[114], P. Wagner[119], H. Wahlen[172], J. Walbersloh[42], J. Walder[71], R. Walker[98], W. Walkowiak[140], R. Wall[173], C. Wang[44], H. Wang[170], J. Wang[55], J.C. Wang[137], S.M. Wang[149], C.P. Ward[27], M. Warsinsky[48], R. Wastie[117], P.M. Watkins[17], A.T. Watson[17], M.F. Watson[17], G. Watts[137], S. Watts[82], A.T. Waugh[148], B.M. Waugh[77], M. Webel[48], J. Weber[42], M.D. Weber[16], M. Weber[128], M.S. Weber[16], P. Weber[58a], A.R. Weidberg[117], J. Weingarten[54], C. Weiser[48], H. Wellenstein[22], P.S. Wells[29], M. Wen[47], T. Wenaus[24], S. Wendler[122], T. Wengler[82], S. Wenig[29], N. Wermes[20], M. Werner[48], P. Werner[29], M. Werth[161], U. Werthenbach[140], M. Wessels[58a], K. Whalen[28], S.J. Wheeler-Ellis[161], S.P. Whitaker[21], A. White[7], M.J. White[27], S. White[24], D. Whiteson[161], D. Whittington[61], F. Wicek[114], D. Wicke[81], F.J. Wickens[128], W. Wiedenmann[170], M. Wielers[128], P. Wienemann[20], C. Wiglesworth[73], L.A.M. Wiik[48], A. Wildauer[165], M.A. Wildt[41,q], I. Wilhelm[125], H.G. Wilkens[29], E. Williams[34], H.H. Williams[119], W. Willis[34], S. Willocq[84], J.A. Wilson[17], M.G. Wilson[142], A. Wilson[87], I. Wingerter-Seez[4], F. Winklmeier[29], M. Wittgen[142], M.W. Wolter[38], H. Wolters[123b], B.K. Wosiek[38], J. Wotschack[29], M.J. Woudstra[84], K. Wraight[53], C. Wright[53], D. Wright[142], B. Wrona[73], S.L. Wu[170], X. Wu[49], E. Wulf[34], S. Xella[35], S. Xie[48], Y. Xie[32], D. Xu[138], N. Xu[170], M. Yamada[158], A. Yamamoto[66], S. Yamamoto[153], T. Yamamura[153], K. Yamanaka[64], J. Yamaoka[44], T. Yamazaki[153], Y. Yamazaki[67], Z. Yan[21], H. Yang[87], U.K. Yang[82], Y. Yang[32], Z. Yang[144], W.-M. Yao[14], Y. Yao[14], Y. Yasu[66], J. Ye[39], S. Ye[24], M. Yilmaz[3,ac], R. Yoosoofmiya[122], K. Yorita[168], R. Yoshida[5], C. Young[142], S.P. Youssef[21], D. Yu[24], J. Yu[7], M. Yu[58c], X. Yu[32], J. Yuan[99], L. Yuan[78], A. Yurkewicz[146], R. Zaidan[63], A.M. Zaitsev[127], Z. Zajacova[29], V. Zambrano[47], L. Zanello[131a,131b], P. Zarzhitsky[39], A. Zaytsev[106], C. Zeitnitz[172], M. Zeller[173], P.F. Zema[29], A. Zemla[38], C. Zendler[20], O. Zenin[127], T. Zenis[143], Z. Zenonos[121a,121b], S. Zenz[14], D. Zerwas[114], G. Zevi della Porta[57], Z. Zhan[32], H. Zhang[83], J. Zhang[5], Q. Zhang[5], X. Zhang[32], L. Zhao[107], T. Zhao[137], Z. Zhao[32], A. Zhemchugov[65], S. Zheng[32], J. Zhong[149,ad], B. Zhou[87], N. Zhou[34], Y. Zhou[149], C.G. Zhu[32], H. Zhu[41], Y. Zhu[170], X. Zhuang[98], V. Zhuravlov[99], B. Zilka[143], R. Zimmermann[20], S. Zimmermann[20], S. Zimmermann[48], M. Ziolkowski[140], R. Zitoun[4], L. Živković[34], V.V. Zmouchko[127,*], G. Zobernig[170], A. Zoccoli[19a,19b], M. zur Nedden[15], V. Zutshi[5]

*CERN, 1211 Genève 23, Switzerland

[1] University at Albany, 1400 Washington Ave, Albany, NY 12222, United States of America

[2] University of Alberta, Department of Physics, Centre for Particle Physics, Edmonton, AB T6G 2G7, Canada

[3] Ankara University, Faculty of Sciences, Department of Physics, TR 061000 Tandogan, Ankara, Turkey

[4] LAPP, Université de Savoie, CNRS/IN2P3, Annecy-le-Vieux, France

[5] Argonne National Laboratory, High Energy Physics Division, 9700 S. Cass Avenue, Argonne IL 60439, United States of America

[6] University of Arizona, Department of Physics, Tucson, AZ 85721, United States of America

[7] The University of Texas at Arlington, Department of Physics, Box 19059, Arlington, TX 76019, United States of America

[8] University of Athens, Nuclear & Particle Physics, Department of Physics, Panepistimiopouli, Zografou, GR 15771 Athens, Greece

[9] National Technical University of Athens, Physics Department, 9-Iroon Polytechniou, GR 15780 Zografou, Greece

[10] Institute of Physics, Azerbaijan Academy of Sciences, H. Javid Avenue 33, AZ 143 Baku, Azerbaijan

[11] Institut de Física d'Altes Energies, IFAE, Edifici Cn, Universitat Autònoma de Barcelona, ES-08193 Bellaterra (Barcelona), Spain

[12] University of Belgrade(a), Institute of Physics, P.O. Box 57, 11001 Belgrade; Vinca Institute of Nuclear Sciences(b), Mihajla Petrovica Alasa 12-14, 11001 Belgrade, Serbia

[13] University of Bergen, Department for Physics and Technology, Allegaten 55, NO-5007 Bergen, Norway

[14] Lawrence Berkeley National Laboratory and University of California, Physics Division, MS50B-6227, 1 Cyclotron Road, Berkeley, CA 94720, United States of America

[15] Humboldt University, Institute of Physics, Berlin, Newtonstr. 15, D-12489 Berlin, Germany

[16] University of Bern

[17] Albert Einstein Center for Fundamental Physics, Laboratory for High Energy Physics, Sidlerstrasse 5, CH-3012 Bern, Switzerland University of Birmingham, School of Physics and Astronomy, Edgbaston, Birmingham B15 2TT, United Kingdom

[18] Bogazici University, Faculty of Sciences, Department of Physics, TR-80815 Bebek-Istanbul, Turkey

[19] INFN Sezione di Bologna(a); Università di Bologna, Dipartimento di Fisica(b), viale C. Berti Pichat, 6/2, IT-40127 Bologna, Italy

[20] University of Bonn, Physikalisches Institut, Nussallee 12, D-53115 Bonn, Germany

[21] Boston University, Department of Physics, 590 Commonwealth Avenue, Boston, MA 02215, United States of America

[22] Brandeis University, Department of Physics, MS057, 415 South Street, Waltham, MA 02454, United States of America

[23] Universidade Federal do Rio De Janeiro, Instituto de Fisica[a], Caixa Postal 68528, Ilha do Fundao, BR-21945-970 Rio de Janeiro; [b] Universidade de Sao Paulo, Instituto de Fisica, R.do Matao Trav. R.187, Sao Paulo-SP, 05508-900, Brazil

[24] Brookhaven National Laboratory, Physics Department, Bldg. 510A, Upton, NY 11973, United States of America

[25] National Institute of Physics and Nuclear Engineering[a], Bucharest-Magurele, Str. Atomistilor 407, P.O. Box MG-6, R-077125, Romania; [b] University Politehnica Bucharest, Rectorat-AN 001, 313 Splaiul Independentei, sector 6, 060042 Bucuresti; [c] West University in Timisoara, Bd. Vasile Parvan 4, Timisoara, Romania

[26] Universidad de Buenos Aires, FCEyN, Dto. Fisica, Pab I-C. Universitaria, 1428 Buenos Aires, Argentina

[27] University of Cambridge, Cavendish Laboratory, J J Thomson Avenue, Cambridge CB3 0HE, United Kingdom

[28] Carleton University, Department of Physics, 1125 Colonel By Drive, Ottawa ON K1S 5B6, Canada

[29] CERN, CH-1211 Geneva 23, Switzerland

[30] University of Chicago, Enrico Fermi Institute, 5640 S. Ellis Avenue, Chicago, IL 60637, United States of America

[31] Pontificia Universidad Católica de Chile, Facultad de Fisica, Departamento de Fisica[a], Avda. Vicuna Mackenna 4860, San Joaquin, Santiago; Universidad Técnica Federico Santa María, Departamento de Física[b], Avda. España 1680, Casilla 110-V, Valparaíso, Chile

[32] Institute of HEP, Chinese Academy of Sciences, P.O. Box 918, CN-100049 Beijing; USTC, Department of Modern Physics, Hefei, CN-230026 Anhui; Nanjing University, Department of Physics, CN-210093 Nanjing; Shandong University, HEP Group, CN-250100 Shadong, China

[33] Laboratoire de Physique Corpusculaire, CNRS-IN2P3, Université Blaise Pascal, FR-63177 Aubiere Cedex, France

[34] Columbia University, Nevis Laboratory, 136 So. Broadway, Irvington, NY 10533, United States of America

[35] University of Copenhagen, Niels Bohr Institute, Blegdamsvej 17, DK-2100 Kobenhavn 0, Denmark

[36] INFN Gruppo Collegato di Cosenza[a]; Università della Calabria, Dipartimento di Fisica[b], IT-87036 Arcavacata di Rende, Italy

[37] Faculty of Physics and Applied Computer Science of the AGH-University of Science and Technology, (FPACS, AGH-UST), al. Mickiewicza 30, PL-30059 Cracow, Poland

[38] The Henryk Niewodniczanski Institute of Nuclear Physics, Polish Academy of Sciences, ul. Radzikowskiego 152, PL-31342 Krakow, Poland

[39] Southern Methodist University, Physics Department, 106 Fondren Science Building, Dallas, TX 75275-0175, United States of America

[40] University of Texas at Dallas, 800 West Campbell Road, Richardson, TX 75080-3021, United States of America

[41] DESY, Notkestr. 85, D-22603 Hamburg , Germany and Platanenalle 6, D-15738 Zeuthen, Germany

[42] TU Dortmund, Experimentelle Physik IV, DE-44221 Dortmund, Germany

[43] Technical University Dresden, Institut fuer Kern- und Teilchenphysik, Zellescher Weg 19, D-01069 Dresden, Germany

[44] Duke University, Department of Physics, Durham, NC 27708, United States of America

[45] University of Edinburgh, School of Physics & Astronomy, James Clerk Maxwell Building, The Kings Buildings, Mayfield Road, Edinburgh EH9 3JZ, United Kingdom

[46] Fachhochschule Wiener Neustadt; Johannes Gutenbergstrasse 3 AT-2700 Wiener Neustadt, Austria

[47] INFN Laboratori Nazionali di Frascati, via Enrico Fermi 40, IT-00044 Frascati, Italy

[48] Albert-Ludwigs-Universität, Fakultät für Mathematik und Physik, Hermann-Herder Str. 3, D-79104 Freiburg i.Br., Germany

[49] Université de Genève, Section de Physique, 24 rue Ernest Ansermet, CH-1211 Geneve 4, Switzerland

[50] INFN Sezione di Genova[a]; Università di Genova, Dipartimento di Fisica[b], via Dodecaneso 33, IT-16146 Genova, Italy

[51] Institute of Physics of the Georgian Academy of Sciences, 6 Tamarashvili St., GE-380077 Tbilisi; Tbilisi State University, HEP Institute, University St. 9, GE-380086 Tbilisi, Georgia

[52] Justus-Liebig-Universitaet Giessen, II Physikalisches Institut, Heinrich-Buff Ring 16, D-35392 Giessen, Germany

[53] University of Glasgow, Department of Physics and Astronomy, Glasgow G12 8QQ, United Kingdom

[54] Georg-August-Universitat, II. Physikalisches Institut, Friedrich-Hund Platz 1, D-37077 Goettingen, Germany

[55] Laboratoire de Physique Subatomique et de Cosmologie, CNRS/IN2P3, Université Joseph Fourier, INPG, 53 avenue des Martyrs, FR-38026 Grenoble Cedex, France

[56] Hampton University, Department of Physics, Hampton, VA 23668, United States of America

[57] Harvard University, Laboratory for Particle Physics and Cosmology, 18 Hammond Street, Cambridge, MA 02138, United States of America

[58] Ruprecht-Karls-Universitaet Heidelberg, Kirchhoff-Institut fuer Physik[a], Im Neuenheimer Feld 227, DE-69120 Heidelberg; [b] Physikalisches Institut, Philosophenweg 12, D-69120 Heidelberg; ZITI Ruprecht-Karls-University Heidelberg[c], Lehrstuhl fuer Informatik V, B6, 23-29, DE-68131 Mannheim, Germany

[59] Hiroshima University, Faculty of Science, 1-3-1 Kagamiyama, Higashihiroshima-shi, JP-Hiroshima 739-8526, Japan

[60] Hiroshima Institute of Technology, Faculty of Applied Information Science, 2-1-1 Miyake Saeki-ku, Hiroshima-shi, JP-Hiroshima 731-5193, Japan

[61] Indiana University, Department of Physics, Swain Hall West 117, Bloomington, IN 47405-7105, United States of America

[62] Institut fuer Astro- und Teilchenphysik, Technikerstrasse 25, A-6020 Innsbruck, Austria

[63] University of Iowa, 203 Van Allen Hall, Iowa City, IA 52242-1479, United States of America

[64] Iowa State University, Department of Physics and Astronomy, Ames High Energy Physics Group, Ames, IA 50011-3160, United States of America

[65] Joint Institute for Nuclear Research, JINR Dubna, RU-141 980 Moscow Region, Russia

[66] KEK, High Energy Accelerator Research Organization, 1-1 Oho, Tsukuba-shi, Ibaraki-ken 305-0801, Japan

[67] Kobe University, Graduate School of Science, 1-1 Rokkodai-cho, Nada-ku, JP Kobe 657-8501, Japan

[68] Kyoto University, Faculty of Science, Oiwake-cho, Kitashirakawa, Sakyou-ku, Kyoto-shi, JP-Kyoto 606-8502, Japan

[69] Kyoto University of Education, 1 Fukakusa, Fujimori, fushimi-ku, Kyoto-shi, JP-Kyoto 612-8522, Japan

[70] Universidad Nacional de La Plata, FCE, Departamento de Física, IFLP (CONICET-UNLP), C.C. 67, 1900 La Plata, Argentina

[71] Lancaster University, Physics Department, Lancaster LA1 4YB, United Kingdom

[72] INFN Sezione di Lecce[a]; Università del Salento, Dipartimento di Fisica[b] Via Arnesano IT-73100 Lecce, Italy

[73] University of Liverpool, Oliver Lodge Laboratory, P.O. Box 147, Oxford Street, Liverpool L69 3BX, United Kingdom

[74] Jožef Stefan Institute and University of Ljubljana, Department of Physics, SI-1000 Ljubljana, Slovenia

[75] Queen Mary University of London, Department of Physics, Mile End Road, London E1 4NS, United Kingdom

[76] Royal Holloway, University of London, Department of Physics, Egham Hill, Egham, Surrey TW20 0EX, United Kingdom

[77] University College London, Department of Physics and Astronomy, Gower Street, London WC1E 6BT, United Kingdom

[78] Laboratoire de Physique Nucléaire et de Hautes Energies, Université Pierre et Marie Curie (Paris 6), Université Denis Diderot (Paris-7), CNRS/IN2P3, Tour 33, 4 place Jussieu, FR-75252 Paris Cedex 05, France

[79] Lunds universitet, Naturvetenskapliga fakulteten, Fysiska institutionen, Box 118, SE-221 00 Lund, Sweden

[80] Universidad Autonoma de Madrid, Facultad de Ciencias, Departamento de Fisica Teorica, ES-28049 Madrid, Spain

[81] Universitaet Mainz, Institut fuer Physik, Staudinger Weg 7, DE-55099 Mainz, Germany

[82] University of Manchester, School of Physics and Astronomy, Manchester M13 9PL, United Kingdom

[83] CPPM, Aix-Marseille Université, CNRS/IN2P3, Marseille, France

[84] University of Massachusetts, Department of Physics, 710 North Pleasant Street, Amherst, MA 01003, United States of America

[85] McGill University, High Energy Physics Group, 3600 University Street, Montreal, Quebec H3A 2T8, Canada

[86] University of Melbourne, School of Physics, AU-Parkville, Victoria 3010, Australia

[87] The University of Michigan, Department of Physics, 2477 Randall Laboratory, 500 East University, Ann Arbor, MI 48109-1120, United States of America

[88] Michigan State University, Department of Physics and Astronomy, High Energy Physics Group, East Lansing, MI 48824-2320, United States of America

[89] INFN Sezione di Milano[a]; Università di Milano, Dipartimento di Fisica[b], via Celoria 16, IT-20133 Milano, Italy

[90] B.I. Stepanov Institute of Physics, National Academy of Sciences of Belarus, Independence Avenue 68, Minsk 220072, Republic of Belarus

[91] National Scientific & Educational Centre for Particle & High Energy Physics, NC PHEP BSU, M. Bogdanovich St. 153, Minsk 220040, Republic of Belarus

[92] Massachusetts Institute of Technology, Department of Physics, Room 24-516, Cambridge, MA 02139, United States of America

[93] University of Montreal, Group of Particle Physics, C.P. 6128, Succursale Centre-Ville, Montreal, Quebec, H3C 3J7 , Canada

[94] P.N. Lebedev Institute of Physics, Academy of Sciences, Leninsky pr. 53, RU-117 924 Moscow, Russia

[95] Institute for Theoretical and Experimental Physics (ITEP), B. Cheremushkinskaya ul. 25, RU 117 218 Moscow, Russia

[96] Moscow Engineering & Physics Institute (MEPhI), Kashirskoe Shosse 31, RU-115409 Moscow, Russia

[97] Lomonosov Moscow State University Skobeltsyn Institute of Nuclear Physics (MSU SINP), 1(2), Leninskie gory, GSP-1, Moscow 119991 Russian Federation, Russia

[98] Ludwig-Maximilians-Universität München, Fakultät für Physik, Am Coulombwall 1, DE-85748 Garching, Germany

[99] Max-Planck-Institut für Physik, (Werner-Heisenberg-Institut), Föhringer Ring 6, 80805 München, Germany

[100] Nagasaki Institute of Applied Science, 536 Aba-machi, JP Nagasaki 851-0193, Japan

[101] Nagoya University, Graduate School of Science, Furo-Cho, Chikusa-ku, Nagoya, 464-8602, Japan

[102] INFN Sezione di Napoli[a]; Università di Napoli, Dipartimento di Scienze Fisiche[b], Complesso Universitario di Monte Sant'Angelo, via Cinthia, IT-80126 Napoli, Italy

[103] University of New Mexico, Department of Physics and Astronomy, MSC07 4220, Albuquerque, NM 87131 USA, United States of America

[104] Radboud University Nijmegen/NIKHEF, Department of Experimental High Energy Physics, Toernooiveld 1, NL-6525 ED Nijmegen, Netherlands

[105] Nikhef National Institute for Subatomic Physics, and University of Amsterdam, Science Park 105, 1098 XG Amsterdam, Netherlands

[106] Budker Institute of Nuclear Physics (BINP), RU-Novosibirsk 630 090, Russia

[107] New York University, Department of Physics, 4 Washington Place, New York NY 10003, USA, United States of America

[108] Ohio State University, 191 West Woodruff Ave, Columbus, OH 43210-1117, United States of America

[109] Okayama University, Faculty of Science, Tsushimanaka 3-1-1, Okayama 700-8530, Japan

[110] University of Oklahoma, Homer L. Dodge Department of Physics and Astronomy, 440 West Brooks, Room 100, Norman, OK 73019-0225, United States of America

[111] Oklahoma State University, Department of Physics, 145 Physical Sciences Building, Stillwater, OK 74078-3072, United States of America

[112] Palacký University, 17.listopadu 50a, 772 07 Olomouc, Czech Republic

[113] University of Oregon, Center for High Energy Physics, Eugene, OR 97403-1274, United States of America

[114] LAL, Univ. Paris-Sud, IN2P3/CNRS, Orsay, France

[115] Osaka University, Graduate School of Science, Machikaneyama-machi 1-1, Toyonaka, Osaka 560-0043, Japan

[116] University of Oslo, Department of Physics, P.O. Box 1048, Blindern, NO-0316 Oslo 3, Norway

[117] Oxford University, Department of Physics, Denys Wilkinson Building, Keble Road, Oxford OX1 3RH, United Kingdom

[118] INFN Sezione di Pavia[a]; Università di Pavia, Dipartimento di Fisica Nucleare e Teorica[b], Via Bassi 6, IT-27100 Pavia, Italy

[119] University of Pennsylvania, Department of Physics, High Energy Physics Group, 209 S. 33rd Street, Philadelphia, PA 19104, United States of America

[120] Petersburg Nuclear Physics Institute, RU-188 300 Gatchina, Russia

[121] INFN Sezione di Pisa[a]; Università di Pisa, Dipartimento di Fisica E. Fermi[b], Largo B. Pontecorvo 3, IT-56127 Pisa, Italy

[122] University of Pittsburgh, Department of Physics and Astronomy, 3941 O'Hara Street, Pittsburgh, PA 15260, United States of America

[123] Universidad de Granada[a], Departamento de Fisica Teorica y del Cosmos and CAFPE, E-18071 Granada; Laboratorio de Instrumentacao e Fisica Experimental de Particulas-LIP[b], Avenida Elias Garcia 14-1, PT-1000-149 Lisboa, Portugal

[124] Institute of Physics, Academy of Sciences of the Czech Republic, Na Slovance 2, CZ-18221 Praha 8, Czech Republic

[125] Charles University in Prague, Faculty of Mathematics and Physics, Institute of Particle and Nuclear Physics, V Holesovickach 2, CZ-18000 Praha 8, Czech Republic

[126]Czech Technical University in Prague, Zikova 4, CZ-166 35 Praha 6, Czech Republic

[127]State Research Center Institute for High Energy Physics, Moscow Region, 142281, Protvino, Pobeda street, 1, Russia

[128]Rutherford Appleton Laboratory, Science and Technology Facilities Council, Harwell Science and Innovation Campus, Didcot OX11 0QX, United Kingdom

[129]University of Regina, Physics Department, Canada

[130]Ritsumeikan University, Noji Higashi 1 chome 1-1, JP-Kusatsu, Shiga 525-8577, Japan

[131]INFN Sezione di Roma I[(a)]; Università La Sapienza, Dipartimento di Fisica[(b)], Piazzale A. Moro 2, IT-00185 Roma, Italy

[132]INFN Sezione di Roma Tor Vergata[(a)]; Università di Roma Tor Vergata, Dipartimento di Fisica[(b)], via della Ricerca Scientifica, IT-00133 Roma, Italy

[133]INFN Sezione di Roma Tre[(a)]; Università Roma Tre, Dipartimento di Fisica[(b)], via della Vasca Navale 84, IT-00146 Roma, Italy

[134]Université Hassan II, Faculté des Sciences Ain Chock[(a)], B.P. 5366, MA-Casablanca; Centre National de l'Energie des Sciences Techniques Nucleaires (CNESTEN)[(b)], B.P. 1382 R.P. 10001 Rabat 10001; Université Mohamed Premier[(c)], LPTPM, Faculté des Sciences, B.P.717. Bd. Mohamed VI, 60000, Oujda ; Université Mohammed V, Faculté des Sciences[(d)], LPNR, BP 1014, 10000 Rabat, Morocco

[135]CEA, DSM/IRFU, Centre d'Etudes de Saclay, FR-91191 Gif-sur-Yvette, France

[136]University of California Santa Cruz, Santa Cruz Institute for Particle Physics (SCIPP), Santa Cruz, CA 95064, United States of America

[137]University of Washington, Seattle, Department of Physics, Box 351560, Seattle, WA 98195-1560, United States of America

[138]University of Sheffield, Department of Physics & Astronomy, Hounsfield Road, Sheffield S3 7RH, United Kingdom

[139]Shinshu University, Department of Physics, Faculty of Science, 3-1-1 Asahi, Matsumoto-shi, JP-Nagano 390-8621, Japan

[140]Universitaet Siegen, Fachbereich Physik, D 57068 Siegen, Germany

[141]Simon Fraser University, Department of Physics, 8888 University Drive, CA-Burnaby, BC V5A 1S6, Canada

[142]SLAC National Accelerator Laboratory, Stanford, California 94309, United States of America

[143]Comenius University, Faculty of Mathematics, Physics & Informatics, Mlynska dolina F2, SK-84248 Bratislava; Institute of Experimental Physics of the Slovak Academy of Sciences, Dept. of Subnuclear Physics, Watsonova 47, SK-04353 Kosice, Slovak Republic

[144]Stockholm University, Department of Physics, AlbaNova, SE-106 91 Stockholm, Sweden

[145]Royal Institute of Technology (KTH), Physics Department, SE-106 91 Stockholm, Sweden

[146]Stony Brook University, Department of Physics and Astronomy, Nicolls Road, Stony Brook, NY 11794-3800, United States of America

[147]University of Sussex, Department of Physics and Astronom

[148]Pevensey 2 Building, Falmer, Brighton BN1 9QH, United Kingdom University of Sydney, School of Physics, AU-Sydney NSW 2006, Australia

[149]Insitute of Physics, Academia Sinica, TW-Taipei 11529, Taiwan

[150]Technion, Israel Inst. of Technology, Department of Physics, Technion City, IL-Haifa 32000, Israel

[151]Tel Aviv University, Raymond and Beverly Sackler School of Physics and Astronomy, Ramat Aviv, IL-Tel Aviv 69978, Israel

[152]Aristotle University of Thessaloniki, Faculty of Science, Department of Physics, Division of Nuclear & Particle Physics, University Campus, GR-54124, Thessaloniki, Greece

[153]The University of Tokyo, International Center for Elementary Particle Physics and Department of Physics, 7-3-1 Hongo, Bunkyo-ku, JP-Tokyo 113-0033, Japan

[154]Tokyo Metropolitan University, Graduate School of Science and Technology, 1-1 Minami-Osawa, Hachioji, Tokyo 192-0397, Japan

[155]Tokyo Institute of Technology, 2-12-1-H-34 O-Okayama, Meguro, Tokyo 152-8551, Japan

[156]University of Toronto, Department of Physics, 60 Saint George Street, Toronto M5S 1A7, Ontario, Canada

[157]TRIUMF[(a)], 4004 Wesbrook Mall, Vancouver, B.C. V6T 2A3; [(b)]York University, Department of Physics and Astronomy, 4700 Keele St., Toronto, Ontario, M3J 1P3, Canada

[158]University of Tsukuba, Institute of Pure and Applied Sciences, 1-1-1 Tennoudai, Tsukuba-shi, JP-Ibaraki 305-8571, Japan

[159]Tufts University, Science & Technology Center, 4 Colby Street, Medford, MA 02155, United States of America

[160]Universidad Antonio Narino, Centro de Investigaciones, Cra 3 Este No.47A-15, Bogota, Colombia

[161]University of California, Irvine, Department of Physics & Astronomy, CA 92697-4575, United States of America

[162]INFN Gruppo Collegato di Udine[(a)]; ICTP[(b)], Strada Costiera 11, IT-34014, Trieste; Università di Udine, Dipartimento di Fisica[(c)], via delle Scienze 208, IT-33100 Udine, Italy

[163]University of Illinois, Department of Physics, 1110 West Green Street, Urbana, Illinois 61801, United States of America

[164]University of Uppsala, Department of Physics and Astronomy, P.O. Box 516, SE -751 20 Uppsala, Sweden

[165]Instituto de Física Corpuscular (IFIC) Centro Mixto UVEG-CSIC, Apdo. 22085 ES-46071 Valencia, Dept. Física At. Mol. y Nuclear; Univ. of Valencia, and Instituto de Microelectrónica de Barcelona (IMB-CNM-CSIC) 08193 Bellaterra Barcelona, Spain

[166]University of British Columbia, Department of Physics, 6224 Agricultural Road, CA-Vancouver, B.C. V6T 1Z1, Canada

[167]University of Victoria, Department of Physics and Astronomy, P.O. Box 3055, Victoria B.C., V8W 3P6, Canada

[168]Waseda University, WISE, 3-4-1 Okubo, Shinjuku-ku, Tokyo, 169-8555, Japan

[169]The Weizmann Institute of Science, Department of Particle Physics, P.O. Box 26, IL-76100 Rehovot, Israel

[170]University of Wisconsin, Department of Physics, 1150 University Avenue, WI 53706 Madison, Wisconsin, United States of America

[171]Julius-Maximilians-University of Würzburg, Physikalisches Institut, Am Hubland, 97074 Wuerzburg, Germany

[172]Bergische Universitaet, Fachbereich C, Physik, Postfach 100127, Gauss-Strasse 20, D- 42097 Wuppertal, Germany

[173]Yale University, Department of Physics, P.O. Box 208121, New Haven CT, 06520-8121, United States of America

[174]Yerevan Physics Institute, Alikhanian Brothers Street 2, AM-375036 Yerevan, Armenia

[175]ATLAS-Canada Tier-1 Data Centre 4004 Wesbrook Mall, Vancouver, BC, V6T 2A3, Canada

[176]GridKA Tier-1 FZK, Forschungszentrum Karlsruhe GmbH, Steinbuch Centre for Computing (SCC), Hermann-von-Helmholtz-Platz 1, 76344 Eggenstein-Leopoldshafen, Germany

[177]Port d'Informacio Cientifica (PIC), Universitat Autonoma de Barcelona (UAB), Edifici D, E-08193 Bellaterra, Spain

[178]Centre de Calcul CNRS/IN2P3, Domaine scientifique de la Doua, 27 bd du 11 Novembre 1918, 69622 Villeurbanne Cedex, France

[179] INFN-CNAF, Viale Berti Pichat 6/2, 40127 Bologna, Italy

[180] Nordic Data Grid Facility, NORDUnet A/S, Kastruplundgade 22, 1, DK-2770 Kastrup, Denmark

[181] SARA Reken- en Netwerkdiensten, Science Park 121, 1098 XG Amsterdam, Netherlands

[182] Academia Sinica Grid Computing, Institute of Physics, Academia Sinica, No. 128, Sect. 2, Academia Rd., Nankang, Taipei, Taiwan 11529, Taiwan

[183] UK-T1-RAL Tier-1, Rutherford Appleton Laboratory, Science and Technology Facilities Council, Harwell Science and Innovation Campus, Didcot OX11 0QX, United Kingdom

[184] RHIC and ATLAS Computing Facility, Physics Department, Building 510, Brookhaven National Laboratory, Upton, New York 11973, United States of America

[a] Also at CPPM, Marseille.

[b] Also at TRIUMF, 4004 Wesbrook Mall, Vancouver, B.C. V6T 2A3, Canada.

[c] Also at Gaziantep University, Turkey.

[d] Also at Faculty of Physics and Applied Computer Science of the AGH-University of Science and Technology, (FPACS, AGH-UST), al. Mickiewicza 30, PL-30059 Cracow, Poland.

[e] Also at Institute for Particle Phenomenology, Ogden Centre for Fundamental Physics, Department of Physics, University of Durham, Science Laboratories, South Rd, Durham DH1 3LE, United Kingdom.

[f] Currently at Dogus University, Kadik.

[g] Also at Università di Napoli Parthenope, via A. Acton 38, IT-80133 Napoli, Italy.

[h] Also at Institute of Particle Physics (IPP), Canada.

[i] Louisiana Tech University, 305 Wisteria Street, P.O. Box 3178, Ruston, LA 71272, United States of America.

[j] Currently at Dumlupinar University, Kutahya, Turkey.

[k] Currently at Department of Physics, University of Helsinki, P.O. Box 64, FI-00014, Finland.

[l] At Department of Physics, California State University, Fresno, 2345 E. San Ramon Avenue, Fresno, CA 93740-8031, United States of America.

[m] Also at TRIUMF, 4004 Wesbrook Mall, Vancouver, B.C. V6T 2A3, Canada.

[n] Currently at Istituto Universitario di Studi Superiori IUSS, V.le Lungo Ticino Sforza 56, 27100 Pavia, Italy.

[o] Also at California Institute of Technology, Physics Department, Pasadena, CA 91125, United States of America.

[p] Also at University of Montreal.

[q] Also at Institut für Experimentalphysik, Universität Hamburg, Luruper Chaussee 149, 22761 Hamburg, Germany.

[r] Also at Petersburg Nuclear Physics Institute, RU-188 300 Gatchina, Russia.

[s] Also at school of physics and engineering, Sun Yat-sen University.

[t] Also at school of physics, Shandong university, Jinan.

[u] Also at Rutherford Appleton Laboratory, Science and Technology Facilities Council, Harwell Science and Innovation Campus, Didcot OX11 0QX.

[v] Also at Rutherford Appleton Laboratory, Science and Technology Facilities Council, Harwell Science and Innovation Campus, Didcot OX11 0QX, United Kingdom.

[w] University of South Carolina, Dept. of Physics and Astronomy, 700 S. Main St, Columbia, SC 29208, United States of America.

[x] Also at KFKI Research Institute for Particle and Nuclear Physics, Budapest, Hungary.

[y] Also at Institute of Physics, Jagiellonian University, Cracow, Poland.

[z] Also at school of physics and engineering, Sun Yat-sen University.

[aa] Currently at TOBB University, Ankara, Turkey.

[ab] Currently at TAEA, Ankara, Turkey.

[ac] Currently at Gazi University, Ankara, Turkey.

[ad] Also at Dept of Physics, Nanjing University.

[*] Deceased.

Abstract The ATLAS liquid argon calorimeter has been operating continuously since August 2006. At this time, only part of the calorimeter was readout, but since the beginning of 2008, all calorimeter cells have been connected to the ATLAS readout system in preparation for LHC collisions. This paper gives an overview of the liquid argon calorimeter performance measured in situ with random triggers, calibration data, cosmic muons, and LHC beam splash events. Results on the detector operation, timing performance, electronics noise, and gain stability are presented. High energy deposits from radiative cosmic muons and beam splash events allow to check the intrinsic constant term of the energy resolution. The uniformity of the electromagnetic barrel calorimeter response along η (averaged over ϕ) is measured at the percent level using minimum ionizing cosmic muons. Finally, studies of electromagnetic showers from radiative muons have been used to cross-check the Monte Carlo simulation. The performance results obtained using the ATLAS readout, data acquisition, and reconstruction software indicate that the liquid argon calorimeter is well-prepared for collisions at the dawn of the LHC era.

[**] e-mail: atlas.secretariat@cern.ch

1 Introduction

Installation of the liquid argon (LAr) calorimeter in the AT-LAS [1] experimental hall was completed in early 2008. Until recently, the expected performance of the LAr calorimeter was extrapolated from intensive testing of a few modules with electron and pion beams from 1998 to 2003 [2–10], and in 2004 of a complete ATLAS detector slice [11–13]. The 20 months separating the completion of the installation from the first LHC collisions have been used to commission the LAr calorimeter. This paper reviews the first in situ measurements of the electronics stability, the quality of the energy reconstruction, the calorimeter response uniformity and the agreement between data and the Monte Carlo simulation of electromagnetic shower shapes. The measurements are performed using calibration triggers, cosmic muons, and the first LHC beam events collected during this 20 months period. The results and the experience gained in the operation of the LAr calorimeter provide the foundation for a more rapid understanding of the experimental signatures of the first LHC collisions, involving electrons, photons, missing transverse energy (E_T^{miss}), jets, and τs where the LAr calorimeter plays a central role.

This paper is organized as follows. Section 2 gives the present hardware status of the LAr calorimeter. Section 3 details the level of understanding of the ingredients entering the cell energy reconstruction: pedestals, noise, electronic gains, timing, and the quality of the signal pulse shape predictions. The current understanding of the first level trigger energy computation is also discussed. Section 4 describes the in situ performance of the electromagnetic LAr calorimeter using ionizing and radiating cosmic muons. Lastly, Sect. 5 draws the conclusions.

2 LAr calorimeter hardware status and data taking conditions

The LAr calorimeter is composed of electromagnetic and hadronic sub-detectors of which the main characteristics are described in Sect. 2.1. During the detector and electronics construction and installation, regular and stringent quality tests were performed, resulting in a fully functional LAr calorimeter. The operational stability of the cryostats since March 2008 is discussed in Sect. 2.2. The current status of the high voltage and the cell readout are discussed in Sects. 2.3 and 2.4 respectively. Finally, the general data taking conditions are given in Sect. 2.5. In ATLAS, the positive x-axis is defined as pointing from the interaction point to the center of the LHC ring, the positive y-axis is defined as pointing upwards, and the positive z-axis corresponds to protons running anti-clockwise. The polar angle θ is measured from the beam axis (z-axis), the azimuthal angle ϕ is measured in the transverse (xy)-plane, and the pseudorapidity is defined as $\eta = -\ln\tan(\theta/2)$.

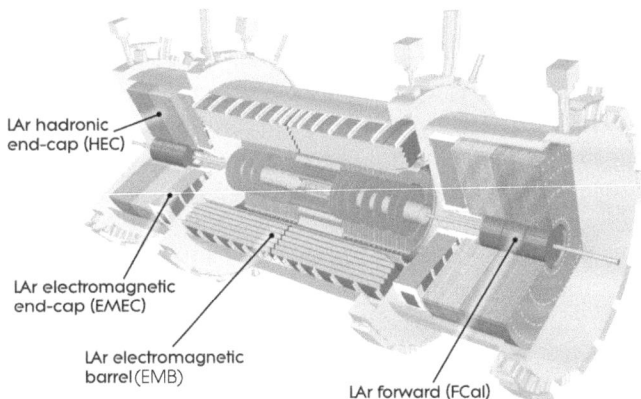

Fig. 1 Cut-away view of the LAr calorimeter, 17 m long (barrel + endcaps) and 4 m of diameter

2.1 Main characteristics of the LAr calorimeter

The LAr calorimeter [1], shown in Fig. 1, is composed of sampling detectors with full azimuthal symmetry, housed in one barrel and two endcap cryostats. More specifically, a highly granular electromagnetic (EM) calorimeter with accordion-shaped electrodes and lead absorbers in liquid argon covers the pseudorapidity range $|\eta| < 3.2$, and contains a barrel part (EMB [14], $|\eta| < 1.475$) and an endcap part (EMEC [15], $1.375 < |\eta| < 3.2$). For $|\eta| < 1.8$, a presampler (PS [15, 16]), consisting of an active LAr layer and installed directly in front of the EM calorimeters, provides a measurement of the energy lost upstream. Located behind the EMEC is a copper-liquid argon hadronic endcap calorimeter (HEC [17], $1.5 < |\eta| < 3.2$), and a copper/tungsten-liquid argon forward calorimeter (FCal [18]) covers the region closest to the beam at $3.1 < |\eta| < 4.9$. An hadronic Tile calorimeter ($|\eta| < 1.7$) surrounding the LAr cryostats completes the ATLAS calorimetry.

All the LAr detectors are segmented transversally and divided in three or four layers in depth, and correspond to a total of 182,468 readout cells, i.e. 97.2% of the full ATLAS calorimeter readout.

The relative energy resolution of the LAr calorimeter is usually parameterized by:

$$\frac{\sigma_E}{E} = \frac{a}{\sqrt{E}} \oplus \frac{b}{E} \oplus c, \tag{1}$$

where (a) is the stochastic term, (b) the noise term and (c) the constant term. The target values for these terms are respectively $a \simeq 10\%$, $b \simeq 170$ MeV (without pile-up) and $c = 0.7\%$.

2.2 Cryostat operation

Variations of the liquid argon temperature have a direct impact on the readout signal, and consequently on the energy scale, partly through the effect on the argon density, but mostly through the effect on the ionization electron drift velocity in the LAr. Overall, a $-2\%/K$ signal variation is expected [19]. The need to keep the corresponding contribution to the constant term of the energy resolution (1) negligible (i.e. well below 0.2%) imposes a temperature uniformity requirement of better than 100 mK in each cryostat. In the liquid, ~500 temperature probes (PT100 platinum resistors) are fixed on the LAr detector components and read out every minute. In 2008–2009, installation activities in the ATLAS cavern prevented a stable cryostat temperature. A quiet period of ten days around the 2008 Christmas break, representative of what is expected during LHC collisions, allowed a check of the temperature stability in the absence of these external factors. The average dispersion (RMS) of the measurements of each temperature probe over this period is 1.6 mK (5 mK maximum), showing that no significant local temperature variation in time is observed in the three cryostats. Over this period, the temperature uniformity (RMS of all probes per cyostat) is illustrated for the barrel in Fig. 2 and gives 59 mK. Results for the two endcap cryostats are also in the range 50–70 mK, below the required level of 100 mK. The average cryostat temperatures are slightly different for the barrel (88.49 K) and the two endcaps (88.67 and 88.45 K) because they are independently regulated. An energy scale correction per cryostat will therefore be applied.

To measure the effects of possible out-gassing of calorimeter materials under irradiation, which has been minimized by careful screening of components, 30 purity monitors measuring the energy deposition of radioactive sources in the LAr are installed in each cryostat and read every 15 minutes. The contribution to the constant term of the energy resolution is negligible for a level of electronegative impurities below 1000 ppb O_2 equivalent. All argon purity

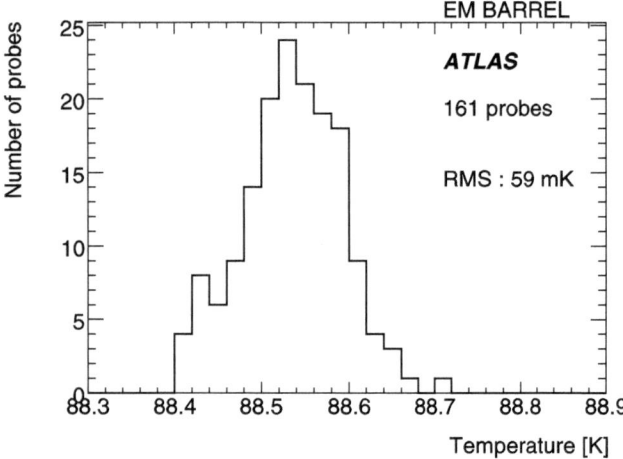

Fig. 2 Distribution of barrel cryostat probe temperatures averaged over a period of ten days

measurements over a period of two years are stable, in the range 200 ± 100 ppb O_2 equivalent, well below this requirement.

In summary, measurements of the liquid argon temperature and purity demonstrate that the stability of the operation of the three LAr cryostats is in the absence of proton beams within the required limits ensuring a negligible contribution to the energy resolution constant term.

2.3 High voltage status

The electron/ion drift speed in the LAr gap depends on the electric field, typically 1 kV/mm. Sub-detector-specific high voltage (HV) settings are applied. In the EM barrel, the high voltage is constant along η, while in the EMEC, where the gap varies continuously with radius, it is adjusted in steps along η. The HV supply granularity is typically in sectors of $\Delta\eta \times \Delta\phi = 0.2 \times 0.2$. For redundancy, each side of an EM electrode, which is in the middle of the LAr gap, is powered separately. In the HEC, each sub-gap is serviced by one of four different HV lines, while for the FCal each of the four electrode groups forming a normal readout channel is served by an independent HV line.

For HV sectors with non-optimal behavior, solutions were implemented in order to recover the corresponding region. For example, in the EM calorimeter, faulty electrodes were connected to separate HV lines during the assembly phase at room temperature while, if the defect was identified during cryostat cold testing, the high voltage sector was divided into two in ϕ, each connected separately. The effect of zero voltage on one side of an electrode was studied in beam tests proving that with offline corrections the energy can still be measured, with only a small loss in accuracy. Finally, for HV sectors with a permanent short-circuit, high voltage modules permitting large DC current draws of up to 3 mA (more than three orders of magnitude above the nominal limit) are used in order to operate the faulty sector at 1000 V or above.

As a result, 93.9% of readout cells are operating under nominal conditions and the rest sees a reduced high voltage. However, even with a reduced high voltage, signals can be well reconstructed by using a correction scale factor. Figure 3 shows the distribution of all HV correction factors for the EM, HEC and FCal cells as of the end of September 2009. Since the beginning of 2008, no changes have been observed. The largest correction occurs if one side of an EM electrode is not powered, and only half of the signal is collected. For the faulty cells, this correction factor is applied online at the energy reconstruction level. A similar correction is currently being implemented at the first level (L1) trigger.

In conclusion, since the beginning of 2008, all 182,468 readout cells are powered with high voltage, and no dead

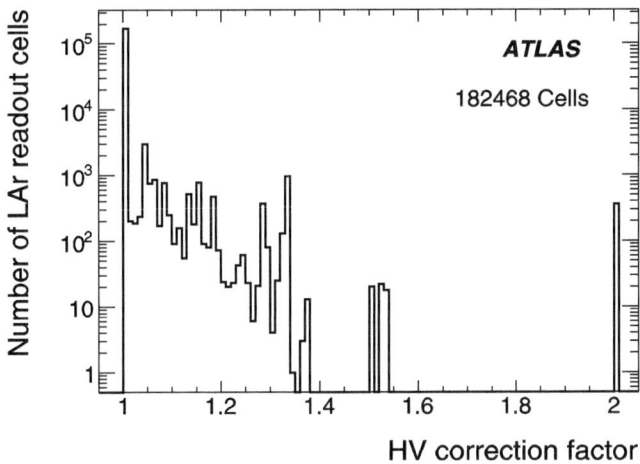

Fig. 3 High voltage correction factors for all LAr cells at the end of September 2009

region exists. Signals from regions with non-nominal high voltage are easily corrected and their impact on physics is negligible.

2.4 Readout cell status

The cell signals are read out through 1524 Front-End Boards (FEBs [20, 21]) with 128 channels each, which sit inside front-end crates that are located around the periphery of the cryostats. The FEBs perform analog processing (amplification and shaping—except for the HEC where the amplification is done inside the cryostat), store the signal while waiting for the L1 trigger decision, and digitize the accepted signals. The FEBs also perform fast analog summing of cell signals in predefined projective "towers" for the L1 trigger.

The digitized signals are transmitted via optical fibers to the Readout Drivers (RODs) [22] located in the counting room 70 m away. The cell energy is reconstructed online in the ROD modules up to a nominal maximum L1 rate of 75 kHz. The cell and trigger tower energy reconstruction is described in detail in Sect. 3.

The response of the 182,468 readout cells is regularly monitored using 122 calibration boards [23] located in the front-end crates. These boards inject calibrated current pulses through high-precision resistors to simulate energy deposits in the calorimeters. At the end of September 2009, 1.3% of cells have problems. The majority of them, i.e. 1.2% of the total number of cells, are not read-out because they are connected to 17 non-functioning FEBs. On these FEBs, the active part (VCSEL) of the optical transmitter to the ROD has failed. This failure, occurring at a rate of two or three devices per month, is under intensive investigation and are expected to be fixed during the next LHC shutdown. The remaining 0.1% of cells with problems can be split in three sub-types: incurable cells, i.e. cells not responding to the input pulse (0.02%), or which are permanently (0.03%) or sporadically (0.07%) very noisy. The first two types are always masked in the event reconstruction (121 cells), while the sporadically very noisy cells, not yet well understood, are masked on an event by event basis. For cells which do not receive calibration signals (0.3%) average calibration constants computed among neighboring cells are used. For cells with non-nominal high voltage (6.1%) a software correction factor is applied. Both have very limited impact on the energy reconstruction.

In total, 180,128 cells, representing 98.7% of the total number of cells in the LAr calorimeter, are used for event reconstruction at the end of September 2009. The number of inactive cells (1.3%) is dominated by the cells lost due to faulty optical drivers (1.2%): apart from these, the number of inactive cells has been stable in time.

2.5 Data taking conditions

The results presented here focus on the period starting in September 2008 when all the ATLAS sub-detectors were completed and integrated into the data acquisition. Apart from regular electronics calibration runs, two interesting types of data are used to commission the LAr calorimeter: the beam splash events and the cosmic muons. The first type corresponds to LHC events of September 10th 2008 when the first LHC beam hit the collimators located 200 m upstream of the ATLAS interaction point. A cascade of pions and muons parallel to the beam axis fired the beam related trigger, illuminated the whole ATLAS detector and deposited several PeV per event in the LAr calorimeter. The second type corresponds to long cosmic muon runs acquired on September–October 2008 and on June–July 2009 where more than 300 million events were recorded, corresponding to more than 500 TB of data.

For the LAr commissioning, L1 calorimeter triggers are used to record radiative energy losses from cosmic muons while the first level muon spectrometer and second level inner detector triggers are used to study pseudo-projective minimum ionizing muons. In most of the runs analyzed, the toroidal and solenoidal magnetic fields were at the nominal value.

3 Electronic performance and quality of cell energy reconstruction

The robustness of the LAr calorimeter energy reconstruction has been studied in detail using calibration and randomly triggered events, cosmic muons and beam splash events. Section 3.1 briefly describes the energy reconstruction method in the trigger towers and in the cells, as well as a validation study of the trigger. The time stability of the electronics is discussed in Sect. 3.2. The status of the electronics timing for the first LHC collisions is presented in Sect. 3.3, and the quality of the LAr calorimeter energy reconstruction is assessed in Sect. 3.4.

3.1 Energy reconstruction in the LAr calorimeter

When charged particles cross the LAr gap between electrodes and absorbers, they ionize the liquid argon. Under the influence of the electric field, the ionization electrons drift towards the electrode inducing a current. The initial current is proportional to the energy deposited in the liquid argon. The calorimeter signals are then used to compute the energy per trigger tower or per cell as discussed in this section.

3.1.1 Energy reconstruction at the first level calorimeter trigger

The timing requirements for the L1 trigger latency can only be met with fast analogue summing in coarse granularity. In the EM part, the pre-summation of analog signals per layer on the FEBs serves as input to tower builder boards where the final trigger tower signal sum and shaping is performed. In the HEC and FCal, the summation is performed on the FEBs and transmitted to tower driver boards where only shaping is done. The tower sizes are $\Delta\eta \times \Delta\phi = 0.1 \times 0.1$ for $|\eta| < 2.5$ and go up to $\Delta\eta \times \Delta\phi = 0.4 \times 0.4$ for $3.1 < |\eta| < 4.9$. The analog trigger sum signals are sent to receiver modules in the service cavern. The main function of these modules is to compensate for the differences in energy calibration and signal attenuation over the long cables using programmable amplifier gains (g_R). The outputs are sent to L1 trigger pre-processor boards which perform the sampling at 40 MHz and the digitization of five samples. At this stage, both the transverse energy and bunch crossing are determined using a finite impulse response filter, in order to maximize the signal-to-noise ratio and bunch crossing identification efficiency. During ATLAS operation, the output $g_R A^{L1}$ of the filter, which uses optimal filtering, is passed to a look-up table where pedestal (P in ADC counts) subtraction, noise suppression and the conversion from ADC counts to transverse energy in GeV ($F_{ADC \to GeV}^{L1}$) is performed in order to extract the final transverse energy value (E_T^{L1}) for each trigger tower:

$$E_T^{L1} = F_{ADC \to GeV}^{L1}\left(g_R A^{L1} - g_R P\right). \tag{2}$$

Arrays (in $\eta - \phi$) of these E_T^{L1} energies, merged with similar information coming from the Tile calorimeter, are subsequently used to trigger on electrons, photons, jets, τs and events with large missing transverse energy.

3.1.2 Energy reconstruction at cell level

At the cell level, the treatment of the analog signal is also performed in the front-end electronics. After shaping, the signal is sampled at 40 MHz and digitized if the event was selected by the L1 trigger. The reconstruction of the cell energy, performed in the ROD, is based on an optimal filtering algorithm applied to the samples s_j [24]. The amplitude A, in ADC counts, is computed as:

$$A = \sum_{j=1}^{N_{samples}} a_j(s_j - p), \tag{3}$$

where p is the ADC pedestal (Sect. 3.2.1). The Optimal Filtering Coefficients (OFCs) a_j are computed per cell from the predicted ionization pulse shape and the measured noise autocorrelation to minimize the noise and pile-up contributions to A. For cells with sufficient signal, the difference (Δt in ns) between the digitization time and the chosen phase is obtained from:

$$\Delta t = \frac{1}{A} \sum_{j=1}^{N_{samples}} b_j(s_j - p), \tag{4}$$

where b_j are time-OFCs. For a perfectly timed detector and in-time particles $|\Delta t|$ must be close to zero, while larger values indicate the need for better timing or the presence of out-of-time particles in the event.

The default number of samples used for A and Δt computation is $N_{samples} = 5$, but for some specific analyses more samples, up to a maximum of 32, are recorded. Finally, including the relevant electronic calibration constants, the deposited energy (in MeV) is extracted with:

$$E_{cell} = F_{\mu A \to MeV} \times F_{DAC \to \mu A} \times \frac{1}{\frac{Mphys}{Mcali}} \times G \times A, \tag{5}$$

where the various constants are linked to the calibration system: the cell gain G (to cover energies ranging from a maximum of 3 TeV down to noise level, three linear gains are used: low, medium and high with ratios $\sim 1/10/100$) is computed by injecting a known calibration signal and reconstructing the corresponding cell response; the factor $1/\frac{Mphys}{Mcali}$ quantifies the ratio of response to a calibration pulse and an ionization pulse corresponding to the same input current; the factor $F_{DAC \to \mu A}$ converts digital-to-analog converter (DAC) counts set on the calibration board to μA; finally, the factor $F_{\mu A \to MeV}$ is estimated from simulations and beam test results, and includes high voltage corrections for non-nominal settings (see Sect. 2.3). Note that the crosstalk bias in the finely segmented first layer of the electromagnetic calorimeter is corrected for in the gain G [4].

3.1.3 Check of the first level tower trigger energy computation

The trigger decision is of utmost importance for ATLAS during LHC collisions since the data-taking rate is at maximum 200 Hz because of bandwidth limitations, i.e. a factor

2×10^5 smaller than the 40 MHz LHC clock. It is therefore important to check that no systematic bias is introduced in the computation of the L1 trigger energy and that the trigger energy resolution is not too degraded with respect to the offline reconstruction. In the following, this check is performed with the most granular part of the LAr calorimeter, the barrel part of the EM calorimeter, where 60 cell signals are summed per trigger tower.

Since cosmic muon events occur asynchronously with respect to the LHC clock, and the electronics for both the trigger and the standard readout is loaded with one set of filtering coefficients (corresponding to beam crossing), the reconstructed energy is biased by up to 10%, depending on the phase. For the study presented here, A^{L1} is recomputed offline by fitting a second-order polynomial to the three highest samples transmitted through the processors. The most critical part in the trigger energy computation is then to calibrate the individual receiver gains g_R. For that purpose, a common linearly increasing calibration pulse is sent to both the L1 trigger and the normal cell circuits: the inverse receiver gain $1/g_R$ is obtained by fitting the correlation between the L1 calorimeter transverse energy (E_T^{L1}) and the sum of cell transverse energies in the same trigger tower, later called *offline* trigger tower (E_T^{LAr}). In cosmic muon runs, receiver gains are set to 1.0 and are recomputed offline with dedicated calibration runs. As a cross check, the gain was also extracted using LHC beam splash event data which covers the full detector. In both cases, the L1 transverse energy is computed as in (2).

In the EM calorimeter, radiating cosmic muons may produce a local energy deposit of a few GeV, and fire the EM calorimeter trigger condition EM3 that requires a transverse energy greater than 3 GeV in a sum of four adjacent EM trigger towers. To mimic an electron coming from the interaction point, only those events that contain a track reconstructed with strict projectivity cuts are considered. Here, the L1 calorimeter transverse energy is computed using the gains determined with calibration runs. Figure 4 shows the correlation between E_T^{L1} and E_T^{LAr}. Computing the ratio of E_T^{L1} and E_T^{LAr} gives a Gaussian distribution with a mean of 1.015 ± 0.002, showing the very good correspondence between these two quantities, especially at low energy. This also shows that the trigger energy is well calibrated and almost unbiased with respect to the LAr readout.

Figure 5 shows the corresponding resolution computed as the relative difference of E_T^{L1} and E_T^{LAr}. At low energy, the difference is dominated by electronic noise since the two readout paths have only part of their electronics in common. The ATLAS specification of 5% of L1 transverse energy resolution is reached for energies greater than 10 GeV. The L1 transverse energy resolution reaches around 3% at high energy.

As a crosscheck, a similar study was performed with gains computed from the beam splash events, without the

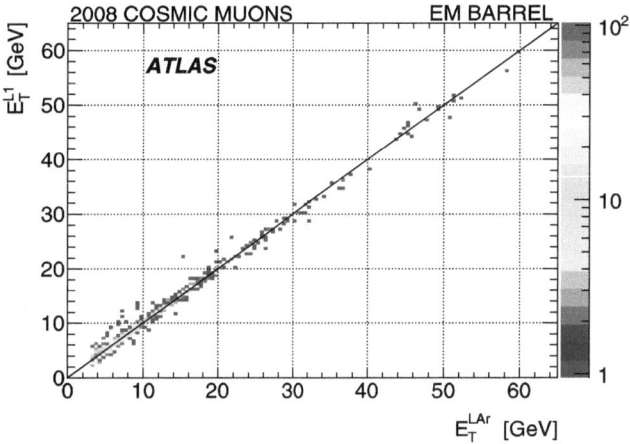

Fig. 4 L1 transverse energy (E_T^{L1}) computed with the receiver gains extracted from calibration runs versus the sum of cell transverse energies in the same trigger tower (E_T^{LAr})

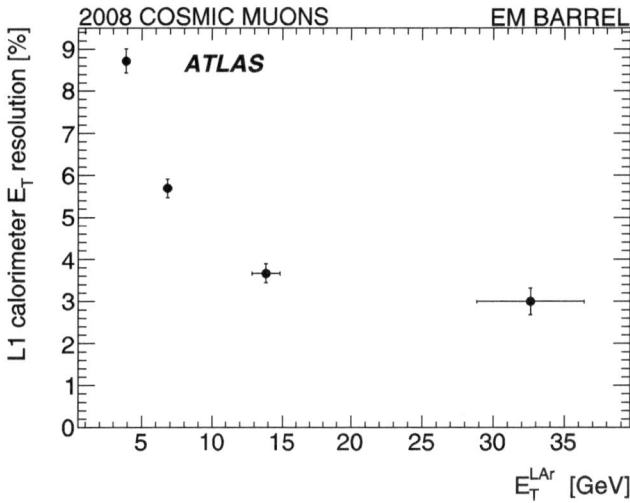

Fig. 5 Relative difference of E_T^{L1} and E_T^{LAr} (L1 Calorimeter E_T resolution) as a function of E_T^{LAr}. Strict projectivity cuts for the track pointing to the EM shower are applied. Horizontal error bars reflect the RMS of E_T^{LAr} in each bin

projectivity cut. A slight degradation of the resolution is observed at high energy, but not at low energy where the noise dominates. Taking advantage of the higher statistics, it is possible to compute the 5 GeV "turn-on curve", i.e. the relative efficiency for an offline trigger tower to meet the requirement $E_T^{L1} \geq 5$ GeV as a function of E_T^{LAr}. This is not the absolute efficiency as the calorimeter trigger condition EM3 is used to trigger the events. The efficiency is shown in Fig. 6, where a sharp variation around a $E_T^{L1} = 5$ GeV energy threshold is observed.

These results give confidence that EM showers (electrons and photons) will be triggered efficiently in LHC events. After this study, the gains g_R were extracted from dedicated calibration runs and loaded into the receivers to be used for the first LHC collisions.

3.2 Electronic stability

Hundreds of millions of randomly triggered and calibration events can be used for a study of the stability of the properties of each readout channel, such as the pedestal, noise and gain. The first two quantities are computed for each cell as the mean (pedestal) of the signal samples s_j in ADC counts, and the width (noise) of the energy distribution. The gain is extracted by fitting the output pulse amplitudes against calibration pulses with increasing amplitudes.

3.2.1 Pedestal

The stability of the pedestals is monitored by measuring variations with respect to a reference pedestal value for each cell. For each FEB, an average over the 128 channels is computed.

As an example, Fig. 7 illustrates the results for the 48 HEC FEBs over a period of six months in 2009. A slight drift

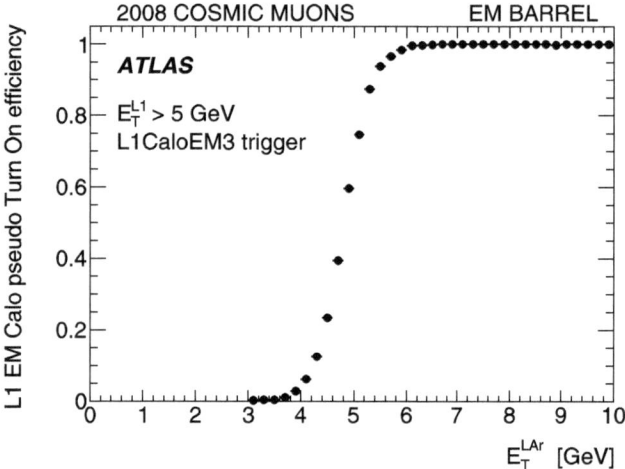

Fig. 6 Turn-on curve efficiency for $E_T^{L1} > 5$ GeV requirement obtained with events triggered by the EM3 L1 Calorimeter trigger

of the pedestal with time, uncorrelated with the FEB temperature and/or magnetic field configurations, is observed. Overall, the FEB pedestal variations follow a Gaussian distribution with a standard deviation of 0.02 ADC counts, i.e. below 2 MeV. The same checks have been performed on all other FEBs, and give typical variations of around 1 (0.1) MeV and 10 (1) MeV in the EM and FCal calorimeters respectively, in medium (high) gain. These variations are much lower for the EM and HEC or at the same level for the FCal than the numerical precision of the energy computation, which is 8 (1) MeV in medium (high) gain.

During the LHC running, it is foreseen to acquire pedestal and calibration runs between fills, thus it will be possible to correct for any small time dependence such as observed in Fig. 7. In the same spirit, random triggers collected during physics runs can be used to track any pedestal variations during an LHC fill.

3.2.2 Noise

Figure 8 shows the noise measured in randomly triggered events at the cell level as a function of η for all layers of the LAr calorimeters. In all layers, a good agreement with the expected noise [1] is observed. Noise values are symmetric with respect to $\eta = 0$ and uniformly in ϕ within few percents. In the EM calorimeters, the noise ranges from 10 to 50 MeV, while it is typically a factor of 10 greater in the hadronic endcap and forward calorimeters where the granularity is 20 times coarser and the sampling fractions are lower. It should be noted that these results are obtained using five samples in (3) and (5), i.e. the noise is reduced by a factor varying from 1.5 to 1.8, depending on η, with respect to the single-sample noise value.

The coherent noise over the many cells used to measure electron and photon energies in the EM calorimeters should be kept below 5% [25] of the incoherent noise (i.e. the quadratic sum of all channel noise). For the second layer of the EM calorimeter, the contribution from the coherent noise has been estimated to 2%, by studying simultaneous increase of noise in a group of channels.

Fig. 7 Average FEB pedestal variations in ADC counts, in medium gain, for the HEC during 6 months of data taking in 2009. The crosses indicate the mean value for each time slice

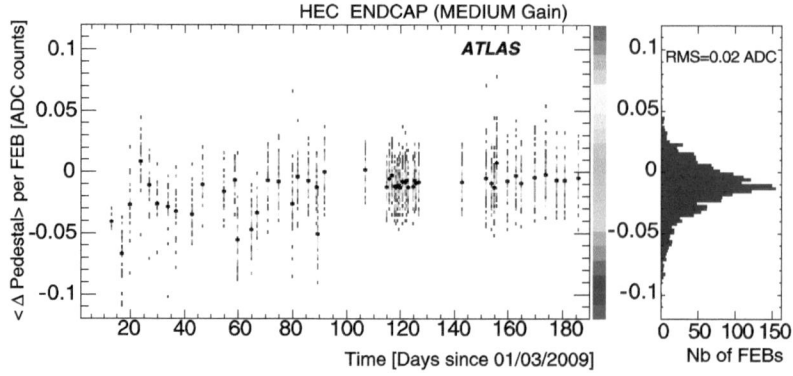

Systematic studies of noise stability have been pursued: all noise variations are typically within ± 1 keV, 0.1 MeV and 1 MeV for EM, HEC and FCal, respectively. No correlations with the FEB temperature and/or changes of magnetic field conditions have been observed.

3.2.3 Gain

The calibration pulse is an exponential signal (controlled by two parameters, f_{step} and τ_{cali}) which emulates the triangular ionization signal. It is injected on the detector as close as possible to the electrodes, except for the FCal where it is applied at the base-plane of the front-end crates [18]. Thus, the analog cell response is treated by the FEBs in the same way as an ionization signal, but it is typically averaged over 100 triggers in the RODs and transmitted offline where the average signal peak height is computed. The cell gain is extracted as the inverse ratio of the response signal in ADC counts to the injected calibration signal in DAC counts.

The stability of the cell gain is monitored by looking at the relative gain difference averaged over 128 FEB channels. This is illustrated in Fig. 9 for the 1448 FEBs of the EM calorimeter, in high gain. All variations are within $\pm 0.3\%$

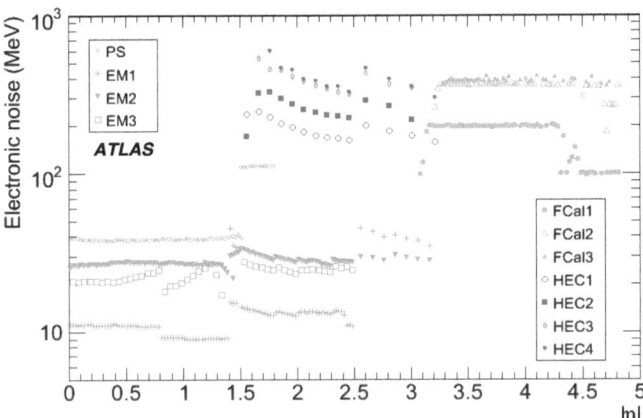

Fig. 8 Electronic noise (σ_{noise}) in randomly triggered events at the EM scale in individual cells for each layer of the calorimeter as a function of $|\eta|$. Results are averaged over ϕ

and similar results are obtained for medium and low gains. An effect of 0.2% on the gains has recently been identified as coming from a particular setting of the FEBs. The two populations are most probably coming from this effect. Regular update of calibration database take account of the variations. Similar results are obtained for the HEC, and variations within $\pm 0.1\%$ are measured for the FCal.

In conclusion, results presented for the pedestals, noise, and gains illustrate the stability of the LAr electronics over several months of data taking. Values are stored in the ATLAS calibration database and are used for online and offline reconstruction.

3.2.4 Global check with E_T^{miss} variable

Another way to investigate the level of understanding of pedestals and noise in the LAr calorimeter is to compute global quantities in randomly triggered events with the calorimeter, such as the vector sum of transverse cell energies. The calorimetric missing transverse energy E_T^{miss} is defined as:

$$E_x^{miss} = -\sum_{i=1}^{N_{cell}} E_i \sin\theta_i \cos\phi_i,$$

$$E_y^{miss} = -\sum_{i=1}^{N_{cell}} E_i \sin\theta_i \sin\phi_i, \qquad (6)$$

$$E_T^{miss} = \sqrt{\left(E_x^{miss}\right)^2 + \left(E_y^{miss}\right)^2},$$

where E_i is the cell energy, θ_i its polar angle and ϕ_i its azimuthal angle. Because of the high granularity of the LAr calorimeter, it is crucial to suppress noise contributions to E_T^{miss}, i.e. limit the number of cells, N_{cell}, used in the sum. In ATLAS, this is done with two methods: (i) a cell-based method in which only cells above a noise threshold of two standard deviations ($|E_i| > 2\sigma_{noise}$) are kept; (ii) a cluster-based method which uses only cells belonging to three-dimensional topological clusters [26]. These clusters are built around $|E_i| > 4\sigma_{noise}$ seeds by iteratively

Fig. 9 Average FEB (high) gain variations during 6 months of 2009 data taking, in the EM part of the calorimeter. The crosses indicate the mean value for each time slice

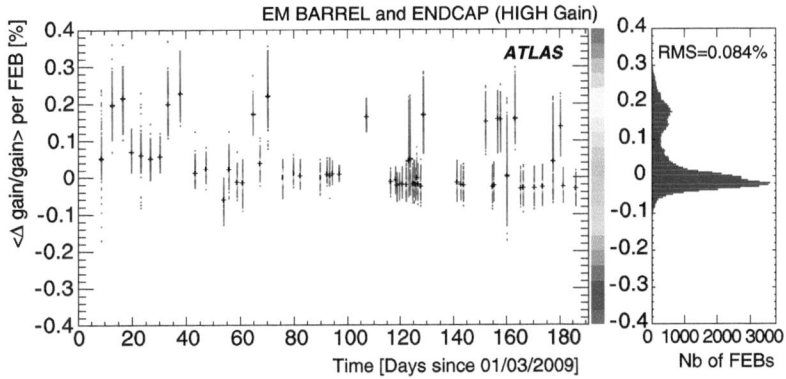

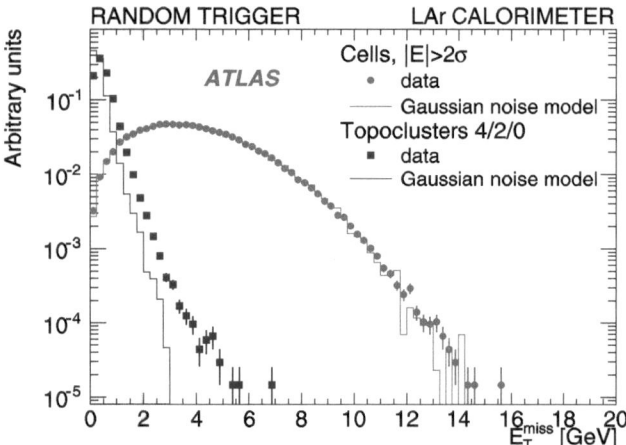

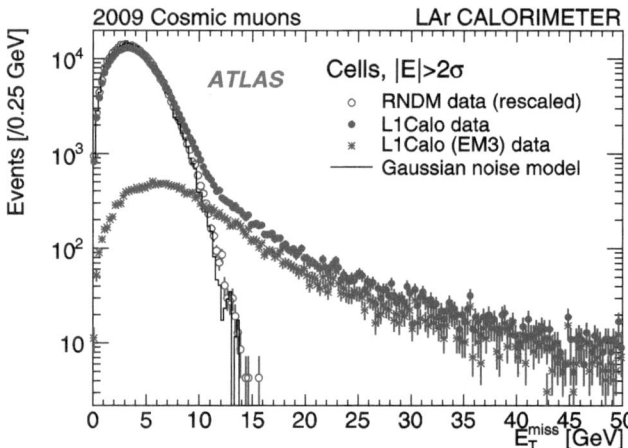

Fig. 10 E_T^{miss} distribution with LAr calorimeter cells for 135,000 randomly triggered events in June 2009. The *dots* (*squares*) show the cell-based (cluster-based) methods in the data, and the histograms show the equivalent distributions for the Gaussian noise model (see text)

Fig. 11 E_T^{miss} distribution with LAr calorimeter cells for 300,000 L1 calorimeter (L1Calo) triggers reconstructed with the cell-based method. Results for EM3 trigger conditions (Sect. 3.1.3) from the same run are superimposed on the same plot and the results from randomly triggered events are again overlaid (*open symbols* and *histogram*)

gathering neighboring cells with $|E_i| > 2\sigma_{\mathrm{noise}}$ and, in a final step, adding all direct neighbors of these accumulated secondary cells (Topocluster 4/2/0). In randomly triggered events, about 8500 and 500 LAr cells, respectively, are selected with these two noise-suppression methods.

The distributions of E_x^{miss} and E_y^{miss} should be Gaussian and centered on zero in randomly triggered events. The measurements are compared with a Gaussian noise model, where no pedestal shift or coherent noise is present, obtained by randomizing the cell energy according to a Gaussian model for the cell noise. For this E_T^{miss} computation, cells with very high noise (see Sect. 2.4) are removed from the computation.

Figure 10 shows the E_T^{miss} distributions for a randomly triggered data sample acquired in 15 hours. The two noise suppression methods are compared to the corresponding Gaussian noise model. For the cell-based method, a good agreement is observed between the data and the simple model. Because of the lower number of cells kept in the cluster-based method, a smaller noise contribution to E_T^{miss} is observed. The agreement between the data and the model is not as good as for the cell-based method, reflecting the higher sensitivity of the cluster-based method to the noise description. In both cases, no E_T^{miss} tails are present, reflecting the absence of large systematic pedestal shifts or abnormal noise.

Using E_T^{miss} it was possible to spot, in 2008, a high coherent noise due to the defective grounding of a barrel presampler HV cable and sporadic noise in a few preamplifiers. These two problems were repaired prior to the 2009 runs. The time stability of E_T^{miss} is regularly monitored using randomly triggered events by observing the mean and width of the E_x^{miss} and E_y^{miss} distributions. With the cluster-based method, the variation of all quantities was measured to be

± 0.1 GeV over 1.5 months. This variation is small compared to the expected E_T^{miss} resolution ($\simeq 5$ GeV for $W \to e\nu$ events) and can be controlled further by more frequent updates of the calibration constants.

A similar analysis was performed with L1 calorimeter triggered events, corresponding to radiative energy losses from cosmic muons, from the same run as used above. The L1 calorimeter trigger (L1calo) triggers events when either the sum of adjacent trigger tower transverse energies is above 3 GeV in the EM calorimeter (EM3) or 5 GeV when summing EM and hadronic towers [27]. The results are illustrated in Fig. 11 for the cell-based noise suppression method. Most of these events are triggered by energy losses in the Tile calorimeter that do not spill in the LAr calorimeter, which therefore mainly records noise, leading to a E_T^{miss} distribution similar to the one obtained with random triggers. However, in few cases, events are triggered by the LAr calorimeter such as the EM3 trigger. The peak at 3 GeV is then shifted upwards to 6 GeV and the proportion of events with E_T^{miss} above 15 GeV is greatly enhanced.

3.3 LAr calorimeter timing

The energy reconstruction in each cell relies on the fact that in the standard (five samples) physics data acquisition mode, the third sample is located close to the signal maximum: this implies an alignment of the timing of all calorimeter cells to within a few ns.

Several parameters determine each cell timing: the first contribution comes from FEB internal delays which induce a cell timing variation of ± 2 ns within each FEB. This is accounted for when computing the optimal filtering coefficients. The second contribution concerns FEB to FEB variations due to different cable lengths to reach a given FEB:

this relative FEB timing can vary by up to ± 10 ns and can be corrected for by setting an adjustable delay on each FEB.

The study presented here aims at predicting (using calibration data and additional hardware inputs) and measuring (using cosmic muons and beam splash data) this relative FEB timing in order to derive timing alignment delays for each FEB.

3.3.1 Timing prediction

The time of the signal maximum is different in a calibration run (t_{calib}) and in a physics run (t_{phys}). The main contribution to this time is the delay T_0 before the pulse starts to rise (the difference between the calibration and physics pulse widths is much smaller than this T_0 delay variation). This delay is driven by cable lengths which are different in these two configurations and additional delays in physics runs because of the particle time of flight, and the Timing, Trigger and Control (TTC) system configurations.

In a calibration run, a signal is injected from the calibration board through the calibration cables, and is then read out through the physics signal cables. The value of the delay T_0^{calib} with respect to the signal injection can thus be computed for each FEB using the various cable lengths (L_{calib}, L_{physics}) and signal propagation speeds (v_{calib}, v_{physics}):

$$T_0^{\text{calib}} = \frac{L_{\text{calib}}}{v_{\text{calib}}} + \frac{L_{\text{physics}}}{v_{\text{physics}}}. \tag{7}$$

The above prediction is compared with the measured value in calibration runs. The measurement corresponds to the time at which the calibration pulse exceeds three standard deviations above the noise; it is found to agree with the prediction to within ± 2 ns, ignoring the variations within each FEB.

The time of the signal maximum t_{calib} is obtained by fitting the peak of the pulse of cells in a given FEB with a third order polynomial. As the cable length is a function of the cell position along the beam axis (z, η), the cell times are averaged per FEBs in a given layer (except for the HEC where layers are mixed inside a FEB) and a given η-bin in order to align the FEBs in time.

The time of the ionization pulse in each cell can then be predicted from the calibration time using the following formula:

$$t_{\text{phys}} = t_{\text{calib}} - \frac{L_{\text{calib}}}{v_{\text{calib}}} + t_{\text{flight}} + \Delta t_{\text{TTC}}, \tag{8}$$

where t_{calib} was defined in the previous paragraph; t_{flight} is the time of flight of an incident particle from the interaction point to the cell, which varies from 5 ns for a presampler cell at $\eta = 0$, to 19 ns for a back cell in the HEC; and Δt_{TTC} is a global correction for the six partitions due to the cabling

of the TTC system which is needed to align all FEBs at the crate level. This predicted ionization pulse time is compared with the corresponding measurement in the next section.

3.3.2 Timing measurement

The ionization pulse time has been measured in beam splash and cosmic muon events. The time is reconstructed using optimal filtering coefficients. Since the arrival time of the particle is not known, one does not know in advance to which samples the time OFCs b_i should be applied (since these OFCs were computed for a particular set of samples around the pulse maximum). Therefore, an iterative procedure is used until the obtained Δt (see (4)) is less than 3 ns.

The time is then corrected for two effects: first, the time-of-flight difference between the beam splash or cosmic muon configurations and the collision configuration, and second, the asynchronicity of the beam splash and cosmic muon events, where arrival times vary with respect to the TTC clock.

The comparison between the measured and the predicted (8) ionization pulse time is shown in Fig. 12 for the C-side ($\eta < 0$) of each LAr sub-detector.

This comparison is performed for each "slot" corresponding to a group of FEBs in a given layer and η-range, averaged over all calorimeter modules over ϕ. As mentionned in the introduction, the relative timing of each group of FEBs varies by ± 10 ns due to the different corresponding cable lengthes.

On the plots, the error bars correspond to the RMS of values for all modules in a slot: in the FCal, there is only one module per slot, so no error bars are shown (also note that slot 8 is empty in the FCal). In some regions, the cosmic data statistics was not sufficient to extract the time: the corresponding bins are thus empty. The agreement between the prediction and the two measurements is within ± 2 ns (and at worst ± 5 ns for two slots of the FCal).

Finally, a set of FEB timing alignment delays is obtained from these well understood measured relative times. These delays will be used at the LHC startup and updated once the phase between the beam and the machine clock will be measured and shown to be stable. The desired precision of ± 1 ns should be reached then.

3.4 Signal reconstruction studies and impact on intrinsic global energy resolution constant term

The main ingredient for accurate energy and time reconstruction of signals from LHC collisions is the prediction of the ionization signal shape, from which the optimal filtering coefficients used in (3) are computed. After recalling the basics of the method used to predict the shape in Sect. 3.4.1, an estimate of the signal prediction quality with three samples

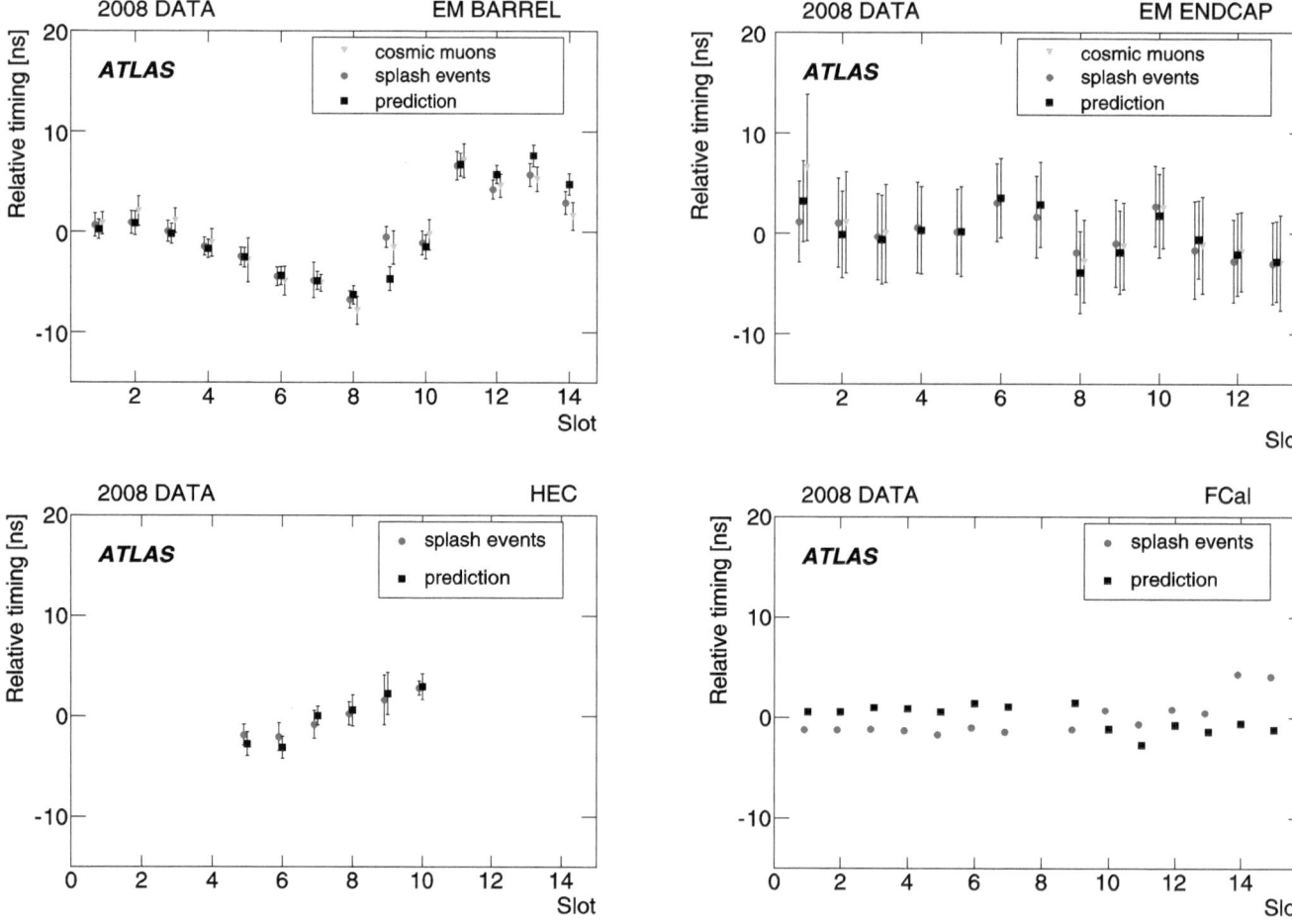

Fig. 12 Relative predicted and measured FEB times in the electromagnetic barrel (*top left*), electromagnetic endcap (*top right*), HEC (*bottom left*) and FCal (*bottom right*) calorimeters, for the C-side ($\eta < 0$). The

x-axis ("Slot") corresponds to a group of FEBs in a given layer (or a group of layers in the HEC) and η-range. The *error bars* show the width of the distributions in each slot

in the EM calorimeter is presented in Sect. 3.4.2. The full 32 samples shape prediction is used to determine the ionization electron drift time needed for the OFC computation in the EM calorimeter (Sect. 3.4.3). Finally, from these two studies an estimate of the main contributions to the constant term in the global energy resolution of the EM calorimeter is given in Sect. 3.4.4.

3.4.1 Prediction of the ionization pulse shape

The standard ATLAS method for prediction of the ionization pulse shape in the EM and the HEC relies on the calibration system. A precisely known calibration signal is sent through the same path as seen by the ionization pulses thus probing the actual electrical and readout properties of each calorimeter cell. In both the EM and the HEC, the calibration pulse properties are parameterized using two variables, f_{step} and τ_{cali}, which have been measured for all calibration boards [23] and are routinely extracted from calibration signals [28].

The predicted ionization shapes are calculated from the calibration pulses by modeling each readout cell as a resonant RLC circuit, where C is the cell capacitance, L the inductive path of ionization signal, and R the contact resistance between the cell electrode and the readout line. The effective LC and RC have been estimated from a frequency analysis of the output calibration pulse shape [28]. They were also measured with a network analyzer during the long validation period of the three cryostats [29–31]. For the HEC, calibration pulses are transformed into ionization signal predictions using a semi-analytical model of the readout electronics, with a functional form with zeros and poles accounting for the cable and pre-amplifier transfer functions [32, 33]. The prediction of both the EM and HEC ionization pulses requires the knowledge of the electron drift time in liquid argon (T_{drift}), which can be inferred from the calorimeter properties or directly measured from data (see Sect. 3.4.3).

To illustrate the good quality of the pulse shape prediction, radiating cosmic muons depositing few GeV in a cell

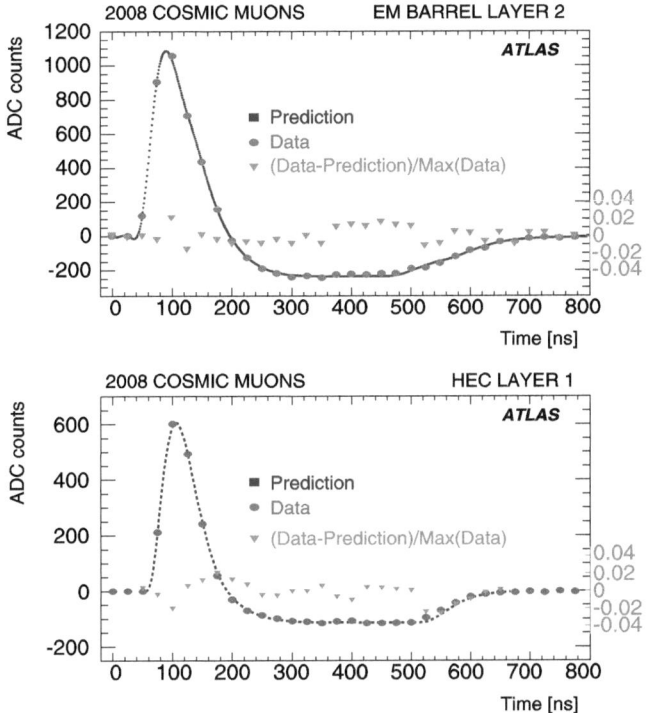

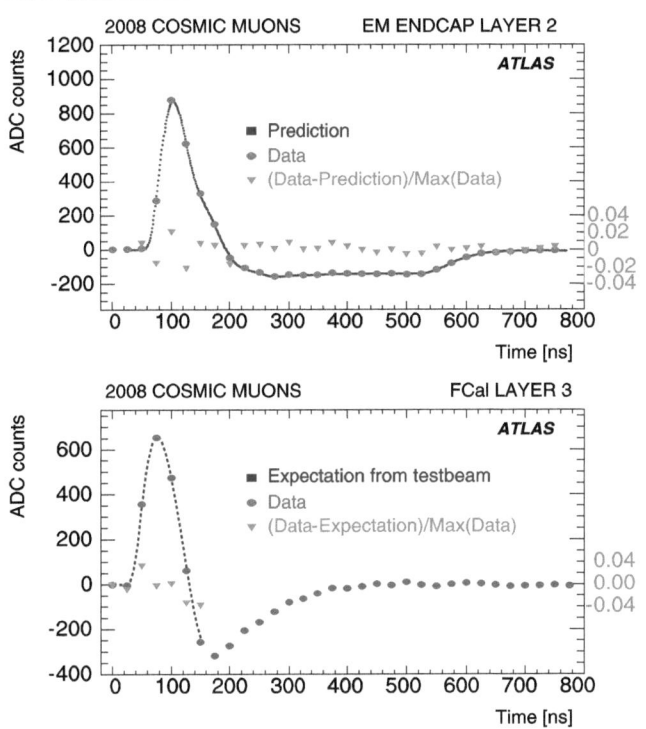

Fig. 13 Typical pulse shapes, recorded during the cosmic ray campaign, for a given cell in the second layer for the barrel (*top left*) and the endcap (*top right*) of the EM calorimeter, as well as in the first layer of the HEC (*bottom left*) and in the third layer of the FCal (*bottom right*). The relative difference between data and prediction is indicated by triangles on the right scale

have been used. Figure 13 shows a typical 32-sample pulse recorded in the barrel (top left) and the endcap (top right) of the EM calorimeter, as well as in the HEC (bottom left). In each case, the pulse shape prediction, scaled to the measured cell energy, agrees at the few percent level with the measured pulse.

As already mentioned, in the FCal the calibration pulse is injected at the base-plane of the front-end crates, and therefore the response to a calibration signal differs significantly from the response to an ionization pulse, preventing the use of methods described above. Instead, seven sample pulse shapes recorded during the beam test campaign [9, 10] have been averaged to obtain a normalized reference pulse shape for each layer. Figure 13 (bottom right) shows a typical example where the agreement between the reference pulse shape and the data is at the 4% level.

3.4.2 Quality of signal reconstruction in the EM calorimeter

Several PeV were deposited in the full calorimeter in LHC beam splash events. As an example, Fig. 14 shows the energy deposited in the second layer of the EM calorimeter. The structure in ϕ reflects the material encountered by the particle flux before hitting the calorimeter, such as the endcap toroid. In this layer, a total of 5×10^5 five sample signal shapes with at least 5 GeV of deposited energy were recorded. These events were used to estimate the quality of the pulse shape prediction for every cell.

For this purpose, a Q^2-estimator is defined as :

$$Q^2 = \frac{1}{N_{\text{dof}}} \sum_{j=1}^{N_{\text{samples}}} \frac{(s_j - A g_j^{\text{phys}})^2}{\sigma_{\text{noise}}^2 + (kA)^2}, \qquad (9)$$

where the amplitude A (3) is computed with a number of samples $N_{\text{samples}} = 3$ (because the timing was not yet adjusted everywhere for the beam splash events, not all samples can be used), s_j is the amplitude of each sample j, in ADC counts, g_j^{phys} is the normalized predicted ionization shape and k is a factor quantifying the relative accuracy of the amplitude A. Assuming an accuracy of around 1%, with the 5 GeV energy cut applied one has $\sigma_{\text{noise}}^2 < (kA)^2$. In this regime, it is possible to fit a χ^2 function with 3 degree of freedom on the $Q^2 \times N_{\text{dof}}$ distribution over cells in the central region (where the Q^2 variation is small). Therefore, $N_{\text{dof}} = 3$. A given value of Q^2 can be interpreted as a precision on the amplitude at the level kQ.

Figure 15 shows the Q^2-estimator in the second layer of the EM calorimeter averaged over ϕ, assuming $k = 1.5\%$ corresponding to $Q^2 \sim 1$ for $\eta \sim 0$. The accuracy is degraded by at most a factor of ~ 2 (i.e. $Q^2 \sim 4$) in some endcap regions. This shows that these data can be described with a reasonable precision.

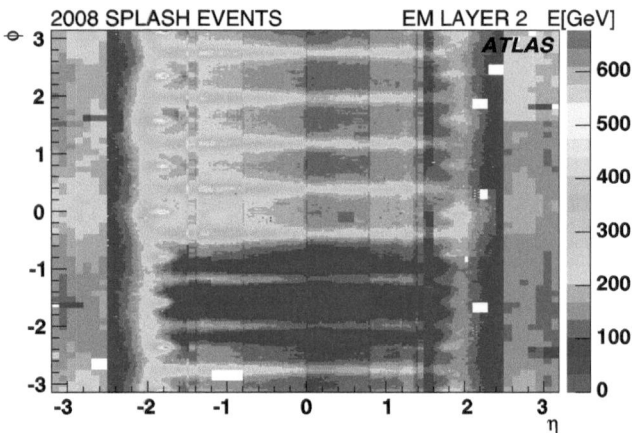

Fig. 14 Total energy deposited in the LHC beam splash events in every cell of the EM calorimeter second layer. Empty bins are due to non functioning electronics

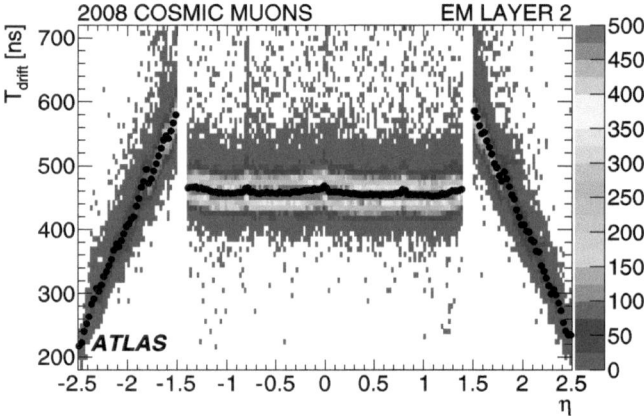

Fig. 16 Drift time measurement in the cells of the EM calorimeter second layer with $E > 1$ GeV for the 2008 cosmic muon run. The *dots* correspond to drift time values averaged in ϕ

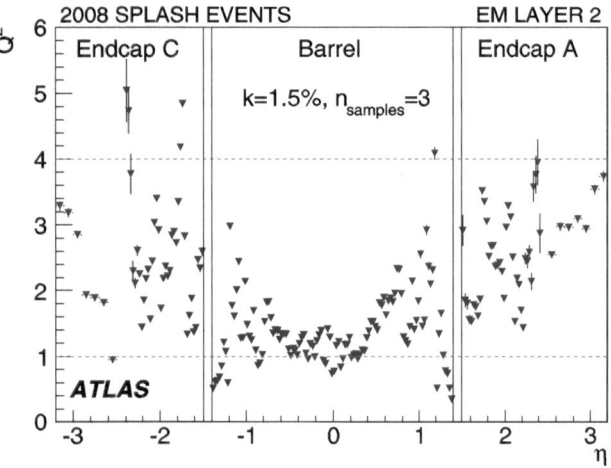

Fig. 15 Estimator Q^2 (defined in the text) as a function of η for 5×10^5 pulse shapes with $E > 5$ GeV in the EM calorimeter second layer cells. Q^2 is defined in (9) with $k = 1.5\%$

3.4.3 Ionization electron drift time measurement in the EM calorimeter

During the 2008 cosmic runs, half a million pulses with 32 samples were recorded in the EM calorimeter from cells in which at least 1 GeV was reconstructed. Given the good accuracy of the predicted signal undershoot (see Fig. 13), the drift time can be extracted from a fit to the measured signal [34].

Figure 16 shows the fitted drift time for all selected cells in the second layer using the standard pulse shape prediction method (Sect. 3.4.1). In the EMB, the drift time has also been measured with a method in which the shape is computed using a more analytical model and LC and RC extracted from network analyzer measurements [30]. The drift times extracted from the two methods are in excellent agreement, giving confidence in the results: a constant

value around the expected 460 ns is obtained, except near the electrode edges ($|\eta| = 0, 0.8$ and 1.4) where the electric field is lower. The decrease of the drift time in the EM endcap ($1.5 < |\eta| < 2.5$) reflects the decrease of the gap size with $|\eta|$. Similar results are obtained for the first and third layers of the EM calorimeters.

3.4.4 Impact on the global energy resolution constant term of the EM calorimeter

When five of the production EM calorimeter modules were tested individually in electron beams, the global constant term c of the energy resolution formula was measured to be $c \sim 0.5\%$ in the EM barrel and 0.7% in the EM endcap [4]. The main contributors are the signal reconstruction accuracy, the LAr gap uniformity, and the electronics calibration system. The first two contributions c_{SR} and c_{gap} can be investigated using results presented in Sects. 3.4.2 and 3.4.3, considering only the second layer of the EM calorimeter where most of the electromagnetic shower energy is deposited.

From Fig. 15, one finds that $\langle Q^2 \rangle \sim 1.4$ in the EM barrel and 2.6 in endcap, and hence $\langle k \rangle = 1.8\%$ and 2.4% respectively. This corresponds to residuals between the predicted and measured pulses of 1 to 2% of the pulse amplitude (see Fig. 13 for illustration), for samples around the signal maximum. Similar residuals were obtained in the electron beam test analysis [28]. At this time, the contribution of the signal reconstruction to the constant term was estimated to be $c_{SR} = 0.25\%$. Given the measured accuracy with beam splash events, the beam test result seems to be reachable with LHC collisions.

The drift time measured in Sect. 3.4.3 is a function of the gap thickness (w_{gap}) and the high voltage (V):

$$T_{drift} \sim \frac{w_{gap}^{\alpha+1}}{V^\alpha} \qquad (10)$$

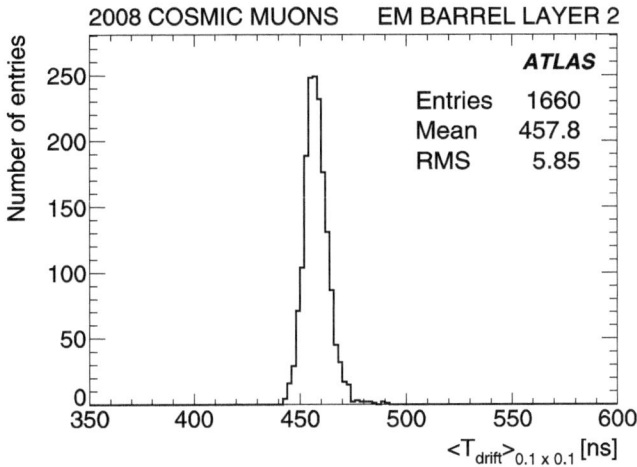

Fig. 17 Distribution of the local average drift time values in $\Delta\eta \times \Delta\phi = 0.1 \times 0.1$ bins, for the middle layer of the EM barrel

where $\alpha \simeq 0.3$ is empirically determined from measurements [19]. In the EM barrel, the electric field is constant, except in transition regions, and thus the drift time uniformity directly measures the LAr gap variations. To reduce statistical fluctuations, the measured drift time values are averaged over regions of $\Delta\eta \times \Delta\phi = 0.1 \times 0.1$. The distribution of the average drift time is shown in Fig. 17 for the second layer of the EM barrel calorimeter.

The drift time uniformity, estimated as the ratio of the RMS of this distribution to its mean value, is $1.28 \pm 0.03\%$. Using the relation between the drift time and the gap from (10) and the fact that the signal amplitude is proportional to the initial ionization current ($I \simeq \frac{\rho \cdot w_{\text{gap}}}{T_{\text{drift}}} \simeq w_{\text{gap}}^{-\alpha}$ where ρ is the linear density of charge), one can relate the relative variation of the drift time to the one of the amplitude applying a factor $\alpha/(1+\alpha)$ to the above result. Therefore, the drift time uniformity leads to a dispersion of response due to the barrel calorimeter gap variations of $(0.29^{+0.05}_{-0.04})\%$ where the systematic uncertainties are included. This represents an upper bound on the corresponding constant term c_{gap}.

For comparison, during the EM calorimeter barrel module construction, the LAr gap thickness was measured, yielding an estimate of the constant term due to gap size variations of $c_{\text{gap}} = 0.16\%$ [14]. The measurement of the gap size uniformity presented here takes into account further effects like deformations in the assembled wheels and possible systematic uncertainties from the in situ cosmic muon analysis.

4 In situ EM calorimeter performance with cosmic muons

In the previous sections, we demonstrated the good performance of the electronics operation and the good understanding of the energy reconstruction. The cosmic ray events can

therefore now be used to validate the Monte Carlo simulation that will be used for the first collisions.

Two such analyses are presented in this section: the first study aims to investigate the electromagnetic barrel calorimeter uniformity using ionization signals from quasi-projective cosmic muons, and the second aims to reconstruct electromagnetic showers from radiative cosmic muons and to compare the measured shower shapes with simulation.

4.1 Monte Carlo simulation

The ATLAS Monte Carlo [35, 36] simulates the interaction of particles produced during LHC collisions or from cosmic muons within the ATLAS sub-detectors. It is based on the Geant4 toolkit [37] that provides the physics lists, geometry description and tracking tools. For cosmic muons, the material between the ground level and the ATLAS cavern is also simulated, i.e. the overburden and the two access shafts. The simulated cosmic ray spectrum corresponds to what was measured at sea level [38]. Air showers are not simulated but have a negligible effect on the analyses presented here. In order to save CPU time, the generated events are filtered before entering the full Geant4 simulation by requiring that the particles cross a specific detector volume (in the following analyses, typically inner detector volumes).

An important use of the simulation, amongst many others, is to validate the selection criteria on shower-shape for high-level trigger and offline algorithms, as well as to derive the electron and photon energy calibrations.

It is important to note that, thanks to the digitization step of the calorimeter simulation which emulates the behavior of the electronics, the standard energy reconstruction procedure can be applied to the simulated events. The special procedure used for asynchronous cosmic muon data, which uses an iterative determination of the event time, is however not applied to the Monte Carlo data.

4.2 Uniformity of the electromagnetic barrel calorimeter

4.2.1 Goals and means of the analysis

Any non-uniformity in the response of the calorimeter has a direct impact on the constant term in the energy resolution (see Sect. 3.4.4); great care was taken during the construction to limit all sources of non-uniformity to the minimum achievable, aiming for a global constant term below 0.7%. The default ATLAS Monte Carlo simulation emulates the effect of the constant term, but for the present analysis, this emulation was turned off.

The uniformity of the calorimeter was measured for three barrel production modules using electrons during beam test campaigns [4]. Cosmic muons provide a unique opportunity to measure the calorimeter uniformity in situ over a larger number of modules, unfortunately limited to the barrel calorimeter due to both the topology of the cosmic muon events and the choice of triggers. The scope of this analysis is nevertheless quite different than in the beam test. First, muons behave very differently from electrons: in most events, they deposit only a minimum ionization energy in the liquid argon and they are much less sensitive to upstream material. The result can therefore not be easily extrapolated to the electron and photon response. Second, the cosmic run statistics are limited, so uniformity cannot be studied with cell-level granularity. The goal of this cosmic muon analysis is rather to quantify the agreement between data and Monte Carlo, and to exclude the presence of any significant non-uniformity in the calorimeter response.

A previous uniformity analysis using cosmic muons [39] from 2006 and 2007 relied on the hadronic Tile calorimeter to trigger events and to measure the muon sample purity. For the 2008 data discussed here, both the muon spectrometer and inner detector were operating and were used for triggering and event selection. The data sample consists of filtered events requiring a reconstructed track in the inner detector with at least one hit in the silicon tracker. The tracks are also selected to be reasonably projective by requiring that their transverse ($|d_0|$) and longitudinal ($|z_0|$) impact parameters, with respect to the center of the coordinate system be smaller than 300 mm.

4.2.2 Signal reconstruction

In the first step, a muon track is reconstructed in the inner detector. For that purpose, a dedicated algorithm looks for a single track crossing both the top and bottom hemispheres. This single track is then extrapolated both downward and upward into the calorimeter.

Around the two track impact positions in the calorimeter, a rectangle of cells (the cell road) is selected in the first and second layers (the signal to noise ratio for muons is too low in the third layer). The cells of the first layer have a size of $\Delta\eta \times \Delta\phi = 0.003 \times 0.1$ and 12×3 such cells are kept. Similarly, the cells of the second layer have a size of $\Delta\eta \times \Delta\phi = 0.025 \times 0.025$, and 5×5 such cells are kept.

To reconstruct the energy of the selected cells, the muon timing is obtained via an iterative procedure that is usually only applied to cells with an ADC signal at least four times the noise level. Since most muons are minimum ionizing particles, the muon signal is small, typically 150 MeV is deposited in the most energetic cell in the second layer, only five times the noise, and many cells do not pass this threshold. Therefore, an alternative reconstruction is used in this analysis: in the first pass, the iteration threshold is lowered to zero so that the timing is computed for most of the cells. In the second pass, the timing of the most energetic cell determined in the first pass is applied to all the other cells of

the road. The cell energy is reconstructed at the electron energy scale and thus does not represent the true energy loss of the muon. Finally, clusters are formed in each layer to reconstruct the muon energy loss. The criteria used to decide on the cluster size are described below.

4.2.3 Optimization of the uniformity measurement

In order to perform the most accurate evaluation of the calorimeter uniformity, the measurement granularity, the cluster size and the selection cuts have been optimized. The granularity chosen is a compromise between the need for high statistics (large binning) and the need for high precision. The cluster size optimizes the signal to noise ratio while the selection cuts reduce the biases while keeping high statistics.

The binning is determined by requiring a minimum of 500 events per unit. In the η direction, this corresponds to bins of 0.025 (equal to the second layer cell width) up to $|\eta| = 0.7$ and wider bins above.

In the first layer, the muon energy loss is measured using a $\Delta\eta \times \Delta\phi = 2 \times 1$ (in first layer cell unit) cluster, which contains most of the deposited energy. Adding an additional cell brings more noise than signal. In the second layer, a 1×3 (in second layer cell unit) cluster is used: it suffers less from noise than a 3×3 cluster, but requires the removal of non-projective events which leak outside the cluster along the η direction.

This projectivity cut is based on the centrality of the muon in the second layer cell: when the muon passes close to the edge of the cell, a very small non-projectivity induces a large energy leakage into the neighboring cell. Therefore, for each second layer cell, eight bins corresponding to the eight first layer cells located in front of it were defined, and in each bin a cut is applied on the beam impact parameter z_0 of the track, such that the muon is geometrically contained in the second layer cell. The remaining statistics after this projectivity cut is 76 k events in the data sample and 113 k events in the Monte Carlo sample. The events are mainly located under the cavern shafts leading to a coverage of around 20% of the full electromagnetic barrel calorimeter.

A comparison of the energy reconstructed in the first and second layers between data and Monte Carlo events is shown in Fig. 18. Because the muon energy loss is mostly η-dependent, both distributions are shown for all events (top), showing a large width due to the variation of the energy response over η, and for a single η-bin (bottom).

The agreement between the data and Monte Carlo distributions is very good, both for the shape and for the absolute energy scale which differs by only 2% in the front layer and 1% in the second layer. Part of the difference comes from the slight difference in acceptance for data and Monte Carlo, as well as from the difference in energy reconstruction. This overall energy scale difference is corrected for in the MC in the rest of the study.

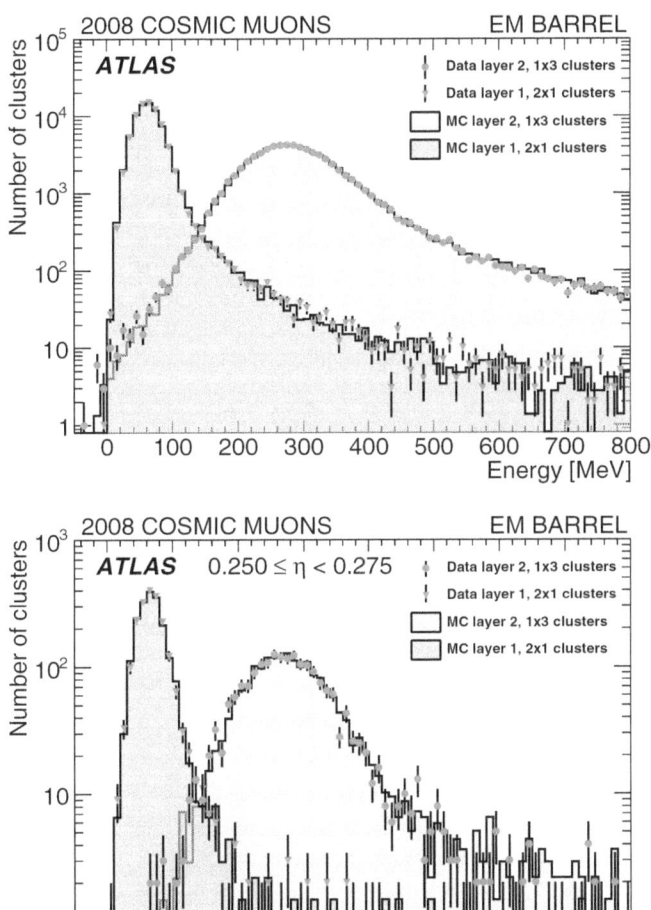

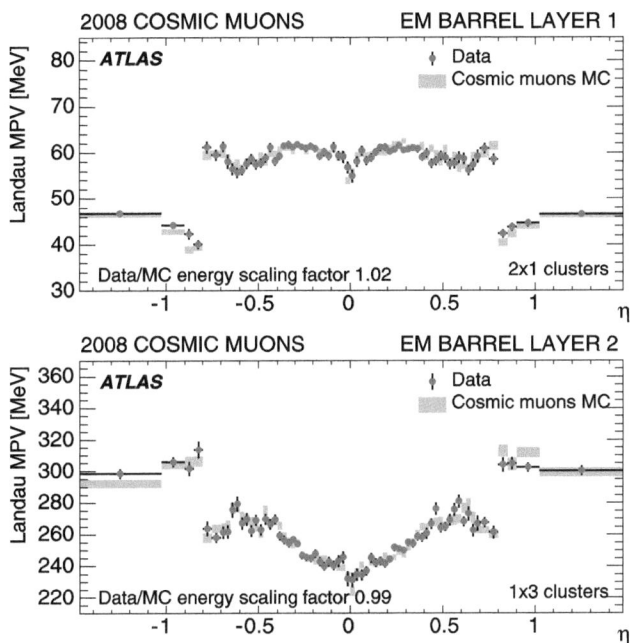

Fig. 19 Landau MPV as a function of η in the first (*top*) and second (*bottom*) layers for the data (*red points*) and Monte Carlo (*grey bands*)

Fig. 18 Energy in a 2×1 cluster in the first layer (*histogram* for Monte Carlo and *triangles* for data) and in a 1×3 cluster in the second layer (*histogram* for Monte Carlo and *full circles* for data) for all events (*top*) and a single η-bin (*bottom*)

4.2.4 Calorimeter uniformity along η

Given the limited statistics of the projective cosmic muon data, the uniformity of the response in η cannot be estimated at the cell level. A natural choice of cell combination is to integrate clusters in ϕ since the response should not vary along this direction due to the ϕ symmetry of the calorimeter. The response along the η direction for cosmic muons depends on the variation of the amount of liquid argon seen by the muon. In particular, a transition occurs at $|\eta| = 0.8$ where the lead thickness goes from 1.53 mm to 1.13 mm.

The estimation of the muon energy in each η-bin is done with a fit of the cluster energy distribution using a Landau function convoluted with a Gaussian. The Landau function accounts for fluctuations of the energy deposition in the ionization process and the Gaussian accounts for the electronic noise and possible remaining fluctuations. In particular, a 10% difference is observed between the width of the Gaussian expected from the electronic noise and the width of

the fitted Gaussian. Mostly this bias comes from remaining cluster non-containment effects which are found to be η-independent and thus do not produce any artificial non-uniformity. The most probable value (MPV) of the Landau distribution estimates the energy deposition.

Distributions of data and Monte Carlo MPVs along the η direction for the first and second layers are shown in Fig. 19.

In the first layer, the MPVs are roughly constant along η, except around $\eta = 0$ where some cells are physically missing in the detector, and around $|\eta| = 0.6$ where the cell depth is varying. In the second layer, the response follows a typical "V-shape" corresponding to the variation of the cell depth along η that rises up to $|\eta| = 0.6$. Again, the agreement between the data and Monte Carlo is very good, showing that the contribution of systematic effects due to the energy reconstruction method or the non-projectivity of the tracks is small.

The response uniformity U_{meas} is given by the RMS of the normalized differences between the data and Monte Carlo MPVs in each η-bin:

$$U_{\text{meas}} = \sqrt{\frac{\sum_{i=1}^{N_b} (U_{i,\text{meas}} - \langle U_{i,\text{meas}} \rangle)^2}{N_b}}, \quad (11)$$

with:

$$U_{i,\text{meas}} = \frac{\text{MPV}_{i,\text{Data}} - \text{MPV}_{i,\text{MC}}}{\text{MPV}_{i,\text{Data}}}, \quad (12)$$

where $U_{i,\text{meas}}$ is averaged over ϕ, N_b is the number of bins in η, and $\langle U_{i,\text{meas}} \rangle = 0$ since the global energy scale difference was corrected by rescaling the MC.

The measured uniformity should be compared to the expected uniformity $U_{\exp}$, which is obtained similarly to (11) with $U_{i,\exp}$ given by:

$$U_{i,\exp} = \frac{\mathrm{MPV}_{i,\mathrm{MC}}}{\mathrm{MPV}_{i,\mathrm{Data}}} \sqrt{U_{i,\mathrm{Data}}^2 + U_{i,\mathrm{MC}}^2} \qquad (13)$$

with:

$$U_{i,\mathrm{Data(MC)}} = \frac{\sigma(\mathrm{MPV}_{i,\mathrm{Data(MC)}})}{\mathrm{MPV}_{i,\mathrm{Data(MC)}}}, \qquad (14)$$

where $\sigma(\mathrm{MPV}_{i,\mathrm{Data(MC)}})$ is the statistical uncertainty on the measured Landau MPV. This uncertainty is due to the finite statistics of the data and Monte Carlo samples in each bin, the Landau dispersion of the ionization, and the electronic noise.

The measured uniformity U_{meas} should agree with the expected uniformity $U_{\exp}$ if the Monte Carlo simulation reproduces the data well: the key ingredients are the acceptance, the muon spectrum, and the energy reconstruction method. A significant departure of the measured uniformity from the expected one would be a measurement of additional non-uniformities U_Δ ($U_\Delta^2 = U_{\mathrm{meas}}^2 - U_{\exp}^2$).

The measured and expected uniformities for the two EM layers are shown in Fig. 20.

The fluctuations of the measured energies are large: the RMS of the corresponding distribution is $2.4 \pm 0.2\%$ in the first layer and $1.7 \pm 0.1\%$ in the second layer, showing that the statistical power of the analysis is limited given the available data and Monte Carlo statistics. The fluctuations mostly remain within the limits of the band representing the expected values. The RMS of the latter distribution is 2.2% in the first layer and 1.6% in the second layer. This demonstrates that no significant additional non-uniformity (U_Δ) is present in the data. An upper limit is derived and yields $U_\Delta < 1.7\%$ @ 95% CL in the first layer, and $U_\Delta < 1.1\%$ @ 95% CL in the second layer.

The calorimeter response uniformity along η (averaged over ϕ) is thus consistent at the percent level with the Monte Carlo simulation and shows no significant non-uniformity.

4.3 Electromagnetic shower studies

The second analysis aims at validating the Monte Carlo simulation of the distribution of some key calorimeter variables used in the ATLAS electron/photon identification. This is done using radiative cosmic muons that can give rise to electromagnetic showers in the calorimeter through bremsstrahlung or pair conversions.

4.3.1 Selection of radiative muons

To increase the probability of the presence of a muon in the event, it is requested that at least one track has been reconstructed in the inner detector barrel with $|d_0| < 220$ mm and $p_T > 5$ GeV: these cuts ensure a similar acceptance for data and Monte Carlo.

A radiative energy loss is searched for in the electromagnetic barrel calorimeter by requiring a cluster with an energy greater than 5 GeV. Since the radiation can occur anywhere along the muon path, the corresponding shower is not always fully contained in the electromagnetic calorimeter: this is visible in Fig. 21 which shows the fraction of the cluster

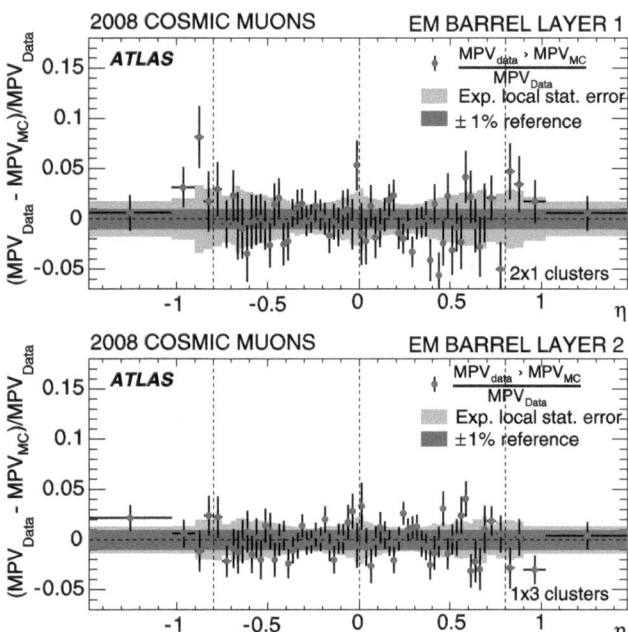

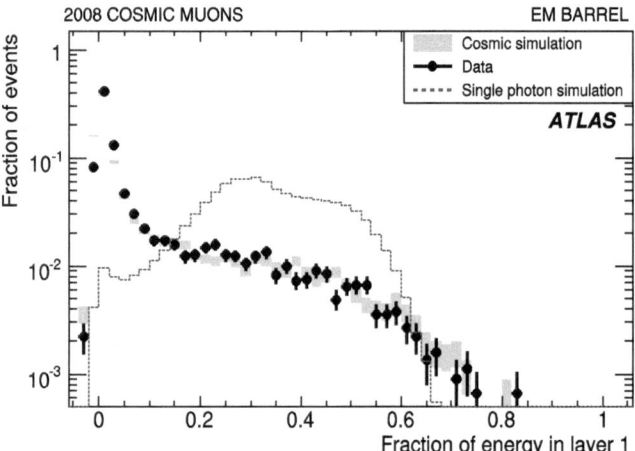

Fig. 20 Measured $U_{i,\mathrm{meas}}$ (*red points*) and expected $U_{i,\exp}$ (*light grey band*) cosmic muon energy dispersions as function of η for the first (*top*) and second (*bottom*) layers of the EM barrel. The *dark grey band* indicates a $\pm 1\%$ strip for reference

Fig. 21 Fraction of cluster energy deposited in the first layer of the electromagnetic barrel calorimeter for cosmic data (*dots*) and Monte Carlo (*rectangles*), as well as for simulated single photons of 5 GeV momentum from interaction vertex (*red histogram*)

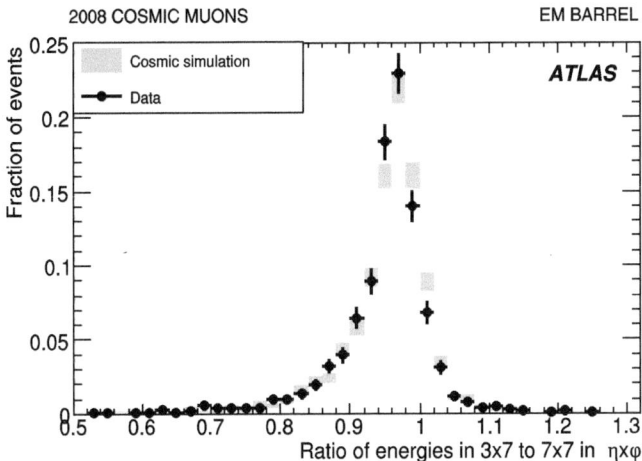

Fig. 22 Lateral shower containment in the second layer of the calorimeter given by the ratio of the energy deposited in a 3×7 cluster to a 7×7 cluster for radiative cosmic muon data (*dots*) and Monte Carlo simulation (*rectangles*)

energy deposited in the first layer for simulated single photons from interaction vertex and for electromagnetic showers from radiating cosmic muons. This shows that the longitudinal shower development of the radiative photons is well reproduced by the Monte Carlo simulation, and that most of the radiating muons deposit very little energy in the first layer. To select "collision-like" showers, this fraction is requested to be greater than 0.1. A total of 1200 candidates remain in the data sample and 2161 in the Monte Carlo after this selection.

4.3.2 Shower shape validation

Various shower shape distributions used for photon identification have been compared with the Monte Carlo simulation: Figs. 22 and 23 show two distributions of variables related to lateral shower containment in the first and second layers of the electromagnetic calorimeter.

Figure 22 shows the ratio of the energy deposited in a $\Delta\eta \times \Delta\phi = 3 \times 7$ (in second layer cell unit) cluster to that in a 7×7 cluster, in the second layer of the barrel calorimeter. In LHC collisions, this variable distinguishes electromagnetic showers, contained in 3 cells in η, from hadronic showers, leaking outside these 3 cells. The contribution from the noise explains that the ratio can be above 1.

Figure 23 shows the variable $F_{side} = (E_{\pm 3} - E_{\pm 1})/E_{\pm 1}$ computed as the ratio of energy within seven central cells in the first layer ($E_{\pm 3}$), outside a core of three central cells ($E_{\pm 1}$), over energy in the three central cells: in LHC collisions, this variable typically separates photons, where little energy is deposited outside the core region, from π^0s, where the two photons produced by the π^0 deposit some energy outside the core region. The agreement between the Monte Carlo simulation and the cosmic ray data is very good in

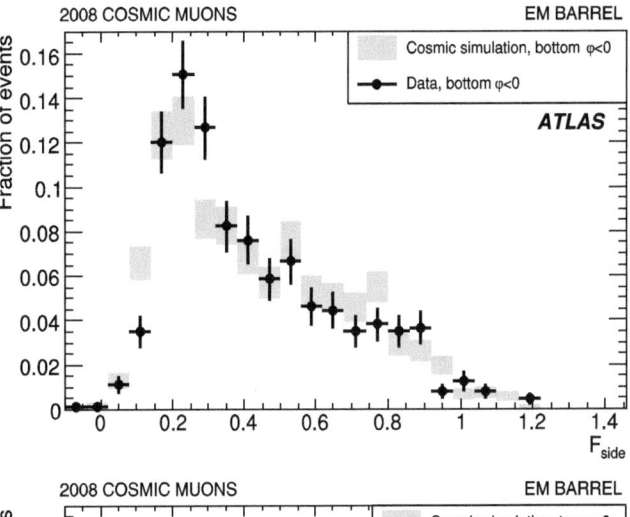

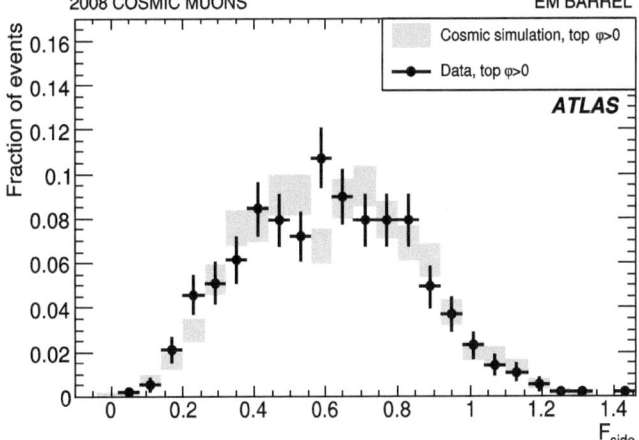

Fig. 23 Lateral shower containment in the first layer for "collision–like" (*top panel*) or "reverse" (*bottom panel*) electromagnetic showers for radiative cosmic muon data (*dots*) and Monte Carlo simulation (*rectangles*). The definition of the F_{side} is given in the text

both the cases where the electromagnetic shower develops in the "collision-like" direction (in the bottom hemisphere) and the case where it develops in the backward direction (in the top hemisphere).

Within the statistics available from data, important calorimeter variables used in the electron/photon identification in ATLAS illustrate the good agreement between the Monte Carlo simulation and electromagnetic showers from radiative cosmic events in the calorimeter. These results, as well as the numerous comparisons done with beam test data [2–6], give confidence that robust photon and electron identification will be available for early data at the LHC.

5 Conclusions and perspectives

The liquid argon calorimeter has been installed, connected and fully readout since the beginning of 2008. Since then, much experience has been gained in operating the system.

Thanks to the very stable cryogenics and electronics operation over this period, first performance studies with the complete LAr calorimeter coverage have been done using several months of cosmic muon data and with LHC beam splash events from September 2008. These data provided a check of the first level trigger energy computation and the timing of the electronics. In the EM calorimeter, detailed studies of the signal shape predictions allow to check that, within the accuracy of the analysis, there is no extra contribution to the dominant contributions to the intrinsic constant term of the energy resolution. This indicates that the reach of a global constant term of 0.7% is achievable. The non-uniformity of the EM barrel calorimeter response to cosmic muons is consistent at the percent level with the simulated response. Finally, the electromagnetic shower profiles are in good agreement with the simulated ones, thus validating the Monte Carlo description. All these results allow for strong confidence in the readiness of the LAr calorimeter for the first LHC collisions.

The ultimate LAr calorimeter performance will be assessed with collision data: this is particularly true for the electromagnetic and hadronic energy scale computation in the ATLAS environment, which is needed for many ATLAS physics analyses.

Acknowledgements We are greatly indebted to all CERN's departments and to the LHC project for their immense efforts not only in building the LHC, but also for their direct contributions to the construction and installation of the ATLAS detector and its infrastructure. We acknowledge equally warmly all our technical colleagues in the collaborating institutions without whom the ATLAS detector could not have been built. Furthermore we are grateful to all the funding agencies which supported generously the construction and the commissioning of the ATLAS detector and also provided the computing infrastructure.

The ATLAS detector design and construction has taken about fifteen years, and our thoughts are with all our colleagues who sadly could not see its final realisation.

We acknowledge the support of ANPCyT, Argentina; Yerevan Physics Institute, Armenia; ARC and DEST, Australia; Bundesministerium für Wissenschaft und Forschung, Austria; National Academy of Sciences of Azerbaijan; State Committee on Science & Technologies of the Republic of Belarus; CNPq and FINEP, Brazil; NSERC, NRC, and CFI, Canada; CERN; NSFC, China; Ministry of Education, Youth and Sports of the Czech Republic, Ministry of Industry and Trade of the Czech Republic, and Committee for Collaboration of the Czech Republic with CERN; Danish Natural Science Research Council; European Commission, through the ARTEMIS Research Training Network; IN2P3-CNRS and Dapnia-CEA, France; Georgian Academy of Sciences; BMBF, DESY, DFG and MPG, Germany; Ministry of Education and Religion, through the EPEAEK program PYTHAGORAS II and GSRT, Greece; ISF, MINERVA, GIF, DIP, and Benoziyo Center, Israel; INFN, Italy; MEXT, Japan; CNRST, Morocco; FOM and NWO, Netherlands; The Research Council of Norway; Ministry of Science and Higher Education, Poland; GRICES and FCT, Portugal; Ministry of Education and Research, Romania; Ministry of Education and Science of the Russian Federation, Russian Federal Agency of Science and Innovations, and Russian Federal Agency of Atomic Energy; JINR; Ministry of Science, Serbia; Department of International Science and Technology Cooperation, Ministry of Education of the Slovak Republic; Slovenian Research Agency, Ministry of Higher Education, Science and Technology, Slovenia; Ministerio de Educación y Ciencia, Spain; The Swedish Research Council, The Knut and Alice Wallenberg Foundation, Sweden; State Secretariat for Education and Science, Swiss National Science Foundation, and Cantons of Bern and Geneva, Switzerland; National Science Council, Taiwan; TAEK, Turkey; The Science and Technology Facilities Council and The Leverhulme Trust, United Kingdom; DOE and NSF, United States of America.

References

1. G. Aad et al., The ATLAS experiment at the CERN large hadron collider. JINST **3**, S08003 (2008)
2. J. Colas et al., Position resolution and particle identification with the ATLAS EM calorimeter. Nucl. Instrum. Methods A **550**, 96 (2005)
3. M. Aharrouche et al., Energy linearity and resolution of the ATLAS electromagnetic barrel calorimeter in an electron test-beam. Nucl. Instrum. Methods A **568**, 601 (2006)
4. M. Aharrouche et al., Response uniformity of the ATLAS liquid argon electromagnetic calorimeter. Nucl. Instrum. Methods A **582**, 429 (2007)
5. M. Aharrouche et al., Time resolution of the ATLAS barrel liquid argon electromagnetic calorimeter. Nucl. Instrum. Methods A **597**, 178 (2008)
6. M. Aharrouche et al., Study of the response of ATLAS electromagnetic liquid argon calorimeters to muons. Nucl. Instrum. Methods A **606**, 419 (2009)
7. B. Dowler et al., Performance of the ATLAS hadronic end-cap calorimeter in beam tests. Nucl. Instrum. Methods A **482**, 96 (2002)
8. A.E. Kiryunin et al., GEANT4 physics evaluation with testbeam data of the ATLAS hadronic end-cap calorimeter. Nucl. Instrum. Methods A **560**, 278 (2006)
9. J.C. Armitage et al., Electron signals in the forward calorimeter prototype for ATLAS. JINST **2**, P11001 (2007)
10. J.P. Archambault et al., Energy calibration of the ATLAS liquid argon forward calorimeter. JINST **3**, P02002 (2008)
11. M. Aharrouche et al., Measurement of the response of the ATLAS liquid argon barrel calorimeter to electrons at the 2004 combined test-beam. Nucl. Instrum. Methods A **614**, 400 (2010)
12. C. Cojocaru et al., Hadronic calibration of the ATLAS liquid argon end-cap calorimeter in the pseudorapidity region $1.6 < \eta < 1.8$ in beam tests. Nucl. Instrum. Methods A **531**, 481 (2004)
13. J. Pinfold et al., Performance of the ATLAS liquid argon endcap calorimeter in the pseudorapidity region $2.5 < \eta < 4.0$ in beam tests. Nucl. Instrum. Methods A **593**, 324 (2008)
14. B. Aubert et al., Construction, assembly and tests of the ATLAS electromagnetic barrel calorimeter. Nucl. Instrum. Methods A **558**, 388 (2006)
15. M. Aleksa et al., Construction, assembly and tests of the ATLAS electromagnetic end-cap calorimeter. JINST **3**, P06002 (2008)
16. M.L. Andrieux et al., Construction and test of the first two sectors of the ATLAS liquid argon presampler. Nucl. Instrum. Methods A **479**, 316 (2002)
17. D.M. Gingrich et al., Construction, assembly and testing of the ATLAS hadronic end-cap calorimeter. JINST **2**, P05005 (2007)
18. A. Artamonov et al., The ATLAS forward calorimeters. JINST **3**, P02010 (2008)
19. C. de La Taille, L. Serin, Temperature dependence of the ATLAS electromagnetic calorimeter signal. Preliminary drift time measurement, ATL-LARG-95-029

20. N.J. Buchanan et al., Design and implementation of the front-end board for the readout of the ATLAS liquid argon calorimeters. JINST **3**, P03004 (2008)

21. N.J. Buchanan et al., ATLAS liquid argon calorimeter front end electronics system. JINST **3**, P09003 (2008)

22. A. Bazan et al., ATLAS liquid argon calorimeter back end electronics. JINST **2**, P06002 (2007)

23. J. Colas et al., Electronics calibration board for the ATLAS liquid argon calorimeters. Nucl. Instrum. Methods A **593**, 269 (2008)

24. W.E. Cleland, E.G. Stern, Signal processing considerations for liquid ionization calorimeter in a high rate environment. Nucl. Instrum. Methods A **338**, 467 (1994)

25. ATLAS collaboration, Liquid argon calorimeter technical design report, CERN-LHCC-96-041, http://cdsweb.cern.ch/record/331061

26. W. Lampl et al., Calorimeter clustering algorithms: Description and performance, ATL-LARG-PUB-2008-002

27. R. Achenbach et al., The ATLAS Level-1 calorimeter trigger. JINST **3**, P03001 (2008)

28. D. Banfi, M. Delmastro, M. Fanti, Cell response equalization of the ATLAS electromagnetic calorimeter without the direct knowledge of the ionization signals. JINST **1**, P08001 (2006)

29. S. Baffioni et al., Electrical measurements on the atlas electromagnetic barrel calorimeter, ATL-LARG-PUB-2007-005

30. C. Collard et al., Prediction of signal amplitude and shape for the ATLAS electromagnetic calorimeter, ATL-LARG-PUB-2007-010

31. C. Gabaldon Ruiz et al., Signal reconstruction in the EM end-cap calorimeter and check with cosmic data in the region $0 < \eta < 3.2$, ATL-LARG-PUB-2008-001

32. H. Brettel et al., Calibration of the ATLAS hadronic end-cap calorimeter, in *Proceedings of the 6th Workshop on Electronics for LHC Experiments*, Krakow, Poland, September 2000, p. 218

33. J. Ban et al., Cold electronics for the liquid argon hadronic end-cap calorimeter of ATLAS. Nucl. Instrum. Methods A **556**, 158 (2006)

34. ATLAS Collaboration, Drift time measurement in the ATLAS liquid argon electromagnetic calorimeter using cosmic muons data. Eur. Phys. J. C (submitted). http://arxiv.org/abs/1002.4189

35. ATLAS Collaboration, The ATLAS Monte Carlo project. Eur. Phys. J. C (submitted). http://arxiv.org/abs/1005.4568

36. J.P. Archambault et al., The simulation of the ATLAS liquid argon calorimetry, ATL-LARG-PUB-2009-001

37. S. Agostinelli et al., GEANT4: a simulation toolkit. Nucl. Instrum. Methods A **506**, 250 (2003)

38. A. Dar, Atmospheric neutrinos, astrophysical neutrons, and proton-decay experiments. Phys. Rev. Lett. **51**, 227–230 (1983)

39. M. Cooke et al., In situ commissioning of the ATLAS electromagnetic calorimeter with cosmic muons, ATL-LARG-PUB-2007-013

The ATLAS Inner Detector commissioning and calibration

The ATLAS Collaboration[*,]**

G. Aad[48], B. Abbott[111], J. Abdallah[11], A.A. Abdelalim[49], A. Abdesselam[118], O. Abdinov[10], B. Abi[112], M. Abolins[88], H. Abramowicz[152], H. Abreu[115], B.S. Acharya[163a,163b], D.L. Adams[24], T.N. Addy[56], J. Adelman[174], C. Adorisio[36a,36b], P. Adragna[75], T. Adye[129], S. Aefsky[22], J.A. Aguilar-Saavedra[124b], M. Aharrouche[81], S.P. Ahlen[21], F. Ahles[48], A. Ahmad[147], M. Ahsan[40], G. Aielli[133a,133b], T. Akdogan[18a], T.P.A. Åkesson[79], G. Akimoto[154], A.V. Akimov[94], A. Aktas[48], M.S. Alam[1], M.A. Alam[76], S. Albrand[55], M. Aleksa[29], I.N. Aleksandrov[65], C. Alexa[25a], G. Alexander[152], G. Alexandre[49], T. Alexopoulos[9], M. Alhroob[20], M. Aliev[15], G. Alimonti[89a], J. Alison[120], M. Aliyev[10], P.P. Allport[73], S.E. Allwood-Spiers[53], J. Almond[82], A. Aloisio[102a,102b], R. Alon[170], A. Alonso[79], M.G. Alviggi[102a,102b], K. Amako[66], C. Amelung[22], A. Amorim[124a], G. Amorós[166], N. Amram[152], C. Anastopoulos[139], T. Andeen[29], C.F. Anders[48], K.J. Anderson[30], A. Andreazza[89a,89b], V. Andrei[58a], X.S. Anduaga[70], A. Angerami[34], F. Anghinolfi[29], N. Anjos[124a], A. Annovi[47], A. Antonaki[8], M. Antonelli[47], S. Antonelli[19a,19b], J. Antos[144b], B. Antunovic[41], F. Anulli[132a], S. Aoun[83], G. Arabidze[8], I. Aracena[143], Y. Arai[66], A.T.H. Arce[14], J.P. Archambault[28], S. Arfaoui[29,a], J.-F. Arguin[14], T. Argyropoulos[9], M. Arik[18a], A.J. Armbruster[87], O. Arnaez[4], C. Arnault[115], A. Artamonov[95], D. Arutinov[20], M. Asai[143], S. Asai[154], R. Asfandiyarov[171], S. Ask[82], B. Åsman[145a,145b], D. Asner[28], L. Asquith[77], K. Assamagan[24], A. Astvatsatourov[52], G. Atoian[174], B. Auerbach[174], K. Augsten[127], M. Aurousseau[4], N. Austin[73], G. Avolio[162], R. Avramidou[9], C. Ay[54], G. Azuelos[93,b], Y. Azuma[154], M.A. Baak[29], A.M. Bach[14], H. Bachacou[136], K. Bachas[29], M. Backes[49], E. Badescu[25a], P. Bagnaia[132a,132b], Y. Bai[32a], T. Bain[157], J.T. Baines[129], O.K. Baker[174], M.D. Baker[24], S. Baker[77], F.Baltasar Dos Santos Pedrosa[29], E. Banas[38], P. Banerjee[93], S. Banerjee[168], D. Banfi[89a,89b], A. Bangert[137], V. Bansal[168], S.P. Baranov[94], A. Barashkou[65], T. Barber[27], E.L. Barberio[86], D. Barberis[50a,50b], M. Barbero[20], D.Y. Bardin[65], T. Barillari[99], M. Barisonzi[173], T. Barklow[143], N. Barlow[27], B.M. Barnett[129], R.M. Barnett[14], A. Baroncelli[134a], A.J. Barr[118], F. Barreiro[80], J. Barreiro Guimarães da Costa[57], P. Barrillon[115], R. Bartoldus[143], D. Bartsch[20], R.L. Bates[53], L. Batkova[144a], J.R. Batley[27], A. Battaglia[16], M. Battistin[29], F. Bauer[136], H.S. Bawa[143], M. Bazalova[125], B. Beare[157], T. Beau[78], P.H. Beauchemin[118], R. Beccherle[50a], P. Bechtle[41], G.A. Beck[75], H.P. Beck[16], M. Beckingham[48], K.H. Becks[173], A.J. Beddall[18c], A. Beddall[18c], V.A. Bednyakov[65], C. Bee[83], M. Begel[24], S. Behar Harpaz[151], P.K. Behera[63], M. Beimforde[99], C. Belanger-Champagne[165], P.J. Bell[49], W.H. Bell[49], G. Bella[152], L. Bellagamba[19a], F. Bellina[29], M. Bellomo[119a], A. Belloni[57], K. Belotskiy[96], O. Beltramello[29], S. Ben Ami[151], O. Benary[152], D. Benchekroun[135a], M. Bendel[81], B.H. Benedict[162], N. Benekos[164], Y. Benhammou[152], D.P. Benjamin[44], M. Benoit[115], J.R. Bensinger[22], K. Benslama[130], S. Bentvelsen[105], M. Beretta[47], D. Berge[29], E. Bergeaas Kuutmann[41], N. Berger[4], F. Berghaus[168], E. Berglund[49], J. Beringer[14], J. Bernabéu[166], P. Bernat[115], R. Bernhard[48], C. Bernius[77], T. Berry[76], A. Bertin[19a,19b], M.I. Besana[89a,89b], N. Besson[136], S. Bethke[99], R.M. Bianchi[48], M. Bianco[72a,72b], O. Biebel[98], J. Biesiada[14], M. Biglietti[132a,132b], H. Bilokon[47], M. Bindi[19a,19b], A. Bingul[18c], C. Bini[132a,132b], C. Biscarat[179], U. Bitenc[48], K.M. Black[57], R.E. Blair[5], J.-B. Blanchard[115], G. Blanchot[29], C. Blocker[22], A. Blondel[49], W. Blum[81], U. Blumenschein[54], G.J. Bobbink[105], A. Bocci[44], M. Boehler[41], J. Boek[173], N. Boelaert[79], S. Böser[77], J.A. Bogaerts[29], A. Bogouch[90,*], C. Bohm[145a], J. Bohm[125], V. Boisvert[76], T. Bold[162,c], V. Boldea[25a], V.G. Bondarenko[96], M. Bondioli[162], M. Boonekamp[136], S. Bordoni[78], C. Borer[16], A. Borisov[128], G. Borissov[71], I. Borjanovic[12a], S. Borroni[132a,132b], K. Bos[105], D. Boscherini[19a], M. Bosman[11], H. Boterenbrood[105], J. Bouchami[93], J. Boudreau[123], E.V. Bouhova-Thacker[71], C. Boulahouache[123], C. Bourdarios[115], A. Boveia[30], J. Boyd[29], I.R. Boyko[65], I. Bozovic-Jelisavcic[12b], J. Bracinik[17], A. Braem[29], P. Branchini[134a], A. Brandt[7], G. Brandt[41], O. Brandt[54], U. Bratzler[155], B. Brau[84], J.E. Brau[114], H.M. Braun[173], B. Brelier[157], J. Bremer[29], R. Brenner[165], S. Bressler[151], D. Britton[53], F.M. Brochu[27], I. Brock[20], R. Brock[88], E. Brodet[152], C. Bromberg[88], G. Brooijmans[34], W.K. Brooks[31b], G. Brown[82], P.A. Bruckman de Renstrom[38], D. Bruncko[144b], R. Bruneliere[48], S. Brunet[41], A. Bruni[19a], G. Bruni[19a], M. Bruschi[19a], F. Bucci[49], J. Buchanan[118], P. Buchholz[141], A.G. Buckley[45], I.A. Budagov[65], B. Budick[108], V. Büscher[81], L. Bugge[117], O. Bulekov[96], M. Bunse[42], T. Buran[117], H. Burckhart[29], S. Burdin[73], T. Burgess[13], S. Burke[129], E. Busato[33], P. Bussey[53], C.P. Buszello[165], F. Butin[29], B. Butler[143], J.M. Butler[21], C.M. Buttar[53], J.M. Butterworth[77], T. Byatt[77], J. Caballero[24], S. Cabrera Urbán[166], D. Caforio[19a,19b], O. Cakir[3a], P. Calafiura[14], G. Calderini[78], P. Calfayan[98], R. Calkins[106],

L.P. Caloba[23a], D. Calvet[33], P. Camarri[133a,133b], D. Cameron[117], S. Campana[29], M. Campanelli[77], V. Canale[102a,102b], F. Canelli[30], A. Canepa[158a], J. Cantero[80], L. Capasso[102a,102b], M.D.M. Capeans Garrido[29], I. Caprini[25a], M. Caprini[25a], M. Capua[36a,36b], R. Caputo[147], C. Caramarcu[25a], R. Cardarelli[133a], T. Carli[29], G. Carlino[102a], L. Carminati[89a,89b], B. Caron[2,d], S. Caron[48], G.D. Carrillo Montoya[171], S. Carron Montero[157], A.A. Carter[75], J.R. Carter[27], J. Carvalho[124a], D. Casadei[108], M.P. Casado[11], M. Cascella[122a,122b], A.M. Castaneda Hernandez[171], E. Castaneda-Miranda[171], V. Castillo Gimenez[166], N.F. Castro[124b], G. Cataldi[72a], A. Catinaccio[29], J.R. Catmore[71], A. Cattai[29], G. Cattani[133a,133b], S. Caughron[34], P. Cavalleri[78], D. Cavalli[89a], M. Cavalli-Sforza[11], V. Cavasinni[122a,122b], F. Ceradini[134a,134b], A.S. Cerqueira[23a], A. Cerri[29], L. Cerrito[75], F. Cerutti[47], S.A. Cetin[18b], A. Chafaq[135a], D. Chakraborty[106], K. Chan[2], J.D. Chapman[27], J.W. Chapman[87], E. Chareyre[78], D.G. Charlton[17], V. Chavda[82], S. Cheatham[71], S. Chekanov[5], S.V. Chekulaev[158a], G.A. Chelkov[65], H. Chen[24], S. Chen[32c], X. Chen[171], A. Cheplakov[65], V.F. Chepurnov[65], R. Cherkaoui El Moursli[135d], V. Tcherniatine[24], D. Chesneanu[25a], E. Cheu[6], S.L. Cheung[157], L. Chevalier[136], F. Chevallier[136], G. Chiefari[102a,102b], L. Chikovani[51], J.T. Childers[58a], A. Chilingarov[71], G. Chiodini[72a], V. Chizhov[65], G. Choudalakis[30], S. Chouridou[137], I.A. Christidi[77], A. Christov[48], D. Chromek-Burckhart[29], M.L. Chu[150], J. Chudoba[125], G. Ciapetti[132a,132b], A.K. Ciftci[3a], R. Ciftci[3a], D. Cinca[33], V. Cindro[74], M.D. Ciobotaru[162], C. Ciocca[19a,19b], A. Ciocio[14], M. Cirilli[87,e], A. Clark[49], P.J. Clark[45], W. Cleland[123], J.C. Clemens[83], B. Clement[55], C. Clement[145a,145b], Y. Coadou[83], M. Cobal[163a,163c], A. Coccaro[50a,50b], J. Cochran[64], J. Coggeshall[164], E. Cogneras[179], A.P. Colijn[105], C. Collard[115], N.J. Collins[17], C. Collins-Tooth[53], J. Collot[55], G. Colon[84], P. Conde Muiño[124a], E. Coniavitis[165], M.C. Conidi[11], M. Consonni[104], S. Constantinescu[25a], C. Conta[119a,119b], F. Conventi[102a,f], M. Cooke[34], B.D. Cooper[75], A.M. Cooper-Sarkar[118], N.J. Cooper-Smith[76], K. Copic[34], T. Cornelissen[50a,50b], M. Corradi[19a], F. Corriveau[85,g], A. Corso-Radu[162], A. Cortes-Gonzalez[164], G. Cortiana[99], G. Costa[89a], M.J. Costa[166], D. Costanzo[139], T. Costin[30], D. Côté[41], R. Coura Torres[23a], L. Courneyea[168], G. Cowan[76], C. Cowden[27], B.E. Cox[82], K. Cranmer[108], J. Cranshaw[5], M. Cristinziani[20], G. Crosetti[36a,36b], R. Crupi[72a,72b], S. Crépé-Renaudin[55], C. Cuenca Almenar[174], T. Cuhadar Donszelmann[139], M. Curatolo[47], C.J. Curtis[17], P. Cwetanski[61], Z. Czyczula[174], S. D'Auria[53], M. D'Onofrio[73], A. D'Orazio[99], C. Da Via[82], W. Dabrowski[37], T. Dai[87], C. Dallapiccola[84], S.J. Dallison[129,*], C.H. Daly[138], M. Dam[35], H.O. Danielsson[29], D. Dannheim[99], V. Dao[49], G. Darbo[50a], G.L. Darlea[25b], W. Davey[86], T. Davidek[126], N. Davidson[86], R. Davidson[71], M. Davies[93], A.R. Davison[77], I. Dawson[139], R.K. Daya[39], K. De[7], R. de Asmundis[102a], S. De Castro[19a,19b], P.E. De Castro Faria Salgado[24], S. De Cecco[78], J. de Graat[98], N. De Groot[104], P. de Jong[105], L. De Mora[71], M. De Oliveira Branco[29], D. De Pedis[132a], A. De Salvo[132a], U. De Sanctis[163a,163c], A. De Santo[148], J.B. De Vivie De Regie[115], S. Dean[77], D.V. Dedovich[65], J. Degenhardt[120], M. Dehchar[118], C. Del Papa[163a,163c], J. Del Peso[80], T. Del Prete[122a,122b], A. Dell'Acqua[29], L. Dell'Asta[89a,89b], M. Della Pietra[102a,h], D. della Volpe[102a,102b], M. Delmastro[29], P.A. Delsart[55], C. Deluca[147], S. Demers[174], M. Demichev[65], B. Demirkoz[11], J. Deng[162], W. Deng[24], S.P. Denisov[128], J.E. Derkaoui[135c], F. Derue[78], P. Dervan[73], K. Desch[20], P.O. Deviveiros[157], A. Dewhurst[129], B. DeWilde[147], S. Dhaliwal[157], R. Dhullipudi[24,i], A. Di Ciaccio[133a,133b], L. Di Ciaccio[4], A. Di Girolamo[29], B. Di Girolamo[29], S. Di Luise[134a,134b], A. Di Mattia[88], R. Di Nardo[133a,133b], A. Di Simone[133a,133b], R. Di Sipio[19a,19b], M.A. Diaz[31a], F. Diblen[18c], E.B. Diehl[87], J. Dietrich[48], T.A. Dietzsch[58a], S. Diglio[115], K. Dindar Yagci[39], J. Dingfelder[48], C. Dionisi[132a,132b], P. Dita[25a], S. Dita[25a], F. Dittus[29], F. Djama[83], R. Djilkibaev[108], T. Djobava[51], M.A.B. do Vale[23a], A. Do Valle Wemans[124a], T.K.O. Doan[4], D. Dobos[29], E. Dobson[29], M. Dobson[162], C. Doglioni[118], T. Doherty[53], J. Dolejsi[126], I. Dolenc[74], Z. Dolezal[126], B.A. Dolgoshein[96], T. Dohmae[154], M. Donega[120], J. Donini[55], J. Dopke[173], A. Doria[102a], A. Dos Anjos[171], A. Dotti[122a,122b], M.T. Dova[70], A. Doxiadis[105], A.T. Doyle[53], Z. Drasal[126], M. Dris[9], J. Dubbert[99], E. Duchovni[170], G. Duckeck[98], A. Dudarev[29], F. Dudziak[115], M. Dührssen[29], L. Duflot[115], M.-A. Dufour[85], M. Dunford[30], H. Duran Yildiz[3b], R. Duxfield[139], M. Dwuznik[37], M. Düren[52], W.L. Ebenstein[44], J. Ebke[98], S. Eckweiler[81], K. Edmonds[81], C.A. Edwards[76], K. Egorov[61], W. Ehrenfeld[41], T. Ehrich[99], T. Eifert[29], G. Eigen[13], K. Einsweiler[14], E. Eisenhandler[75], T. Ekelof[165], M. El Kacimi[4], M. Ellert[165], S. Elles[4], F. Ellinghaus[81], K. Ellis[75], N. Ellis[29], J. Elmsheuser[98], M. Elsing[29], D. Emeliyanov[129], R. Engelmann[147], A. Engl[98], B. Epp[62], A. Eppig[87], J. Erdmann[54], A. Ereditato[16], D. Eriksson[145a], I. Ermoline[88], J. Ernst[1], M. Ernst[24], J. Ernwein[136], D. Errede[164], S. Errede[164], E. Ertel[81], M. Escalier[115], C. Escobar[166], X. Espinal Curull[11], B. Esposito[47], A.I. Etienvre[136], E. Etzion[152], H. Evans[61], L. Fabbri[19a,19b], C. Fabre[29], K. Facius[35], R.M. Fakhrutdinov[128], S. Falciano[132a], Y. Fang[171], M. Fanti[89a,89b], A. Farbin[7], A. Farilla[134a], J. Farley[147], T. Farooque[157], S.M. Farrington[118], P. Farthouat[29], P. Fassnacht[29], D. Fassouliotis[8], B. Fatholahzadeh[157], L. Fayard[115], F. Fayette[54], R. Febbraro[33], P. Federic[144a], O.L. Fedin[121], W. Fedorko[29], L. Feligioni[83], C.U. Felzmann[86], C. Feng[32d], E.J. Feng[30], A.B. Fenyuk[128], J. Ferencei[144b], J. Ferland[93], B. Fernandes[124a], W. Fernando[109], S. Ferrag[53], J. Ferrando[118],

V. Ferrara[41], A. Ferrari[165], P. Ferrari[105], R. Ferrari[119a], A. Ferrer[166], M.L. Ferrer[47], D. Ferrere[49], C. Ferretti[87], M. Fiascaris[118], F. Fiedler[81], A. Filipčič[74], A. Filippas[9], F. Filthaut[104], M. Fincke-Keeler[168], M.C.N. Fiolhais[124a], L. Fiorini[11], A. Firan[39], G. Fischer[41], M.J. Fisher[109], M. Flechl[48], I. Fleck[141], J. Fleckner[81], P. Fleischmann[172], S. Fleischmann[20], T. Flick[173], L.R. Flores Castillo[171], M.J. Flowerdew[99], T. Fonseca Martin[76], A. Formica[136], A. Forti[82], D. Fortin[158a], D. Fournier[115], A.J. Fowler[44], K. Fowler[137], H. Fox[71], P. Francavilla[122a,122b], S. Franchino[119a,119b], D. Francis[29], M. Franklin[57], S. Franz[29], M. Fraternali[119a,119b], S. Fratina[120], J. Freestone[82], S.T. French[27], R. Froeschl[29], D. Froidevaux[29], J.A. Frost[27], C. Fukunaga[155], E. Fullana Torregrosa[5], J. Fuster[166], C. Gabaldon[80], O. Gabizon[170], T. Gadfort[24], S. Gadomski[49], G. Gagliardi[50a,50b], P. Gagnon[61], C. Galea[98], E.J. Gallas[118], V. Gallo[16], B.J. Gallop[129], P. Gallus[125], E. Galyaev[40], K.K. Gan[109], Y.S. Gao[143,j], A. Gaponenko[14], M. Garcia-Sciveres[14], C. García[166], J.E. García Navarro[49], R.W. Gardner[30], N. Garelli[29], H. Garitaonandia[105], V. Garonne[29], C. Gatti[47], G. Gaudio[119a], V. Gautard[136], P. Gauzzi[132a,132b], I.L. Gavrilenko[94], C. Gay[167], G. Gaycken[20], E.N. Gazis[9], P. Ge[32d], C.N.P. Gee[129], Ch. Geich-Gimbel[20], K. Gellerstedt[145a,145b], C. Gemme[50a], M.H. Genest[98], S. Gentile[132a,132b], F. Georgatos[9], S. George[76], A. Gershon[152], H. Ghazlane[135d], N. Ghodbane[33], B. Giacobbe[19a], S. Giagu[132a,132b], V. Giakoumopoulou[8], V. Giangiobbe[122a,122b], F. Gianotti[29], B. Gibbard[24], A. Gibson[157], S.M. Gibson[118], L.M. Gilbert[118], M. Gilchriese[14], V. Gilewsky[91], D.M. Gingrich[2,k], J. Ginzburg[152], N. Giokaris[8], M.P. Giordani[163a,163c], R. Giordano[102a,102b], F.M. Giorgi[15], P. Giovannini[99], P.F. Giraud[29], P. Girtler[62], D. Giugni[89a], P. Giusti[19a], B.K. Gjelsten[117], L.K. Gladilin[97], C. Glasman[80], A. Glazov[41], K.W. Glitza[173], G.L. Glonti[65], J. Godfrey[142], J. Godlewski[29], M. Goebel[41], T. Göpfert[43], C. Goeringer[81], C. Gössling[42], T. Göttfert[99], V. Goggi[119a,119b,l], S. Goldfarb[87], D. Goldin[39], T. Golling[174], A. Gomes[124a], L.S. Gomez Fajardo[41], R. Gonçalo[76], L. Gonella[20], C. Gong[32b], S. González de la Hoz[166], M.L. Gonzalez Silva[26], S. Gonzalez-Sevilla[49], J.J. Goodson[147], L. Goossens[29], H.A. Gordon[24], I. Gorelov[103], G. Gorfine[173], B. Gorini[29], E. Gorini[72a,72b], A. Gorišek[74], E. Gornicki[38], B. Gosdzik[41], M. Gosselink[105], M.I. Gostkin[65], I. Gough Eschrich[162], M. Gouighri[135a], D. Goujdami[135a], M.P. Goulette[49], A.G. Goussiou[138], C. Goy[4], I. Grabowska-Bold[162,m], P. Grafström[29], K.-J. Grahn[146], S. Grancagnolo[15], V. Grassi[147], V. Gratchev[121], N. Grau[34], H.M. Gray[34,n], J.A. Gray[147], E. Graziani[134a], B. Green[76], T. Greenshaw[73], Z.D. Greenwood[24,o], I.M. Gregor[41], P. Grenier[143], E. Griesmayer[46], J. Griffiths[138], N. Grigalashvili[65], A.A. Grillo[137], K. Grimm[147], S. Grinstein[11], Y.V. Grishkevich[97], M. Groh[99], M. Groll[81], E. Gross[170], J. Grosse-Knetter[54], J. Groth-Jensen[79], K. Grybel[141], C. Guicheney[33], A. Guida[72a,72b], T. Guillemin[4], H. Guler[85,p], J. Gunther[125], B. Guo[157], Y. Gusakov[65], A. Gutierrez[93], P. Gutierrez[111], N. Guttman[152], O. Gutzwiller[171], C. Guyot[136], C. Gwenlan[118], C.B. Gwilliam[73], A. Haas[143], S. Haas[29], C. Haber[14], H.K. Hadavand[39], D.R. Hadley[17], P. Haefner[99], Z. Hajduk[38], H. Hakobyan[175], J. Haller[41,q], K. Hamacher[173], A. Hamilton[49], S. Hamilton[160], L. Han[32b], K. Hanagaki[116], M. Hance[120], C. Handel[81], P. Hanke[58a], J.R. Hansen[35], J.B. Hansen[35], J.D. Hansen[35], P.H. Hansen[35], T. Hansl-Kozanecka[137], P. Hansson[143], K. Hara[159], G.A. Hare[137], T. Harenberg[173], R.D. Harrington[21], O.M. Harris[138], K. Harrison[17], J. Hartert[48], F. Hartjes[105], A. Harvey[56], S. Hasegawa[101], Y. Hasegawa[140], S. Hassani[136], S. Haug[16], M. Hauschild[29], R. Hauser[88], M. Havranek[125], C.M. Hawkes[17], R.J. Hawkings[29], T. Hayakawa[67], H.S. Hayward[73], S.J. Haywood[129], S.J. Head[82], V. Hedberg[79], L. Heelan[28], S. Heim[88], B. Heinemann[14], S. Heisterkamp[35], L. Helary[4], M. Heller[115], S. Hellman[145a,145b], C. Helsens[11], T. Hemperek[20], R.C.W. Henderson[71], M. Henke[58a], A. Henrichs[54], A.M. Henriques Correia[29], S. Henrot-Versille[115], C. Hensel[54], T. Henß[173], Y. Hernández Jiménez[166], A.D. Hershenhorn[151], G. Herten[48], R. Hertenberger[98], L. Hervas[29], N.P. Hessey[105], E. Higón-Rodriguez[166], J.C. Hill[27], K.H. Hiller[41], S. Hillert[145a,145b], S.J. Hillier[17], I. Hinchliffe[14], E. Hines[120], M. Hirose[116], F. Hirsch[42], D. Hirschbuehl[173], J. Hobbs[147], N. Hod[152], M.C. Hodgkinson[139], P. Hodgson[139], A. Hoecker[29], M.R. Hoeferkamp[103], J. Hoffman[39], D. Hoffmann[83], M. Hohlfeld[81], T. Holy[127], J.L. Holzbauer[88], Y. Homma[67], T. Horazdovsky[127], T. Hori[67], C. Horn[143], S. Horner[48], S. Horvat[99], J.-Y. Hostachy[55], S. Hou[150], A. Hoummada[135a], T. Howe[39], J. Hrivnac[115], T. Hryn'ova[4], P.J. Hsu[174], S.-C. Hsu[14], G.S. Huang[111], Z. Hubacek[127], F. Hubaut[83], F. Huegging[20], T.B. Huffman[118], E.W. Hughes[34], G. Hughes[71], M. Hurwitz[30], U. Husemann[41], N. Huseynov[10], J. Huston[88], J. Huth[57], G. Iacobucci[102a], G. Iakovidis[9], I. Ibragimov[141], L. Iconomidou-Fayard[115], J. Idarraga[158b], P. Iengo[4], O. Igonkina[105], Y. Ikegami[66], M. Ikeno[66], Y. Ilchenko[39], D. Iliadis[153], T. Ince[20], P. Ioannou[8], M. Iodice[134a], A. Irles Quiles[166], A. Ishikawa[67], M. Ishino[66], R. Ishmukhametov[39], T. Isobe[154], C. Issever[118], S. Istin[18a], Y. Itoh[101], A.V. Ivashin[128], W. Iwanski[38], H. Iwasaki[66], J.M. Izen[40], V. Izzo[102a], B. Jackson[120], J.N. Jackson[73], P. Jackson[143], M.R. Jaekel[29], V. Jain[61], K. Jakobs[48], S. Jakobsen[35], J. Jakubek[127], D.K. Jana[111], E. Jankowski[157], E. Jansen[77], A. Jantsch[99], M. Janus[48], G. Jarlskog[79], L. Jeanty[57], I. Jen-La Plante[30], P. Jenni[29], P. Jez[35], S. Jézéquel[4], W. Ji[79], J. Jia[147], Y. Jiang[32b], M. Jimenez Belenguer[29], S. Jin[32a], O. Jinnouchi[156], D. Joffe[39], M. Johansen[145a,145b], K.E. Johansson[145a],

P. Johansson[139], S. Johnert[41], K.A. Johns[6], K. Jon-And[145a,145b], G. Jones[82], R.W.L. Jones[71], T.J. Jones[73], P.M. Jorge[124a], J. Joseph[14], V. Juranek[125], P. Jussel[62], V.V. Kabachenko[128], M. Kaci[166], A. Kaczmarska[38], M. Kado[115], H. Kagan[109], M. Kagan[57], S. Kaiser[99], E. Kajomovitz[151], S. Kalinin[173], L.V. Kalinovskaya[65], S. Kama[41], N. Kanaya[154], M. Kaneda[154], V.A. Kantserov[96], J. Kanzaki[66], B. Kaplan[174], A. Kapliy[30], J. Kaplon[29], D. Kar[43], M. Karagounis[20], M. Karagoz Unel[118], M. Karnevskiy[41], V. Kartvelishvili[71], A.N. Karyukhin[128], L. Kashif[57], A. Kasmi[39], R.D. Kass[109], A. Kastanas[13], M. Kastoryano[174], M. Kataoka[4], Y. Kataoka[154], E. Katsoufis[9], J. Katzy[41], V. Kaushik[6], K. Kawagoe[67], T. Kawamoto[154], G. Kawamura[81], M.S. Kayl[105], F. Kayumov[94], V.A. Kazanin[107], M.Y. Kazarinov[65], J.R. Keates[82], R. Keeler[168], P.T. Keener[120], R. Kehoe[39], M. Keil[54], G.D. Kekelidze[65], M. Kelly[82], M. Kenyon[53], O. Kepka[125], N. Kerschen[29], B.P. Kerševan[74], S. Kersten[173], K. Kessoku[154], M. Khakzad[28], F. Khalil-zada[10], H. Khandanyan[164], A. Khanov[112], D. Kharchenko[65], A. Khodinov[147], A. Khomich[58a], G. Khoriauli[20], N. Khovanskiy[65], V. Khovanskiy[95], E. Khramov[65], J. Khubua[51], H. Kim[7], M.S. Kim[2], P.C. Kim[143], S.H. Kim[159], O. Kind[15], P. Kind[173], B.T. King[73], J. Kirk[129], G.P. Kirsch[118], L.E. Kirsch[22], A.E. Kiryunin[99], D. Kisielewska[37], T. Kittelmann[123], H. Kiyamura[67], E. Kladiva[144b], M. Klein[73], U. Klein[73], K. Kleinknecht[81], M. Klemetti[85], A. Klier[170], A. Klimentov[24], R. Klingenberg[42], E.B. Klinkby[44], T. Klioutchnikova[29], P.F. Klok[104], S. Klous[105], E.-E. Kluge[58a], T. Kluge[73], P. Kluit[105], M. Klute[54], S. Kluth[99], N.S. Knecht[157], E. Kneringer[62], B.R. Ko[44], T. Kobayashi[154], M. Kobel[43], B. Koblitz[29], M. Kocian[143], A. Kocnar[113], P. Kodys[126], K. Köneke[41], A.C. König[104], S. Koenig[81], L. Köpke[81], F. Koetsveld[104], P. Koevesarki[20], T. Koffas[29], E. Koffeman[105], F. Kohn[54], Z. Kohout[127], T. Kohriki[66], H. Kolanoski[15], V. Kolesnikov[65], I. Koletsou[4], J. Koll[88], D. Kollar[29], S. Kolos[162,r], S.D. Kolya[82], A.A. Komar[94], J.R. Komaragiri[142], T. Kondo[66], T. Kono[41,s], R. Konoplich[108], S.P. Konovalov[94], N. Konstantinidis[77], S. Koperny[37], K. Korcyl[38], K. Kordas[153], A. Korn[14], I. Korolkov[11], E.V. Korolkova[139], V.A. Korotkov[128], O. Kortner[99], P. Kostka[41], V.V. Kostyukhin[20], S. Kotov[99], V.M. Kotov[65], K.Y. Kotov[107], C. Kourkoumelis[8], A. Koutsman[105], R. Kowalewski[168], H. Kowalski[41], T.Z. Kowalski[37], W. Kozanecki[136], A.S. Kozhin[128], V. Kral[127], V.A. Kramarenko[97], G. Kramberger[74], M.W. Krasny[78], A. Krasznahorkay[108], J. Kraus[88], A. Kreisel[152], F. Krejci[127], J. Kretzschmar[73], N. Krieger[54], P. Krieger[157], K. Kroeninger[54], H. Kroha[99], J. Kroll[120], J. Kroseberg[20], J. Krstic[12a], U. Kruchonak[65], H. Krüger[20], Z.V. Krumshteyn[65], T. Kubota[154], S. Kuehn[48], A. Kugel[58c], T. Kuhl[173], D. Kuhn[62], V. Kukhtin[65], Y. Kulchitsky[90], S. Kuleshov[31b], C. Kummer[98], M. Kuna[83], J. Kunkle[120], A. Kupco[125], H. Kurashige[67], M. Kurata[159], Y.A. Kurochkin[90], V. Kus[125], R. Kwee[15], A. La Rosa[29], L. La Rotonda[36a,36b], J. Labbe[4], C. Lacasta[166], F. Lacava[132a,132b], H. Lacker[15], D. Lacour[78], V.R. Lacuesta[166], E. Ladygin[65], R. Lafaye[4], B. Laforge[78], T. Lagouri[80], S. Lai[48], M. Lamanna[29], C.L. Lampen[6], W. Lampl[6], E. Lancon[136], U. Landgraf[48], M.P.J. Landon[75], J.L. Lane[82], A.J. Lankford[162], F. Lanni[24], K. Lantzsch[29], A. Lanza[119a], S. Laplace[4], C. Lapoire[83], J.F. Laporte[136], T. Lari[89a], A. Larner[118], M. Lassnig[29], P. Laurelli[47], W. Lavrijsen[14], P. Laycock[73], A.B. Lazarev[65], A. Lazzaro[89a,89b], O. Le Dortz[78], E. Le Guirriec[83], E. Le Menedeu[136], A. Lebedev[64], C. Lebel[93], T. LeCompte[5], F. Ledroit-Guillon[55], H. Lee[105], J.S.H. Lee[149], S.C. Lee[150], M. Lefebvre[168], M. Legendre[136], B.C. LeGeyt[120], F. Legger[98], C. Leggett[14], M. Lehmacher[20], G. Lehmann Miotto[29], X. Lei[6], R. Leitner[126], D. Lellouch[170], J. Lellouch[78], V. Lendermann[58a], K.J.C. Leney[73], T. Lenz[173], G. Lenzen[173], B. Lenzi[136], K. Leonhardt[43], C. Leroy[93], J.-R. Lessard[168], C.G. Lester[27], A. Leung Fook Cheong[171], J. Levêque[83], D. Levin[87], L.J. Levinson[170], M. Leyton[15], H. Li[171], X. Li[87], Z. Liang[39], Z. Liang[150,t], B. Liberti[133a], P. Lichard[29], M. Lichtnecker[98], K. Lie[164], W. Liebig[105], J.N. Lilley[17], A. Limosani[86], M. Limper[63], S.C. Lin[150], J.T. Linnemann[88], E. Lipeles[120], L. Lipinsky[125], A. Lipniacka[13], T.M. Liss[164], D. Lissauer[24], A. Lister[49], A.M. Litke[137], C. Liu[28], D. Liu[150,u], H. Liu[87], J.B. Liu[87], M. Liu[32b], T. Liu[39], Y. Liu[32b], M. Livan[119a,119b], A. Lleres[55], S.L. Lloyd[75], E. Lobodzinska[41], P. Loch[6], W.S. Lockman[137], S. Lockwitz[174], T. Loddenkoetter[20], F.K. Loebinger[82], A. Loginov[174], C.W. Loh[167], T. Lohse[15], K. Lohwasser[48], M. Lokajicek[125], R.E. Long[71], L. Lopes[124a], D. Lopez Mateos[34,v], M. Losada[161], P. Loscutoff[14], X. Lou[40], A. Lounis[115], K.F. Loureiro[109], L. Lovas[144a], J. Love[21], P.A. Love[71], A.J. Lowe[61], F. Lu[32a], H.J. Lubatti[138], C. Luci[132a,132b], A. Lucotte[55], A. Ludwig[43], D. Ludwig[41], I. Ludwig[48], F. Luehring[61], D. Lumb[48], L. Luminari[132a], E. Lund[117], B. Lund-Jensen[146], B. Lundberg[79], J. Lundberg[29], J. Lundquist[35], D. Lynn[24], J. Lys[14], E. Lytken[79], H. Ma[24], L.L. Ma[171], J.A. Macana Goia[93], G. Maccarrone[47], A. Macchiolo[99], B. Maček[74], J. Machado Miguens[124a], R. Mackeprang[35], R.J. Madaras[14], W.F. Mader[43], R. Maenner[58c], T. Maeno[24], P. Mättig[173], S. Mättig[41], P.J. Magalhaes Martins[124a], E. Magradze[51], Y. Mahalalel[152], K. Mahboubi[48], A. Mahmood[1], C. Maiani[132a,132b], C. Maidantchik[23a], A. Maio[124a], S. Majewski[24], Y. Makida[66], M. Makouski[128], N. Makovec[115], Pa. Malecki[38], P. Malecki[38], V.P. Maleev[121], F. Malek[55], U. Mallik[63], D. Malon[5], S. Maltezos[9], V. Malyshev[107], S. Malyukov[65], M. Mambelli[30], R. Mameghani[98], J. Mamuzic[41], L. Mandelli[89a], I. Mandić[74], R. Mandrysch[15], J. Maneira[124a],

P.S. Mangeard[88], I.D. Manjavidze[65], P.M. Manning[137], A. Manousakis-Katsikakis[8], B. Mansoulie[136], A. Mapelli[29], L. Mapelli[29], L. March[80], J.F. Marchand[4], F. Marchese[133a,133b], G. Marchiori[78], M. Marcisovsky[125], C.P. Marino[61], F. Marroquim[23a], Z. Marshall[34,v], S. Marti-Garcia[166], A.J. Martin[75], A.J. Martin[174], B. Martin[29], B. Martin[88], F.F. Martin[120], J.P. Martin[93], T.A. Martin[17], B. Martin dit Latour[49], M. Martinez[11], V. Martinez Outschoorn[57], A.C. Martyniuk[82], F. Marzano[132a], A. Marzin[136], L. Masetti[20], T. Mashimo[154], R. Mashinistov[96], J. Masik[82], A.L. Maslennikov[107], I. Massa[19a,19b], N. Massol[4], A. Mastroberardino[36a,36b], T. Masubuchi[154], P. Matricon[115], H. Matsunaga[154], T. Matsushita[67], C. Mattravers[118,w], S.J. Maxfield[73], A. Mayne[139], R. Mazini[150], M. Mazur[48], J. Mc Donald[85], S.P. Mc Kee[87], A. McCarn[164], R.L. McCarthy[147], N.A. McCubbin[129], K.W. McFarlane[56], H. McGlone[53], G. Mchedlidze[51], S.J. McMahon[129], R.A. McPherson[168,g], A. Meade[84], J. Mechnich[105], M. Mechtel[173], M. Medinnis[41], R. Meera-Lebbai[111], T.M. Meguro[116], S. Mehlhase[41], A. Mehta[73], K. Meier[58a], B. Meirose[48], C. Melachrinos[30], B.R. Mellado Garcia[171], L. Mendoza Navas[161], Z. Meng[150,x], S. Menke[99], E. Meoni[11], P. Mermod[118], L. Merola[102a,102b], C. Meroni[89a], F.S. Merritt[30], A.M. Messina[29], J. Metcalfe[103], A.S. Mete[64], J.-P. Meyer[136], J. Meyer[172], J. Meyer[54], T.C. Meyer[29], W.T. Meyer[64], J. Miao[32d], S. Michal[29], L. Micu[25a], R.P. Middleton[129], S. Migas[73], L. Mijović[74], G. Mikenberg[170], M. Mikestikova[125], M. Mikuž[74], D.W. Miller[143], W.J. Mills[167], C.M. Mills[57], A. Milov[170], D.A. Milstead[145a,145b], D. Milstein[170], A.A. Minaenko[128], M. Miñano[166], I.A. Minashvili[65], A.I. Mincer[108], B. Mindur[37], M. Mineev[65], Y. Ming[130], L.M. Mir[11], G. Mirabelli[132a], S. Misawa[24], A. Misiejuk[76], J. Mitrevski[137], V.A. Mitsou[166], P.S. Miyagawa[82], J.U. Mjörnmark[79], T. Moa[145a,145b], S. Moed[57], V. Moeller[27], K. Mönig[41], N. Möser[20], W. Mohr[48], S. Mohrdieck-Möck[99], R. Moles-Valls[166], J. Molina-Perez[29], J. Monk[77], E. Monnier[83], S. Montesano[89a,89b], F. Monticelli[70], R.W. Moore[2], C. Mora Herrera[49], A. Moraes[53], A. Morais[124a], J. Morel[54], G. Morello[36a,36b], D. Moreno[161], M. Moreno Llácer[166], P. Morettini[50a], M. Morii[57], A.K. Morley[86], G. Mornacchi[29], S.V. Morozov[96], J.D. Morris[75], H.G. Moser[99], M. Mosidze[51], J. Moss[109], R. Mount[143], E. Mountricha[136], S.V. Mouraviev[94], E.J.W. Moyse[84], M. Mudrinic[12b], F. Mueller[58a], J. Mueller[123], K. Mueller[20], T.A. Müller[98], D. Muenstermann[42], A. Muir[167], Y. Munwes[152], R. Murillo Garcia[162], W.J. Murray[129], I. Mussche[105], E. Musto[102a,102b], A.G. Myagkov[128], M. Myska[125], J. Nadal[11], K. Nagai[159], K. Nagano[66], Y. Nagasaka[60], A.M. Nairz[29], K. Nakamura[154], I. Nakano[110], H. Nakatsuka[67], G. Nanava[20], A. Napier[160], M. Nash[77,y], N.R. Nation[21], T. Nattermann[20], T. Naumann[41], G. Navarro[161], S.K. Nderitu[20], H.A. Neal[87], E. Nebot[80], P. Nechaeva[94], A. Negri[119a,119b], G. Negri[29], A. Nelson[64], T.K. Nelson[143], S. Nemecek[125], P. Nemethy[108], A.A. Nepomuceno[23a], M. Nessi[29], M.S. Neubauer[164], A. Neusiedl[81], R.M. Neves[108], P. Nevski[24], F.M. Newcomer[120], R.B. Nickerson[118], R. Nicolaidou[136], L. Nicolas[139], G. Nicoletti[47], B. Nicquevert[29], F. Niedercorn[115], J. Nielsen[137], A. Nikiforov[15], K. Nikolaev[65], I. Nikolic-Audit[78], K. Nikolopoulos[8], H. Nilsen[48], P. Nilsson[7], A. Nisati[132a], T. Nishiyama[67], R. Nisius[99], L. Nodulman[5], M. Nomachi[116], I. Nomidis[153], M. Nordberg[29], B. Nordkvist[145a,145b], D. Notz[41], J. Novakova[126], M. Nozaki[66], M. Nožička[41], I.M. Nugent[158a], A.-E. Nuncio-Quiroz[20], G. Nunes Hanninger[20], T. Nunnemann[98], E. Nurse[77], D.C. O'Neil[142], V. O'Shea[53], F.G. Oakham[28,d], H. Oberlack[99], A. Ochi[67], S. Oda[154], S. Odaka[66], J. Odier[83], H. Ogren[61], A. Oh[82], S.H. Oh[44], C.C. Ohm[145a,145b], T. Ohshima[101], H. Ohshita[140], T. Ohsugi[59], S. Okada[67], H. Okawa[162], Y. Okumura[101], T. Okuyama[154], A.G. Olchevski[65], M. Oliveira[124a], D. Oliveira Damazio[24], E. Oliver Garcia[166], D. Olivito[120], A. Olszewski[38], J. Olszowska[38], C. Omachi[67,z], A. Onofre[124a], P.U.E. Onyisi[30], C.J. Oram[158a], M.J. Oreglia[30], Y. Oren[152], D. Orestano[134a,134b], I. Orlov[107], C. Oropeza Barrera[53], R.S. Orr[157], E.O. Ortega[130], B. Osculati[50a,50b], R. Ospanov[120], C. Osuna[11], J.P. Ottersbach[105], F. Ould-Saada[117], A. Ouraou[136], Q. Ouyang[32a], M. Owen[82], S. Owen[139], A. Oyarzun[31b], V.E. Ozcan[77], K. Ozone[66], N. Ozturk[7], A. Pacheco Pages[11], C. Padilla Aranda[11], E. Paganis[139], C. Pahl[63], F. Paige[24], K. Pajchel[117], S. Palestini[29], D. Pallin[33], A. Palma[124a], J.D. Palmer[17], Y.B. Pan[171], E. Panagiotopoulou[9], B. Panes[31a], N. Panikashvili[87], S. Panitkin[24], D. Pantea[25a], M. Panuskova[125], V. Paolone[123], Th.D. Papadopoulou[9], S.J. Park[54], W. Park[24,aa], M.A. Parker[27], F. Parodi[50a,50b], J.A. Parsons[34], U. Parzefall[48], E. Pasqualucci[132a], A. Passeri[134a], F. Pastore[134a,134b], Fr. Pastore[29], G. Pásztor[49,ab], S. Pataraia[99], J.R. Pater[82], S. Patricelli[102a,102b], T. Pauly[29], L.S. Peak[149], M. Pecsy[144a], M.I. Pedraza Morales[171], S.V. Peleganchuk[107], H. Peng[171], A. Penson[34], J. Penwell[61], M. Perantoni[23a], K. Perez[34,v], E. Perez Codina[11], M.T. Pérez García-Estañ[166], V. Perez Reale[34], L. Perini[89a,89b], H. Pernegger[29], R. Perrino[72a], S. Persembe[3a], P. Perus[115], V.D. Peshekhonov[65], B.A. Petersen[29], T.C. Petersen[35], E. Petit[83], C. Petridou[153], E. Petrolo[132a], F. Petrucci[134a,134b], D. Petschull[41], M. Petteni[142], R. Pezoa[31b], A. Phan[86], A.W. Phillips[27], P.W. Phillips[129], G. Piacquadio[29], M. Piccinini[19a,19b], R. Piegaia[26], J.E. Pilcher[30], A.D. Pilkington[82], J. Pina[124a], M. Pinamonti[163a,163c], J.L. Pinfold[2], B. Pinto[124a], C. Pizio[89a,89b], R. Placakyte[41], M. Plamondon[168], M.-A. Pleier[24], A. Poblaguev[174], S. Poddar[58a], F. Podlyski[33], L. Poggioli[115], M. Pohl[49], F. Polci[55], G. Polesello[119a], A. Policicchio[138], A. Polini[19a], J. Poll[75], V. Polychronakos[24], D. Pomeroy[22], K. Pommès[29], P. Ponsot[136],

L. Pontecorvo[132a], B.G. Pope[88], G.A. Popeneciu[25a], D.S. Popovic[12a], A. Poppleton[29], J. Popule[125], X. Portell Bueso[48], R. Porter[162], G.E. Pospelov[99], S. Pospisil[127], M. Potekhin[24], I.N. Potrap[99], C.J. Potter[148], C.T. Potter[85], K.P. Potter[82], G. Poulard[29], J. Poveda[171], R. Prabhu[20], P. Pralavorio[83], S. Prasad[57], R. Pravahan[7], L. Pribyl[29], D. Price[61], L.E. Price[5], P.M. Prichard[73], D. Prieur[123], M. Primavera[72a], K. Prokofiev[29], F. Prokoshin[31b], S. Protopopescu[24], J. Proudfoot[5], X. Prudent[43], H. Przysiezniak[4], S. Psoroulas[20], E. Ptacek[114], J. Purdham[87], M. Purohit[24,ac], P. Puzo[115], Y. Pylypchenko[117], M. Qi[32c], J. Qian[87], W. Qian[129], Z. Qin[41], A. Quadt[54], D.R. Quarrie[14], W.B. Quayle[171], F. Quinonez[31a], M. Raas[104], V. Radeka[24], V. Radescu[58b], B. Radics[20], T. Rador[18a], F. Ragusa[89a,89b], G. Rahal[179], A.M. Rahimi[109], S. Rajagopalan[24], M. Rammensee[48], M. Rammes[141], F. Rauscher[98], E. Rauter[99], M. Raymond[29], A.L. Read[117], D.M. Rebuzzi[119a,119b], A. Redelbach[172], G. Redlinger[24], R. Reece[120], K. Reeves[40], E. Reinherz-Aronis[152], A. Reinsch[114], I. Reisinger[42], D. Reljic[12a], C. Rembser[29], Z.L. Ren[150], P. Renkel[39], S. Rescia[24], M. Rescigno[132a], S. Resconi[89a], B. Resende[136], P. Reznicek[126], R. Rezvani[157], A. Richards[77], R. Richter[99], E. Richter-Was[38,ad], M. Ridel[78], M. Rijpstra[105], M. Rijssenbeek[147], A. Rimoldi[119a,119b], L. Rinaldi[19a], R.R. Rios[39], I. Riu[11], F. Rizatdinova[112], E. Rizvi[75], D.A. Roa Romero[161], S.H. Robertson[85,g], A. Robichaud-Veronneau[49], D. Robinson[27], J.E.M. Robinson[77], M. Robinson[114], A. Robson[53], J.G. Rocha de Lima[106], C. Roda[122a,122b], D. Roda Dos Santos[29], D. Rodriguez[161], Y. Rodriguez Garcia[15], S. Roe[29], O. Røhne[117], V. Rojo[1], S. Rolli[160], A. Romaniouk[96], V.M. Romanov[65], G. Romeo[26], D. Romero Maltrana[31a], L. Roos[78], E. Ros[166], S. Rosati[138], G.A. Rosenbaum[157], L. Rosselet[49], V. Rossetti[11], L.P. Rossi[50a], M. Rotaru[25a], J. Rothberg[138], D. Rousseau[115], C.R. Royon[136], A. Rozanov[83], Y. Rozen[151], X. Ruan[115], B. Ruckert[98], N. Ruckstuhl[105], V.I. Rud[97], G. Rudolph[62], F. Rühr[58a], F. Ruggieri[134a], A. Ruiz-Martinez[64], L. Rumyantsev[65], Z. Rurikova[48], N.A. Rusakovich[65], J.P. Rutherfoord[6], C. Ruwiedel[20], P. Ruzicka[125], Y.F. Ryabov[121], P. Ryan[88], G. Rybkin[115], S. Rzaeva[10], A.F. Saavedra[149], H.F.-W. Sadrozinski[137], R. Sadykov[65], F. Safai Tehrani[132a,132b], H. Sakamoto[154], G. Salamanna[105], A. Salamon[133a], M.S. Saleem[111], D. Salihagic[99], A. Salnikov[143], J. Salt[166], B.M. Salvachua Ferrando[5], D. Salvatore[36a,36b], F. Salvatore[148], A. Salvucci[47], A. Salzburger[29], D. Sampsonidis[153], B.H. Samset[117], H. Sandaker[13], H.G. Sander[81], M.P. Sanders[98], M. Sandhoff[173], P. Sandhu[157], R. Sandstroem[105], S. Sandvoss[173], D.P.C. Sankey[129], B. Sanny[173], A. Sansoni[47], C. Santamarina Rios[85], C. Santoni[33], R. Santonico[133a,133b], J.G. Saraiva[124a], T. Sarangi[171], E. Sarkisyan-Grinbaum[7], F. Sarri[122a,122b], O. Sasaki[66], N. Sasao[68], I. Satsounkevitch[90], G. Sauvage[4], P. Savard[157,d], A.Y. Savine[6], V. Savinov[123], L. Sawyer[24,ae], D.H. Saxon[53], L.P. Says[33], C. Sbarra[19a,19b], A. Sbrizzi[19a,19b], D.A. Scannicchio[29], J. Schaarschmidt[43], P. Schacht[99], U. Schäfer[81], S. Schaetzel[58b], A.C. Schaffer[115], D. Schaile[98], R.D. Schamberger[147], A.G. Schamov[107], V. Scharf[58a], V.A. Schegelsky[121], D. Scheirich[87], M. Schernau[162], M.I. Scherzer[14], C. Schiavi[50a,50b], J. Schieck[99], M. Schioppa[36a,36b], S. Schlenker[29], E. Schmidt[48], K. Schmieden[20], C. Schmitt[81], M. Schmitz[20], A. Schönig[58b], M. Schott[29], D. Schouten[142], J. Schovancova[125], M. Schram[85], A. Schreiner[63], C. Schroeder[81], N. Schroer[58c], M. Schroers[173], J. Schultes[173], H.-C. Schultz-Coulon[58a], J.W. Schumacher[43], M. Schumacher[48], B.A. Schumm[137], Ph. Schune[136], C. Schwanenberger[82], A. Schwartzman[143], Ph. Schwemling[78], R. Schwienhorst[88], R. Schwierz[43], J. Schwindling[136], W.G. Scott[129], J. Searcy[114], E. Sedykh[121], E. Segura[11], S.C. Seidel[103], A. Seiden[137], F. Seifert[43], J.M. Seixas[23a], G. Sekhniaidze[102a], D.M. Seliverstov[121], B. Sellden[145a], N. Semprini-Cesari[19a,19b], C. Serfon[98], L. Serin[115], R. Seuster[99], H. Severini[111], M.E. Sevior[86], A. Sfyrla[164], E. Shabalina[54], M. Shamim[114], L.Y. Shan[32a], J.T. Shank[21], Q.T. Shao[86], M. Shapiro[14], P.B. Shatalov[95], K. Shaw[139], D. Sherman[29], P. Sherwood[77], A. Shibata[108], M. Shimojima[100], T. Shin[56], A. Shmeleva[94], M.J. Shochet[30], M.A. Shupe[6], P. Sicho[125], A. Sidoti[15], F. Siegert[77], J. Siegrist[14], Dj. Sijacki[12a], O. Silbert[170], J. Silva[124a], Y. Silver[152], D. Silverstein[143], S.B. Silverstein[145a], V. Simak[127], Lj. Simic[12a], S. Simion[115], B. Simmons[77], M. Simonyan[35], P. Sinervo[157], N.B. Sinev[114], V. Sipica[141], G. Siragusa[81], A.N. Sisakyan[65], S.Yu. Sivoklokov[97], J. Sjoelin[145a,145b], T.B. Sjursen[13], K. Skovpen[107], P. Skubic[111], M. Slater[17], T. Slavicek[127], K. Sliwa[160], J. Sloper[29], V. Smakhtin[170], S.Yu. Smirnov[96], Y. Smirnov[24], L.N. Smirnova[97], O. Smirnova[79], B.C. Smith[57], D. Smith[143], K.M. Smith[53], M. Smizanska[71], K. Smolek[127], A.A. Snesarev[94], S.W. Snow[82], J. Snow[111], J. Snuverink[105], S. Snyder[24], M. Soares[124a], R. Sobie[168,g], J. Sodomka[127], A. Soffer[152], C.A. Solans[166], M. Solar[127], J. Solc[127], E. Solfaroli Camillocci[132a,132b], A.A. Solodkov[128], O.V. Solovyanov[128], J. Sondericker[24], V. Sopko[127], B. Sopko[127], M. Sosebee[7], A. Soukharev[107], S. Spagnolo[72a,72b], F. Spanò[34], R. Spighi[19a], G. Spigo[29], F. Spila[132a,132b], R. Spiwoks[29], M. Spousta[126], T. Spreitzer[142], B. Spurlock[7], R.D.St. Denis[53], T. Stahl[141], J. Stahlman[120], R. Stamen[58a], S.N. Stancu[162], E. Stanecka[29], R.W. Stanek[5], C. Stanescu[134a], S. Stapnes[117], E.A. Starchenko[128], J. Stark[55], P. Staroba[125], P. Starovoitov[91], J. Stastny[125], P. Stavina[144a], G. Steele[53], P. Steinbach[43], P. Steinberg[24], I. Stekl[127], B. Stelzer[142], H.J. Stelzer[41], O. Stelzer-Chilton[158a], H. Stenzel[52], K. Stevenson[75], G.A. Stewart[53], M.C. Stockton[29], K. Stoerig[48], G. Stoicea[25a], S. Stonjek[99], P. Strachota[126],

A.R. Stradling[7], A. Straessner[43], J. Strandberg[87], S. Strandberg[14], A. Strandlie[117], M. Strauss[111], P. Strizenec[144b],
R. Ströhmer[172], D.M. Strom[114], R. Stroynowski[39], J. Strube[129], B. Stugu[13], P. Sturm[173], D.A. Soh[150,af], D. Su[143],
Y. Sugaya[116], T. Sugimoto[101], C. Suhr[106], M. Suk[126], V.V. Sulin[94], S. Sultansoy[3d], T. Sumida[29], X.H. Sun[32d],
J.E. Sundermann[48], K. Suruliz[163a,163b], S. Sushkov[11], G. Susinno[36a,36b], M.R. Sutton[139], T. Suzuki[154], Y. Suzuki[66],
I. Sykora[144a], T. Sykora[126], T. Szymocha[38], J. Sánchez[166], D. Ta[20], K. Tackmann[29], A. Taffard[162], R. Tafirout[158a],
A. Taga[117], Y. Takahashi[101], H. Takai[24], R. Takashima[69], H. Takeda[67], T. Takeshita[140], M. Talby[83], A. Talyshev[107],
M.C. Tamsett[76], J. Tanaka[154], R. Tanaka[115], S. Tanaka[131], S. Tanaka[66], S. Tapprogge[81], D. Tardif[157], S. Tarem[151],
F. Tarrade[24], G.F. Tartarelli[89a], P. Tas[126], M. Tasevsky[125], E. Tassi[36a,36b], M. Tatarkhanov[14], C. Taylor[77],
F.E. Taylor[92], G.N. Taylor[86], R.P. Taylor[168], W. Taylor[158b], P. Teixeira-Dias[76], H. Ten Kate[29], P.K. Teng[150],
Y.D. Tennenbaum-Katan[151], S. Terada[66], K. Terashi[154], J. Terron[80], M. Terwort[41,q], M. Testa[47], R.J. Teuscher[157,g],
J. Therhaag[20], M. Thioye[174], S. Thoma[48], J.P. Thomas[17], E.N. Thompson[84], P.D. Thompson[17], P.D. Thompson[157],
R.J. Thompson[82], A.S. Thompson[53], E. Thomson[120], R.P. Thun[87], T. Tic[125], V.O. Tikhomirov[94], Y.A. Tikhonov[107],
P. Tipton[174], F.J. Tique Aires Viegas[29], S. Tisserant[83], B. Toczek[37], T. Todorov[4], S. Todorova-Nova[160],
B. Toggerson[162], J. Tojo[66], S. Tokár[144a], K. Tokushuku[66], K. Tollefson[88], L. Tomasek[125], M. Tomasek[125],
M. Tomoto[101], L. Tompkins[14], K. Toms[103], A. Tonoyan[13], C. Topfel[16], N.D. Topilin[65], I. Torchiani[29], E. Torrence[114],
E. Torró Pastor[166], J. Toth[83,ab], F. Touchard[83], D.R. Tovey[139], T. Trefzger[172], L. Tremblet[29], A. Tricoli[29],
I.M. Trigger[158a], S. Trincaz-Duvoid[78], T.N. Trinh[78], M.F. Tripiana[70], N. Triplett[64], W. Trischuk[157], A. Trivedi[24,ag],
B. Trocmé[55], C. Troncon[89a], A. Trzupek[38], C. Tsarouchas[9], J.C.-L. Tseng[118], M. Tsiakiris[105], P.V. Tsiareshka[90],
D. Tsionou[139], G. Tsipolitis[9], V. Tsiskaridze[51], E.G. Tskhadadze[51], I.I. Tsukerman[95], V. Tsulaia[123], J.-W. Tsung[20],
S. Tsuno[66], D. Tsybychev[147], J.M. Tuggle[30], D. Turecek[127], I. Turk Cakir[3e], E. Turlay[105], P.M. Tuts[34],
M.S. Twomey[138], M. Tylmad[145a,145b], M. Tyndel[129], K. Uchida[116], I. Ueda[154], R. Ueno[28], M. Ugland[13],
M. Uhlenbrock[20], M. Uhrmacher[54], F. Ukegawa[159], G. Unal[29], A. Undrus[24], G. Unel[162], Y. Unno[66], D. Urbaniec[34],
E. Urkovsky[152], P. Urquijo[49,ah], P. Urrejola[31a], G. Usai[7], M. Uslenghi[119a,119b], L. Vacavant[83], V. Vacek[127],
B. Vachon[85], S. Vahsen[14], P. Valente[132a], S. Valentinetti[19a,19b], S. Valkar[126], E. Valladolid Gallego[166], S. Vallecorsa[151],
J.A. Valls Ferrer[166], R. Van Berg[120], H. van der Graaf[105], E. van der Kraaij[105], E. van der Poel[105], D. van der Ster[29],
N. van Eldik[84], P. van Gemmeren[5], Z. van Kesteren[105], I. van Vulpen[105], W. Vandelli[29], A. Vaniachine[5], P. Vankov[73],
F. Vannucci[78], R. Vari[132a], E.W. Varnes[6], D. Varouchas[14], A. Vartapetian[7], K.E. Varvell[149], L. Vasilyeva[94],
V.I. Vassilakopoulos[56], F. Vazeille[33], C. Vellidis[8], F. Veloso[124a], S. Veneziano[132a], A. Ventura[72a,72b], D. Ventura[138],
M. Venturi[48], N. Venturi[16], V. Vercesi[119a], M. Verducci[172], W. Verkerke[105], J.C. Vermeulen[105], M.C. Vetterli[142,d],
I. Vichou[164], T. Vickey[118], G.H.A. Viehhauser[118], M. Villa[19a,19b], E.G. Villani[129], M. Villaplana Perez[166],
E. Vilucchi[47], M.G. Vincter[28], E. Vinek[29], V.B. Vinogradov[65], S. Viret[33], J. Virzi[14], A. Vitale[19a,19b], O. Vitells[170],
I. Vivarelli[48], F. Vives Vaque[11], S. Vlachos[9], M. Vlasak[127], N. Vlasov[20], A. Vogel[20], P. Vokac[127], M. Volpi[11],
H. von der Schmitt[99], J. von Loeben[99], H. von Radziewski[48], E. von Toerne[20], V. Vorobel[126], V. Vorwerk[11],
M. Vos[166], R. Voss[29], T.T. Voss[173], J.H. Vossebeld[73], N. Vranjes[12a], M. Vranjes Milosavljevic[12a], V. Vrba[125],
M. Vreeswijk[105], T. Vu Anh[81], D. Vudragovic[12a], R. Vuillermet[29], I. Vukotic[115], P. Wagner[120], J. Walbersloh[42],
J. Walder[71], R. Walker[98], W. Walkowiak[141], R. Wall[174], C. Wang[44], H. Wang[171], J. Wang[55], S.M. Wang[150],
A. Warburton[85], C.P. Ward[27], M. Warsinsky[48], R. Wastie[118], P.M. Watkins[17], A.T. Watson[17], M.F. Watson[17],
G. Watts[138], S. Watts[82], A.T. Waugh[149], B.M. Waugh[77], M.D. Weber[16], M. Weber[129], M.S. Weber[16], P. Weber[58a],
A.R. Weidberg[118], J. Weingarten[54], C. Weiser[48], H. Wellenstein[22], P.S. Wells[29], T. Wenaus[24], S. Wendler[123],
T. Wengler[82], S. Wenig[29], N. Wermes[20], M. Werner[48], P. Werner[29], M. Werth[162], U. Werthenbach[141], M. Wessels[58a],
K. Whalen[28], A. White[7], M.J. White[27], S. White[24], S.R. Whitehead[118], D. Whiteson[162], D. Whittington[61],
F. Wicek[115], D. Wicke[81], F.J. Wickens[129], W. Wiedenmann[171], M. Wielers[129], P. Wienemann[20], C. Wiglesworth[73],
L.A.M. Wiik[48], A. Wildauer[166], M.A. Wildt[41,q], H.G. Wilkens[29], E. Williams[34], H.H. Williams[120], S. Willocq[84],
J.A. Wilson[17], M.G. Wilson[143], A. Wilson[87], I. Wingerter-Seez[4], F. Winklmeier[29], M. Wittgen[143], M.W. Wolter[38],
H. Wolters[124a], B.K. Wosiek[38], J. Wotschack[29], M.J. Woudstra[84], K. Wraight[53], C. Wright[53], D. Wright[143],
B. Wrona[73], S.L. Wu[171], X. Wu[49], E. Wulf[34], B.M. Wynne[45], L. Xaplanteris[9], S. Xella[35], S. Xie[48], D. Xu[139], N. Xu[171],
M. Yamada[159], A. Yamamoto[66], K. Yamamoto[64], S. Yamamoto[154], T. Yamamura[154], J. Yamaoka[44], T. Yamazaki[154],
Y. Yamazaki[67], Z. Yan[21], H. Yang[87], U.K. Yang[82], Z. Yang[145a,145b], W.-M. Yao[14], Y. Yao[14], Y. Yasu[66], J. Ye[39], S. Ye[24],
M. Yilmaz[3c], R. Yoosoofmiya[123], K. Yorita[169], R. Yoshida[5], C. Young[143], S.P. Youssef[21], D. Yu[24], J. Yu[7], L. Yuan[78],
A. Yurkewicz[147], R. Zaidan[63], A.M. Zaitsev[128], Z. Zajacova[29], V. Zambrano[47], L. Zanello[132a,132b], A. Zaytsev[107],
C. Zeitnitz[173], M. Zeller[174], A. Zemla[38], C. Zendler[20], O. Zenin[128], T. Zenis[144a], Z. Zenonos[122a,122b], S. Zenz[14],
D. Zerwas[115], G. Zevi della Porta[57], Z. Zhan[32d], H. Zhang[83], J. Zhang[5], Q. Zhang[5], X. Zhang[32d], L. Zhao[108],

T. Zhao[138], **Z. Zhao**[32b], **A. Zhemchugov**[65], **J. Zhong**[150,ai], **B. Zhou**[87], **N. Zhou**[34], **Y. Zhou**[150], **C.G. Zhu**[32d], **H. Zhu**[41], **Y. Zhu**[171], **X. Zhuang**[98], **V. Zhuravlov**[99], **R. Zimmermann**[20], **S. Zimmermann**[20], **S. Zimmermann**[48], **M. Ziolkowski**[141], **L. Živković**[34], **G. Zobernig**[171], **A. Zoccoli**[19a,19b], **M. zur Nedden**[15], **V. Zutshi**[106]

*CERN, 1211 Geneva 23, Switzerland

[1]University at Albany, 1400 Washington Ave, Albany, NY 12222, United States of America

[2]University of Alberta, Department of Physics, Centre for Particle Physics, Edmonton, AB T6G 2G7, Canada

[3]Ankara University[a], Faculty of Sciences, Department of Physics, TR 061000 Tandogan, Ankara; Dumlupinar University[b], Faculty of Arts and Sciences, Department of Physics, Kutahya; Gazi University[c], Faculty of Arts and Sciences, Department of Physics, 06500, Teknikokullar, Ankara; TOBB University of Economics and Technology[d], Faculty of Arts and Sciences, Division of Physics, 06560, Sogutozu, Ankara; Turkish Atomic Energy Authority[e], 06530, Lodumlu, Ankara, Turkey

[4]LAPP, Université de Savoie, CNRS/IN2P3, Annecy-le-Vieux, France

[5]Argonne National Laboratory, High Energy Physics Division, 9700 S. Cass Avenue, Argonne IL 60439, United States of America

[6]University of Arizona, Department of Physics, Tucson, AZ 85721, United States of America

[7]The University of Texas at Arlington, Department of Physics, Box 19059, Arlington, TX 76019, United States of America

[8]University of Athens, Nuclear & Particle Physics, Department of Physics, Panepistimiopouli, Zografou, GR 15771 Athens, Greece

[9]National Technical University of Athens, Physics Department, 9-Iroon Polytechniou, GR 15780 Zografou, Greece

[10]Institute of Physics, Azerbaijan Academy of Sciences, H. Javid Avenue 33, AZ 143 Baku, Azerbaijan

[11]Institut de Física d'Altes Energies, IFAE, Edifici Cn, Universitat Autònoma de Barcelona, ES-08193 Bellaterra (Barcelona), Spain

[12]University of Belgrade[a], Institute of Physics, P.O. Box 57, 11001 Belgrade; Vinca Institute of Nuclear Sciences[b], Mihajla Petrovica Alasa 12-14, 11001 Belgrade, Serbia

[13]University of Bergen, Department for Physics and Technology, Allegaten 55, NO-5007 Bergen, Norway

[14]Lawrence Berkeley National Laboratory and University of California, Physics Division, MS50B-6227, 1 Cyclotron Road, Berkeley, CA 94720, United States of America

[15]Humboldt University, Institute of Physics, Berlin, Newtonstr. 15, D-12489 Berlin, Germany

[16]University of Bern, Albert Einstein Center for Fundamental Physics, Laboratory for High Energy Physics, Sidlerstrasse 5, CH-3012 Bern, Switzerland

[17]University of Birmingham, School of Physics and Astronomy, Edgbaston, Birmingham B15 2TT, United Kingdom

[18]Bogazici University[a], Faculty of Sciences, Department of Physics, TR-80815 Bebek-Istanbul; Dogus University[b], Faculty of Arts and Sciences, Department of Physics, 34722, Kadikoy, Istanbul; [c]Gaziantep University, Faculty of Engineering, Department of Physics Engineering, 27310, Sehitkamil, Gaziantep, Turkey; Istanbul Technical University[d], Faculty of Arts and Sciences, Department of Physics, 34469, Maslak, Istanbul, Turkey

[19]INFN Sezione di Bologna[a]; Università di Bologna, Dipartimento di Fisica[b], viale C. Berti Pichat, 6/2, IT-40127 Bologna, Italy

[20]University of Bonn, Physikalisches Institut, Nussallee 12, D-53115 Bonn, Germany

[21]Boston University, Department of Physics, 590 Commonwealth Avenue, Boston, MA 02215, United States of America

[22]Brandeis University, Department of Physics, MS057, 415 South Street, Waltham, MA 02454, United States of America

[23]Universidade Federal do Rio De Janeiro, COPPE/EE/IF [a], Caixa Postal 68528, Ilha do Fundao, BR-21945-970 Rio de Janeiro; [b]Universidade de Sao Paulo, Instituto de Fisica, R.do Matao Trav. R.187, Sao Paulo-SP, 05508-900, Brazil

[24]Brookhaven National Laboratory, Physics Department, Bldg. 510A, Upton, NY 11973, United States of America

[25]National Institute of Physics and Nuclear Engineering[a], Bucharest-Magurele, Str. Atomistilor 407, P.O. Box MG-6, R-077125, Romania; University Politehnica Bucharest[b], Rectorat-AN 001, 313 Splaiul Independentei, sector 6, 060042 Bucuresti; West University[c] in Timisoara, Bd. Vasile Parvan 4, Timisoara, Romania

[26]Universidad de Buenos Aires, FCEyN, Dto. Fisica, Pab I-C. Universitaria, 1428 Buenos Aires, Argentina

[27]University of Cambridge, Cavendish Laboratory, J J Thomson Avenue, Cambridge CB3 0HE, United Kingdom

[28]Carleton University, Department of Physics, 1125 Colonel By Drive, Ottawa ON K1S 5B6, Canada

[29]CERN, CH-1211 Geneva 23, Switzerland

[30]University of Chicago, Enrico Fermi Institute, 5640 S. Ellis Avenue, Chicago, IL 60637, United States of America

[31]Pontificia Universidad Católica de Chile, Facultad de Fisica, Departamento de Fisica[a], Avda. Vicuna Mackenna 4860, San Joaquin, Santiago; Universidad Técnica Federico Santa María, Departamento de Física[b], Avda. España 1680, Casilla 110-V, Valparaíso, Chile

[32]Institute of High Energy Physics, Chinese Academy of Sciences[a], P.O. Box 918, 19 Yuquan Road, Shijing Shan District, CN-Beijing 100049; University of Science & Technology of China (USTC), Department of Modern Physics[b], Hefei, CN-Anhui 230026; Nanjing University, Department of Physics[c], 22 Hankou Road, Nanjing, 210093; Shandong University, High Energy Physics Group[d], Jinan, CN-Shandong 250100, China

[33]Laboratoire de Physique Corpusculaire, Clermont Université, Université Blaise Pascal, CNRS/IN2P3, FR-63177 Aubiere Cedex, France

[34]Columbia University, Nevis Laboratory, 136 So. Broadway, Irvington, NY 10533, United States of America

[35]University of Copenhagen, Niels Bohr Institute, Blegdamsvej 17, DK-2100 Kobenhavn 0, Denmark

[36]INFN Gruppo Collegato di Cosenza[a]; Università della Calabria, Dipartimento di Fisica[b], IT-87036 Arcavacata di Rende, Italy

[37]Faculty of Physics and Applied Computer Science of the AGH-University of Science and Technology (FPACS, AGH-UST), al. Mickiewicza 30, PL-30059 Cracow, Poland

[38]The Henryk Niewodniczanski Institute of Nuclear Physics, Polish Academy of Sciences, ul. Radzikowskiego 152, PL-31342 Krakow, Poland

[39]Southern Methodist University, Physics Department, 106 Fondren Science Building, Dallas, TX 75275-0175, United States of America

[40]University of Texas at Dallas, 800 West Campbell Road, Richardson, TX 75080-3021, United States of America

[41]DESY, Notkestr. 85, D-22603 Hamburg and Platanenallee 6, D-15738 Zeuthen, Germany

[42]TU Dortmund, Experimentelle Physik IV, DE-44221 Dortmund, Germany

[43]Technical University Dresden, Institut für Kern- und Teilchenphysik, Zellescher Weg 19, D-01069 Dresden, Germany

[44] Duke University, Department of Physics, Durham, NC 27708, United States of America

[45] University of Edinburgh, School of Physics & Astronomy, James Clerk Maxwell Building, The Kings Buildings, Mayfield Road, Edinburgh EH9 3JZ, United Kingdom

[46] Fachhochschule Wiener Neustadt; Johannes Gutenbergstrasse 3 AT-2700 Wiener Neustadt, Austria

[47] INFN Laboratori Nazionali di Frascati, via Enrico Fermi 40, IT-00044 Frascati, Italy

[48] Albert-Ludwigs-Universität, Fakultät für Mathematik und Physik, Hermann-Herder Str. 3, D-79104 Freiburg i.Br., Germany

[49] Université de Genève, Section de Physique, 24 rue Ernest Ansermet, CH-1211 Geneve 4, Switzerland

[50] INFN Sezione di Genova[a]; Università di Genova, Dipartimento di Fisica[b], via Dodecaneso 33, IT-16146 Genova, Italy

[51] Institute of Physics of the Georgian Academy of Sciences, 6 Tamarashvili St., GE-380077 Tbilisi; Tbilisi State University, HEP Institute, University St. 9, GE-380086 Tbilisi, Georgia

[52] Justus-Liebig-Universität Giessen, II Physikalisches Institut, Heinrich-Buff Ring 16, D-35392 Giessen, Germany

[53] University of Glasgow, Department of Physics and Astronomy, Glasgow G12 8QQ, United Kingdom

[54] Georg-August-Universität, II. Physikalisches Institut, Friedrich-Hund Platz 1, D-37077 Göttingen, Germany

[55] Laboratoire de Physique Subatomique et de Cosmologie, CNRS/IN2P3, Université Joseph Fourier, INPG, 53 avenue des Martyrs, FR-38026 Grenoble Cedex, France

[56] Hampton University, Department of Physics, Hampton, VA 23668, United States of America

[57] Harvard University, Laboratory for Particle Physics and Cosmology, 18 Hammond Street, Cambridge, MA 02138, United States of America

[58] Ruprecht-Karls-Universität Heidelberg: Kirchhoff-Institut für Physik[a], Im Neuenheimer Feld 227, D-69120 Heidelberg; Physikalisches Institut[b], Philosophenweg 12, D-69120 Heidelberg; ZITI Ruprecht-Karls-University Heidelberg[c], Lehrstuhl für Informatik V, B6, 23-29, DE-68131 Mannheim, Germany

[59] Hiroshima University, Faculty of Science, 1-3-1 Kagamiyama, Higashihiroshima-shi, JP-Hiroshima 739-8526, Japan

[60] Hiroshima Institute of Technology, Faculty of Applied Information Science, 2-1-1 Miyake Saeki-ku, Hiroshima-shi, JP-Hiroshima 731-5193, Japan

[61] Indiana University, Department of Physics, Swain Hall West 117, Bloomington, IN 47405-7105, United States of America

[62] Institut für Astro- und Teilchenphysik, Technikerstrasse 25, A-6020 Innsbruck, Austria

[63] University of Iowa, 203 Van Allen Hall, Iowa City, IA 52242-1479, United States of America

[64] Iowa State University, Department of Physics and Astronomy, Ames High Energy Physics Group, Ames, IA 50011-3160, United States of America

[65] Joint Institute for Nuclear Research, JINR Dubna, RU-141 980 Moscow Region, Russia

[66] KEK, High Energy Accelerator Research Organization, 1-1 Oho, Tsukuba-shi, Ibaraki-ken 305-0801, Japan

[67] Kobe University, Graduate School of Science, 1-1 Rokkodai-cho, Nada-ku, JP Kobe 657-8501, Japan

[68] Kyoto University, Faculty of Science, Oiwake-cho, Kitashirakawa, Sakyou-ku, Kyoto-shi, JP-Kyoto 606-8502, Japan

[69] Kyoto University of Education, 1 Fukakusa, Fujimori, fushimi-ku, Kyoto-shi, JP-Kyoto 612-8522, Japan

[70] Universidad Nacional de La Plata, FCE, Departamento de Física, IFLP (CONICET-UNLP), C.C. 67, 1900 La Plata, Argentina

[71] Lancaster University, Physics Department, Lancaster LA1 4YB, United Kingdom

[72] INFN Sezione di Lecce[a]; Università del Salento, Dipartimento di Fisica[b] Via Arnesano IT-73100 Lecce, Italy

[73] University of Liverpool, Oliver Lodge Laboratory, P.O. Box 147, Oxford Street, Liverpool L69 3BX, United Kingdom

[74] Jožef Stefan Institute and University of Ljubljana, Department of Physics, SI-1000 Ljubljana, Slovenia

[75] Queen Mary University of London, Department of Physics, Mile End Road, London E1 4NS, United Kingdom

[76] Royal Holloway, University of London, Department of Physics, Egham Hill, Egham, Surrey TW20 0EX, United Kingdom

[77] University College London, Department of Physics and Astronomy, Gower Street, London WC1E 6BT, United Kingdom

[78] Laboratoire de Physique Nucléaire et de Hautes Energies, Université Pierre et Marie Curie (Paris 6), Université Denis Diderot (Paris-7), CNRS/IN2P3, Tour 33, 4 place Jussieu, FR-75252 Paris Cedex 05, France

[79] Lunds universitet, Naturvetenskapliga fakulteten, Fysiska institutionen, Box 118, SE-221 00 Lund, Sweden

[80] Universidad Autonoma de Madrid, Facultad de Ciencias, Departamento de Fisica Teorica, ES-28049 Madrid, Spain

[81] Universität Mainz, Institut für Physik, Staudinger Weg 7, DE-55099 Mainz, Germany

[82] University of Manchester, School of Physics and Astronomy, Manchester M13 9PL, United Kingdom

[83] CPPM, Aix-Marseille Université, CNRS/IN2P3, Marseille, France

[84] University of Massachusetts, Department of Physics, 710 North Pleasant Street, Amherst, MA 01003, United States of America

[85] McGill University, High Energy Physics Group, 3600 University Street, Montreal, Quebec H3A 2T8, Canada

[86] University of Melbourne, School of Physics, AU-Parkville, Victoria 3010, Australia

[87] The University of Michigan, Department of Physics, 2477 Randall Laboratory, 500 East University, Ann Arbor, MI 48109-1120, United States of America

[88] Michigan State University, Department of Physics and Astronomy, High Energy Physics Group, East Lansing, MI 48824-2320, United States of America

[89] INFN Sezione di Milano[a]; Università di Milano, Dipartimento di Fisica[b], via Celoria 16, IT-20133 Milano, Italy

[90] B.I. Stepanov Institute of Physics, National Academy of Sciences of Belarus, Independence Avenue 68, Minsk 220072, Republic of Belarus

[91] National Scientific & Educational Centre for Particle & High Energy Physics, NC PHEP BSU, M. Bogdanovich St. 153, Minsk 220040, Republic of Belarus

[92] Massachusetts Institute of Technology, Department of Physics, Room 24-516, Cambridge, MA 02139, United States of America

[93] University of Montreal, Group of Particle Physics, C.P. 6128, Succursale Centre-Ville, Montreal, Quebec, H3C 3J7, Canada

[94] P.N. Lebedev Institute of Physics, Academy of Sciences, Leninsky pr. 53, RU-117 924 Moscow, Russia

[95] Institute for Theoretical and Experimental Physics (ITEP), B. Cheremushkinskaya ul. 25, RU 117 218 Moscow, Russia

[96] Moscow Engineering & Physics Institute (MEPhI), Kashirskoe Shosse 31, RU-115409 Moscow, Russia

[97] Lomonosov Moscow State University Skobeltsyn Institute of Nuclear Physics (MSU SINP), 1(2), Leninskie gory, GSP-1, Moscow 119991, Russian Federation, Russia

[98] Ludwig-Maximilians-Universität München, Fakultät für Physik, Am Coulombwall 1, DE-85748 Garching, Germany

[99] Max-Planck-Institut für Physik, (Werner-Heisenberg-Institut), Föhringer Ring 6, 80805 München, Germany

[100] Nagasaki Institute of Applied Science, 536 Aba-machi, JP Nagasaki 851-0193, Japan

[101] Nagoya University, Graduate School of Science, Furo-Cho, Chikusa-ku, Nagoya, 464-8602, Japan

[102] INFN Sezione di Napoli[a]; Università di Napoli, Dipartimento di Scienze Fisiche[b], Complesso Universitario di Monte Sant'Angelo, via Cinthia, IT-80126 Napoli, Italy

[103] University of New Mexico, Department of Physics and Astronomy, MSC07 4220, Albuquerque, NM 87131, United States of America

[104] Radboud University Nijmegen/NIKHEF, Department of Experimental High Energy Physics, Heyendaalseweg 135, NL-6525 AJ, Nijmegen, Netherlands

[105] Nikhef National Institute for Subatomic Physics, and University of Amsterdam, Science Park 105, 1098 XG Amsterdam, Netherlands

[106] Department of Physics, Northern Illinois University, LaTourette Hall Normal Road, DeKalb, IL 60115, United States of America

[107] Budker Institute of Nuclear Physics (BINP), RU-Novosibirsk 630 090, Russia

[108] New York University, Department of Physics, 4 Washington Place, New York NY 10003, United States of America

[109] Ohio State University, 191 West Woodruff Ave, Columbus, OH 43210-1117, United States of America

[110] Okayama University, Faculty of Science, Tsushimanaka 3-1-1, Okayama 700-8530, Japan

[111] University of Oklahoma, Homer L. Dodge Department of Physics and Astronomy, 440 West Brooks, Room 100, Norman, OK 73019-0225, United States of America

[112] Oklahoma State University, Department of Physics, 145 Physical Sciences Building, Stillwater, OK 74078-3072, United States of America

[113] Palacký University, 17.listopadu 50a, 772 07 Olomouc, Czech Republic

[114] University of Oregon, Center for High Energy Physics, Eugene, OR 97403-1274, United States of America

[115] LAL, Univ. Paris-Sud, IN2P3/CNRS, Orsay, France

[116] Osaka University, Graduate School of Science, Machikaneyama-machi 1-1, Toyonaka, Osaka 560-0043, Japan

[117] University of Oslo, Department of Physics, P.O. Box 1048, Blindern, NO-0316 Oslo 3, Norway

[118] Oxford University, Department of Physics, Denys Wilkinson Building, Keble Road, Oxford OX1 3RH, United Kingdom

[119] INFN Sezione di Pavia[a]; Università di Pavia, Dipartimento di Fisica Nucleare e Teorica[b], Via Bassi 6, IT-27100 Pavia, Italy

[120] University of Pennsylvania, Department of Physics, High Energy Physics Group, 209 S. 33rd Street, Philadelphia, PA 19104, United States of America

[121] Petersburg Nuclear Physics Institute, RU-188 300 Gatchina, Russia

[122] INFN Sezione di Pisa[a]; Università di Pisa, Dipartimento di Fisica E. Fermi[b], Largo B. Pontecorvo 3, IT-56127 Pisa, Italy

[123] University of Pittsburgh, Department of Physics and Astronomy, 3941 O'Hara Street, Pittsburgh, PA 15260, United States of America

[124] Laboratorio de Instrumentacao e Fisica Experimental de Particulas-LIP[a], Avenida Elias Garcia 14-1, PT-1000-149 Lisboa, Portugal; Universidad de Granada, Departamento de Fisica Teorica y del Cosmos and CAFPE[b], E-18071 Granada, Spain

[125] Institute of Physics, Academy of Sciences of the Czech Republic, Na Slovance 2, CZ-18221 Praha 8, Czech Republic

[126] Charles University in Prague, Faculty of Mathematics and Physics, Institute of Particle and Nuclear Physics, V Holesovickach 2, CZ-18000 Praha 8, Czech Republic

[127] Czech Technical University in Prague, Zikova 4, CZ-166 35 Praha 6, Czech Republic

[128] State Research Center Institute for High Energy Physics, Moscow Region, 142281, Protvino, Pobeda street, 1, Russia

[129] Rutherford Appleton Laboratory, Science and Technology Facilities Council, Harwell Science and Innovation Campus, Didcot OX11 0QX, United Kingdom

[130] University of Regina, Physics Department, Canada

[131] Ritsumeikan University, Noji Higashi 1 chome 1-1, JP-Kusatsu, Shiga 525-8577, Japan

[132] INFN Sezione di Roma I[a]; Università La Sapienza, Dipartimento di Fisica[b], Piazzale A. Moro 2, IT- 00185 Roma, Italy

[133] INFN Sezione di Roma Tor Vergata[a]; Università di Roma Tor Vergata, Dipartimento di Fisica[b], via della Ricerca Scientifica, IT-00133 Roma, Italy

[134] INFN Sezione di Roma Tre[a]; Università Roma Tre, Dipartimento di Fisica[b], via della Vasca Navale 84, IT-00146 Roma, Italy

[135] Réseau Universitaire de Physique des Hautes Energies (RUPHE): Université Hassan II, Faculté des Sciences Ain Chock[a], B.P. 5366, MA-Casablanca; Centre National de l'Energie des Sciences Techniques Nucleaires (CNESTEN)[b], B.P. 1382 R.P. 10001 Rabat 10001; Université Mohamed Premier[c], LPTPM, Faculté des Sciences, B.P.717. Bd. Mohamed VI, 60000, Oujda; Université Mohammed V, Faculté des Sciences[d] 4 Avenue Ibn Battouta, BP 1014 RP, 10000 Rabat, Morocco

[136] CEA, DSM/IRFU, Centre d'Etudes de Saclay, FR-91191 Gif-sur-Yvette, France

[137] University of California Santa Cruz, Santa Cruz Institute for Particle Physics (SCIPP), Santa Cruz, CA 95064, United States of America

[138] University of Washington, Seattle, Department of Physics, Box 351560, Seattle, WA 98195-1560, United States of America

[139] University of Sheffield, Department of Physics & Astronomy, Hounsfield Road, Sheffield S3 7RH, United Kingdom

[140] Shinshu University, Department of Physics, Faculty of Science, 3-1-1 Asahi, Matsumoto-shi, JP-Nagano 390-8621, Japan

[141] Universität Siegen, Fachbereich Physik, D 57068 Siegen, Germany

[142] Simon Fraser University, Department of Physics, 8888 University Drive, CA-Burnaby, BC V5A 1S6, Canada

[143] SLAC National Accelerator Laboratory, Stanford, California 94309, United States of America

[144] Comenius University, Faculty of Mathematics, Physics & Informatics[a], Mlynska dolina F2, SK-84248 Bratislava; Institute of Experimental Physics of the Slovak Academy of Sciences, Dept. of Subnuclear Physics[b], Watsonova 47, SK-04353 Kosice, Slovak Republic

[145] Stockholm University: Department of Physics[a]; The Oskar Klein Centre[b], AlbaNova, SE-106 91 Stockholm, Sweden

[146] Royal Institute of Technology (KTH), Physics Department, SE-106 91 Stockholm, Sweden

[147] Stony Brook University, Department of Physics and Astronomy, Nicolls Road, Stony Brook, NY 11794-3800, United States of America

[148] University of Sussex, Department of Physics and Astronomy Pevensey 2 Building, Falmer, Brighton BN1 9QH, United Kingdom

[149] University of Sydney, School of Physics, AU-Sydney NSW 2006, Australia

[150] Insitute of Physics, Academia Sinica, TW-Taipei 11529, Taiwan

[151] Technion, Israel Inst. of Technology, Department of Physics, Technion City, IL-Haifa 32000, Israel

[152] Tel Aviv University, Raymond and Beverly Sackler School of Physics and Astronomy, Ramat Aviv, IL-Tel Aviv 69978, Israel

[153] Aristotle University of Thessaloniki, Faculty of Science, Department of Physics, Division of Nuclear & Particle Physics, University Campus, GR-54124, Thessaloniki, Greece

[154] The University of Tokyo, International Center for Elementary Particle Physics and Department of Physics, 7-3-1 Hongo, Bunkyo-ku, JP-Tokyo 113-0033, Japan

[155] Tokyo Metropolitan University, Graduate School of Science and Technology, 1-1 Minami-Osawa, Hachioji, Tokyo 192-0397, Japan

[156] Tokyo Institute of Technology, 2-12-1-H-34 O-Okayama, Meguro, Tokyo 152-8551, Japan

[157] University of Toronto, Department of Physics, 60 Saint George Street, Toronto M5S 1A7, Ontario, Canada

[158] TRIUMF[a], 4004 Wesbrook Mall, Vancouver, B.C. V6T 2A3; [b] York University, Department of Physics and Astronomy, 4700 Keele St., Toronto, Ontario, M3J 1P3, Canada

[159] University of Tsukuba, Institute of Pure and Applied Sciences, 1-1-1 Tennoudai, Tsukuba-shi, JP-Ibaraki 305-8571, Japan

[160] Tufts University, Science & Technology Center, 4 Colby Street, Medford, MA 02155, United States of America

[161] Universidad Antonio Narino, Centro de Investigaciones, Cra 3 Este No.47A-15, Bogota, Colombia

[162] University of California, Irvine, Department of Physics & Astronomy, CA 92697-4575, United States of America

[163] INFN Gruppo Collegato di Udine[a]; ICTP[b], Strada Costiera 11, IT-34014, Trieste; Università di Udine, Dipartimento di Fisica[c], via delle Scienze 208, IT-33100 Udine, Italy

[164] University of Illinois, Department of Physics, 1110 West Green Street, Urbana, Illinois 61801, United States of America

[165] University of Uppsala, Department of Physics and Astronomy, P.O. Box 516, SE-751 20 Uppsala, Sweden

[166] Instituto de Física Corpuscular (IFIC) Centro Mixto UVEG-CSIC, Apdo. 22085 ES-46071 Valencia, Dept. Física At. Mol. y Nuclear; Univ. of Valencia, and Instituto de Microelectrónica de Barcelona (IMB-CNM-CSIC) 08193 Bellaterra Barcelona, Spain

[167] University of British Columbia, Department of Physics, 6224 Agricultural Road, CA-Vancouver, B.C. V6T 1Z1, Canada

[168] University of Victoria, Department of Physics and Astronomy, P.O. Box 3055, Victoria B.C., V8W 3P6, Canada

[169] Waseda University, WISE, 3-4-1 Okubo, Shinjuku-ku, Tokyo, 169-8555, Japan

[170] The Weizmann Institute of Science, Department of Particle Physics, P.O. Box 26, IL-76100 Rehovot, Israel

[171] University of Wisconsin, Department of Physics, 1150 University Avenue, WI 53706 Madison, Wisconsin, United States of America

[172] Julius-Maximilians-University of Würzburg, Physikalisches Institut, Am Hubland, 97074 Würzburg, Germany

[173] Bergische Universität, Fachbereich C, Physik, Postfach 100127, Gauss-Strasse 20, D-42097 Wuppertal, Germany

[174] Yale University, Department of Physics, PO Box 208121, New Haven CT, 06520-8121, United States of America

[175] Yerevan Physics Institute, Alikhanian Brothers Street 2, AM-375036 Yerevan, Armenia

[176] ATLAS-Canada Tier-1 Data Centre, TRIUMF, 4004 Wesbrook Mall, Vancouver, BC, V6T 2A3, Canada

[177] GridKA Tier-1 FZK, Forschungszentrum Karlsruhe GmbH, Steinbuch Centre for Computing (SCC), Hermann-von-Helmholtz-Platz 1, 76344 Eggenstein-Leopoldshafen, Germany

[178] Port d'Informacio Cientifica (PIC), Universitat Autonoma de Barcelona (UAB), Edifici D, E-08193 Bellaterra, Spain

[179] Centre de Calcul CNRS/IN2P3, Domaine scientifique de la Doua, 27 bd du 11 Novembre 1918, 69622 Villeurbanne Cedex, France

[180] INFN-CNAF, Viale Berti Pichat 6/2, 40127 Bologna, Italy

[181] Nordic Data Grid Facility, NORDUnet A/S, Kastruplundgade 22, 1, DK-2770 Kastrup, Denmark

[182] SARA Reken- en Netwerkdiensten, Science Park 121, 1098 XG Amsterdam, Netherlands

[183] Academia Sinica Grid Computing, Institute of Physics, Academia Sinica, No.128, Sec. 2, Academia Rd., Nankang, Taipei, Taiwan 11529, Taiwan

[184] UK-T1-RAL Tier-1, Rutherford Appleton Laboratory, Science and Technology Facilities Council, Harwell Science and Innovation Campus, Didcot OX11 0QX, United Kingdom

[185] RHIC and ATLAS Computing Facility, Physics Department, Building 510, Brookhaven National Laboratory, Upton, New York 11973, United States of America

[a] Also at CPPM, Marseille, France.

[b] Also at TRIUMF, Vancouver, Canada.

[c] Also at FPACS, AGH-UST, Cracow, Poland.

[d] Also at TRIUMF, Vancouver, Canada.

[e] Now at CERN.

[f] Also at Università di Napoli Parthenope, Napoli, Italy.

[g] Also at Institute of Particle Physics (IPP), Canada.

[h] Also at Università di Napoli Parthenope, via A. Acton 38, IT-80133 Napoli, Italy.

[i] Louisiana Tech University, 305 Wisteria Street, P.O. Box 3178, Ruston, LA 71272, United States of America.

[j] At California State University, Fresno, USA.

[k] Also at TRIUMF, 4004 Wesbrook Mall, Vancouver, B.C. V6T 2A3, Canada.

[l] Currently at Istituto Universitario di Studi Superiori IUSS, Pavia, Italy.

[m] Also at FPACS, AGH-UST, Cracow, Poland.

[n] Also at California Institute of Technology, Pasadena, USA.

[o] Louisiana Tech University, Ruston, USA.

[p] Also at University of Montreal, Montreal, Canada.

[q] Also at Institut für Experimentalphysik, Universität Hamburg, Hamburg, Germany.

[r] Also at Petersburg Nuclear Physics Institute, Gatchina, Russia.

[s] Also at Institut für Experimentalphysik, Universität Hamburg, Luruper Chaussee 149, 22761 Hamburg, Germany.

[t] Also at School of Physics and Engineering, Sun Yat-sen University, China.

[u] Also at School of Physics, Shandong University, Jinan, China.

[v] Also at California Institute of Technology, Pasadena, USA.

[w] Also at Rutherford Appleton Laboratory, Didcot, UK.

[x] Also at School of Physics, Shandong University, Jinan.

[y] Also at Rutherford Appleton Laboratory, Didcot, UK.

[z] Now at KEK.

[aa] University of South Carolina, Columbia, USA.

[ab] Also at KFKI Research Institute for Particle and Nuclear Physics, Budapest, Hungary.

[ac] University of South Carolina, Dept. of Physics and Astronomy, 700 S. Main St, Columbia, SC 29208, United States of America.

[ad] Also at Institute of Physics, Jagiellonian University, Cracow, Poland.

[ae] Louisiana Tech University, Ruston, USA.

[af] Also at School of Physics and Engineering, Sun Yat-sen University, Taiwan.

[ag] University of South Carolina, Columbia, USA.

[ah] Transfer to LHCb 31.01.2010.

[ai] Also at Nanjing University, China.

[*] Deceased.

Abstract The ATLAS Inner Detector is a composite tracking system consisting of silicon pixels, silicon strips and straw tubes in a 2 T magnetic field. Its installation was completed in August 2008 and the detector took part in data-taking with single LHC beams and cosmic rays. The initial detector operation, hardware commissioning and in-situ calibrations are described. Tracking performance has been measured with 7.6 million cosmic-ray events, collected using a tracking trigger and reconstructed with modular pattern-recognition and fitting software. The intrinsic hit efficiency and tracking trigger efficiencies are close to 100%. Lorentz angle measurements for both electrons and holes, specific energy-loss calibration and transition radiation turn-on measurements have been performed. Different alignment techniques have been used to reconstruct the detector geometry. After the initial alignment, a transverse impact parameter resolution of 22.1 ± 0.9 μm and a relative momentum resolution $\sigma_p/p = (4.83 \pm 0.16) \times 10^{-4}$ GeV^{-1} $\times p_T$ have been measured for high momentum tracks.

1 Introduction

The ATLAS detector [1] is one of two large general-purpose detectors designed to probe new physics at the unprecedented energies and luminosities available at the Large Hadron Collider at CERN [2]. ATLAS is divided into three major regions: a large toroidal-field high-precision muon spectrometer surrounding a set of high-granularity calorimeters which, in turn, surround an optimized, multi-technology tracker situated in a 2 T magnetic field provided by a solenoid.

** e-mail: atlas.secretariat@cern.ch

This central tracking detector is referred to as the Inner Detector (ID). This paper describes the commissioning and calibration of the Inner Detector from its final installation in August of 2008 through cosmic-ray data-taking until the end of the year. In this period the full tracking system operated for the first time. The aim of this commissioning phase was to prepare the detector for LHC collisions which took place in 2009. The necessary steps were:

- to operate all the services and controls,
- to perform an in-situ calibration of the detector,
- to synchronise all sub-detectors,
- to measure efficiency and noise occupancy for each sub-detector in combined operation,
- to test the reconstruction software and the tracking triggers on real data,
- to perform an initial alignment of the detector.

A significant component of the commissioning involved setting up the hardware and software infrastructure needed to operate the detector. This included the calibration procedures, which will be repeated regularly during proton-proton data-taking periods. The most relevant aspects are therefore described here.

Cosmic-ray events were used to perform a preliminary alignment and to commission the track reconstruction. They mostly consist of a single muon traversing the whole detector, and have a hard momentum spectrum. Their kinematics makes them particularly suitable for some specific measurements, for example intrinsic detector efficiency, track resolution and study of detector response to ionisation as a function of momentum and incident angle.

The layout of the paper is as follows. The main components of the ID are briefly described in Sect. 2. The operating modes and conditions during the different data-taking periods, the reconstruction software and the tracking triggers are described in Sect. 3. The synchronisation of the sub-detectors is presented in Sect. 4 and the calibration procedures and results in Sect. 5. Section 6 describes the alignment, while Sect. 7 presents measurements of the detector performance: intrinsic efficiency, the Lorentz angle in silicon for both electrons and holes, resolution of tracking parameters, the specific energy loss for particle identification at low momentum and the observation of transition radiation turn-on.

In the following, the ATLAS coordinate system will be used. The nominal interaction point is defined as the origin of a right-handed coordinate system. The beam direction defines the z-axis and the x–y plane is transverse to it. The positive x-axis is defined as pointing from the interaction point to the centre of the LHC ring and the positive y-axis points upwards. Cylindrical coordinates R and ϕ are often used in the transverse plane. The pseudorapidity η is defined in terms of the polar angle θ: $\eta = -\ln\tan(\theta/2)$.

Tracks are described using the parameters of a helical trajectory at the point of closest approach to the z-axis: the transverse impact parameter, d_0, the z coordinate, z_0, the angles of the momentum direction, ϕ_0 and θ, and the inverse of the particle momentum multiplied by the charge, q/p.

2 The ATLAS Inner Detector

The layout of the Inner Detector is shown in Fig. 1. The acceptance in pseudorapidity is $|\eta| < 2.5$ for particles coming from the LHC beam-interaction region, with full coverage in ϕ. The detector has been designed to provide a transverse momentum resolution, in the plane perpendicular to the beam axis, of $\sigma_{p_T}/p_T = 0.05\% p_T$ GeV $\oplus$ 1% and a transverse impact parameter resolution of 10 μm for high momentum particles in the central η region [1]. The Inner Detector comprises three complementary sub-detectors: the Pixel Detector, the SemiConductor Tracker and the Transition Radiation Tracker. Relevant features are described briefly below; full details can be found in [1].

The Pixel Detector sensitive elements cover radial distances between 50.5 mm and 150 mm. The detector consists of 1 744 silicon pixel modules [3] arranged in three concentric barrel layers and two endcaps of three disks each. It provides typically three measurement points for particles originating in the beam-interaction region. Each module covers an active area of 16.4 mm×60.8 mm and contains 47 232 pixels, most of size 50 μm × 400 μm. The direction of the shorter pitch defines the local x-coordinate on the module and corresponds to the high-precision position measurement in the $R\phi$ plane. The longer pitch, corresponding to the local y-coordinate, is oriented approximately along the z direction in the barrel and along R in the endcaps. A module is read out by 16 radiation-hard front-end chips [4] bump-bonded to the sensor; the total number of readout channels is ∼80.4 million. Hits in a pixel are read out if the signal exceeds a tunable threshold. The pulse height is measured using the Time-over-Threshold (ToT) technique.

The SemiConductor Tracker (SCT) sensitive elements span radial distances from 299 mm to 560 mm. The detector consists of 4 088 modules of silicon-strip detectors arranged in four concentric barrels and two endcaps of nine disks each. It provides typically eight strip measurements (four space-points) for particles originating in the beam-interaction region. The strips in the barrel are approximately parallel to the solenoid field and beam axis, and have a constant pitch of 80 μm, while in the endcaps the strip direction is radial and of variable pitch. Most modules [5, 6] consist of four silicon-strip sensors [7]; two sensors on each side are daisy-chained together to give 768 strips of approximately 12 cm in length. A second pair of identical sensors is glued

Fig. 1 Cut-away image of the ATLAS Inner Detector

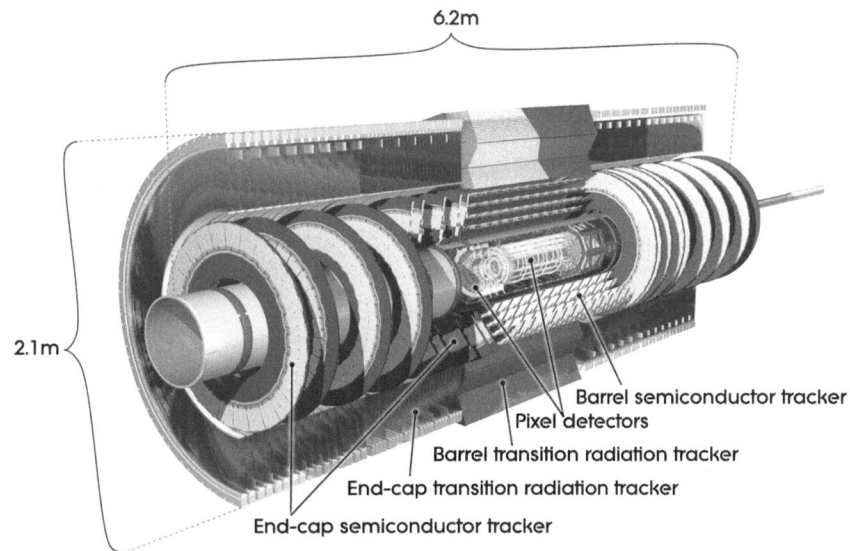

6.2m

2.1m

Barrel semiconductor tracker
Pixel detectors
Barrel transition radiation tracker
End-cap transition radiation tracker
End-cap semiconductor tracker

back-to-back with the first pair at a stereo angle of 40 mrad to provide space points. The strips are read out by radiation-hard front-end readout chips [8], each chip reading out 128 channels; the total number of readout channels is ∼6.3 million. The hit information is binary: a hit is registered if the pulse height in a channel exceeds a preset threshold, normally corresponding to a charge of 1 fC.

Measurements in the silicon detectors often perform a selection on the angle of a track incident on a module. The angle between a track and the normal to the plane of a sensor is called α. The angle between a track and the normal to the sensor in the plane defined by the normal to the sensor and the local x-axis (i.e. the axis in the plane of the sensor corresponding to the high-precision measurement in the Pixel Detector or perpendicular to the strip direction in the SCT) is termed ϕ_{local}.

The Transition Radiation Tracker (TRT) sensitive volume covers radial distances from 563 mm to 1 066 mm. The detector consists of 298 304 proportional drift tubes (straws), 4 mm in diameter, read out by 350 848 channels of electronics. The straws in the barrel region are arranged in three cylindrical layers and 32 ϕ sectors; they have split anodes and are read out from each side [9]. The straws in the endcap regions are radially oriented and arranged in 80 wheel-like modular structures [10]. The TRT straw layout is designed so that charged particles with transverse momentum $p_T > 0.5$ GeV and with pseudorapidity $|\eta| < 2.0$ cross typically more than 30 straws. The TRT provides electron identification via transition radiation from polypropylene fibres (barrel) or foils (endcaps) interleaved between the straws. The much higher energy of the transition radiation photons (∼6 keV compared with the few hundred eV deposited by an ionising particle in the Xe, CO_2, O_2 gas) is detected by a second, high-threshold, discriminator in the radiation-hard front-end electronics [11].

The Beam Conditions Monitor (BCM) [12] is designed to monitor the rate of background particles and to protect the silicon trackers from instantaneous high radiation doses caused by LHC beam incidents. The BCM consists of two stations, forward and backward, each with four modules located at a radius of 5.5 cm and at a distance of ±1.84 m from the interaction point. Each module has two pCVD diamond sensors of 1×1 cm^2 surface area and 500 µm thickness mounted back-to-back. The 1 ns signal rise-time allows the discrimination of particle hits due to collisions (in-time) from background (out-of-time). The BCM signal provides both trigger information and an instantaneous hit-rate used as input to a beam-abort signal.

Readout systems The Pixel and SCT detectors' readout systems use optical transmission for the outgoing module data and the incoming timing, trigger and control data. The

transmission is based on VCSELs operating at a wavelength of 850 nm and radiation-hard fibres [13, 14]. For each SCT module, there are two optical links operating at 40 Mbits/s for the data readout. Redundancy is implemented to allow for the loss of one optical link, without significant loss of data. For the cosmic-ray data-taking, the Pixel Detector links also operated at 40 MBits/s. The TRT uses shielded twisted-pair lines to transfer data to a patch panel inside the muon spectrometer, where up to 31 lines are multiplexed [15] into one 1.6 Gbits/s optical link.

The off-detector readout electronics is based on custom-made Read-Out Driver (ROD) modules [16, 17]. The RODs gather the data belonging to a single trigger into one packet (and in the case of the TRT perform data compression) and transmit the data to the ATLAS readout system using optical links operating at 1.6 Gbits/s [15]. The RODs also perform monitoring and calibration tasks [18].

Cooling The silicon detectors are cooled with a bi-phase evaporative system [19] which is designed to deliver C_3F_8 fluid at −25 °C in the low-mass cooling structures on the detector. The target temperature for the silicon sensors after irradiation is 0 °C for the Pixel Detector and −7 °C for the SCT; these values were chosen to mitigate the effects of radiation damage. In the commissioning phase in 2008 both detectors limited the coolant temperature to −10 °C in the circuits cooling their sensors. The resulting sensor temperatures were in the range −7 °C to +5 °C, depending on layer and module type. In 2009 the coolant temperature was reduced. Sensor temperatures were in the range −17 °C to −7 °C for the Pixel Detector and −7 °C to −2 °C for the SCT.

In contrast to the silicon detectors, the TRT operates at room temperature. The electronics is cooled by a monophase-liquid cooling loop separate from the Pixel and SCT bi-phase system.

3 Data samples and operation conditions

3.1 Data-taking periods

In 2008 the Inner Detector participated in three main data-taking periods:

– Single-beam LHC running. Particularly relevant were the so called *beam-splash* events, where the LHC beams were directed into the tertiary collimators located 150 m from the interaction point in order to provide secondary particles crossing the whole cross-section of the ATLAS detector. Since the incident particles had a direction almost parallel to the beam axis, they crossed many detector elements and were used for synchronization of the individual TRT readout units (see Sect. 4). For reasons of detector safety, during this period the Pixel Detector and SCT barrel were switched off and the SCT endcaps were operated at a reduced bias voltage of 20 V instead of 150 V, with the readout threshold increased to 1.2 fC to reduce the data volume.

– Combined ATLAS cosmic-ray run. Data were taken by the full ATLAS detector with different magnetic field combinations: toroid and solenoid switched on and off independently.

– Standalone ID cosmic-ray run. Only the Inner Detector took part in this run, which used a newly introduced Level-1 tracking trigger (see Sect. 3.4). All data taken during this period were with the solenoid off.

Cosmic rays come predominantly from the vertical direction. They were therefore particularly useful for studying the barrel region of the detector, where they resemble particles from collisions.

In the time between the combined and standalone cosmic-ray data-taking periods, a complete tuning and calibration of the detectors was performed as detailed in Sect. 5.

A summary of the numbers of reconstructed tracks in the 2008 cosmic-ray data-taking periods is shown in Table 1. Similar data-taking periods in 2009 have been used to confirm the performance achieved in the 2008 commissioning period.

3.2 Operating conditions

Most of the detector was operational during the cosmic-ray data-taking periods. Loss of coverage was mainly due to issues with the recently-commissioned evaporative cooling system and the optical links. The fractions of non-operational channels in each sub-detector are summarised in Table 2.

In the Pixel Detector three cooling loops, each serving 12 modules, showed apparent leaks, two on the positive-z endcap and one on the negative-z endcap. For safety, these loops were disabled in 2008, but were operated successfully in 2009, after the installation of a leak-monitoring system during the winter shutdown. In the SCT, 36 modules in the negative-z endcap were turned off because of problems in two cooling loops. One of these loops was repaired after the end of 2008 operation, resulting in the recovery of 23 modules.

Table 1 Number of tracks collected during the 2008 cosmic-ray runs. Numbers are given for all reconstructed Inner Detector tracks, those having at least one SCT hit and those having at least one Pixel hit

Detector	Solenoid off	Solenoid on
All	4 940 000	2 670 000
≥1 SCT hit	1 150 000	880 000
≥1 Pixel hit	230 000	190 000

A major problem with the optical links for the SCT and Pixel detectors was the failure of VCSEL arrays in the off-detector electronics. The loss of data for the SCT was reduced because of the redundancy system, but the problem prevented the read-out of 35 pixel modules in the combined run. These were recovered by replacing the defective VCSEL arrays with spare parts between the combined and standalone data-taking periods. The VCSEL failures are believed to be due to Electro Static Discharge (ESD) damage. During the 2008–2009 shutdown all VCSEL arrays in the off-detector electronics were replaced with new components produced with much tighter ESD controls. A very low rate of problems was observed in 2009.

Remaining inactive parts in the Pixel Detector and SCT were mainly due to failure in high- or low-voltage connections.

In the TRT barrel 1.6% of the straws were inactive due to mechanical problems in the detector which had been noted prior to installation and 0.7% were inactive due to scattered electronics problems at the board and chip level after installation. In the endcaps about 1.6% of the electronics channels were inactive, largely due to high- and low-voltage power connection problems, while only 0.3% of the straws had known mechanical problems. The mechanical defects were always straw cathodes that had been deformed during module or wheel construction so that they would not reliably hold high-voltage, and in these cases the anode wires were removed. These numbers remained essentially constant throughout the 2008 and 2009 data-taking periods.

The detector conditions were supervised and monitored by a Detector Control System [20], which monitored high-voltage and low-voltage values, temperatures and other environmental parameters. In particular the applied bias voltage on the silicon detectors was used to compute the Lorentz angle (Sect. 7.2) during track reconstruction, and the detector status was used to assess the data quality.

Table 2 Fraction of non-operational channels for each sub-detector in the 2008 cosmic-ray run and at the beginning of LHC collisions in 2009. For the Pixel Detector in 2008 the first numbers correspond to the earlier combined run, the second to the later standalone run

Detector	Reason	2008	2009
Pixel	Cooling	2.1%	0.0%
	Optical links	2.0%–0.0%	0.3%
	Other	1.9%	2.4%
	Total	6.0%–4.0%	2.7%
SCT	Cooling	0.9%	0.3%
	Optical links	0.4%	0.0%
	Other	0.8%	0.7%
	Total	2.1%	1.0%
TRT	Total	2.0%	2.0%

Monitoring software [21] running within the ATLAS Athena framework [22] was used to analyse data and to reconstruct tracks as described in Sect. 3.3, both online during the physics run and during offline reconstruction. The lightweight online monitoring ran on a limited subset of data, while the offline monitoring provided more in-depth analysis over larger samples of data.

3.3 Track reconstruction

Data were reconstructed using ATLAS software in the Athena framework [22]. In a first step, groups of contiguous pixels (in the Pixel Detector) or strips (in the SCT) with a hit were grouped into clusters. Channels which were noisy, as determined from either online calibration data or offline monitoring, were rejected at this stage. The one-dimensional strip clusters from the two sides of an SCT module were combined into three-dimensional space-points using knowledge of the stereo angle and the radial (longitudinal) positions of the barrel (endcap) modules; in the case of pixel clusters, only the knowledge of the radial (longitudinal) position was necessary to construct a barrel (endcap) space-point. The construction of TRT drift circles, i.e. the radial distance of the particle trajectory to the wire in a tube, required knowledge of the time of the cosmic ray passing through, which was determined using the iterative procedure described in Sect. 4. The three-dimensional space-points, in the Pixel Detector and SCT, and the drift circles, in the TRT, formed the input to the pattern-recognition algorithms.

The track reconstruction [23] started the pattern recognition by using space-points from the silicon detectors. In cosmic-ray data, these track candidates were allowed to span the central beam-axis region, and no cut was placed on the transverse impact parameter d_0. These silicon-only tracks were extended in both directions into the TRT, and refitted using all associated space-points from the silicon and TRT detectors. As shown in Table 1, a significant fraction of tracks from cosmic rays do not pass through the silicon detectors, and these were found by running a TRT stand-alone track-finding algorithm on the remaining measurements. At all stages, the track fitting was performed using the global χ^2 fitter described in [24].

To measure the resolution of the track parameters the cosmic-ray tracks which traverse the ATLAS detector from top to bottom were split into two halves. This was done by fitting two new tracks, each containing the hits in the upper or lower half of the detector only. These new tracks are referred to as split tracks. Figure 2 shows the momentum and angular distributions of the split tracks as measured in data. The shapes of the ϕ_0 and θ distributions reflect the fact that particles could enter the ATLAS cavern through the access shafts more easily than through the rock. The range of ϕ_0 is always negative as the split tracks in both the upper and

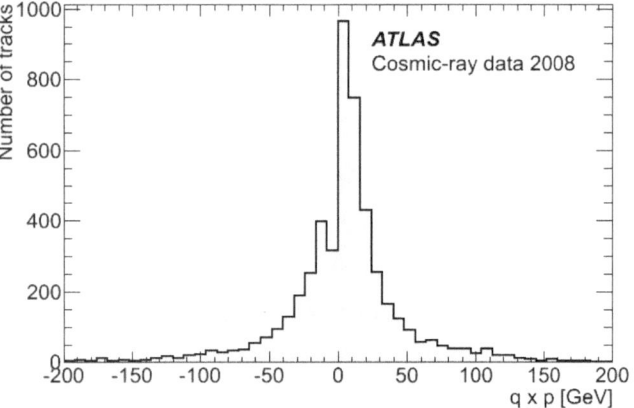

(a) Momentum distribution

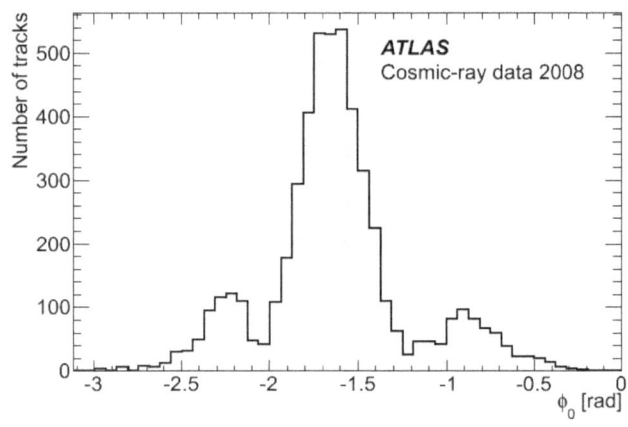

(b) ϕ_0 distribution

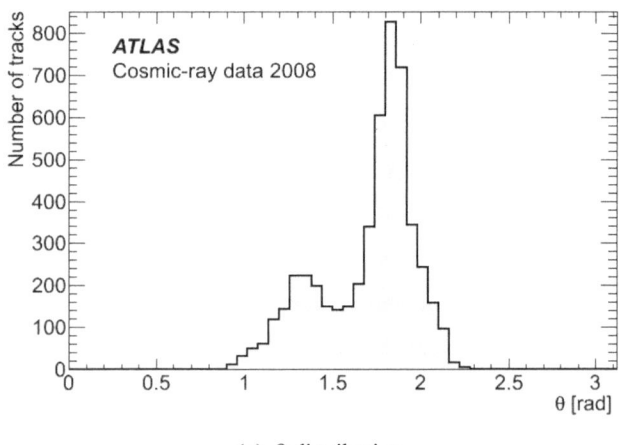

(c) θ distribution

Fig. 2 Distribution of split-track parameters for a set of cosmic-ray data with solenoid on: (**a**) particle charge multiplied by momentum ($q \times p$), (**b**) azimuthal (ϕ_0) and (**c**) polar (θ) angles

lower halves of the detector are reconstructed from top to bottom. The high μ^+/μ^- asymmetry in the low momentum bins in Fig. 2(a) is due to the toroid deflecting μ^- coming from the shafts away from the ID. The resolution results are presented in Sect. 7.3.

3.4 Tracking triggers

The ATLAS trigger system has a three-level architecture: Level-1, Level-2 and Event Filter. Level-2 and Event Filter together form the High Level Trigger (HLT) [1].

The trigger for cosmic-ray events was provided by the muon or calorimeter systems at Level-1. For the ID standalone data-taking, a Level-1 TRT trigger was added, based on a fast digital OR of groups of approximately 200 TRT straws [25].

Three Inner Detector tracking algorithms were run at Level-2. One algorithm was specifically designed for cosmic-ray running and used only barrel TRT information. It reconstructed tracks in a search window of up to about 45° to the vertical in azimuthal angle. The other two algorithms [26] were designed for collisions but were adapted for cosmic-ray running in order to exercise the algorithms online and also to complement the coverage of the TRT trigger. These algorithms started with track reconstruction in the silicon detectors and then extrapolated tracks to the TRT. As a consequence of being designed for collisions, the cosmic-particle trajectory was reconstructed as two tracks: one going upwards and the other downwards. The two algorithms used a common input consisting of space-points formed from clusters of hits in the pixel layers and from associated stereo-layer hits in the SCT. They shared common tools for track fitting and extrapolation to the TRT, but differed in the initial track-finding step:

- SiTrack was based on a combinatorial method. It first looked for pairs of space-points in the inner layers consistent with beam-line constraints, then combined these pairs with space-points in other layers to form triplets and finally merged triplets to form track candidates. In order to achieve good efficiency in cosmic-ray data-taking, the beam-line constraints were relaxed compared with those used for collision data.
- IDSCAN used a three-stage histogramming method to first determine the z-coordinate (position along the beam) of the interaction point in collision events, and then look for track candidates consistent with this interaction point. For cosmic-ray data-taking a first step was introduced which shifted the space-points in the direction transverse to the beam-axis, so that the shifted points lay on a trajectory passing close to the nominal beam position.

The efficiency of the Level-2 ID cosmic-ray trigger was determined using events triggered by the Level-1 muon trigger and containing an offline ID track. In Fig. 3 the efficiency is shown as a function of the transverse impact parameter of the offline track, d_0, for each of the three different algorithms as well as for the combined trigger. The efficiency was calculated for the sample of offline tracks with 3+3 space-points on the upper+lower track segments in the

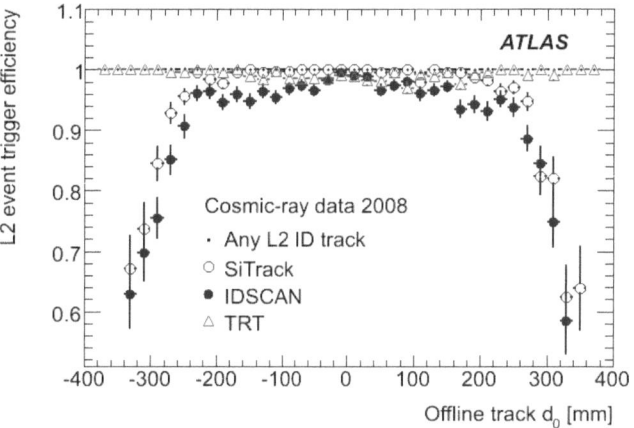

Fig. 3 Efficiency of Level-2 tracking algorithms in cosmic-ray events, as a function of d_0; the efficiency drop for the silicon based algorithms at about 300 mm corresponds to the acceptance of the first SCT barrel layer

silicon barrel. The track was also required to be within the TRT readout time window. The efficiency for IDSCAN and SiTrack falls off for tracks with d_0 approaching the radius of the first SCT layer (300 mm). The space-point shifting step that precedes IDSCAN fails for high curvature tracks, and this is reflected in a lower efficiency for IDSCAN. The combined efficiency is $(99.96 \pm 0.02)\%$.

3.5 Simulation

Cosmic-ray events were simulated by a sequence which first generated single particles at the surface above ATLAS, then filtered them for acceptance in the detector and finally ran the standard detector simulation, digitisation and reconstruction.

The generator used the flux calculations in [27] and a standard cosmic-ray momentum spectrum [28]. Muons pointing to a sphere representing the inside of the experimental cavern were propagated through the rock, cavern structures and the detector using simulation software based on GEANT4 [29, 30]. To increase the acceptance, only events with at least one hit in a given volume inside the detector were submitted to the digitization algorithms and the event reconstruction. The digitisation was adapted to reproduce the timing properties of cosmic-ray muons (see Sect. 4), and tracks were reconstructed as described in Sect. 3.3.

4 Detector timing

All sub-detectors use a common clock signal, with a 25 ns period corresponding to the spacing of LHC bunch-crossings (BC). This is either an ATLAS internal clock or one provided by the LHC and synchronised to the bunch-crossing. A delay to this signal is then applied by each detector component in order to account for signal propagation times.

A major difference between cosmic-ray running and detector operation with LHC collisions is that cosmic-ray events occur evenly distributed in the interval between two clock edges. In order to properly treat cosmic-ray events, it is therefore necessary to measure for each event the time difference between the clock edge and the passage of the cosmic-ray particle. This time difference is then an input to the track reconstruction and analysis. The TRT timing determines the precision of this measurement, because the granularity of its leading-edge measurement is 3.125 ns (1/8 of a BC) instead of one BC as for the silicon detectors. It is therefore used as a reference. The broader readout window of the Pixel Detector helped in verifying the coarse selection of beam clock offsets for both the TRT and SCT, and in understanding the trigger time offsets for the various triggers used in cosmic-ray data-taking.

4.1 TRT timing

TRT timing requirements are set by the constraint that both the leading-edge and trailing-edge transitions of a signal must be within the 75 ns (three BC) readout window. About 50 ns are required to cover the range of electron drift times at the full 2 T magnetic field. Propagation time differences within a front-end board are about 5 ns and, combined with small cabling and time-of-flight effects, imply that a time offset bigger than 10 ns would result in acceptance losses. The readout timing was initially synchronized across the detector using measured cable lengths, which gave a spread of ±5 ns in the barrel, and within one bunch-crossing in the endcaps.

In the barrel region, the time offset T_0 for each Trigger, Timing and Control unit [11] was improved using cosmic-ray tracks, and the corresponding corrections were applied to the hardware settings. These offsets were validated using the LHC beam-splash events. In these events many particles passed through the detector at the same time. Almost every TRT straw was hit multiple times and, apart from time-of-flight effects, different parts of the detector were hit simultaneously. Figure 4(a) shows T_0 settings which were estimated with a single beam-splash event. Since the readout timing before beam-splash events had already been adjusted using cosmic-ray events, the systematic effect due to time-of-flight in cosmic-ray data can clearly be seen. Apart from this, the measured time is uniform, with variations of about 1 ns. These settings were monitored in the subsequent running periods and they have remained stable.

In the endcap regions very few cosmic-ray events had been collected by September 2008. The initial correction

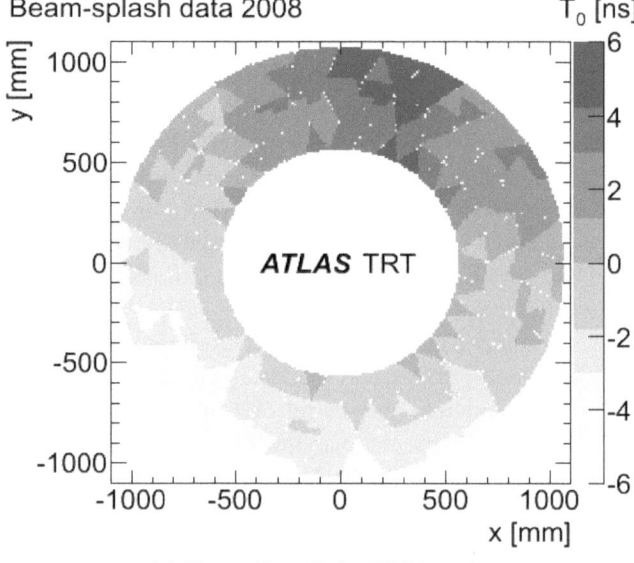

(a) Time offsets T_0 for TRT barrel A

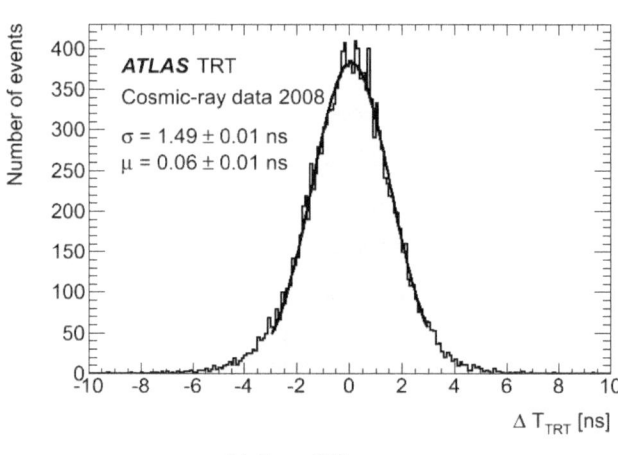

(b) T_{TRT} difference

Fig. 4 (**a**) Validation of TRT T_0 hardware settings in TRT barrel A with September 2008 beam-splash data. (**b**) Difference between the T_{TRT} value obtained from the upper and lower parts of a split track for a sample of cosmic-ray tracks

was derived from beam-splash data. This adjustment was validated using cosmic-ray data and, after subtracting the time-of-flight, the measured T_0 constants in the endcap showed an accuracy of 1.3 ns.

In the cosmic-ray run the TRT time measurement was used to determine the time, T_{TRT}, of a cosmic ray passing through the ID. This was determined by the average of measured TRT leading-edge times for all hits on a track, corrected for electron drift time and offline T_0 calibration constants (see Sect. 5.3). Since the estimated electron drift time depends on the track trajectory, the track was first fit using only the position of the centre of each hit wire, without using the drift-time information. These track parameters were then used to estimate T_{TRT} and this estimate was used to correct the position of TRT hits and to repeat the track fit.

The accuracy of this T_{TRT} measurement procedure was studied by splitting the cosmic-ray track into upper and lower parts and fitting T_{TRT} separately for each. The time difference between the two segments is shown in Fig. 4(b). The resolution is estimated as the spread of this difference, divided by two. This factor assumes a statistical error only, and is a combination of a $\sqrt{2}$ due to both upper and lower T_{TRT} uncertainties contributing to the spread, and another factor of $\sqrt{2}$ because split tracks have half the number of hits. The accuracy of T_{TRT} for barrel tracks in the 2008 cosmic-ray data was shown to be better than 1 ns.

4.2 Pixel Detector timing

The Pixel Detector front-end electronics can read out up to 16 consecutive BC for each trigger [4]. Each recorded hit includes the number of the BC in which it occurred.

At luminosities higher than 10^{32} cm^{-2} s^{-1}, the expected occupancy will only permit read-out of a single BC per trigger. In cosmic-ray data-taking the low trigger rate allows a broader time window. In the 2008 commissioning run, eight BC were read out per trigger.

The BC distribution for hits from cosmic-ray muons is shown in Fig. 5(a). The spread is due to the convolution of the front-end electronics timewalk, which results in low pulse-height hits being assigned to a late BC, and to the uniform time distribution of cosmic rays.

The distribution of hits among bunch crossings can be used to improve the detector timing relative to the corrections computed from measured signal delays in cables and read-out electronics.

Module-to-module synchronization in the barrel was assessed averaging the BC, corrected for T_{TRT}, of clusters with a pulse height greater than 15 000 e. The subtraction of T_{TRT} reduces the spread due to the event time and the requirement on pulse height removes the timewalk effect. The measured values are shown in Fig. 5(b) and indicate a time variation of 0.17 BC, equivalent to 4.25 ns without any specific module-to-module tuning. This is sufficient to obtain full efficiency in the readout window used for detector commissioning. To reduce the spread and extend the tuning to the endcap region, the higher statistics from collision events will be needed.

4.3 SCT timing

The readout of the SCT needs to be synchronized with the bunch-crossing time to ensure that the signal is sampled at the peak of the charge-response curve. In cosmic-ray data-taking, a strip is read out if the signal is above threshold in any one of three 25 ns time-bins centred on the triggered bunch-crossing.

Prior to cosmic-ray data-taking, the timing of each module was adjusted to compensate for differences in the lengths

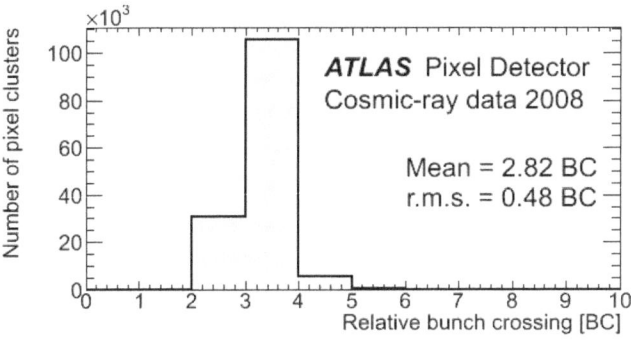

(a) Timing of pixel clusters

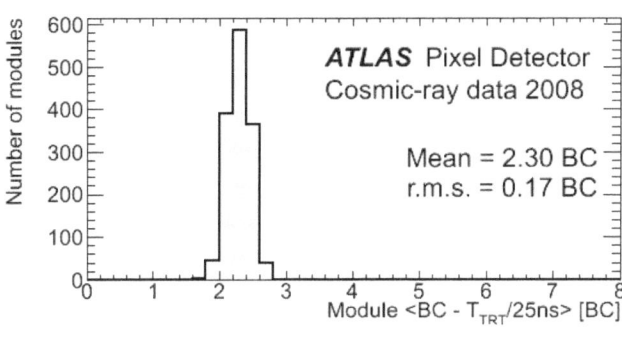

(b) Average pixel module timing

Fig. 5 Pixel Detector BC distributions for individual clusters on track (**a**) and per-module average BC relative to the T_{TRT} in units of 25 ns (**b**). The dispersion in (**a**) is due to timewalk and event time spread, while in (**b**) is the module-to-module synchronization

of the optical fibres used for data transmission to and from the modules. During data-taking, the overall timing of the SCT was adjusted in steps of 25 ns until a peak in occupancy associated with tracks was observed. No attempt was made to refine this timing using finer adjustments, and no corrections for time-of-flight were applied.

The degree of synchronisation of the SCT was studied using the cosmic-ray timing derived from the TRT. Figure 6 shows the fraction of in-time clusters on a track as a function of T_{TRT} for barrel modules. The clusters were required to contain at least two strips, all from the same BC, to reduce the effect of variations in the charge-collection time. The distribution has a flat top with a width of about 25 ns and can be fitted to a step function convolved with two Gaussian functions. The peak time of the charge response corresponds to the mid-point of the step function. Separate fits have been performed for the SCT barrel modules served by a single optical-fibre 'harness' (each harness serves six modules on a barrel at the same azimuthal angle). Most of the barrel harnesses are well synchronised: the r.m.s. width of the distribution is 1.8 ns.

4.4 BCM timing

Even though the BCM acceptance for cosmic rays is very limited, during the November 2008 operation, a total of 131 events had muons passing through this detector. These allowed the relative timing between the BCM signal and the trigger to be measured. From the timing distributions, an offset of 19.5 ± 0.4 BC was observed for triggers based on the muon system and of 19.4 ± 0.1 BC for the events triggered by the TRT Fast-Or, as shown in Fig. 7. These observed time offsets agree well with the expectation of 19 BC from the estimation of propagation time along cables and optical fibres.

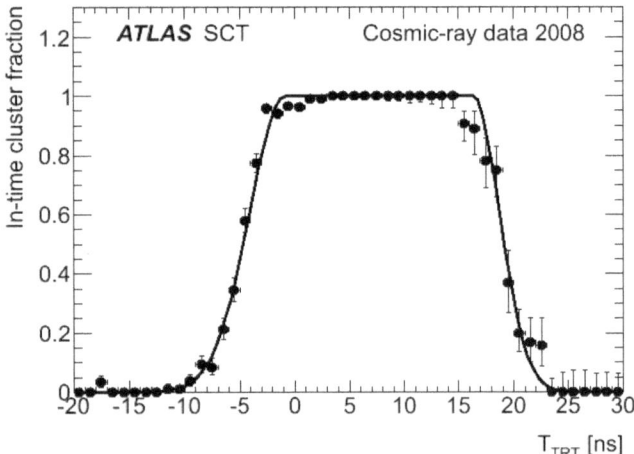

Fig. 6 Fraction of in-time clusters on track as a function of T_{TRT} for SCT barrel modules. The curve shows a fit to a step function convolved with two Gaussian functions. The peak time of the response curve is assumed to be at the centre of the step function

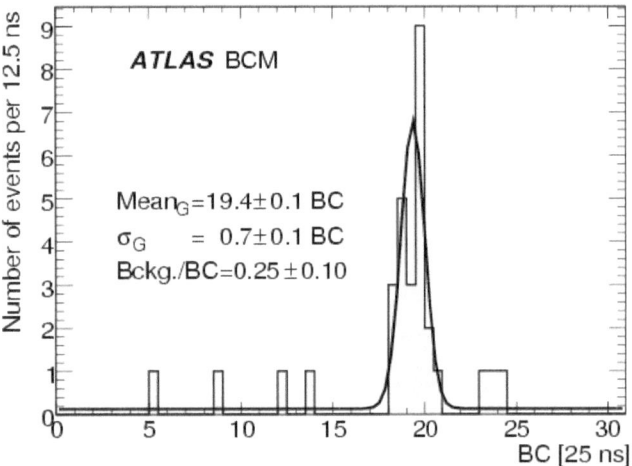

Fig. 7 Timing distribution of BCM events triggered by the TRT Fast-Or. The data are fitted with a Gaussian over a flat background

5 Sub-detector calibration

To be prepared for data-taking, each sub-detector performs a set of calibrations necessary to provide a uniform response, to map defective channels and to ensure an acceptable noise rate. Offline calibrations are then obtained during normal data-taking. They consist of additional noise suppression and, for the Pixel Detector and TRT, corrections to the position measurement of reconstructed tracks.

During collision data-taking, it is planned that offline calibrations will be performed on a subset of the data and the bulk processing of most data will start only after these calibrations have been validated. This model could not be applied during the 2008 data-taking, since the rate of events with tracks, especially in the silicon detectors, is many orders of magnitude lower than in LHC collisions. Therefore offline calibration results were only used in the reprocessing at the end of the data-taking period.

5.1 Pixel Detector calibration

The calibration of the Pixel Detector consists in tuning the optical communication links and adjusting the front-end electronics to provide uniform thresholds and response to injected charge. Suppression of noisy channels is also done at this time. Data for these calibrations are acquired in special runs. The quality of the calibration is then verified using measurements of noise rate, charge collection and timing in normal ATLAS runs. The cluster reconstruction algorithm, which uses the pulse height to improve the accuracy of the position measurement is also calibrated.

The optical data-links contain arrays of 8 or 16 VCSEL devices [14, 31]. The bias voltage which controls optical power can only be adjusted for the data-link as a whole. Due to the spread in the device characteristics, the optical power for a setting is not uniform and a scan of the bias voltage is performed to determine a suitable value for all devices in the data-link. A bit-error rate of $< 2.7 \times 10^{-8}$ with a confidence level of 99% was measured for the two bandwidth configurations, 40 and 80 Mbits/s, which will be used for operation up to a luminosity of 10^{33} cm^{-2} s^{-1}. At higher luminosity, the innermost layer will be operated at a readout speed of 160 Mbits/s, by using two 80 Mbits/s channels for each module.

Threshold calibration of the front-end electronics is performed by injecting known amplitude signals into the input of the electronics chain. The fraction of observed hits as a function of the injected charge is fitted with an error function, providing the threshold, defined as the 50% efficiency point, and the electronic noise. An 8-bit DAC is used to adjust the threshold to the target value. The distributions of threshold and noise for the whole detector are shown in Fig. 8. At the nominal working point, corresponding to a $4\,000\ e$ threshold, a uniformity of $40\ e$ r.m.s. is

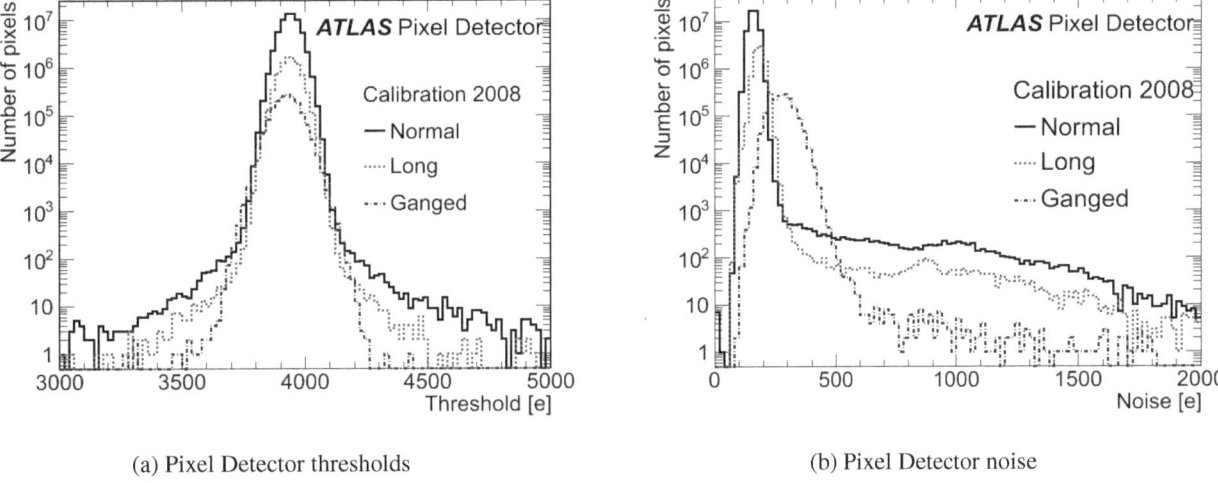

(a) Pixel Detector thresholds   (b) Pixel Detector noise

Fig. 8 Pixel Detector threshold (**a**) and noise (**b**) distributions, as obtained from in-situ calibrations based on charge injection

achieved after tuning. In these conditions the average noise level is $160\ e$ for most pixels, and slightly higher for pixels of $600\ \mu m$ size (long pixels) or for pairs of interconnected pixels (ganged pixels), which are used to cover the otherwise dead area between front-end chips [32]. The tails in Fig. 8 correspond to 4×10^{-5} of channels differing by more than $250\ e$ from the nominal threshold and 1.3×10^{-4} of channels with noise greater than $600\ e$, which may give high noise occupancy during operation.

Due to the finite electronics rise-time, low-amplitude pulses may be assigned to a BC later than the one in which the signal is generated [4]. Therefore the *in-time* threshold is also measured. This is the minimal signal for which the hit is located in the same BC as the particle crossing. For the reference $4\,000\ e$ threshold, the *in-time* threshold is $5\,400\ e$, with a r.m.s. spread of $240\ e$.

Due to the high threshold-to-noise ratio, random noise occupancy, i.e. the probability for a channel to give a noise hit per BC, is extremely low. Dedicated standalone runs with random triggers are used to find and mask the small fraction of channels that show an anomalous occupancy, greater than 10^{-6} hits/BC. Random triggers during normal data-taking runs are used for monitoring additional noisy channels which are not used in reconstruction if they have an occupancy greater than 10^{-5} hits/BC.

The actual fraction of noisy pixels was below 2.2×10^{-4} for all the 2008 data-taking. After masking these channels, the noise occupancy was $\sim 10^{-10}$ hits/BC, corresponding to less then one noise hit per event in the Pixel Detector.

The pulse height is measured using the Time-over-Threshold (ToT) method. The relationship between amplitude and ToT is calibrated with charge injection and the resulting calibration curve is used to reconstruct the energy deposited in the detector by charged particles. The absolute scale of the ToT calibration can be estimated by comparing

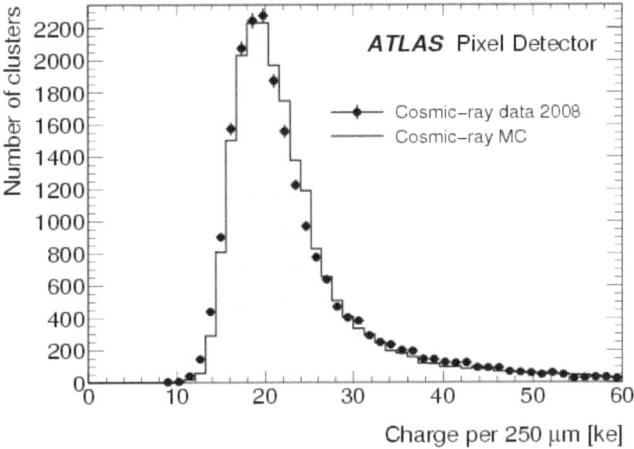

Fig. 9 Spectrum of charge release by cosmic-ray muons in the Pixel Detector, as obtanied from the Time-over-Threshold measurement

the observed spectrum of collected charge with the expectation obtained by combining the theoretical model of energy loss in silicon [33], the average energy needed to create an electron-hole pair, $W = 3.68 \pm 0.02$ eV/pair [34], and the effect of losses of collected charge due to the finite threshold of pixels (Fig. 9). For this study two methods were used. The first selected two-pixel clusters on tracks with incident angle $\alpha < 25°$: for these clusters the losses due to threshold effects are negligible and the most probable value could be directly compared to theoretical predictions. The second compared the pulse height of one-pixel and two-pixel clusters in data and Monte Carlo as a function of α in the range $\alpha < 30°$. Both methods agreed, providing a calibration factor for the charge scale of 0.986 ± 0.002 (stat.) ± 0.030 (syst.), consistent with unity. The largest systematic uncertainties are 2.4% from the spread of the measured values of W [34–37] and 2% from the theoretical modelling of energy loss in silicon.

Pulse-height measurements improve the accuracy of the position measurement, in both the local x and y coordinates, for clusters consisting of more than one pixel. The charge-sharing ratios, Ω_x and Ω_y, between the signals collected on the first and last row or column in the cluster

$$\Omega_x = \frac{Q_{\text{last row}}}{Q_{\text{last row}} + Q_{\text{first row}}},$$

$$\Omega_y = \frac{Q_{\text{last column}}}{Q_{\text{last column}} + Q_{\text{first column}}}$$

are used to correct the geometrical centre-of-cluster positions (x_c, y_c) with a linear function

$$(x_c, y_c) \rightarrow \left[x_c + \Delta_x \left(\Omega_x - \frac{1}{2} \right), y_c + \Delta_y \left(\Omega_y - \frac{1}{2} \right) \right], \quad (1)$$

with weights, Δ_x and Δ_y, depending on the particle incident angle and cluster size [38].

Cosmic rays with transverse momenta $p_T > 5$ GeV provided a calibration of Δ_x for two- and three-pixel clusters and $\phi_{\text{local}} < 45°$ (Fig. 10), a range much wider than expected for particles from proton-proton collisions. Along the beam direction, the limited range of cosmic-ray polar angles (Fig. 2(c)) only allowed the Δ_y calibration for two-pixel clusters up to $|\eta| < 1$; collisions are needed to cover the full acceptance in pseudorapidity. This calibrated position-reconstruction algorithm is expected to provide a measurement accuracy of 6 μm in the transverse plane for two-pixel clusters.

5.2 SCT calibration

Good front-end calibration is essential to the operation of the SCT because of the binary readout employed. The channel thresholds must be set to provide good efficiency ($>99\%$)

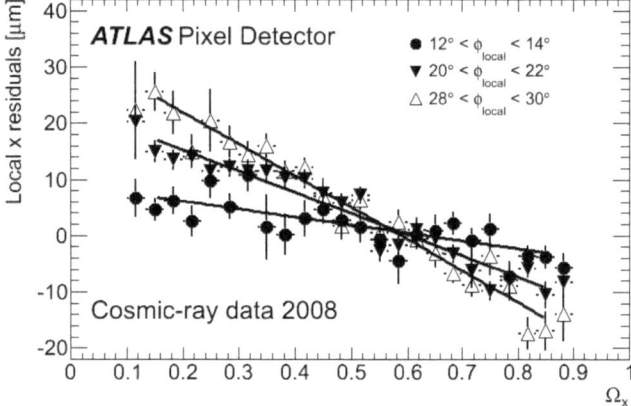

Fig. 10 Residual between track extrapolation and the centre-of-cluster position in the Pixel Detector for two-pixel clusters in the local x direction and different incident angles. The measured slopes are used to improve the position resolution with respect to the purely binary readout according to (1)

and uniformity of response while keeping the noise occupancy below 5×10^{-4} hits/BC. The calibration procedure is described in [18] and it follows a sequence similar to the one described for the Pixel Detector. Calibration runs are performed with the SCT data-acquisition system in a standalone mode, and the data analysed online. As a first step the parameters of the optical data links [13] are tuned to ensure reliable communication to and from the modules.

Threshold calibration is performed by injecting known charges into the front-end of each readout channel and measuring the occupancy as a function of threshold. For each input charge the dependence is parameterized using a complementary error function. The threshold at which the occupancy is 50% (V_{t50}) corresponds to the median of the injected charge while the sigma gives the noise after amplification. Channel gains are extracted from the dependence of V_{t50} on the input charge, and are used to set the discriminator thresholds. Channel-to-channel variations are compensated using a 4-bit DAC (TrimDAC). The TrimDAC steps can themselves be set to one of four different values to allow uniformity of response to be maintained when uncorrected channel-to-channel variations increase after irradiation. The achieved uniformity of response is shown in Fig. 11(a), which shows the distribution of the r.m.s. spread of V_{t50} values on a chip. Distributions are shown separately for chips in each TrimDAC range; most of the chips are configured in the finest setting, with a small spread. After irradiation it is expected that coarser settings will become necessary. The uniformity at the nominal threshold of 1 fC, corresponding to a signal of 54–58 mV, is $\sim 4\%$. The corresponding noise level, shown in Fig. 11(b), is between 900 and 1 700 e, depending on the strip length.

Threshold scans with no injected charge are used to measure the noise occupancy and strips with occupancy greater than 5×10^{-4} hits/BC are disabled. Figure 12 shows the occupancy values measured in calibration mode after removing the $\sim 0.2\%$ of noisy strips. Normal data-taking runs are used for the identification of noisy channels which escape detection during the calibration runs. Strips with an occupancy above 5×10^{-3} hits/BC are subsequently removed during reconstruction. The number of such strips never exceeds 0.1% of the channels. The noise occupancy in cosmic-ray data was calculated as the number of hits per event not associated to a track, per channel and BC. This rate was found to be of order of 10^{-5}, in good agreement with the calibration-mode data.

5.3 TRT calibration

As for the other sub-systems, the first step in calibrating the TRT is to adjust the data-links to provide reliable communication. There are separate steps for adjusting, on one hand, the phasing of the clock and the trigger and control lines

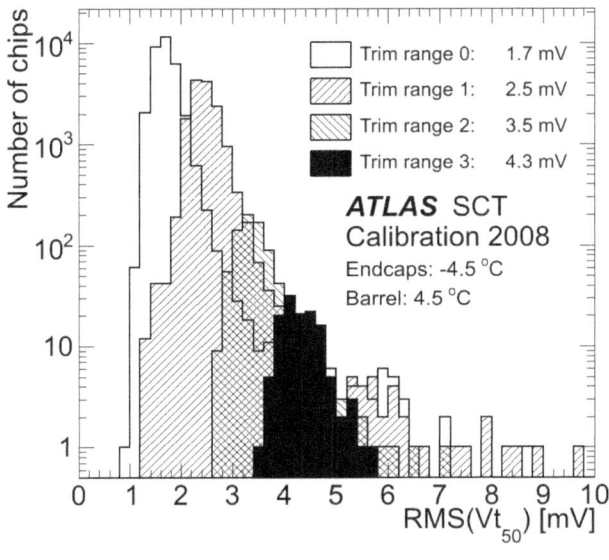

(a) SCT threshold dispersion

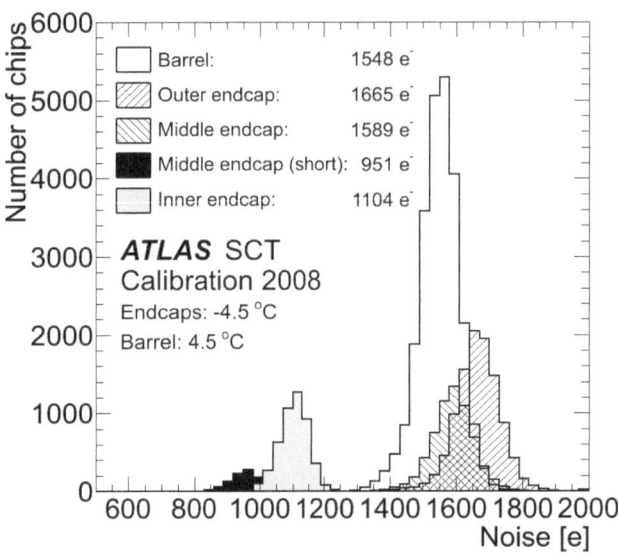

(b) SCT noise

Fig. 11 SCT threshold dispersion and noise from calibrations at 2 fC threshold based on charge injection. (**a**) Distribution of the r.m.s. spread of the threshold V_{t50} for each chip. The average values for each trim range are given. (**b**) Distribution of the input noise values for each chip as obtained in response curve tests. The average values for each detector region are given. The average SCT sensor temperatures for barrel and endcap modules as estimated from the operation conditions are also given

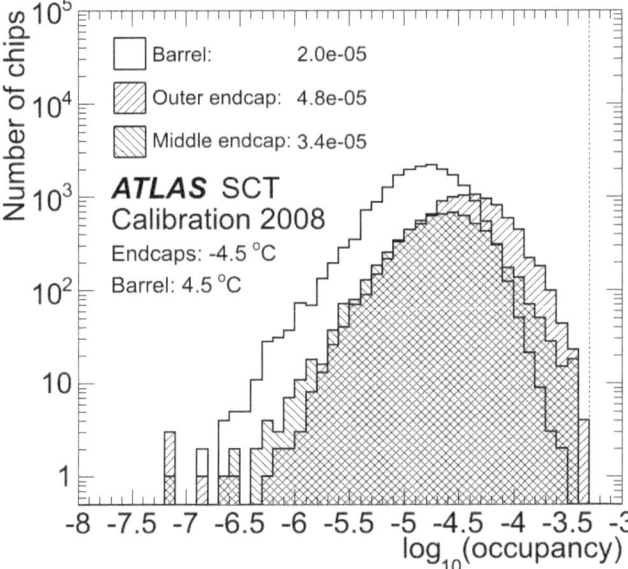

Fig. 12 The SCT noise occupancy per channel measured in calibration mode at 1 fC threshold for barrel and endcap modules in 2008 data. The dotted line is the specification value of 5×10^{-4}. A fraction of 0.2% of strips with occupancy above specification are excluded. The average noise occupancies and operational temperatures are shown

and, on the other hand, the phasing of the data lines from the front end into the optical links going to the TRT RODs. Noise data are then acquired in special calibration runs and

are used for the high-uniformity tuning of detector thresholds.

The effective gain and inherent noise of the front-end chips were measured during production by injecting each channel with known amplitude signals at multiple threshold settings. At the board, module and detector level, thresholds were set to give a noise occupancy corresponding to the desired threshold in fC. The uniformity of the random noise occupancy (or rate) for different detector elements at the same effective threshold gives a measure of element-to-element matching.

The TRT low (tracking) threshold is set to about 2 fC, corresponding to 250 eV of deposited ionization energy. This setting gives an average noise occupancy of about 2% for the three bunch-crossings sampled by each trigger. This calibration process achieves a uniform response to particles across the detector, correcting, for example, for the effect on the physical thresholds of ground offsets in the low voltage levels supplied to the front-end electronics. Figure 13 shows the TRT low threshold noise occupancy in 2008 cosmic-ray data. The occupancy is uniform with a r.m.s. spread across the detector of 0.5%. The ~2% permanently dead straws and the handful with 100% occupancy are discarded.

Normal data-taking runs are used for the identification of noisy channels and measurement of random noise. These runs are also used to compute parameters needed to optimize the determination of the particle crossing point. The parameters consist of the T_0 for each 16-straw time-measuring chip and the global time-distance relationship, R–T, shown in

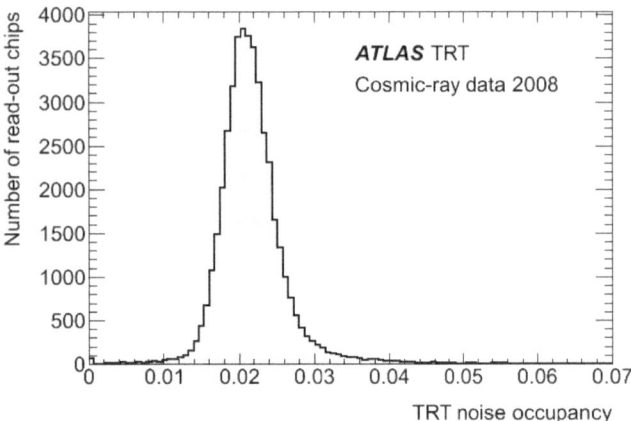

Fig. 13 TRT low threshold noise occupancy for 2008 cosmic-ray data averaged over each group of eight straws

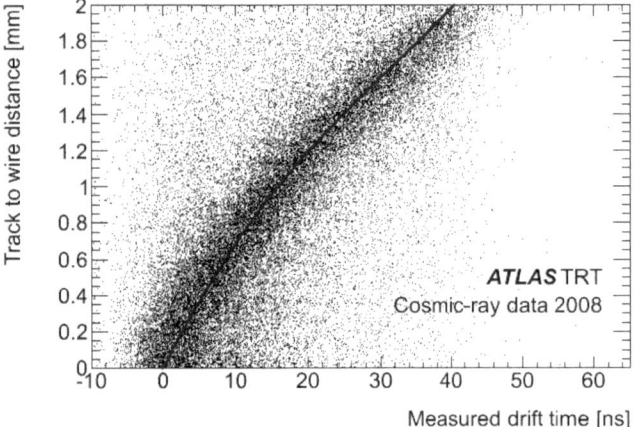

Fig. 14 Measured time–distance (R–T) relationship for the TRT barrel with solenoid field on

Fig. 14. The R–T relationship is obtained by fitting a third-order polynomial to the distance of the reconstructed track from the centre of the straw as a function of the time of the leading-edge, corrected by T_{TRT}.

6 Alignment

The accuracy with which particle tracks can be reconstructed is limited by how precisely the positions and orientations of the ID sensor modules and wires are known. The requirement on the alignment quality is that the resolution of track parameters is to be degraded by no more than 20% with respect to the intrinsic resolution [39]. The silicon pixel and strip modules must be aligned with a precision of respectively 7 μm and 12 μm in the sensitive $R\phi$ direction. In the z (R for the endcap) direction of silicon modules and for the TRT, the alignment precision is required to be of several tens of micrometres. In addition, the alignment should

have minimal systematic effects which could bias the track-parameter determination.

The alignment is specified by a set of constants, six for each individual module or assembly structure (barrel layer, endcap disk, etc.) corresponding to the six degrees-of-freedom of a rigid body: three translations T_x, T_y and T_z with respect to the nominal position and three rotations R_x, R_y and R_z with respect to the nominal axis orientations.

Track-based alignment algorithms were used to determine alignment constants using the cosmic-ray data collected in 2008. The algorithms use the tracking residual distributions of the modules; a residual is defined as the distance between the position of the measurement and the intersection of the fitted track with that module. The alignment constants can be determined via a minimisation of the following χ^2 function:

$$\chi^2 = \sum_{tracks} \mathbf{r}^T V^{-1} \mathbf{r} \tag{2}$$

where the sum is over all tracks in a given event sample, $\mathbf{r}$ is the vector of residuals for a given track and V is the covariance matrix of those residuals. In general, $\mathbf{r}$ is a function of both the track parameters,

$$\boldsymbol{\tau} = (d_0, z_0, \phi_0, \theta, q/p), \tag{3}$$

and of the alignment constants,

$$\mathbf{a} = (T_x, T_y, T_z, R_x, R_y, R_z), \tag{4}$$

of those modules with hits contributing to the track fit. The alignment was determined using the Global χ^2 algorithm [40]. In this algorithm the χ^2 given by (2) was simultaneously minimised with respect to $\boldsymbol{\tau}$ and $\mathbf{a}$ to determine the alignment constants.

The results were cross-checked using two alternative algorithms, which gave consistent results. In the Local χ^2 algorithm [41, 42] the minimisation was done only with respect to $\mathbf{a}$. In the Robust algorithm [43], used only for silicon detectors, the alignment corrections were calculated directly from the size of the residual bias. In all cases, an iterative procedure was used.

The 7.6 million tracks reconstructed in the Inner Detector during the 2008 cosmic-ray data-taking period were used to perform a preliminary alignment of the tracking system which significantly improved the tracking performance.

Because cosmic rays come from above and not from the centre of the ATLAS detector, more hits were recorded in silicon modules in the top and bottom quadrants of the barrel than the side quadrants or the endcaps. In addition, the large incidence angles in the side and endcap modules result in poor-resolution large or fragmented clusters. This limits the precision to which these regions of the Pixel Detector

Table 3 Alignment levels used with cosmic-ray data for the Inner Detector subsystems. Naming, brief description, number of structures and the total number of degrees of freedom to be aligned at each level are given. The six degrees of freedom per structure in (4) are used, unless otherwise indicated

Level	Brief description		Structures	Degrees of freedom
0		Total:	7	41
	Whole Pixel detector		1	6
	SCT barrel and 2 endcaps		3	18
	TRT barrel (except T_z) and 2 endcaps		3	17
1		Total:	14	84
	Pixel barrel layers split into upper and lower halves plus 2 endcaps		6+2	48
	SCT barrel split into 4 layers plus 2 endcaps		4+2	24
2		Total:		2 472
	Pixel barrel layers split into staves plus 2 endcaps		112+2	684
	SCT barrel layers split into staves plus 2 endcaps		176+2	1 068
	TRT barrel modules (except T_z)		96	480
	TRT endcap wheels (only T_x, T_y and R_z)		40 × 2	240
3		Total:	3 568	7 136
	Pixel barrel modules (only T_x and R_z)		1 456	2 912
	SCT barrel modules (only T_x and R_z)		2 112	4 224

and SCT can be aligned. Due to its structure and larger acceptance, the TRT is less sensitive to this anisotropy and its alignment precision was more uniform.

6.1 Global alignment

The alignment proceeds in stages from larger structures to the individual module level, as detailed in Table 3. At each stage more degrees of freedom are introduced, but the expected sizes of the corrections are smaller.

In the first step, the Level 0 alignment, the SCT barrel and two endcaps are aligned relative to the entire Pixel Detector, followed by the TRT alignment with respect to the silicon detectors. In aligning the TRT barrel, only 5 degrees of freedom are used; the T_z is not considered because the TRT barrel modules are almost 1 m long and do not measure the z coordinate.

Cosmic-ray simulation studies with a misaligned geometry showed that, using solenoid-on tracks for the silicon detectors' Level 0 alignment, may lead to corrections being underestimated. The presence of a misalignment between the sub-detectors could lead to a bias in reconstructed track momentum, with part of the misalignment being absorbed into the curvature. Therefore these alignment corrections were derived using only solenoid-off data. The simulation tests also showed that the solenoid-off data were able to estimate the Level 0 misalignments with a precision better than 100 μm. This precision is limited by misalignments of the internal structures and by multiple Coulomb scattering effects.

Table 4 Level 0 alignment parameters, translations (T_x, T_y and T_z) and rotation (R_z only), of the SCT and TRT barrel, endcap A (positive z) and endcap C (negative z). The statistical errors were much smaller than the last digit

Structure	T_x [mm]	T_y [mm]	T_z [mm]	R_z [mrad]
SCT barrel	0.9	0.6	0.5	−1.8
SCT endcap A	−1.8	0.5	0.0	−1.3
SCT endcap C	−0.4	0.6	1.0	−1.3
TRT barrel	0.2	−0.1	N/A	0.0
TRT endcap A	−1.5	0.2	−3.4	−7.0
TRT endcap C	−1.0	1.7	2.1	6.4

For the TRT instead, both a solenoid-on and a solenoid-off sets of tracks were used. The results were compared and found consistent within the uncertainties.

Shifts from the nominal positions of up to 2 mm were observed, with rotations R_z of several mrad, as shown in Table 4; the rotations R_x and R_y were all consistent with zero.

6.2 Local alignment of the Pixel Detector and SCT

After the initial alignment of the detector components as a whole, the subsequent alignment levels consider smaller structures.

Due to the low statistics the endcaps were aligned globally, but no attempt was made to align individual disks or modules. The initial geometry for the alignment was based on the nominal position of the modules.

The first stage in the internal alignment of the Pixel Detector and SCT (Level 1) was the alignment of the pixel half-shell barrel layers, the full SCT barrel layers and the four endcap structures (two for each of the Pixel Detector and the SCT). The SCT barrel layers were considered to be rigid cylinders, whilst the pixel half-shells were considered rigid half-cylinders. For all the structures, the full set of 6 degrees of freedom was considered in the alignment. This level was aligned combining both solenoid-on and solenoid-off cosmic-ray data. The computed alignment corrections were of the order of hundreds of micrometres in all T_x, T_y and T_z, with in particular a rotation of the first pixel upper half shell of almost 2 mrad with respect to the other layers.

The next step was the alignment of the Pixel Detector and SCT stave-by-stave (Level 2). The pixel staves are real structures, composed of 13 modules in the same ϕ position, which were assembled and surveyed. The SCT was not assembled in staves but the modules were individually mounted on the support cylinder. Nevertheless, for alignment purposes the SCT barrel was also split into rows of 12 modules. The staves were considered a rigid body and all 6 degrees of freedom were used. The alignment corrections for the translations of the staves were of the order of tens of micrometres.

Once the staves were aligned the alignment at module-to-module level (Level 3) was performed. The positions of pixel modules mounted within the staves were surveyed just after assembly [44]. This survey information was used as a starting point for the internal alignment of the pixel modules, but not to constrain the alignment corrections, because the deformation of staves after the survey was expected to be significantly larger than survey errors. This step was performed in the local coordinate system described in Sect. 2 for individual silicon modules.

The number of hits per module was much smaller than for the larger structures, and thus the statistical precision of the alignment becomes a significant consideration. Therefore the number of degrees of freedom was reduced to just two per module, T_x and R_z. These two parameters were chosen because they were appropriate to describe the lateral bending along the pixel staves, the largest deformation observed in the residuals, with an amplitude reaching 500 μm for the worst case.

Pixel Detector and SCT residual distributions before and after the alignment procedure are shown in Fig. 15 for tracks with $p_T > 2$ GeV and $|d_0| < 50$ mm. These are compared to distributions obtained using a perfectly-aligned Monte Carlo simulation of cosmic rays. Before alignment the residual distributions are very wide compared to the Monte Carlo simulation and also biased. After alignment their widths were substantially reduced and the means are consistent with zero to within a few micrometres.

The residuals cannot be used to quote the point resolution, because their errors include a contribution from extrap-

olation uncertainties larger than the point resolution. This contribution also depends on the track momentum and silicon layer, resulting in strongly non-Gaussian distributions. By comparing the width of the aligned residual distributions to the simulation, and assuming that the only contribution to the increased width is from misalignments, the size of the remaining module-level misalignments is estimated to be approximately 20 μm.

6.3 Local alignment of the TRT

The second step of the TRT barrel alignment internally aligned the 96 individual TRT barrel modules (three layers of 32 ϕ-sectors each). Although the straw anodes inside the barrel modules are physically separated at $z = 0$, no such distinction exists at the module level. As for the Level 0 barrel alignment, only five degrees of freedom were used, T_z being non-measurable. The internal alignment was determined separately for different periods of cosmic-ray data taking, which could either be solenoid on or solenoid off. This internal alignment used TRT stand-alone tracks, giving high statistics because of the larger acceptance of the TRT volume. The size of the translation alignment corrections was of the order of 200–300 μm with respect to the nominal position of the modules.

In each endcap, the 40 wheels were aligned in three degrees of freedom: T_x, T_y, and R_z. The corrections for the translations were of the order of 100 μm and the rotations were tenths of a milliradian.

Figure 15(d) shows the residual distribution for tracks with $p_T > 2$ GeV in the barrel modules, both before and after alignment. The distributions are compared to those obtained using a perfectly aligned cosmic-ray Monte Carlo simulation. Again the width and bias of the residual distribution were improved after alignment.

6.4 Summary and perspectives

The cosmic-ray alignment significantly improved the track reconstruction and the track-parameter resolutions, presented in Sect. 7.3. The achieved level of precision, about 20 μm, ensures that track reconstruction efficiency with early LHC data will not be significantly affected by residual misalignments.

Local alignment with cosmic rays is statistically limited by the small acceptance of individual detector modules, especially in the endcap region. Therefore it was not possible to perform a Level 3 alignment in the endcaps. In addition, a reduced set of degrees of freedom was used in the barrel region. That not all possible misalignments can be recovered using only cosmic-ray data partially explains why the nominal Monte Carlo resolution has not yet been achieved.

In order to reach the design granularity, a high statistics sample of tracks from proton-proton collisions is needed.

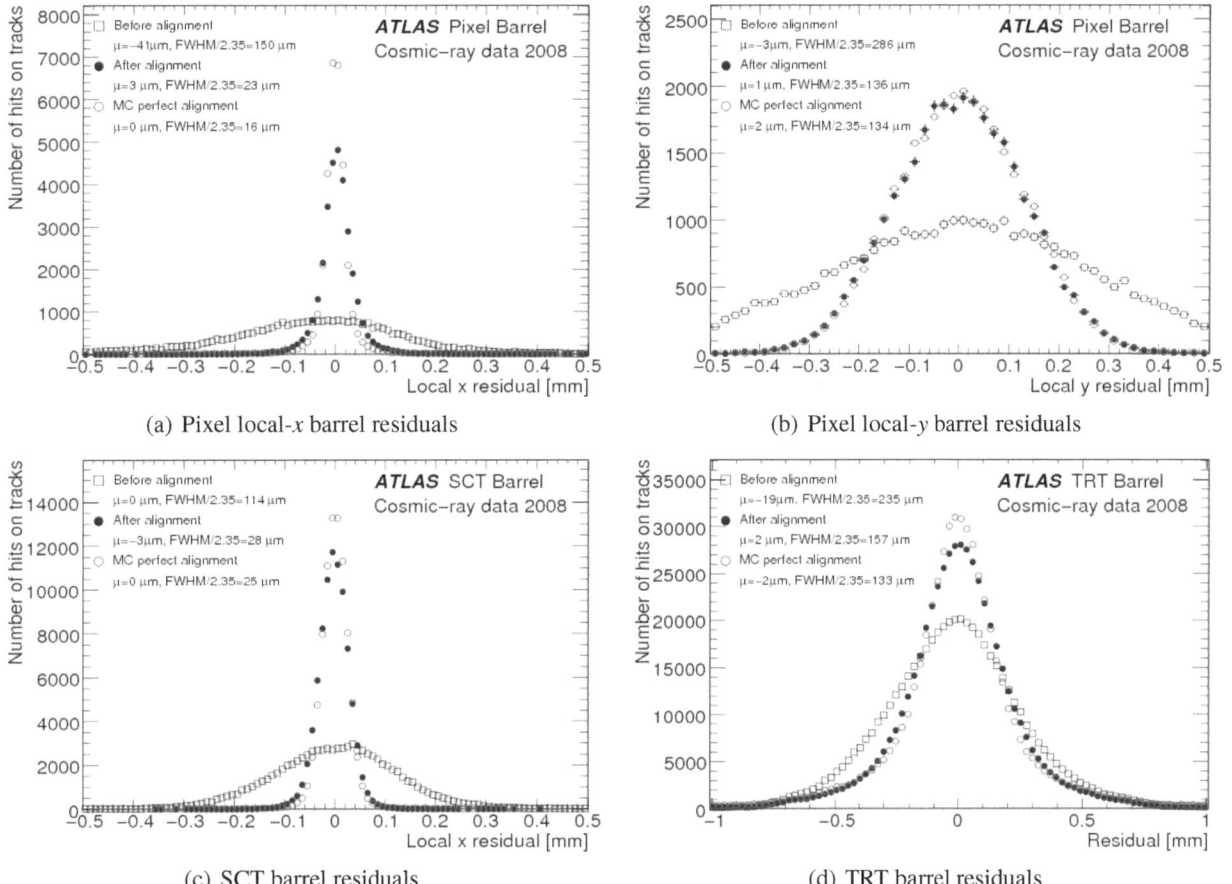

(a) Pixel local-*x* barrel residuals

(b) Pixel local-*y* barrel residuals

(c) SCT barrel residuals

(d) TRT barrel residuals

Fig. 15 Residual distributions in the local reference frame for hits in barrel regions for all ID sub-detectors. The plots show the results for 2008 cosmic-ray tracks before and after alignment and a comparison with a perfectly aligned cosmic-ray Monte Carlo simulation. Tracks are selected requiring $p_T > 2$ GeV

When this has been collected, all 1 744 and 4 088 Pixel Detector and SCT modules will be aligned with the full set of degrees of freedom in (4). Individual TRT wires will also be aligned with the two more sensitive degrees of freedom: the translation along the ϕ direction and the rotation about the R or z directions in the barrel and endcap regions, respectively.

7 Detector performance

7.1 Intrinsic detector efficiency

The intrinsic detector efficiency measures the probability of a hit being registered in an operational detector element when a charged particle traverses the sensitive part of the element. Both a high intrinsic efficiency and a low non-operational fraction are essential to ensure good-quality tracking.

The intrinsic efficiencies of the Pixel and SCT detectors are measured by extrapolating well-reconstructed tracks through the detector and counting the numbers of hits (clusters) on the track and 'holes' where a hit would be expected but is not found. The track extrapolation uses the full track fit described in Sect. 3.3 to compute the intersections of the track with all modules along its trajectory. If a module (module side for the SCT) does not have a cluster associated to the track and the intersection point is more than 3σ from the edge of the sensitive area the absence is called a hole. The efficiency, ε, is defined as the ratio of the number of clusters found to the number expected:

$$\varepsilon = \frac{N_{\text{clusters}}}{N_{\text{clusters}} + N_{\text{holes}}} \tag{5}$$

where N_{clusters} is the number of clusters found and N_{holes} is the number of holes.

Pixel efficiencies are determined using tracks with at least 30 TRT hits (40 for the data with solenoid off), at least 12 SCT hits and $\sin \alpha < 0.7$. There must be only one track passing these cuts in the event. Tracks used to measure the SCT efficiency must have at least 30 TRT hits or 7 SCT hits, a hit both before and after the module side under investigation

and $|\phi_{local}| < 40°$. A run-dependent cut on T_{TRT} is applied
to ensure good timing. The angular cuts are applied because
the tracking algorithm does not function as well at high
incidence angle; charge sharing among many channels
combined with the readout threshold may result in multiple
clusters and reduced apparent efficiency.

The track extrapolation does not predict holes near the
sensor edges or ambiguously mapped pixels, so these areas
are excluded from the efficiency calculation. For the Pixel
detector, clusters or holes within 0.6 mm of ganged pixels in
the ϕ direction, or within 1.0 mm of the sensor edge in the ϕ
or z direction, are excluded. Similarly, for the SCT the intersection of the track with the sensor is required to be at least
2 mm from the edge in ϕ and at least 3 mm in z. To reduce
the bias due to the track fitting and pattern recognition criteria, which are affected by residual misalignments, clusters
not already associated to a track but close to an intersection are included in $N_{clusters}$ in (5) and removed from N_{holes}.
Due to the low noise occupancy (Sect. 5), it is likely that
these result from track reconstruction inefficiencies rather
than noise. The inclusion of these clusters improves the efficiency by 0.04% in the Pixel barrel and 0.2% in the SCT
barrel. Varying the distance for inclusion of non-associated
clusters between 2 mm and 10 mm changes the efficiencies
by at most 0.002% and 0.004% for Pixel Detector and SCT
respectively, and is included in the systematic uncertainties.

Non-functioning detector elements (Sect. 3.2) are not
included in the calculation of the intrinsic efficiency. In
the SCT, complete module sides and chips are excluded;
these amount to ~2% of the detector. The measured inefficiency contains a contribution from isolated dead strips for
which no correction is applied. For the Pixel detector, non-
operational modules and front-end chips amount to 4–6% of
the detector.

The measured efficiency of each barrel layer is shown
for the Pixels and SCT in Fig. 16(a) for data taken with
solenoid on. Efficiencies measured with solenoid off are typically ~0.2% lower, indicating some residual inefficiencies
arising from track reconstruction when the particle momentum is unknown. The overall efficiency of the Pixel barrel
is $(99.974 \pm 0.004(stat.) \pm 0.003(syst.))\%$ and of the SCT
barrel is $(99.78 \pm 0.01(stat.) \pm 0.01(syst.))\%$; the systematic error in each case is determined by varying the track selection criteria. Of the remaining 0.026% pixel inefficiency,
$(0.017 \pm 0.004)\%$ is the contribution due to known defective
channels observed during detector construction.

The efficiency of the TRT is determined in a similar manner to that of the silicon detectors, excluding the 2% non-
functioning channels. Tracks are extrapolated through the
TRT in a series of steps. To reduce tracking biases, at each
point all straws in a region containing up to the third near-
est neighbour are considered. The efficiency is determined
by dividing the number of hit straws by the total number of

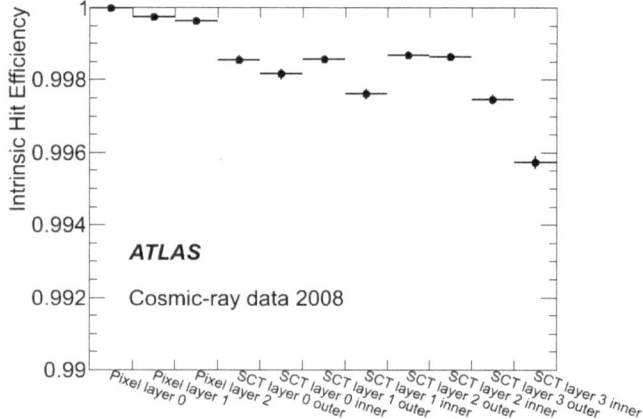

(a) Pixel and SCT barrel efficiencies

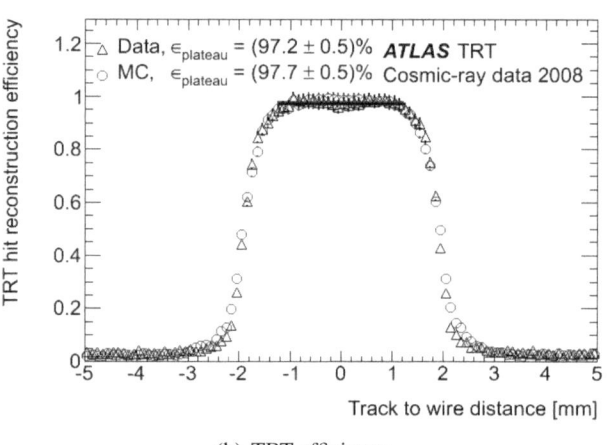

(b) TRT efficiency

Fig. 16 (a) Intrinsic efficiency of each Pixel Detector and SCT barrel
layer. (b) TRT efficiency as a function of distance from the wire

straws within the region. The efficiency depends on the path
length of a track inside a straw, and is therefore determined
as a function of the distance of a track from the wire. Tracks
are required to have at least 20 TRT hits, at least 6 SCT hits,
T_{TRT} between 5 ns and 25 ns and an angle to the vertical of
less than 15°. The efficiency of the TRT barrel, for data with
solenoid on, is shown in Fig. 16(b). The overall efficiency
over the plateau region is $(97.2 \pm 0.5)\%$.

7.2 Lorentz angle measurement

The charge carriers in the silicon detectors are subject to
the electric field **E**, generated by the bias voltage and oriented normal to the module plane, and the solenoid magnetic
field **B**. In the endcaps the fields are nearly parallel and the
charge carriers drift directly towards the electrodes. In the
barrel modules these fields are perpendicular and the charge
carriers drift at the Lorentz angle, θ_L, with respect to the normal to the sensor plane. The Lorentz angle depends on the
charge carrier mobility, which in turn depends on the bias

voltage, the thickness of the depleted region and the temperature [45]. For fully-depleted modules, the average shift in collected charge is approximately 30 µm for the Pixel Detector and 10 µm for the SCT, in both cases not negligible with respect to the detector resolution and alignment precision. Measurements of the Lorentz angle for the ATLAS sensors have already been performed in test beams [38, 46], but in conditions different from the actual operation in ATLAS.

The Lorentz angle is measured from the dependence of the cluster size on the incident angle of the particle. When the incident angle equals the Lorentz angle, all the charge carriers generated by the particle drift along the particle direction and, apart from charge diffusion, are collected at the same point on the sensor surface, giving a minimum cluster size.

The dependence of the cluster size on the incident angle ϕ_{local} is shown for the Pixel Detector and SCT in Fig. 17. Data are fitted using the convolution of the function:

$$f(\phi_{\text{local}}) = a|\tan\phi_{\text{local}} - \tan\theta_L| + b \qquad (6)$$

with a Gaussian distribution. Fit parameters are the Lorentz angle θ_L, the shape parameters a, b and the width of the Gaussian. For the Pixel Detector an improvement of the fit quality was observed by replacing the second term in (6) by $b/\sqrt{\cos\phi_{\text{local}}}$, which is a phenomenological attempt to describe the bigger relative weight of diffusion effects for tracks at high incident angle.

The measured values are $11.77° \pm 0.03°$ and $-3.93° \pm 0.03°$ for the Pixel Detector and SCT respectively, where the errors are statistical only. The values differ by a factor of three due to the different mobility of the charge carriers which provide the dominant signal: electrons in the Pixel Detector, holes in the SCT.

As a cross check for systematic effects, the same measurement was performed for data with no magnetic field, giving values of $0.09° \pm 0.03°$ and $0.05° \pm 0.05°$ for the Pixel Detector and SCT respectively. Since for the Pixel Detector the disagreement with respect to the expected null value is statistically significant, it is used as a component of the systematic uncertainty. The other dominant source of systematic uncertainty is the fit range, which has been estimated to give a contribution of $0.07°$ for the Pixel Detector and $0.10°$ for the SCT. The measured values of the

Lorentz angle in the 2 T magnetic field are shown in Table 5 where they are compared with the expectation from the model in [45]. The measurements are compatible with the model predictions within the uncertainties on the predictions arising from the values of charge-carrier mobilities.

Since Pixel Detector modules operated with different temperature ranges in 2008 and 2009, it was possible to measure the dependence of the Lorentz angle on the silicon

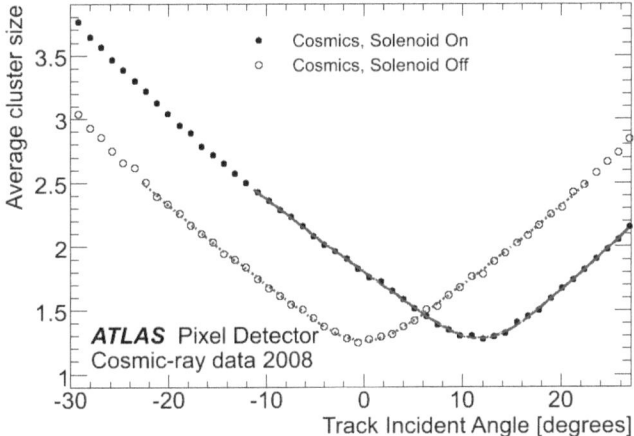

(a) Pixel Detector mean cluster width

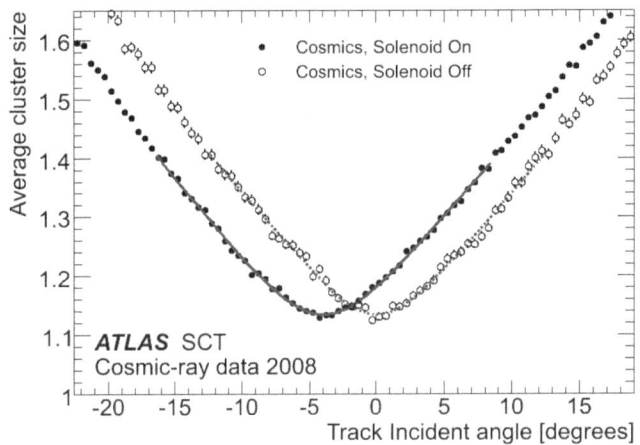

(b) SCT mean cluster width

Fig. 17 Cluster-size dependence on the particle incident angle for the Pixel Detector (**a**) and the SCT (**b**). The displacement of the minimum for the data with solenoid on is a measurement of the Lorentz angle θ_L

Table 5 Measured values of the Lorentz angle in 2 T magnetic field at the average operational temperature in 2008, compared with model expectations [45]. For the measurements, the first error is statistical and the second systematic. The error on the model prediction arises from uncertainties in the charge-carrier mobility

Detector	T [°C]	Measured θ_L [°]	Model θ_L [°]
Pixel (electrons)	-3	$11.77 \pm 0.03 \pm^{0.13}_{0.23}$	12.89 ± 1.55
SCT (holes)	5	$-3.93 \pm 0.03 \pm 0.10$	-3.69 ± 0.26

temperature. The resulting dependence

$$d\theta_L/dT = (-0.042 \pm 0.003)°/K \qquad (7)$$

is in agreement with the model expectation of $-0.042°/K$.

7.3 Track parameter resolution

The expected resolution of the perigee parameters d_0, z_0, ϕ_0, θ and q/p of a particle emerging from proton-proton collisions in the LHC can be predicted using reconstructed and split tracks from cosmic-ray data. Since particles coming from cosmic-ray showers mostly traverse the detector from top to bottom, the resolutions can only be derived for the ATLAS barrel detectors.

In order to select tracks with good quality, the split tracks are each required to have at least 2, 6 and 25 hits in the barrel of the Pixel, SCT and TRT detectors respectively, and a transverse momentum of more than 1 GeV. The $|d_0|$ impact parameter has to be less than 40 mm to guarantee that the split tracks originate in the interaction region inside the beam pipe.

The perigee parameters T_{up} and T_{down}, where T is any of the five parameters, of each split-track pair are compared to each other to extract the overall track parameter resolutions. Since both tracks come from the same particle, their difference $\Delta\tau = T_{\tau,\text{up}} - T_{\tau,\text{down}}$ for each perigee parameter τ must have a variance $\sigma^2(\Delta\tau)$ which is two times the variance $\sigma^2(T_\tau)$ of the parameters of each track. The resolution of the track parameter τ is therefore given by the root mean square of the $\Delta\tau$ distribution divided by $\sqrt{2}$. This method has been used to study the resolution of the perigee parameters of Inner Detector tracks. The variances were calculated excluding the outermost 0.3% of events in each distribution.

The measured resolution is compared to the Monte Carlo expectation for a perfectly-aligned detector. The difference in performance is attributed to the remaining misalignment

after the procedure in Sect. 6. In addition, the refit of the split-track pair can be restricted to a subset of measurements in the Inner Detector. This has been done to study the perigee parameter resolutions of silicon-only tracks (Pixel and SCT) and compare them to resolutions of the same tracks which have been fitted using the full Inner Detector.

A summary of the measured track-parameter resolutions for $p_T > 30$ GeV, where the multiple-scattering contribution can be neglected, is given in Table 6.

Impact parameter resolution Figure 18 shows the transverse and longitudinal impact parameter resolutions as determined from the data using the track-splitting method. They are displayed as a function of transverse momentum. At low momenta the resolution is governed by multiple scattering in the beam pipe and first pixel layers. For higher momenta, above about 10 GeV, the impact parameter resolutions rapidly approach an asymptotic limit which is given by the intrinsic detector resolution and residual misalignments.

Resolutions as a function of η are constant and symmetric around $\eta = 0$, as shown in Fig. 19. Both Figs. 18 and 19 compare the resolution obtained for Inner Detector tracks with that from a fit to solely the silicon part. The d_0 resolution is slightly more precise for full tracks, as the TRT

Table 6 Track parameter resolution for tracks with $p_T > 30$ GeV in cosmic-ray data and simulation

Parameter	Asymptotic resolution	
	Cosmic-ray data 2008	Monte Carlo
d_0 [μm]	22.1 ± 0.9	14.3 ± 0.2
z_0 [μm]	112 ± 4	101 ± 1
ϕ_0 [mrad]	0.147 ± 0.006	0.115 ± 0.001
θ [mrad]	0.88 ± 0.03	0.794 ± 0.006
q/p [GeV^{-1}]	$(4.83 \pm 0.16) \times 10^{-4}$	$(3.28 \pm 0.03) \times 10^{-4}$

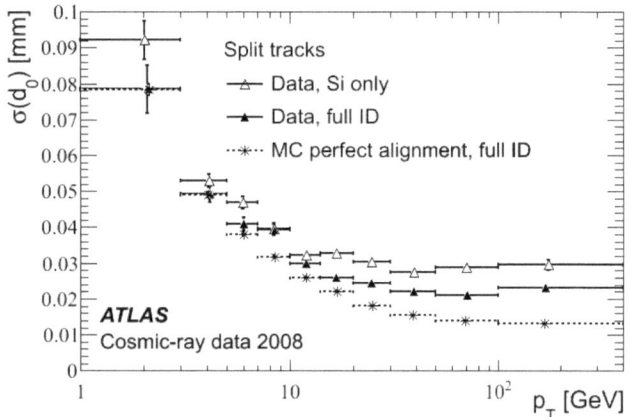 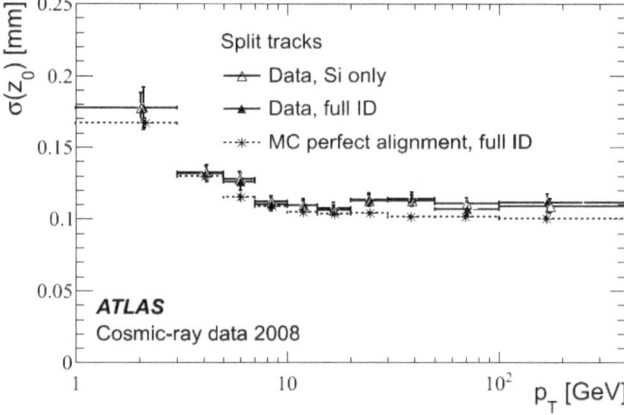

Fig. 18 Impact parameter resolution determined from data for the track impact parameters as a function of transverse momentum. Resolutions of full ID (*solid triangles*) and silicon-only (*open triangles*) tracks are compared to those from full tracks in MC simulation (*stars*)

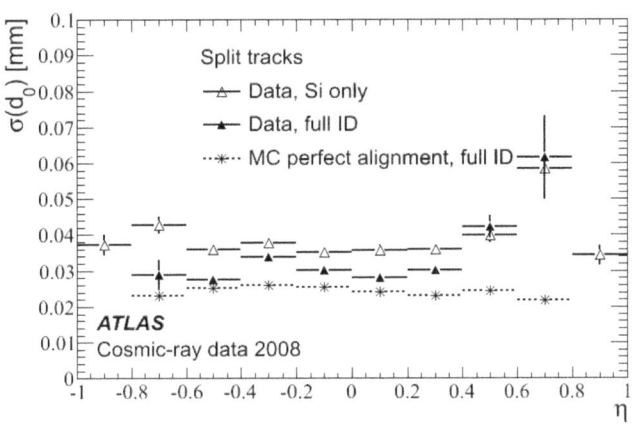
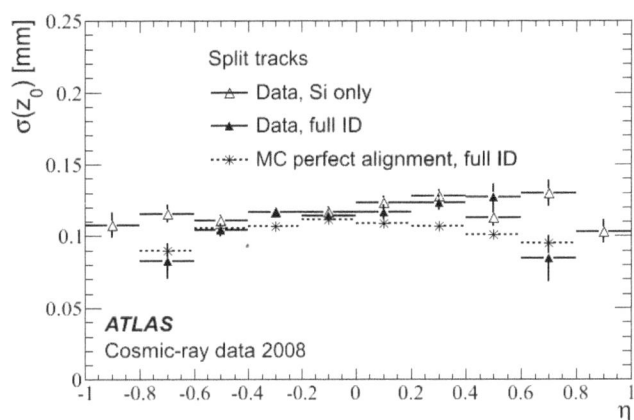

Fig. 19 Impact parameter resolution determined from data for tracks with $p_T > 1$ GeV, as a function of pseudorapidity η. The resolutions are shown for full ID tracks (*solid triangles*), silicon-only tracks (*open triangles*) and simulated full ID tracks (*stars*)

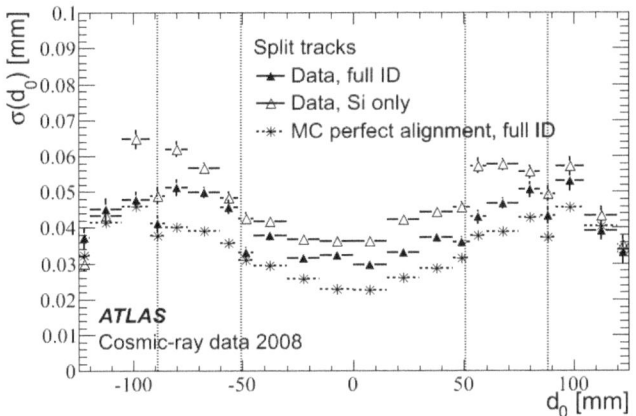
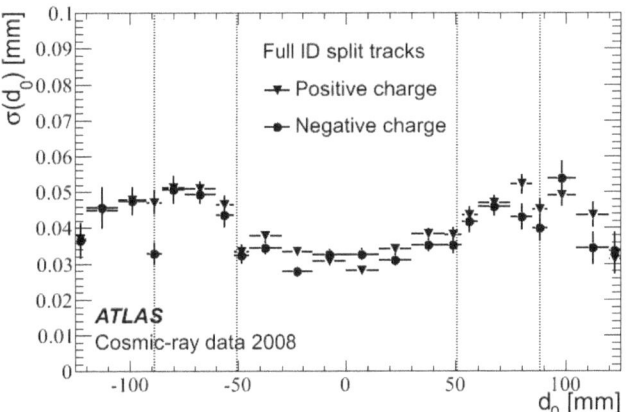

Fig. 20 Transverse impact parameter resolution as a function of transverse impact parameter for tracks with $p_T > 1$ GeV. As for the previous figures, the *left plot* compares resolutions of full ID tracks, silicon-only tracks and simulated full ID tracks. In the *right plot* resolutions are compared for full Inner Detector tracks with positive (*circles*) and negative charge (*squares*). The *vertical lines* indicate the positions of the pixel barrel layers

measurements add to the momentum resolution and thus to the precision of the track extrapolation to the perigee point.

The d_0 resolution has also been studied as a function of d_0 on a sample without the cut on $|d_0|$. The results are presented in Fig. 20 and show a worsening in resolution towards larger $|d_0|$, which corresponds to tracks crossing pixel layers at high incident angle. Pixel clusters from such tracks are wider and possibly fragmented due to a geometrically reduced charge deposition per pixel. This effect degrades the resolution, as does the smaller number of pixel layers crossed. The resolution of full ID tracks at d_0 values near to the radii of pixel layers (about 50, 90 and 120 mm) improves because of the reduction in the extrapolation length between the closest measurement and the perigee of the track.

A dependence on the charge of the reconstructed tracks has also been investigated as shown in Fig. 20 (right plot). Small differences appear in some bins, but do not allow for a conclusive result. A dependence of the resolutions on z_0

and ϕ_0 has been checked as well, and none was found. This means that the impact parameter resolutions follow the symmetries in the barrel part of the Inner Detector.

Angular resolution A precise and reliable reconstruction of the track direction contributes to the knowledge of the momentum vector and thus is vital for finding decay vertices and matching with signals from other detectors. A precision on the track angles below 1 mrad is achieved, as shown in Figs. 21 and 22.

The angular resolutions have been found to be independent of other track parameters, except for an expected small worsening at $|d_0| > 50$ mm.

Momentum resolution A precise momentum determination of high-energy particles is a key ingredient for any physics analysis. In Fig. 23 the relative momentum resolution $p \times \sigma(q/p)$ is shown as a function of p_T (left plot)

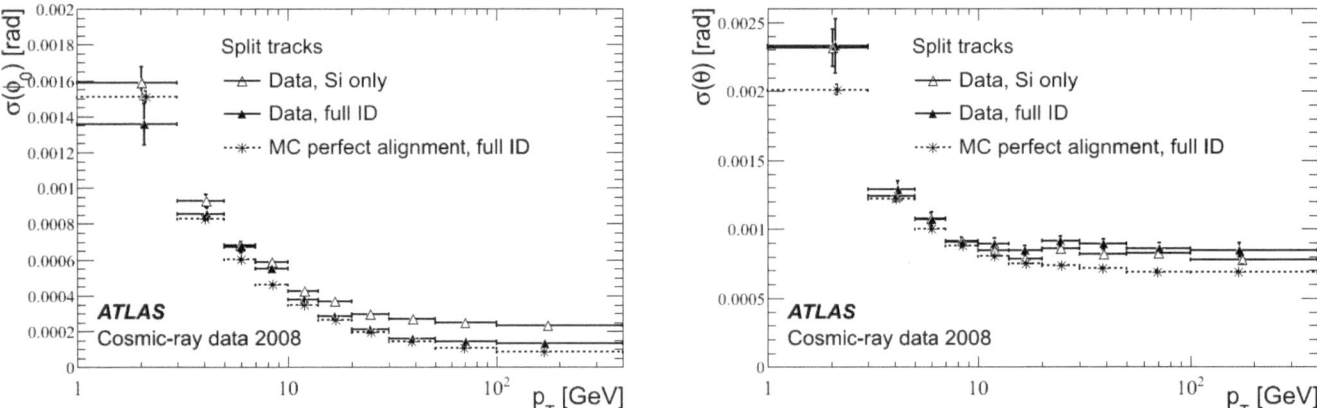

Fig. 21 Angular resolution determined from data as a function of transverse momentum. The resolutions are shown for full ID tracks (*solid triangles*), silicon-only tracks (*open triangles*) and simulated full ID tracks (*stars*)

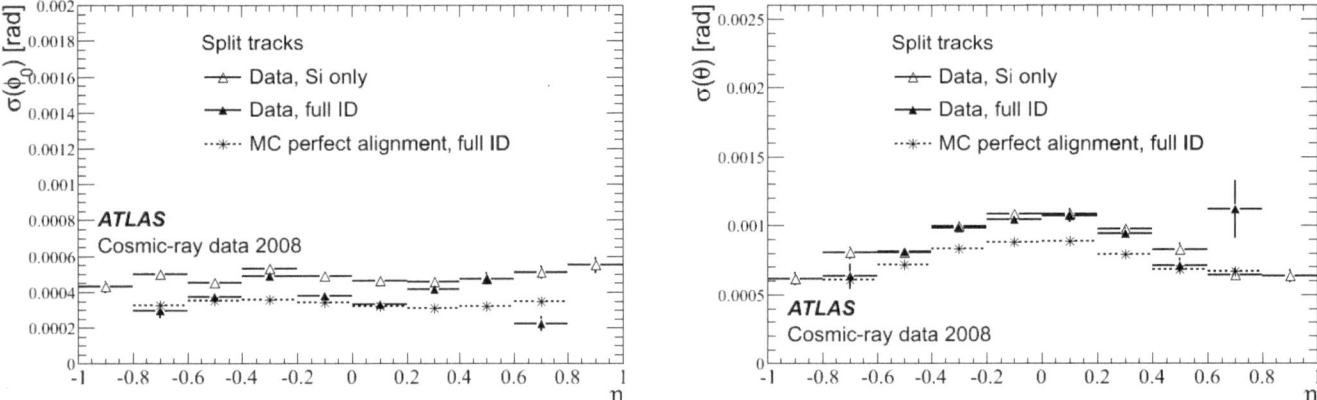

Fig. 22 Angular resolution determined from data for tracks with $p_T > 1$ GeV as a function of pseudorapidity η. The resolutions are shown for full ID tracks (*solid triangles*), silicon-only tracks (*open triangles*) and simulated full ID tracks (*stars*)

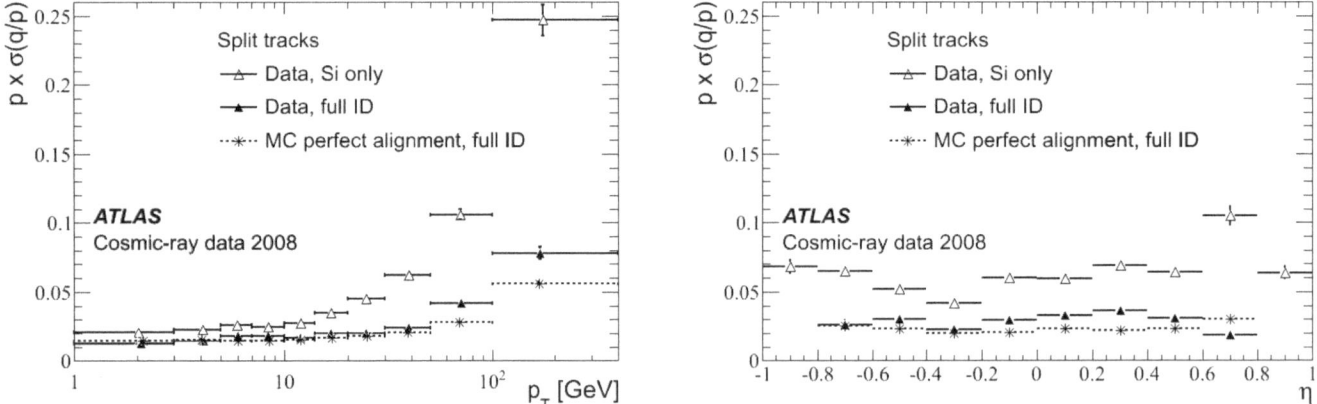

Fig. 23 Momentum resolutions determined from data as a function of transverse momentum and η. The resolutions are shown for full ID tracks (*solid triangles*), silicon-only tracks (*open triangles*) and simulated full ID tracks (*stars*)

and η (right plot). While the resolution is flat in η, it shows the expected degradation at higher transverse momenta. In this region, the contribution of the TRT to the momentum resolution becomes clearly visible.

7.4 Energy-loss measurement

The average specific energy loss of charged particle dE/dx is described by the Bethe-Bloch function [28]. The specific energy loss, sensitive to the particle speed $\beta = v/c$, can be combined with the momentum measurement to provide particle identification. Because of the energy loss tails (see Fig. 9) a truncated mean can be used to reduce the variance of the estimation.

Split tracks from cosmic-ray muons have been used to measure the resolution on dE/dx of the Pixel Detector. Tracks are required to have a transverse momentum $p_T > 0.5$ GeV and relative momentum resolution $\sigma(p_T)/p_T < 20\%$. In addition a cut on the distance of closest approach to the beam axis, $|d_0| < 10$ mm, is made in order to select tracks similar to the ones generated by LHC collisions.

The specific energy loss in a Pixel Detector module is derived from the cluster charge, Q, taking into account the average energy needed to create an electron-hole pair W (Sect. 5.1) and the path in silicon $d/\cos\alpha$ where d is the detector sensitive thickness (250 μm):

$$\frac{dE}{dx} = \frac{Q}{e}\frac{W\cos\alpha}{d}. \tag{8}$$

At high incident angle particles cross several pixel cells; the signal released in some of them may be below threshold and the energy loss underestimated. To reduce this effect, only clusters with $\cos\alpha > 0.6$ and $|\phi_{local}| < 0.5$ rad are used. The correct association of clusters to the reconstructed track is ensured by requiring position residuals to be less than 300 μm in the local x coordinate and less than 900 μm in local y.

Figure 24 shows the most probable dE/dx value of individual clusters in the barrel region as a function of the track momentum. The relativistic rise and its saturation due to the density effect are clearly visible and there is a good agreement between the $7.2 \pm 0.4\%$ rise observed in data from 0.5 GeV to 20 GeV in p_T, and the $7.5 \pm 0.4\%$ estimated from the simulation. For tracks with at least three

clusters, a global dE/dx estimation is made by averaging all the individual measurements after the exclusion of the cluster with the maximum $Q\cos\alpha$. This procedure has been verified to produce an almost Gaussian estimator on the relativistic plateau, $p_T > 20$ GeV, with a resolution of 15%. This would allow a limited particle identification capability, with a 2σ separation between K and π for $p < 500$ MeV.

7.5 Transition radiation measurement

The large spread of momenta of the cosmic rays recorded has allowed a validation of the transition-radiation performance of the TRT by measuring the percentage of high-threshold hits on tracks at different momenta. The probability of producing a transition radiation photon at each material boundary is dependent upon the Lorentz gamma factor of the particle. Since the threshold for producing transition radiation is $E/m \sim 1\,000$, in LHC collision events transition radiation is essentially limited to electrons. However, the mean p_T of recorded cosmic-ray muons was 60 GeV with a significant tail to almost 1 TeV (see Fig. 2(a)). The high-momentum muons produce enough transition-radiation photons to allow an initial calibration of the TRT as a transition radiation detector.

The transition radiation study used 20 000 nearly-vertical tracks in the barrel TRT. The tracks were required to have at least four SCT hits and at least 20 TRT hits, a fit $\chi^2/$Ndof < 10.0, $\sigma(p_T)/p_T < 3.0$ and $0.5 < p_T < 1\,000$ GeV. The track angle to the vertical, measured using hits in the SCT, was restricted to be less than 15°. Tracks were assigned to (logarithmic) momentum bins, and the high-threshold hit probability calculated as a simple ratio in each bin.

Figure 25 shows the probability of seeing a high-threshold hit on a muon track in the TRT barrel as a function of the Lorentz gamma factor of the particle; the probability is averaged over positively and negatively charged muons. The

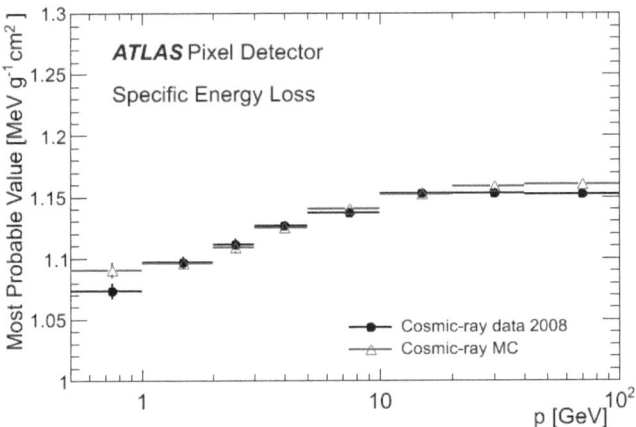

Fig. 24 Most probable value of the specific energy loss dE/dx in the Pixel Detector as a function of muon momentum in the relativistic rise region. Monte Carlo points are scaled according to the absolute charge calibration determined in Sect. 5.1

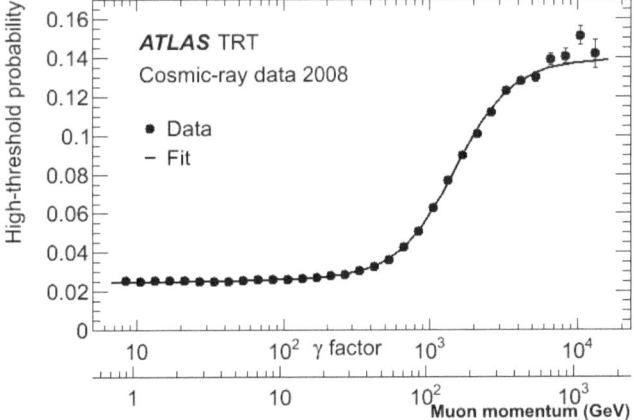

Fig. 25 High-threshold hit probability as a function of muon Lorentz γ factor for selected tracks in the October 2008 cosmic-ray data. The *line* shows a sigmoid fit to the data

fitted curve shown in Fig. 25 is consistent with the result obtained in the 2004 ATLAS combined test beam run and confirms the design of the TRT electron identification capabilities.

8 Conclusions

The final installation of the ATLAS Inner Detector in August 2008 was followed by a period of commissioning and calibration. During this period the detector took data with high efficiency with both LHC single beams and cosmic rays. These data allowed full tests of trigger, data-acquisition and monitoring systems, and of offline track reconstruction. Some problems with the newly-installed evaporative cooling system and the optical links of the silicon detectors were exposed. These were addressed before data-taking with LHC beams in 2009, when more than 98% of the detector was operational.

Detector gains were calibrated and thresholds adjusted to give good uniformity of response. The components of the detector were timed-in with a precision of 1–2 ns. Many detector performance properties were measured. The average noise occupancies were $\sim 10^{-10}$ hit/channel/BC for the Pixel Detector and $\sim 3 \times 10^{-5}$ hit/channel/BC for the SCT, well within specifications. The intrinsic efficiencies of the silicon detectors were measured to be close to 100% and of the TRT to be 97.2±0.5%. The Lorentz angle in the silicon detectors in the 2 T magnetic field was found to be consistent with model expectations. Energy loss in the Pixel Detector and transition radiation were measured and found to be in agreement with expectations from test beams.

A new Level-1 track trigger based on a fast OR of TRT signals was commissioned. The Level-2 trigger tracking-algorithms were modified for cosmic rays, resulting in a trigger efficiency of 99.6±0.02% for tracks reconstructed offline. The cosmic-ray data were used to perform an initial detector alignment. The resolution of track parameters was measured by comparing two segments of a cosmic-ray track. After detector alignment, the impact parameter resolutions for high-momentum tracks were found to be 22.1 ± 0.9 μm and 112 ± 4 μm in the transverse and longitudinal directions, respectively. In this asymptotic limit, the relative momentum resolution was measured to be $\sigma_p/p = (4.83 \pm 0.16) \times 10^{-4} \text{ GeV}^{-1} \times p_T$.

The observed performance on this early data showed the ATLAS Inner Detector to be fully operational and providing high-quality tracking before the first LHC collisions.

Acknowledgements We are greatly indebted to all CERN's departments and to the LHC project for their immense efforts not only in building the LHC, but also for their direct contributions to the construction and installation of the ATLAS detector and its infrastructure. We acknowledge equally warmly all our technical colleagues in the collaborating Institutions without whom the ATLAS detector could not have been built. Furthermore we are grateful to all the funding agencies which supported generously the construction and the commissioning of the ATLAS detector and also provided the computing infrastructure.

The ATLAS detector design and construction has taken about fifteen years, and our thoughts are with all our colleagues who sadly could not see its final realisation.

We acknowledge the support of ANPCyT, Argentina; Yerevan Physics Institute, Armenia; ARC and DEST, Australia; Bundesministerium für Wissenschaft und Forschung, Austria; National Academy of Sciences of Azerbaijan; State Committee on Science & Technologies of the Republic of Belarus; CNPq and FINEP, Brazil; NSERC, NRC, and CFI, Canada; CERN; CONICYT, Chile; NSFC, China; COLCIENCIAS, Colombia; Ministry of Education, Youth and Sports of the Czech Republic, Ministry of Industry and Trade of the Czech Republic, and Committee for Collaboration of the Czech Republic with CERN; Danish Natural Science Research Council and the Lundbeck Foundation; European Commission, through the ARTEMIS Research Training Network; IN2P3-CNRS and CEA-DSM/IRFU, France; Georgian Academy of Sciences; BMBF, DFG, HGF and MPG, Germany; Ministry of Education and Religion, through the EPEAEK program PYTHAGORAS II and GSRT, Greece; ISF, MINERVA, GIF, DIP, and Benoziyo Center, Israel; INFN, Italy; MEXT, Japan; CNRST, Morocco; FOM and NWO, Netherlands; The Research Council of Norway; Ministry of Science and Higher Education, Poland; GRICES and FCT, Portugal; Ministry of Education and Research, Romania; Ministry of Education and Science of the Russian Federation and State Atomic Energy Corporation ROSATOM; JINR; Ministry of Science, Serbia; Department of International Science and Technology Cooperation, Ministry of Education of the Slovak Republic; Slovenian Research Agency, Ministry of Higher Education, Science and Technology, Slovenia; Ministerio de Educación y Ciencia, Spain; The Swedish Research Council, The Knut and Alice Wallenberg Foundation, Sweden; State Secretariat for Education and Science, Swiss National Science Foundation, and Cantons of Bern and Geneva, Switzerland; National Science Council, Taiwan; TAEK, Turkey; The Science and Technology Facilities Council and The Leverhulme Trust, United Kingdom; DOE and NSF, United States of America.

References

1. G. Aad et al., The ATLAS experiment at the CERN large hadron collider. J. Nucl. Sci. Technol. **3**, S08003 (2008). doi:10.1088/1748-0221/3/08/S08003
2. L. Evans, P. Bryant (eds.), LHC machine. J. Nucl. Sci. Technol. **3**, S08001 (2008). doi:10.1088/1748-0221/3/08/S08001
3. G. Aad et al., ATLAS pixel detector electronics and sensors. J. Nucl. Sci. Technol. **3**, P07007 (2008). doi:10.1088/1748-0221/3/08/P07007
4. I. Peric et al., The FEI3 readout chip for the ATLAS pixel detector. Nucl. Instrum. Methods A **565**, 178–187 (2006). doi:10.1016/j.nima.2006.05.032
5. A. Abdesselam et al., The barrel modules of the ATLAS semiconductor tracker. Nucl. Instrum. Methods A **568**, 642–671 (2006). doi:10.1016/j.nima.2006.08.036
6. A. Abdesselam et al., The ATLAS semiconductor tracker end-cap module. Nucl. Instrum. Methods A **575**, 353–389 (2007). doi:10.1016/j.nima.2007.02.019

7. A. Ahmad et al., The Silicon microstrip sensors of the ATLAS semiconductor tracker. Nucl. Instrum. Methods A **578**, 98–118 (2007). doi:10.1016/j.nima.2007.04.157

8. F. Campabadal et al., Design and performance of the ABCD3TA ASIC for readout of silicon strip detectors in the ATLAS semiconductor tracker. Nucl. Instrum. Methods A **552**, 292–328 (2005). doi:10.1016/j.nima.2005.07.002

9. E. Abat et al., The ATLAS TRT barrel detector. J. Nucl. Sci. Technol. **3**, P02014 (2008). doi:10.1088/1748-0221/3/02/P02014

10. E. Abat et al., The ATLAS TRT end-cap detectors. J. Nucl. Sci. Technol. **3**, P10003 (2008). doi:10.1088/1748-0221/3/10/P10003

11. E. Abat et al., The ATLAS TRT electronics. J. Nucl. Sci. Technol. **3**, P06007 (2008). doi:10.1088/1748-0221/3/08/S08003

12. V. Cindro et al., The ATLAS beam conditions monitor. J. Nucl. Sci. Technol. **3**, P02004 (2008). doi:10.1088/1748-0221/3/02/P02004

13. A. Abdesselam et al., The optical links of the ATLAS semiconductor tracker. J. Nucl. Sci. Technol. **2**, P09003 (2007). doi:10.1088/1748-0221/2/09/P09003

14. K.K. Gan et al., Optical link of the ATLAS pixel detector. Nucl. Instrum. Methods A **570**, 292–294 (2007). doi:10.1016/j.nima.2006.09.042

15. P. Moreira et al., G-Link and Gigabit Ethernet compliant serializer for LHC data transmission. IEEE Nucl. Sci. Symp. **2**, 96–99 (2000). doi:10.1109/NSSMIC.2000.949860

16. M.L. Chu et al., The off-detector opto-electronics for the optical links of the ATLAS semiconductor tracker and pixel detector. Nucl. Instrum. Methods A **530**, 293–310 (2004). doi:10.1016/j.nima.2004.04.228

17. P. Lichard et al., Evolution of the TRT backend and the new TRT-TTC board, in *Proceedings of the 2005 LECC*, Heidelberg. CERN-LHCC-2005-038 (CERN, Geneva, 2005), p. 253

18. A. Abdesselam et al., The Data acquisition and calibration system for the ATLAS semiconductor tracker. J. Nucl. Sci. Technol. **3**, P01003 (2008). doi:10.1088/1748-0221/3/01/P01003

19. D. Attree et al., The evaporative cooling system for the ATLAS inner detector. J. Nucl. Sci. Technol. **3**, P07003 (2008). doi:10.1088/1748-0221/3/07/P07003

20. A. Barriuso Poy et al., The detector control system of the ATLAS experiment. J. Nucl. Sci. Technol. **3**, P05006 (2008). doi:10.1088/1748-0221/3/05/P05006

21. M. White, Data quality monitor for the ATLAS inner detector, in *17th International Workshop on Vertex detectors*. Proceedings of Science, 2008. PoS (Vertex 2008) 044

22. ATLAS Collaboration, The athena framework, in *ATLAS Computing Technical Design Report*, CERN-LHCC-2005-022 (CERN, Geneva, 2005) p. 27

23. T. Cornelissen et al., *Concepts, Design and Implementation of the ATLAS New Tracking (NEWT)*. ATLAS Note ATL-SOFT-PUB-2007-007

24. T. Cornelissen et al., The global χ^2 track fitter in ATLAS. J. Phys.: Conf. Ser. **119**, 032013 (2008). doi:10.1088/1742-6596/119/3/032013

25. A. Fratina et al., *The TRT Fast-OR Trigger*. ATLAS Note ATL-INDET-PUB-2009-002

26. G. Aad et al., HLT track reconstruction performance, in *Expected Performance of the ATLAS Experiment: Detector, Trigger and Physics*. CERN-OPEN-2008-020 (CERN, Geneva, 2009), p. 565

27. A. Dar, Atm. neutrinos, astrophysical neutrons, and proton-decay experiments. Phys. Rev. Lett. **51**, 227 (1983). doi:10.1103/PhysRevLett.51.22

28. C. Amsler et al., The review of particle physics. Phys. Lett. B **667**, 1 (2008). doi:10.1016/j.physletb.2008.07.018

29. S. Agostinelli et al., Geant4—a simulation Toolkit. Nucl. Instrum. Methods A **506**, 250–303 (2003). doi:10.1016/S0168-9002(03)01368-8

30. J. Allison et al., Geant4 developments and applications. IEEE Trans. Nucl. Sci. **53**(1), 270–278 (2006). doi:10.1109/TNS.2006.869826

31. K.E. Arms et al., ATLAS pixel opto-electronics. Nucl. Instrum. Methods A **554**, 458–468 (2005). doi:10.1016/j.nima.2005.07.070

32. M.S. Alam et al., The ATLAS silicon pixel sensors. Nucl. Instrum. Methods A **456**, 217–232 (2001). doi:10.1016/S0168-9002(00)00574-X

33. H. Bichsel, Straggling in thin silicon detectors. Rev. Mod. Phys. **60**, 663–669 (1988). doi:10.1103/RevModPhys.60.663

34. R.D. Ryan, Precision measurements of the ionization energy and its temperature variation in high purity silicon radiation detectors. IEEE Trans. Nucl. Sci. **20**(1), 473–480 (1973). doi:10.1109/TNS.1973.4326950

35. P. Christmas, Average energy required to produce an ion pair. Tech. Rep. Report 31 (ICRU, 1979)

36. R.H. Pehl et al., Accurate determination of the ionization energy in semiconductor devices. Nucl. Instrum. Methods **59**, 45–55 (1968)

37. F. Scholze et al., Determination of the electron-hole pair creation energy for semiconductors from the spectral responsivity of photodiodes. Nucl. Instrum. Methods A **439**, 208–215 (2000). doi:10.1016/S0168-9002(99)00937-7

38. I. Gorelov et al., A measurement of Lorentz angle and spatial resolution of radiation hard silicon pixel sensors. Nucl. Instrum. Methods A **481**, 204–221 (2002). doi:10.1016/S0168-9002(01)01413-9

39. ATLAS Collaboration, Alignment requirements, in *ATLAS Inner Detector Technical Design Report*, vol. I. CERN-LHCC-1997-016 (CERN, Geneva, 1997), p. 215

40. P. Brükman, A. Hicheur, S.J. Haywood, *Global χ^2 Approach to the Alignment of the ATLAS Silicon Tracking Detectors*. ATLAS Note ATL-INDET-PUB-2005-002

41. R. Härtel, Iterative local χ^2 alignment approach for the ATLAS SCT detector. Master's thesis, MPI Munich (2005)

42. T. Göttfert, Iterative local χ^2 alignment algorithm for the ATLAS Pixel detector. Master's thesis, Universität Würzburg and MPI Munich (2006)

43. F. Heinemann, *Track Based Alignment of the ATLAS Silicon Detectors with the Robust Alignment Algorithm*. ATLAS Note ATL-INDET-PUB-2007-011

44. A. Andreazza, V. Kostyukhin, R.J. Madaras, *Survey of the ATLAS Pixel Detector Components*. ATLAS Note ATL-INDET-PUB-2008-012

45. C. Jacoboni et al., A review of some charge transport properties of silicon. Solid State Electron. **20**, 77–89 (1977). doi:10.1016/0038-1101(77)90054-5

46. F. Campabadal et al., Beam tests of ATLAS SCT silicon strip detector modules. Nucl. Instrum. Methods A **538**, 384–407 (2005). doi:10.1016/j.nima.2004.08.133

Drift Time Measurement in the ATLAS Liquid Argon Electromagnetic Calorimeter using Cosmic Muons

The ATLAS Collaboration[*,**]

G. Aad[48], B. Abbott[110], J. Abdallah[11], A.A. Abdelalim[49], A. Abdesselam[117], O. Abdinov[10], B. Abi[111], M. Abolins[88], H. Abramowicz[151], H. Abreu[114], B.S. Acharya[162a,162b], D.L. Adams[24], T.N. Addy[56], J. Adelman[173], C. Adorisio[36a,36b], P. Adragna[75], T. Adye[128], S. Aefsky[22], J.A. Aguilar-Saavedra[123a], M. Aharrouche[81], S.P. Ahlen[21], F. Ahles[48], A. Ahmad[146], H. Ahmed[2], M. Ahsan[40], G. Aielli[132a,132b], T. Akdogan[18], T.P.A. Åkesson[79], G. Akimoto[153], A.V. Akimov[94], A. Aktas[48], M.S. Alam[1], M.A. Alam[76], J. Albert[167], S. Albrand[55], M. Aleksa[29], I.N. Aleksandrov[65], F. Alessandria[89a], C. Alexa[25a], G. Alexander[151], G. Alexandre[49], T. Alexopoulos[9], M. Alhroob[20], M. Aliev[15], G. Alimonti[89a], J. Alison[119], M. Aliyev[10], P.P. Allport[73], S.E. Allwood-Spiers[53], J. Almond[82], A. Aloisio[102a,102b], R. Alon[169], A. Alonso[79], M.G. Alviggi[102a,102b], K. Amako[66], C. Amelung[22], V.V. Ammosov[127], A. Amorim[123b], G. Amorós[165], N. Amram[151], C. Anastopoulos[138], T. Andeen[29], C.F. Anders[48], K.J. Anderson[30], A. Andreazza[89a,89b], V. Andrei[58a], X.S. Anduaga[70], A. Angerami[34], F. Anghinolfi[29], N. Anjos[123b], A. Antonaki[8], M. Antonelli[47], S. Antonelli[19a,19b], J. Antos[143], B. Antunovic[41], F. Anulli[131a], S. Aoun[83], G. Arabidze[8], I. Aracena[142], Y. Arai[66], A.T.H. Arce[14], J.P. Archambault[28], S. Arfaoui[29,a], J.-F. Arguin[14], T. Argyropoulos[9], E. Arik[18,*], M. Arik[18], A.J. Armbruster[87], O. Arnaez[4], C. Arnault[114], A. Artamonov[95], D. Arutinov[20], M. Asai[142], S. Asai[153], R. Asfandiyarov[170], S. Ask[82], B. Åsman[144], D. Asner[28], L. Asquith[77], K. Assamagan[24], A. Astbury[167], A. Astvatsatourov[52], G. Atoian[173], B. Auerbach[173], E. Auge[114], K. Augsten[126], M. Aurousseau[4], N. Austin[73], G. Avolio[161], R. Avramidou[9], D. Axen[166], C. Ay[54], G. Azuelos[93,b], Y. Azuma[153], M.A. Baak[29], C. Bacci[133a,133b], A. Bach[14], H. Bachacou[135], K. Bachas[29], M. Backes[49], E. Badescu[25a], P. Bagnaia[131a,131b], Y. Bai[32], D.C. Bailey[156], T. Bain[156], J.T. Baines[128], O.K. Baker[173], M.D. Baker[24], S. Baker[77], F. Baltasar Dos Santos Pedrosa[29], E. Banas[38], P. Banerjee[93], S. Banerjee[167], D. Banfi[89a,89b], A. Bangert[136], V. Bansal[167], S.P. Baranov[94], S. Baranov[65], A. Barashkou[65], T. Barber[27], E.L. Barberio[86], D. Barberis[50a,50b], M. Barbero[20], D.Y. Bardin[65], T. Barillari[99], M. Barisonzi[172], T. Barklow[142], N. Barlow[27], B.M. Barnett[128], R.M. Barnett[14], S. Baron[29], A. Baroncelli[133a], A.J. Barr[117], F. Barreiro[80], J. Barreiro Guimarães da Costa[57], P. Barrillon[114], N. Barros[123b], R. Bartoldus[142], D. Bartsch[20], J. Bastos[123b], R.L. Bates[53], L. Batkova[143], J.R. Batley[27], A. Battaglia[16], M. Battistin[29], F. Bauer[135], H.S. Bawa[142], M. Bazalova[124], B. Beare[156], T. Beau[78], P.H. Beauchemin[117], R. Beccherle[50a], N. Becerici[18], P. Bechtle[41], G.A. Beck[75], H.P. Beck[16], M. Beckingham[48], K.H. Becks[172], I. Bedajanek[126], A.J. Beddall[18,c], A. Beddall[18,c], P. Bednár[143], V.A. Bednyakov[65], C. Bee[83], M. Begel[24], S. Behar Harpaz[150], P.K. Behera[63], M. Beimforde[99], C. Belanger-Champagne[164], P.J. Bell[82], W.H. Bell[49], G. Bella[151], L. Bellagamba[19a], F. Bellina[29], M. Bellomo[118a], A. Belloni[57], K. Belotskiy[96], O. Beltramello[29], S.Ben Ami[150], O. Benary[151], D. Benchekroun[134a], M. Bendel[81], B.H. Benedict[161], N. Benekos[163], Y. Benhammou[151], G.P. Benincasa[123b], D.P. Benjamin[44], M. Benoit[114], J.R. Bensinger[22], K. Benslama[129], S. Bentvelsen[105], M. Beretta[47], D. Berge[29], E. Bergeaas Kuutmann[144], N. Berger[4], F. Berghaus[167], E. Berglund[49], J. Beringer[14], K. Bernardet[83], P. Bernat[114], R. Bernhard[48], C. Bernius[77], T. Berry[76], A. Bertin[19a,19b], M.I. Besana[89a,89b], N. Besson[135], S. Bethke[99], R.M. Bianchi[48], M. Bianco[72a,72b], O. Biebel[98], J. Biesiada[14], M. Biglietti[131a,131b], H. Bilokon[47], M. Bindi[19a,19b], S. Binet[114], A. Bingul[18,c], C. Bini[131a,131b], C. Biscarat[178], U. Bitenc[48], K.M. Black[57], R.E. Blair[5], J.-B. Blanchard[114], G. Blanchot[29], C. Blocker[22], J. Blocki[38], A. Blondel[49], W. Blum[81], U. Blumenschein[54], G.J. Bobbink[105], A. Bocci[44], M. Boehler[41], J. Boek[172], N. Boelaert[79], S. Böser[77], J.A. Bogaerts[29], A. Bogouch[90,*], C. Bohm[144], J. Bohm[124], V. Boisvert[76], T. Bold[161,d], V. Boldea[25a], A. Boldyrev[97], V.G. Bondarenko[96], M. Bondioli[161], M. Boonekamp[135], S. Bordoni[78], C. Borer[16], A. Borisov[127], G. Borissov[71], I. Borjanovic[72a], S. Borroni[131a,131b], K. Bos[105], D. Boscherini[19a], M. Bosman[11], M. Bosteels[29], H. Boterenbrood[105], J. Bouchami[93], J. Boudreau[122], E.V. Bouhova-Thacker[71], C. Boulahouache[122], C. Bourdarios[114], J. Boyd[29], I.R. Boyko[65], I. Bozovic-Jelisavcic[12b], J. Bracinik[17], A. Braem[29], P. Branchini[133a], G.W. Brandenburg[57], A. Brandt[7], G. Brandt[41], O. Brandt[54], U. Bratzler[154], B. Brau[84], J.E. Brau[113], H.M. Braun[172], B. Brelier[156], J. Bremer[29], R. Brenner[164], S. Bressler[150], D. Breton[114], D. Britton[53], F.M. Brochu[27], I. Brock[20], R. Brock[88], T.J. Brodbeck[71], E. Brodet[151], F. Broggi[89a], C. Bromberg[88], G. Brooijmans[34], W.K. Brooks[31b],

G. Brown[82], E. Brubaker[30], P.A. Bruckman de Renstrom[38], D. Bruncko[143], R. Bruneliere[48], S. Brunet[41], A. Bruni[19a], G. Bruni[19a], M. Bruschi[19a], T. Buanes[13], F. Bucci[49], J. Buchanan[117], P. Buchholz[140], A.G. Buckley[45,e], I.A. Budagov[65], B. Budick[107], V. Büscher[81], L. Bugge[116], O. Bulekov[96], M. Bunse[42], T. Buran[116], H. Burckhart[29], S. Burdin[73], T. Burgess[13], S. Burke[128], E. Busato[33], P. Bussey[53], C.P. Buszello[164], F. Butin[29], B. Butler[142], J.M. Butler[21], C.M. Buttar[53], J.M. Butterworth[77], T. Byatt[77], J. Caballero[24], S. Cabrera Urbán[165], D. Caforio[19a,19b], O. Cakir[3], P. Calafiura[14], G. Calderini[78], P. Calfayan[98], R. Calkins[5], L.P. Caloba[23a], R. Caloi[131a,131b], D. Calvet[33], P. Camarri[132a,132b], M. Cambiaghi[118a,118b], D. Cameron[116], F. Campabadal Segura[165], S. Campana[29], M. Campanelli[77], V. Canale[102a,102b], F. Canelli[30], A. Canepa[157a], J. Cantero[80], L. Capasso[102a,102b], M.D.M. Capeans Garrido[29], I. Caprini[25a], M. Caprini[25a], M. Capua[36a,36b], R. Caputo[146], D. Caracinha[123b], C. Caramarcu[25a], R. Cardarelli[132a], T. Carli[29], G. Carlino[102a], L. Carminati[89a,89b], B. Caron[2,b], S. Caron[48], G.D. Carrillo Montoya[170], S. Carron Montero[156], A.A. Carter[75], J.R. Carter[27], J. Carvalho[123b], D. Casadei[107], M.P. Casado[11], M. Cascella[121a,121b], C. Caso[50a,50b,*], A.M. Castaneda Hernadez[170], E. Castaneda-Miranda[170], V. Castillo Gimenez[165], N. Castro[123a], G. Cataldi[72a], A. Catinaccio[29], J.R. Catmore[71], A. Cattai[29], G. Cattani[132a,132b], S. Caughron[34], D. Cauz[162a,162c], P. Cavalleri[78], D. Cavalli[89a], M. Cavalli-Sforza[11], V. Cavasinni[121a,121b], F. Ceradini[133a,133b], A.S. Cerqueira[23a], A. Cerri[29], L. Cerrito[75], F. Cerutti[47], S.A. Cetin[18,f], F. Cevenini[102a,102b], A. Chafaq[134a], D. Chakraborty[5], K. Chan[2], J.D. Chapman[27], J.W. Chapman[87], E. Chareyre[78], D.G. Charlton[17], V. Chavda[82], S. Cheatham[71], S. Chekanov[5], S.V. Chekulaev[157a], G.A. Chelkov[65], H. Chen[24], S. Chen[32], T. Chen[32], X. Chen[170], S. Cheng[32], A. Cheplakov[65], V.F. Chepurnov[65], R. Cherkaoui El Moursli[134d], V. Tcherniatine[24], D. Chesneanu[25a], E. Cheu[6], S.L. Cheung[156], L. Chevalier[135], F. Chevallier[135], V. Chiarella[47], G. Chiefari[102a,102b], L. Chikovani[51], J.T. Childers[58a], A. Chilingarov[71], G. Chiodini[72a], M. Chizhov[65], G. Choudalakis[30], S. Chouridou[136], I.A. Christidi[152], A. Christov[48], D. Chromek-Burckhart[29], M.L. Chu[149], J. Chudoba[124], G. Ciapetti[131a,131b], A.K. Ciftci[3], R. Ciftci[3], D. Cinca[33], V. Cindro[74], M.D. Ciobotaru[161], C. Ciocca[19a,19b], A. Ciocio[14], M. Cirilli[87], M. Citterio[89a], A. Clark[49], W. Cleland[122], J.C. Clemens[83], B. Clement[55], C. Clement[144], Y. Coadou[83], M. Cobal[162a,162c], A. Coccaro[50a,50b], J. Cochran[64], S. Coelli[89a], J. Coggeshall[163], E. Cogneras[16], C.D. Cojocaru[28], J. Colas[4], B. Cole[34], A.P. Colijn[105], C. Collard[114], N.J. Collins[17], C. Collins-Tooth[53], J. Collot[55], G. Colon[84], P. Conde Muiño[123b], E. Coniavitis[164], M. Consonni[104], S. Constantinescu[25a], C. Conta[118a,118b], F. Conventi[102a,g], J. Cook[29], M. Cooke[34], B.D. Cooper[75], A.M. Cooper-Sarkar[117], N.J. Cooper-Smith[76], K. Copic[34], T. Cornelissen[50a,50b], M. Corradi[19a], F. Corriveau[85,h], A. Corso-Radu[161], A. Cortes-Gonzalez[163], G. Cortiana[99], G. Costa[89a], M.J. Costa[165], D. Costanzo[138], T. Costin[30], D. Côté[41], R. Coura Torres[23a], L. Courneyea[167], G. Cowan[76], C. Cowden[27], B.E. Cox[82], K. Cranmer[107], J. Cranshaw[5], M. Cristinziani[20], G. Crosetti[36a,36b], R. Crupi[72a,72b], S. Crépé-Renaudin[55], C. Cuenca Almenar[173], T. Cuhadar Donszelmann[138], M. Curatolo[47], C.J. Curtis[17], P. Cwetanski[61], Z. Czyczula[173], S. D'Auria[53], M. D'Onofrio[73], A. D'Orazio[99], P.V.M. Da Silva[23a], C. Da Via[82], W. Dabrowski[37], T. Dai[87], C. Dallapiccola[84], S.J. Dallison[128,*], C.H. Daly[137], M. Dam[35], H.O. Danielsson[29], D. Dannheim[99], V. Dao[49], G. Darbo[50a], G.L. Darlea[25a], W. Davey[86], T. Davidek[125], N. Davidson[86], R. Davidson[71], M. Davies[93], A.R. Davison[77], I. Dawson[138], J.W. Dawson[5], R.K. Daya[39], K. De[7], R. de Asmundis[102a], S. De Castro[19a,19b], P.E. De Castro Faria Salgado[24], S. De Cecco[78], J. de Graat[98], N. De Groot[104], P. de Jong[105], E. De La Cruz-Burelo[87], C. De La Taille[114], L. De Mora[71], M. De Oliveira Branco[29], D. De Pedis[131a], A. De Salvo[131a], U. De Sanctis[162a,162c], A. De Santo[147], J.B. De Vivie De Regie[114], G. De Zorzi[131a,131b], S. Dean[77], H. Deberg[163], G. Dedes[99], D.V. Dedovich[65], P.O. Defay[33], J. Degenhardt[119], M. Dehchar[117], C. Del Papa[162a,162c], J. Del Peso[80], T. Del Prete[121a,121b], A. Dell'Acqua[29], L. Dell'Asta[89a,89b], M. Della Pietra[102a,g], D. della Volpe[102a,102b], M. Delmastro[29], N. Delruelle[29], P.A. Delsart[55], C. Deluca[146], S. Demers[173], M. Demichev[65], B. Demirkoz[11], J. Deng[161], W. Deng[24], S.P. Denisov[127], C. Dennis[117], J.E. Derkaoui[134c], F. Derue[78], P. Dervan[73], K. Desch[20], P.O. Deviveiros[156], A. Dewhurst[71], B. DeWilde[146], S. Dhaliwal[156], R. Dhullipudi[24,i], A. Di Ciaccio[132a,132b], L. Di Ciaccio[4], A. Di Domenico[131a,131b], A. Di Girolamo[29], B. Di Girolamo[29], S. Di Luise[133a,133b], A. Di Mattia[88], R. Di Nardo[132a,132b], A. Di Simone[132a,132b], R. Di Sipio[19a,19b], M.A. Diaz[31a], F. Diblen[18], E.B. Diehl[87], J. Dietrich[48], T.A. Dietzsch[58a], S. Diglio[114], K. Dindar Yagci[39], D.J. Dingfelder[48], C. Dionisi[131a,131b], P. Dita[25a], S. Dita[25a], F. Dittus[29], F. Djama[83], R. Djilkibaev[107], T. Djobava[51], M.A.B. do Vale[23a], A. Do Valle Wemans[123b], T.K.O. Doan[4], M. Dobbs[85], D. Dobos[29], E. Dobson[29], M. Dobson[161], J. Dodd[34], T. Doherty[53], Y. Doi[66], J. Dolejsi[125], I. Dolenc[74], Z. Dolezal[125], B.A. Dolgoshein[96], T. Dohmae[153], M. Donega[119], J. Donini[55], J. Dopke[172], A. Doria[102a], A. Dos Anjos[170], A. Dotti[121a,121b], M.T. Dova[70], A. Doxiadis[105], A.T. Doyle[53], Z. Drasal[125], C. Driouichi[35], M. Dris[9], J. Dubbert[99], E. Duchovni[169], G. Duckeck[98], A. Dudarev[29], F. Dudziak[114], M. Dührssen[48], L. Duflot[114], M.-A. Dufour[85], M. Dunford[30], A. Duperrin[83], H.Duran Yildiz[3,j], A. Dushkin[22], R. Duxfield[138], M. Dwuznik[37],

M. Düren[52], W.L. Ebenstein[44], J. Ebke[98], S. Eckert[48], S. Eckweiler[81], K. Edmonds[81], C.A. Edwards[76], P. Eerola[79,k],
K. Egorov[61], W. Ehrenfeld[41], T. Ehrich[99], T. Eifert[29], G. Eigen[13], K. Einsweiler[14], E. Eisenhandler[75], T. Ekelof[164],
M. El Kacimi[4], M. Ellert[164], S. Elles[4], F. Ellinghaus[81], K. Ellis[75], N. Ellis[29], J. Elmsheuser[98], M. Elsing[29], R. Ely[14],
D. Emeliyanov[128], R. Engelmann[146], A. Engl[98], B. Epp[62], A. Eppig[87], V.S. Epshteyn[95], A. Ereditato[16],
D. Eriksson[144], I. Ermoline[88], J. Ernst[1], M. Ernst[24], J. Ernwein[135], D. Errede[163], S. Errede[163], E. Ertel[81],
M. Escalier[114], C. Escobar[165], X. Espinal Curull[11], B. Esposito[47], F. Etienne[83], A.I. Etienvre[135], E. Etzion[151],
H. Evans[61], L. Fabbri[19a,19b], C. Fabre[29], K. Facius[35], R.M. Fakhrutdinov[127], S. Falciano[131a], A.C. Falou[114],
Y. Fang[170], M. Fanti[89a,89b], A. Farbin[7], A. Farilla[133a], J. Farley[146], T. Farooque[156], S.M. Farrington[117],
P. Farthouat[29], F. Fassi[165], P. Fassnacht[29], D. Fassouliotis[8], B. Fatholahzadeh[156], L. Fayard[114], F. Fayette[54],
R. Febbraro[33], P. Federic[143], O.L. Fedin[120], I. Fedorko[29], W. Fedorko[29], L. Feligioni[83], C.U. Felzmann[86], C. Feng[32],
E.J. Feng[30], A.B. Fenyuk[127], J. Ferencei[143], J. Ferland[93], B. Fernandes[123b], W. Fernando[108], S. Ferrag[53],
J. Ferrando[117], A. Ferrari[164], P. Ferrari[105], R. Ferrari[118a], A. Ferrer[165], M.L. Ferrer[47], D. Ferrere[49], C. Ferretti[87],
M. Fiascaris[117], F. Fiedler[81], A. Filipčič[74], A. Filippas[9], F. Filthaut[104], M. Fincke-Keeler[167], M.C.N. Fiolhais[123b],
L. Fiorini[11], A. Firan[39], G. Fischer[41], M.J. Fisher[108], M. Flechl[164], I. Fleck[140], J. Fleckner[81], P. Fleischmann[171],
S. Fleischmann[20], T. Flick[172], L.R. Flores Castillo[170], M.J. Flowerdew[99], F. Föhlisch[58a], M. Fokitis[9],
T. Fonseca Martin[76], D.A. Forbush[137], A. Formica[135], A. Forti[82], D. Fortin[157a], J.M. Foster[82], D. Fournier[114],
A. Foussat[29], A.J. Fowler[44], K. Fowler[136], H. Fox[71], P. Francavilla[121a,121b], S. Franchino[118a,118b], D. Francis[29],
M. Franklin[57], S. Franz[29], M. Fraternali[118a,118b], S. Fratina[119], J. Freestone[82], S.T. French[27], R. Froeschl[29],
D. Froidevaux[29], J.A. Frost[27], C. Fukunaga[154], E. Fullana Torregrosa[5], J. Fuster[165], C. Gabaldon[80], O. Gabizon[169],
T. Gadfort[24], S. Gadomski[49], G. Gagliardi[50a,50b], P. Gagnon[61], C. Galea[98], E.J. Gallas[117], M.V. Gallas[29], V. Gallo[16],
B.J. Gallop[128], P. Gallus[124], E. Galyaev[40], K.K. Gan[108], Y.S. Gao[142,l], A. Gaponenko[14], M. Garcia-Sciveres[14],
C. García[165], J.E. García Navarro[49], R.W. Gardner[30], N. Garelli[29], H. Garitaonandia[105], V. Garonne[29], C. Gatti[47],
G. Gaudio[118a], O. Gaumer[49], P. Gauzzi[131a,131b], I.L. Gavrilenko[94], C. Gay[166], G. Gaycken[20], J.-C. Gayde[29],
E.N. Gazis[9], P. Ge[32], C.N.P. Gee[128], Ch. Geich-Gimbel[20], K. Gellerstedt[144], C. Gemme[50a], M.H. Genest[98],
S. Gentile[131a,131b], F. Georgatos[9], S. George[76], P. Gerlach[172], A. Gershon[151], C. Geweniger[58a], H. Ghazlane[134d],
P. Ghez[4], N. Ghodbane[33], B. Giacobbe[19a], S. Giagu[131a,131b], V. Giakoumopoulou[8], V. Giangiobbe[121a,121b],
F. Gianotti[29], B. Gibbard[24], A. Gibson[156], S.M. Gibson[117], L.M. Gilbert[117], M. Gilchriese[14], V. Gilewsky[91],
A.R. Gillman[128], D.M. Gingrich[2,b], J. Ginzburg[151], N. Giokaris[8], M.P. Giordani[162a,162c], R. Giordano[102a,102b],
P. Giovannini[99], P.F. Giraud[29], P. Girtler[62], D. Giugni[89a], P. Giusti[19a], B.K. Gjelsten[116], L.K. Gladilin[97],
C. Glasman[80], A. Glazov[41], K.W. Glitza[172], G.L. Glonti[65], J. Godfrey[141], J. Godlewski[29], M. Goebel[41], T. Göpfert[43],
C. Goeringer[81], C. Gössling[42], T. Göttfert[99], V. Goggi[118a,118b,,m], S. Goldfarb[87], D. Goldin[39], T. Golling[173],
N.P. Gollub[29], A. Gomes[123b], L.S. Gomez Fajardo[41], R. Gonçalo[76], L. Gonella[20], C. Gong[32],
S. González de la Hoz[165], M.L. Gonzalez Silva[26], S. Gonzalez-Sevilla[49], J.J. Goodson[146], L. Goossens[29],
P.A. Gorbounov[156], H.A. Gordon[24], I. Gorelov[103], G. Gorfine[172], B. Gorini[29], E. Gorini[72a,72b], A. Gorišek[74],
E. Gornicki[38], V.N. Goryachev[127], B. Gosdzik[41], M. Gosselink[105], M.I. Gostkin[65], I. Gough Eschrich[161],
M. Gouighri[134a], D. Goujdami[134a], M.P. Goulette[49], A.G. Goussiou[137], C. Goy[4], I. Grabowska-Bold[161,d],
P. Grafström[29], K.-J. Grahn[145], L. Granado Cardoso[123b], F. Grancagnolo[72a], S. Grancagnolo[15], V. Grassi[89a],
V. Gratchev[120], N. Grau[34], H.M. Gray[34,n], J.A. Gray[146], E. Graziani[133a], B. Green[76], T. Greenshaw[73],
Z.D. Greenwood[24,i], I.M. Gregor[41], P. Grenier[142], E. Griesmayer[46], J. Griffiths[137], N. Grigalashvili[65], A.A. Grillo[136],
K. Grimm[146], S. Grinstein[11], Y.V. Grishkevich[97], L.S. Groer[156], J. Grognuz[29], M. Groh[99], M. Groll[81], E. Gross[169],
J. Grosse-Knetter[54], J. Groth-Jensen[79], K. Grybel[140], V.J. Guarino[5], C. Guicheney[33], A. Guida[72a,72b], T. Guillemin[4],
H. Guler[85,o], J. Gunther[124], B. Guo[156], A. Gupta[30], Y. Gusakov[65], A. Gutierrez[93], P. Gutierrez[110], N. Guttman[151],
O. Gutzwiller[29], C. Guyot[135], C. Gwenlan[117], C.B. Gwilliam[73], A. Haas[142], S. Haas[29], C. Haber[14], R. Hackenburg[24],
H.K. Hadavand[39], D.R. Hadley[17], P. Haefner[99], R. Härtel[99], Z. Hajduk[38], H. Hakobyan[174], J. Haller[41,p],
K. Hamacher[172], A. Hamilton[49], S. Hamilton[159], H. Han[32], L. Han[32], K. Hanagaki[115], M. Hance[119], C. Handel[81],
P. Hanke[58a], J.R. Hansen[35], J.B. Hansen[35], J.D. Hansen[35], P.H. Hansen[35], T. Hansl-Kozanecka[136], P. Hansson[142],
K. Hara[158], G.A. Hare[136], T. Harenberg[172], R.D. Harrington[21], O.M. Harris[137], K. Harrison[17], J. Hartert[48],
F. Hartjes[105], T. Haruyama[66], A. Harvey[56], S. Hasegawa[101], Y. Hasegawa[139], K. Hashemi[22], S. Hassani[135],
M. Hatch[29], F. Haug[29], S. Haug[16], M. Hauschild[29], R. Hauser[88], M. Havranek[124], C.M. Hawkes[17], R.J. Hawkings[29],
D. Hawkins[161], T. Hayakawa[67], H.S. Hayward[73], S.J. Haywood[128], M. He[32], S.J. Head[82], V. Hedberg[79], L. Heelan[28],
S. Heim[88], B. Heinemann[14], S. Heisterkamp[35], L. Helary[4], M. Heller[114], S. Hellman[144], C. Helsens[11],
T. Hemperek[20], R.C.W. Henderson[71], M. Henke[58a], A. Henrichs[54], A.M. Henriques Correia[29], S. Henrot-Versille[114],

C. Hensel[54], T. Henß[172], Y. Hernández Jiménez[165], A.D. Hershenhorn[150], G. Herten[48], R. Hertenberger[98], L. Hervas[29], N.P. Hessey[105], A. Hidvegi[144], E. Higón-Rodriguez[165], D. Hill[5,*], J.C. Hill[27], K.H. Hiller[41], S. Hillert[144], S.J. Hillier[17], I. Hinchliffe[14], E. Hines[119], M. Hirose[115], F. Hirsch[42], D. Hirschbuehl[172], J. Hobbs[146], N. Hod[151], M.C. Hodgkinson[138], P. Hodgson[138], A. Hoecker[29], M.R. Hoeferkamp[103], J. Hoffman[39], D. Hoffmann[83], M. Hohlfeld[81], S.O. Holmgren[144], T. Holy[126], J.L. Holzbauer[88], Y. Homma[67], P. Homola[126], T. Horazdovsky[126], T. Hori[67], C. Horn[142], S. Horner[48], S. Horvat[99], J.-Y. Hostachy[55], S. Hou[149], M.A. Houlden[73], A. Hoummada[134a], T. Howe[39], J. Hrivnac[114], T. Hryn'ova[4], P.J. Hsu[173], S.-C. Hsu[14], G.S. Huang[110], Z. Hubacek[126], F. Hubaut[83], F. Huegging[20], E.W. Hughes[34], G. Hughes[71], R.E. Hughes-Jones[82], P. Hurst[57], M. Hurwitz[30], U. Husemann[41], N. Huseynov[10], J. Huston[88], J. Huth[57], G. Iacobucci[102a], G. Iakovidis[9], I. Ibragimov[140], L. Iconomidou-Fayard[114], J. Idarraga[157b], P. Iengo[4], O. Igonkina[105], Y. Ikegami[66], M. Ikeno[66], Y. Ilchenko[39], D. Iliadis[152], Y. Ilyushenka[65], M. Imori[153], T. Ince[167], P. Ioannou[8], M. Iodice[133a], A. Irles Quiles[165], A. Ishikawa[67], M. Ishino[66], R. Ishmukhametov[39], T. Isobe[153], V. Issakov[173,*], C. Issever[117], S. Istin[18], Y. Itoh[101], A.V. Ivashin[127], H. Iwasaki[66], J.M. Izen[40], V. Izzo[102a], B. Jackson[119], J.N. Jackson[73], P. Jackson[142], M. Jaekel[29], M. Jahoda[124], V. Jain[61], K. Jakobs[48], S. Jakobsen[29], J. Jakubek[126], D. Jana[110], E. Jansen[104], A. Jantsch[99], M. Janus[48], R.C. Jared[170], G. Jarlskog[79], P. Jarron[29], L. Jeanty[57], I. Jen-La Plante[30], P. Jenni[29], P. Jez[35], S. Jézéquel[4], W. Ji[79], J. Jia[146], Y. Jiang[32], M. Jimenez Belenguer[29], G. Jin[32], S. Jin[32], O. Jinnouchi[155], D. Joffe[39], M. Johansen[144], K.E. Johansson[144], P. Johansson[138], S. Johnert[41], K.A. Johns[6], K. Jon-And[144], G. Jones[82], R.W.L. Jones[71], T.W. Jones[77], T.J. Jones[73], O. Jonsson[29], D. Joos[48], C. Joram[29], P.M. Jorge[123b], V. Juranek[124], P. Jussel[62], V.V. Kabachenko[127], S. Kabana[16], M. Kaci[165], A. Kaczmarska[38], M. Kado[114], H. Kagan[108], M. Kagan[57], S. Kaiser[99], E. Kajomovitz[150], S. Kalinin[172], L.V. Kalinovskaya[65], A. Kalinowski[129], S. Kama[41], N. Kanaya[153], M. Kaneda[153], V.A. Kantserov[96], J. Kanzaki[66], B. Kaplan[173], A. Kapliy[30], J. Kaplon[29], M. Karagounis[20], M. Karagoz Unel[117], V. Kartvelishvili[71], A.N. Karyukhin[127], L. Kashif[57], A. Kasmi[39], R.D. Kass[108], A. Kastanas[13], M. Kastoryano[173], M. Kataoka[4], Y. Kataoka[153], E. Katsoufis[9], J. Katzy[41], V. Kaushik[6], K. Kawagoe[67], T. Kawamoto[153], G. Kawamura[81], M.S. Kayl[105], F. Kayumov[94], V.A. Kazanin[106], M.Y. Kazarinov[65], S.I. Kazi[86], J.R. Keates[82], R. Keeler[167], P.T. Keener[119], R. Kehoe[39], M. Keil[49], G.D. Kekelidze[65], M. Kelly[82], J. Kennedy[98], M. Kenyon[53], O. Kepka[124], N. Kerschen[29], B.P. Kerševan[74], S. Kersten[172], K. Kessoku[153], M. Khakzad[28], F. Khalil-zada[10], H. Khandanyan[163], A. Khanov[111], D. Kharchenko[65], A. Khodinov[146], A.G. Kholodenko[127], A. Khomich[58a], G. Khoriauli[20], N. Khovanskiy[65], V. Khovanskiy[95], E. Khramov[65], J. Khubua[51], G. Kilvington[76], H. Kim[7], M.S. Kim[2], P.C. Kim[142], S.H. Kim[158], O. Kind[15], P. Kind[172], B.T. King[73], J. Kirk[128], G.P. Kirsch[117], L.E. Kirsch[22], A.E. Kiryunin[99], D. Kisielewska[37], T. Kittelmann[122], H. Kiyamura[67], E. Kladiva[143], M. Klein[73], U. Klein[73], K. Kleinknecht[81], M. Klemetti[85], A. Klier[169], A. Klimentov[24], R. Klingenberg[42], E.B. Klinkby[44], T. Klioutchnikova[29], P.F. Klok[104], S. Klous[105], E.-E. Kluge[58a], T. Kluge[73], P. Kluit[105], M. Klute[54], S. Kluth[99], N.S. Knecht[156], E. Kneringer[62], B.R. Ko[44], T. Kobayashi[153], M. Kobel[43], B. Koblitz[29], M. Kocian[142], A. Kocnar[112], P. Kodys[125], K. Köneke[41], A.C. König[104], L. Köpke[81], F. Koetsveld[104], P. Koevesarki[20], T. Koffas[29], E. Koffeman[105], F. Kohn[54], Z. Kohout[126], T. Kohriki[66], T. Kokott[20], H. Kolanoski[15], V. Kolesnikov[65], I. Koletsou[4], J. Koll[88], D. Kollar[29], S. Kolos[161,q], S.D. Kolya[82], A.A. Komar[94], J.R. Komaragiri[141], T. Kondo[66], T. Kono[41,p], A.I. Kononov[48], R. Konoplich[107], S.P. Konovalov[94], N. Konstantinidis[77], S. Koperny[37], K. Korcyl[38], K. Kordas[16], V. Koreshev[127], A. Korn[14], I. Korolkov[11], E.V. Korolkova[138], V.A. Korotkov[127], O. Kortner[99], P. Kostka[41], V.V. Kostyukhin[20], M.J. Kotamäki[29], S. Kotov[99], V.M. Kotov[65], K.Y. Kotov[106], Z. Koupilova[125], C. Kourkoumelis[8], A. Koutsman[105], R. Kowalewski[167], H. Kowalski[41], T.Z. Kowalski[37], W. Kozanecki[135], A.S. Kozhin[127], V. Kral[126], V.A. Kramarenko[97], G. Kramberger[74], M.W. Krasny[78], A. Krasznahorkay[107], A. Kreisel[151], F. Krejci[126], A. Krepouri[152], J. Kretzschmar[73], P. Krieger[156], G. Krobath[98], K. Kroeninger[54], H. Kroha[99], J. Kroll[119], J. Kroseberg[20], J. Krstic[12a], U. Kruchonak[65], H. Krüger[20], Z.V. Krumshteyn[65], T. Kubota[153], S. Kuehn[48], A. Kugel[58c], T. Kuhl[172], D. Kuhn[62], V. Kukhtin[65], Y. Kulchitsky[90], S. Kuleshov[31b], C. Kummer[98], M. Kuna[83], J. Kunkle[119], A. Kupco[124], H. Kurashige[67], M. Kurata[158], L.L. Kurchaninov[157a], Y.A. Kurochkin[90], V. Kus[124], E. Kuznetsova[131a,131b], O. Kvasnicka[124], R. Kwee[15], L. La Rotonda[36a,36b], L. Labarga[80], J. Labbe[4], C. Lacasta[165], F. Lacava[131a,131b], H. Lacker[15], D. Lacour[78], V.R. Lacuesta[165], E. Ladygin[65], R. Lafaye[4], B. Laforge[78], T. Lagouri[80], S. Lai[48], M. Lamanna[29], C.L. Lampen[6], W. Lampl[6], E. Lancon[135], U. Landgraf[48], M.P.J. Landon[75], J.L. Lane[82], A.J. Lankford[161], F. Lanni[24], K. Lantzsch[29], A. Lanza[118a], S. Laplace[4], C. Lapoire[83], J.F. Laporte[135], T. Lari[89a], A.V. Larionov[127], A. Larner[117], C. Lasseur[29], M. Lassnig[29], P. Laurelli[47], W. Lavrijsen[14], P. Laycock[73], A.B. Lazarev[65], A. Lazzaro[89a,89b], O. Le Dortz[78], E. Le Guirriec[83], C. Le Maner[156], E. Le Menedeu[135], M. Le Vine[24], M. Leahu[29], A. Lebedev[64], C. Lebel[93], T. LeCompte[5], F. Ledroit-Guillon[55], H. Lee[105], J.S.H. Lee[148], S.C. Lee[149],

M. Lefebvre[167], M. Legendre[135], B.C. LeGeyt[119], F. Legger[98], C. Leggett[14], M. Lehmacher[20], G. Lehmann Miotto[29], X. Lei[6], R. Leitner[125], D. Lelas[167], D. Lellouch[169], J. Lellouch[78], M. Leltchouk[34], V. Lendermann[58a], K.J.C. Leney[73], T. Lenz[172], G. Lenzen[172], B. Lenzi[135], K. Leonhardt[43], C. Leroy[93], J.-R. Lessard[167], C.G. Lester[27], A. Leung Fook Cheong[170], J. Levêque[83], D. Levin[87], L.J. Levinson[169], M.S. Levitski[127], S. Levonian[41], M. Lewandowska[21], M. Leyton[14], H. Li[170], J. Li[7], S. Li[41], X. Li[87], Z. Liang[39], Z. Liang[149,r], B. Liberti[132a], P. Lichard[29], M. Lichtnecker[98], K. Lie[163], W. Liebig[105], D. Liko[29], J.N. Lilley[17], H. Lim[5], A. Limosani[86], M. Limper[63], S.C. Lin[149], S.W. Lindsay[73], V. Linhart[126], J.T. Linnemann[88], A. Liolios[152], E. Lipeles[119], L. Lipinsky[124], A. Lipniacka[13], T.M. Liss[163], D. Lissauer[24], A. Lister[49], A.M. Litke[136], C. Liu[28], D. Liu[149,s], H. Liu[87], J.B. Liu[87], M. Liu[32], S. Liu[2], T. Liu[39], Y. Liu[32], M. Livan[118a,118b], A. Lleres[55], S.L. Lloyd[75], E. Lobodzinska[41], P. Loch[6], W.S. Lockman[136], S. Lockwitz[173], T. Loddenkoetter[20], F.K. Loebinger[82], A. Loginov[173], C.W. Loh[166], T. Lohse[15], K. Lohwasser[48], M. Lokajicek[124], J. Loken[117], L. Lopes[123b], D. Lopez Mateos[34,n], M. Losada[160], P. Loscutoff[14], M.J. Losty[157a], X. Lou[40], A. Lounis[114], K.F. Loureiro[108], L. Lovas[143], J. Love[21], P. Love[71], A.J. Lowe[61], F. Lu[32], J. Lu[2], H.J. Lubatti[137], C. Luci[131a,131b], A. Lucotte[55], A. Ludwig[43], D. Ludwig[41], I. Ludwig[48], J. Ludwig[48], F. Luehring[61], L. Luisa[162a,162c], D. Lumb[48], L. Luminari[131a], E. Lund[116], B. Lund-Jensen[145], B. Lundberg[79], J. Lundberg[29], J. Lundquist[35], G. Lutz[99], D. Lynn[24], J. Lys[14], E. Lytken[79], H. Ma[24], L.L. Ma[170], J.A. Macana Goia[93], G. Maccarrone[47], A. Macchiolo[99], B. Maček[74], J. Machado Miguens[123b], R. Mackeprang[29], R.J. Madaras[14], W.F. Mader[43], R. Maenner[58c], T. Maeno[24], P. Mättig[172], S. Mättig[41], P.J. Magalhaes Martins[123b], E. Magradze[51], C.A. Magrath[104], Y. Mahalalel[151], K. Mahboubi[48], A. Mahmood[1], G. Mahout[17], C. Maiani[131a,131b], C. Maidantchik[23a], A. Maio[123b], S. Majewski[24], Y. Makida[66], M. Makouski[127], N. Makovec[114], Pa. Malecki[38], P. Malecki[38], V.P. Maleev[120], F. Malek[55], U. Mallik[63], D. Malon[5], S. Maltezos[9], V. Malyshev[106], S. Malyukov[65], M. Mambelli[30], R. Mameghani[98], J. Mamuzic[41], A. Manabe[66], L. Mandelli[89a], I. Mandić[74], R. Mandrysch[15], J. Maneira[123b], P.S. Mangeard[88], I.D. Manjavidze[65], P.M. Manning[136], A. Manousakis-Katsikakis[8], B. Mansoulie[135], A. Mapelli[29], L. Mapelli[29], L. March[80], J.F. Marchand[4], F. Marchese[132a,132b], G. Marchiori[78], M. Marcisovsky[124], C.P. Marino[61], C.N. Marques[123b], F. Marroquim[23a], R. Marshall[82], Z. Marshall[34,n], F.K. Martens[156], S. Marti i Garcia[165], A.J. Martin[75], A.J. Martin[173], B. Martin[29], B. Martin[88], F.F. Martin[119], J.P. Martin[93], T.A. Martin[17], B. Martin dit Latour[49], M. Martinez[11], V. Martinez Outschoorn[57], A. Martini[47], A.C. Martyniuk[82], T. Maruyama[158], F. Marzano[131a], A. Marzin[135], L. Masetti[20], T. Mashimo[153], R. Mashinistov[96], J. Masik[82], A.L. Maslennikov[106], G. Massaro[105], N. Massol[4], A. Mastroberardino[36a,36b], T. Masubuchi[153], M. Mathes[20], P. Matricon[114], H. Matsunaga[153], T. Matsushita[67], C. Mattravers[117,t], S.J. Maxfield[73], E.N. May[5], A. Mayne[138], R. Mazini[149], M. Mazur[48], M. Mazzanti[89a], P. Mazzanti[19a], J. Mc Donald[85], S.P. Mc Kee[87], A. McCarn[163], R.L. McCarthy[146], N.A. McCubbin[128], K.W. McFarlane[56], H. McGlone[53], G. Mchedlidze[51], R.A. McLaren[29], S.J. McMahon[128], T.R. McMahon[76], R.A. McPherson[167,h], A. Meade[84], J. Mechnich[105], M. Mechtel[172], M. Medinnis[41], R. Meera-Lebbai[110], T.M. Meguro[115], R. Mehdiyev[93], S. Mehlhase[41], A. Mehta[73], K. Meier[58a], B. Meirose[48], C. Melachrinos[30], A. Melamed-Katz[169], B.R. Mellado Garcia[170], Z. Meng[149,u], S. Menke[99], E. Meoni[11], D. Merkl[98], P. Mermod[117], L. Merola[102a,102b], C. Meroni[89a], F.S. Merritt[30], A.M. Messina[29], I. Messmer[48], J. Metcalfe[103], A.S. Mete[64], J.-P. Meyer[135], J. Meyer[171], J. Meyer[54], T.C. Meyer[29], W.T. Meyer[64], J. Miao[32], S. Michal[29], L. Micu[25a], R.P. Middleton[128], S. Migas[73], L. Mijović[74], G. Mikenberg[169], M. Mikuž[74], D.W. Miller[142], W.J. Mills[166], C.M. Mills[57], A. Milov[169], D.A. Milstead[144], A.A. Minaenko[127], M. Miñano[165], I.A. Minashvili[65], A.I. Mincer[107], B. Mindur[37], M. Mineev[65], Y. Ming[129], L.M. Mir[11], G. Mirabelli[131a], S. Misawa[24], S. Miscetti[47], A. Misiejuk[76], J. Mitrevski[136], V.A. Mitsou[165], P.S. Miyagawa[82], J.U. Mjörnmark[79], D. Mladenov[22], T. Moa[144], S. Moed[57], V. Moeller[27], K. Mönig[41], N. Möser[20], B. Mohn[13], W. Mohr[48], S. Mohrdieck-Möck[99], R. Moles-Valls[165], J. Molina-Perez[29], G. Moloney[86], J. Monk[77], E. Monnier[83], S. Montesano[89a,89b], F. Monticelli[70], R.W. Moore[2], C. Mora Herrera[49], A. Moraes[53], A. Morais[123b], J. Morel[4], G. Morello[36a,36b], D. Moreno[160], M. Moreno Llácer[165], P. Morettini[50a], M. Morii[57], A.K. Morley[86], G. Mornacchi[29], S.V. Morozov[96], J.D. Morris[75], H.G. Moser[99], M. Mosidze[51], J. Moss[108], R. Mount[142], E. Mountricha[9], S.V. Mouraviev[94], E.J.W. Moyse[84], M. Mudrinic[12b], F. Mueller[58a], J. Mueller[122], K. Mueller[20], T.A. Müller[98], D. Muenstermann[42], A. Muir[166], Y. Munwes[151], R. Murillo Garcia[161], W.J. Murray[128], I. Mussche[105], E. Musto[102a,102b], A.G. Myagkov[127], M. Myska[124], J. Nadal[11], K. Nagai[24], K. Nagano[66], Y. Nagasaka[60], A.M. Nairz[29], K. Nakamura[153], I. Nakano[109], H. Nakatsuka[67], G. Nanava[20], A. Napier[159], M. Nash[77,v], N.R. Nation[21], T. Nattermann[20], T. Naumann[41], G. Navarro[160], S.K. Nderitu[20], H.A. Neal[87], E. Nebot[80], P. Nechaeva[94], A. Negri[118a,118b], G. Negri[29], A. Nelson[64], T.K. Nelson[142], S. Nemecek[124], P. Nemethy[107], A.A. Nepomuceno[23a], M. Nessi[29], M.S. Neubauer[163], A. Neusiedl[81], R.N. Neves[123b], P. Nevski[24], F.M. Newcomer[119], R.B. Nickerson[117], R. Nicolaidou[135], L. Nicolas[138], G. Nicoletti[47],

F. Niedercorn[114], J. Nielsen[136], A. Nikiforov[15], K. Nikolaev[65], I. Nikolic-Audit[78], K. Nikolopoulos[8], H. Nilsen[48], P. Nilsson[7], A. Nisati[131a], T. Nishiyama[67], R. Nisius[99], L. Nodulman[5], M. Nomachi[115], I. Nomidis[152], M. Nordberg[29], B. Nordkvist[144], D. Notz[41], J. Novakova[125], M. Nozaki[66], M. Nožička[41], I.M. Nugent[157a], A.-E. Nuncio-Quiroz[20], G. Nunes Hanninger[20], T. Nunnemann[98], E. Nurse[77], D.C. O'Neil[141], V. O'Shea[53], F.G. Oakham[28,b], H. Oberlack[99], A. Ochi[67], S. Oda[153], S. Odaka[66], J. Odier[83], G.A. Odino[50a,50b], H. Ogren[61], A. Oh[82], S.H. Oh[44], C.C. Ohm[144], T. Ohshima[101], H. Ohshita[139], T. Ohsugi[59], S. Okada[67], H. Okawa[153], Y. Okumura[101], M. Olcese[50a], A.G. Olchevski[65], M. Oliveira[123b], D. Oliveira Damazio[24], J. Oliver[57], E. Oliver Garcia[165], D. Olivito[119], A. Olszewski[38], J. Olszowska[38], C. Omachi[67], A. Onofre[123b], P.U.E. Onyisi[30], C.J. Oram[157a], G. Ordonez[104], M.J. Oreglia[30], Y. Oren[151], D. Orestano[133a,133b], I. Orlov[106], C. Oropeza Barrera[53], R.S. Orr[156], E.O. Ortega[129], B. Osculati[50a,50b], R. Ospanov[119], C. Osuna[11], R. Otec[126], J. P Ottersbach[105], F. Ould-Saada[116], A. Ouraou[135], Q. Ouyang[32], M. Owen[82], S. Owen[138], A. Oyarzun[31b], V.E. Ozcan[77], K. Ozone[66], N. Ozturk[7], A. Pacheco Pages[11], S. Padhi[170], C. Padilla Aranda[11], E. Paganis[138], C. Pahl[63], F. Paige[24], K. Pajchel[116], S. Palestini[29], D. Pallin[33], A. Palma[123b], J.D. Palmer[17], Y.B. Pan[170], E. Panagiotopoulou[9], B. Panes[31a], N. Panikashvili[87], S. Panitkin[24], D. Pantea[25a], M. Panuskova[124], V. Paolone[122], Th.D. Papadopoulou[9], S.J. Park[54], W. Park[24,w], M.A. Parker[27], S.I. Parker[14], F. Parodi[50a,50b], J.A. Parsons[34], U. Parzefall[48], E. Pasqualucci[131a], G. Passardi[29], A. Passeri[133a], F. Pastore[133a,133b], Fr. Pastore[29], G. Pásztor[49,x], S. Pataraia[99], J.R. Pater[82], S. Patricelli[102a,102b], A. Patwa[24], T. Pauly[29], L.S. Peak[148], M. Pecsy[143], M.I. Pedraza Morales[170], S.V. Peleganchuk[106], H. Peng[170], A. Penson[34], J. Penwell[61], M. Perantoni[23a], K. Perez[34,n], E. Perez Codina[11], M.T. Pérez García-Estañ[165], V. Perez Reale[34], L. Perini[89a,89b], H. Pernegger[29], R. Perrino[72a], P. Perrodo[4], S. Persembe[3], P. Perus[114], V.D. Peshekhonov[65], B.A. Petersen[29], J. Petersen[29], T.C. Petersen[35], E. Petit[83], C. Petridou[152], E. Petrolo[131a], F. Petrucci[133a,133b], D. Petschull[41], M. Petteni[141], R. Pezoa[31b], B. Pfeifer[48], A. Phan[86], A.W. Phillips[27], G. Piacquadio[48], M. Piccinini[19a,19b], R. Piegaia[26], J.E. Pilcher[30], A.D. Pilkington[82], J. Pina[123b], M. Pinamonti[162a,162c], J.L. Pinfold[2], J. Ping[32], B. Pinto[123b], C. Pizio[89a,89b], R. Placakyte[41], M. Plamondon[167], W.G. Plano[82], M.-A. Pleier[24], A. Poblaguev[173], S. Poddar[58a], F. Podlyski[33], P. Poffenberger[167], L. Poggioli[114], M. Pohl[49], F. Polci[55], G. Polesello[118a], A. Policicchio[137], A. Polini[19a], J. Poll[75], V. Polychronakos[24], D.M. Pomarede[135], D. Pomeroy[22], K. Pommès[29], L. Pontecorvo[131a], B.G. Pope[88], D.S. Popovic[12a], A. Poppleton[29], J. Popule[124], X. Portell Bueso[48], R. Porter[161], G.E. Pospelov[99], P. Pospichal[29], S. Pospisil[126], M. Potekhin[24], I.N. Potrap[99], C.J. Potter[147], C.T. Potter[85], K.P. Potter[82], G. Poulard[29], J. Poveda[170], R. Prabhu[20], P. Pralavorio[83], S. Prasad[57], R. Pravahan[7], T. Preda[25a], K. Pretzl[16], L. Pribyl[29], D. Price[61], L.E. Price[5], P.M. Prichard[73], D. Prieur[122], M. Primavera[72a], K. Prokofiev[29], F. Prokoshin[31b], S. Protopopescu[24], J. Proudfoot[5], X. Prudent[43], H. Przysiezniak[4], S. Psoroulas[20], E. Ptacek[113], C. Puigdengoles[11], J. Purdham[87], M. Purohit[24,w], P. Puzo[114], Y. Pylypchenko[116], M. Qi[32], J. Qian[87], W. Qian[128], Z. Qian[83], Z. Qin[41], D. Qing[157a], A. Quadt[54], D.R. Quarrie[14], W.B. Quayle[170], F. Quinonez[31a], M. Raas[104], V. Radeka[24], V. Radescu[58b], B. Radics[20], T. Rador[18], F. Ragusa[89a,89b], G. Rahal[178], A.M. Rahimi[108], D. Rahm[24], S. Rajagopalan[24], M. Rammes[140], P.N. Ratoff[71], F. Rauscher[98], E. Rauter[99], M. Raymond[29], A.L. Read[116], D.M. Rebuzzi[118a,118b], A. Redelbach[171], G. Redlinger[24], R. Reece[119], K. Reeves[40], E. Reinherz-Aronis[151], A. Reinsch[113], I. Reisinger[42], D. Reljic[12a], C. Rembser[29], Z.L. Ren[149], P. Renkel[39], S. Rescia[24], M. Rescigno[131a], S. Resconi[89a], B. Resende[105], P. Reznicek[125], R. Rezvani[156], A. Richards[77], R.A. Richards[88], R. Richter[99], E. Richter-Was[38,y], M. Ridel[78], S. Rieke[81], M. Rijpstra[105], M. Rijssenbeek[146], A. Rimoldi[118a,118b], L. Rinaldi[19a], R.R. Rios[39], I. Riu[11], G. Rivoltella[89a,89b], F. Rizatdinova[111], E.R. Rizvi[75], D.A. Roa Romero[160], S.H. Robertson[85,h], A. Robichaud-Veronneau[49], D. Robinson[27], J. Robinson[77], M. Robinson[113], A. Robson[53], J.G. Rocha de Lima[5], C. Roda[121a,121b], D. Roda Dos Santos[29], D. Rodriguez[160], Y. Rodriguez Garcia[15], S. Roe[29], O. Røhne[116], V. Rojo[1], S. Rolli[159], A. Romaniouk[96], V.M. Romanov[65], G. Romeo[26], D. Romero Maltrana[31a], L. Roos[78], E. Ros[165], S. Rosati[131a,131b], G.A. Rosenbaum[156], E.I. Rosenberg[64], L. Rosselet[49], V. Rossetti[11], L.P. Rossi[50a], M. Rotaru[25a], J. Rothberg[137], I. Rottländer[20], D. Rousseau[114], C.R. Royon[135], A. Rozanov[83], Y. Rozen[150], X. Ruan[114], B. Ruckert[98], N. Ruckstuhl[105], V.I. Rud[97], G. Rudolph[62], F. Rühr[58a], F. Ruggieri[133a], A. Ruiz-Martinez[64], L. Rumyantsev[65], N.A. Rusakovich[65], J.P. Rutherfoord[6], C. Ruwiedel[20], P. Ruzicka[124], Y.F. Ryabov[120], V. Ryadovikov[127], P. Ryan[88], G. Rybkin[114], S. Rzaeva[10], A.F. Saavedra[148], H.F.-W. Sadrozinski[136], R. Sadykov[65], H. Sakamoto[153], G. Salamanna[105], A. Salamon[132a], M. Saleem[110], D. Salihagic[99], A. Salnikov[142], J. Salt[165], B.M. Salvachua Ferrando[5], D. Salvatore[36a,36b], F. Salvatore[147], A. Salvucci[47], A. Salzburger[29], D. Sampsonidis[152], B.H. Samset[116], M.A. Sanchis Lozano[165], H. Sandaker[13], H.G. Sander[81], M.P. Sanders[98], M. Sandhoff[172], R. Sandstroem[105], S. Sandvoss[172], D.P.C. Sankey[128], B. Sanny[172], A. Sansoni[47], C. Santamarina Rios[85], L. Santi[162a,162c], C. Santoni[33], R. Santonico[132a,132b], J. Santos[123b], J.G. Saraiva[123b], T. Sarangi[170],

E. Sarkisyan-Grinbaum[7], F. Sarri[121a,121b], O. Sasaki[66], T. Sasaki[66], N. Sasao[68], I. Satsounkevitch[90], G. Sauvage[4], P. Savard[156,b], A.Y. Savine[6], V. Savinov[122], L. Sawyer[24,i], D.H. Saxon[53], L.P. Says[33], C. Sbarra[19a,19b], A. Sbrizzi[19a,19b], D.A. Scannicchio[29], J. Schaarschmidt[43], P. Schacht[99], U. Schäfer[81], S. Schaetzel[58b], A.C. Schaffer[114], D. Schaile[98], R.D. Schamberger[146], A.G. Schamov[106], V.A. Schegelsky[120], D. Scheirich[87], M. Schernau[161], M.I. Scherzer[14], C. Schiavi[50a,50b], J. Schieck[99], M. Schioppa[36a,36b], S. Schlenker[29], J.L. Schlereth[5], P. Schmid[62], K. Schmieden[20], C. Schmitt[81], M. Schmitz[20], M. Schott[29], D. Schouten[141], J. Schovancova[124], M. Schram[85], A. Schreiner[63], C. Schroeder[81], N. Schroer[58c], M. Schroers[172], G. Schuler[29], J. Schultes[172], H.-C. Schultz-Coulon[58a], J.W. Schumacher[43], M. Schumacher[48], B.A. Schumm[136], Ph. Schune[135], C. Schwanenberger[82], A. Schwartzman[142], Ph. Schwemling[78], R. Schwienhorst[88], R. Schwierz[43], J. Schwindling[135], W.G. Scott[128], J. Searcy[113], E. Sedykh[120], E. Segura[11], S.C. Seidel[103], A. Seiden[136], F. Seifert[43], J.M. Seixas[23a], G. Sekhniaidze[102a], D.M. Seliverstov[120], B. Sellden[144], M. Seman[143], N. Semprini-Cesari[19a,19b], C. Serfon[98], L. Serin[114], R. Seuster[99], H. Severini[110], M.E. Sevior[86], A. Sfyrla[163], E. Shabalina[54], M. Shamim[113], L.Y. Shan[32], J.T. Shank[21], Q.T. Shao[86], M. Shapiro[14], P.B. Shatalov[95], L. Shaver[6], K. Shaw[138], D. Sherman[29], P. Sherwood[77], A. Shibata[107], M. Shimojima[100], T. Shin[56], A. Shmeleva[94], M.J. Shochet[30], M.A. Shupe[6], P. Sicho[124], A. Sidoti[15], A. Siebel[172], F. Siegert[77], J. Siegrist[14], Dj. Sijacki[12a], O. Silbert[169], J. Silva[123b], Y. Silver[151], D. Silverstein[142], S.B. Silverstein[144], V. Simak[126], Lj. Simic[12a], S. Simion[114], B. Simmons[77], M. Simonyan[4], P. Sinervo[156], N.B. Sinev[113], V. Sipica[140], G. Siragusa[81], A.N. Sisakyan[65], S.Yu. Sivoklokov[97], J. Sjoelin[144], T.B. Sjursen[13], P. Skubic[110], N. Skvorodnev[22], M. Slater[17], T. Slavicek[126], K. Sliwa[159], J. Sloper[29], T. Sluka[124], V. Smakhtin[169], S.Yu. Smirnov[96], Y. Smirnov[24], L.N. Smirnova[97], O. Smirnova[79], B.C. Smith[57], D. Smith[142], K.M. Smith[53], M. Smizanska[71], K. Smolek[126], A.A. Snesarev[94], S.W. Snow[82], J. Snow[110], J. Snuverink[105], S. Snyder[24], M. Soares[123b], R. Sobie[167,h], J. Sodomka[126], A. Soffer[151], C.A. Solans[165], M. Solar[126], J. Solc[126], E. Solfaroli Camillocci[131a,131b], A.A. Solodkov[127], O.V. Solovyanov[127], R. Soluk[2], J. Sondericker[24], V. Sopko[126], B. Sopko[126], M. Sosebee[7], V.V. Sosnovtsev[96], L. Sospedra Suay[165], A. Soukharev[106], S. Spagnolo[72a,72b], F. Spanò[34], P. Speckmayer[29], E. Spencer[136], R. Spighi[19a], G. Spigo[29], F. Spila[131a,131b], R. Spiwoks[29], M. Spousta[125], T. Spreitzer[141], B. Spurlock[7], R.D.St. Denis[53], T. Stahl[140], J. Stahlman[119], R. Stamen[58a], S.N. Stancu[161], E. Stanecka[29], R.W. Stanek[5], C. Stanescu[133a], S. Stapnes[116], E.A. Starchenko[127], J. Stark[55], P. Staroba[124], P. Starovoitov[91], J. Stastny[124], A. Staude[98], P. Stavina[143], G. Stavropoulos[14], G. Steele[53], P. Steinbach[43], P. Steinberg[24], I. Stekl[126], B. Stelzer[141], H.J. Stelzer[41], O. Stelzer-Chilton[157a], H. Stenzel[52], K. Stevenson[75], G. Stewart[53], M.C. Stockton[29], K. Stoerig[48], G. Stoicea[25a], S. Stonjek[99], P. Strachota[125], A. Stradling[7], A. Straessner[43], J. Strandberg[87], S. Strandberg[14], A. Strandlie[116], M. Strauss[110], P. Strizenec[143], R. Ströhmer[98], D.M. Strom[113], J.A. Strong[76,*], R. Stroynowski[39], J. Strube[128], B. Stugu[13], I. Stumer[24,*], D.A. Soh[149,r], D. Su[142], S.I. Suchkov[96], Y. Sugaya[115], T. Sugimoto[101], C. Suhr[5], M. Suk[125], V.V. Sulin[94], S. Sultansoy[3,z], T. Sumida[29], X. Sun[32], J.E. Sundermann[48], K. Suruliz[162a,162b], S. Sushkov[11], G. Susinno[36a,36b], M.R. Sutton[138], T. Suzuki[153], Y. Suzuki[66], Yu.M. Sviridov[127], I. Sykora[143], T. Sykora[125], T. Szymocha[38], J. Sánchez[165], D. Ta[20], K. Tackmann[29], A. Taffard[161], R. Tafirout[157a], A. Taga[116], Y. Takahashi[101], H. Takai[24], R. Takashima[69], H. Takeda[67], T. Takeshita[139], M. Talby[83], A. Talyshev[106], M.C. Tamsett[76], J. Tanaka[153], R. Tanaka[114], S. Tanaka[130], S. Tanaka[66], G.P. Tappern[29], S. Tapprogge[81], D. Tardif[156], S. Tarem[150], F. Tarrade[24], G.F. Tartarelli[89a], P. Tas[125], M. Tasevsky[124], E. Tassi[36a,36b], M. Tatarkhanov[14], C. Taylor[77], F.E. Taylor[92], G.N. Taylor[86], R.P. Taylor[167], W. Taylor[157b], P. Teixeira-Dias[76], H. Ten Kate[29], P.K. Teng[149], Y.D. Tennenbaum-Katan[150], S. Terada[66], K. Terashi[153], J. Terron[80], M. Terwort[41,p], M. Testa[47], R.J. Teuscher[156,h], C.M. Tevlin[82], J. Thadome[172], R. Thananuwong[49], M. Thioye[173], S. Thoma[48], J.P. Thomas[17], T.L. Thomas[103], E.N. Thompson[84], P.D. Thompson[17], P.D. Thompson[156], R.J. Thompson[82], A.S. Thompson[53], E. Thomson[119], R.P. Thun[87], T. Tic[124], V.O. Tikhomirov[94], Y.A. Tikhonov[106], C.J.W.P. Timmermans[104], P. Tipton[173], F.J. Tique Aires Viegas[29], S. Tisserant[83], J. Tobias[48], B. Toczek[37], T. Todorov[4], S. Todorova-Nova[159], B. Toggerson[161], J. Tojo[66], S. Tokár[143], K. Tokushuku[66], K. Tollefson[88], L. Tomasek[124], M. Tomasek[124], F. Tomasz[143], M. Tomoto[101], D. Tompkins[6], L. Tompkins[14], K. Toms[103], G. Tong[32], A. Tonoyan[13], C. Topfel[16], N.D. Topilin[65], E. Torrence[113], E. Torró Pastor[165], J. Toth[83,x], F. Touchard[83], D.R. Tovey[138], S.N. Tovey[86], T. Trefzger[171], L. Tremblet[29], A. Tricoli[29], I.M. Trigger[157a], S. Trincaz-Duvoid[78], T.N. Trinh[78], M.F. Tripiana[70], N. Triplett[64], W. Trischuk[156], A. Trivedi[24,w], B. Trocmé[55], C. Troncon[89a], A. Trzupek[38], C. Tsarouchas[9], J.C.-L. Tseng[117], I. Tsiafis[152], M. Tsiakiris[105], P.V. Tsiareshka[90], D. Tsionou[138], G. Tsipolitis[9], V. Tsiskaridze[51], E.G. Tskhadadze[51], I.I. Tsukerman[95], V. Tsulaia[122], J.-W. Tsung[20], S. Tsuno[66], D. Tsybychev[146], M. Turala[38], D. Turecek[126], I. Turk Cakir[3,aa], E. Turlay[105], P.M. Tuts[34], M.S. Twomey[137], M. Tylmad[144], M. Tyndel[128], G. Tzanakos[8], K. Uchida[115], I. Ueda[153], M. Ugland[13], M. Uhlenbrock[20], M. Uhrmacher[54], F. Ukegawa[158], G. Unal[29], D.G. Underwood[5], A. Undrus[24],

G. Unel[161], Y. Unno[66], D. Urbaniec[34], E. Urkovsky[151], P. Urquijo[49], P. Urrejola[31a], G. Usai[7], M. Uslenghi[118a,118b], L. Vacavant[83], V. Vacek[126], B. Vachon[85], S. Vahsen[14], J. Valenta[124], P. Valente[131a], S. Valentinetti[19a,19b], S. Valkar[125], E. Valladolid Gallego[165], S. Vallecorsa[150], J.A. Valls Ferrer[165], R. Van Berg[119], H. van der Graaf[105], E. van der Kraaij[105], E. van der Poel[105], D. Van Der Ster[29], N. van Eldik[84], P. van Gemmeren[5], Z. van Kesteren[105], I. van Vulpen[105], W. Vandelli[29], G. Vandoni[29], A. Vaniachine[5], P. Vankov[73], F. Vannucci[78], F. Varela Rodriguez[29], R. Vari[131a], E.W. Varnes[6], D. Varouchas[14], A. Vartapetian[7], K.E. Varvell[148], L. Vasilyeva[94], V.I. Vassilakopoulos[56], F. Vazeille[33], G. Vegni[89a,89b], J.J. Veillet[114], C. Vellidis[8], F. Veloso[123b], R. Veness[29], S. Veneziano[131a], A. Ventura[72a,72b], D. Ventura[137], M. Venturi[48], N. Venturi[16], V. Vercesi[118a], M. Verducci[171], W. Verkerke[105], J.C. Vermeulen[105], M.C. Vetterli[141,b], I. Vichou[163], T. Vickey[170], G.H.A. Viehhauser[117], M. Villa[19a,19b], E.G. Villani[128], M. Villaplana Perez[165], J. Villate[123b], E. Vilucchi[47], M.G. Vincter[28], E. Vinek[29], V.B. Vinogradov[65], S. Viret[33], J. Virzi[14], A. Vitale[19a,19b], O.V. Vitells[169], I. Vivarelli[48], F. Vives Vaques[11], S. Vlachos[9], M. Vlasak[126], N. Vlasov[20], A. Vogel[20], P. Vokac[126], M. Volpi[11], G. Volpini[89a], H. von der Schmitt[99], J. von Loeben[99], H. von Radziewski[48], E. von Toerne[20], V. Vorobel[125], A.P. Vorobiev[127], V. Vorwerk[11], M. Vos[165], R. Voss[29], T.T. Voss[172], J.H. Vossebeld[73], N. Vranjes[12a], M. Vranjes Milosavljevic[12a], V. Vrba[124], M. Vreeswijk[105], T. Vu Anh[81], D. Vudragovic[12a], R. Vuillermet[29], I. Vukotic[114], P. Wagner[119], H. Wahlen[172], J. Walbersloh[42], J. Walder[71], R. Walker[98], W. Walkowiak[140], R. Wall[173], C. Wang[44], H. Wang[170], J. Wang[55], J.C. Wang[137], S.M. Wang[149], C.P. Ward[27], M. Warsinsky[48], R. Wastie[117], P.M. Watkins[17], A.T. Watson[17], M.F. Watson[17], G. Watts[137], S. Watts[82], A.T. Waugh[148], B.M. Waugh[77], M. Webel[48], J. Weber[42], M.D. Weber[16], M. Weber[128], M.S. Weber[16], P. Weber[58a], A.R. Weidberg[117], J. Weingarten[54], C. Weiser[48], H. Wellenstein[22], P.S. Wells[29], M. Wen[47], T. Wenaus[24], S. Wendler[122], T. Wengler[82], S. Wenig[29], N. Wermes[20], M. Werner[48], P. Werner[29], M. Werth[161], U. Werthenbach[140], M. Wessels[58a], K. Whalen[28], S.J. Wheeler-Ellis[161], S.P. Whitaker[21], A. White[7], M.J. White[27], S. White[24], D. Whiteson[161], D. Whittington[61], F. Wicek[114], D. Wicke[81], F.J. Wickens[128], W. Wiedenmann[170], M. Wielers[128], P. Wienemann[20], C. Wiglesworth[73], L.A.M. Wiik[48], A. Wildauer[165], M.A. Wildt[41,p], I. Wilhelm[125], H.G. Wilkens[29], E. Williams[34], H.H. Williams[119], W. Willis[34], S. Willocq[84], J.A. Wilson[17], M.G. Wilson[142], A. Wilson[87], I. Wingerter-Seez[4], F. Winklmeier[29], M. Wittgen[142], M.W. Wolter[38], H. Wolters[123b], B.K. Wosiek[38], J. Wotschack[29], M.J. Woudstra[84], K. Wraight[53], C. Wright[53], D. Wright[142], B. Wrona[73], S.L. Wu[170], X. Wu[49], E. Wulf[34], S. Xella[35], S. Xie[48], Y. Xie[32], D. Xu[138], N. Xu[170], M. Yamada[158], A. Yamamoto[66], S. Yamamoto[153], T. Yamamura[153], K. Yamanaka[64], J. Yamaoka[44], T. Yamazaki[153], Y. Yamazaki[67], Z. Yan[21], H. Yang[87], U.K. Yang[82], Y. Yang[32], Z. Yang[144], W.-M. Yao[14], Y. Yao[14], Y. Yasu[66], J. Ye[39], S. Ye[24], M. Yilmaz[3,ab], R. Yoosoofmiya[122], K. Yorita[168], R. Yoshida[5], C. Young[142], S.P. Youssef[21], D. Yu[24], J. Yu[7], M. Yu[58c], X. Yu[32], J. Yuan[99], L. Yuan[78], A. Yurkewicz[146], R. Zaidan[63], A.M. Zaitsev[127], Z. Zajacova[29], V. Zambrano[47], L. Zanello[131a,131b], P. Zarzhitsky[39], A. Zaytsev[106], C. Zeitnitz[172], M. Zeller[173], P.F. Zema[29], A. Zemla[38], C. Zendler[20], O. Zenin[127], T. Zenis[143], Z. Zenonos[121a,121b], S. Zenz[14], D. Zerwas[114], G. Zevi della Porta[57], Z. Zhan[32], H. Zhang[83], J. Zhang[5], Q. Zhang[5], X. Zhang[32], L. Zhao[107], T. Zhao[137], Z. Zhao[32], A. Zhemchugov[65], S. Zheng[32], J. Zhong[149,ac], B. Zhou[87], N. Zhou[34], Y. Zhou[149], C.G. Zhu[32], H. Zhu[41], Y. Zhu[170], X. Zhuang[98], V. Zhuravlov[99], R. Zimmermann[20], S. Zimmermann[20], S. Zimmermann[48], M. Ziolkowski[140], R. Zitoun[4], L. Živković[34], V.V. Zmouchko[127,*], G. Zobernig[170], A. Zoccoli[19a,19b], M. zur Nedden[15], V. Zutshi[5]

*CERN, 1211 Geneva 23, Switzerland
[1]University at Albany, 1400 Washington Ave, Albany, NY 12222, United States of America
[2]University of Alberta, Department of Physics, Centre for Particle Physics, Edmonton, AB T6G 2G7, Canada
[3]Ankara University, Faculty of Sciences, Department of Physics, TR 061000 Tandogan, Ankara, Turkey
[4]LAPP, Université de Savoie, CNRS/IN2P3, Annecy-le-Vieux, France
[5]Argonne National Laboratory, High Energy Physics Division, 9700 S. Cass Avenue, Argonne IL 60439, United States of America
[6]University of Arizona, Department of Physics, Tucson, AZ 85721, United States of America
[7]The University of Texas at Arlington, Department of Physics, Box 19059, Arlington, TX 76019, United States of America
[8]University of Athens, Nuclear & Particle Physics, Department of Physics, Panepistimiopouli, Zografou, GR 15771 Athens, Greece
[9]National Technical University of Athens, Physics Department, 9-Iroon Polytechniou, GR 15780 Zografou, Greece
[10]Institute of Physics, Azerbaijan Academy of Sciences, H. Javid Avenue 33, AZ 143 Baku, Azerbaijan
[11]Institut de Física d'Altes Energies, IFAE, Edifici Cn, Universitat Autònoma de Barcelona, ES-08193 Bellaterra (Barcelona), Spain
[12](a)University of Belgrade, Institute of Physics, P.O. Box 57, 11001 Belgrade; Vinca Institute of Nuclear Sciences(b), Mihajla Petrovica Alasa 12-14, 11001 Belgrade, Serbia
[13]University of Bergen, Department for Physics and Technology, Allegaten 55, NO-5007 Bergen, Norway
[14]Lawrence Berkeley National Laboratory and University of California, Physics Division, MS50B-6227, 1 Cyclotron Road, Berkeley, CA 94720, United States of America
[15]Humboldt University, Institute of Physics, Berlin, Newtonstr. 15, D-12489 Berlin, Germany
[16]University of Bern, Albert Einstein Center for Fundamental Physics, Laboratory for High Energy Physics, Sidlerstrasse 5, CH-3012 Bern, Switzerland

[17]University of Birmingham, School of Physics and Astronomy, Edgbaston, Birmingham B15 2TT, United Kingdom

[18]Bogazici University, Faculty of Sciences, Department of Physics, TR-80815 Bebek-Istanbul, Turkey

[19]INFN Sezione di Bologna[a]; Università di Bologna, Dipartimento di Fisica[b], viale C. Berti Pichat, 6/2, IT-40127 Bologna, Italy

[20]University of Bonn, Physikalisches Institut, Nussallee 12, D-53115 Bonn, Germany

[21]Boston University, Department of Physics, 590 Commonwealth Avenue, Boston, MA 02215, United States of America

[22]Brandeis University, Department of Physics, MS057, 415 South Street, Waltham, MA 02454, United States of America

[23]Universidade Federal do Rio De Janeiro, Instituto de Fisica[a], Caixa Postal 68528, Ilha do Fundao, BR-21945-970 Rio de Janeiro; [b]Universidade de Sao Paulo, Instituto de Fisica, R.do Matao Trav. R.187, Sao Paulo-SP, 05508-900, Brazil

[24]Brookhaven National Laboratory, Physics Department, Bldg. 510A, Upton, NY 11973, United States of America

[25]National Institute of Physics and Nuclear Engineering[a], Bucharest-Magurele, Str. Atomistilor 407, P.O. Box MG-6, R-077125, Romania; [b]University Politehnica Bucharest, Rectorat-AN 001, 313 Splaiul Independentei, sector 6, 060042 Bucuresti; [c]West University in Timisoara, Bd. Vasile Parvan 4, Timisoara, Romania

[26]Universidad de Buenos Aires, FCEyN, Dto. Fisica, Pab I-C. Universitaria, 1428 Buenos Aires, Argentina

[27]University of Cambridge, Cavendish Laboratory, J J Thomson Avenue, Cambridge CB3 0HE, United Kingdom

[28]Carleton University, Department of Physics, 1125 Colonel By Drive, Ottawa ON K1S 5B6, Canada

[29]CERN, CH-1211 Geneva 23, Switzerland

[30]University of Chicago, Enrico Fermi Institute, 5640 S. Ellis Avenue, Chicago, IL 60637, United States of America

[31]Pontificia Universidad Católica de Chile, Facultad de Fisica, Departamento de Fisica[a], Avda. Vicuna Mackenna 4860, San Joaquin, Santiago; Universidad Técnica Federico Santa María, Departamento de Física[b], Avda. España 1680, Casilla 110-V, Valparaíso, Chile

[32]Institute of HEP, Chinese Academy of Sciences, P.O. Box 918, CN-100049 Beijing; USTC, Department of Modern Physics, Hefei, CN-230026 Anhui; Nanjing University, Department of Physics, CN-210093 Nanjing; Shandong University, HEP Group, CN-250100 Shadong, China

[33]Laboratoire de Physique Corpusculaire, CNRS-IN2P3, Université Blaise Pascal, FR-63177 Aubiere Cedex, France

[34]Columbia University, Nevis Laboratory, 136 So. Broadway, Irvington, NY 10533, United States of America

[35]University of Copenhagen, Niels Bohr Institute, Blegdamsvej 17, DK-2100 Kobenhavn 0, Denmark

[36]INFN Gruppo Collegato di Cosenza[a]; Università della Calabria, Dipartimento di Fisica[b], IT-87036 Arcavacata di Rende, Italy

[37]Faculty of Physics and Applied Computer Science of the AGH-University of Science and Technology (FPACS, AGH-UST), al. Mickiewicza 30, PL-30059 Cracow, Poland

[38]The Henryk Niewodniczanski Institute of Nuclear Physics, Polish Academy of Sciences, ul. Radzikowskiego 152, PL-31342 Krakow, Poland

[39]Southern Methodist University, Physics Department, 106 Fondren Science Building, Dallas, TX 75275-0175, United States of America

[40]University of Texas at Dallas, 800 West Campbell Road, Richardson, TX 75080-3021, United States of America

[41]DESY, Notkestr. 85, D-22603 Hamburg, Germany and Platanenallee 6, D-15738 Zeuthen, Germany

[42]TU Dortmund, Experimentelle Physik IV, DE-44221 Dortmund, Germany

[43]Technical University Dresden, Institut fuer Kern- und Teilchenphysik, Zellescher Weg 19, D-01069 Dresden, Germany

[44]Duke University, Department of Physics, Durham, NC 27708, United States of America

[45]University of Edinburgh, School of Physics & Astronomy, James Clerk Maxwell Building, The Kings Buildings, Mayfield Road, Edinburgh EH9 3JZ, United Kingdom

[46]Fachhochschule Wiener Neustadt; Johannes Gutenbergstrasse 3 AT-2700 Wiener Neustadt, Austria

[47]INFN Laboratori Nazionali di Frascati, via Enrico Fermi 40, IT-00044 Frascati, Italy

[48]Albert-Ludwigs-Universität, Fakultät für Mathematik und Physik, Hermann-Herder Str. 3, D-79104 Freiburg i.Br., Germany

[49]Université de Genève, Section de Physique, 24 rue Ernest Ansermet, CH-1211 Geneve 4, Switzerland

[50]INFN Sezione di Genova[a]; Università di Genova, Dipartimento di Fisica[b], via Dodecaneso 33, IT-16146 Genova, Italy

[51]Institute of Physics of the Georgian Academy of Sciences, 6 Tamarashvili St., GE-380077 Tbilisi; Tbilisi State University, HEP Institute, University St. 9, GE-380086 Tbilisi, Georgia

[52]Justus-Liebig-Universitaet Giessen, II Physikalisches Institut, Heinrich-Buff Ring 16, D-35392 Giessen, Germany

[53]University of Glasgow, Department of Physics and Astronomy, Glasgow G12 8QQ, United Kingdom

[54]Georg-August-Universitat, II. Physikalisches Institut, Friedrich-Hund Platz 1, D-37077 Goettingen, Germany

[55]Laboratoire de Physique Subatomique et de Cosmologie, CNRS/IN2P3, Université Joseph Fourier, INPG, 53 avenue des Martyrs, FR-38026 Grenoble Cedex, France

[56]Hampton University, Department of Physics, Hampton, VA 23668, United States of America

[57]Harvard University, Laboratory for Particle Physics and Cosmology, 18 Hammond Street, Cambridge, MA 02138, United States of America

[58]Ruprecht-Karls-Universitaet Heidelberg, Kirchhoff-Institut fuer Physik[a], Im Neuenheimer Feld 227, D-69120 Heidelberg; [b]Physikalisches Institut, Philosophenweg 12, D-69120 Heidelberg; ZITI Ruprecht-Karls-University Heidelberg[c], Lehrstuhl fuer Informatik V, B6, 23-29, DE-68131 Mannheim, Germany

[59]Hiroshima University, Faculty of Science, 1-3-1 Kagamiyama, Higashihiroshima-shi, JP-Hiroshima 739-8526, Japan

[60]Hiroshima Institute of Technology, Faculty of Applied Information Science, 2-1-1 Miyake Saeki-ku, Hiroshima-shi, JP-Hiroshima 731-5193, Japan

[61]Indiana University, Department of Physics, Swain Hall West 117, Bloomington, IN 47405-7105, United States of America

[62]Institut fuer Astro- und Teilchenphysik, Technikerstrasse 25, A-6020 Innsbruck, Austria

[63]University of Iowa, 203 Van Allen Hall, Iowa City, IA 52242-1479, United States of America

[64]Iowa State University, Department of Physics and Astronomy, Ames High Energy Physics Group, Ames, IA 50011-3160, United States of America

[65]Joint Institute for Nuclear Research, JINR Dubna, RU-141 980 Moscow Region, Russia

[66]KEK, High Energy Accelerator Research Organization, 1-1 Oho, Tsukuba-shi, Ibaraki-ken 305-0801, Japan

[67]Kobe University, Graduate School of Science, 1-1 Rokkodai-cho, Nada-ku, JP Kobe 657-8501, Japan

[68]Kyoto University, Faculty of Science, Oiwake-cho, Kitashirakawa, Sakyou-ku, Kyoto-shi, JP-Kyoto 606-8502, Japan

[69] Kyoto University of Education, 1 Fukakusa, Fujimori, fushimi-ku, Kyoto-shi, JP-Kyoto 612-8522, Japan

[70] Universidad Nacional de La Plata, FCE, Departamento de Física, IFLP (CONICET-UNLP), C.C. 67, 1900 La Plata, Argentina

[71] Lancaster University, Physics Department, Lancaster LA1 4YB, United Kingdom

[72] INFN Sezione di Lecce[a]; Università del Salento, Dipartimento di Fisica[b] Via Arnesano IT-73100 Lecce, Italy

[73] University of Liverpool, Oliver Lodge Laboratory, P.O. Box 147, Oxford Street, Liverpool L69 3BX, United Kingdom

[74] Jožef Stefan Institute and University of Ljubljana, Department of Physics, SI-1000 Ljubljana, Slovenia

[75] Queen Mary University of London, Department of Physics, Mile End Road, London E1 4NS, United Kingdom

[76] Royal Holloway, University of London, Department of Physics, Egham Hill, Egham, Surrey TW20 0EX, United Kingdom

[77] University College London, Department of Physics and Astronomy, Gower Street, London WC1E 6BT, United Kingdom

[78] Laboratoire de Physique Nucléaire et de Hautes Energies, Université Pierre et Marie Curie (Paris 6), Université Denis Diderot (Paris-7), CNRS/IN2P3, Tour 33, 4 place Jussieu, FR-75252 Paris Cedex 05, France

[79] Lunds universitet, Naturvetenskapliga fakulteten, Fysiska institutionen, Box 118, SE-221 00 Lund, Sweden

[80] Universidad Autonoma de Madrid, Facultad de Ciencias, Departamento de Fisica Teorica, ES-28049 Madrid, Spain

[81] Universitaet Mainz, Institut fuer Physik, Staudinger Weg 7, DE-55099 Mainz, Germany

[82] University of Manchester, School of Physics and Astronomy, Manchester M13 9PL, United Kingdom

[83] CPPM, Aix-Marseille Université, CNRS/IN2P3, Marseille, France

[84] University of Massachusetts, Department of Physics, 710 North Pleasant Street, Amherst, MA 01003, United States of America

[85] McGill University, High Energy Physics Group, 3600 University Street, Montreal, Quebec H3A 2T8, Canada

[86] University of Melbourne, School of Physics, AU-Parkville, Victoria 3010, Australia

[87] The University of Michigan, Department of Physics, 2477 Randall Laboratory, 500 East University, Ann Arbor, MI 48109-1120, United States of America

[88] Michigan State University, Department of Physics and Astronomy, High Energy Physics Group, East Lansing, MI 48824-2320, United States of America

[89] INFN Sezione di Milano[a]; Università di Milano, Dipartimento di Fisica[b], via Celoria 16, IT-20133 Milano, Italy

[90] B.I. Stepanov Institute of Physics, National Academy of Sciences of Belarus, Independence Avenue 68, Minsk 220072, Republic of Belarus

[91] National Scientific & Educational Centre for Particle & High Energy Physics, NC PHEP BSU, M. Bogdanovich St. 153, Minsk 220040, Republic of Belarus

[92] Massachusetts Institute of Technology, Department of Physics, Room 24-516, Cambridge, MA 02139, United States of America

[93] University of Montreal, Group of Particle Physics, C.P. 6128, Succursale Centre-Ville, Montreal, Quebec, H3C 3J7, Canada

[94] P.N. Lebedev Institute of Physics, Academy of Sciences, Leninsky pr. 53, RU-117 924 Moscow, Russia

[95] Institute for Theoretical and Experimental Physics (ITEP), B. Cheremushkinskaya ul. 25, RU 117 218 Moscow, Russia

[96] Moscow Engineering & Physics Institute (MEPhI), Kashirskoe Shosse 31, RU-115409 Moscow, Russia

[97] Lomonosov Moscow State University Skobeltsyn Institute of Nuclear Physics (MSU SINP), 1(2), Leninskie gory, GSP-1, Moscow 119991 Russian Federation, Russia

[98] Ludwig-Maximilians-Universität München, Fakultät für Physik, Am Coulombwall 1, DE-85748 Garching, Germany

[99] Max-Planck-Institut für Physik, (Werner-Heisenberg-Institut), Föhringer Ring 6, 80805 München, Germany

[100] Nagasaki Institute of Applied Science, 536 Aba-machi, JP Nagasaki 851-0193, Japan

[101] Nagoya University, Graduate School of Science, Furo-Cho, Chikusa-ku, Nagoya, 464-8602, Japan

[102] INFN Sezione di Napoli[a]; Università di Napoli, Dipartimento di Scienze Fisiche[b], Complesso Universitario di Monte Sant'Angelo, via Cinthia, IT-80126 Napoli, Italy

[103] University of New Mexico, Department of Physics and Astronomy, MSC07 4220, Albuquerque, NM 87131 USA, United States of America

[104] Radboud University Nijmegen/NIKHEF, Department of Experimental High Energy Physics, Toernooiveld 1, NL-6525 ED Nijmegen, Netherlands

[105] Nikhef National Institute for Subatomic Physics, and University of Amsterdam, Science Park 105, 1098 XG Amsterdam, Netherlands

[106] Budker Institute of Nuclear Physics (BINP), RU-Novosibirsk 630 090, Russia

[107] New York University, Department of Physics, 4 Washington Place, New York NY 10003, USA, United States of America

[108] Ohio State University, 191 West Woodruff Ave, Columbus, OH 43210-1117, United States of America

[109] Okayama University, Faculty of Science, Tsushimanaka 3-1-1, Okayama 700-8530, Japan

[110] University of Oklahoma, Homer L. Dodge Department of Physics and Astronomy, 440 West Brooks, Room 100, Norman, OK 73019-0225, United States of America

[111] Oklahoma State University, Department of Physics, 145 Physical Sciences Building, Stillwater, OK 74078-3072, United States of America

[112] Palacký University, 17.listopadu 50a, 772 07 Olomouc, Czech Republic

[113] University of Oregon, Center for High Energy Physics, Eugene, OR 97403-1274, United States of America

[114] LAL, Univ. Paris-Sud, IN2P3/CNRS, Orsay, France

[115] Osaka University, Graduate School of Science, Machikaneyama-machi 1-1, Toyonaka, Osaka 560-0043, Japan

[116] University of Oslo, Department of Physics, P.O. Box 1048, Blindern, NO-0316 Oslo 3, Norway

[117] Oxford University, Department of Physics, Denys Wilkinson Building, Keble Road, Oxford OX1 3RH, United Kingdom

[118] INFN Sezione di Pavia[a]; Università di Pavia, Dipartimento di Fisica Nucleare e Teorica[b], Via Bassi 6, IT-27100 Pavia, Italy

[119] University of Pennsylvania, Department of Physics, High Energy Physics Group, 209 S. 33rd Street, Philadelphia, PA 19104, United States of America

[120] Petersburg Nuclear Physics Institute, RU-188 300 Gatchina, Russia

[121] INFN Sezione di Pisa[a]; Università di Pisa, Dipartimento di Fisica E. Fermi[b], Largo B. Pontecorvo 3, IT-56127 Pisa, Italy

[122] University of Pittsburgh, Department of Physics and Astronomy, 3941 O'Hara Street, Pittsburgh, PA 15260, United States of America

[123] [a] Universidad de Granada, Departamento de Fisica Teorica y del Cosmos and CAFPE, E-18071 Granada; Laboratorio de Instrumentacao e Fisica Experimental de Particulas-LIP[b], Avenida Elias Garcia 14-1, PT-1000-149 Lisboa, Portugal

[124]Institute of Physics, Academy of Sciences of the Czech Republic, Na Slovance 2, CZ-18221 Praha 8, Czech Republic

[125]Charles University in Prague, Faculty of Mathematics and Physics, Institute of Particle and Nuclear Physics, V Holesovickach 2, CZ-18000 Praha 8, Czech Republic

[126]Czech Technical University in Prague, Zikova 4, CZ-166 35 Praha 6, Czech Republic

[127]State Research Center Institute for High Energy Physics, Moscow Region, 142281, Protvino, Pobeda street, 1, Russia

[128]Rutherford Appleton Laboratory, Science and Technology Facilities Council, Harwell Science and Innovation Campus, Didcot OX11 0QX, United Kingdom

[129]University of Regina, Physics Department, Canada

[130]Ritsumeikan University, Noji Higashi 1 chome 1-1, JP-Kusatsu, Shiga 525-8577, Japan

[131]INFN Sezione di Roma I[(a)]; Università La Sapienza, Dipartimento di Fisica[(b)], Piazzale A. Moro 2, IT- 00185 Roma, Italy

[132]INFN Sezione di Roma Tor Vergata[(a)]; Università di Roma Tor Vergata, Dipartimento di Fisica[(b)], via della Ricerca Scientifica, IT-00133 Roma, Italy

[133]INFN Sezione di Roma Tre[(a)]; Università Roma Tre, Dipartimento di Fisica[(b)], via della Vasca Navale 84, IT-00146 Roma, Italy

[134]Université Hassan II, Faculté des Sciences Ain Chock[(a)], B.P. 5366, MA-Casablanca; Centre National de l'Energie des Sciences Techniques Nucleaires (CNESTEN)[(b)], B.P. 1382 R.P. 10001 Rabat 10001; Université Mohamed Premier[(c)], LPTPM, Faculté des Sciences, B.P.717. Bd. Mohamed VI, 60000, Oujda; Université Mohammed V, Faculté des Sciences[(d)], LPNR, BP 1014, 10000 Rabat, Morocco

[135]CEA, DSM/IRFU, Centre d'Etudes de Saclay, FR-91191 Gif-sur-Yvette, France

[136]University of California Santa Cruz, Santa Cruz Institute for Particle Physics (SCIPP), Santa Cruz, CA 95064, United States of America

[137]University of Washington, Seattle, Department of Physics, Box 351560, Seattle, WA 98195-1560, United States of America

[138]University of Sheffield, Department of Physics & Astronomy, Hounsfield Road, Sheffield S3 7RH, United Kingdom

[139]Shinshu University, Department of Physics, Faculty of Science, 3-1-1 Asahi, Matsumoto-shi, JP-Nagano 390-8621, Japan

[140]Universitaet Siegen, Fachbereich Physik, D 57068 Siegen, Germany

[141]Simon Fraser University, Department of Physics, 8888 University Drive, CA-Burnaby, BC V5A 1S6, Canada

[142]SLAC National Accelerator Laboratory, Stanford, California 94309, United States of America

[143]Comenius University, Faculty of Mathematics, Physics & Informatics, Mlynska dolina F2, SK-84248 Bratislava; Institute of Experimental Physics of the Slovak Academy of Sciences, Dept. of Subnuclear Physics, Watsonova 47, SK-04353 Kosice, Slovak Republic

[144]Stockholm University, Department of Physics, AlbaNova, SE-106 91 Stockholm, Sweden

[145]Royal Institute of Technology (KTH), Physics Department, SE-106 91 Stockholm, Sweden

[146]Stony Brook University, Department of Physics and Astronomy, Nicolls Road, Stony Brook, NY 11794-3800, United States of America

[147]University of Sussex, Department of Physics and Astronomy Pevensey 2 Building, Falmer, Brighton BN1 9QH, United Kingdom

[148]University of Sydney, School of Physics, AU-Sydney NSW 2006, Australia

[149]Insitute of Physics, Academia Sinica, TW-Taipei 11529, Taiwan

[150]Technion, Israel Inst. of Technology, Department of Physics, Technion City, IL-Haifa 32000, Israel

[151]Tel Aviv University, Raymond and Beverly Sackler School of Physics and Astronomy, Ramat Aviv, IL-Tel Aviv 69978, Israel

[152]Aristotle University of Thessaloniki, Faculty of Science, Department of Physics, Division of Nuclear & Particle Physics, University Campus, GR-54124, Thessaloniki, Greece

[153]The University of Tokyo, International Center for Elementary Particle Physics and Department of Physics, 7-3-1 Hongo, Bunkyo-ku, JP-Tokyo 113-0033, Japan

[154]Tokyo Metropolitan University, Graduate School of Science and Technology, 1-1 Minami-Osawa, Hachioji, Tokyo 192-0397, Japan

[155]Tokyo Institute of Technology, 2-12-1-H-34 O-Okayama, Meguro, Tokyo 152-8551, Japan

[156]University of Toronto, Department of Physics, 60 Saint George Street, Toronto M5S 1A7, Ontario, Canada

[157]TRIUMF[(a)], 4004 Wesbrook Mall, Vancouver, B.C. V6T 2A3; [(b)]York University, Department of Physics and Astronomy, 4700 Keele St., Toronto, Ontario, M3J 1P3, Canada

[158]University of Tsukuba, Institute of Pure and Applied Sciences, 1-1-1 Tennoudai, Tsukuba-shi, JP-Ibaraki 305-8571, Japan

[159]Tufts University, Science & Technology Center, 4 Colby Street, Medford, MA 02155, United States of America

[160]Universidad Antonio Narino, Centro de Investigaciones, Cra 3 Este No. 47A-15, Bogota, Colombia

[161]University of California, Irvine, Department of Physics & Astronomy, CA 92697-4575, United States of America

[162]INFN Gruppo Collegato di Udine[(a)]; ICTP[(b)], Strada Costiera 11, IT-34014, Trieste; Università di Udine, Dipartimento di Fisica[(c)], via delle Scienze 208, IT-33100 Udine, Italy

[163]University of Illinois, Department of Physics, 1110 West Green Street, Urbana, Illinois 61801, United States of America

[164]University of Uppsala, Department of Physics and Astronomy, P.O. Box 516, SE -751 20 Uppsala, Sweden

[165]Instituto de Física Corpuscular (IFIC) Centro Mixto UVEG-CSIC, Apdo. 22085 ES-46071 Valencia, Dept. Física At. Mol. y Nuclear; Univ. of Valencia, and Instituto de Microelectrónica de Barcelona (IMB-CNM-CSIC) 08193 Bellaterra Barcelona, Spain

[166]University of British Columbia, Department of Physics, 6224 Agricultural Road, CA-Vancouver, B.C. V6T 1Z1, Canada

[167]University of Victoria, Department of Physics and Astronomy, P.O. Box 3055, Victoria B.C., V8W 3P6, Canada

[168]Waseda University, WISE, 3-4-1 Okubo, Shinjuku-ku, Tokyo, 169-8555, Japan

[169]The Weizmann Institute of Science, Department of Particle Physics, P.O. Box 26, IL-76100 Rehovot, Israel

[170]University of Wisconsin, Department of Physics, 1150 University Avenue, WI 53706 Madison, Wisconsin, United States of America

[171]Julius-Maximilians-University of Würzburg, Physikalisches Institut, Am Hubland, 97074 Wuerzburg, Germany

[172]Bergische Universitaet, Fachbereich C, Physik, Postfach 100127, Gauss-Strasse 20, D-42097 Wuppertal, Germany

[173]Yale University, Department of Physics, P.O. Box 208121, New Haven CT, 06520-8121, United States of America

[174]Yerevan Physics Institute, Alikhanian Brothers Street 2, AM-375036 Yerevan, Armenia

[175]ATLAS-Canada Tier-1 Data Centre 4004 Wesbrook Mall, Vancouver, BC, V6T 2A3, Canada

[176]GridKA Tier-1 FZK, Forschungszentrum Karlsruhe GmbH, Steinbuch Centre for Computing (SCC), Hermann-von-Helmholtz-Platz 1, 76344 Eggenstein-Leopoldshafen, Germany

[177] Port d'Informacio Cientifica (PIC), Universitat Autonoma de Barcelona (UAB), Edifici D, E-08193 Bellaterra, Spain

[178] Centre de Calcul CNRS/IN2P3, Domaine scientifique de la Doua, 27 bd du 11 Novembre 1918, 69622 Villeurbanne Cedex, France

[179] INFN-CNAF, Viale Berti Pichat 6/2, 40127 Bologna, Italy

[180] Nordic Data Grid Facility, NORDUnet A/S, Kastruplundgade 22, 1, DK-2770 Kastrup, Denmark

[181] SARA Reken- en Netwerkdiensten, Science Park 121, 1098 XG Amsterdam, Netherlands

[182] Academia Sinica Grid Computing, Institute of Physics, Academia Sinica, No. 128, Sec. 2, Academia Rd., Nankang, Taipei, Taiwan 11529, Taiwan

[183] UK-T1-RAL Tier-1, Rutherford Appleton Laboratory, Science and Technology Facilities Council, Harwell Science and Innovation Campus, Didcot OX11 0QX, United Kingdom

[184] RHIC and ATLAS Computing Facility, Physics Department, Building 510, Brookhaven National Laboratory, Upton, New York 11973, United States of America

[a] Also at CPPM, Marseille, France.

[b] Also at TRIUMF, 4004 Wesbrook Mall, Vancouver, B.C. V6T 2A3, Canada.

[c] Also at Gaziantep University, Turkey.

[d] Also at Faculty of Physics and Applied Computer Science of the AGH-University of Science and Technology (FPACS, AGH-UST), al. Mickiewicza 30, PL-30059 Cracow, Poland.

[e] Also at Institute for Particle Phenomenology, Ogden Centre for Fundamental Physics, Department of Physics, University of Durham, Science Laboratories, South Rd, Durham DH1 3LE, United Kingdom.

[f] Currently at Dogus University, Kadik.

[g] Also at Università di Napoli Parthenope, via A. Acton 38, IT-80133 Napoli, Italy.

[h] Also at Institute of Particle Physics (IPP), Canada.

[i] Louisiana Tech University, 305 Wisteria Street, P.O. Box 3178, Ruston, LA 71272, United States of America.

[j] Currently at Dumlupinar University, Kutahya, Turkey.

[k] Currently at Department of Physics, University of Helsinki, P.O. Box 64, FI-00014, Finland.

[l] At Department of Physics, California State University, Fresno, 2345 E. San Ramon Avenue, Fresno, CA 93740-8031, United States of America.

[n] Also at California Institute of Technology, Physics Department, Pasadena, CA 91125, United States of America.

[o] Also at University of Montreal, Canada.

[p] Also at Institut für Experimentalphysik, Universität Hamburg, Luruper Chaussee 149, 22761 Hamburg, Germany.

[q] Also at Petersburg Nuclear Physics Institute, RU-188 300 Gatchina, Russia.

[r] Also at School of Physics and Engineering, Sun Yat-sen University, Taiwan.

[s] Also at School of Physics, Shandong University, Jinan, China.

[t] Also at Rutherford Appleton Laboratory, Science and Technology Facilities Council, Harwell Science and Innovation Campus, Didcot OX11, United Kingdom.

[u] Also at school of physics, Shandong university, Jinan.

[v] Also at Rutherford Appleton Laboratory, Science and Technology Facilities Council, Harwell Science and Innovation Campus, Didcot OX11 0QX, United Kingdom.

[w] University of South Carolina, Dept. of Physics and Astronomy, 700 S. Main St, Columbia, SC 29208, United States of America.

[x] Also at KFKI Research Institute for Particle and Nuclear Physics, Budapest, Hungary.

[y] Also at Institute of Physics, Jagiellonian University, Cracow, Poland.

[z] Currently at TOBB University, Ankara, Turkey.

[aa] Currently at TAEA, Ankara, Turkey.

[ab] Currently at Gazi University, Ankara, Turkey.

[ac] Also at Dept of Physics, Nanjing University, China.

[*] Deceased.

[**] e-mail: atlas.secretariat@cern.ch

Abstract The ionization signals in the liquid argon of the ATLAS electromagnetic calorimeter are studied in detail using cosmic muons. In particular, the drift time of the ionization electrons is measured and used to assess the intrinsic uniformity of the calorimeter gaps and estimate its impact on the constant term of the energy resolution. The drift times of electrons in the cells of the second layer of the calorimeter are uniform at the level of 1.3% in the barrel and 2.8% in the endcaps. This leads to an estimated contribution to the constant term of $(0.29^{+0.05}_{-0.04})\%$ in the barrel and $(0.54^{+0.06}_{-0.04})\%$ in the endcaps. The same data are used to measure the drift velocity of ionization electrons in liquid argon, which is found to be 4.61 ± 0.07 mm/μs at 88.5 K and 1 kV/mm.

1 Introduction

The ATLAS liquid argon (LAr) calorimeter [1] is composed of sampling detectors with full azimuthal[1] symmetry and is housed in one barrel and two endcap cryostats.

---

[1] The azimuthal angle ϕ is measured in the plane transverse to the beam axis. Positive ϕ is in the up direction. The pseudorapidity is defined as $\eta = -\ln(\tan(\theta/2))$, where θ is the polar angle from the beam axis. Positive η is for the proton beam circulating anticlockwise.

A highly granular electromagnetic (EM) calorimeter with accordion–shaped electrodes and lead absorbers covers the pseudorapidity range $|\eta| < 3.2$, and contains a barrel part ($|\eta| < 1.475$) [2] made of two half-barrels joined at $\eta = 0$ and two endcap parts ($1.375 < |\eta| < 3.2$) [3]. Each section is segmented in depth in three layers (denoted as layer 1, 2, 3). For $|\eta| < 1.8$, a presampler (PS) [3, 4], installed in the cryostat in front of the EM calorimeter, provides a measurement of the energy lost upstream.

The EM calorimeter plays a crucial role during the operation of the LHC, since physics channels involving electrons and photons in the final state form a crucial part of the ATLAS physics program. Achieving the required precision and discovery reach places stringent requirements on the performance of the calorimeter. The uniformity of the calorimeter response over a large acceptance is particularly important for the overall resolution. This drives several design choices for the calorimeter: lead-liquid argon calorimetry provides a good energy resolution and homogeneity even in the presence of strong radiation; the accordion geometry (see Fig. 1) avoids readout cracks between calorimeter modules, thus also providing good uniformity.

In order to equalize the gains of different calorimeter channels, a calibration procedure involving electronic charge injection is used. This is however not sensitive to intrinsic characteristics of the ionization gaps in the liquid argon system, such as variations in gap sizes and LAr temperature changes. Such non-uniformities can be measured from the ionization signals created by charged particles. The calorimeter energy response to this ionization is not the best quantity for this purpose, because it requires a knowledge

of the energy of the incoming particle. However the electron drift time in LAr, which can be obtained from the signal pulse shape resulting from ionizing particles that deposit sufficient energy above the intrinsic noise level in a calorimeter cell, is a powerful monitoring tool. As explained in Sect. 2, the drift time is also about four times more sensitive to changes in the LAr gap size than is the energy response. Cosmic muons have been used to this end as part of the calorimeter commissioning before the LHC start-up.

The EM calorimeter installation in the ATLAS cavern was completed at the end of 2006. Before LHC start-up, the main challenge was to commission the associated electronics and automate all of the calibration steps for the full 173,312 channels. Cosmic muon data have been taken regularly for commissioning purposes since 2006. At the end of the summer and during autumn of 2008 stable cosmic muon runs were taken with the detector fully operational and using various trigger menus. In normal data taking only 5 samples around the pulse peak at 25 ns intervals are taken, but in order to accurately measure the drift time 32 samples are needed. The pulse height is also relevant, since larger pulses are less affected by electronic noise. A summary of the detector performance obtained from calibration data, cosmic muons and beam splash events is detailed in [5].

Measurements of the drift time (T_{drift}) in the ATLAS EM calorimeter using cosmic muon data are presented in this paper. These drift times, which are independent of the amplitude of the pulses used for their determination, can be compared from one calorimeter region to another, and thus allow a measurement of the uniformity of the calorimeter.

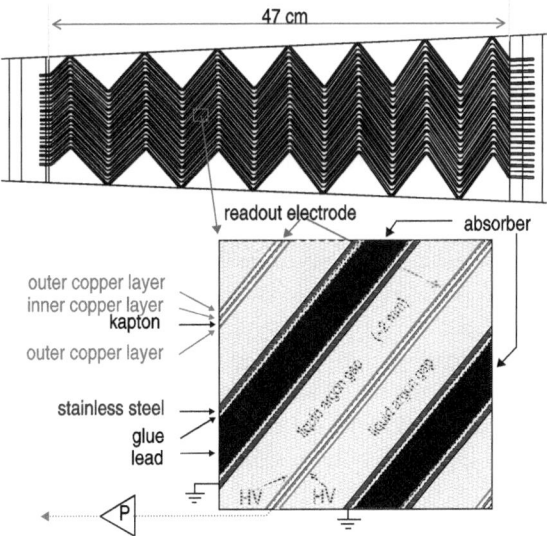

Fig. 1 Accordion structure of the barrel. The *top figure* is a view of a small sector of the barrel calorimeter in a plane transverse to the LHC beams. Honeycomb spacers, in the liquid argon gap, position the electrodes between the lead absorber plates

2 Ionization signal in the calorimeter

The current resulting from the passage of a charged particle through a liquid argon gap has the typical ionization-chamber triangular shape, with a short rise time (smaller than 1 ns) which is neglected in the rest of this note, followed by a linear decay for the duration of the maximum drift time

$$T_{\mathrm{drift}} = w_{\mathrm{gap}}/V_{\mathrm{drift}}, \tag{1}$$

where w_{gap} is the LAr gap width and V_{drift} the electron drift velocity [6]. The ionization current, I, is then modeled as:

$$I(t; I_0, T_{\mathrm{drift}}) = I_0 \left(1 - \frac{t}{T_{\mathrm{drift}}} \right) \quad \text{for } 0 < t < T_{\mathrm{drift}} \tag{2}$$

where I_0 is the current at $t = 0$. The peak current amplitude $I_0 = \rho \cdot V_{\mathrm{drift}}$ is proportional to the drift velocity and to the negative linear charge density ρ along the direction perpendicular to the readout electrode, which varies with the lead

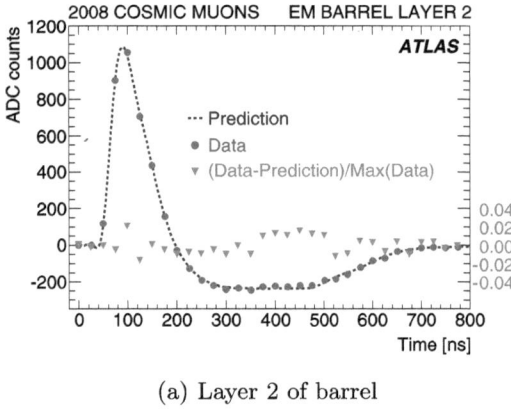

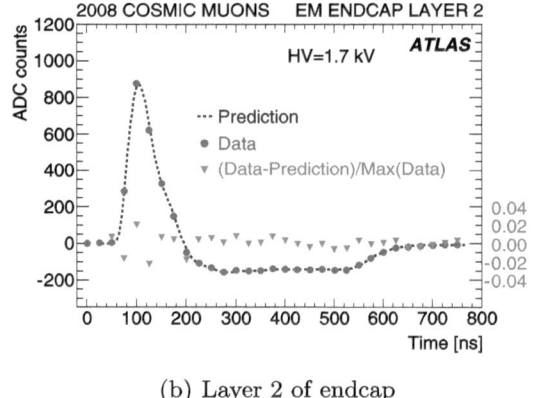

(a) Layer 2 of barrel

(b) Layer 2 of endcap

Fig. 2 Typical single ionization pulse in a cell of layer 2 of the barrel (**a**) and endcap (**b**) of the calorimeter. The *large red dots* show the data samples, the *small blue dots* the prediction and the *grey triangles* the

relative difference (data (S) – prediction (g))/S_{max}, on the scale shown on the right side of the plot (normalized to the data)

thickness.[2] Since the determination of the energy is based on the measurement of I_0, it is crucial to be able to precisely evaluate and monitor V_{drift}. While the LAr gap thickness is mechanically constrained, the drift velocity depends on the actual conditions of the detector: the LAr temperature and density, and the local high voltage. Uniform response in a calorimeter with constant lead thickness requires uniform drift velocity in the gaps.

At this point it is appropriate to recall that each liquid argon electronic cell is built out of several gaps connected in parallel: for layers 2 and 3, there are 4(3) double-gaps in parallel in the barrel (endcap) respectively; there are four times as many gaps per cell in layer 1, given the coarser granularity of the readout in the azimuthal direction [1]. The parameters measured represent an average of the local gaps, both in depth along the cell, and in between the gaps forming a cell.

At the end of the readout chain the triangular signal is amplified, shaped and passed through a switched capacitor array which samples the signal every 25 ns. The shaping function (see Sect. 3) includes one integration and two derivatives. Their net effect is to transform the triangular signal in a positive spike, followed by a flat undershoot, the length of the undershoot being equal to the drift time. The net area of the pulse, except for small fluctuations due to noise, being equal to 0. Upon Level 1 trigger decision, the samples are then digitized using a fast-ADC and recorded [7, 8]. Figure 2 shows two typical digitized signal shapes, one for the barrel and the other for the endcap. The data samples in each plot correspond to a single cosmic muon event in a single cell, and fluctuations of the amplitude in each sample due to noise can be observed. The pulses shown pass the analysis

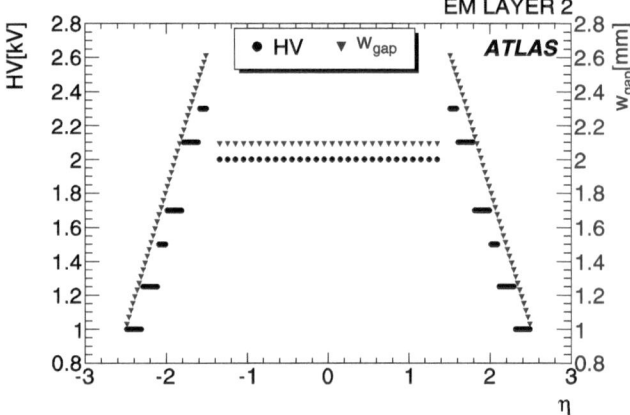

Fig. 3 Nominal HV (*black dots*) and nominal gap width w_{gap} (*blue triangles*) versus η in the 2nd layer of the EM calorimeter

criteria described in Sect. 4. The prediction is obtained by modeling the readout chain as described in Sect. 3. In the barrel section, the nominal gap size is constant (2.09 mm); in the endcap the gap size changes with pseudorapidity (see Fig. 3), so that at larger values of η smaller gaps lead to a shorter pulse undershoot.

In the ideal case, an electrode is surrounded by two identical gaps, one on each side (see Fig. 4). Any modification of one of the gaps by a relative fraction x will break the symmetry, leading to two different values of drift time T_{Di} ($i = 1, 2$) ((4) and (5)). Figure 5 demonstrates this effect by showing the total collected current versus time in the case where the electrode is at the nominal position ($\delta_{gap} = 0$ μm) or shifted by 100 μm and 200 μm. This affects the rise at the end of the pulse (between 450 and 650 ns on Fig. 2(a) for example) which is sensitive to changes in the gap size over the charge collection area. The variation of the drift time inside the cell arises in part from the slight opening of the gaps along the accordion folds (see Fig. 1), but the bulk of the ef-

---

[2]If the LAr gap increases (as in the endcap) ρ increases slightly on average due to showering in LAr. This is accounted for using detector simulation.

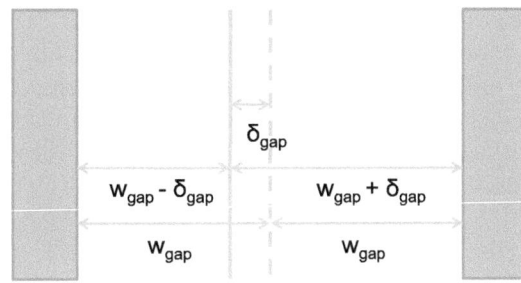

Fig. 4 Schematic view of a LAr gap. The nominal position of the readout electrode (*dashed line*) is exactly equidistant from the lead absorbers. Any shift with respect to the nominal position (*solid line*) causes an increase of the gap width on one side of the electrode, and a decrease on the other side

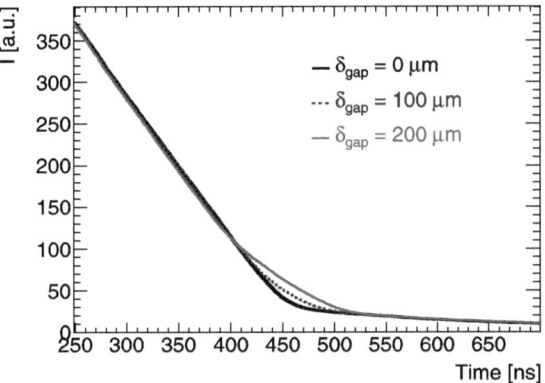

Fig. 5 Current as a function of time for a perfect centering of the electrode ($\delta_{\text{gap}} = 0$ µm), a shift of $\delta_{\text{gap}} = 100$ µm and $\delta_{\text{gap}} = 200$ µm

fect is caused by random or systematic displacements of the electrodes away from the gap center. Both effects are parametrized by the shift parameter $\delta_{\text{gap}} = x \cdot w_{\text{gap}}$. This shift parameter is limited to a maximum of 400 µm due to the honeycomb filling the gaps, however, some modifications of electrical field lines (like edge effects) can contribute to local enlargements.

Beside the gap width, w_{gap}, the model of the signal takes into account the electrode shift parameter as well as possible variations in high voltage on both sides (neglecting in a first description the bend parts). The total signal can be expressed as a sum of two triangular signals, one for each side of the gap, each described by a drift time T_{Di} and peak current $f_i \cdot I_0$ ($i = 1, 2$). Since the drift velocity V_{drift} in liquid argon follows, for the range of electric fields relevant for this study, a power-law dependence on the electric field value [9, 10], with an exponent denoted here by α

$$V_{\text{drift}} = K \cdot \left[\frac{HV}{w_{\text{gap}}} \right]^{\alpha} \qquad (3)$$

the drift time and peak current fraction are given by:

$$T_{D1} = T_{\text{drift}}(1 - x)^{1+\alpha}(HV_1/HV_{\text{nom}})^{-\alpha},$$

$$f_1 = \frac{f_{\text{nom}}}{2}(1 - x)^{-\alpha}(HV_1/HV_{\text{nom}})^{\alpha}, \qquad (4)$$

$$T_{D2} = T_{\text{drift}}(1 + x)^{1+\alpha}(HV_2/HV_{\text{nom}})^{-\alpha},$$

$$f_2 = \frac{f_{\text{nom}}}{2}(1 + x)^{-\alpha}(HV_2/HV_{\text{nom}})^{\alpha}, \qquad (5)$$

where T_{drift} and f_{nom} ($f_{\text{nom}} = 1$ when the bend parts are neglected) are respectively the drift time value and the fraction of current corresponding to the nominal high voltage HV_{nom}, and HV_i corresponds to the actual high voltage applied on side i. In the barrel the nominal high voltage is 2 kV; in the endcap, the high voltage varies with η (see Fig. 3) to cope with the varying gap, ensuring in principle a constant drift velocity by keeping the electric field constant. For the high voltage distribution, electrodes are grouped by sectors of $\Delta\eta \times \Delta\phi = 0.2 \times 0.2$ and for redundancy separated supplies are used for each side of the electrodes. While in the vast majority of the sectors the high voltage has the nominal value, a few of them are operated at lower values, to prevent accidental sparking or excess noise.

Both in the barrel and in the endcap, the nominal operating field is close to 1 kV/mm. The range of variation of x (up to typically 20%) induces a corresponding variation of the operating field of $\pm20\%$. In this reduced range, and for a fixed value of the liquid argon temperature, 88.5 K, the variation of the drift velocity with the field is well described [9, 10] by a power law (3). Fitting the data published in [11] with such law gives $\alpha_1 = 0.316 \pm 0.030$. Additional information was obtained with our own data comparing a group of sectors in the barrel operated at 1600 V, to the bulk operated at 2000 V. The ratio of the velocity values obtained, taking into account small position dependence (see Sect. 6), gives: $\alpha_2 = 0.295 \pm 0.020$. Considering these two values, and given the low sensitivity of our results to the exact value of α (see Sect. 9) we decided to use $\alpha = 0.3$ with a systematic uncertainty of $^{+0.04}_{-0.02}$.

In the accordion geometry, the electric field in the bent sections has a lower value than in the flat parts. This leads to another contribution to the ionization signal in the form of two smaller triangular signals with a longer time constant and smaller f_{bend}. The sum of the current fractions ($f_{\text{nom}} + f_{\text{bend}}$) must be equal to 1; the main contribution on Fig. 5 is related to the drift time in flat sections, the tail at large time ($t > 500$ ns) is due to the larger gap width in the bent sections of the accordion. These triangular shapes are parametrized (neglecting the electrode shift effect) by

$$T_{D3} = T_{\text{bend}}(HV_1/HV_{\text{nom}})^{-\alpha},$$

$$f_3 = \frac{f_{\text{bend}}}{2}(HV_1/HV_{\text{nom}})^{\alpha}, \qquad (6)$$

$$T_{D4} = T_{\mathrm{bend}}(HV_2/HV_{\mathrm{nom}})^{-\alpha},$$

$$f_4 = \frac{f_{\mathrm{bend}}}{2}(HV_2/HV_{\mathrm{nom}})^{\alpha}. \tag{7}$$

In the barrel, the T_{bend} and f_{bend} contributions per layer are estimated using the GEANT4 simulation of a uniform charge density in the gap. These values are given in Table 1 for layers 1 to 3 (there are no bent sections in the presampler).

In the endcaps, for practical reasons a different simulation was used, MC GAMMA, where 10 GeV electromagnetic showers have been simulated to predict the drift time T_{drift} and to estimate T_{bend} and f_{bend}. A photon simulation was chosen since the signals relevant to this study originate from electromagnetic showers produced by cosmic muons. The simulated photons were generated with a flight direction originating from the ATLAS Interaction Point. This differs from cosmic muons which cross the calorimeter in a quasi-vertical direction. Both T_{drift} and T_{bend} are plotted in Fig. 6 as a function of pseudorapidity for the three layers. These quantities are obtained from the distribution of the local drift time where the contributions from straight and bent sections of the accordion are clearly distinguished. Figure 6 shows that both quantities decrease with increasing η, following the reduction of the gap size. The difference observed between the layers is due to the depth variation of the gap size: the gap grows continuously from layer 1 to 3 due to the pro-

jective geometry of the cells. The values for layer 2 lie closer to those of layer 1. This is explained by the fact that at the energy of the simulated showers (10 GeV), the shower maximum is closer to layer 1 than to layer 3. The current fraction f_{bend} is also estimated from the simulation for every η cell, with values ranging from 5% to 20% depending on pseudorapidity.

3 Prediction of the ionization pulse shape

The LAr calorimeters are equipped with a calibration system to inject an exponential pulse of precisely known amplitude onto intermediate "mother" boards located on the front face (for layer 1) and back face (for layers 2 and 3) of the calorimeter. The exponential decay time of these calibration signals has been trimmed to mimic the triangular ionization pulse shape as closely as possible. Since the readout path of the calibration signals is identical to that of the ionization pulses, the gain and pulse response of the electronics can be measured with the calibration system over the full range of signal amplitudes and time delays. The exponential calibration pulse properties are analytically modeled via two parameters τ_{cali} (inverse of the exponential slope) and f_{step} (relative amplitude of a voltage step coming together with the main exponential signal).

The main ingredient needed for accurate energy and time reconstruction in the LAr EM calorimeter is the precise knowledge of the ionization signal shape in each readout cell, from which the optimal filtering coefficients [12] are computed. This knowledge of the ionization pulse shape is also necessary to accurately equalize the response across cells to account for its difference in shape and amplitude with respect to the calibration pulse. The difference between the two pulses is due to the slightly different shape of the

Table 1 T_{bend} and f_{bend} values for the different layers in the barrel

Layer	T_{bend} (ns)	f_{bend} (%)
Layer 1	820	4.9
Layer 2	898	7.1
Layer 3	941	8.5

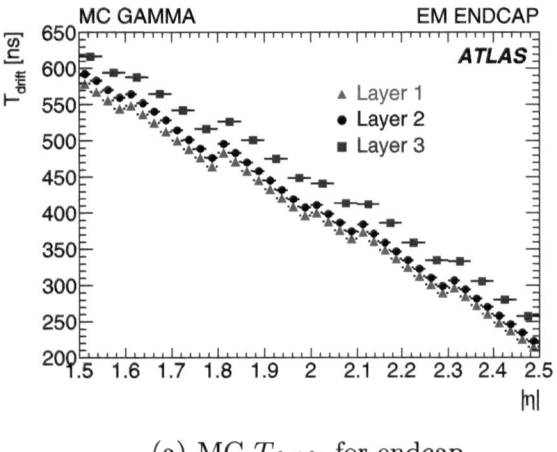

(a) MC T_{drift} for endcap

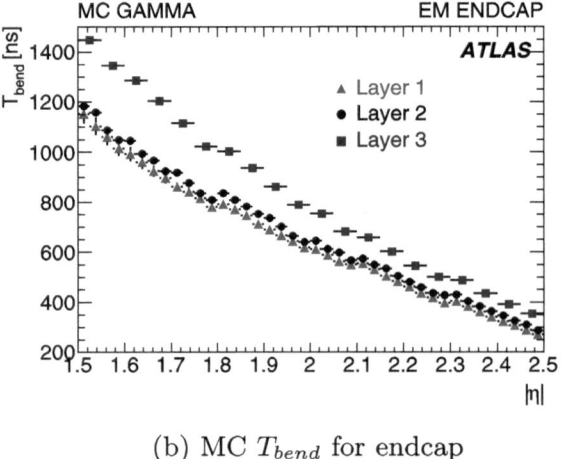

(b) MC T_{bend} for endcap

Fig. 6 Monte Carlo simulation for (**a**) T_{drift} and (**b**) T_{bend} versus η for the three endcap layers: layer 1 (*red triangles*), layer 2 (*black dots*) and layer 3 (*blue squares*)

induced current (triangle versus exponential) and the different injection point for the currents (electrode versus mother board).

The prediction of the ionization pulse shape relies on the modeling of each readout cell as a resonant RLC circuit (where C corresponds to the cell capacitance, L to the inductive path of the ionization signal and R to the contact resistance between the detector cell and the readout line) and on the description of the signal propagation including reflections, amplification and shaping by the readout electronics.

In the standard ATLAS pulse shape prediction method, Response Transformation Method (RTM) [13], calibration pulses are used to determine the description of the signal propagation and the response of the readout electronics, as well as the parameters describing the electrical properties of the readout cell, $(LC$ and $RC)$ and the calibration signal $(\tau_{\text{cali}}$ and $f_{\text{step}})$.

A second method has been developed for the EM barrel, First Principles Method (FPM) [14], where the signal propagation and the response of the readout electronics are analytically described, and the goodness of the analytical description is tuned using the measured calibration pulses.

Both methods need, as an input parameter, the value of the drift time in each cell, which can be either inferred from the local geometry of the detector along with the actual LAr temperature and high voltage, or measured from data pulses as described in this work. Details on the two methods, which describe the ionization pulse equally well, are given below.

3.1 RTM method

The properties of the signal propagation and of the electronic response of the readout of the LAr EM calorimeter cells are probed by the calibration system and can be determined from the measured calibration pulses. The two underlying assumptions behind the RTM [13] are that:

- The ionization pulse (g_{phys}) can be numerically predicted from the corresponding calibration pulse (g_{cali}) by means of time domain convolution with two simple functions, parameterizing respectively the shape difference between the exponential and triangular currents, and their different injection points in the detector, see [13]:

$$
\begin{aligned}
g_{\text{phys}}(t) \\
= g_{\text{cali}}(t) \\
* \mathcal{L}^{-1}\left\{\frac{(1 + s\tau_{\text{cali}})(sT_{\text{drift}} - 1 + e^{-sT_{\text{drift}}})}{sT_{\text{drift}}(f_{\text{step}} + s\tau_{\text{cali}})}\right\} \\
* \mathcal{L}^{-1}\left\{\frac{1}{1 + s^2LC + sRC}\right\} \quad (8)
\end{aligned}
$$

where $\mathcal{L}^{-1}$ denotes an inverse Laplace transform, with s being the variable in the frequency space. The first

time-domain convolution corrects for the different signal shapes through the calibration pulse parameters τ_{cali} and f_{step} and the drift time T_{drift}, while the second convolution accounts for the different injection points on the detector cell, modeled as a lumped RLC electrical circuit.

- All parameters $(\tau_{\text{cali}}, f_{\text{step}}, LC, RC)$ used in the convolution functions, apart from the drift time, are directly extracted from the measured calibration pulses by numerical analysis [13].

3.2 FPM method

In the FPM method, the signal generation is based on "first principles" of signal propagation [14]. All the calculations are made in the frequency domain, and when the signal at the output of the final shaping amplifier is obtained, it is transformed to the time domain by using a fast Fourier transform [15].

After generation at the detector level, a signal is propagated along the signal cable, taking into account its impedance, propagation velocity, and absorption by the skin effect [14]. A small fraction of this signal is reflected at the signal cable-feedthrough transition, while the rest is transmitted. A second reflection takes place at the feedthrough-preamplifier transition. In this model, the feedthrough is modeled as a single cable section, with its own impedance, skin effect absorption constant, propagation velocity and length. The preamplifier is described by a complex impedance, the real part and the imaginary part (Re[Z_{PA}], Im[Z_{PA}]) being both functions of the frequency ω. The last element of the chain is the $CR - RC^2$ shaping amplifier, described by the transfer function:

$$
F_{\text{sh}}(\omega) = \omega \cdot \tau_{\text{sh}}/\left(1 + (\omega \cdot \tau_{\text{sh}})^2\right)^{3/2} \quad (9)
$$

where τ_{sh} is the RC time constant of this element. The model accounts for both the directly transmitted signal and the reflections up to the second order (i.e. two forward-backward reflections and two backward-forward reflections).

Parameters are taken from construction (cable lengths, f_{step} and τ_{cali}, which were measured for all calibration boards [16]), from direct measurements channel-by-channel (resonance frequency $\omega_0 = 1/\sqrt{LC}$ and R) [17], and from measurements on representative samples (Re[Z_{PA}], Im[Z_{PA}], propagation velocity and skin effect constants). The signal cable impedance Z_{S} and the shaper time constant τ_{sh} were left as free parameters and fitted channel-by-channel on calibration pulses [14]. The values obtained for Z_{S} and τ_{sh} came out close to expectations, giving confidence in the method which describes calibration pulses to 1% or better. The relative timing of all calibration signals was also reproduced with an accuracy of about 1 ns.

Table 2 Cut values for the most energetic sample of the data pulse. The approximate electronic cell noise (σ_{noise}) averaged over layer and the approximate multiplicative conversion factor from ADC counts to MeV (F) are given as well

	Layer	S_{max} (ADC count) lower limit	σ_{noise} (ADC count)	F (MeV/ADC count)
Barrel	Presampler	200	8.0	7.0
	Layer 1	500	8.0	2.5
	Layer 2 ($|\eta| \leq 0.8$)	160	5.0	10.0
	Layer 2 ($|\eta| > 0.8$)	100	3.5	17.0
	Layer 3	160	5.0	7.0
Endcap	Layer 1	500	7.0	3.0
	Layer 2	160	4.0	14.0
	Layer 3	160	2.0	7.0

This method was not extended to the EM endcap because not all the necessary parameters have been measured with the required precision due to a more complex geometry.

4 Description of the data

Cosmic muon runs from the data-taking period of September–November 2008 are used in this analysis, corresponding to a period where the LAr data acquisition system transmitted and saved 32 samples of the readout signals. The events of interest are those where muons lose a substantial fraction of their energy by radiation (the energy lost by dE/dx in layer 2 is in average about 300 MeV [5]). These events were triggered using calorimeter trigger towers over the full calorimeter depth, of size $\Delta\eta \times \Delta\phi = 0.1 \times 0.1$ for $|\eta| < 2.5$, 0.2×0.2 for $2.5 < |\eta| < 3.2$, and up to 0.4×0.4 for $3.2 < |\eta| < 4.9$. Since the data were collected from cosmic muons instead of LHC collisions, trigger thresholds were adjusted accordingly. For technical reasons, only cells which were readout in high gain (LAr readout has three gains with ratio $\sim$100/10/1) are selected for this analysis. This has a very small impact on the selected sample as the energy deposits are typically in the high gain range (energies below 20 GeV).

Despite the small rate of cosmic muons depositing significant electromagnetic energy, the number of events recorded during the run period ensured sufficient statistics for most of the calorimeter regions, with the exception of the high-η region of the endcaps. The pseudorapidity range in this study is hence restricted to $|\eta| < 2.5$.

To minimize distortion of the signal shape, the energy deposited in a cell is required to be well above its typical noise value. This is particularly important since the drift time is obtained on an event-by-event basis. The quantity S_{max} is defined as the amplitude of the most energetic sample of the data pulse. The minimal required values for S_{max} are given in Table 2 for the different layers of the calorimeter; these values correspond to about 1–2 GeV. The average noise is also quoted, representing between 1 and 4% of the minimal

value for S_{max}. Unless differently stated, all ADC values are pedestal subtracted. The difference of thresholds between the $|\eta| < 0.8$ and $0.8 < |\eta| < 1.4$ regions in layer 2 of the barrel is required by a difference in gain. To correct for this effect, the normalized variable $S_{\text{max}}^{\text{gain}}$ is used for the selection, defined as $S_{\text{max}}^{\text{gain}} = 1.6 \cdot S_{\text{max}}$ for $0.8 < |\eta| < 1.4$, and $S_{\text{max}}^{\text{gain}} = S_{\text{max}}$ everywhere else. An upper limit of 3900 ADC counts for $S_{\text{max}}^{\text{gain}}$ plus pedestal is also required to avoid saturation.

As a small fraction of the ionization pulses are distorted and their drift times cannot be determined accurately, a set of cuts has been defined to select good quality pulses:

– The data should have a negative undershoot in the pulse shape. This is ensured by requiring that at least 5 samples after the peak have a negative amplitude.
– In order to prevent pulses with too short an undershoot (as can be the case for signals resulting from crosstalk for instance), a condition requires that the pulse does not contain more than 12 samples around 0 ADC counts at the end of the pulse. This condition cannot be applied to the endcap where such shapes occur due to smaller drift-time values at high pseudorapidity.

For a small fraction (6%) of the LAr EM calorimeter the high voltage cannot be safely set to the nominal value. The cells belonging to these regions are excluded in the follow-

Table 3 Approximate number of cosmic muon induced pulses in each layer after quality cuts

	Layer	# pulses
Barrel	Presampler	20 k
	Layer 1	43 k
	Layer 2	331 k
	Layer 3	79 k
Endcap	Layer 1	13 k
	Layer 2	45 k
	Layer 3	18 k

ing. The numbers of pulses per layer after quality cuts are given in Table 3.

5 Extraction of the drift time

The 32 data samples S_i of each calorimeter cell selected by the criteria given in Sect. 4 are fitted using the pulse predictions described in Sect. 3, scaled by an amplitude $A_{\max}$ and shifted in time by an offset t_0:

$$g_{\mathrm{fit}}(t; A_{\max}, t_0, T_{\mathrm{drift}}, x)$$

$$= A_{\max} \cdot g_{\mathrm{phys}}(t; f_{\mathrm{nom}}, T_{\mathrm{drift}}, x, f_{\mathrm{bend}}, T_{\mathrm{bend}})$$

for $t > t_0$. $\qquad(10)$

Four parameters are left free in this procedure: the drift time (T_{drift}), the associated shift of the electrode estimated as $\delta_{\mathrm{gap}} = x \cdot w_{\mathrm{gap}}$ which is in fact only sensitive to the absolute value of x when the high voltage is the same on both sides of electrodes, the global normalization factor $A_{\max}$ and the timing adjustment t_0. The optimal set of these four parameters is estimated using the least squares method to minimize the quantity:

$$Q_0^2 = \frac{1}{n - N_p} \sum_{i=1}^{n} \frac{(S_i - g_{\mathrm{fit}}(t_i; A_{\max}, t_0, T_{\mathrm{drift}}, x))^2}{\sigma_{\mathrm{noise}}^2} \qquad(11)$$

where n is the total number of data samples used in the fit (usually $n = 32$), N_p the number of free parameters ($N_p = 4$), and σ_{noise} is given in Table 2. This minimization is performed using the MINUIT package [18].

Figure 7(a) presents the variation of Q_0^2 with $S_{\max}^{\mathrm{gain}}$ for layer 2 of the barrel. An increase of the Q_0^2 value is observed when $S_{\max}^{\mathrm{gain}}$ is larger. The same behavior is observed in the other calorimeter layers, as expected. In order to be able to apply a global selection to the fit quality independently of the data amplitude, a "normalized" Q_0^2, called Q^2, has been used:

$$Q^2 = \frac{1}{n - N_p} \sum_{i=1}^{n} \frac{(S_i - g_{\mathrm{fit}}(t_i; A_{\max}, t_0, T_{\mathrm{drift}}, x))^2}{\sigma_{\mathrm{noise}}^2 + (k S_{\max})^2} \qquad(12)$$

where k is chosen such that Q^2 is independent of $S_{\max}$, as represented in Fig. 7(b). The denominator in (12) is the quadratic sum of the noise and of the relative inaccuracy of the predicted shape. It represents the numerator uncertainty. The values of k are given in Table 4 for the different layers of barrel (two methods) and endcap.

For the measurement of the drift time, the last data samples corresponding to the end of the pulse are very important. It was noticed that for a small fraction of pulses ($\sim$0.6% for the layer 2) the fit converges successfully but the predicted pulse does not succeed in describing the rise at the

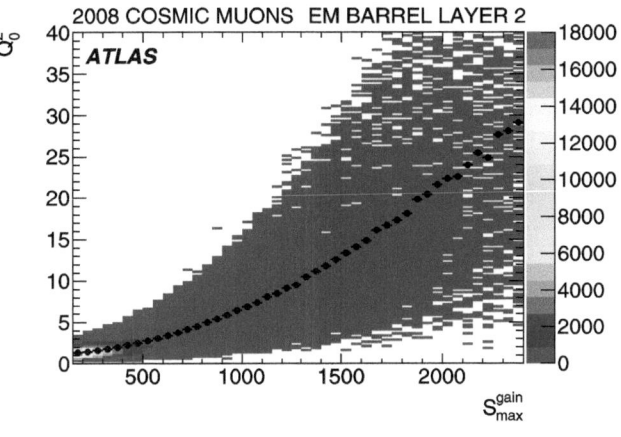

(a) Q_0^2 versus S_{max}^{gain}

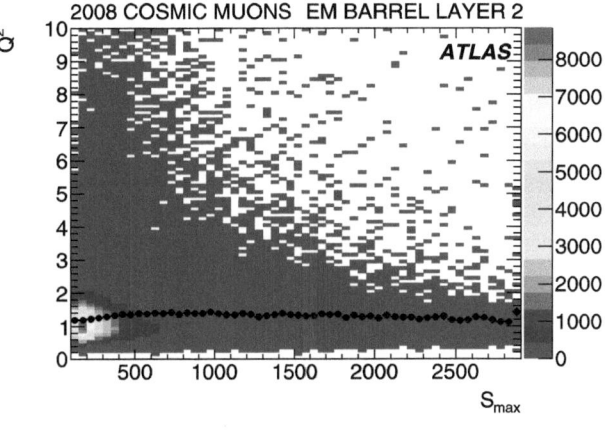

(b) Q^2 versus S_{max}

Fig. 7 (a) Q_0^2 versus $S_{\max}^{\mathrm{gain}}$ and (b) Q^2 versus $S_{\max}$ in layer 2 of the barrel. The *black points* correspond to the mean value

end of the pulse. This implies an incorrect estimate of the drift time. To specifically quantify the quality of the fit at the end of the pulse, the variable Δ_{last7} has been defined, based only on the last 7 samples:

$$\Delta_{\mathrm{last7}} = \sum_{i=26}^{32} \frac{S_i - g_{\mathrm{fit}}(t_i; A_{\max}, t_0, T_{\mathrm{drift}}, x)}{S_{\max}}. \qquad(13)$$

Large values of $|\Delta_{\mathrm{last7}}|$ single out pulses with erroneous fitted drift times. This effect is also observed with a toy simulation, and therefore seems to be an intrinsic feature of the fitted function, with a large peak followed by a flat tail.

To remove events for which the end of the pulse is badly described by the model, a cleaning selection requiring $|\Delta_{\mathrm{last7}}| < 0.15$ and $Q^2 < 2.5$ (3) in the barrel (endcap) is imposed.

An additional set of cuts on the maximum relative residual over all samples is applied for presampler cells, where pick-up of oscillatory signals was in a few places observed (3% of the pulses):

Table 4 k values for the different methods in the different regions of the EM calorimeter

Layer	k_{FPM} in barrel	k_{RTM} in barrel	k_{RTM} in endcap
Presampler	0.9%		
Layer 1	1.1%	0.8%	0.9%
Layer 2	0.8%	1.0%	1.4%
Layer 3	0.75%	1.0%	1.3%

- $|S_i - g_{\mathrm{fit}}(t_i)|_{\max}/S_{\max} < 10\%$
- If the residual is small ($|S_i - g_{\mathrm{fit}}(t_i)|_{\max} < 20$ ADC counts), the cut is relaxed to $|S_i - g_{\mathrm{fit}}(t_i)|_{\max}/S_{\max} < 20\%$

After these selections, the fit parameters are examined in more detail. Figure 8 presents the distribution of the absolute value of the shift parameter, $\delta_{\mathrm{gap}} = x w_{\mathrm{gap}}$, as a function of the drift time.

The region in Fig. 8(a) with a drift time T_{drift} comprised between 380 and 550 ns corresponds to the expected range for the drift time in the barrel given the resolution of the measurement. The low drift time region $T_{\mathrm{drift}} < 380$ ns of Fig. 8(a) (0.05% of the pulses) is dominated by low-amplitude pulses distributed evenly in the calorimeter. A closer examination shows that in about 80% of the cases for the layer 2 barrel, signals in excess of $S_{\max} = 1500$ ADC counts or cells sampled at medium gain are found as first neighbors which corroborates a crosstalk hypothesis.

In the region $T_{\mathrm{drift}} > 550$ ns of Fig. 8(a) (0.25% of the pulses), some pulses are still significantly negative, more than 700 ns after the time of signal maximum. A possible explanation is that the energy deposit originates from a photon emitted along a bent section, thus having an abnormally enhanced f_{bend} contribution. Unfortunately the runs taken with 32 samples do not contain information from the inner tracker which would have allowed this hypothesis to be validated by a projectivity study. Aside from these extremely large drift time pulses, there is a larger class of pulses which are only somewhat longer than normal. They are distributed

along specific η and ϕ directions: in the transition regions at $|\eta| = 0.8$ and between the two half-barrels at $\eta = 0$ (see Sect. 6.1.1) where a slight dilution or leakage of the electric field lines yields a larger drift time (this is also observed in layer 1 of the barrel); in the intermodular regions in ϕ in the upper part of the detector, where mechanical assembly tolerances allow for a slightly increased gap at the interface between modules due to gravity effects (this is not seen in barrel layers 1 and 3 which are much closer to the mechanical fixed points).

In the endcap, the cloud of points corresponding to the expected T_{drift} is broader than in the barrel, as can be seen in Fig. 8(b): it ranges from 300 to 600 ns as a consequence of the gap size variation with η of the endcap design. The fact that the dispersion of $|\delta_{\mathrm{gap}}|$ is larger at higher values of T_{drift} is explained as a consequence of the larger gap size: the larger the gap width, the larger the displacement of the electrode can be. A few events (0.9% of the pulses) are observed at very high values of both T_{drift} and $|\delta_{\mathrm{gap}}|$. They are located at low $|\eta|$ where the drift time is very large by construction (see Fig. 6(a)). Their pulse shape cannot be completely readout using 32 samples, and in particular the rise following the undershoot is partially absent, which leads to unphysical values of the shift above 400 μm.

A distinctive aspect of the fit, which is clearly visible in Fig. 9, is that it yields a peak at $|\delta_{\mathrm{gap}}| = 0$. This is mainly explained by noise fluctuations. The superposition of two triangles of ionization current with unequal length due to an electrode shift (see Fig. 5) can only lead to a softening

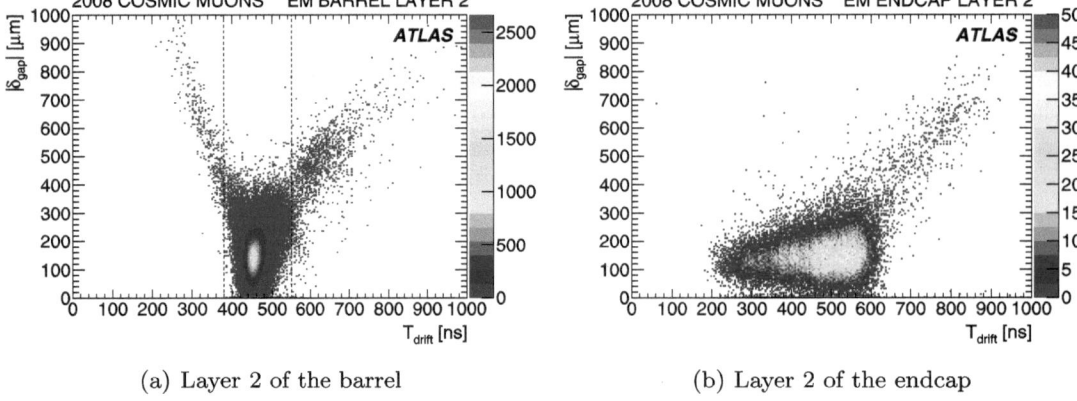

(a) Layer 2 of the barrel   (b) Layer 2 of the endcap

Fig. 8 Absolute value of the shift parameter as a function of the drift time in the barrel (**a**) and in the endcap (**b**), for layer 2

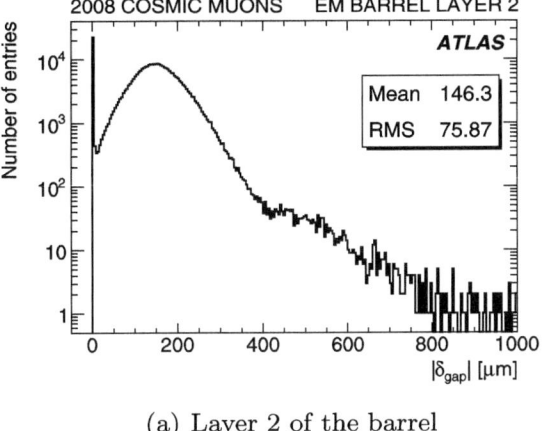

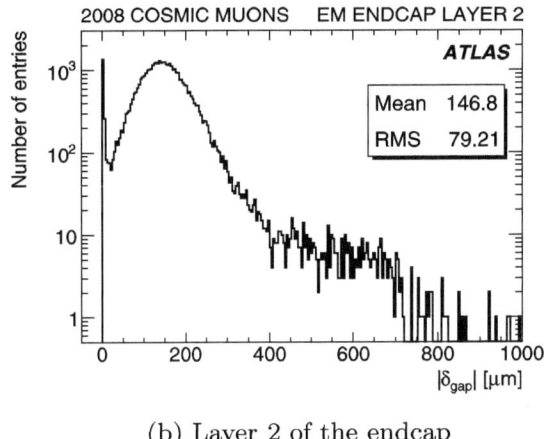

(a) Layer 2 of the barrel   (b) Layer 2 of the endcap

Fig. 9 Distribution of the absolute value of the shift parameter in layer 2 of the barrel (**a**) and endcap (**b**)

of the rise at the end of the pulse, compared to the single-triangle case. If, due to noise, the rise is steeper than for a single-triangular shape, the fit forces δ_{gap} to 0. In order to improve the statistical significance of high-amplitude signals and minimize the impact of noise fluctuations, it has been decided to weight the events by $(S_{max}^{gain})^2$. The results in the following sections of this note are produced with this weighting factor.

6 Results in the calorimeter barrel

Two parallel analyses have been performed for this part of the calorimeter using the two pulse shape prediction methods described in Sect. 3. The analyses agree at the level of 0.3%, which provides a good check of the robustness of these results. In this section the measurement of the drift time is presented, along with its implications for the calorimeter response uniformity and an estimation of the electrode shift.

6.1 Drift time measurement
in pseudorapidity and azimuthal angle

Results in layer 2 are presented first because the statistical uncertainties are lower in this layer (see Table 3). More refined comparisons between the two methods are then possible. The following subsection reports on the results in the other layers. The presampler is discussed separately due to its different structure.

6.1.1 Layer 2 of the barrel

Figure 10 presents the drift time T_{drift} extracted from the fit as a function of η. The results of the two methods differ by 0.1 ns on average with an RMS of 1.3 ns. The full purple line

illustrates the prediction from absorber thickness measurements made during the calorimeter construction [2]. This prediction is based on the fact that the mechanical structure of the calorimeter ensures that the pitch (with nominal values shown in parentheses) is constant to within about 5 μm:

$$\text{Absorber}(2.2 \text{ mm}) + w_{gap}(2.09 \text{ mm})$$
$$+ \text{Electrode}(0.280 \text{ mm}) + w_{gap}(2.09 \text{ mm})$$
$$= 6.66 \text{ mm} = (2\pi/1024) \cdot R_i \cos\theta_i \qquad (14)$$

where R_i and θ_i are the average radius and the local angle of the 1024 accordion-shaped absorbers with respect to the radial direction. So if the thickness of the absorber varies with η, the gap will also vary in the opposite direction. As the drift time T_{drift} is directly related to the gap by:

$$T_{drift} = T_{D0}(w_{gap}/w_{gap0})^{1+\alpha} \qquad (15)$$

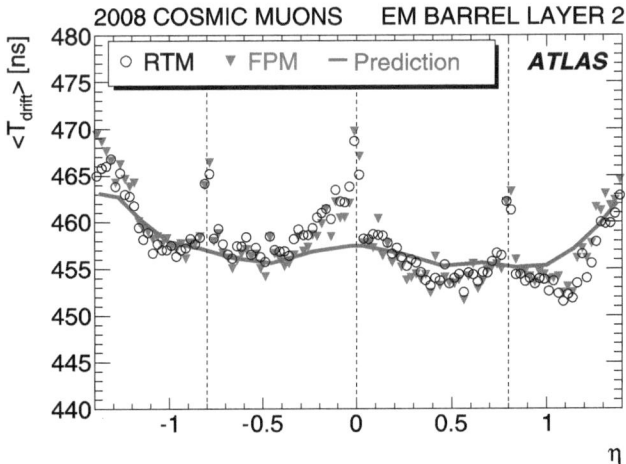

Fig. 10 Drift time as a function of η in layer 2 of the barrel: using the RTM method (*open dots*), the FPM method (*red triangles*) and the prediction described in the text (*purple line*)

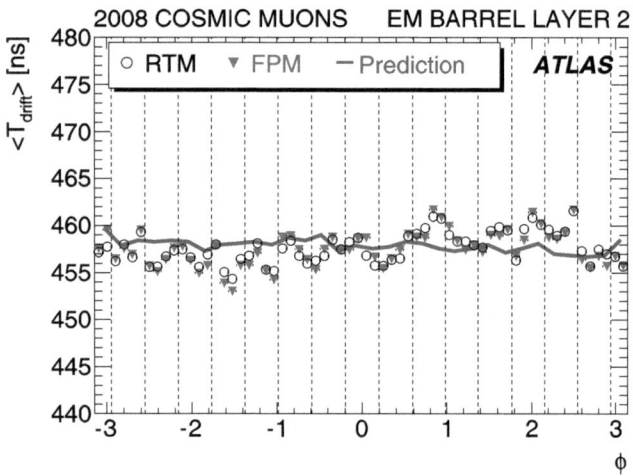

Fig. 11 Drift time as a function of ϕ in layer 2 of the barrel: using the RTM method (*open dots*), the FPM method (*red triangles*) and the prediction described in the text (*purple line*)

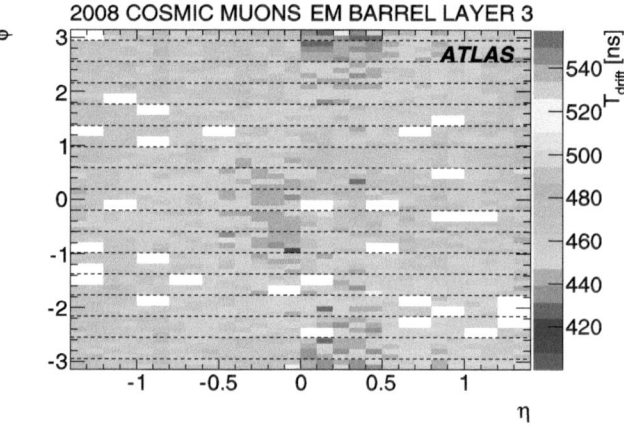

Fig. 12 2D map of T_{drift} in (η, ϕ) for layer 3. The empty bins correspond to sectors with non nominal HV

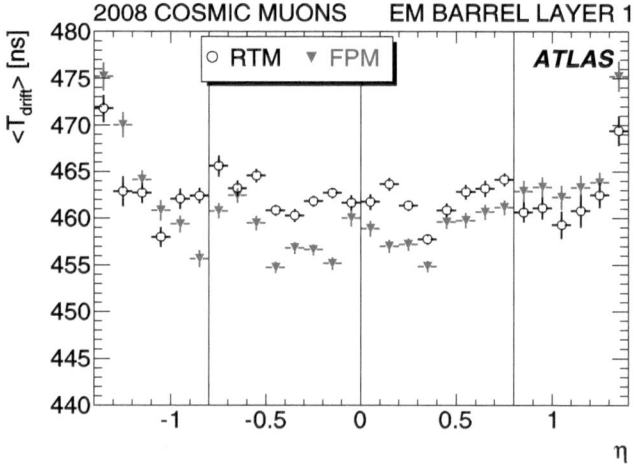

Fig. 13 Drift time as a function of η in layer 1 of the barrel: using the RTM method (*open dots*) and the FPM method (*red triangles*)

a prediction can be derived for the drift time from the variations around the nominal gap size ($w_{\text{gap0}} = 2.09$ mm) associated with $T_{D0} = \langle T_{\text{drift}} \rangle = 457.9$ ns.

The agreement between the prediction coming from precision mechanical probe measurements of the absorber thickness and the data is rather good, except in the transition regions around $\eta = 0$, ± 0.8 and -1.4, where the lower field induces a larger T_{drift}. To quantify the agreement between the drift time measurements from the fit and the estimate from the measurement of the absorbers, the RMS of the difference between the data points and the prediction is computed. This yields a value of 2.9 ns, as compared to an RMS deviation with respect to a constant value of 3.7 ns, excluding the data points around the transition region in each case. Comparing bin by bin the drift times obtained (Fig. 10) for the negative and positive values of η, one gets a distribution with a mean of 3.4 ± 0.2 ns and RMS of 1.7 ns. The predicted value is 1.5 ± 0.2 ns.

The T_{drift} distribution as a function of ϕ is presented in Fig. 11, for both methods. There is a small difference between the $\phi < 0$ ((456.8 ± 0.3) ns) and $\phi > 0$ ((458.3 ± 0.3) ns) regions: a $(0.3 \pm 0.1)\%$ relative effect consistent with sagging and pear shape deformation of the calorimeter. No significant variations are observed in the absorber thickness measurements. The distribution of the results is also rather uniform when looking at the two half-barrels separately. The RMS of the ϕ distribution is smaller (1.8 ns) when the two half-barrels are combined, than for the $\eta < 0$ (2.8 ns) and $\eta > 0$ (3.1 ns) half-barrels separately. This may be due to the existence of small ϕ modulations with opposite phases in the two half-barrels that appear to be more visible in layer 3 (see Fig. 12).

6.1.2 Other layers of the barrel

The distribution of T_{drift} as a function of η for layer 1 is displayed in Fig. 13. The results of the two methods differ by 1.3 ns on average, with an RMS of 4 ns, and at some points by up to 7 ns. The front layer is particularly intricate because of the large relative variations of the cell depths which present a discontinuity at $|\eta| = 0.8$, inducing a corresponding variation of the cell capacitance and bent-to-flat ratio. Given that the two methods differ in their estimation of the cell capacitance, such a difference is not unexpected.

In Fig. 12, a drift time modulation with ϕ is clearly visible for $|\eta| < 0.5$ (and equally present in both methods) in both half barrels of layer 3. While the source of the modulation is so far unexplained, the fact that the modulations in the two half-barrels are opposite in phase is expected, since one of the half-barrels was rotated by 180 degrees about the vertical direction for final integration.

6.1.3 Presampler

The presampler is constructed differently from the other layers of the calorimeter. It is made of narrow flat electrodes. The size of the gaps is slightly smaller than elsewhere, leading to values of T_{drift} lower than in the rest of the calorimeter. In addition, this gap varies with η; the values for the 4 regions are given in Table 5. The effect on the fitted drift time can be immediately seen in Fig. 14. The prediction superimposed on the measured distribution is normalized to the region $0.8 < |\eta| < 1.2$. Good agreement within 1% is observed between the measured and expected drift times as a function of η. As there are no bent sections in the presampler, the pulse description is simpler than in the case of the other layers. While the variations in η are large, the ϕ dependence of the drift time is negligible.

6.2 Response uniformity

The reconstructed value of the energy deposited in the calorimeter by an electron or photon should be independent of the position of its impact on the calorimeter. The non-uniformity coming from local variations of the response due to gap fluctuations can be determined using the drift time

Table 5 Gap values in presampler

η region	w_{gap} (in mm)
$|\eta| < 0.4$	1.966
$0.4 \leq |\eta| < 0.8$	1.936
$0.8 \leq |\eta| < 1.2$	2.006
$1.2 \leq |\eta|$	1.906

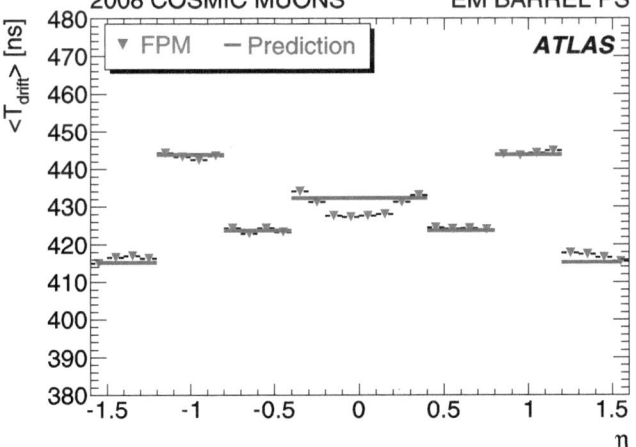

Fig. 14 Drift time as a function of η in the presampler barrel using the FPM method (*red triangles*). The *full purple line* represents the prediction normalized to the region $0.8 < |\eta| < 1.2$, using (15) and the gap values given in Table 5. The *empty bins* correspond to sectors with non nominal HV

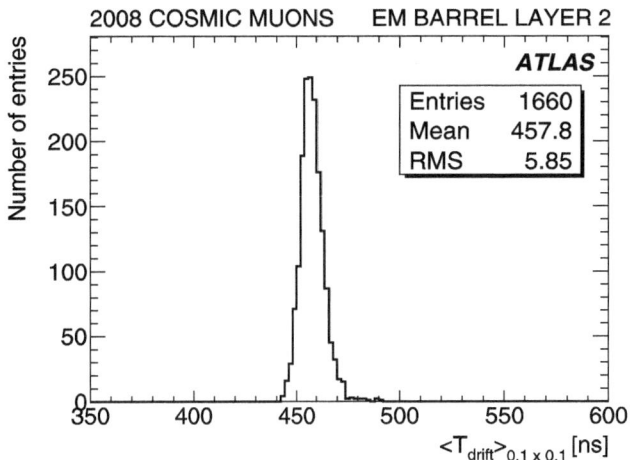

Fig. 15 Drift time uniformity between groups of 4×4 cells ($\Delta\eta \times \Delta\phi = 0.1 \times 0.1$) for barrel layer 2

measurements. This study is done only for layer 2, which is the main contributor to the energy response of the detector as it collects $\sim$70% of the total electromagnetic signal in the calorimeter.

Figure 15 shows the distribution of the drift time averaged over groups of 4×4 cells corresponding to an area of 0.1×0.1 in $\Delta\eta \times \Delta\phi$ plane. This area represent a typical transverse size of a single particle shower. The average of the statistical uncertainties on T_{drift} obtained for pulses within the various 4×4 groups is 1.25 ns, well below the dispersion of the determined T_{drift} values of the groups (the RMS is 5.85 ns). From the measurement of drift times, the systematic dispersion of the gaps can be estimated and its impact on the calorimeter energy response can be assessed.

The drift time uniformity, corresponding to the ratio of the RMS and the mean value of the local T_{drift} distribution amounts to $(5.85 \pm 0.14)/457.8 = (1.28 \pm 0.03)\%$. From the relation between the drift time and the drift velocity (1), the latter being proportional to the energy response, together with (15), it follows, that the drift time uniformity leads to a dispersion of the response due to the gap variations of $(1.28 \pm 0.03)\% \cdot (\alpha/(1 + \alpha)) = (0.29 \pm 0.01)\%$. Excluding transition regions in η and in ϕ, the gap variations amount to $5.7/457.4 = 1.25\%$ and the impact on the response is 0.28%. Taking into account small variations observed in the result when changing the weighting, the fit strategy (see Sect. 9) or the pulse reconstruction method, a systematic error of 0.03% is obtained. The uncertainty on α (see Sect. 2), treated here as external parameter, contributes with a systematic uncertainty of $^{+0.04}_{-0.02}$. Grouping all errors together in quadrature gives as the final result: $(0.29^{+0.05}_{-0.04})\%$.

6.3 Electrode shift

As presented in Sect. 2, there is some freedom for the electrodes to be displaced with respect to their nominal positions

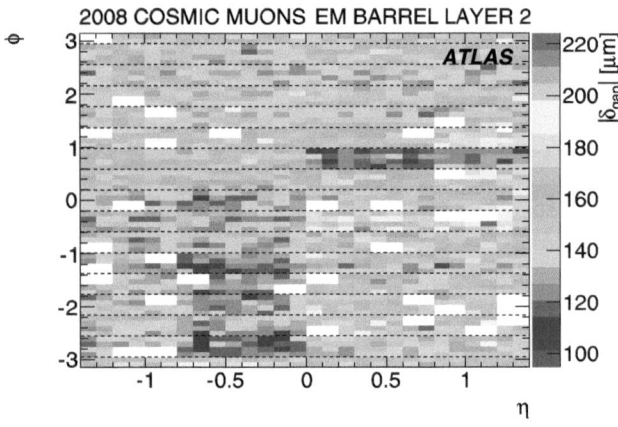

Fig. 16 (η, ϕ) map in which $|\delta_{gap}|$ is plotted per bin of 0.1×0.1

equidistant between two neighboring absorbers. This displacement is expected to be less than 400 μm except perhaps in the transition regions between modules.

The electrode shift is left as a free parameter in the fit to the data, which yields one value per calorimeter cell. Only the average of the absolute value of the displacement can be observed. Since a cell consists of several electrodes, an effective value is obtained which is a combination of the individual movements of each electrode within a cell.

The local average value for the shift parameter per bin of 0.1×0.1 is shown in Fig. 16 for layer 2. It indicates that the bottom half (negative ϕ) of the negative-η half-barrel has shift parameter values somewhat lower than average. Similarly the module azimuthally located between $4\pi/16$ and $5\pi/16$ in the $\eta > 0$ half-barrel presents lower shift values. These variations given their distribution throughout the detector, are likely to be due to mechanical construction issues.

The shift parameter also covers local variations of the "double-gap" within a cell, for example, by the slight opening of gaps along an accordion fold. This latter variation is in general much smaller than the off-centering of electrodes between absorbers.

Smaller values of the shift parameter are expected for the presampler compared to the accordion layers, due to mechanical constraints on the electrodes which are individually glued in between two precision structural frames [2]. The mean value of the shift in the presampler is found to be $\langle|\delta_{gap}|\rangle = 66.5$ μm, as compared to 146 μm in the accordion section.

7 Results in the calorimeter endcap

As was done for the barrel, the endcap results are grouped in three different parts: drift time measurements, calorimeter response uniformity and electrode shift determination.

7.1 Drift time measurement in pseudorapidity and azimuthal angle

The drift time T_{drift} averaged over ϕ is studied as a function of η for each of the three layers of the endcap (see Fig. 17). The two endcaps, A ($\eta > 0$) and C ($\eta < 0$), are combined in the figure. A general decrease of T_{drift} with increasing pseudorapidity is observed, as expected from the corresponding reduction of the design gap size. Fewer fluctuations are observed in layer 2, which offers a larger cross section to cosmic muon-induced electromagnetic showers. In all layers regular steps are observed, corresponding to the locations of the boundaries between high voltage regions.

The data are compared to the Monte Carlo calculation described in Section 2. Good agreement is observed at high η, however the Monte Carlo is slightly above the data at low values of η ($\sim$1–3%), which is a more difficult region to simulate.

In Fig. 18, for a comparison, the data points from the three distributions of Fig. 17 are super-imposed on the same plot. An increase of the drift time with the cell gap size at fixed η is clearly observed, with T_{drift} being smallest for layer 1 and highest for layer 3 (see Sect. 2 and Fig. 6(a)). The drift time for layer 2 lies half way between layers 1 and 3 in contrast to the Monte Carlo simulation (Fig. 6(a)) where the values for layer 2 are closer to the values of the layer 1. This difference reflects the fact that cosmic muons are randomly distributed within the depth of layer 2, while the photons of the simulation develop there shower closer to layer 1.

Figure 19 shows the drift time T_{drift} as a function of azimuthal angle for layer 2 for the two endcaps. The values of T_{drift} for each given pseudorapidity bin have been normalized to the average in order to mask the dependence on η. Vertical dashed lines indicate the boundaries between modules. An asymmetry is visible on Fig. 19 between positive and negative values of ϕ: $T_{drift}(\phi > 0)$ is larger (0.996 ± 0.002) than $T_{drift}(\phi < 0)$ (0.980 ± 0.002). Since $\phi < 0$ is the lower half of the calorimeter, we associate this effect to the greater gravitational compression of this part leading to slightly smaller gaps than in the upper half $\phi > 0$.

7.2 Response uniformity

An estimate of the intrinsic uniformity of the endcap can be made in a similar manner as presented for the barrel in Sect. 6.2. The average drift time across a region of size 0.1×0.1 on the (η, ϕ) plane is computed, with special care to take into account the varying gap thickness.

Figure 20 represents the distribution of $T_{drift}/\langle T_0 \rangle$ for layer 2. The normalization $\langle T_0 \rangle$ corresponds to the value (per η cell) predicted from a first order polynomial fit to the data

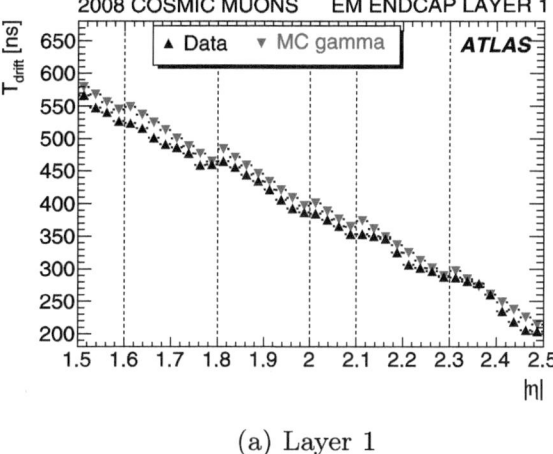

(a) Layer 1

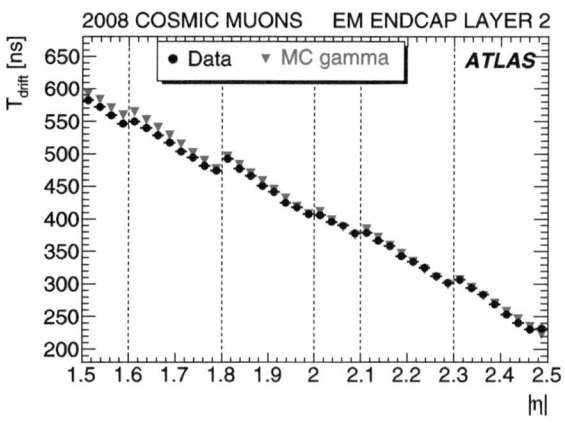

(b) Layer 2

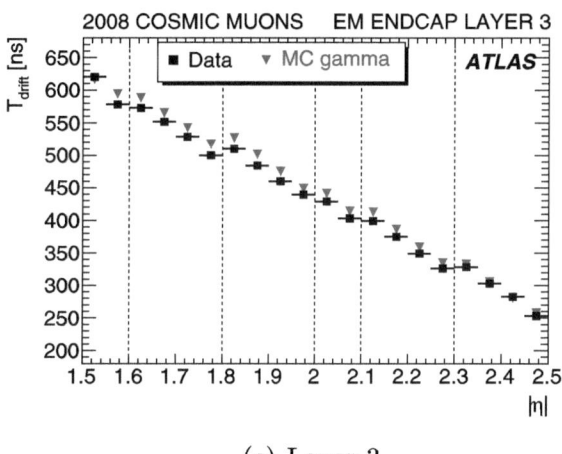

(c) Layer 3

Fig. 17 Drift time versus pseudorapidity for layer 1 (**a**), layer 2 (**b**), and layer 3 (**c**) cells of the endcap. *Black points* are the data and *red triangles* Monte Carlo predictions for photons. The *vertical dashed lines* show the boundaries between different high voltage regions

T_drift in each high voltage region. This normalization cancels out the change of the drift time due to the nominal design gap size variation with η. The study is carried out only

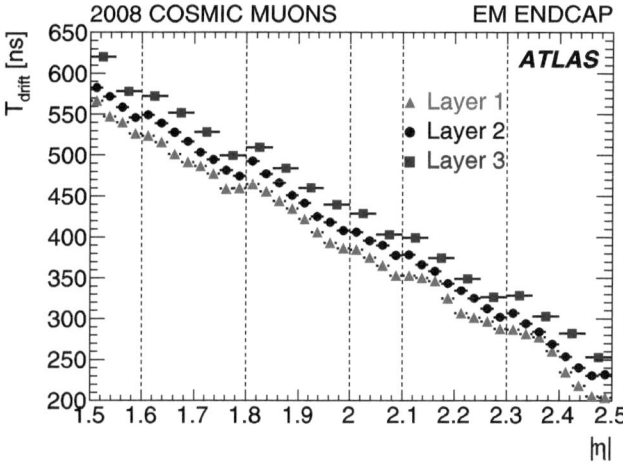

Fig. 18 Drift time versus pseudorapidity for the three layers of the endcap: layer 1 (*red triangles*), layer 2 (*black dots*), layer 3 (*blue squares*). The vertical dashed lines show the boundaries between different high voltage regions

for layer 2 since it contains most of the shower energy of typical LHC electrons and photons. In addition, more events have been recorded in layer 2 than in the other layers, which increases the statistical accuracy of the measurement.

The drift time uniformity of the T_drift (0.1×0.1) distribution has an RMS of $(2.8 \pm 0.1)\%$. To get the pure systematic non-uniformity between the 0.1×0.1 cells, the dispersion within the 0.1×0.1 cells, which in this case is not negligible, $(1.5 \pm 0.1)\%$, is quadratically subtracted. These numbers translate to a uniformity of the endcap calorimeter response due to intrinsic gap variations of $(0.54 \pm 0.02)\%$. Systematic effects as discussed in Sect. 9 and the uncertainty on α (see Sect. 2) increase the error to $(0.54^{+0.06}_{-0.04})\%$.

7.3 Electrode shift

The distribution of the electrode shift as a function of the azimuthal angle is presented in Fig. 21 for layer 2. A rather flat behavior is observed. Vertical dashed lines correspond to the boundaries between consecutive modules. With a finer binning no particular increase of the shift is observed at these transitions, even when extending the scale to 1000 μm. The average of about 146m is independent of the layer.

8 Drift time and velocity measurements

To quantify the consistency of the drift time measurements, the drift velocity (V_drift) is studied more closely. The drift velocity can be extracted from drift time measurements if the local gap values are accurately known (see (1)). Both w_gap and T_drift are designed to be constant for the barrel, but varying with pseudorapidity for the endcap. The variation of the drift time T_drift (see Fig. 22(a)) does not compensate for

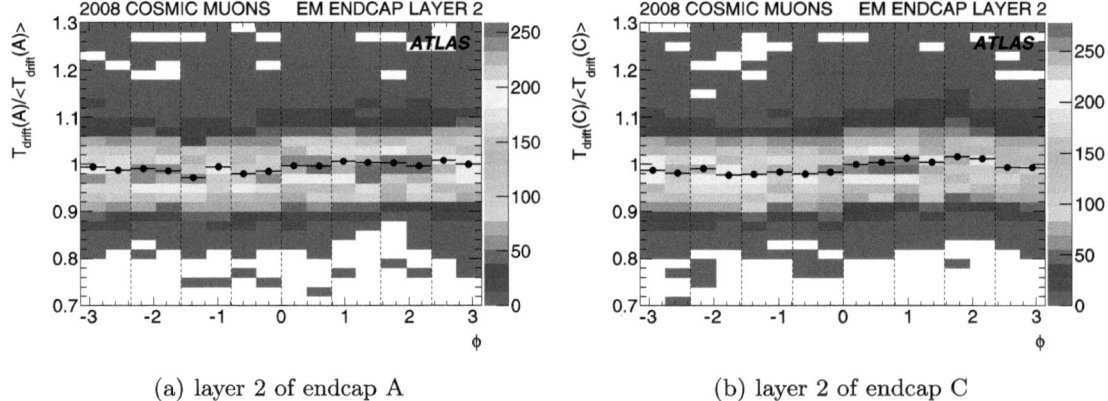

(a) layer 2 of endcap A   (b) layer 2 of endcap C

Fig. 19 Drift time normalized to the average value versus ϕ for layer 2 of the $\eta > 0$ (**a**) and $\eta < 0$ (**b**) endcap wheels. The *black dots* are the average per ϕ bin and the *vertical dashed lines* show the boundaries between different modules

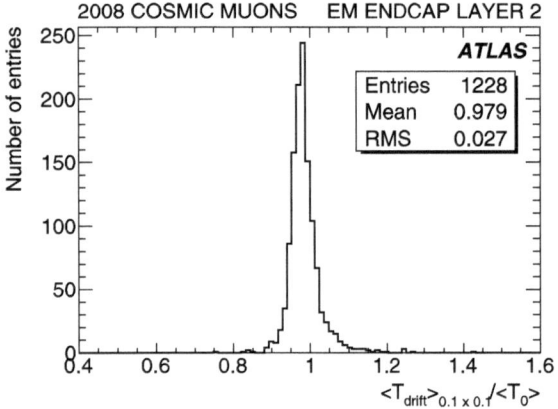

Fig. 20 Drift time uniformity between groups of 4×4 cells ($\Delta\eta \times \Delta\phi = 0.1 \times 0.1$) for endcap layer 2. The normalization $\langle T_0 \rangle$ is obtained as a fit to the data using a first order polynomial in each HV region to cancel out the influence of the gap variation with η

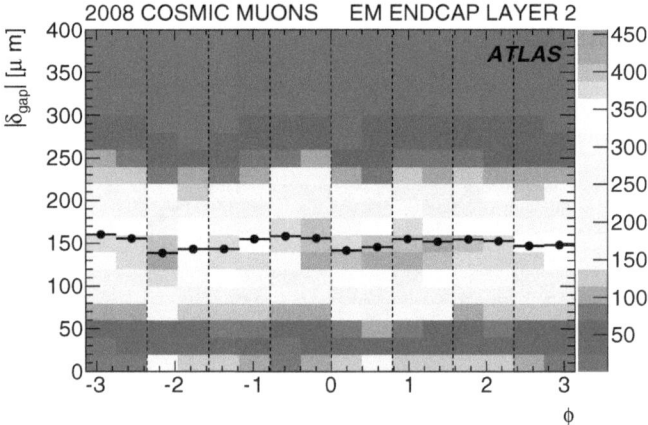

Fig. 21 Electrode shift as function of ϕ for layer 2 of the endcap. The *black dots* are the average per ϕ bin and the *vertical dashed lines* show the boundaries between different modules

the variation of w_{gap} because $T_{\text{drift}} \sim w_{\text{gap}}^{1+\alpha}$. In addition, the different high voltage regions in the endcap introduce steps in the behavior of the drift velocity as a function of η.

In order to compare accurately the drift velocities between barrel and endcap and for each calorimeter layer, they are scaled to a reference field of 1 kV/mm:

$$V_{\text{drift}}(1 \text{ kV/mm}) = \frac{w_{\text{gap}}}{T_{\text{drift}}} \left(\frac{2000 \text{ V} \cdot w_{\text{gap}}}{HV_{\text{nom}} \cdot 2 \text{ mm}} \right)^{\alpha} \qquad (16)$$

where HV_{nom} is the nominal high voltage value, w_{gap} is taken from the design value and α is the exponent introduced in Sect. 2. Figure 22(a) shows the drift velocity at the same field 1 kV/mm for layer 2 of the entire calorimeter as a function of η. As expected, a rather constant behavior is observed over the entire calorimeter. The deviations from a perfect horizontal line is explained by local non-uniformities. Deviations are observed at the transition regions at $\eta = 0$ and

$|\eta| = 0.8$ and in the crack region between barrel and endcap at $|\eta| = 1.4$, where the field is lower.

The temperature in the endcap A ($\eta > 0$) is slightly higher (by about 0.3 K) than the temperatures of the barrel (88.5 K) and endcap C (88.4 K). This can explain the larger drift velocity measured in endcap C ($\eta < 0$) with respect to endcap A, by $\sim 0.6\%$ (see Fig. 22(b)), the expected difference being approximately 0.5%.

Figure 23 shows the comparison of V_{drift} for the different layers of the barrel and endcaps. The mean values of the distributions are also quoted. The errors on these means, given the large number of pulses averaged and the random nature of the noise dominating the error on single measurements, are much smaller than the systematic uncertainties (see Sect. 9). According to (16), the uncertainty in the drift velocity depends on uncertainties in both the gap size and the drift time. The former can be extracted from an azimuthal and pseudorapidity uniformity study, giving values smaller or equal to 1% and 2% for the barrel and endcap

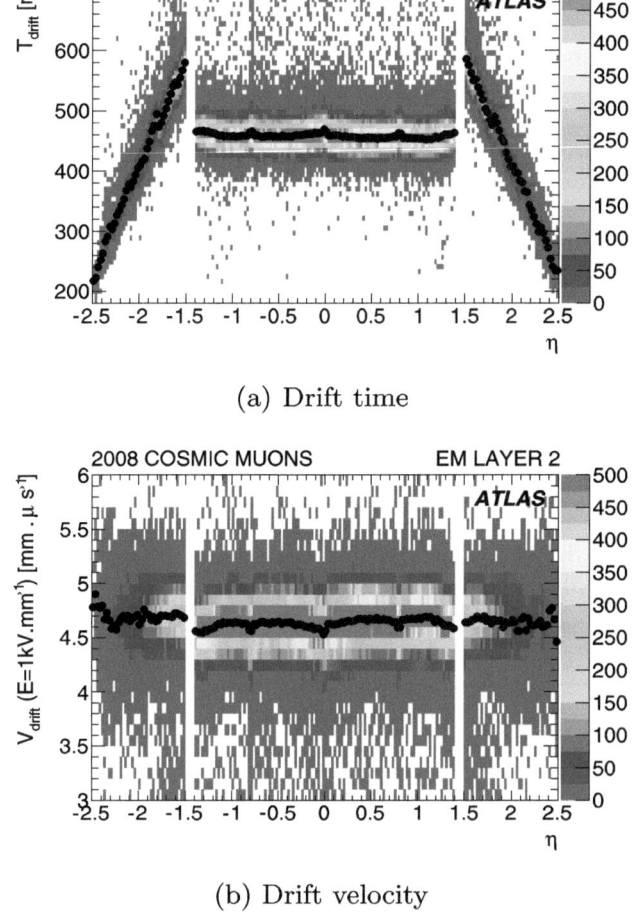

(a) Drift time

(b) Drift velocity

Fig. 22 (a) Drift time and (b) Drift velocity (at $E = 1$ kV/mm) versus η in layer 2. The *black dots* are the average per η bin

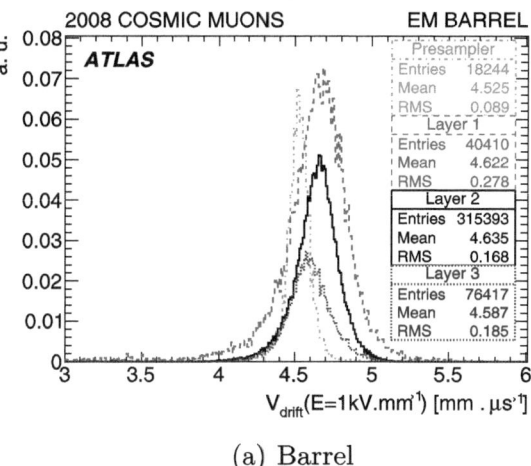

(a) Barrel

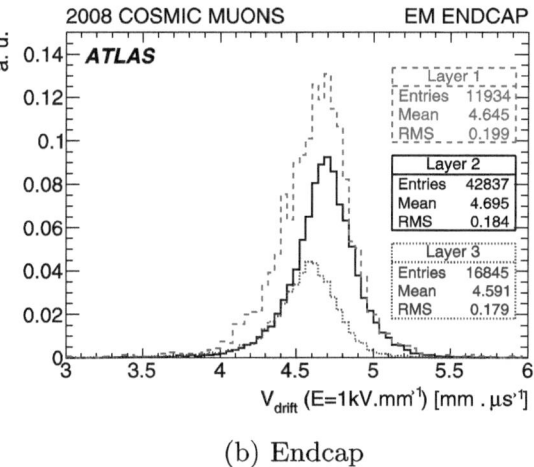

(b) Endcap

Fig. 23 Drift velocity distribution for the barrel (a) and endcap (b)

respectively. The latter receives contributions from several sources (see Sect. 9). The mean values of the drift velocity for the different layers of the barrel and endcap are given in Table 6. They are all compatible within errors, although the barrel presampler is somewhat below the average.

These results can be compared with the measurements from [11] which give (4.65 ± 0.12) mm/µs for a LAr tem-

perature of 88.5 K and provides good agreement with the present measurement.

9 Systematic uncertainties

The different sources of systematic uncertainties affecting the measurement of the drift time which have been studied are discussed below. The resulting systematic uncertainties

Table 6 Drift velocity at $E = 1$ kV/mm in the different layers of the calorimeter

	Layer	Drift velocity (in mm/µs at 1 kV/mm)
Barrel	Presampler	4.52 ± 0.001 (stat) $^{+0.11}_{-0.07}$ (syst)
	Layer 1	4.62 ± 0.003 (stat) $^{+0.06}_{-0.14}$ (syst)
	Layer 2	4.63 ± 0.002 (stat) $^{+0.06}_{-0.14}$ (syst)
	Layer 3	4.59 ± 0.002 (stat) $^{+0.06}_{-0.14}$ (syst)
Endcap	Layer 1	4.65 ± 0.002 (stat) $^{+0.10}_{-0.14}$ (syst)
	Layer 2	4.69 ± 0.001 (stat) $^{+0.10}_{-0.14}$ (syst)
	Layer 3	4.59 ± 0.002 (stat) $^{+0.10}_{-0.14}$ (syst)

on the velocity are given in Table 6, and in Sects. 6.2 and 7.2 for what concerns the uniformity of response.

9.1 Comparison of the results obtained in the barrel with the two prediction methods

Two pulse shape prediction methods have been used for the barrel. Their results are compared to give an estimate of the systematic uncertainty on the prediction. For layers 2 and 3, the mean value of the difference between the predicted distributions is ~0.2 ns and the RMS in the η direction is ~1.2 ns which is of the order of the precision of the measurement for both methods: hence no significant difference is observed for these layers. For layer 1, which also suffers from low statistics, the mean value of the difference (1.3 ns) (see Sect. 6.1.2) is taken as an estimate of the systematic uncertainty associated with the prediction.

9.2 Different fit strategies

In addition to the fit procedure described in this paper, another approach was also followed in layer 2 of the barrel: the cell-based fit. The method consists of fitting simultaneously all the (N) pulses collected in a given cell, using a single value for each of the drift time and the shift parameter, and N global normalization factors and timing adjustments (one of each per pulse). This yields results that are similar but not identical to those obtained from a weighted average of the individual fits with the weight $(S_{\max}^{\text{gain}})^2$. For instance the average drift time in the case of the cell-based fit is 1.2 ns (i.e. 0.3%) lower due to a somewhat reduced effect of pulses with large T_{drift}. With the cell-based method, which has more statistical power for a given fit, the spike at zero visible in Fig. 9 is very much reduced, confirming that it originates from statistical fluctuations of the noise leading to a rising slope around 550 ns steeper than for a single triangle.

9.3 Variation of parameters of the cell

The effect of the uncertainty on the capacitance in layer 2 of the barrel on the FPM determination of the drift time is studied as follows: the capacitance is varied by $\pm5\%$ based on measurements, and a new set of the parameters τ_{sh} and Z_S (defined in Sect. 3) are recalculated from the FPM calibration fits and used in new fits of the cosmic muon data. A small change in the overall drift time scale is observed, but no significant variation in the drift time dependence on η. It should nevertheless be noted that when varying the capacitance in either direction, the drift times increased by about 3 ns. As discussed in [14], an increase (decrease) in the value of the capacitance is partially compensated by a smaller (larger) value of the shaper time constant τ_{sh}, which leads to only minor variations in the pulse shape.

For the RTM method, the estimated uncertainty for the determination of LC and τ_{cali} is less than 3%. The τ_{cali} uncertainty induces an uncertainty of about 0.5% on T_{drift}, with an additional contribution of less than 0.1% coming from the LC uncertainty.

9.4 Effect of electron attachment

In the presence of impurities in the LAr medium, drift electrons may attach to the impurities with an associated lifetime T_{live}, and the signal shape is no longer triangular but has the form:

$$I(t) = \frac{Q_0}{T_{\text{drift}}} e^{-t/T_{\text{live}}} \left(1 - \frac{t}{T_{\text{drift}}}\right) \quad \text{for } 0 < t < T_{\text{drift}}. \quad (17)$$

Using the Fourier transform of $I(t)$, the pulse shape is derived by convolution of the various factors affecting pulse formation and propagation (see [14] for the general case). The data are then fitted with the additional parameter T_{live}.

Although this new parametrization allows to reduce the size of residuals, the values obtained for T_{live} have a large dispersion (about 6 µs for both the average and RMS of the distribution). Another weak point of this description is that the effect is totally absent in the presampler, which is in the same liquid bath as the calorimeter.

A systematic uncertainty of $^{+1.5}_{-0}\%$ in the drift time is conservatively estimated from the difference between the cases of including or not the T_{live} parameter. The η dependence of T_{drift} remains essentially the same in both cases. The drift velocity remains unchanged in the presampler, but is reduced by 1.5 to 2% in the other layers, which would make the presampler and the rest of the barrel more compatible.

While this study was made only in the barrel, the estimated systematic uncertainty is also used for the endcap.

9.5 Variation of the bent triangle contribution

The amount of energy deposited in the bent sections of the calorimeter is estimated using the simulation. To account for possible differences between data and simulation, a systematic uncertainty related to the estimate of the fraction of signal collected in the bent sections is assessed by varying the contribution of the triangle associated with the bends f_{bend} by $\pm20\%$ based on Table 1. This test was done in a limited section of layer 2 of the barrel. The resulting systematic variation of the drift time is ∓3 ns, as if T_{drift} were compensating the absence of the bent triangles. It should be noted that the variations of the drift time with the relative amplitude of the third and fourth triangle (see (4) to (7)) are constant throughout the detector; uncertainties on the contribution from bent sections should therefore not affect the estimate of the intrinsic uniformity, except in layer 1 (see Sect. 6.1.2).

The procedure to estimate T_{bend} and f_{bend} in the endcap requires that the contributions from the bent and straight parts of the accordion can be separated using the local drift time distribution of simulated 10 GeV photon showers. The uncertainty induced by this procedure is propagated to the final T_{drift} value, leading to a 0.2% variation that is compatible with the precision of the measurement.

9.6 Variation of the parameter α

The effect of the uncertainty on the exponent α in the determination of the drift time in the endcap was studied by varying α in the range from 0.30 to 0.39 larger than the range determined in Sect. 2 (0.28 to 0.34). This larger range was initially motivated by a previous measurement of this exponent during the beam test of the endcap prototype using 120 GeV electrons, where a value of 0.39 seemed to describe the data better, however over a larger electrical field range than relevant here. The effect of this difference (0.30 to 0.39) on the drift time is approximately 1 ns or about 0.2%, again at the level of precision of the measurement. The effect on the retained range (0.28 to 0.34) would be even smaller. The arithmetic effect of the uncertainty on α on the uniformity was considered in Sects. 6.2 and 7.2.

9.7 Summary of the systematic uncertainties

The systematic uncertainties discussed above apply to the drift time measurements and can be translated into drift velocity through (1). The drift velocities for each layer are summarized in Table 6.

Averaging over the presampler and layer 2 (barrel and endcap) values, for which most of the systematics are uncorrelated, gives as the final result for the reference field of 1 kV/mm and a temperature $T = 88.5$ K:

$$V_{\text{ref}} = (4.61 \pm 0.07) \text{ mm/μs}. \tag{18}$$

10 Direct determination of local gap and drift velocity at operating point

Taking advantage of the studies presented above, a somewhat more global treatment of the data is presented below, which allows:

- To unify the comparison of the local measured gaps, and their reference value from construction in both the barrel and the endcaps.
- To obtain for the whole calorimeter the values of the drift velocity at the local operating points.

If the drift velocity were to be fully saturated, i.e. independent of the electric field, a measurement of the drift time would trivially give the associated local gap using (1). In the situation analyzed here, the drift velocity depends weakly on

the electric field, with a power law already given in Sect. 2 (see (3)).

Using (1) and (3) rewritten below as

$$V_{\text{drift}} = V_{\text{ref}} \cdot \left[\frac{HV}{HV_0} \cdot \frac{w_{\text{gap0}}}{w_{\text{gap}}} \right]^{\alpha} \tag{19}$$

it is possible to express both the local velocity and the local gap, as functions of the measured T_{drift}:

$$w_{\text{gap}} = [A \cdot T_{\text{drift}}]^{1/(1+\alpha)}, \tag{20}$$

$$V_{\text{drift}} = \frac{A^{1/(1+\alpha)}}{T_{\text{drift}}^{\alpha/(1+\alpha)}} \tag{21}$$

with $A = V_{\text{ref}} \cdot [\frac{HV}{HV_0}]^{\alpha} \cdot w_{\text{gap0}}^{\alpha}$. The analysis presented below uses: $\alpha = 0.3$, $w_{\text{gap0}} = 2$ mm, $HV_0 = 2$ kV and normalizes the drift velocity at 1 kV/mm to the average $V_{\text{ref}} = 4.61$ mm/μs, as reported in Sect. 9. The effect of the shift ($x \sim 0.1$) was estimated to bias the above analysis by less than 0.2% on the extracted gap value, and is therefore neglected. Data for the endcaps have been corrected for the temperature difference, and rescaled to 88.5 K.

The additional information yielded by this analysis shows directly how the ratio of the measured gap to the designed gap varies as a function of position in the detector. Figure 24 shows the relative difference between the calculated and design values. The average difference is not exactly 0. This comes from the fact that the average velocity value used for the normalization includes presampler data, while the gap calculation presented in Fig. 24 contains only layer 2.

One can see that the ratio between calculated and design values, spanning a gap range between 1 mm and 2.5 mm, has an RMS of 0.83%, i.e. typically 16 μm. In the presampler, the corresponding dispersion is 7 μm, reflecting a more rigid fixing of the electrodes defining the gaps. In the barrel part one recognizes the systematic effects in the results discussed in Sect. 6.1.1 (see in particular Fig. 10) associated with the

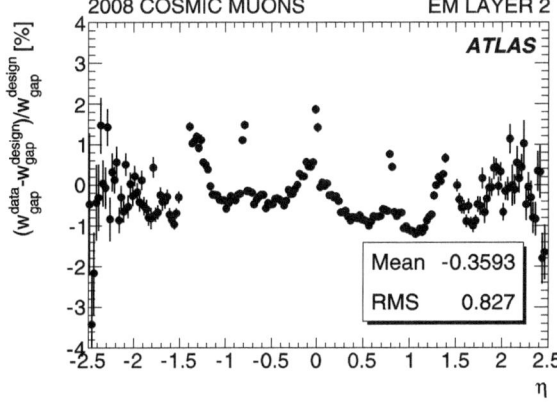

Fig. 24 Relative difference between the design gap values and the values extracted from T_{drift} measurements

slight bulging of the absorbers, and the "transition regions" at $\eta = 0$, ± 0.8 and ± 1.4. Strictly speaking these transitions areas, for which additional effects enter into play, should be corrected for in the calculation of the RMS. In the endcaps the statistical power is unfortunately lower giving rise to larger fluctuations, but no significant trend is observed.

Figure 25 shows the drift velocity obtained using (21) as a function of pseudorapidity and the same normalization as above. As opposed to Fig. 22, which gave the velocity at a reference field of 1 kV/mm, Fig. 25 shows the drift velocity at the local operating field, which is directly related to the peak current (see Sect. 2) associated with an energy deposition.

In the barrel region, the drift velocity is essentially flat, with a slight modulation reflecting the variation of the absorber thickness with pseudorapidity. Taking the average value of the drift velocity in sectors of $\Delta\eta \times \Delta\phi = 0.1 \times 0.1$, as was done in Sect. 6.2 for T_{drift}, one obtains a distribution with an RMS of 0.29% exactly equal to what was derived in Sect. 6.2 from the RMS of the T_{drift} distribution, showing the expected consistency of the analyses using T_{drift} or V_{drift}.

In the endcap region, one observes the 6 sawteeth on each side resulting from the finite granularity of the HV distribution (see Fig. 3). Corrections are made in the energy reconstruction to normalize the response of each strip in pseudorapidity to the response of the strip in the center of the HV sector, using the power law dependence. Beside these modulations, one observes that:

– The average velocity in the endcaps is smaller than in the barrel. In the energy reconstruction this is accounted for by correction factors (which also take into account the fact that the lead thicknesses are different) determined from test beam and implemented in the detailed Monte Carlo simulation of the full ATLAS detector.
– The measured velocity averaged over an HV sector somewhat diminishes with increasing pseudorapidity. This effect goes in the same direction (lowering the response)

as the reduced contribution of liquid argon to showering/conversion effects at large pseudorapidities (small gaps). Both effects are qualitatively counterbalanced by the fact that the relative contribution of bends as compared to flat parts is lower at high pseudorapidity, resulting in an increased response. As already mentioned, detailed Monte Carlo simulations normalized with test beam scans were used to determine the HV values optimizing the uniformity of response of the endcaps. This will be cross checked when enough $Z^0 \rightarrow e^+ e^-$ decays become available.

11 Conclusions

We have shown in this paper that sufficient amounts of ionization data (~ 0.5 million pulses of energy larger than ~ 1 GeV) can be used for a precision measurement of the average electron drift time in each cell of the highly granular LAr electromagnetic calorimeter of ATLAS that has been readout with fast electronics, in the current mode. In this regime, the recorded energy is directly proportional to the drift velocity of ionization electrons, which is readily obtained from the drift time measurement. Furthermore, the drift velocity and thus the recorded energy are ~ 4 times less sensitive to gap variations than the drift time.

Taking advantage of these facts, we derived an estimate of the calorimeter non-uniformity of response due to gap size variations, of $(0.29^{+0.05}_{-0.04})$ and $(0.54^{+0.06}_{-0.04})\%$ respectively for the barrel and the endcaps. The other main contribution to the intrinsic non-uniformity of the calorimeter is the dispersion of the thickness of the lead absorbers which contributes 0.18% for both barrel and endcaps [2, 3].

The drift time is also an input needed in order to reconstruct the signal amplitude by optimal filtering. An examination of the tails of the drift time distributions singles out "transition areas" of the calorimeter, in both azimuthal or pseudorapidity angle, where the electrical field is lower than average due to "edge effects". Some modulations in the third layer of the barrel have also been observed.

The analysis method used to derive the drift time provides as another parameter the average absolute value of the amount the electrodes are off center between their two neighboring absorbers. The values obtained are around 146 µm for both barrel and endcap accordion layers, and are substantially smaller for the presampler (66.5 µm) as expected from its design.

The drift velocity, rescaled to a field of 1 kV/mm, is obtained from the drift time measurements leading to an average of (4.61 ± 0.07) mm/µs. This value is compatible with previously published measurements at the same operating temperature of 88.5 K.

The measurements presented in this paper illustrate the accuracy achieved with this method even using cosmic

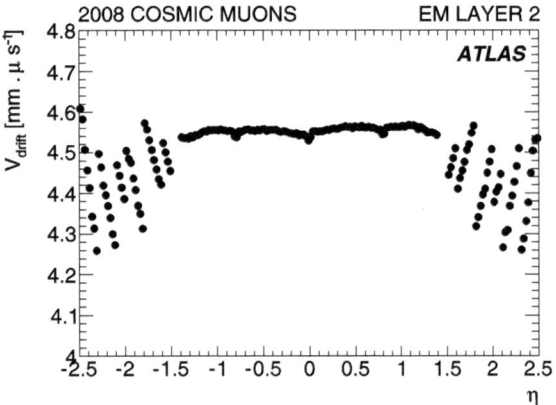

Fig. 25 Drift velocity versus η in the layer 2 at the operating point extracted from T_{drift} measurements

muon data, thus demonstrating that it can be used to correct for the measured gap variations in order to eventually reduce the constant term of the energy resolution, especially if the measurements are repeated with collision data. It is therefore important, in the quest to improve the energy resolution constant term, that in the future these measurements be done with LHC collision data.

Acknowledgements We are greatly indebted to all CERN's departments and to the LHC project for their immense efforts not only in building the LHC, but also for their direct contributions to the construction and installation of the ATLAS detector and its infrastructure. We acknowledge equally warmly all our technical colleagues in the collaborating institutions without whom the ATLAS detector could not have been built. Furthermore we are grateful to all the funding agencies which supported generously the construction and the commissioning of the ATLAS detector and also provided the computing infrastructure.

The ATLAS detector design and construction has taken about fifteen years, and our thoughts are with all our colleagues who sadly could not see its final realization.

We acknowledge the support of ANPCyT, Argentina; Yerevan Physics Institute, Armenia; ARC and DEST, Australia; Bundesministerium für Wissenschaft und Forschung, Austria; National Academy of Sciences of Azerbaijan; State Committee on Science & Technologies of the Republic of Belarus; CNPq and FINEP, Brazil; NSERC, NRC, and CFI, Canada; CERN; NSFC, China; Ministry of Education, Youth and Sports of the Czech Republic, Ministry of Industry and Trade of the Czech Republic, and Committee for Collaboration of the Czech Republic with CERN; Danish Natural Science Research Council; European Commission, through the ARTEMIS Research Training Network; IN2P3-CNRS and Dapnia-CEA, France; Georgian Academy of Sciences; BMBF, DESY, DFG and MPG, Germany; Ministry of Education and Religion, through the EPEAEK program PYTHAGORAS II and GSRT, Greece; ISF, MINERVA, GIF, DIP, and Benoziyo Center, Israel; INFN, Italy; MEXT, Japan; CNRST, Morocco; FOM and NWO, Netherlands; The Research Council of Norway; Ministry of Science and Higher Education, Poland; GRICES and FCT, Portugal; Ministry of Education and Research, Romania; Ministry of Education and Science of the Russian Federation, Russian Federal Agency of Science and Innovations, and Russian Federal Agency of Atomic Energy; JINR; Ministry of Science, Serbia; Department of International Science and Technology Cooperation, Ministry of Education of the Slovak Republic; Slovenian Research Agency, Ministry of Higher Education, Science and Technology, Slovenia; Ministerio de Educación y Ciencia, Spain; The Swedish Research Council, The Knut and Alice Wallenberg Foundation, Sweden; State Secretariat for Education and Science, Swiss National Science Foundation, and Cantons of Bern and Geneva, Switzerland; National Science Council, Taiwan; TAEK, Turkey; The Science and Technology Facilities Council and The Leverhulme Trust, United Kingdom; DOE and NSF, United States of America.

References

1. ATLAS Collaboration, The ATLAS experiment at the CERN large hadron collider. JINST **3**, S08003 (2008)
2. B. Aubert et al., Construction, assembly and tests of the ATLAS electromagnetic barrel calorimeter. Nucl. Instrum. Methods A **558**, 388 (2006)
3. M. Aleksa et al., Construction, assembly and tests of the ATLAS electromagnetic end-cap calorimeter. JINST **3**, P06002 (2008)
4. M.L. Andrieux et al., Construction and test of the first two sectors of the ATLAS liquid argon presampler. Nucl. Instrum. Methods A **479**, 316 (2002)
5. ATLAS Collaboration, Readiness of the ATLAS liquid argon calorimeter for LHC collisions, submitted for publication in EPJC. arXiv:0912.2642v2
6. ATLAS Collaboration, ATLAS liquid argon calorimeter. Technical design report, ATLAS-TDR-002, CERN-LHCC-96-41
7. N.J. Buchanan et al., ATLAS liquid argon calorimeter front end electronics. JINST **3**, P02010 (2008)
8. A. Bazan et al., ATLAS liquid argon calorimeter back end electronics (RODs). JINST **2**, P06002 (2007)
9. L.S. Miller et al., Charge transport in solid and liquid Ar, Kr and Xe. Phys. Rev. **166**, 871–878 (1968)
10. K. Yoshino et al., Effect of molecular solutes on the electron drift velocity in liquid Ar, Kr, and Xe. Phys. Rev. A **14**, 438–444 (1976)
11. W. Walkowiak, Drift velocity of free electrons in liquid argon. Nucl. Instrum. Methods A **449**, 288–294 (2000)
12. W.E. Cleland, E.G. Stern, Signal processing considerations for liquid ionization calorimeter in a high rate environment. Nucl. Instrum. Methods A **338**, 467 (1994)
13. D. Banfi, M. Delmastro, M. Fanti, Cell response equalization of the ATLAS electromagnetic calorimeter without the direct knowledge of the ionization signals. JINST **1**, P08001 (2006)
14. C. Collard, D. Fournier, S. Henrot-Versillé, L. Serin, Prediction of signal amplitude and shape for the ATLAS electromagnetic calorimeter. ATL-LARG-PUB-2007-010
15. M. Frigo, S.G. Johnson, The design and implementation of FFTW3. Proc. IEEE 93 **2**, 216–231 (2005)
16. J. Colas et al., Electronics calibration board for the ATLAS liquid argon calorimeters. Nucl. Instrum. Methods A **593**, 269 (2008)
17. S. Baffioni et al., Electrical measurements on the ATLAS electromagnetic barrel calorimeter. ATL-LARG-PUB-2007-005
18. F. James, M. Roos, Minuit: a system for function minimization and analysis of the parameter errors and correlations. Comput. Phys. Commun. **10**, 343–367 (1975)

Commissioning of the ATLAS Muon Spectrometer with cosmic rays

The ATLAS Collaboration[*,]**

G. Aad[48], B. Abbott[111], J. Abdallah[11], A.A. Abdelalim[49], A. Abdesselam[118], O. Abdinov[10], B. Abi[112], M. Abolins[88], H. Abramowicz[152], H. Abreu[115], B.S. Acharya[163a,163b], D.L. Adams[24], T.N. Addy[56], J. Adelman[174], C. Adorisio[36a,36b], P. Adragna[75], T. Adye[129], S. Aefsky[22], J.A. Aguilar-Saavedra[124b], M. Aharrouche[81], S.P. Ahlen[21], F. Ahles[48], A. Ahmad[147], H. Ahmed[2], M. Ahsan[40], G. Aielli[133a,133b], T. Akdogan[18a], T.P.A. Åkesson[79], G. Akimoto[154], A.V. Akimov[94], A. Aktas[48], M.S. Alam[1], M.A. Alam[76], S. Albrand[55], M. Aleksa[29], I.N. Aleksandrov[65], C. Alexa[25a], G. Alexander[152], G. Alexandre[49], T. Alexopoulos[9], M. Alhroob[20], M. Aliev[15], G. Alimonti[89a], J. Alison[120], M. Aliyev[10], P.P. Allport[73], S.E. Allwood-Spiers[53], J. Almond[82], A. Aloisio[102a,102b], R. Alon[170], A. Alonso[79], M.G. Alviggi[102a,102b], K. Amako[66], C. Amelung[22], A. Amorim[124a], G. Amorós[166], N. Amram[152], C. Anastopoulos[139], T. Andeen[29], C.F. Anders[48], K.J. Anderson[30], A. Andreazza[89a,89b], V. Andrei[58a], X.S. Anduaga[70], A. Angerami[34], F. Anghinolfi[29], N. Anjos[124a], A. Annovi[47], A. Antonaki[8], M. Antonelli[47], S. Antonelli[19a,19b], J. Antos[144b], B. Antunovic[41], F. Anulli[132a], S. Aoun[83], G. Arabidze[8], I. Aracena[143], Y. Arai[66], A.T.H. Arce[14], J.P. Archambault[28], S. Arfaoui[29,a], J.-F. Arguin[14], T. Argyropoulos[9], M. Arik[18a], A.J. Armbruster[87], O. Arnaez[4], C. Arnault[115], A. Artamonov[95], D. Arutinov[20], M. Asai[143], S. Asai[154], R. Asfandiyarov[171], S. Ask[82], B. Åsman[145a,145b], D. Asner[28], L. Asquith[77], K. Assamagan[24], A. Astbury[168], A. Astvatsatourov[52], G. Atoian[174], B. Auerbach[174], K. Augsten[127], M. Aurousseau[4], N. Austin[73], G. Avolio[162], R. Avramidou[9], D. Axen[167], C. Ay[54], G. Azuelos[93,b], Y. Azuma[154], M.A. Baak[29], A.M. Bach[14], H. Bachacou[136], K. Bachas[29], M. Backes[49], E. Badescu[25a], P. Bagnaia[132a,132b], Y. Bai[32a], T. Bain[157], J.T. Baines[129], O.K. Baker[174], M.D. Baker[24], S. Baker[77], F. Baltasar Dos Santos Pedrosa[29], E. Banas[38], P. Banerjee[93], S. Banerjee[168], D. Banfi[89a,89b], A. Bangert[137], V. Bansal[168], S.P. Baranov[94], S. Baranov[65], A. Barashkou[65], T. Barber[27], E.L. Barberio[86], D. Barberis[50a,50b], M. Barbero[20], D.Y. Bardin[65], T. Barillari[99], M. Barisonzi[173], T. Barklow[143], N. Barlow[27], B.M. Barnett[129], R.M. Barnett[14], A. Baroncelli[134a], A.J. Barr[118], F. Barreiro[80], J. Barreiro Guimarães da Costa[57], P. Barrillon[115], R. Bartoldus[143], D. Bartsch[20], R.L. Bates[53], L. Batkova[144a], J.R. Batley[27], A. Battaglia[16], M. Battistin[29], F. Bauer[136], H.S. Bawa[143], M. Bazalova[125], B. Beare[157], T. Beau[78], P.H. Beauchemin[118], R. Beccherle[50a], N. Becerici[18a], P. Bechtle[41], G.A. Beck[75], H.P. Beck[16], M. Beckingham[48], K.H. Becks[173], A.J. Beddall[18c], A. Beddall[18c], V.A. Bednyakov[65], C. Bee[83], M. Begel[24], S. Behar Harpaz[151], P.K. Behera[63], M. Beimforde[99], C. Belanger-Champagne[165], P.J. Bell[49], W.H. Bell[49], G. Bella[152], L. Bellagamba[19a], F. Bellina[29], M. Bellomo[119a], A. Belloni[57], K. Belotskiy[96], O. Beltramello[29], S. Ben Ami[151], O. Benary[152], D. Benchekroun[135a], M. Bendel[81], B.H. Benedict[162], N. Benekos[164], Y. Benhammou[152], G.P. Benincasa[124a], D.P. Benjamin[44], M. Benoit[115], J.R. Bensinger[22], K. Benslama[130], S. Bentvelsen[105], M. Beretta[47], D. Berge[29], E. Bergeaas Kuutmann[41], N. Berger[4], F. Berghaus[168], E. Berglund[49], J. Beringer[14], P. Bernat[115], R. Bernhard[48], C. Bernius[77], T. Berry[76], A. Bertin[19a,19b], M.I. Besana[89a,89b], N. Besson[136], S. Bethke[99], R.M. Bianchi[48], M. Bianco[72a,72b], O. Biebel[98], J. Biesiada[14], M. Biglietti[132a,132b], H. Bilokon[47], M. Bindi[19a,19b], S. Binet[115], A. Bingul[18c], C. Bini[132a,132b], C. Biscarat[179], U. Bitenc[48], K.M. Black[57], R.E. Blair[5], J.-B. Blanchard[115], G. Blanchot[29], C. Blocker[22], A. Blondel[49], W. Blum[81], U. Blumenschein[54], G.J. Bobbink[105], A. Bocci[44], M. Boehler[41], J. Boek[173], N. Boelaert[79], S. Böser[77], J.A. Bogaerts[29], A. Bogouch[90,†], C. Bohm[145a], J. Bohm[125], V. Boisvert[76], T. Bold[162,c], V. Boldea[25a], V.G. Bondarenko[96], M. Bondioli[162], M. Boonekamp[136], S. Bordoni[78], C. Borer[16], A. Borisov[128], G. Borissov[71], I. Borjanovic[72a], S. Borroni[132a,132b], K. Bos[105], D. Boscherini[19a], M. Bosman[11], H. Boterenbrood[105], J. Bouchami[93], J. Boudreau[123], E.V. Bouhova-Thacker[71], C. Boulahouache[123], C. Bourdarios[115], A. Boveia[30], J. Boyd[29], I.R. Boyko[65], I. Bozovic-Jelisavcic[12b], J. Bracinik[17], A. Braem[29], P. Branchini[134a], G.W. Brandenburg[57], A. Brandt[7], G. Brandt[41], O. Brandt[54], U. Bratzler[155], B. Brau[84], J.E. Brau[114], H.M. Braun[173], B. Brelier[157], J. Bremer[29], R. Brenner[165], S. Bressler[151], D. Britton[53], F.M. Brochu[27], I. Brock[20], R. Brock[88], E. Brodet[152], C. Bromberg[88], G. Brooijmans[34], W.K. Brooks[31b], G. Brown[82], P.A. Bruckman de Renstrom[38], D. Bruncko[144b], R. Bruneliere[48], S. Brunet[41], A. Bruni[19a], G. Bruni[19a], M. Bruschi[19a], F. Bucci[49], J. Buchanan[118], P. Buchholz[141], A.G. Buckley[45], I.A. Budagov[65], B. Budick[108],

V. Büscher[81], L. Bugge[117], O. Bulekov[96], M. Bunse[42], T. Buran[117], H. Burckhart[29], S. Burdin[73], T. Burgess[13], S. Burke[129], E. Busato[33], P. Bussey[53], C.P. Buszello[165], F. Butin[29], B. Butler[143], J.M. Butler[21], C.M. Buttar[53], J.M. Butterworth[77], T. Byatt[77], J. Caballero[24], S. Cabrera Urbán[166], D. Caforio[19a,19b], O. Cakir[3a], P. Calafiura[14], G. Calderini[78], P. Calfayan[98], R. Calkins[106a], L.P. Caloba[23a], D. Calvet[33], P. Camarri[133a,133b], D. Cameron[117], S. Campana[29], M. Campanelli[77], V. Canale[102a,102b], F. Canelli[30], A. Canepa[158a], J. Cantero[80], L. Capasso[102a,102b], M.D.M. Capeans Garrido[29], I. Caprini[25a], M. Caprini[25a], M. Capua[36a,36b], R. Caputo[147], C. Caramarcu[25a], R. Cardarelli[133a], T. Carli[29], G. Carlino[102a], L. Carminati[89a,89b], B. Caron[2,b], S. Caron[48], G.D. Carrillo Montoya[171], S. Carron Montero[157], A.A. Carter[75], J.R. Carter[27], J. Carvalho[124a], D. Casadei[108], M.P. Casado[11], M. Cascella[122a,122b], A.M. Castaneda Hernandez[171], E. Castaneda-Miranda[171], V. Castillo Gimenez[166], N.F. Castro[124b], G. Cataldi[72a], A. Catinaccio[29], J.R. Catmore[71], A. Cattai[29], G. Cattani[133a,133b], S. Caughron[34], D. Cauz[163a,163c], P. Cavalleri[78], D. Cavalli[89a], M. Cavalli-Sforza[11], V. Cavasinni[122a,122b], F. Ceradini[134a,134b], A.S. Cerqueira[23a], A. Cerri[29], L. Cerrito[75], F. Cerutti[47], S.A. Cetin[18b], A. Chafaq[135a], D. Chakraborty[106a], K. Chan[2], J.D. Chapman[27], J.W. Chapman[87], E. Chareyre[78], D.G. Charlton[17], V. Chavda[82], S. Cheatham[71], S. Chekanov[5], S.V. Chekulaev[158a], G.A. Chelkov[65], H. Chen[24], S. Chen[32c], X. Chen[171], A. Cheplakov[65], V.F. Chepurnov[65], R. Cherkaoui El Moursli[135d], V. Tcherniatine[24], D. Chesneanu[25a], E. Cheu[6], S.L. Cheung[157], L. Chevalier[136], F. Chevallier[136], V. Chiarella[47], G. Chiefari[102a,102b], L. Chikovani[51], J.T. Childers[58a], A. Chilingarov[71], G. Chiodini[72a], V. Chizhov[65], G. Choudalakis[30], S. Chouridou[137], I.A. Christidi[77], A. Christov[48], D. Chromek-Burckhart[29], M.L. Chu[150], J. Chudoba[125], G. Ciapetti[132a,132b], A.K. Ciftci[3a], R. Ciftci[3a], D. Cinca[33], V. Cindro[74], M.D. Ciobotaru[162], C. Ciocca[19a,19b], A. Ciocio[14], M. Cirilli[87], M. Citterio[89a], A. Clark[49], P.J. Clark[45], W. Cleland[123], J.C. Clemens[83], B. Clement[55], C. Clement[145a,145b], Y. Coadou[83], M. Cobal[163a,163c], A. Coccaro[50a,50b], J. Cochran[64], J. Coggeshall[164], E. Cogneras[179], A.P. Colijn[105], C. Collard[115], N.J. Collins[17], C. Collins-Tooth[53], J. Collot[55], G. Colon[84], P. Conde Muiño[124a], E. Coniavitis[165], M. Consonni[104], S. Constantinescu[25a], C. Conta[119a,119b], F. Conventi[102a,d], M. Cooke[34], B.D. Cooper[75], A.M. Cooper-Sarkar[118], N.J. Cooper-Smith[76], K. Copic[34], T. Cornelissen[50a,50b], M. Corradi[19a], F. Corriveau[85,e], A. Corso-Radu[162], A. Cortes-Gonzalez[164], G. Cortiana[99], G. Costa[89a], M.J. Costa[166], D. Costanzo[139], T. Costin[30], D. Côté[41], R. Coura Torres[23a], L. Courneyea[168], G. Cowan[76], C. Cowden[27], B.E. Cox[82], K. Cranmer[108], J. Cranshaw[5], M. Cristinziani[20], G. Crosetti[36a,36b], R. Crupi[72a,72b], S. Crépe-Renaudin[55], C .Cuenca Almenar[174], T. Cuhadar Donszelmann[139], M. Curatolo[47], C.J. Curtis[17], P. Cwetanski[61], Z. Czyczula[174], S. D'Auria[53], M. D'Onofrio[73], A. D'Orazio[99], C. Da Via[82], W. Dabrowski[37], T. Dai[87], C. Dallapiccola[84], S.J. Dallison[129,*], C.H. Daly[138], M. Dam[35], H.O. Danielsson[29], D. Dannheim[99], V. Dao[49], G. Darbo[50a], G.L. Darlea[25b], W. Davey[86], T. Davidek[126], N. Davidson[86], R. Davidson[71], M. Davies[93], A.R. Davison[77], I. Dawson[139], R.K. Daya[39], K. De[7], R. de Asmundis[102a], S. De Castro[19a,19b], P.E. De Castro Faria Salgado[24], S. De Cecco[78], J. de Graat[98], N. De Groot[104], P. de Jong[105], L. De Mora[71], M. De Oliveira Branco[29], D. De Pedis[132a], A. De Salvo[132a], U. De Sanctis[163a,163c], A. De Santo[148], J.B. De Vivie De Regie[115], G. De Zorzi[132a,132b], S. Dean[77], D.V. Dedovich[65], J. Degenhardt[120], M. Dehchar[118], C. Del Papa[163a,163c], J. Del Peso[80], T. Del Prete[122a,122b], A. Dell'Acqua[29], L. Dell'Asta[89a,89b], M. Della Pietra[102a,d], D. della Volpe[102a,102b], M. Delmastro[29], P.A. Delsart[55], C. Deluca[147], S. Demers[174], M. Demichev[65], B. Demirkoz[11], J. Deng[162], W. Deng[24], S.P. Denisov[128], J.E. Derkaoui[135c], F. Derue[78], P. Dervan[73], K. Desch[20], P.O. Deviveiros[157], A. Dewhurst[129], B. DeWilde[147], S. Dhaliwal[157], R. Dhullipudi[24,f], A. Di Ciaccio[133a,133b], L. Di Ciaccio[4], A. Di Domenico[132a,132b], A. Di Girolamo[29], B. Di Girolamo[29], S. Di Luise[134a,134b], A. Di Mattia[88], R. Di Nardo[133a,133b], A. Di Simone[133a,133b], R. Di Sipio[19a,19b], M.A. Diaz[31a], F. Diblen[18c], E.B. Diehl[87], J. Dietrich[48], T.A. Dietzsch[58a], S. Diglio[115], K. Dindar Yagci[39], J. Dingfelder[48], C. Dionisi[132a,132b], P. Dita[25a], S. Dita[25a], F. Dittus[29], F. Djama[83], R. Djilkibaev[108], T. Djobava[51], M.A.B. do Vale[23a], A. Do Valle Wemans[124a], T.K.O. Doan[4], D. Dobos[29], E. Dobson[29], M. Dobson[162], C. Doglioni[118], T. Doherty[53], J. Dolejsi[126], I. Dolenc[74], Z. Dolezal[126], B.A. Dolgoshein[96], T. Dohmae[154], M. Donega[120], J. Donini[55], J. Dopke[173], A. Doria[102a], A. Dos Anjos[171], A. Dotti[122a,122b], M.T. Dova[70], A. Doxiadis[105], A.T. Doyle[53], Z. Drasal[126], M. Dris[9], J. Dubbert[99], E. Duchovni[170], G. Duckeck[98], A. Dudarev[29], F. Dudziak[115], M. Dührssen[29], L. Duflot[115], M.-A. Dufour[85], M. Dunford[30], H. Duran Yildiz[3b], A. Dushkin[22], R. Duxfield[139], M. Dwuznik[37], M. Düren[52], W.L. Ebenstein[44], J. Ebke[98], S. Eckweiler[81], K. Edmonds[81], C.A. Edwards[76], K. Egorov[61], W. Ehrenfeld[41], T. Ehrich[99], T. Eifert[29], G. Eigen[13], K. Einsweiler[14], E. Eisenhandler[75], T. Ekelof[165], M. El Kacimi[4], M. Ellert[165], S. Elles[4], F. Ellinghaus[81], K. Ellis[75], N. Ellis[29], J. Elmsheuser[98], M. Elsing[29], D. Emeliyanov[129], R. Engelmann[147], A. Engl[98], B. Epp[62], A. Eppig[87], J. Erdmann[54], A. Ereditato[16], D. Eriksson[145a], I. Ermoline[88], J. Ernst[1], M. Ernst[24], J. Ernwein[136], D. Errede[164], S. Errede[164], E. Ertel[81], M. Escalier[115], C. Escobar[166], X. Espinal Curull[11], B. Esposito[47], A.I. Etienvre[136], E. Etzion[152],

H. Evans[61], L. Fabbri[19a,19b], C. Fabre[29], K. Facius[35], R.M. Fakhrutdinov[128], S. Falciano[132a], Y. Fang[171],
M. Fanti[89a,89b], A. Farbin[7], A. Farilla[134a], J. Farley[147], T. Farooque[157], S.M. Farrington[118], P. Farthouat[29],
P. Fassnacht[29], D. Fassouliotis[8], B. Fatholahzadeh[157], L. Fayard[115], F. Fayette[54], R. Febbraro[33], P. Federic[144a],
O.L. Fedin[121], W. Fedorko[29], L. Feligioni[83], C.U. Felzmann[86], C. Feng[32d], E.J. Feng[30], A.B. Fenyuk[128],
J. Ferencei[144b], J. Ferland[93], B. Fernandes[124a], W. Fernando[109], S. Ferrag[53], J. Ferrando[118], V. Ferrara[41],
A. Ferrari[165], P. Ferrari[105], R. Ferrari[119a], A. Ferrer[166], M.L. Ferrer[47], D. Ferrere[49], C. Ferretti[87], M. Fiascaris[118],
F. Fiedler[81], A. Filipčič[74], A. Filippas[9], F. Filthaut[104], M. Fincke-Keeler[168], M.C.N. Fiolhais[124a], L. Fiorini[11],
A. Firan[39], G. Fischer[41], M.J. Fisher[109], M. Flechl[165], I. Fleck[141], J. Fleckner[81], P. Fleischmann[172],
S. Fleischmann[20], T. Flick[173], L.R. Flores Castillo[171], M.J. Flowerdew[99], T. Fonseca Martin[76], A. Formica[136],
A. Forti[82], D. Fortin[158a], D. Fournier[115], A.J. Fowler[44], K. Fowler[137], H. Fox[71], P. Francavilla[122a,122b],
S. Franchino[119a,119b], D. Francis[29], M. Franklin[57], S. Franz[29], M. Fraternali[119a,119b], S. Fratina[120], J. Freestone[82],
S.T. French[27], R. Froeschl[29], D. Froidevaux[29], J.A. Frost[27], C. Fukunaga[155], E. Fullana Torregrosa[5], J. Fuster[166],
C. Gabaldon[80], O. Gabizon[170], T. Gadfort[24], S. Gadomski[49], G. Gagliardi[50a,50b], P. Gagnon[61], C. Galea[98],
E.J. Gallas[118], V. Gallo[16], B.J. Gallop[129], P. Gallus[125], P. Galyaev[40], K.K. Gan[109], Y.S. Gao[143,g], A. Gaponenko[14],
M. Garcia-Sciveres[14], C. García[166], J.E. García Navarro[49], R.W. Gardner[30], N. Garelli[29], H. Garitaonandia[105],
V. Garonne[29], C. Gatti[47], G. Gaudio[119a], V. Gautard[136], P. Gauzzi[132a,132b], I.L. Gavrilenko[94], C. Gay[167],
G. Gaycken[20], E.N. Gazis[9], P. Ge[32d], C.N.P. Gee[129], Ch. Geich-Gimbel[20], K. Gellerstedt[145a,145b], C. Gemme[50a],
M.H. Genest[98], S. Gentile[132a,132b], F. Georgatos[9], S. George[76], A. Gershon[152], H. Ghazlane[135d], N. Ghodbane[33],
B. Giacobbe[19a], S. Giagu[132a,132b], V. Giakoumopoulou[8], V. Giangiobbe[122a,122b], F. Gianotti[29], B. Gibbard[24],
A. Gibson[157], S.M. Gibson[118], L.M. Gilbert[118], M. Gilchriese[14], V. Gilewsky[91], D.M. Gingrich[2,b], J. Ginzburg[152],
N. Giokaris[8], M.P. Giordani[163a,163c], R. Giordano[102a,102b], F.M. Giorgi[15], P. Giovannini[99], P.F. Giraud[29], P. Girtler[62],
D. Giugni[89a], P. Giusti[19a], B.K. Gjelsten[117], L.K. Gladilin[97], C. Glasman[80], A. Glazov[41], K.W. Glitza[173],
G.L. Glonti[65], J. Godfrey[142], J. Godlewski[29], M. Goebel[41], T. Göpfert[43], C. Goeringer[81], C. Gössling[42], T. Göttfert[99],
V. Goggi[119a,119b,h], S. Goldfarb[87], D. Goldin[39], T. Golling[174], A. Gomes[124a], L.S. Gomez Fajardo[41], R. Gonçalo[76],
L. Gonella[20], C. Gong[32b], S. González de la Hoz[166], M.L. Gonzalez Silva[26], S. Gonzalez-Sevilla[49], J.J. Goodson[147],
L. Goossens[29], H.A. Gordon[24], I. Gorelov[103], G. Gorfine[173], B. Gorini[29], E. Gorini[72a,72b], A. Gorišek[74],
E. Gornicki[38], B. Gosdzik[41], M. Gosselink[105], M.I. Gostkin[65], I. Gough Eschrich[162], M. Gouighri[135a],
D. Goujdami[135a], M.P. Goulette[49], A.G. Goussiou[138], C. Goy[4], I. Grabowska-Bold[162,c], P. Grafström[29],
K.-J. Grahn[146], S. Grancagnolo[15], V. Grassi[147], V. Gratchev[121], N. Grau[34], H.M. Gray[34,i], J.A. Gray[147],
E. Graziani[134a], B. Green[76], T. Greenshaw[73], Z.D. Greenwood[24,f], I.M. Gregor[41], P. Grenier[143], E. Griesmayer[46],
J. Griffiths[138], N. Grigalashvili[65], A.A. Grillo[137], K. Grimm[147], S. Grinstein[11], Y.V. Grishkevich[97], M. Groh[99],
M. Groll[81], E. Gross[170], J. Grosse-Knetter[54], J. Groth-Jensen[79], K. Grybel[141], C. Guicheney[33], A. Guida[72a,72b],
T. Guillemin[4], H. Guler[85,j], J. Gunther[125], B. Guo[157], A. Gupta[30], Y. Gusakov[65], A. Gutierrez[93], P. Gutierrez[111],
N. Guttman[152], O. Gutzwiller[171], C. Guyot[136], C. Gwenlan[118], C.B. Gwilliam[73], A. Haas[143], S. Haas[29], C. Haber[14],
H.K. Hadavand[39], D.R. Hadley[17], P. Haefner[99], R. Härtel[99], Z. Hajduk[38], H. Hakobyan[175], J. Haller[41,k],
K. Hamacher[173], A. Hamilton[49], S. Hamilton[160], L. Han[32b], K. Hanagaki[116], M. Hance[120], C. Handel[81], P. Hanke[58a],
J.R. Hansen[35], J.B. Hansen[35], J.D. Hansen[35], P.H. Hansen[35], T. Hansl-Kozanecka[137], P. Hansson[143], K. Hara[159],
G.A. Hare[137], T. Harenberg[173], R.D. Harrington[21], O.M. Harris[138], K. Harrison[17], J. Hartert[48], F. Hartjes[105],
A. Harvey[56], S. Hasegawa[101], Y. Hasegawa[140], K. Hashemi[22], S. Hassani[136], S. Haug[16], M. Hauschild[29], R. Hauser[88],
M. Havranek[125], C.M. Hawkes[17], R.J. Hawkings[29], T. Hayakawa[67], H.S. Hayward[73], S.J. Haywood[129], S.J. Head[82],
V. Hedberg[79], L. Heelan[28], S. Heim[88], B. Heinemann[14], S. Heisterkamp[35], L. Helary[4], M. Heller[115],
S. Hellman[145a,145b], C. Helsens[11], T. Hemperek[20], R.C.W. Henderson[71], M. Henke[58a], A. Henrichs[54],
A.M. Henriques Correia[29], S. Henrot-Versille[115], C. Hensel[54], T. Henß[173], Y. Hernández Jiménez[166],
A.D. Hershenhorn[151], G. Herten[48], R. Hertenberger[98], L. Hervas[29], N.P. Hessey[105], E. Higón-Rodriguez[166],
J.C. Hill[27], K.H. Hiller[41], S. Hillert[145a,145b], S.J. Hillier[17], I. Hinchliffe[14], E. Hines[120], M. Hirose[116], F. Hirsch[42],
D. Hirschbuehl[173], J. Hobbs[147], N. Hod[152], M.C. Hodgkinson[139], P. Hodgson[139], A. Hoecker[29], M.R. Hoeferkamp[103],
J. Hoffman[39], D. Hoffmann[83], M. Hohlfeld[81], T. Holy[127], J.L. Holzbauer[88], Y. Homma[67], T. Horazdovsky[127],
T. Hori[67], C. Horn[143], S. Horner[48], S. Horvat[99], J.-Y. Hostachy[55], S. Hou[150], A. Hoummada[135a], T. Howe[39],
J. Hrivnac[115], T. Hryn'ova[4], P.J. Hsu[174], S.-C. Hsu[14], G.S. Huang[111], Z. Hubacek[127], F. Hubaut[83], F. Huegging[20],
E.W. Hughes[34], G. Hughes[71], M. Hurwitz[30], U. Husemann[41], N. Huseynov[10], J. Huston[88], J. Huth[57],
G. Iacobucci[102a], G. Iakovidis[9], I. Ibragimov[141], L. Iconomidou-Fayard[115], J. Idarraga[158b], P. Iengo[4],
O. Igonkina[105], Y. Ikegami[66], M. Ikeno[66], Y. Ilchenko[39], D. Iliadis[153], T. Ince[168], P. Ioannou[8], M. Iodice[134a],

A. Irles Quiles[166], A. Ishikawa[67], M. Ishino[66], R. Ishmukhametov[39], T. Isobe[154], V. Issakov[174,*], C. Issever[118], S. Istin[18a], Y. Itoh[101], A.V. Ivashin[128], W. Iwanski[38], H. Iwasaki[66], J.M. Izen[40], V. Izzo[102a], B. Jackson[120], J.N. Jackson[73], P. Jackson[143], M.R. Jaekel[29], V. Jain[61], K. Jakobs[48], S. Jakobsen[35], J. Jakubek[127], D.K. Jana[111], E. Jansen[104], A. Jantsch[99], M. Janus[48], R.C. Jared[171], G. Jarlskog[79], L. Jeanty[57], I. Jen-La Plante[30], P. Jenni[29], P. Jez[35], S. Jézéquel[4], W. Ji[79], J. Jia[147], Y. Jiang[32b], M. Jimenez Belenguer[29], S. Jin[32a], O. Jinnouchi[156], D. Joffe[39], M. Johansen[145a,145b], K.E. Johansson[145a], P. Johansson[139], S. Johnert[41], K.A. Johns[6], K. Jon-And[145a,145b], G. Jones[82], R.W.L. Jones[71], T.J. Jones[73], P.M. Jorge[124a], J. Joseph[14], V. Juranek[125], P. Jussel[62], V.V. Kabachenko[128], M. Kaci[166], A. Kaczmarska[38], M. Kado[115], H. Kagan[109], M. Kagan[57], S. Kaiser[99], E. Kajomovitz[151], S. Kalinin[173], L.V. Kalinovskaya[65], A. Kalinowski[130], S. Kama[41], N. Kanaya[154], M. Kaneda[154], V.A. Kantserov[96], J. Kanzaki[66], B. Kaplan[174], A. Kapliy[30], J. Kaplon[29], D. Kar[43], M. Karagounis[20], M. Karagoz Unel[118], V. Kartvelishvili[71], A.N. Karyukhin[128], L. Kashif[57], A. Kasmi[39], R.D. Kass[109], A. Kastanas[13], M. Kastoryano[174], M. Kataoka[4], Y. Kataoka[154], E. Katsoufis[9], J. Katzy[41], V. Kaushik[6], K. Kawagoe[67], T. Kawamoto[154], G. Kawamura[81], M.S. Kayl[105], F. Kayumov[94], V.A. Kazanin[107], M.Y. Kazarinov[65], J.R. Keates[82], R. Keeler[168], P.T. Keener[120], R. Kehoe[39], M. Keil[54], G.D. Kekelidze[65], M. Kelly[82], M. Kenyon[53], O. Kepka[125], N. Kerschen[29], B.P. Kerševan[74], S. Kersten[173], K. Kessoku[154], M. Khakzad[28], F. Khalil-zada[10], H. Khandanyan[164], A. Khanov[112], D. Kharchenko[65], A. Khodinov[147], A. Khomich[58a], G. Khoriauli[20], N. Khovanskiy[65], V. Khovanskiy[95], E. Khramov[65], J. Khubua[51], H. Kim[7], M.S. Kim[2], P.C. Kim[143], S.H. Kim[159], O. Kind[15], P. Kind[173], B.T. King[73], J. Kirk[129], G.P. Kirsch[118], L.E. Kirsch[22], A.E. Kiryunin[99], D. Kisielewska[37], T. Kittelmann[123], H. Kiyamura[67], E. Kladiva[144b], M. Klein[73], U. Klein[73], K. Kleinknecht[81], M. Klemetti[85], A. Klier[170], A. Klimentov[24], R. Klingenberg[42], E.B. Klinkby[44], T. Klioutchnikova[29], P.F. Klok[104], S. Klous[105], E.-E. Kluge[58a], T. Kluge[73], P. Kluit[105], M. Klute[54], S. Kluth[99], N.S. Knecht[157], E. Kneringer[62], B.R. Ko[44], T. Kobayashi[154], M. Kobel[43], B. Koblitz[29], M. Kocian[143], A. Kocnar[113], P. Kodys[126], K. Köneke[41], A.C. König[104], S. Koenig[81], L. Köpke[81], F. Koetsveld[104], P. Koevesarki[20], T. Koffas[29], E. Koffeman[105], F. Kohn[54], Z. Kohout[127], T. Kohriki[66], H. Kolanoski[15], V. Kolesnikov[65], I. Koletsou[4], J. Koll[88], D. Kollar[29], S. Kolos[162,l], S.D. Kolya[82], A.A. Komar[94], J.R. Komaragiri[142], T. Kondo[66], T. Kono[41,k], R. Konoplich[108], S.P. Konovalov[94], N. Konstantinidis[77], S. Koperny[37], K. Korcyl[38], K. Kordas[153], A. Korn[14], I. Korolkov[11], E.V. Korolkova[139], V.A. Korotkov[128], O. Kortner[99], P. Kostka[41], V.V. Kostyukhin[20], S. Kotov[99], V.M. Kotov[65], K.Y. Kotov[107], C. Kourkoumelis[8], A. Koutsman[105], R. Kowalewski[168], H. Kowalski[41], T.Z. Kowalski[37], W. Kozanecki[136], A.S. Kozhin[128], V. Kral[127], V.A. Kramarenko[97], G. Kramberger[74], M.W. Krasny[78], A. Krasznahorkay[108], A. Kreisel[152], F. Krejci[127], J. Kretzschmar[73], N. Krieger[54], P. Krieger[157], K. Kroeninger[54], H. Kroha[99], J. Kroll[120], J. Kroseberg[20], J. Krstic[12a], U. Kruchonak[65], H. Krüger[20], Z.V. Krumshteyn[65], T. Kubota[154], S. Kuehn[48], A. Kugel[58c], T. Kuhl[173], D. Kuhn[62], V. Kukhtin[65], Y. Kulchitsky[90], S. Kuleshov[31b], C. Kummer[98], M. Kuna[83], J. Kunkle[120], A. Kupco[125], H. Kurashige[67], M. Kurata[159], L.L. Kurchaninov[158a], Y.A. Kurochkin[90], V. Kus[125], R. Kwee[15], L. La Rotonda[36a,36b], J. Labbe[4], C. Lacasta[166], F. Lacava[132a,132b], H. Lacker[15], D. Lacour[78], V.R. Lacuesta[166], E. Ladygin[65], R. Lafaye[4], B. Laforge[78], T. Lagouri[80], S. Lai[48], M. Lamanna[29], C.L. Lampen[6], W. Lampl[6], E. Lancon[136], U. Landgraf[48], M.P.J. Landon[75], J.L. Lane[82], A.J. Lankford[162], F. Lanni[24], K. Lantzsch[29], A. Lanza[119a], S. Laplace[4], C. Lapoire[83], J.F. Laporte[136], T. Lari[89a], A. Larner[118], M. Lassnig[29], P. Laurelli[47], W. Lavrijsen[14], P. Laycock[73], A.B. Lazarev[65], A. Lazzaro[89a,89b], O. Le Dortz[78], E. Le Guirriec[83], E. Le Menedeu[136], M. Le Vine[24], A. Lebedev[64], C. Lebel[93], T. LeCompte[5], F. Ledroit-Guillon[55], H. Lee[105], J.S.H. Lee[149], S.C. Lee[150], M. Lefebvre[168], M. Legendre[136], B.C. LeGeyt[120], F. Legger[98], C. Leggett[14], M. Lehmacher[20], G. Lehmann Miotto[29], X. Lei[6], R. Leitner[126], D. Lellouch[170], J. Lellouch[78], V. Lendermann[58a], K.J.C. Leney[73], T. Lenz[173], G. Lenzen[173], B. Lenzi[136], K. Leonhardt[43], C. Leroy[93], J.-R. Lessard[168], C.G. Lester[27], A. Leung Fook Cheong[171], J. Levêque[83], D. Levin[87], L.J. Levinson[170], M. Leyton[14], H. Li[171], S. Li[41], X. Li[87], Z. Liang[39], Z. Liang[150,m], B. Liberti[133a], P. Lichard[29], M. Lichtnecker[98], K. Lie[164], W. Liebig[105], J.N. Lilley[17], H. Lim[5], A. Limosani[86], M. Limper[63], S.C. Lin[150], J.T. Linnemann[88], E. Lipeles[120], L. Lipinsky[125], A. Lipniacka[13], T.M. Liss[164], D. Lissauer[24], A. Lister[49], A.M. Litke[137], C. Liu[28], D. Liu[150,n], H. Liu[87], J.B. Liu[87], M. Liu[32b], T. Liu[39], Y. Liu[32b], M. Livan[119a,119b], A. Lleres[55], S.L. Lloyd[75], E. Lobodzinska[41], P. Loch[6], W.S. Lockman[137], S. Lockwitz[174], T. Loddenkoetter[20], F.K. Loebinger[82], A. Loginov[174], C.W. Loh[167], T. Lohse[15], K. Lohwasser[48], M. Lokajicek[125], R.E. Long[71], L. Lopes[124a], D. Lopez Mateos[34,i], M. Losada[161], P. Loscutoff[14], X. Lou[40], A. Lounis[115], K.F. Loureiro[109], L. Lovas[144a], J. Love[21], P.A. Love[71], A.J. Lowe[61], F. Lu[32a], H.J. Lubatti[138], C. Luci[132a,132b], A. Lucotte[55], A. Ludwig[43], D. Ludwig[41], I. Ludwig[48], F. Luehring[61], L. Luisa[163a,163c], D. Lumb[48], L. Luminari[132a], E. Lund[117], B. Lund-Jensen[146], B. Lundberg[79], J. Lundberg[29], J. Lundquist[35], D. Lynn[24], J. Lys[14], E. Lytken[79], H. Ma[24], L.L. Ma[171], J.A. Macana Goia[93], G. Maccarrone[47], A. Macchiolo[99], B. Maček[74],

J. Machado Miguens[124a], R. Mackeprang[35], R.J. Madaras[14], W.F. Mader[43], R. Maenner[58c], T. Maeno[24],
P. Mättig[173], S. Mättig[41], P.J. Magalhaes Martins[124a], E. Magradze[51], Y. Mahalalel[152], K. Mahboubi[48],
A. Mahmood[1], C. Maiani[132a,132b], C. Maidantchik[23a], A. Maio[124a], S. Majewski[24], Y. Makida[66], M. Makouski[128],
N. Makovec[115], Pa. Malecki[38], P. Malecki[38], V.P. Maleev[121], F. Malek[55], U. Mallik[63], D. Malon[5], S. Maltezos[9],
V. Malyshev[107], S. Malyukov[65], M. Mambelli[30], R. Mameghani[98], J. Mamuzic[41], L. Mandelli[89a], I. Mandić[74],
R. Mandrysch[15], J. Maneira[124a], P.S. Mangeard[88], I.D. Manjavidze[65], P.M. Manning[137], A. Manousakis-Katsikakis[8],
B. Mansoulie[136], A. Mapelli[29], L. Mapelli[29], L. March[80], J.F. Marchand[4], F. Marchese[133a,133b], G. Marchiori[78],
M. Marcisovsky[125], C.P. Marino[61], F. Marroquim[23a], Z. Marshall[34,i], S. Marti-Garcia[166], A.J. Martin[75],
A.J. Martin[174], B. Martin[29], B. Martin[88], F.F. Martin[120], J.P. Martin[93], T.A. Martin[17], B. Martin dit Latour[49],
M. Martinez[11], V. Martinez Outschoorn[57], A. Martini[47], A.C. Martyniuk[82], F. Marzano[132a], A. Marzin[136],
L. Masetti[20], T. Mashimo[154], R. Mashinistov[96], J. Masik[82], A.L. Maslennikov[107], I. Massa[19a,19b], N. Massol[4],
A. Mastroberardino[36a,36b], T. Masubuchi[154], P. Matricon[115], H. Matsunaga[154], T. Matsushita[67], C. Mattravers[118,o],
S.J. Maxfield[73], A. Mayne[139], R. Mazini[150], M. Mazur[48], M. Mazzanti[89a], J. Mc Donald[85], S.P. Mc Kee[87],
A. McCarn[164], R.L. McCarthy[147], N.A. McCubbin[129], K.W. McFarlane[56], H. McGlone[53], G. Mchedlidze[51],
S.J. McMahon[129], R.A. McPherson[168,e], A. Meade[84], J. Mechnich[105], M. Mechtel[173], M. Medinnis[41],
R. Meera-Lebbai[111], T.M. Meguro[116], S. Mehlhase[41], A. Mehta[73], K. Meier[58a], B. Meirose[48], C. Melachrinos[30],
B.R. Mellado Garcia[171], L. Mendoza Navas[161], Z. Meng[150,p], S. Menke[99], E. Meoni[11], P. Mermod[118],
L. Merola[102a,102b], C. Meroni[89a], F.S. Merritt[30], A.M. Messina[29], J. Metcalfe[103], A.S. Mete[64], J.-P. Meyer[136],
J. Meyer[172], J. Meyer[54], T.C. Meyer[29], W.T. Meyer[64], J. Miao[32d], S. Michal[29], L. Micu[25a], R.P. Middleton[129],
S. Migas[73], L. Mijović[74], G. Mikenberg[170], M. Mikestikova[125], M. Mikuž[74], D.W. Miller[143], W.J. Mills[167],
C.M. Mills[57], A. Milov[170], D.A. Milstead[145a,145b], D. Milstein[170], A.A. Minaenko[128], M. Miñano[166], I.A. Minashvili[65],
A.I. Mincer[108], B. Mindur[37], M. Mineev[65], Y. Ming[130], L.M. Mir[11], G. Mirabelli[132a], S. Misawa[24], S. Miscetti[47],
A. Misiejuk[76], J. Mitrevski[137], V.A. Mitsou[166], P.S. Miyagawa[82], J.U. Mjörnmark[79], D. Mladenov[22], T. Moa[145a,145b],
S. Moed[57], V. Moeller[27], K. Mönig[41], N. Möser[20], W. Mohr[48], S. Mohrdieck-Möck[99], R. Moles-Valls[166],
J. Molina-Perez[29], J. Monk[77], E. Monnier[83], S. Montesano[89a,89b], F. Monticelli[70], R.W. Moore[2], C. Mora Herrera[49],
A. Moraes[53], A. Morais[124a], J. Morel[54], G. Morello[36a,36b], D. Moreno[161], M. Moreno Llácer[166], P. Morettini[50a],
M. Morii[57], A.K. Morley[86], G. Mornacchi[29], S.V. Morozov[96], J.D. Morris[75], H.G. Moser[99], M. Mosidze[51], J. Moss[109],
R. Mount[143], E. Mountricha[136], S.V. Mouraviev[94], E.J.W. Moyse[84], M. Mudrinic[12b], F. Mueller[58a], J. Mueller[123],
K. Mueller[20], T.A. Müller[98], D. Muenstermann[42], A. Muir[167], Y. Munwes[152], R. Murillo Garcia[162], W.J. Murray[129],
I. Mussche[105], E. Musto[102a,102b], A.G. Myagkov[128], M. Myska[125], J. Nadal[11], K. Nagai[159], K. Nagano[66],
Y. Nagasaka[60], A.M. Nairz[29], K. Nakamura[154], I. Nakano[110], H. Nakatsuka[67], G. Nanava[20], A. Napier[160],
M. Nash[77,q], N.R. Nation[21], T. Nattermann[20], T. Naumann[41], G. Navarro[161], S.K. Nderitu[20], H.A. Neal[87], E. Nebot[80],
P. Nechaeva[94], A. Negri[119a,119b], G. Negri[29], A. Nelson[64], T.K. Nelson[143], S. Nemecek[125], P. Nemethy[108],
A.A. Nepomuceno[23a], M. Nessi[29], M.S. Neubauer[164], A. Neusiedl[81], R.M. Neves[108], P. Nevski[24], F.M. Newcomer[120],
R.B. Nickerson[118], R. Nicolaidou[136], L. Nicolas[139], G. Nicoletti[47], B. Nicquevert[29], F. Niedercorn[115], J. Nielsen[137],
A. Nikiforov[15], K. Nikolaev[65], I. Nikolic-Audit[78], K. Nikolopoulos[8], H. Nilsen[48], P. Nilsson[7], A. Nisati[132a],
T. Nishiyama[67], R. Nisius[99], L. Nodulman[5], M. Nomachi[116], I. Nomidis[153], M. Nordberg[29], B. Nordkvist[145a,145b],
D. Notz[41], J. Novakova[126], M. Nozaki[66], M. Nožička[41], I.M. Nugent[158a], A.-E. Nuncio-Quiroz[20],
G. Nunes Hanninger[20], T. Nunnemann[98], E. Nurse[77], D.C. O'Neil[142], V. O'Shea[53], F.G. Oakham[28,b], H. Oberlack[99],
A. Ochi[67], S. Oda[154], S. Odaka[66], J. Odier[83], H. Ogren[61], A. Oh[82], S.H. Oh[44], C.C. Ohm[145a,145b], T. Ohshima[101],
H. Ohshita[140], T. Ohsugi[59], S. Okada[67], H. Okawa[162], Y. Okumura[101], T. Okuyama[154], A.G. Olchevski[65],
M. Oliveira[124a], D. Oliveira Damazio[24], J. Oliver[57], E. Oliver Garcia[166], D. Olivito[120], A. Olszewski[38],
J. Olszowska[38], C. Omachi[67,r], A. Onofre[124a], P.U.E. Onyisi[30], C.J. Oram[158a], M.J. Oreglia[30], Y. Oren[152],
D. Orestano[134a,134b], I. Orlov[107], C. Oropeza Barrera[53], R.S. Orr[157], E.O. Ortega[130], B. Osculati[50a,50b],
R. Ospanov[120], C. Osuna[11], J.P. Ottersbach[105], F. Ould-Saada[117], A. Ouraou[136], Q. Ouyang[32a], M. Owen[82],
S. Owen[139], A. Oyarzun[31b], V.E. Ozcan[77], K. Ozone[66], N. Ozturk[7], A. Pacheco Pages[11], C. Padilla Aranda[11],
E. Paganis[139], C. Pahl[63], F. Paige[24], K. Pajchel[117], S. Palestini[29], D. Pallin[33], A. Palma[124a], J.D. Palmer[17],
Y.B. Pan[171], E. Panagiotopoulou[9], B. Panes[31a], N. Panikashvili[87], S. Panitkin[24], D. Pantea[25a], M. Panuskova[125],
V. Paolone[123], Th.D. Papadopoulou[9], S.J. Park[54], W. Park[24,s], M.A. Parker[27], S.I. Parker[14], F. Parodi[50a,50b],
J.A. Parsons[34], U. Parzefall[48], E. Pasqualucci[132a], A. Passeri[134a], F. Pastore[134a,134b], Fr. Pastore[29], G. Pásztor[49,t],
S. Pataraia[99], J.R. Pater[82], S. Patricelli[102a,102b], A. Patwa[24], T. Pauly[29], L.S. Peak[149], M. Pecsy[144a],
M.I. Pedraza Morales[171], S.V. Peleganchuk[107], H. Peng[171], A. Penson[34], J. Penwell[61], M. Perantoni[23a], K. Perez[34,i],

E. Perez Codina[11], M.T. Pérez García-Estañ[166], V. Perez Reale[34], L. Perini[89a,89b], H. Pernegger[29], R. Perrino[72a], S. Persembe[3a], P. Perus[115], V.D. Peshekhonov[65], B.A. Petersen[29], T.C. Petersen[35], E. Petit[83], C. Petridou[153], E. Petrolo[132a], F. Petrucci[134a,134b], D. Petschull[41], M. Petteni[142], R. Pezoa[31b], A. Phan[86], A.W. Phillips[27], G. Piacquadio[29], M. Piccinini[19a,19b], R. Piegaia[26], J.E. Pilcher[30], A.D. Pilkington[82], J. Pina[124a], M. Pinamonti[163a,163c], J.L. Pinfold[2], B. Pinto[124a], C. Pizio[89a,89b], R. Placakyte[41], M. Plamondon[168], M.-A. Pleier[24], A. Poblaguev[174], S. Poddar[58a], F. Podlyski[33], P. Poffenberger[168], L. Poggioli[115], M. Pohl[49], F. Polci[55], G. Polesello[119a], A. Policicchio[138], A. Polini[19a], J. Poll[75], V. Polychronakos[24], D. Pomeroy[22], K. Pommès[29], P. Ponsot[136], L. Pontecorvo[132a], B.G. Pope[88], G.A. Popeneciu[25a], D.S. Popovic[12a], A. Poppleton[29], J. Popule[125], X. Portell Bueso[48], R. Porter[162], G.E. Pospelov[99], S. Pospisil[127], M. Potekhin[24], I.N. Potrap[99], C.J. Potter[148], C.T. Potter[85], K.P. Potter[82], G. Poulard[29], J. Poveda[171], R. Prabhu[20], P. Pralavorio[83], S. Prasad[57], R. Pravahan[7], L. Pribyl[29], D. Price[61], L.E. Price[5], P.M. Prichard[73], D. Prieur[123], M. Primavera[72a], K. Prokofiev[29], F. Prokoshin[31b], S. Protopopescu[24], J. Proudfoot[5], X. Prudent[43], H. Przysiezniak[4], S. Psoroulas[20], E. Ptacek[114], C. Puigdengoles[11], J. Purdham[87], M. Purohit[24,s], P. Puzo[115], Y. Pylypchenko[117], M. Qi[32c], J. Qian[87], W. Qian[129], Z. Qin[41], A. Quadt[54], D.R. Quarrie[14], W.B. Quayle[171], F. Quinonez[31a], M. Raas[104], V. Radeka[24], V. Radescu[58b], B. Radics[20], T. Rador[18a], F. Ragusa[89a,89b], G. Rahal[179], A.M. Rahimi[109], S. Rajagopalan[24], M. Rammensee[48], M. Rammes[141], F. Rauscher[98], E. Rauter[99], M. Raymond[29], A.L. Read[117], D.M. Rebuzzi[119a,119b], A. Redelbach[172], G. Redlinger[24], R. Reece[120], K. Reeves[40], E. Reinherz-Aronis[152], A. Reinsch[114], I. Reisinger[42], D. Reljic[12a], C. Rembser[29], Z.L. Ren[150], P. Renkel[39], S. Rescia[24], M. Rescigno[132a], S. Resconi[89a], B. Resende[136], P. Reznicek[126], R. Rezvani[157], A. Richards[77], R.A. Richards[88], R. Richter[99], E. Richter-Was[38,u], M. Ridel[78], M. Rijpstra[105], M. Rijssenbeek[147], A. Rimoldi[119a,119b], L. Rinaldi[19a], R.R. Rios[39], I. Riu[11], F. Rizatdinova[112], E. Rizvi[75], D.A. Roa Romero[161], S.H. Robertson[85,e], A. Robichaud-Veronneau[49], D. Robinson[27], J.E.M. Robinson[77], M. Robinson[114], A. Robson[53], J.G. Rocha de Lima[106a], C. Roda[122a,122b], D. Roda Dos Santos[29], D. Rodriguez[161], Y. Rodriguez Garcia[15], S. Roe[29], O. Røhne[117], V. Rojo[1], S. Rolli[160], A. Romaniouk[96], V.M. Romanov[65], G. Romeo[26], D. Romero Maltrana[31a], L. Roos[78], E. Ros[166], S. Rosati[138], G.A. Rosenbaum[157], L. Rosselet[49], V. Rossetti[11], L.P. Rossi[50a], M. Rotaru[25a], J. Rothberg[138], D. Rousseau[115], C.R. Royon[136], A. Rozanov[83], Y. Rozen[151], X. Ruan[115], B. Ruckert[98], N. Ruckstuhl[105], V.I. Rud[97], G. Rudolph[62], F. Rühr[58a], F. Ruggieri[134a], A. Ruiz-Martinez[64], L. Rumyantsev[65], Z. Rurikova[48], N.A. Rusakovich[65], J.P. Rutherfoord[6], C. Ruwiedel[20], P. Ruzicka[125], Y.F. Ryabov[121], P. Ryan[88], G. Rybkin[115], S. Rzaeva[10], A.F. Saavedra[149], H.F.-W. Sadrozinski[137], R. Sadykov[65], H. Sakamoto[154], G. Salamanna[105], A. Salamon[133a], M.S. Saleem[111], D. Salihagic[99], A. Salnikov[143], J. Salt[166], B.M. Salvachua Ferrando[5], D. Salvatore[36a,36b], F. Salvatore[148], A. Salvucci[47], A. Salzburger[29], D. Sampsonidis[153], B.H. Samset[117], H. Sandaker[13], H.G. Sander[81], M.P. Sanders[98], M. Sandhoff[173], P. Sandhu[157], R. Sandstroem[105], S. Sandvoss[173], D.P.C. Sankey[129], B. Sanny[173], A. Sansoni[47], C. Santamarina Rios[85], C. Santoni[33], R. Santonico[133a,133b], J.G. Saraiva[124a], T. Sarangi[171], E. Sarkisyan-Grinbaum[7], F. Sarri[122a,122b], O. Sasaki[66], N. Sasao[68], I. Satsounkevitch[90], G. Sauvage[4], P. Savard[157,b], A.Y. Savine[6], V. Savinov[123], L. Sawyer[24,f], D.H. Saxon[53], L.P. Says[33], C. Sbarra[19a,19b], A. Sbrizzi[19a,19b], D.A. Scannicchio[29], J. Schaarschmidt[43], P. Schacht[99], U. Schäfer[81], S. Schaetzel[58b], A.C. Schaffer[115], D. Schaile[98], R.D. Schamberger[147], A.G. Schamov[107], V.A. Schegelsky[121], D. Scheirich[87], M. Schernau[162], M.I. Scherzer[14], C. Schiavi[50a,50b], J. Schieck[99], M. Schioppa[36a,36b], S. Schlenker[29], K. Schmieden[20], C. Schmitt[81], M. Schmitz[20], M. Schott[29], D. Schouten[142], J. Schovancova[125], M. Schram[85], A. Schreiner[63], C. Schroeder[81], N. Schroer[58c], M. Schroers[173], J. Schultes[173], H.-C. Schultz-Coulon[58a], J.W. Schumacher[43], M. Schumacher[48], B.A. Schumm[137], Ph. Schune[136], C. Schwanenberger[82], A. Schwartzman[143], Ph. Schwemling[78], R. Schwienhorst[88], R. Schwierz[43], J. Schwindling[136], W.G. Scott[129], J. Searcy[114], E. Sedykh[121], E. Segura[11], S.C. Seidel[103], A. Seiden[137], F. Seifert[43], J.M. Seixas[23a], G. Sekhniaidze[102a], D.M. Seliverstov[121], B. Sellden[145a], N. Semprini-Cesari[19a,19b], C. Serfon[98], L. Serin[115], R. Seuster[99], H. Severini[111], M.E. Sevior[86], A. Sfyrla[164], E. Shabalina[54], M. Shamim[114], L.Y. Shan[32a], J.T. Shank[21], Q.T. Shao[86], M. Shapiro[14], P.B. Shatalov[95], K. Shaw[139], D. Sherman[29], P. Sherwood[77], A. Shibata[108], M. Shimojima[100], T. Shin[56], A. Shmeleva[94], M.J. Shochet[30], M.A. Shupe[6], P. Sicho[125], A. Sidoti[15], F. Siegert[77], J. Siegrist[14], Dj. Sijacki[12a], O. Silbert[170], J. Silva[124a], Y. Silver[152], D. Silverstein[143], S.B. Silverstein[145a], V. Simak[127], Lj. Simic[12a], S. Simion[115], B. Simmons[77], M. Simonyan[35], P. Sinervo[157], N.B. Sinev[114], V. Sipica[141], G. Siragusa[81], A.N. Sisakyan[65], S.Yu. Sivoklokov[97], J. Sjoelin[145a,145b], T.B. Sjursen[13], K. Skovpen[107], P. Skubic[111], M. Slater[17], T. Slavicek[127], K. Sliwa[160], J. Sloper[29], T. Sluka[125], V. Smakhtin[170], S.Yu. Smirnov[96], Y. Smirnov[24], L.N. Smirnova[97], O. Smirnova[79], B.C. Smith[57], D. Smith[143], K.M. Smith[53], M. Smizanska[71], K. Smolek[127], A.A. Snesarev[94], S.W. Snow[82], J. Snow[111], J. Snuverink[105], S. Snyder[24], M. Soares[124a], R. Sobie[168,e], J. Sodomka[127], A. Soffer[152], C.A. Solans[166], M. Solar[127],

J. Solc[127], E. Solfaroli Camillocci[132a,132b], A.A. Solodkov[128], O.V. Solovyanov[128], R. Soluk[2], J. Sondericker[24], V. Sopko[127], B. Sopko[127], M. Sosebee[7], A. Soukharev[107], S. Spagnolo[72a,72b], F. Spanò[34], E. Spencer[137], R. Spighi[19a], G. Spigo[29], F. Spila[132a,132b], R. Spiwoks[29], M. Spousta[126], T. Spreitzer[142], B. Spurlock[7], R.D. St. Denis[53], T. Stahl[141], J. Stahlman[120], R. Stamen[58a], S.N. Stancu[162], E. Stanecka[29], R.W. Stanek[5], C. Stanescu[134a], S. Stapnes[117], E.A. Starchenko[128], J. Stark[55], P. Staroba[125], P. Starovoitov[91], J. Stastny[125], P. Stavina[144a], G. Steele[53], P. Steinbach[43], P. Steinberg[24], I. Stekl[127], B. Stelzer[142], H.J. Stelzer[41], O. Stelzer-Chilton[158a], H. Stenzel[52], K. Stevenson[75], G.A. Stewart[53], M.C. Stockton[29], K. Stoerig[48], G. Stoicea[25a], S. Stonjek[99], P. Strachota[126], A.R. Stradling[7], A. Straessner[43], J. Strandberg[87], S. Strandberg[14], A. Strandlie[117], M. Strauss[111], P. Strizenec[144b], R. Ströhmer[172], D.M. Strom[114], R. Stroynowski[39], J. Strube[129], B. Stugu[13], D.A. Soh[150,v], D. Su[143], Y. Sugaya[116], T. Sugimoto[101], C. Suhr[106a], M. Suk[126], V.V. Sulin[94], S. Sultansoy[3d], T. Sumida[29], X.H. Sun[32d], J.E. Sundermann[48], K. Suruliz[163a,163b], S. Sushkov[11], G. Susinno[36a,36b], M.R. Sutton[139], T. Suzuki[154], Y. Suzuki[66], I. Sykora[144a], T. Sykora[126], T. Szymocha[38], J. Sánchez[166], D. Ta[20], K. Tackmann[29], A. Taffard[162], R. Tafirout[158a], A. Taga[117], Y. Takahashi[101], H. Takai[24], R. Takashima[69], H. Takeda[67], T. Takeshita[140], M. Talby[83], A. Talyshev[107], M.C. Tamsett[76], J. Tanaka[154], R. Tanaka[115], S. Tanaka[131], S. Tanaka[66], S. Tapprogge[81], D. Tardif[157], S. Tarem[151], F. Tarrade[24], G.F. Tartarelli[89a], P. Tas[126], M. Tasevsky[125], E. Tassi[36a,36b], M. Tatarkhanov[14], C. Taylor[77], F.E. Taylor[92], G.N. Taylor[86], R.P. Taylor[168], W. Taylor[158b], P. Teixeira-Dias[76], H. Ten Kate[29], P.K. Teng[150], Y.D. Tennenbaum-Katan[151], S. Terada[66], K. Terashi[154], J. Terron[80], M. Terwort[41,k], M. Testa[47], R.J. Teuscher[157,e], M. Thioye[174], S. Thoma[48], J.P. Thomas[17], E.N. Thompson[84], P.D. Thompson[17], P.D. Thompson[157], R.J. Thompson[82], A.S. Thompson[53], E. Thomson[120], R.P. Thun[87], T. Tic[125], V.O. Tikhomirov[94], Y.A. Tikhonov[107], P. Tipton[174], F.J. Tique Aires Viegas[29], S. Tisserant[83], B. Toczek[37], T. Todorov[4], S. Todorova-Nova[160], B. Toggerson[162], J. Tojo[66], S. Tokár[144a], K. Tokushuku[66], K. Tollefson[88], L. Tomasek[125], M. Tomasek[125], M. Tomoto[101], L. Tompkins[14], K. Toms[103], A. Tonoyan[13], C. Topfel[16], N.D. Topilin[65], E. Torrence[114], E. Torró Pastor[166], J. Toth[83,t], F. Touchard[83], D.R. Tovey[139], T. Trefzger[172], L. Tremblet[29], A. Tricoli[29], I.M. Trigger[158a], S. Trincaz-Duvoid[78], T.N. Trinh[78], M.F. Tripiana[70], N. Triplett[64], W. Trischuk[157], A. Trivedi[24,s], B. Trocmé[55], C. Troncon[89a], A. Trzupek[38], C. Tsarouchas[9], J.C.-L. Tseng[118], M. Tsiakiris[105], P.V. Tsiareshka[90], D. Tsionou[139], G. Tsipolitis[9], V. Tsiskaridze[51], E.G. Tskhadadze[51], I.I. Tsukerman[95], V. Tsulaia[123], J.-W. Tsung[20], S. Tsuno[66], D. Tsybychev[147], J.M. Tuggle[30], D. Turecek[127], I. Turk Cakir[3e], E. Turlay[105], P.M. Tuts[34], M.S. Twomey[138], M. Tylmad[145a,145b], M. Tyndel[129], K. Uchida[116], I. Ueda[154], M. Ugland[13], M. Uhlenbrock[20], M. Uhrmacher[54], F. Ukegawa[159], G. Unal[29], A. Undrus[24], G. Unel[162], Y. Unno[66], D. Urbaniec[34], E. Urkovsky[152], P. Urquijo[49,w], P. Urrejola[31a], G. Usai[7], M. Uslenghi[119a,119b], L. Vacavant[83], V. Vacek[127], B. Vachon[85], S. Vahsen[14], P. Valente[132a], S. Valentinetti[19a,19b], S. Valkar[126], E. Valladolid Gallego[166], S. Vallecorsa[151], J.A. Valls Ferrer[166], R. Van Berg[120], H. van der Graaf[105], E. van der Kraaij[105], E. van der Poel[105], D. van der Ster[29], N. van Eldik[84], P. van Gemmeren[5], Z. van Kesteren[105], I. van Vulpen[105], W. Vandelli[29], A. Vaniachine[5], P. Vankov[73], F. Vannucci[78], R. Vari[132a], E.W. Varnes[6], D. Varouchas[14], A. Vartapetian[7], K.E. Varvell[149], L. Vasilyeva[94], V.I. Vassilakopoulos[56], F. Vazeille[33], C. Vellidis[8], F. Veloso[124a], S. Veneziano[132a], A. Ventura[72a,72b], D. Ventura[138], M. Venturi[48], N. Venturi[16], V. Vercesi[119a], M. Verducci[172], W. Verkerke[105], J.C. Vermeulen[105], M.C. Vetterli[142,b], I. Vichou[164], T. Vickey[118], G.H.A. Viehhauser[118], M. Villa[19a,19b], E.G. Villani[129], M. Villaplana Perez[166], E. Vilucchi[47], M.G. Vincter[28], E. Vinek[29], V.B. Vinogradov[65], S. Viret[33], J. Virzi[14], A. Vitale[19a,19b], O. Vitells[170], I. Vivarelli[48], F. Vives Vaque[11], S. Vlachos[9], M. Vlasak[127], N. Vlasov[20], A. Vogel[20], P. Vokac[127], M. Volpi[11], H. von der Schmitt[99], J. von Loeben[99], H. von Radziewski[48], E. von Toerne[20], V. Vorobel[126], V. Vorwerk[11], M. Vos[166], R. Voss[29], T.T. Voss[173], J.H. Vossebeld[73], N. Vranjes[12a], M. Vranjes Milosavljevic[12a], V. Vrba[125], M. Vreeswijk[105], T. Vu Anh[81], D. Vudragovic[12a], R. Vuillermet[29], I. Vukotic[115], P. Wagner[120], J. Walbersloh[42], J. Walder[71], R. Walker[98], W. Walkowiak[141], R. Wall[174], C. Wang[44], H. Wang[171], J. Wang[55], S.M. Wang[150], A. Warburton[85], C.P. Ward[27], M. Warsinsky[48], R. Wastie[118], P.M. Watkins[17], A.T. Watson[17], M.F. Watson[17], G. Watts[138], S. Watts[82], A.T. Waugh[149], B.M. Waugh[77], M.D. Weber[16], M. Weber[129], M.S. Weber[16], P. Weber[58a], A.R. Weidberg[118], J. Weingarten[54], C. Weiser[48], H. Wellenstein[22], P.S. Wells[29], M. Wen[47], T. Wenaus[24], S. Wendler[123], T. Wengler[82], S. Wenig[29], N. Wermes[20], M. Werner[48], P. Werner[29], M. Werth[162], U. Werthenbach[141], M. Wessels[58a], K. Whalen[28], A. White[7], M.J. White[27], S. White[24], S.R. Whitehead[118], D. Whiteson[162], D. Whittington[61], F. Wicek[115], D. Wicke[81], F.J. Wickens[129], W. Wiedenmann[171], M. Wielers[129], P. Wienemann[20], C. Wiglesworth[73], L.A.M. Wiik[48], A. Wildauer[166], M.A. Wildt[41,k], H.G. Wilkens[29], E. Williams[34], H.H. Williams[120], S. Willocq[84], J.A. Wilson[17], M.G. Wilson[143], A. Wilson[87], I. Wingerter-Seez[4], F. Winklmeier[29], M. Wittgen[143], M.W. Wolter[38], H. Wolters[124a], B.K. Wosiek[38], J. Wotschack[29], M.J. Woudstra[84], K. Wraight[53], C. Wright[53], D. Wright[143], B. Wrona[73], S.L. Wu[171],

X. Wu[49], E. Wulf[34], B.M. Wynne[45], L. Xaplanteris[9], S. Xella[35], S. Xie[48], D. Xu[139], N. Xu[171], M. Yamada[159], A. Yamamoto[66], K. Yamamoto[64], S. Yamamoto[154], T. Yamamura[154], J. Yamaoka[44], T. Yamazaki[154], Y. Yamazaki[67], Z. Yan[21], H. Yang[87], U.K. Yang[82], Z. Yang[145a,145b], W.-M. Yao[14], Y. Yao[14], Y. Yasu[66], J. Ye[39], S. Ye[24], M. Yilmaz[3c], R. Yoosoofmiya[123], K. Yorita[169], R. Yoshida[5], C. Young[143], S.P. Youssef[21], D. Yu[24], J. Yu[7], L. Yuan[78], A. Yurkewicz[147], R. Zaidan[63], A.M. Zaitsev[128], Z. Zajacova[29], V. Zambrano[47], L. Zanello[132a,132b], A. Zaytsev[107], C. Zeitnitz[173], M. Zeller[174], A. Zemla[38], C. Zendler[20], O. Zenin[128], T. Zenis[144a], Z. Zenonos[122a,122b], S. Zenz[14], D. Zerwas[115], G. Zevi della Porta[57], Z. Zhan[32d], H. Zhang[83], J. Zhang[5], Q. Zhang[5], X. Zhang[32d], L. Zhao[108], T. Zhao[138], Z. Zhao[32b], A. Zhemchugov[65], J. Zhong[150,x], B. Zhou[87], N. Zhou[34], Y. Zhou[150], C.G. Zhu[32d], H. Zhu[41], Y. Zhu[171], X. Zhuang[98], V. Zhuravlov[99], R. Zimmermann[20], S. Zimmermann[20], S. Zimmermann[48], M. Ziolkowski[141], L. Živković[34], G. Zobernig[171], A. Zoccoli[19a,19b], M. zur Nedden[15], V. Zutshi[106a]

*CERN, 1211 Genève 23, Switzerland

[1] University at Albany, 1400 Washington Ave, Albany, NY 12222, United States of America

[2] University of Alberta, Department of Physics, Centre for Particle Physics, Edmonton, AB T6G 2G7, Canada

[3] Ankara University[(a)], Faculty of Sciences, Department of Physics, TR 061000 Tandogan, Ankara; Dumlupinar University[(b)], Faculty of Arts and Sciences, Department of Physics, Kutahya; Gazi University[(c)], Faculty of Arts and Sciences, Department of Physics, 06500, Teknikokullar, Ankara; TOBB University of Economics and Technology[(d)], Faculty of Arts and Sciences, Division of Physics, 06560, Sogutozu, Ankara; Turkish Atomic Energy Authority[(e)], 06530, Lodumlu, Ankara, Turkey

[4] LAPP, Université de Savoie, CNRS/IN2P3, Annecy-le-Vieux, France

[5] Argonne National Laboratory, High Energy Physics Division, 9700 S. Cass Avenue, Argonne IL 60439, United States of America

[6] University of Arizona, Department of Physics, Tucson, AZ 85721, United States of America

[7] The University of Texas at Arlington, Department of Physics, Box 19059, Arlington, TX 76019, United States of America

[8] University of Athens, Nuclear & Particle Physics, Department of Physics, Panepistimiopouli, Zografou, GR 15771 Athens, Greece

[9] National Technical University of Athens, Physics Department, 9-Iroon Polytechniou, GR 15780 Zografou, Greece

[10] Institute of Physics, Azerbaijan Academy of Sciences, H. Javid Avenue 33, AZ 143 Baku, Azerbaijan

[11] Institut de Física d'Altes Energies, IFAE, Edifici Cn, Universitat Autònoma de Barcelona, ES-08193 Bellaterra (Barcelona), Spain

[12] University of Belgrade[(a)], Institute of Physics, P.O. Box 57, 11001 Belgrade; Vinca Institute of Nuclear Sciences[(b)], Mihajla Petrovica Alasa 12-14, 11001 Belgrade, Serbia

[13] University of Bergen, Department for Physics and Technology, Allegaten 55, NO-5007 Bergen, Norway

[14] Lawrence Berkeley National Laboratory and University of California, Physics Division, MS50B-6227, 1 Cyclotron Road, Berkeley, CA 94720, United States of America

[15] Humboldt University, Institute of Physics, Berlin, Newtonstr. 15, D-12489 Berlin, Germany

[16] University of Bern, Albert Einstein Center for Fundamental Physics, Laboratory for High Energy Physics, Sidlerstrasse 5, CH-3012 Bern, Switzerland

[17] University of Birmingham, School of Physics and Astronomy, Edgbaston, Birmingham B15 2TT, United Kingdom

[18] Bogazici University[(a)], Faculty of Sciences, Department of Physics, TR-80815 Bebek-Istanbul; Dogus University[(b)], Faculty of Arts and Sciences, Department of Physics, 34722, Kadikoy, Istanbul; [(c)]Gaziantep University, Faculty of Engineering, Department of Physics Engineering, 27310, Sehitkamil, Gaziantep, Turkey; Istanbul Technical University[(d)], Faculty of Arts and Sciences, Department of Physics, 34469, Maslak, Istanbul, Turkey

[19] INFN Sezione di Bologna[(a)]; Università di Bologna, Dipartimento di Fisica[(b)], viale C. Berti Pichat, 6/2, IT - 40127 Bologna, Italy

[20] University of Bonn, Physikalisches Institut, Nussallee 12, D-53115 Bonn, Germany

[21] Boston University, Department of Physics, 590 Commonwealth Avenue, Boston, MA 02215, United States of America

[22] Brandeis University, Department of Physics, MS057, 415 South Street, Waltham, MA 02454, United States of America

[23] Universidade Federal do Rio De Janeiro, COPPE/EE/IF [(a)], Caixa Postal 68528, Ilha do Fundao, BR-21945-970 Rio de Janeiro; [(b)]Universidade de Sao Paulo, Instituto de Fisica, R.do Matao Trav. R.187, Sao Paulo-SP, 05508-900, Brazil

[24] Brookhaven National Laboratory, Physics Department, Bldg. 510A, Upton, NY 11973, United States of America

[25] National Institute of Physics and Nuclear Engineering[(a)], Bucharest-Magurele, Str. Atomistilor 407, P.O. Box MG-6, R-077125, Romania; University Politehnica Bucharest[(b)], Rectorat–AN 001, 313 Splaiul Independentei, sector 6, 060042 Bucuresti; West University[(c)] in Timisoara, Bd. Vasile Parvan 4, Timisoara, Romania

[26] Universidad de Buenos Aires, FCEyN, Dto. Fisica, Pab I–C. Universitaria, 1428 Buenos Aires, Argentina

[27] University of Cambridge, Cavendish Laboratory, J.J. Thomson Avenue, Cambridge CB3 0HE, United Kingdom

[28] Carleton University, Department of Physics, 1125 Colonel By Drive, Ottawa ON K1S 5B6, Canada

[29] CERN, CH-1211 Geneva 23, Switzerland

[30] University of Chicago, Enrico Fermi Institute, 5640 S. Ellis Avenue, Chicago, IL 60637, United States of America

[31] Pontificia Universidad Católica de Chile, Facultad de Fisica, Departamento de Fisica[(a)], Avda. Vicuna Mackenna 4860, San Joaquin, Santiago; Universidad Técnica Federico Santa María, Departamento de Física[(b)], Avda. España 1680, Casilla 110-V, Valparaíso, Chile

[32] Institute of High Energy Physics, Chinese Academy of Sciences[(a)], P.O. Box 918, 19 Yuquan Road, Shijing Shan District, CN-Beijing 100049; University of Science & Technology of China (USTC), Department of Modern Physics[(b)], Hefei, CN-Anhui 230026; Nanjing University, Department of Physics[(c)], 22 Hankou Road, Nanjing, 210093; Shandong University, High Energy Physics Group[(d)], Jinan, CN-Shandong 250100, China

[33] Laboratoire de Physique Corpusculaire, Clermont Université, Université Blaise Pascal, CNRS/IN2P3, FR-63177 Aubiere Cedex, France

[34] Columbia University, Nevis Laboratory, 136 So. Broadway, Irvington, NY 10533, United States of America

[35] University of Copenhagen, Niels Bohr Institute, Blegdamsvej 17, DK-2100 Kobenhavn 0, Denmark

[36] INFN Gruppo Collegato di Cosenza[(a)]; Università della Calabria, Dipartimento di Fisica[(b)], IT-87036 Arcavacata di Rende, Italy

[37] Faculty of Physics and Applied Computer Science of the AGH-University of Science and Technology (FPACS, AGH-UST), al. Mickiewicza 30, PL-30059 Cracow, Poland

[38] The Henryk Niewodniczanski Institute of Nuclear Physics, Polish Academy of Sciences, ul. Radzikowskiego 152, PL-31342 Krakow, Poland

[39] Southern Methodist University, Physics Department, 106 Fondren Science Building, Dallas, TX 75275-0175, United States of America

[40] University of Texas at Dallas, 800 West Campbell Road, Richardson, TX 75080-3021, United States of America

[41] DESY, Notkestr. 85, D-22603 Hamburg and Platanenallee 6, D-15738 Zeuthen, Germany

[42] TU Dortmund, Experimentelle Physik IV, DE-44221 Dortmund, Germany

[43] Technical University Dresden, Institut für Kern- und Teilchenphysik, Zellescher Weg 19, D-01069 Dresden, Germany

[44] Duke University, Department of Physics, Durham, NC 27708, United States of America

[45] University of Edinburgh, School of Physics & Astronomy, James Clerk Maxwell Building, The Kings Buildings, Mayfield Road, Edinburgh EH9 3JZ, United Kingdom

[46] Fachhochschule Wiener Neustadt; Johannes Gutenbergstrasse 3 AT-2700 Wiener Neustadt, Austria

[47] INFN Laboratori Nazionali di Frascati, via Enrico Fermi 40, IT-00044 Frascati, Italy

[48] Albert-Ludwigs-Universität, Fakultät für Mathematik und Physik, Hermann-Herder Str. 3, D-79104 Freiburg i.Br., Germany

[49] Université de Genève, Section de Physique, 24 rue Ernest Ansermet, CH-1211 Geneve 4, Switzerland

[50] INFN Sezione di Genova(a); Università di Genova, Dipartimento di Fisica(b), via Dodecaneso 33, IT-16146 Genova, Italy

[51] Institute of Physics of the Georgian Academy of Sciences, 6 Tamarashvili St., GE-380077 Tbilisi; Tbilisi State University, HEP Institute, University St. 9, GE-380086 Tbilisi, Georgia

[52] Justus-Liebig-Universität Giessen, II Physikalisches Institut, Heinrich-Buff Ring 16, D-35392 Giessen, Germany

[53] University of Glasgow, Department of Physics and Astronomy, Glasgow G12 8QQ, United Kingdom

[54] Georg-August-Universität, II. Physikalisches Institut, Friedrich-Hund Platz 1, D-37077 Göttingen, Germany

[55] Laboratoire de Physique Subatomique et de Cosmologie, CNRS/IN2P3, Université Joseph Fourier, INPG, 53 avenue des Martyrs, FR-38026 Grenoble Cedex, France

[56] Hampton University, Department of Physics, Hampton, VA 23668, United States of America

[57] Harvard University, Laboratory for Particle Physics and Cosmology, 18 Hammond Street, Cambridge, MA 02138, United States of America

[58] Ruprecht-Karls-Universität Heidelberg: Kirchhoff-Institut für Physik(a), Im Neuenheimer Feld 227, D-69120 Heidelberg; Physikalisches Institut(b), Philosophenweg 12, D-69120 Heidelberg; ZITI Ruprecht-Karls-University Heidelberg(c), Lehrstuhl für Informatik V, B6, 23-29, DE-68131 Mannheim, Germany

[59] Hiroshima University, Faculty of Science, 1-3-1 Kagamiyama, Higashihiroshima-shi, JP–Hiroshima 739-8526, Japan

[60] Hiroshima Institute of Technology, Faculty of Applied Information Science, 2-1-1 Miyake Saeki-ku, Hiroshima-shi, JP–Hiroshima 731-5193, Japan

[61] Indiana University, Department of Physics, Swain Hall West 117, Bloomington, IN 47405-7105, United States of America

[62] Institut für Astro- und Teilchenphysik, Technikerstrasse 25, A-6020 Innsbruck, Austria

[63] University of Iowa, 203 Van Allen Hall, Iowa City, IA 52242-1479, United States of America

[64] Iowa State University, Department of Physics and Astronomy, Ames High Energy Physics Group, Ames, IA 50011-3160, United States of America

[65] Joint Institute for Nuclear Research, JINR Dubna, RU-141 980 Moscow Region, Russia

[66] KEK, High Energy Accelerator Research Organization, 1-1 Oho, Tsukuba-shi, Ibaraki-ken 305-0801, Japan

[67] Kobe University, Graduate School of Science, 1-1 Rokkodai-cho, Nada-ku, JP Kobe 657-8501, Japan

[68] Kyoto University, Faculty of Science, Oiwake-cho, Kitashirakawa, Sakyou-ku, Kyoto-shi, JP–Kyoto 606-8502, Japan

[69] Kyoto University of Education, 1 Fukakusa, Fujimori, Fushimi-ku, Kyoto-shi, JP–Kyoto 612-8522, Japan

[70] Universidad Nacional de La Plata, FCE, Departamento de Física, IFLP (CONICET-UNLP), C.C. 67, 1900 La Plata, Argentina

[71] Lancaster University, Physics Department, Lancaster LA1 4YB, United Kingdom

[72] INFN Sezione di Lecce(a); Università del Salento, Dipartimento di Fisica(b) Via Arnesano IT-73100 Lecce, Italy

[73] University of Liverpool, Oliver Lodge Laboratory, P.O. Box 147, Oxford Street, Liverpool L69 3BX, United Kingdom

[74] Jožef Stefan Institute and University of Ljubljana, Department of Physics, SI-1000 Ljubljana, Slovenia

[75] Queen Mary University of London, Department of Physics, Mile End Road, London E1 4NS, United Kingdom

[76] Royal Holloway, University of London, Department of Physics, Egham Hill, Egham, Surrey TW20 0EX, United Kingdom

[77] University College London, Department of Physics and Astronomy, Gower Street, London WC1E 6BT, United Kingdom

[78] Laboratoire de Physique Nucléaire et de Hautes Energies, Université Pierre et Marie Curie (Paris 6), Université Denis Diderot (Paris-7), CNRS/IN2P3, Tour 33, 4 place Jussieu, FR-75252 Paris Cedex 05, France

[79] Lunds Universitet, Naturvetenskapliga Fakulteten, Fysiska Institutionen, Box 118, SE-221 00 Lund, Sweden

[80] Universidad Autonoma de Madrid, Facultad de Ciencias, Departamento de Fisica Teorica, ES-28049 Madrid, Spain

[81] Universität Mainz, Institut für Physik, Staudinger Weg 7, DE-55099 Mainz, Germany

[82] University of Manchester, School of Physics and Astronomy, Manchester M13 9PL, United Kingdom

[83] CPPM, Aix-Marseille Université, CNRS/IN2P3, Marseille, France

[84] University of Massachusetts, Department of Physics, 710 North Pleasant Street, Amherst, MA 01003, United States of America

[85] McGill University, High Energy Physics Group, 3600 University Street, Montreal, Quebec H3A 2T8, Canada

[86] University of Melbourne, School of Physics, AU–Parkville, Victoria 3010, Australia

[87] The University of Michigan, Department of Physics, 2477 Randall Laboratory, 500 East University, Ann Arbor, MI 48109-1120, United States of America

[88] Michigan State University, Department of Physics and Astronomy, High Energy Physics Group, East Lansing, MI 48824-2320, United States of America

[89] INFN Sezione di Milano(a); Università di Milano, Dipartimento di Fisica(b), via Celoria 16, IT-20133 Milano, Italy

[90] B.I. Stepanov Institute of Physics, National Academy of Sciences of Belarus, Independence Avenue 68, Minsk 220072, Republic of Belarus

[91] National Scientific & Educational Centre for Particle & High Energy Physics, NC PHEP BSU, M. Bogdanovich St. 153, Minsk 220040, Republic of Belarus

[92] Massachusetts Institute of Technology, Department of Physics, Room 24-516, Cambridge, MA 02139, United States of America

[93] University of Montreal, Group of Particle Physics, C.P. 6128, Succursale Centre-Ville, Montreal, Quebec, H3C 3J7, Canada

[94] P.N. Lebedev Institute of Physics, Academy of Sciences, Leninsky pr. 53, RU-117 924 Moscow, Russia

[95] Institute for Theoretical and Experimental Physics (ITEP), B. Cheremushkinskaya ul. 25, RU 117 218 Moscow, Russia

[96] Moscow Engineering & Physics Institute (MEPhI), Kashirskoe Shosse 31, RU-115409 Moscow, Russia

[97] Lomonosov Moscow State University Skobeltsyn Institute of Nuclear Physics (MSU SINP), 1(2), Leninskie gory, GSP-1, Moscow 119991 Russian Federation, Russia

[98] Ludwig-Maximilians-Universität München, Fakultät für Physik, Am Coulombwall 1, DE-85748 Garching, Germany

[99] Max-Planck-Institut für Physik (Werner-Heisenberg-Institut), Föhringer Ring 6, 80805 München, Germany

[100] Nagasaki Institute of Applied Science, 536 Aba-machi, JP Nagasaki 851-0193, Japan

[101] Nagoya University, Graduate School of Science, Furo-Cho, Chikusa-ku, Nagoya, 464-8602, Japan

[102] INFN Sezione di Napoli[a]; Università di Napoli, Dipartimento di Scienze Fisiche[b], Complesso Universitario di Monte Sant'Angelo, via Cinthia, IT-80126 Napoli, Italy

[103] University of New Mexico, Department of Physics and Astronomy, MSC07 4220, Albuquerque, NM 87131 USA, United States of America

[104] Radboud University Nijmegen/NIKHEF, Department of Experimental High Energy Physics, Heyendaalseweg 135, NL-6525 AJ, Nijmegen, Netherlands

[105] Nikhef National Institute for Subatomic Physics, and University of Amsterdam, Science Park 105, 1098 XG Amsterdam, Netherlands

[106] [a] DeKalb, Illinois 60115, United States of America

[107] Budker Institute of Nuclear Physics (BINP), RU–Novosibirsk 630 090, Russia

[108] New York University, Department of Physics, 4 Washington Place, New York NY 10003, USA, United States of America

[109] Ohio State University, 191 West Woodruff Ave, Columbus, OH 43210-1117, United States of America

[110] Okayama University, Faculty of Science, Tsushimanaka 3-1-1, Okayama 700-8530, Japan

[111] University of Oklahoma, Homer L. Dodge Department of Physics and Astronomy, 440 West Brooks, Room 100, Norman, OK 73019-0225, United States of America

[112] Oklahoma State University, Department of Physics, 145 Physical Sciences Building, Stillwater, OK 74078-3072, United States of America

[113] Palacký University, 17. listopadu 50a, 772 07 Olomouc, Czech Republic

[114] University of Oregon, Center for High Energy Physics, Eugene, OR 97403-1274, United States of America

[115] LAL, Univ. Paris-Sud, IN2P3/CNRS, Orsay, France

[116] Osaka University, Graduate School of Science, Machikaneyama-machi 1-1, Toyonaka, Osaka 560-0043, Japan

[117] University of Oslo, Department of Physics, P.O. Box 1048, Blindern, NO-0316 Oslo 3, Norway

[118] Oxford University, Department of Physics, Denys Wilkinson Building, Keble Road, Oxford OX1 3RH, United Kingdom

[119] INFN Sezione di Pavia[a]; Università di Pavia, Dipartimento di Fisica Nucleare e Teorica[b], Via Bassi 6, IT-27100 Pavia, Italy

[120] University of Pennsylvania, Department of Physics, High Energy Physics Group, 209 S. 33rd Street, Philadelphia, PA 19104, United States of America

[121] Petersburg Nuclear Physics Institute, RU-188 300 Gatchina, Russia

[122] INFN Sezione di Pisa[a]; Università di Pisa, Dipartimento di Fisica E. Fermi[b], Largo B. Pontecorvo 3, IT-56127 Pisa, Italy

[123] University of Pittsburgh, Department of Physics and Astronomy, 3941 O'Hara Street, Pittsburgh, PA 15260, United States of America

[124] Laboratorio de Instrumentacao e Fisica Experimental de Particulas–LIP[a], Avenida Elias Garcia 14-1, PT-1000-149 Lisboa, Portugal; Universidad de Granada, Departamento de Fisica Teorica y del Cosmos and CAFPE[b], E-18071 Granada, Spain

[125] Institute of Physics, Academy of Sciences of the Czech Republic, Na Slovance 2, CZ-18221 Praha 8, Czech Republic

[126] Charles University in Prague, Faculty of Mathematics and Physics, Institute of Particle and Nuclear Physics, V Holesovickach 2, CZ-18000 Praha 8, Czech Republic

[127] Czech Technical University in Prague, Zikova 4, CZ-166 35 Praha 6, Czech Republic

[128] State Research Center Institute for High Energy Physics, Moscow Region, 142281, Protvino, Pobeda street, 1, Russia

[129] Rutherford Appleton Laboratory, Science and Technology Facilities Council, Harwell Science and Innovation Campus, Didcot OX11 0QX, United Kingdom

[130] University of Regina, Physics Department, Canada

[131] Ritsumeikan University, Noji Higashi 1 chome 1-1, JP–Kusatsu, Shiga 525-8577, Japan

[132] INFN Sezione di Roma I[a]; Università La Sapienza, Dipartimento di Fisica[b], Piazzale A. Moro 2, IT-00185 Roma, Italy

[133] INFN Sezione di Roma Tor Vergata[a]; Università di Roma Tor Vergata, Dipartimento di Fisica[b], via della Ricerca Scientifica, IT-00133 Roma, Italy

[134] INFN Sezione di Roma Tre[a]; Università Roma Tre, Dipartimento di Fisica[b], via della Vasca Navale 84, IT-00146 Roma, Italy

[135] Réseau Universitaire de Physique des Hautes Energies (RUPHE): Université Hassan II, Faculté des Sciences Ain Chock[a], B.P. 5366, MA–Casablanca; Centre National de l'Energie des Sciences Techniques Nucleaires (CNESTEN)[b], B.P. 1382 R.P. 10001 Rabat 10001; Université Mohamed Premier[c], LPTPM, Faculté des Sciences, B.P.717. Bd. Mohamed VI, 60000, Oujda; Université Mohammed V, Faculté des Sciences[d], 4 Avenue Ibn Battouta, BP 1014 RP, 10000 Rabat, Morocco

[136] CEA, DSM/IRFU, Centre d'Etudes de Saclay, FR-91191 Gif-sur-Yvette, France

[137] University of California Santa Cruz, Santa Cruz Institute for Particle Physics (SCIPP), Santa Cruz, CA 95064, United States of America

[138] University of Washington, Seattle, Department of Physics, Box 351560, Seattle, WA 98195-1560, United States of America

[139] University of Sheffield, Department of Physics & Astronomy, Hounsfield Road, Sheffield S3 7RH, United Kingdom

[140] Shinshu University, Department of Physics, Faculty of Science, 3-1-1 Asahi, Matsumoto-shi, JP–Nagano 390-8621, Japan

[141] Universität Siegen, Fachbereich Physik, D 57068 Siegen, Germany

[142] Simon Fraser University, Department of Physics, 8888 University Drive, CA–Burnaby, BC V5A 1S6, Canada

[143] SLAC National Accelerator Laboratory, Stanford, California 94309, United States of America

[144] Comenius University, Faculty of Mathematics, Physics & Informatics[(a)], Mlynska dolina F2, SK-84248 Bratislava; Institute of Experimental Physics of the Slovak Academy of Sciences, Dept. of Subnuclear Physics[(b)], Watsonova 47, SK-04353 Kosice, Slovak Republic

[145] Stockholm University: Department of Physics[(a)]; The Oskar Klein Centre[(b)], AlbaNova, SE-106 91 Stockholm, Sweden

[146] Royal Institute of Technology (KTH), Physics Department, SE-106 91 Stockholm, Sweden

[147] Stony Brook University, Department of Physics and Astronomy, Nicolls Road, Stony Brook, NY 11794-3800, United States of America

[148] University of Sussex, Department of Physics and Astronomy Pevensey 2 Building, Falmer, Brighton BN1 9QH, United Kingdom

[149] University of Sydney, School of Physics, AU–Sydney NSW 2006, Australia

[150] Insitute of Physics, Academia Sinica, TW–Taipei 11529, Taiwan

[151] Technion, Israel Inst. of Technology, Department of Physics, Technion City, IL–Haifa 32000, Israel

[152] Tel Aviv University, Raymond and Beverly Sackler School of Physics and Astronomy, Ramat Aviv, IL–Tel Aviv 69978, Israel

[153] Aristotle University of Thessaloniki, Faculty of Science, Department of Physics, Division of Nuclear & Particle Physics, University Campus, GR-54124, Thessaloniki, Greece

[154] The University of Tokyo, International Center for Elementary Particle Physics and Department of Physics, 7-3-1 Hongo, Bunkyo-ku, JP–Tokyo 113-0033, Japan

[155] Tokyo Metropolitan University, Graduate School of Science and Technology, 1-1 Minami-Osawa, Hachioji, Tokyo 192-0397, Japan

[156] Tokyo Institute of Technology, 2-12-1-H-34 O-Okayama, Meguro, Tokyo 152-8551, Japan

[157] University of Toronto, Department of Physics, 60 Saint George Street, Toronto M5S 1A7, Ontario, Canada

[158] TRIUMF[(a)], 4004 Wesbrook Mall, Vancouver, B.C. V6T 2A3; [(b)] York University, Department of Physics and Astronomy, 4700 Keele St., Toronto, Ontario, M3J 1P3, Canada

[159] University of Tsukuba, Institute of Pure and Applied Sciences, 1-1-1 Tennoudai, Tsukuba-shi, JP–Ibaraki 305-8571, Japan

[160] Tufts University, Science & Technology Center, 4 Colby Street, Medford, MA 02155, United States of America

[161] Universidad Antonio Narino, Centro de Investigaciones, Cra 3 Este No. 47A-15, Bogota, Colombia

[162] University of California, Irvine, Department of Physics & Astronomy, CA 92697-4575, United States of America

[163] INFN Gruppo Collegato di Udine[(a)]; ICTP[(b)], Strada Costiera 11, IT-34014, Trieste; Università di Udine, Dipartimento di Fisica[(c)], via delle Scienze 208, IT-33100 Udine, Italy

[164] University of Illinois, Department of Physics, 1110 West Green Street, Urbana, Illinois 61801, United States of America

[165] University of Uppsala, Department of Physics and Astronomy, P.O. Box 516, SE-51 20 Uppsala, Sweden

[166] Instituto de Física Corpuscular (IFIC) Centro Mixto UVEG-CSIC, Apdo. 22085 ES-46071 Valencia, Dept. Física At. Mol. y Nuclear; Univ. of Valencia, and Instituto de Microelectrónica de Barcelona (IMB-CNM-CSIC) 08193 Bellaterra Barcelona, Spain

[167] University of British Columbia, Department of Physics, 6224 Agricultural Road, CA–Vancouver, B.C. V6T 1Z1, Canada

[168] University of Victoria, Department of Physics and Astronomy, P.O. Box 3055, Victoria B.C., V8W 3P6, Canada

[169] Waseda University, WISE, 3-4-1 Okubo, Shinjuku-ku, Tokyo, 169-8555, Japan

[170] The Weizmann Institute of Science, Department of Particle Physics, P.O. Box 26, IL-76100 Rehovot, Israel

[171] University of Wisconsin, Department of Physics, 1150 University Avenue, WI 53706 Madison, Wisconsin, United States of America

[172] Julius-Maximilians-University of Würzburg, Physikalisches Institute, Am Hubland, 97074 Würzburg, Germany

[173] Bergische Universität, Fachbereich C, Physik, Postfach 100127, Gauss-Strasse 20, D-42097 Wuppertal, Germany

[174] Yale University, Department of Physics, PO Box 208121, New Haven CT, 06520-8121, United States of America

[175] Yerevan Physics Institute, Alikhanian Brothers Street 2, AM-375036 Yerevan, Armenia

[176] ATLAS-Canada Tier-1 Data Centre, TRIUMF, 4004 Wesbrook Mall, Vancouver, BC, V6T 2A3, Canada

[177] GridKA Tier-1 FZK, Forschungszentrum Karlsruhe GmbH, Steinbuch Centre for Computing (SCC), Hermann-von-Helmholtz-Platz 1, 76344 Eggenstein-Leopoldshafen, Germany

[178] Port d'Informacio Cientifica (PIC), Universitat Autonoma de Barcelona (UAB), Edifici D, E-08193 Bellaterra, Spain

[179] Centre de Calcul CNRS/IN2P3, Domaine Scientifique de la Doua, 27 bd du 11 Novembre 1918, 69622 Villeurbanne Cedex, France

[180] INFN-CNAF, Viale Berti Pichat 6/2, 40127 Bologna, Italy

[181] Nordic Data Grid Facility, NORDUnet A/S, Kastruplundgade 22, 1, DK-2770 Kastrup, Denmark

[182] SARA Reken- en Netwerkdiensten, Science Park 121, 1098 XG Amsterdam, Netherlands

[183] Academia Sinica Grid Computing, Institute of Physics, Academia Sinica, No.128, Sec. 2, Academia Rd., Nankang, Taipei, Taiwan 11529, Taiwan

[184] UK-T1-RAL Tier-1, Rutherford Appleton Laboratory, Science and Technology Facilities Council, Harwell Science and Innovation Campus, Didcot OX11 0QX, United Kingdom

[185] RHIC and ATLAS Computing Facility, Physics Department, Building 510, Brookhaven National Laboratory, Upton, New York 11973, United States of America

[a] Also at CPPM, Marseille, France.

[b] Also at TRIUMF, 4004 Wesbrook Mall, Vancouver, B.C. V6T 2A3, Canada

[c] Also at Faculty of Physics and Applied Computer Science of the AGH-University of Science and Technology (FPACS, AGH-UST), al. Mickiewicza 30, PL-30059 Cracow, Poland

[d] Also at Università di Napoli Parthenope, via A. Acton 38, IT-80133 Napoli, Italy

[e] Also at Institute of Particle Physics (IPP), Canada

[f] Louisiana Tech University, 305 Wisteria Street, P.O. Box 3178, Ruston, LA 71272, United States of America

[g] At Department of Physics, California State University, Fresno, 2345 E. San Ramon Avenue, Fresno, CA 93740-8031, United States of America

[h] Currently at Istituto Universitario di Studi Superiori IUSS, V.le Lungo Ticino Sforza 56, 27100 Pavia, Italy

[i] Also at California Institute of Technology, Physics Department, Pasadena, CA 91125, United States of America

[j] Also at University of Montreal, Canada

[k] Also at Institut für Experimentalphysik, Universität Hamburg, Luruper Chaussee 149, 22761 Hamburg, Germany

[l] Also at Petersburg Nuclear Physics Institute, RU-188 300 Gatchina, Russia

[m] Also at School of Physics and Engineering, Sun Yat-sen University, China

[n] Also at School of Physics, Shandong University, Jinan, China

[o] Also at Rutherford Appleton Laboratory, Science and Technology Facilities Council, Harwell Science and Innovation Campus, Didcot OX11, United Kingdom

[p] Also at School of Physics, Shandong University, Jinan

[q] Also at Rutherford Appleton Laboratory, Science and Technology Facilities Council, Harwell Science and Innovation Campus, Didcot OX11 0QX, United Kingdom

[r] Now at KEK

[s] University of South Carolina, Dept. of Physics and Astronomy, 700 S. Main St, Columbia, SC 29208, United States of America

[t] Also at KFKI Research Institute for Particle and Nuclear Physics, Budapest, Hungary

[u] Also at Institute of Physics, Jagiellonian University, Cracow, Poland

[v] Also at School of Physics and Engineering, Sun Yat-sen University, Taiwan

[w] Transfer to LHCb 31.01.2010

[x] Also at Dept of Physics, Nanjing University, China

[†] Deceased.

Abstract The ATLAS detector at the Large Hadron Collider has collected several hundred million cosmic ray events during 2008 and 2009. These data were used to commission the Muon Spectrometer and to study the performance of the trigger and tracking chambers, their alignment, the detector control system, the data acquisition and the analysis programs. We present the performance in the relevant parameters that determine the quality of the muon measurement. We discuss the single element efficiency, resolution and noise rates, the calibration method of the detector response and of the alignment system, the track reconstruction efficiency and the momentum measurement. The results show that the detector is close to the design performance and that the Muon Spectrometer is ready to detect muons produced in high energy proton–proton collisions.

Contents

** e-mail: atlas.secretariat@cern.ch

1 The ATLAS Muon Spectrometer

The ATLAS Muon Spectrometer (MS in the following) is designed to provide a standalone measurement of the muon momentum with an uncertainty in the transverse momentum varying from 3% at 100 GeV to about 10% at 1 TeV, and to provide a trigger for muons with varying transverse momentum thresholds down to a few GeV. A detailed description of the muon spectrometer and of its expected performance can be found in [1–3]. Here only a brief overview is given. The muon momentum is determined by measuring the track curvature in a toroidal magnetic field. The muon trajectory is always normal to the main component of the magnetic field so that the transverse momentum resolution is roughly independent of η over the whole acceptance. The magnetic field is provided by three toroids, one in the "barrel" ($|\eta| < 1.1$) and one for each "end-cap" ($1.1 < |\eta| < 2.7$), with a field integral between 2 and 8 Tm. The muon curvature is measured by means of three precision chamber stations positioned along its trajectory. In order to meet the required precision each muon station should provide a measurement on the muon trajectory with an accuracy of 50 μm. In Fig. 1 a schematic view of the muon spectrometer[1] is given.

For most of the acceptance Monitored Drift Tube (MDT) chambers are deployed [2]. The coordinate in the plane perpendicular to the wires, measured by the MDT, is referred to as the precision, or bending coordinate, being mainly per-

---

[1] The ATLAS reference system is a Cartesian right-handed coordinate system, with the nominal collision point at the origin. The positive x-axis is defined as pointing from the collision point to the center of the LHC ring and the positive y-axis points upwards while the z-axis is tangent to the beam direction at the collision point. The azimuthal angle ϕ is measured around the beam axis, and the polar angle θ is the angle measured with respect to the z-axis. The pseudorapidity is defined as $\eta = -\ln\tan\theta/2$.

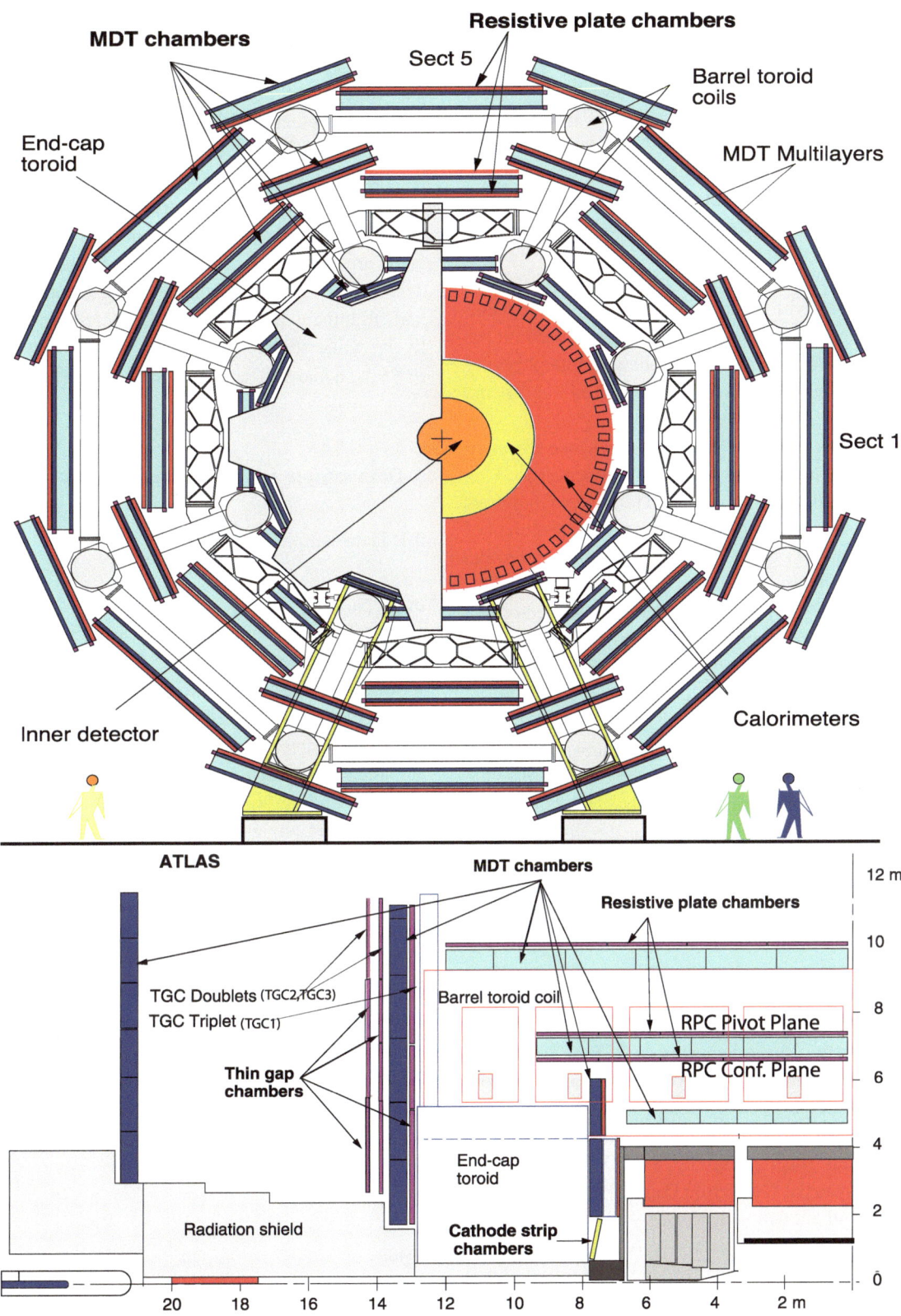

Fig. 1 Schematic view of the muon spectrometer in the $x-y$ (*top*) and $z-y$ (*bottom*) projections. Inner, Middle and Outer chamber stations are denoted BI, BM, BO in the barrel and EI, EM, EO in the end-cap

pendicular to the direction of the toroidal field. In the end-cap inner region, for $|\eta| < 2.0$, Cathode Strip Chambers (CSC) [2] are used because of their capability to cope with higher background rates.

The MDT chambers are composed of two MultiLayers (ML) made of three or four layers of tubes. Each tube is 30 mm in diameter and has an anode wire of 50 µm diameter. The gas mixture used is 93% Ar and 7% CO_2 with a small admixture of water vapor, the drift velocity is not saturated and the total drift time is about 700 ns. The space resolution attainable with a single tube is about 80 µm, measured in a test beam [4, 5]. The CSC chambers are multiwire proportional chambers with cathode strip read out. The cathode planes are equipped with orthogonal strips and the precision coordinate is obtained measuring the charge induced on the strips making the charge interpolation between neighboring strips. Typical resolution obtained with this read-out scheme is about 50 µm.

The trigger system of the MS is based on two different chamber technologies: Resistive Plate Chambers (RPC) [2] instrument the barrel region while Thin Gap Chambers (TGC) [2] are used in the higher background environment of the end-cap regions. Two RPC chambers are attached to the middle barrel chambers providing a low-p_T trigger. A high-p_T trigger is provided by the RPC modules installed on the outer barrel chambers in combination with the low p_T signal provided by the middle chambers. The RPCs also provide the coordinate along the MDT wires that is not measured by the MDT chambers.

Similarly in the end-cap two TGC doublets and one triplet are installed close to the middle station and provide the low-p_T and high-p_T trigger signals. The TGCs also measure the coordinate of the muons in the direction parallel to the MDT wires. This coordinate is referred to as the second, or non-bending coordinate. For this purpose TGC chambers are also installed close to the MDTs in the inner layer of the end-cap (EI).

Some MS naming conventions adopted in this paper are introduced here. The MS is divided in the x–y-plane (also referred to as ϕ-plane) in 16 sectors: Sector 5 being the upper most and Sector 13 the lower most. In both barrel and end-cap regions the MS is divided into 8 'Large' sectors (odd numbered sectors) and 8 'Small' sectors (even numbered sectors), determined by their coverage in ϕ. The muon stations are named 'Inner', 'Middle', and 'Outer', according to the distance from the Interaction Point (IP). The three stations for the barrel are denoted BI, BM, and BO, and for the End-Cap EI, EM, and EO, respectively. Along the z axis, the MS is divided into two sides, called side A (positive z) and C (negative z).

As a complementary source of information, two publications [4, 5] on a detector system-test with a high momentum muon beam can be consulted.

Beginning in September 2008 the ATLAS detector was operated continuously up to November 2008 and then for different periods starting from Spring 2009. The first beams were circulated in the LHC machine in September 2008 but no beam-beam collisions were delivered. During these periods, the ATLAS detector collected mainly cosmic ray data. All muon detector technologies were included in the run with the exception of CSCs for which the Read Out chain was still not yet commissioned and therefore they are not included in the results presented in this paper.

The analyzed data samples and the reconstruction software are described in Sect. 2. The cosmic ray trigger is described in Sect. 3. Studies of data quality, calibration, and alignment are presented in Sects. 4, 5, 6, 7 respectively, while studies on tracking performance are presented in Sects. 8 and 9. The results are summarized in Sect. 10.

2 Data sample and reconstruction software

2.1 Data sample

In preparation for LHC collisions, the ATLAS detector has acquired several hundred million cosmic ray events during several run periods in 2008 and 2009. The analysis of a subset of data corresponding to about 60 M events is presented here. These runs allowed commissioning the ATLAS experiment, the trigger, the data acquisition, the various detectors and the reconstruction software. Most of the cosmic rays reach the underground detectors via the two big shafts. They have incident angles close to the vertical axis and they are mainly triggered by the RPCs. The selected runs, together with the status of the magnetic field in the MS and the number of collected events for the different trigger streams, are listed in Table 1.

Table 1 List of analyzed data runs together with the corresponding trigger stream, statistics and status of the MS magnetic field. All runs were collected in Fall 2008, with the exceptions of run 113860 collected in Spring 2009, and run 121080 in Summer 2009

Run	Trigger	B-field	N of Evts	Period
91060	RPC	Off	17 M	Fall 08
91060	TGC	Off	0.2 M	Fall 08
89106	TGC	Off	0.4 M	Fall 08
89403	TGC	Off	0.4 M	Fall 08
91803	TGC	**On**	50 K	Fall 08
91890	RPC	**On**	16 M	Fall 08
113860	RPC	Off	6 M	Spring 09
121080	RPC	**On**	21 M	Summ 09

2.2 Muon reconstruction software

The data were processed using the complete ATLAS software chain [6]: data decoding, data preparation (which includes calibration and alignment), and track reconstruction. Muon reconstruction has been handled by two independent packages, namely *Moore* [7] and *Muonboy* [8]. The two reconstruction algorithms are similar in design but differ in some details. The general strategy is to reconstruct muon trajectories both at the local (individual chamber), as well as at the global (spectrometer), level. The trajectories reconstructed in individual chambers can be approximated as straight lines over a short distance where bending has little effect and are therefore fit to track *segments*. Full tracks are formed by combining segments from multiple chambers.

Prompt muons produced in proton–proton collisions have trajectories that point back to the Interaction Point (IP). Moreover they are synchronous with the collision since all the detector front end electronics are synchronized with the LHC bunch crossing frequency of 40 MHz. In contrast, cosmic ray muons are "non-pointing" and are asynchronous with the detector clock: they have an additional 25 ns jitter with respect to the clock selected by the trigger. In addition, during commissioning the different trigger detectors were not timed with sufficient precision, leading to variations in timing depending on the region of the detector that originated the trigger. A further difficulty in track reconstruction was due to the lack of precise alignment of the muon detectors during this commissioning phase, as described in Sect. 7.

The reconstruction algorithms were adapted for these "cosmic ray" conditions as described below. Both programs were modified by relaxing the standard tracking requirements and implementing a procedure to accommodate the cosmic ray timing conditions. The tolerance for hit association to form track segments and the uncertainty associated with each hit position were increased. Moreover, a procedure called *global t_0 refit* (Gt$_0$-refit) was developed in both reconstruction algorithms to compensate for the 25 ns time jitter and the imprecise trigger timing. The aim of this procedure is to determine with better precision the time when the cosmic ray crossed the detector by introducing a free global timing parameter (gt_0) in the segment reconstruction. The implementation of the Gt$_0$-refit in the two reconstruction algorithms is briefly described below while the results are presented in Sect. 5.

2.3 Muonboy track reconstruction

The strategy of the *Muonboy* reconstruction algorithm can be summarized in four main steps:

– identification of Regions Of Activity (ROA) in the muon system with the information provided by the RPC/TGC detectors

– reconstruction of local segments in each muon station in the identified ROA

– combination of segments of different muon stations to form muon track candidates using three-dimensional tracking in magnetic field

– global track fit of the muon track candidates through the full system using individual hit information.

The topology of cosmic ray tracks is accommodated by relaxing the Region Of Activity requirement of pointing in a projective geometry when associating hits to form segments, or matching segments to form tracks. Moreover, since cosmic ray events have low occupancy, looser quality criteria were used for the selection of segments and tracks.

The *Muonboy* algorithm for the Gt$_0$-refit consists of a scan of different gt_0 values in steps of 10 ns, doing the full segment reconstruction at each step. The gt_0 value giving the best reconstruction quality factor is kept and a parabolic fit is performed using this best value and the two closer values along the parabola. Then the gt_0 corresponding to the minimum of the quality factor parabola is chosen. In order to obtain high efficiency, the accuracy requirement for the MDT single hit resolution is relaxed by adding in quadrature a 0.5 mm constant smearing to the intrinsic resolution function (described in Sect. 6). This smearing is increased by additional 0.5 mm if the Gt$_0$-refit fails. Moreover, a less demanding track quality factor is required for tracks when hits are missing or are not associated to the track.

2.4 Moore track reconstruction

The *Moore* reconstruction algorithm is built out of several distinct stages:

– identification of global roads throughout the entire spectrometer using all muon detectors (MDT, CSC, RPC and TGC)

– reconstruction of local segments in each muon station seeded by the identified global roads

– combination of segments of different muon stations to form muon track candidates

– global, three-dimensional, tracking and final track fit.

Several modifications to the standard pattern recognition were made to optimize the reconstruction of cosmic ray tracks. In the global road finding step, a straight line *Hough transform* was used to allow for non-pointing tracks. The cuts on distance and direction between the road and the segment were relaxed. In the segment finding no cuts were applied on the number of missing hits (i.e. drift tubes that are expected to be crossed but have no hits).

The Gt$_0$-refit consists in varying simultaneously the global time offset (gt_0) for each segment reconstructed in a chamber. Then all measured times of hits associated to the segment are translated into drift radii after subtraction of the

gt_0. The gt_0 value that minimizes the sum in quadrature of the weighted residuals (corresponding to the segment reconstruction χ^2) is selected.

In this fit the MDT uncertainties are set to twice the test-beam drift tube resolution. If the segment fit is not successful, a straight line fit is performed assuming a constant 1 mm error. Hits are removed if their distance from the segment is greater than 7σ. In the track fit the MDT errors are enlarged to 2 mm to account for uncertainties in the alignment of chamber stations.

3 Trigger configuration during data taking

A more detailed description of the trigger system can be found in [2, 9]. Here only specific issues related to the 2008–2009 cosmic ray data taking are introduced. The muon level-1 trigger is issued by the RPC in the barrel and by the TGC in the end-caps. During cosmic ray data taking most of the statistics were collected using this trigger. Special trigger configurations were adopted with different geometries (e.g. non pointing to the IP) and different timing (e.g. delaying the triggers issued by the upper sectors in order to trigger only in the lower sectors to mimic particles coming from the IP) when commissioning the muon trigger system itself or when selecting cosmic rays for commissioning the other ATLAS sub-systems.

In *beam-collision* configuration, the level-1 muon trigger selects pointing tracks with six different thresholds in transverse momentum and sends information to the Central-Trigger-Processor (CTP). The six thresholds, three low-p_T and three high-p_T, do not distinguish between different detector regions, barrel or end-cap. For cosmic rays, to help commissioning separately the two regions, it was chosen to assign three thresholds to the barrel and three to the end-cap.

3.1 Barrel level-1 trigger

The barrel trigger detectors are arranged in three stations each having a doublet of RPC layers at increasing distances from the IP. In each sector the first two stations are mechanically coupled to the BM MDT while the third is coupled with the BO MDT as shown in Fig. 1.

The trigger algorithm is steered by signals on the middle layers, named Pivot plane. When a hit is found on this plane, the low-p_T trigger logic searches for hits in the inner layers, named Confirm plane, and requires a coincidence in time of three hits over the four layers in a pre-calculated cone. The width of this cone defines the p_T threshold. If hits are also found in a pre-calculated cone of the outermost plane in coincidence with a low-p_T trigger, a high-p_T trigger is issued. Also in this case the p_T threshold is defined by the width of the cone. In addition to the p_T requirement, the

trigger logic also demands the track to be pointing towards the IP both in ϕ and η. In the cosmic ray runs only three of the six thresholds were used in the barrel and were defined as *MU0_LOW*, *MU0_HIGH* and *MU6*. The two thresholds *MU0_LOW*/*HIGH* did not select a physical p_T range; in fact, the *MU0_LOW* was triggered only by the time coincidence of 3 out of 4 hits without any pointing constraint and the *MU0_HIGH* was triggered by the coincidence of a *MU0_LOW* with at least a hit in the corresponding outer plane. The threshold *MU6* required not only a time coincidence but also an IP-pointing constraint in the ϕ-projection only. To emulate the timing expected for beam collisions, and to enhance the illumination of the Inner Detector (ID), the cosmic ray trigger was issued mainly by the bottom sectors. This was achieved by delaying the top sector trigger by 5 BC (125 ns) preventing it from arriving first at the Central Trigger Processor (CTP) and thus forming the trigger.

In the fall 2008 data taking period, the timing of the low-p_T trigger and the data read-out latencies were still under commissioning. This had a large impact on the detector coverage. The situation has largely improved for the runs taken in 2009 both in terms of detector coverage and in trigger timing as shown in Sects. 4.3 and 6.2.

3.2 End-cap level-1 trigger

The level-1 TGC trigger system provided three thresholds, named *MU0_TGC_HALO*, *MU0_TGC* and *MU6_TGC*. The trigger was issued by the coincidence between several TGC layers. The logic was based both on timing (BC identification) and geometry (pointing track). The main difference between the three trigger thresholds is related to the required number of layers and to the degree of pointing to the IP. *MU0_TGC_HALO* required a 3 out of 4 layer coincidence in the two outermost TGC stations, the so-called *Doublet chambers*, in both η (bending) and ϕ (non-bending) projections and a pointing requirement within $20°$. *MU0_TGC* and *MU6_TGC* required in addition a 2 out of 3 layer coincidence in the TGC stations closer to the IP, the so-called *Triplet chambers*, in the η projection only. The pointing requirement of *MU0_TGC* was of $\pm 10°$ degrees while for *MU6_TGC* was of $\pm 5°$.

The trigger was timed for high-momentum muons coming from the IP. All the delays due to different time-of-flight and cable lengths were properly set and cross-checked using a test pulse system achieving a relative timing within 4 ns. For most of the cosmic run period, only the level-1 trigger generated from the TGC bottom sectors was used. This was chosen to ensure good timing of the trigger with the read-out of the ID, since cosmic muons triggered by the TGC bottom sectors and crossing the ID have a time-of-flight similar to muons produced in collisions.

4 Data quality assessment

4.1 Introduction

The data quality assessment consists of several software algorithms working at different levels of the data taking. The Detector Control System (DCS) [11] is the first source of information available during the operation of the detector. Here information on the hardware status of the different sub-detectors and on the settings of Low Voltage (LV) and High Voltage (HV) power supplies and on the gas system is available. The DCS also receives information from the Data Acquisition (DAQ) [11] as soon as problems during the read-out of a chamber appear.

The next stage in the chain of data quality assessment is the on-line monitoring. It receives input from the data acquisition system running in a spectator mode. Once the data are decoded, monitoring histograms are filled showing quantities related to the detector operation. Part of the muon data selected by the level-1 trigger Region Of Interest (ROI) are transferred by the level-2 trigger processors to three dedicated computing farms (referred to as *calibration centers*) to monitor and determine the calibration parameters of the MS chambers. The larger event samples available at the calibration centers allow the analysis of single drift tube responses. The goal of the analysis at the calibration centers is to provide drift tube and trigger chambers calibration constants and to give general feedback on the detector operation within 24 hours, which is the time needed, at high luminosity, to collect enough statistics to calculate new calibration constants.

On a longer time scale, using the full reconstructed ATLAS event information, the off-line data monitoring provides the final information on the data quality. At each step a flag summarizing the data quality at that level is stored in a database.

4.2 MDT chambers

In the fall 2008 period (e.g. Run 91060) only five out of 1110 MDT chambers were not included in the data taking. Of these five chambers, two were not yet connected to services and three had problems with the gas system. Due to the cosmic ray illumination and the trigger coverage not all chambers had sufficient event samples to determine the performance of single drift tubes. The studies reported here were done at different levels of detail, from chamber information down to single drift tube information when the event samples were sufficient. The data survey searched for problems of individual read-out channels as well as of clusters corresponding to hardware related groups of tubes. A screen shot of one of the online monitoring applications used for the MDT chambers is shown in Fig. 2. Here the average number of hits per tube for each MDT is represented in a η–ϕ plot where the higher cosmic illumination on the top and bottom sectors (3–7, 11–15) compared to the vertical sectors (16–2 and 8–10) is clearly seen as well as the larger illumination on the A side of the detector where the larger shaft is present. The five chambers not included in the data acquisition are marked as dark gray boxes. Two more chambers are visible with very low statistics due to problems with the HV supplies. For 32 MDT chambers one of the two multi-layers was disconnected from HV.

A detailed list of hardware problems found in run 91060 is reported in Table 2. The cosmic ray flux was not sufficient for a detailed analysis of single drift tubes for 15 MDT chambers ($\sim$3 K channels). Thus we were able to analyze individually 336 K, out of the working 339 K, drift tubes. To summarize, about 5 K channels, out of 336 K, have shown some problems in run 91,060, corresponding to 1.5%. Most of these channels have been recovered during the 2008–2009 shutdown period. Only a very small fraction of problems, at the level of a few per mill, could not be solved, such as permanently disconnected tubes (broken wires) or chambers with very difficult access.

In addition to monitoring in the DAQ framework (on-line monitoring), the data are also processed with the off-line reconstruction program which produces monitoring histograms. This ensures that the reconstruction works properly and that the correct *conditions data* (calibration and alignment constants) are used in the first processing of the data. The off-line monitoring gathers and presents information on several variables for single drift tubes, e.g. drift time and collected charge distributions, hit occupancy and noise rate. These variables are obtained for individual MDTs or grouped for regions, such as η or ϕ sectors, barrel or end-cap, side A or C. Variables related to segments or tracks are also monitored.

4.3 Barrel trigger chambers: RPC

Commissioning of the RPC detectors progressed continuously and substantial improvements were made during the 2008–2009 shut-down. As an example Fig. 3 shows a two-dimensional distribution of RPC strips requiring a 3 out of 4 majority coincidence for the low-p_T trigger demonstrating that the trigger coverage in Spring 2009 was at the 95% level.

Studies of the trigger performance were made using the data of run 91,060 after implementation of the *trigger roads* [10]. For the low-p_T trigger the four RPC layers in the Middle station are involved, both in η and ϕ projections. Tracks were accepted by the trigger if any strip of the pivot plane was in coincidence with a group of strips of the confirm plane aligned with the IP, realizing a majority combination of 3 out of 4 RPC layers. Figure 4(A) shows the spatial

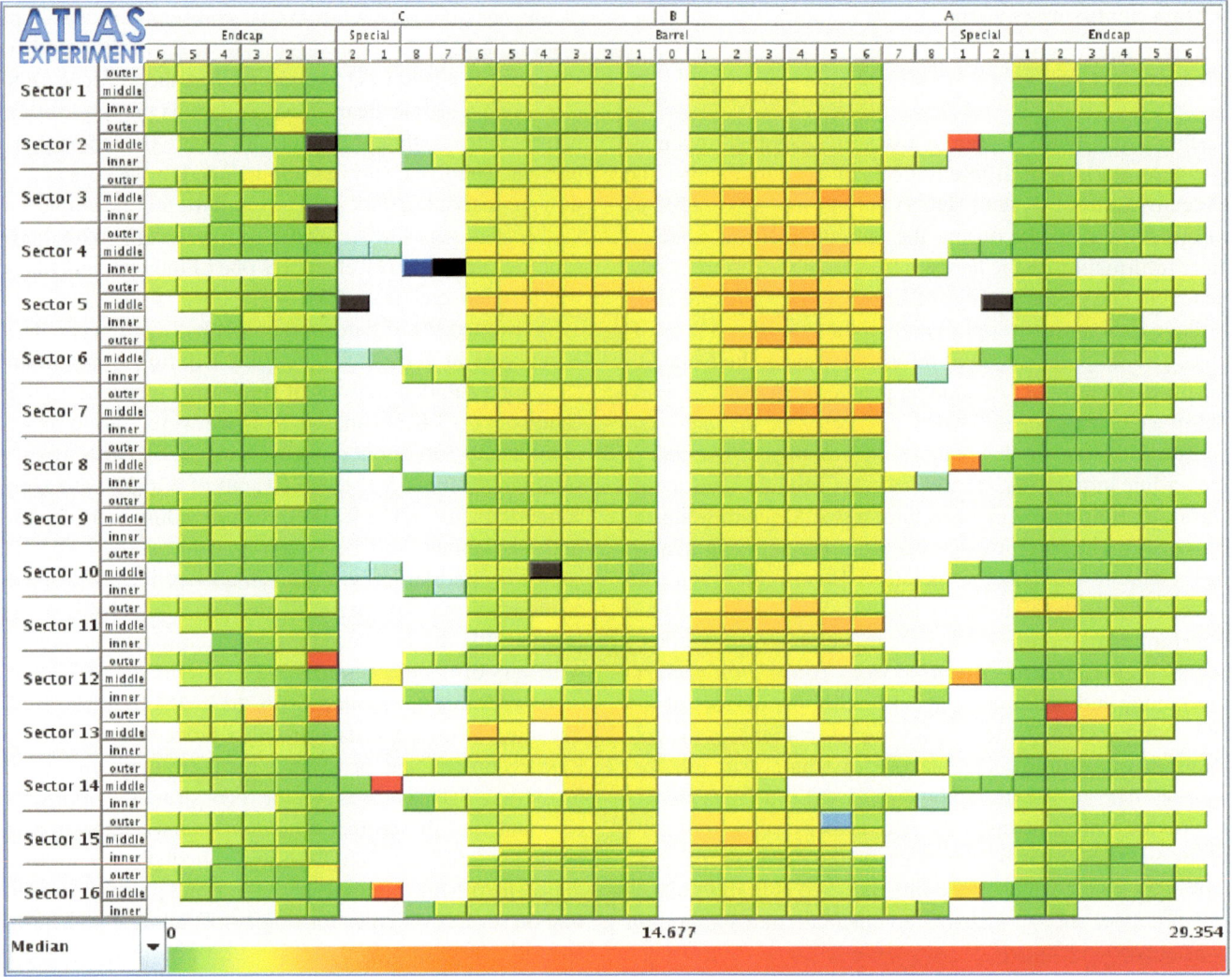

Fig. 2 Screen shot of a monitoring application displaying the MDT hit occupancy for all chambers. Each chamber is represented by *a small box. The color of the box* is related to the average number of raw hits per tube. *The boxes* are arranged in an η–ϕ grid: *a column* represents an η slice, perpendicular to the beam axis; *a row* represents one of the sixteen ϕ sectors. Within each sector chambers of the Inner, Middle, Outer ring are displayed separately

Table 2 List of MDT channels with problems in run 91,060

		Fraction
Number of channels analyzed with sufficient event samples	336,144	
Channels not included in the read-out	936	0.28%
Channels with read-out or initialization problems	744	0.22%
Channels with HV or gas problems	2942	0.88%
Permanently dead channels (broken wires)	323	0.10%
Total problematic channels	4945	1.47%

correlation between ϕ strips in the pivot planes and ϕ strips in the confirm planes. The correlation line is slightly rotated with respect to the diagonal due to the different distance of the confirm and pivot planes with respect to the IP.

A random trigger was used to measure the counting rate for each read-out strip. This is a measurement of the RPC system noise rate. About 310 K strips were analyzed over a total of 350 K working strips. Figure 4(B) shows the distribution of single channel noise rate, normalized to an area of 1 cm^2. For each strip, the noise rate is calculated as the number of hits divided by the number of random triggers and the width of the read-out gate of 200 ns, and is normal-

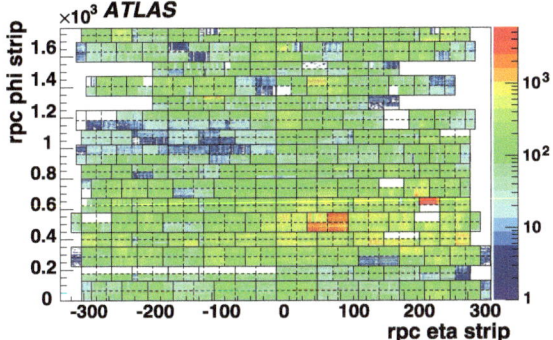

Fig. 3 RPC low-p_T trigger coverage in η–ϕ for Run 113,860 (Spring 2009). Each η and ϕ strip producing a low-p_T trigger corresponds to an entry in the plot. The coverage in Spring 2009 was about 95%

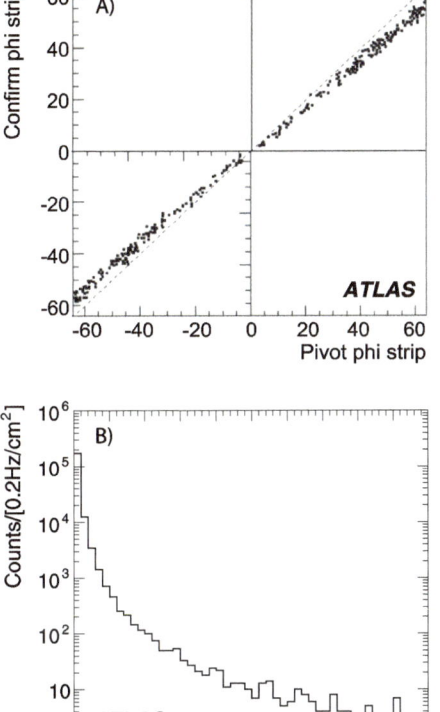

Fig. 4 (**A**): RPC spatial correlation between the pivot strip number and the confirm strip number in the ϕ projection for a programmed trigger road. 128 strips correspond to a RPC plane 3.8 m long. (**B**): Distribution of strip noise rates per unit area measured with a random trigger for 310 K RPC strips. The larger noise present on some strips is probably due to local weaknesses of grounding connections

ized to the area of the strip (typically 550 cm^2 for a BM eta strips and 900 cm^2 for a BO eta strips). Only a few hundred strips showed a counting rate above 10 Hz/cm^2 which is the background rate expected when the LHC will run at high-luminosity. The average noise rate of the RPC was stable during the different running periods.

The fraction of dead channels, considering only the part of the detector included in the read-out in the Fall 2008 runs, was 1.5%, mainly due to problems in the front-end electronics.

4.4 End-cap trigger chambers: TGC

In the end-caps the muon trigger is provided by the TGC chambers installed in three layers that surround the MDT Middle chambers. All together they form the so-called Big Wheels (BW), one in each end-cap. In addition, TGC chambers are also installed close to the EI chambers in the Small Wheels (SW), but these are only used to measure the muon ϕ coordinate. In Fall 2008 all the BW TGC sectors were read-out. Given the installation schedule for the ATLAS detectors, the Inner TGC station were the last chambers installed and they were not fully operational during 2008 runs. For this reason they are not discussed in the following.

Two types of trigger configuration were adopted in Fall 2008. One was optimized to study the end-cap muon detectors with cosmic rays. In this configuration all TGC BW sectors were used in the trigger. The other setting was optimized to provide the trigger for the ID tracking detectors and was used for timing the ID. In order to mimic muons coming from the IP, only the five bottom sectors were used to trigger. The typical detector coverage in these two trigger configurations is shown in Fig. 5 by plotting the coincidence positions in the x–y plane for wire and strip hits for run 91,060 (A) and run 91,803 (B). Only about 0.8% of chambers were not operational due to HV or gas problems. Since for the trigger a majority logic is required these inactive chambers do not produce any dead regions in the trigger acceptance.

The HV and front-end threshold setting, the gate widths for wires and strips, and the trigger sectors are listed in Table 3 for these two runs.

For each trigger issued by the CTP, the TGC Read Out Driver (ROD) sends to the DAQ system the data corresponding to three Bunch Crossings (Previous, Current and Next BC) contained in two separate buffers. Of the two buffers, one is located in the front-end board where the wires and strips providing the low-p_T coincidence are separately recorded. In the second buffer, located in the Sector Logic Board in the service counting room, the coincidence of the wire and strip signals is done. Each buffer has a programmable identifier that has to be adjusted in order to read out the correct (Current) BC data. Figure 6 shows the read-out timing for the front-end and the sector logic buffers for level-1 triggers issued by the TGC. About 98.6% of data in the front-end buffer, and 99.8% of data in the sector logic buffer are read out with the correct timing. The small population in the previous or next BC is due to cosmic ray showers.

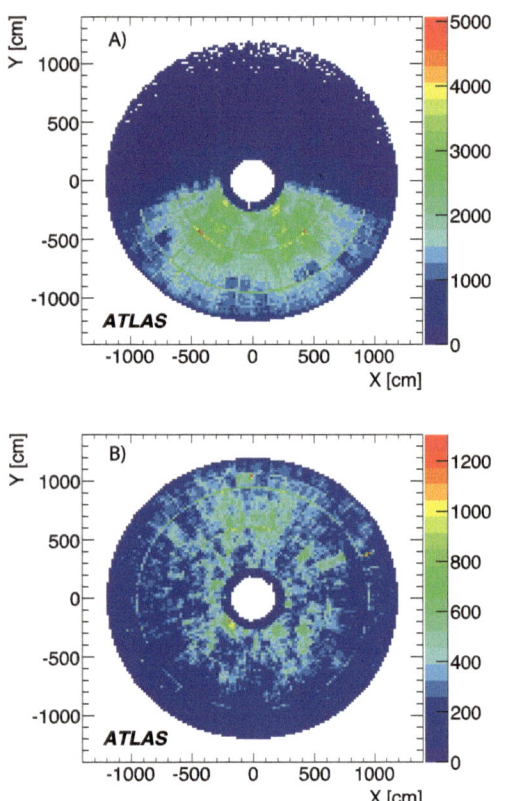

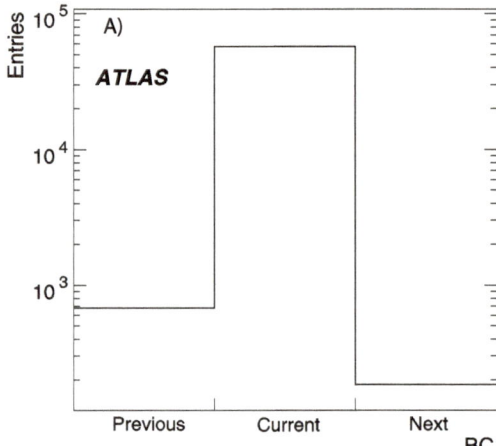

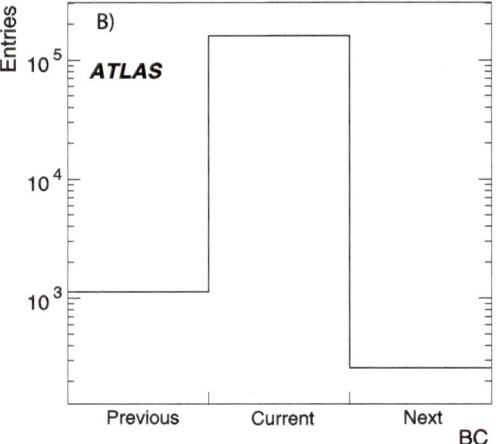

Fig. 5 Map of coincidences of wire and strip hits in the $x-y$ plane. (**A**): The five bottom sectors (sectors 8–12, $195° < \phi < 345°$) used for timing the ID tracking detectors in run 91,060. (**B**): With all sectors participating in the trigger during run 91,803

Table 3 TGC sectors participating in the trigger, high voltage setting, threshold and gate width

Run	Trigger sector	HV	Threshold	Gate widths for wire/strip
91,060	8 to 12	2800 V	100 mV	35/45 ns
91,803	1 to 12	2650 V	80 mV	35/45 ns

5 MDT chamber calibration

5.1 Calibration method

The MDTs require a calibration procedure [12] to precisely convert the measured drift time into a drift distances from the anode wire (drift radius) that is subsequently used in pattern recognition and track fitting. The calibration of the MDT chambers is performed in three steps. In the first step the time offset with respect to the trigger signal, t_0, for each tube or group of tubes is determined; in the second step the drift-time to space relation, $r(t)$ function, is computed; in the third step the spatial resolution is determined.

The calibration constants are loaded in the Conditions Data Base (known as 'COOL') [13] and then retrieved, ac-

Fig. 6 (**A**): TGC front-end and (**B**): sector logic buffers for BC identification. Three BC crossing, previous, current and next are readout

cording to an Interval Of Validity (IOV) mechanism, to be used in the offline reconstruction. The IOV determines for which group of runs the calibration constants are valid. The gas mixture composition varied during the data taking period since the water injection part of the gas system was under commissioning resulting in a not constant admixture of water vapor, as can be seen in Fig. 9. Nonetheless the calibration procedure based on the IOV mechanism was able to provide good calibration constants for all the running period.

The t_0 offset depends on many fixed delays like cable lengths, front-end electronics response, Level-1 trigger latency, time of flight from the IP and has to be determined for each drift tube. The offset is obtained by fitting a Fermi function to the leading edge of the drift time distribution as shown in Fig. 7(A). The precision expected in LHC collision data is better than 1 ns with a dataset of about 10 K muons crossing the drift tube. This uncertainty does not significantly degrade the position resolution of the MDT tubes which corresponds to a time span of about 5 ns. In Fig. 7(B) also the typical spectrum of ADC for all tubes in a chamber is reported. Charge information in each tube is obtained using a Wilkinson ADC [14]. As the MDT chambers are operated at different temperatures depending on their positions in the MS, the $r(t)$ functions differ depending on location and are determined separately. In addition, variations of the toroidal magnetic field along the drift tubes produce different Lorentz angles, thus different $r(t)$ functions. An initial rough estimate of the $r(t)$ function is obtained with an accuracy of 0.5 mm by integrating the drift-time distribution. This is correct under the approximation of a uniform dn/dr distribution, where n is the number of hits at a drift radius r

$$\frac{dn}{dt} = \frac{dn}{dr}\frac{dr}{dt} = \frac{N_{\text{hits}}}{r_{\text{max}}}\frac{dr}{dt} \quad \Rightarrow \quad r(t) = \frac{r_{\text{max}}}{N_{\text{hits}}}\int_0^t \frac{dn}{dt'}dt'.$$

N_{hits} is the total number of hits in the time spectrum and r_{max} is the maximum drift radius (14.4 mm). In cosmic rays this approximation is only good at the level of a few hundred µm mainly because of the production of δ-ray electrons along the track. An $r(t)$ relation with a higher accuracy, of about 20 µm, is obtained from this initial estimate by applying corrections, $\delta r(t)$, which minimize the

residuals of track segment fits with an iterative procedure. This minimization procedure, called *auto-calibration*, takes into account the dependence of the parameters of the fitted segments on the applied corrections $\delta r(t)$ and is mainly based on the geometrical constraints from the precise knowledge of the wire positions. Figure 8 shows a typical residual distribution of a chamber, as a function of the distance of the track segment from the anode wire, after the auto-calibration.

In cosmic ray events additional sources of time jitter, beyond the intrinsic resolution, spoil the MDT measurement. The first cause of time jitter is due to cosmic ray muons crossing the tubes with an arbitrary phase with re-

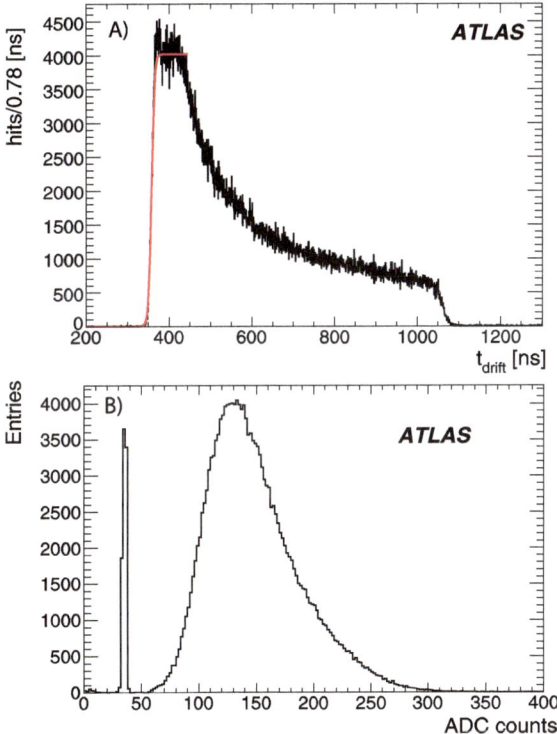

Fig. 7 (**A**): typical drift time spectrum in cosmic ray events for an MDT chamber. The position of the inflection point of the leading edge of the spectrum, t_0, is determined by fitting a Fermi function (shown in *red*) to the beginning of the spectrum. (**B**): Typical spectrum of ADC for all tubes in a chamber. Hits below 50 ADC counts are identified as electronic noise

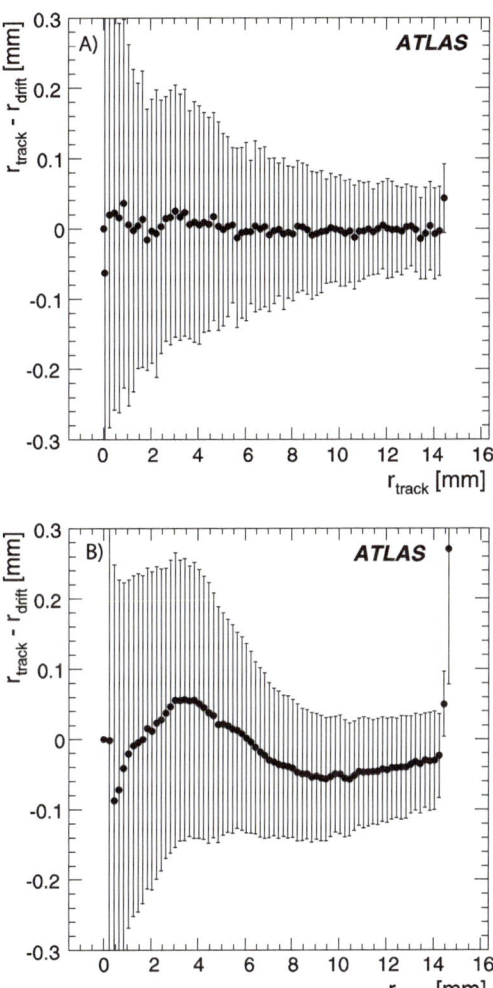

Fig. 8 (**A**): Residuals as a function of the track segment distance from the wire after the $r(t)$ auto-calibration and RPC-time corrections. *The points* correspond to the mean value of the distribution of residuals and the error bars to its RMS value. (**B**): Residuals as a function of the track segment distance from the wire after the $r(t)$ auto-calibration using the Gt_0-refit method. The points correspond to the mean value of the distribution of residuals and the error bars to its RMS value. Residual systematics at the level of 50 µm are present using this correction

spect to the front-end electronics clock [15]. This implies a time jitter corresponding to a 25 ns uniform distribution. The second cause is related to the spread of the trigger time for triggers generated in different parts of the detector (up to about 100 ns due to the initial stage of the trigger timing). Two different methods have been alternatively used to reduce the impact of these effects: the *RPC-time correction* and the MDT Gt_0-*refit*. The achieved performance with both methods are discussed in Sect. 6. In the following a brief description of the former method is given.

The RPC-time correction uses the trigger time measured by the RPC chambers on an event by event basis. This time correction was applied only to the MDT chambers of the BM stations since these chambers are close to the two RPC stations used to issue the trigger and so no corrections due to time of flight and, more importantly, no corrections due to the spread in timing of the trigger signals issued by different parts of the detector are needed. This method cannot be applied to the end-cap region since the TGC do not provide a measurement of the trigger time but rather they select the appropriate BC.

With this correction the time jitter due to the two effects mentioned above is reduced from ∼100 ns to few ns (see Sect. 6). The distribution of the residuals obtained after calibration using the RPC-time correction method is presented in Fig. 8(A). The precision of the auto-calibration is better than ∼20 μm using this correction.

The Gt_0-refit has also been used to improve the single tube resolution, as discussed in Sect. 6. Also the precision of the auto-calibration is much improved with respect to the uncorrected situation. As shown in Fig. 8 a precision of ∼50 μm is obtained for the residuals of the segment fit after auto-calibration with small residual systematics on the auto-calibration.

5.2 End-cap chambers calibration with monitoring chamber

For the end-cap MDT system, due to the limited number of cosmic ray events, a different method to determine the $r(t)$ relation was used. A small MDT chamber installed on the surface of the ATLAS underground hall was set up [16] to monitor continuously the MDT gas composition. One multi-layer is connected to the supply line of the gas recycling system while the other is connected to the return line. This chamber benefits from a very large cosmic ray rate and can therefore determine the $r(t)$ function with high precision in short time intervals. Cosmic ray muons are triggered by scintillator counters mounted on the monitoring chamber. The trigger time is measured and subtracted event-by-event from the tube drift times, in this way the jitter related to the asynchronous front-end clock is automatically removed.

Data from the monitoring chamber are used to derive a $r(t)$ function every 6 hours to monitor the gas drift properties. Figure 9 shows the variation of the maximum drift time (the drift time of muons crossing the drift tube close to its edge) over the period September–October 2008. Two $r(t)$ functions were used to cover the Fall 2008 run period, for each period the $r(t)$ function for each chamber of the MS was corrected to account for the temperature difference using the data measured by the sensors mounted on any particular chamber. The temperature corrections to the $r(t)$ function were derived from the Garfield–MagBoltz simulation program [17–19]. The output of the simulation was validated by several measurements with a muon beam [5]. In the end-cap region, the temperature varies by about 4°C from top to bottom of the MS, resulting in a variation of the maximum drift time of about 10 ns. On the other hand the temperature of the cavern was remarkably stable in time.

Fig. 9 Maximum drift time measured by the gas monitor chamber versus time during September–October 2008. The red points refers to the return line and *the blue points* to the supply line (*green* and *light blue points* are the time average of the supply and return line measurements). The large variation seen between middle of September and 10th of October is due to the change of the quantity of water vapor added to the standard mixture

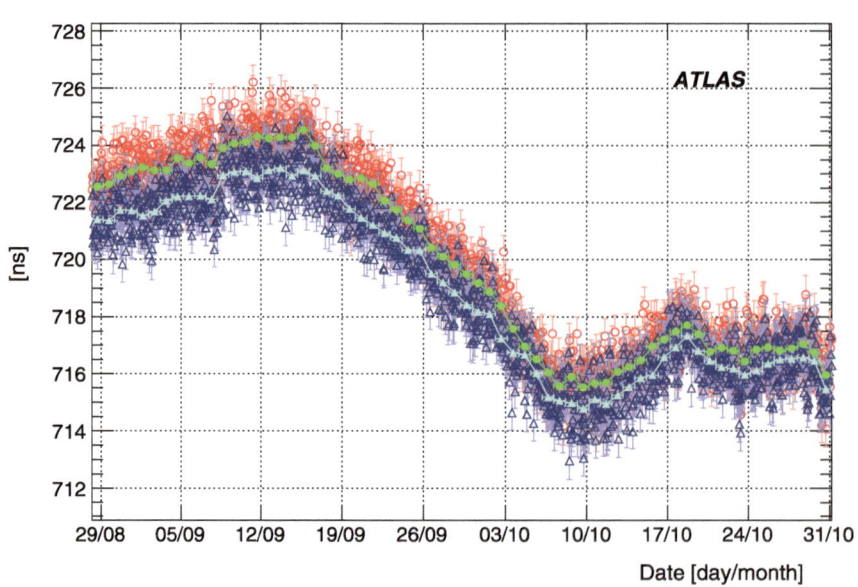

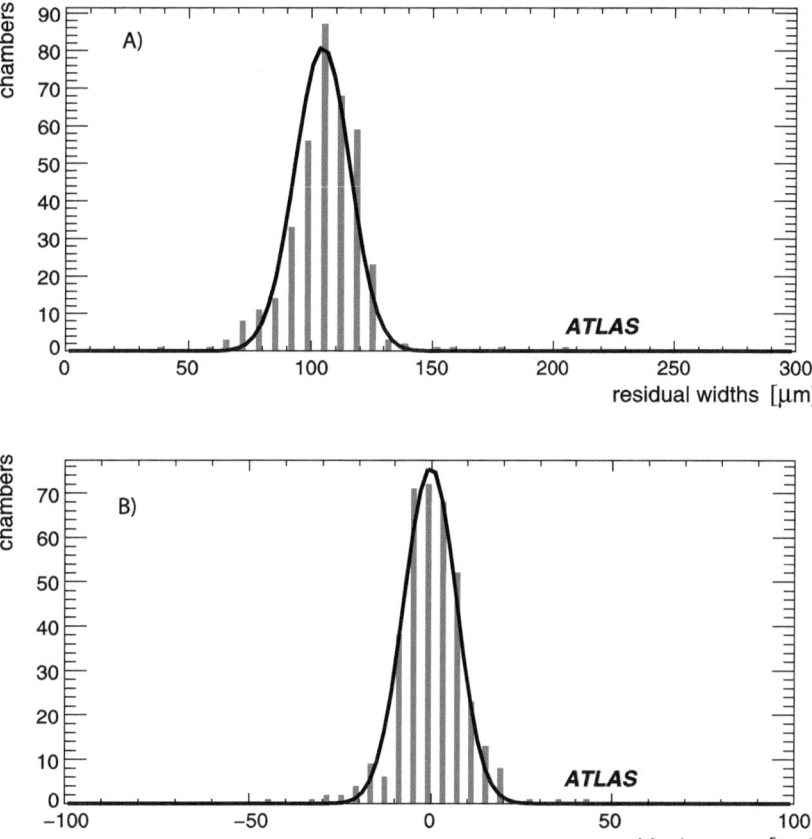

Figure 10 shows the distribution of the mean and RMS
value of the residuals from the fit to track segments in
all end-cap chambers (run 91,060). A Gaussian fit is superimposed. The $r(t)$ function derived from the gas monitor chamber with temperature corrections provides an acceptable calibration for all the MDT chambers of the endcap: the average standard deviation of the residuals is about
100 μm.

6 Detector performance: efficiency and resolution

6.1 MDT

6.1.1 MDT drift time distribution

The behavior of the drift time distributions of individual
tubes is an important quality criterion for the MDT performance. The minimum and maximum drift times, t_0 and t_{max},
respectively, correspond to particles passing very close to
the wire and close to the tube walls, and their stability indicates the stability of the calibration. The number of hits
recorded in a small time window before the rising edge of
the drift time distribution t_0 can be used to estimate the rate
of noise due to hits not correlated with the trigger. A precise

knowledge of t_0 for each tube is essential for high quality
segment and track reconstruction. As explained in Sect. 5,
for cosmic rays some additional time jitter is present and
must be accounted for.

In order to improve the quality of track reconstruction the
Gt_0-refit time correction has been used. The performance of
the Gt_0-refit algorithm has been investigated in the past, both
using simulated data and using data taken with a BIL (Barrel
Inner Large) chamber in a cosmic ray test stand under controlled trigger conditions [20]. The achieved Gt_0 resolution
ranged between 2 and 4 ns depending on the chamber geometry (8 layer chambers have better resolution than 6 layer
chambers) and hit topology. In particular the Gt_0-refit algorithm cannot work if all the hits are on the same side of the
wires, typically for tracks at 30° with respect to the chamber plane. The selection of good quality segments requires a
minimum of five MDT hits and segments with all hits on the
same side of the wires are removed.

In addition to the Gt_0-refit also the RPC-time correction
method was used for the MDT chambers in the middle barrel station (BM) which are located closely to the RPC trigger
chambers. The time measured by these RPC can be used to
correct for a global time offset. An example of the effective-
ness of the method is given in Fig. 7 where the drift time
distribution for a BML chamber is shown after RPC-time

corrections. The steepness of the rising edge, measured as one of the parameters of the Fermi distribution, is improved, passing from 22 ns without correction, to 3 ns with RPC time corrections, a value in agreement with results from muon beam tests [5]. The precision of the RPC-time correction is about 2 ns as explained in Sect. 6.2. This also includes the contribution of the signal propagation time in the RPC strips.

The distribution of the difference between the fitted Gt_0 and the RPC timing correction per segment is shown in Fig. 11 for a BML chamber. The standard deviation of about 4 ns is consistent with an uncertainty of 2 ns from the RPC-time correction added to an uncertainty of 3 ns introduced by the Gt_0-refit method. Tails up to 30 ns are present in the distribution due to bad hit topologies and background hits.

6.1.2 Drift tube spatial resolution

The MDT single tube resolution, as a function of drift distance, was studied using different time corrections. The extraction of the resolution function is based on an iterative method. At the first iteration an approximate input resolution function is assumed. Only segments with a minimum of six hits are considered. These segments are fitted again after removing one hit at the time. Subsequently, the width of the distribution of the residuals for the excluded hit, j, is computed as a function of the drift distance from the wire, $\sigma_{\mathrm{fit},j}(r)$. The errors of the straight line fit (depending on the assumed tube resolution) are then propagated to the excluded hit. The resolution $\sigma_j(r)$ is then computed by quadratically subtracting from the standard deviation of the residuals the fit extrapolation error, $\sigma_{\mathrm{extr},j}(r)$:

$$\sigma_j(r) = \sqrt{\sigma_{\mathrm{fit},j}^2(r) - \sigma_{\mathrm{extr},j}^2(r)}$$

The procedure is iterated using the new resolution function until the input and output resolutions agree within statistical uncertainties; a small number of iterations (two to four) is usually needed.

In Fig. 12 the tube resolution obtained for a BML chamber is shown as the green band. The width of the band accounts for the systematic uncertainty of the method. Also shown (solid line) is the resolution function obtained for an MDT chamber at a high energy muon test beam [5] with well controlled trigger timing. This can be considered as reference for the single-tube resolution. The resolution function measured with cosmic rays is consistent with a time degradation of the reference resolution of about 3 ns. This is in reasonable agreement with the 2 ns time resolution quoted for the RPC-time correction in addition to a small contribution from multiple scattering and individual tube differences in t_0.

The single hit spatial resolution was determined also by applying the Gt_0-refit method to track segments reconstructed in the same chamber. The procedure was similar

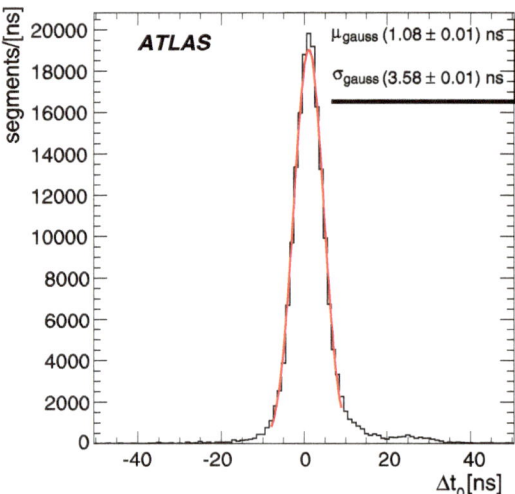

Fig. 11 Difference between the gt_0 obtained with the Gt_0-refit method and with RPC-time correction. The width of the distribution is a convolution of the uncertainties of the RPC-time correction and the Gt_0-refit method

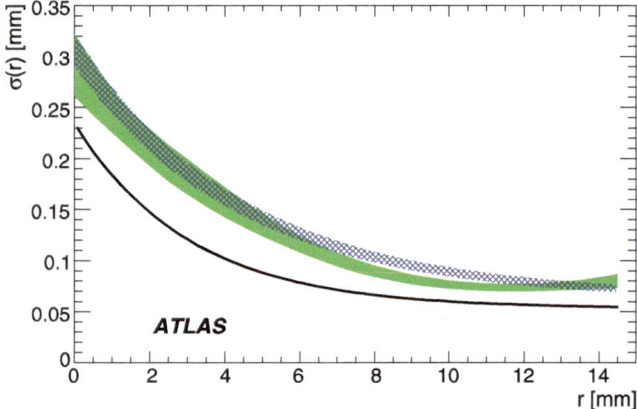

Fig. 12 Drift tube resolution as a function of the radius. *The green shadowed* (RPC correction method) and *the blue* hatched (Gt_0-refit method) bands represent the resolution function measured with cosmic rays with the two different methods described in the text. *The solid line* represents the resolution measured with a high momentum muon beam [5]

to that presented above with the convergence of the method driven by the estimate of the residual pulls. The resolution function is shown as the blue hatched band in Fig. 12. The measured resolution is consistent with the test beam measured resolution provided that an additional time uncertainty of about 2–3 ns is taken into account.

6.1.3 Drift tube noise

The level of noise can be measured in each drift tube by looking at the drift time distribution in a given interval before t_0 where only hits uncorrelated with the trigger are present. The noise rate is obtained by dividing the number of hits normalized to the number of triggers by the chosen time interval. The charge of drift tube signals, at nominal running conditions, is well above the ADC pedestal corresponding to about 50 counts, see Fig. 7. In the reconstruction algorithms only hits with charge above this value are considered. The distribution of noise rate with and without the ADC charge cut is shown in Fig. 13 for all MDT drift tubes. The average noise rate is only 60 Hz without the ADC cut and 13 Hz with ADC cut, the former figure corresponds to an average tube occupancy of less than 10^{-4}.

6.1.4 Drift tube efficiency

The single tube efficiency was studied by reconstructing segments in a chamber using all tubes except the one under observation i.e. excluding one MDT layer at the time in segment reconstruction. Two different types of inefficiencies can be defined: (i) absence of a hit in the tube; (ii) a hit is present but is not associated to the segment because its residual is larger than the association cut. The inefficiency of type (i), referred to as *hardware inefficiency*, is very small, mostly occurring at large drift distances, near the tube wall, where the short track length results in fewer primary electrons or due to the track passing through the dead material between adjacent tubes. The inefficiency of type (ii), referred to as *tracking inefficiency*, is dominated by δ-electrons, produced by the muon itself, which can mask the muon hit if the δ-electron has a smaller drift time than the muon. Tube noise can be an additional source of this type of inefficiency.

Figure 14 shows the distribution of the signed residuals for hits in the tube of one barrel chamber as a function of the distance of the segment from the wire. A large population at small values of the residual, compatible with the spatial resolution, is visible. Large positive residuals are associated with early hits mainly due to δ-electrons. If a hit is not found

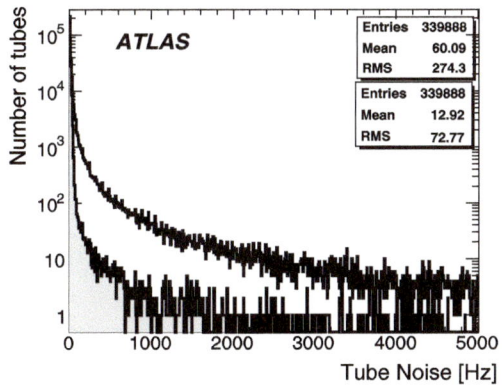
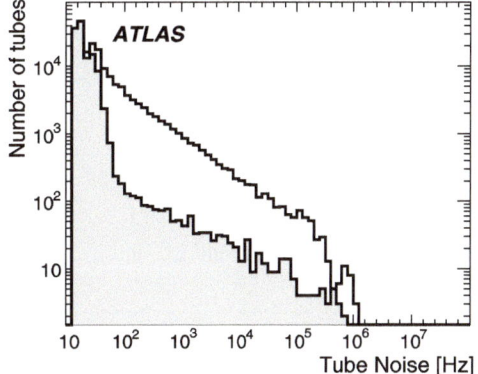

Fig. 13 Distribution of the drift tube noise rate with (*shadowed histogram, bottom statistical box*) and without (*empty histogram, top statistical box*) the ADC cut described in the text. *In the right plot* the logarithmic scale allows observation of the very few noisy tubes.

The tail of the distribution is due to very few noisy tubes that are suffering from pick up of high frequencies through the HV cables or interferences due to the digital clock present in the front end electronics

Fig. 14 Distribution of the hit residuals for tubes excluded in the segment fit, as a function of the distance of the track from the wire. Small residuals are associated with efficient hits. The triangular region is populated by early hits produced by δ-electrons. Missing hits, as explained in the text, are assigned a residual value of 15.5 mm. The histogram on the right represent the projection on the residual axis of the plot on the left pane

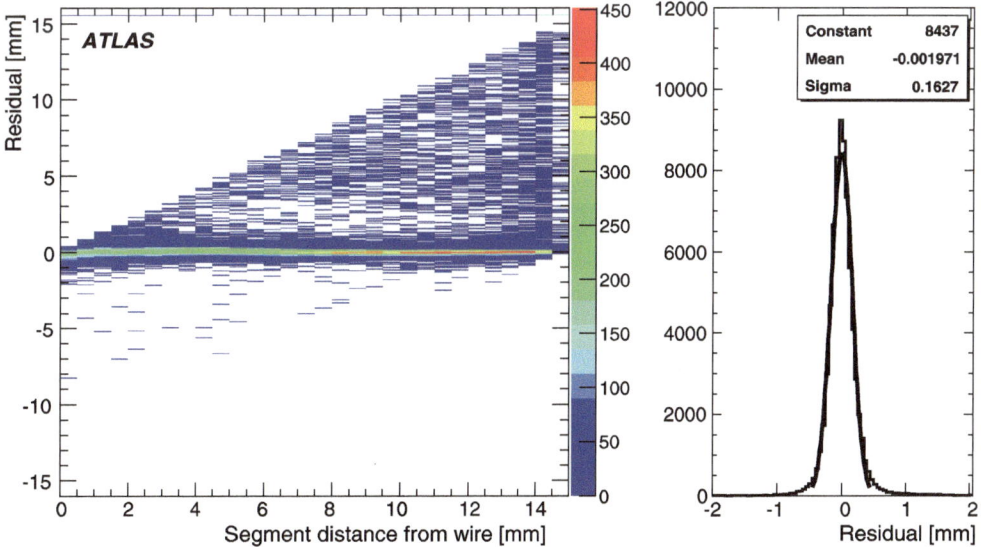

in the tube traversed by the muon (thus a residual cannot be computed) a value of 15.5 mm is assigned, larger than the tube radius of 15 mm. The population of missing hits is visible at the top of Fig. 14 and it peaks close to the tube wall.

The tracking efficiency is defined as the fraction of hits with a distance from the segment smaller than n times its error, this error being a convolution of the tube resolution and the track extrapolation uncertainty. Figure 15 shows the hardware efficiency and the tracking efficiency as a function of the drift radius for $n = 3$, 5, and 10. The tracking efficiency decreases with increasing radius, mainly due to the contribution of δ-electrons. The average tube hardware efficiency is 99.8%; the tracking efficiency is 97.2%, 96.3% and 94.6% for n equals to 10, 5 and 3 respectively.

Figure 16 shows the average value of the tracking efficiency for each tube of a BML chamber for $n = 5$. The average value is about 96%. An efficiency consistent with zero was obtained for two tubes as can be seen in the expanded view on the right plot. These were recognized as tubes with

disconnected wires and were not considered in the average value.

The results of a study on all the barrel chambers with enough cosmic ray illumination to allow the determination of the single tube efficiency is presented in Fig. 17. The distribution of the tracking efficiency for a 5σ hit association cut is shown for about 81 K drift tubes. In addition to about 0.2% of dead channels, less than 1% of tubes have tracking efficiency below 90%, mainly due to calibration constants determined with insufficient precision.

6.2 RPC

In addition to providing the barrel muon trigger, the RPC system is also used to identify the BC of the interaction that produced the muon. This requires a time resolution much better than the bunch crossing period of 25 ns. For this, the time of the strips that form the trigger coincidence is encoded in the front-end with a 3-bit interpolator providing an accuracy of 3.125 ns [10]. The distribution of the time difference between the two layers of a pivot plane in the ϕ projection was used to determine the RPC time resolution. With this method there is no need to correct for the muon time of flight and the signal propagation along the read-out strips. The RMS width of the distribution shown in Fig. 18 is 2.5 ns. From this a time resolution of 1.8 ns is derived for the two RPC layers forming the coincidence. For this measurement only strips associated to a reconstructed muon track and belonging to events with one and only one RPC trigger were considered.

Two other important RPC quantities related to the detector performance are the efficiency and the spatial resolution. In order to determine the RPC efficiency two main issues have to be taken into account. The first one is due to the fact that the RPCs are actually providing the muon trigger thus resulting in a *trigger bias* on the efficiency calculation. The second one is caused by the fact that the RPC hits are

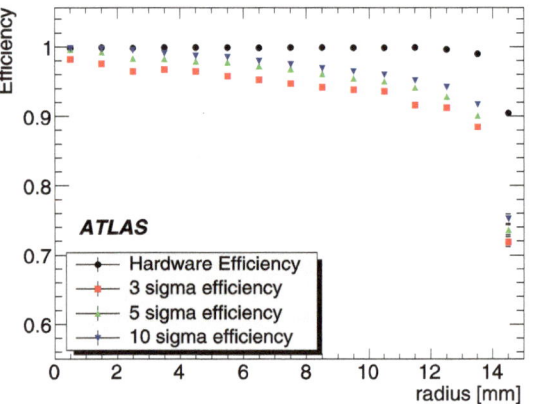

Fig. 15 Tube efficiency as a function of the drift distance averaged over all tubes of a BML chamber. Shown are the hardware efficiency and the tracking efficiency for hit residuals smaller than 3, 5, and 10 times the standard deviation of the distribution

Fig. 16 Single tube tracking efficiencies with a 5σ association cut, as explained in the text, for a BML chamber. *The plot on the right* shows an expanded view in the region where two tubes with the wire disconnected were found

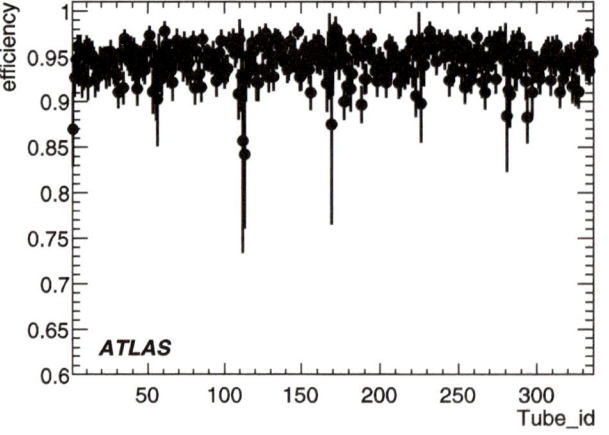
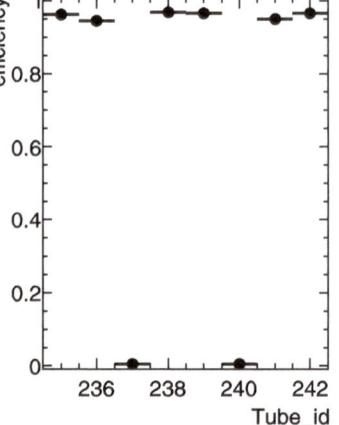

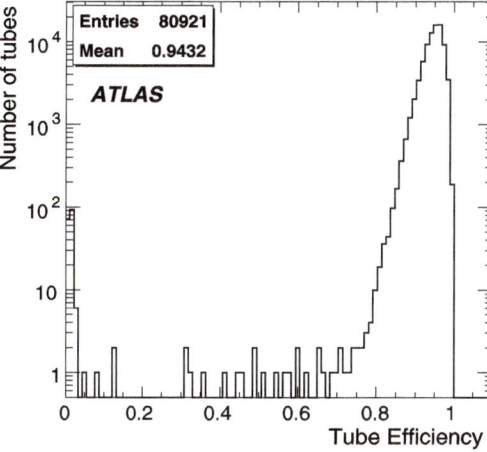

Fig. 17 Distribution of the tracking efficiency, with a 5σ hit association cut, for ~81 K drift tubes in the barrel MDT. About 0.2% of tubes were not working and have efficiency compatible with zero

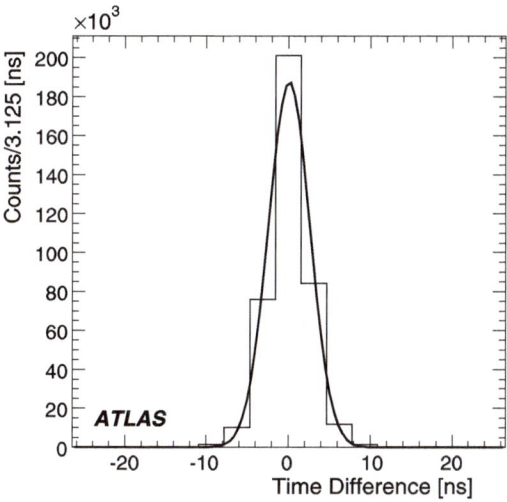

Fig. 18 Distribution of the time difference between the two RPC layers of a pivot plane in the φ projection. The binning of the histogram corresponds to the strip read-out time encoding of 1/8 of BC

also used in the track reconstruction; in particular, they measure the coordinate in the non-bending, φ projection. The second effect has negligible contribution if the efficiency is measured for the η strips, since in this projection the track reconstruction is driven by the MDT. For the efficiency measurement, MDT tracks were extrapolated to the RPC plane and the layer was counted as efficient if at least one η hit was found with a distance of less than 7 cm from the extrapolation. The effect of the *trigger bias* has been removed from the efficiency measurement of an RPC plane by selecting all the events in which the other three planes (in the case of a Middle Station) were producing hits, since the trigger requirement is a 3 over 4 planes majority. The distribution of the efficiency, averaged over each layer, for the RPC chambers in the Middle stations is shown in Fig. 19, the dis-

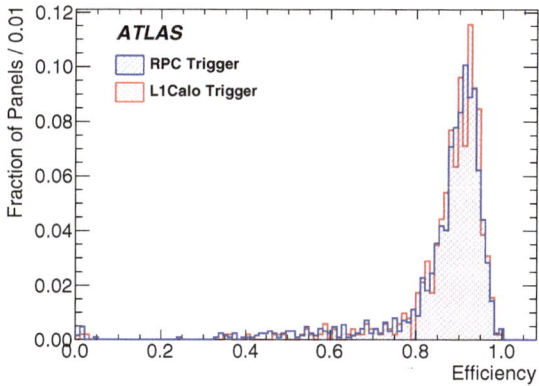

Fig. 19 Distribution of the average efficiency for RPC of the Middle stations for run 91,060. The two distributions refer to two different triggers: RPC trigger (*full line*, 91.33% peak efficiency) and calorimeter trigger (*dashed line*, 92.0% peak efficiency). Both distributions are normalized to unit area. The measured efficiency is lower than expected mainly because the read-out timing was still not optimal

tribution is peaked at an efficiency of 91.3%. To check the remaining impact of the trigger bias on the efficiency measurement, the same analysis was repeated with a sample of cosmic rays selected with a calorimeter trigger (*Level-1Calo* trigger) independently of the RPC trigger response. The result for the efficiency is superimposed in Fig. 19: a good agreement between the two distributions is observed.

The spatial resolution is related to the clusters size, that is the number of strips associated to a muon track. A muon crossing the detector near the center of a readout strip, will in general produce a cluster of size one, while clusters of size two are only observed when muons hit a narrow region at the boundary between two strips. The actual sizes of the regions corresponding to clusters of size one and two depends on the detector operating parameters, but it is in general true that the latter is smaller than the former. This implies that the spatial resolution must be smaller when measured on a subset of data with only clusters of size two. The spatial resolutions of η strips was determined selecting muon tracks reconstructed in the MDT as explained above. For each RPC read out plane, the distribution of the distance from the extrapolated track was obtained separately for clusters of size one and two and then was fitted with a Gaussian. The RMS widths of the fit were divided by the strip pitch (ranging from 27 to 32 mm depending on the chamber type) to allow for comparison between different RPC and are shown in Fig. 20. This technique has been used only for the η panels since the MDT are measuring in the Z–Y plane. On average, clusters of size two give a spatial resolution about half as for clusters of size one, which is below 10 mm as expected.

6.3 TGC

The basic structure of the TGC chambers and their assembly in the MS end-cap wheels is presented elsewhere [2]. Inactive regions due to the gas-gap frame and the wire supports account for a loss of active area varying from 3% to 6% depending on the chamber type. In order to optimize the trigger efficiency these inactive regions are staggered with respect to the trajectory of high momentum muons produced at the IP. In the active area the TGC wires are expected to have an efficiency of more than 98%. For the cosmic ray run 91,060 the trigger logic required a coincidence of 3 out of 4 layers in the doublet chambers (referred to as TGC2 and TGC3 as in Fig. 1). In evaluating the detector efficiency one has to take into account the trigger bias and the fact that cosmic rays are non-pointing to the IP, asynchronous, and do not only consist of single muons but also of extended showers.

To evaluate the efficiency of a layer in the doublet chambers, it is required that there is one and only one hit in each of the other three layers and that these three hits are associated to the current BC. This is intended to remove high multiplicity events (showers) and out-of-time tracks. As a result of this selection, the 3 out of 4 trigger condition is satisfied independently of the presence of a hit in the layer under evaluation. The efficiency of this layer is thus determined in an unbiased way.

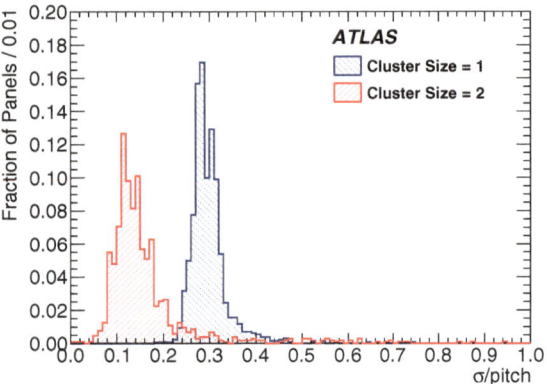

Fig. 20 Distribution of the spatial resolution provided by the η strips for RPC of the Middle stations. The spatial resolution is divided by the strip pitch. The distributions are normalized to unit area

A similar procedure is used for the triplet chambers (TGC1). When evaluating, the efficiency of a layer, it is required (i) that the other two layers satisfy the 2 out of 3 trigger coincidence and (ii) that the line joining the two hits (track) crosses the layer in its active area.

In both cases, the layer under test is considered efficient if there is at least one hit associated to any of the previous, current or next BC. Figure 21 on the left shows an efficiency map in the wire-strip (η–ϕ) plane, and on the right its η projection, i.e. the strip efficiency. Some inactive regions are clearly visible: the bands in Fig. 21—Left indicate the location of the wire supports.

The overall efficiency, including the inactive regions, is evaluated for a fraction of TGC layers (about 40% of TGC doublet layers) by requiring that a muon track crosses the layer under test at least 10 cm away from its edge. The muon track is defined using MDT hits combined with TGC hits in the layers that are not under evaluation. Figure 22 shows the distribution of the wire efficiency for different values of high voltage setting: 2650, 2750, 2800 and 2850 V. The average value of the efficiency, at the nominal voltage of 2800 V, is 92% consistent with the local efficiency measured as explained above and the contribution from inactive-regions.

7 MDT optical alignment

The design transverse momentum resolution at 1 TeV of the MS is about 10%, this translates into a sagitta resolution of 50 μm. The intrinsic resolution of MDT chambers contributes a 40 μm uncertainty to the track sagitta, hence other systematic uncertainties (alignment and calibration) should be kept at the level of 30 μm or smaller. Since long-term mechanical stability in a large structure such as the MS cannot be guaranteed at this level, a continuously running alignment monitoring system [21] has been installed. This system is based on optical and temperature sensors and detects slow chamber displacements, occurring at a timescale of hours or

Fig. 21 *Left*: efficiency map for a TGC chamber layer. *The horizontal axis is the strip channel and the vertical axis is the wire channel*. *Right*: efficiency projection to the strip channels. Observed efficiency drops are consistent with the wire support locations

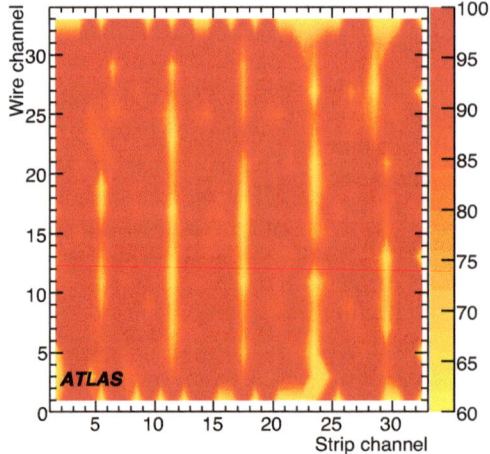

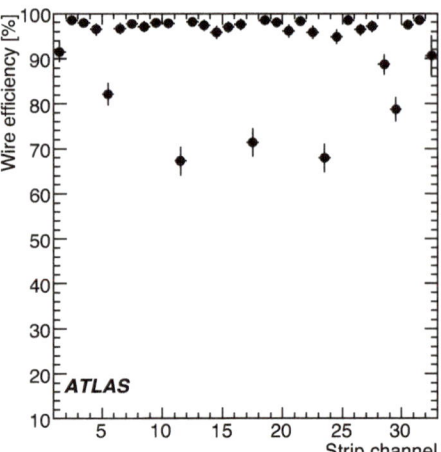

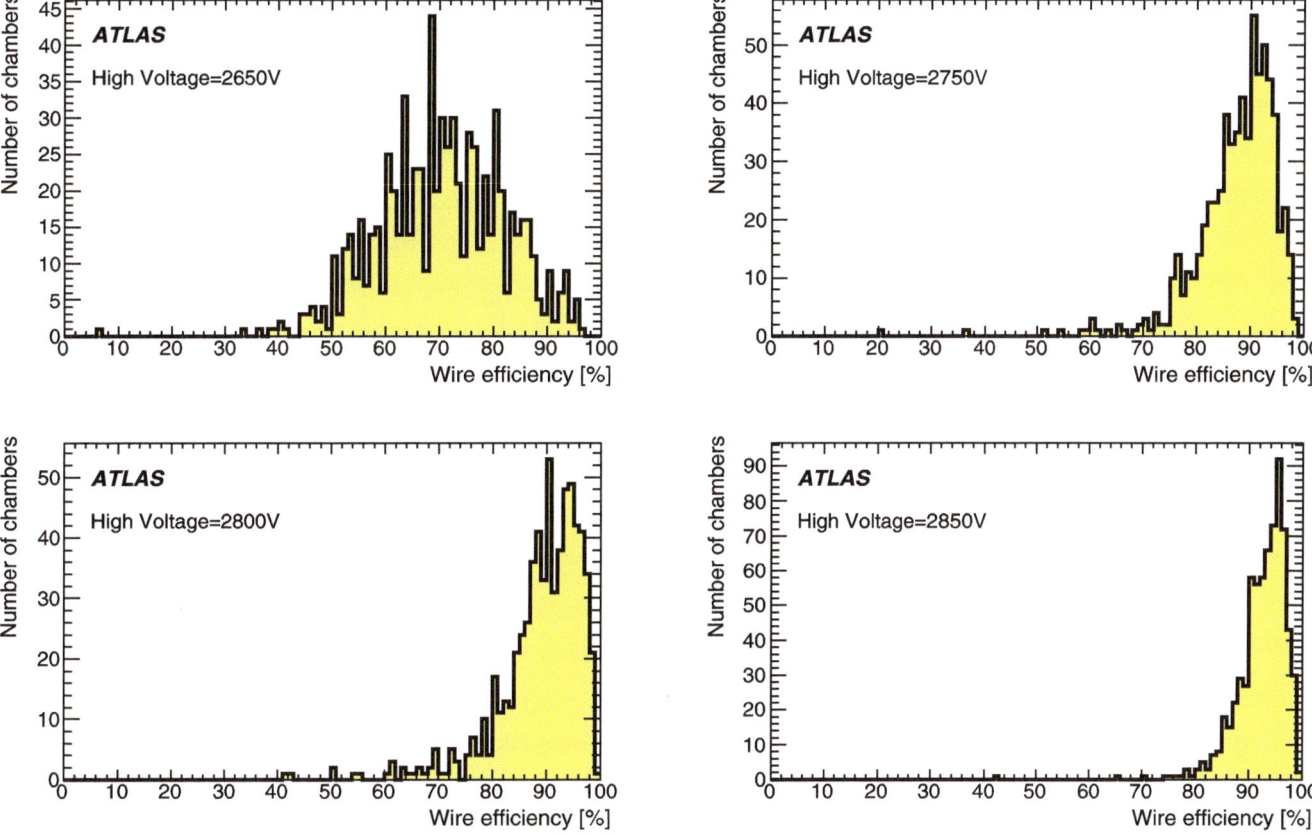

Fig. 22 Distribution of the TGC wire efficiency of individual layers for different high voltage values. The distribution for 2800 V, the nominal voltage in 2008, was obtained with run 91,060

more. The information from the alignment system is used in the offline track reconstruction to correct for the chamber misalignment. No mechanical adjustments were made to the chambers after the initial positioning. The system consists of a variety of optical sensors, all sharing the same design principle: a source of light is imaged through a lens onto an electronic image sensor acting as a screen. In addition to optical position measurements, it is also necessary to determine the thermal expansion of the chambers. In total, there are about 12,000 optical sensors and a similar number of temperature sensors in the system. Optical and temperature sensors were calibrated before the installation such that they can be used to make an *absolute* measurement of the chamber positions in space, rather than only following their movements with time relative to some initial positions.

7.1 End-cap chamber alignment

The end-cap chambers and their alignment system [22] were installed and commissioned during 2005–2008, and continuous alignment data-taking with the complete system started in Summer 2008. After commissioning, more than 99% of all alignment sensors were operational, and only a small number failed during the data-taking in 2008. The effect of

the missing sensors on the final alignment quality is negligible.

The position coordinates, rotation angles, and deformation parameters of the chambers are determined by a global χ^2 minimization procedure. The total χ^2, as well as the contributions of the individual sensor measurements to the χ^2 (pulls) can be used to estimate the alignment quality from the internal consistency of the fit. If the observed sensor resolutions agree with the design values, one expects approximately $\chi^2/\text{ndf} = 1$ and a pull distribution with zero mean and unit RMS width. Figure 23 shows the observed and expected pull distributions in the end-caps, obtained by assuming the design resolutions for all sensor types.

In a second step, the assumed sensor resolutions are adjusted until the observed pull distributions, broken down by sensor type, agree with the expected distribution. This yields the observed sensor resolutions, which are used as input to a Monte Carlo simulation of the alignment system. The simulation predicts a sagitta accuracy due to alignment of about 45 μm, which is close to the design performance.

Validating the alignment as reconstructed from the optical sensor measurements requires an external reference. During chamber installation, surveys of the completed endcap wheels were done using photogrammetry, and the

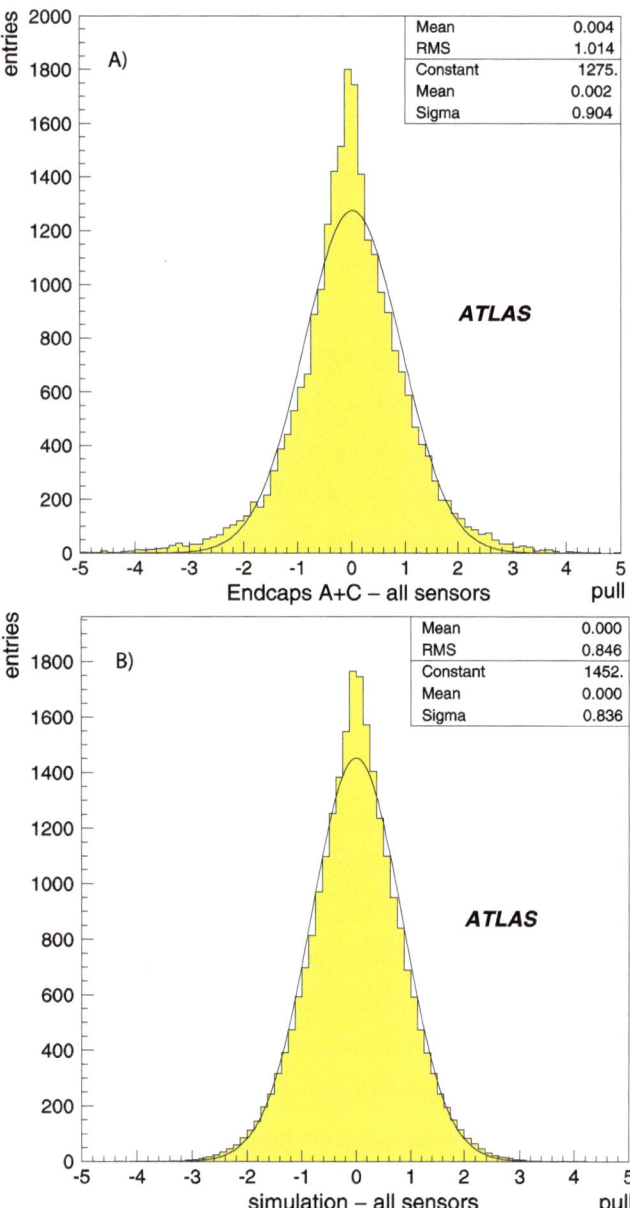

Fig. 23 The observed (**A**, from data) and expected (**B**, from simulation) pull distributions for the end-caps, assuming design resolution for all sensor types. Correlations and weakly constrained degrees of freedom cause the expected pull distribution to have RMS width below unity. The observed χ^2/ndf from the fit on data is 1.4, while the one from simulation is 1.0

chamber positions measured with the alignment system agreed with the survey results within 0.5 mm, the quoted accuracy of the survey. While establishing confidence in the optical system, the full validation of the alignment can only be done with tracks. Thus, cosmic muons recorded during magnet-off running were used to cross-check the alignment provided by the optical system.

For a perfect alignment, the reconstructed sagitta of straight tracks should be zero for each EI-EM-EO measurement tower (note that, when averaged over many towers, the

mean value can be accidentally compatible with zero despite single towers being significantly misaligned). For cosmic muons, the observed width of the sagitta distribution is dominated by multiple scattering. A shifted and/or broadened distribution would indicate imperfections of the alignment. Triplets of track segments were selected in the EI-EM-EO chambers, requiring the three segments to be in the same sector and assigned to the same reconstructed track. Some segment quality cuts were applied for this analysis: (i) $\chi^2/ndf < 10$ and at most one expected hit missing per chamber; (ii) the angle between the segments and the straight line joining the segments in EI and EO was required to be smaller than 5 (50) mrad in the precision (second) coordinate; (iii) at least one trigger hit in the second coordinate was required to be associated to the track. A total of 1700 segment triplets passing the cuts were selected in run 91,060.

Figure 24(A) shows, for the two end-caps, the observed sagitta distribution before and after applying alignment corrections (i.e. the chamber positions, rotations, and deformations as determined by the optical system, as well as a correction for the gravitational sag of the MDT wires). Figure 24(B) shows the corresponding difference in angle in the precision coordinate between each of the segments and the track (the straight line joining the EI and EO segments). For the distribution in Fig. 24(B), the cut at 5 mrad was omitted. The improvement in both variables is clearly visible, the mean value of the corrected sagitta distribution as obtained from the fit with a double-Gaussian function is (-33 ± 42) μm and thus perfectly compatible with zero within the 45 μm error estimated above from the internal consistency of the alignment fit. The width of the corrected sagitta distribution agrees approximately with expectations for the typical energies of triggered cosmic muons. The width of the corrected angle distribution, on the other hand, is about twice as large as expected. This is mainly a consequence of the additional time jitter of MDT measurements described in Sect. 5 which deteriorates the segment resolution.

For the two end-caps separately, the mean value of the sagitta distribution is (-30 ± 61) μm in side A and (-37 ± 57) μm in side C. The sign of the sagitta is defined in such a way that most of the conceivable systematic errors would cause deviations from zero with the same sign in side A and side C. The analysis is limited by statistics even though it uses a significant fraction of the full 2008 data sample. Breaking it down further to the level of sectors, or even to projective towers (where the best sensitivity is obtained) would require significantly more data.

The cross-check with straight tracks confirms that, with the limitations of the analysis, the chamber positions given by the optical alignment system are within the estimated sagitta uncertainties, indicating that the optical system

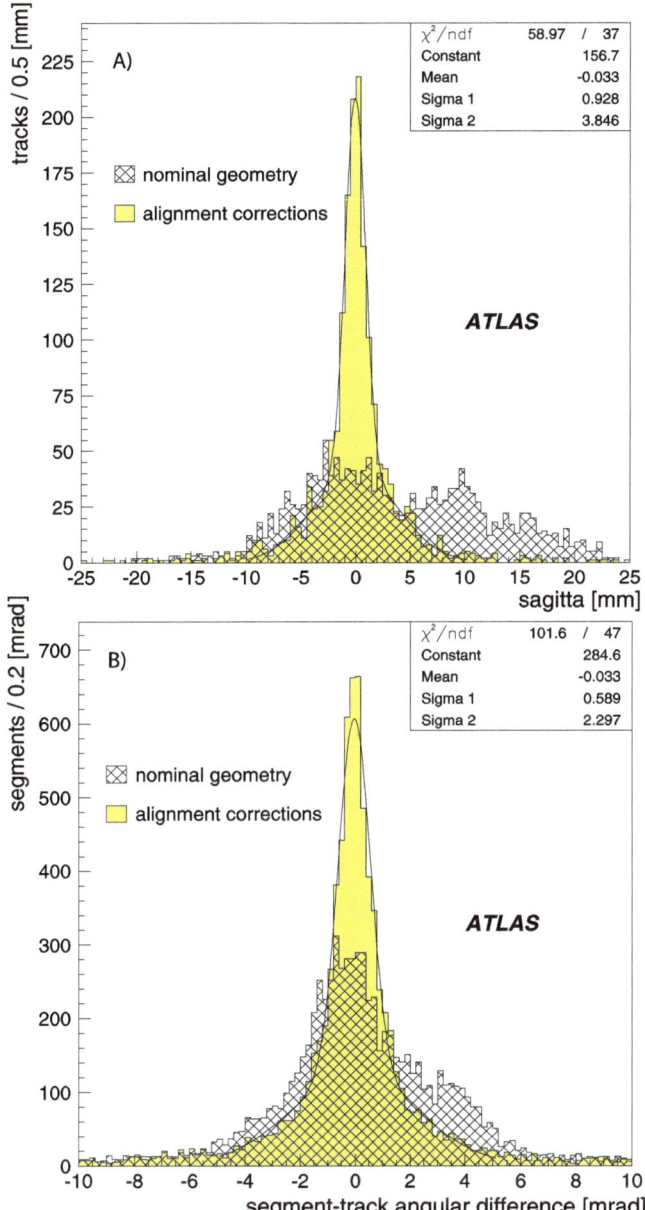

Fig. 24 (**A**): Measured sagitta distribution for the two end-caps. *The cross-hatched histogram* shows the sagitta before alignment corrections, thus reflecting the accuracy of chamber positioning. *The filled histogram* shows the sagitta after applying alignment corrections, *the curve* is the fit of a double-Gaussian function, each Gaussian containing 50% of the events. (**B**): Measured angle in the precision coordinate between the segments and the track to which they are associated

Table 4 Status of the barrel optical system in Fall 2008. No data were recorded during this period from the "broken" sensors. Naming and functions of the different sensors are detailed in reference [23]

Type	Total	Working	Broken
Projective	117	117	0
Axial	1036	1031	5
Praxial	2010	2008	2
Reference	256	253	3
CCC	260	260	0
BIR-BIM	32	32	0
Inplane	2110	2101	9
Total	5817	5798	19
%		99.7	0.3

ing cosmic ray data, the barrel optical system was fully installed and 99.7% of the sensors were functioning correctly. Table 4 summarizes the status of the 5817 installed sensors. The complete system is read out continuously, at a rate of one cycle every 20 minutes. The readout was functioning correctly during the complete period of acquisition of cosmic ray data.

The alignment reconstruction consists in determining the chamber positions and orientations (referred to as "alignment corrections") from the optical sensor measurements. This requires the precise knowledge of the positions of the sensors with respect to the MDT wires. To this purpose, the optical sensors were calibrated before installation and their mechanical supports were glued with precise tools onto the MDT tubes. However, the original design of the barrel optical system suffered from a few errors that eventually degraded the precision of the alignment corrections. Furthermore, the only devices giving projective information in the Small sectors are the CCC sensors which are designed to provide 1 mm accuracy. The alignment of the chambers of the Small sectors is, by design, based on tracks that cross the overlap region between the Small and the Large sectors. However, the statistics obtained in cosmic runs was not sufficient to perform a precise check of this method.

The alignment corrections discussed here cover the nine upper sectors (1 to 9). The complete period of cosmic data taking was divided in intervals of 6 hours, and alignment corrections were reconstructed using the sensor measurements recorded in that interval. This provided data for monitoring of significant movements of the MS, e.g. when the magnetic field in the toroids was switched on.

The barrel alignment reconstruction is based on the minimization of a χ^2, whose inputs are, for each optical sensor i:

– the recorded response $\mathbf{r}_i$
– a model $\mathbf{m}_i(\mathbf{a})$, representing the predicted response of sensor i with respect to the alignment corrections $\mathbf{a}$
– the error σ_i, the estimated uncertainty of the model $\mathbf{m}_i$.

works properly. The design accuracy has nearly been reached in the end-caps. It also shows that the system produces a reliable estimate of the uncertainty of the alignment corrections.

7.2 Barrel chamber alignment

The installation and commissioning of the barrel optical system [23] began in 2005 and continued together with the installation of the chambers until 2008. At the time of record-

The critical part is the model $\mathbf{m}_i$, as it combines all the knowledge of the precise geometry of the optical sensors and their calibration. The free parameters in the fit are the alignment corrections $\mathbf{a}$, and in some cases additional parameters used to model the effect of imprecise sensor positioning or of an incorrect calibration. For all these additional parameters appropriate constraints are included in the fit reflecting the best estimates of the error contributions mentioned above. Overall, 4099 parameters are fit simultaneously. The total reconstruction time for the full barrel is less than one minute.

Given the uncertainties introduced by the additional parameters in the fit procedure, the strategy for alignment in the barrel is slightly different from the one in the end-cap. Dedicated runs without magnetic field in the toroids (but *with* field in the solenoid to tag high momentum tracks) will be used to get initial alignment corrections with a precision of 30 µm. The optical alignment system is then used to monitor movements due to the switching on of the toroidal field and to temperature effects. The mechanical stability of the system, in periods where the magnetic field was constant, is at the level of 100 µm, while movements of the magnet structures at the level of few mm were observed when the magnets were switched on and off. The optical alignment system, which continuously monitors the position of the chambers, is able to follow these movements with the required accuracy. This so-called *relative alignment mode* has already been tested with success in the MS system-test done with a high-energy muon beam [4, 5]. After the minimization, the value obtained for χ^2/ndf is 1.9, which shows that the sensor errors are underestimated.

7.2.1 Performance of the optical alignment in the barrel

Similarly to what is done in the end-cap, an estimate of the contribution to the sagitta error due to the alignment system may be inferred from the χ^2, using the following formula

$$(V^{-1})_{ij} = \frac{1}{2} \frac{\partial^2 \chi^2}{\partial \theta_i \partial \theta_j} \bigg|_{\hat{\theta}}$$

where θ_i are the fitted parameters and V is the global error matrix, of size 4099×4099, of all fitted parameters. To estimate the performance of the alignment system in terms of sagitta measurement, straight tracks, originating in the IP and crossing three layers of chambers, were simulated and the whole fit procedure was applied to these tracks. The sagitta of these pseudo-tracks is a function of some of the alignment corrections, and thus the formula of error propagation may be used to infer the contribution of the alignment to the error of the resulting sagitta. This technique relies on the hypothesis that the errors of the optical sensors are correctly estimated, and thus that the χ^2 is correctly normalized. As this is not the case ($\chi^2/\mathrm{ndf} = 1.9$), the results are

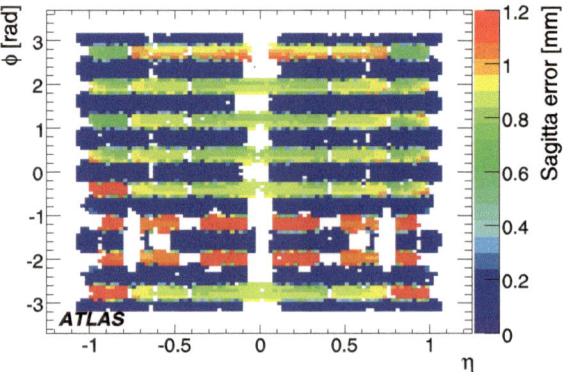

Fig. 25 $\eta \times \phi$ map of the contribution to the sagitta error due to alignment, as estimated with the method described in the text. As expected from the system design, the Small sectors (even sector numbers) are aligned with significantly less precision than the Large sectors (odd sector numbers)

only considered as a rough estimate of the optical alignment performance.

The result is shown in Fig. 25. The Small sectors have a significantly worse alignment than the Large sectors, as explained above. Conservatively, one can conclude that the performance of the optical system, in terms of sagitta precision, is ~200 µm for the Large sectors, and ~1 mm for the Small sectors.

7.2.2 Alignment with straight tracks

Data with the toroidal field off were used to improve the alignment precision in the barrel and to validate the alignment corrections in relative mode. The method is to use straight muon tracks to determine in absolute mode the *initial* spectrometer geometry and, once this geometry is determined, to use the optical alignment system to trace all chamber displacements in a relative mode. The alignment procedure with straight tracks is based on the so-called MILLE-PEDE fitting method [24]. This method uses both alignment and track parameters inside a global fit. As a result, all correlations between alignment and track parameters are taken into account and the alignment algorithm is unbiased.

The track alignment algorithm has been tested with Monte Carlo simulations and with cosmic ray data. The simulation studies show that 10^5 muon tracks with a momentum greater than 20 GeV and pointing to the IP are needed to align the Large sectors with a precision of 30 µm. Small sectors require five times more tracks than Large sectors, due to the multiple scattering in the toroid coils.

Using straight cosmic muon tracks recorded in run 91,060, a set of alignment constants has been produced. A total of 10^7 events were used corresponding to about 3×10^5 cosmic muon tracks in each of the most illuminated barrel sectors. The statistical uncertainty of the sagitta using this track alignment procedure was estimated to be 30 μm for Large sectors.

The data of run 91,060 were processed with the track reconstruction software twice: (i) using the optical alignment corrections and (ii) using the track-based alignment corrections. Both geometries were then tested by measuring the distribution of the track sagitta for muons crossing three chamber stations (Inner, Middle and Outer). Only tracks passing close to the IP in the η projection were chosen. Hits in the Inner and Outer chambers were fit to a straight line, and the distribution of the hit residuals in the Middle chambers was used to evaluate the sagitta. For perfect alignment, the mean value of the sagitta should be zero for straight tracks and, to a good approximation, the mean value of the distribution gives an estimate of the sagitta error.

The results are shown in Fig. 26 for the sets of alignment corrections; on the left for a station in a Large sector, on the right for a station in a Small sector. For reference, the distributions using the design geometry are also shown. The tails of the distributions are due to multiple scattering of muons. In the Large sector station, the two distributions are almost identical, but the distribution with optical alignment is centered at ∼100 μm. In the Small sector station, the distribution with the optical alignment is centered around

1 mm. To compare the results obtained in different stations, Fig. 27 shows the mean values of the sagitta distribution for the Large upper sectors (3, 5 and 7). One Small sector is also presented, sector 4, since this was illuminated with enough events during the same run to produce a meaningful distribution.

The results show that the optical alignment system alone provides a precision at the level of 200 μm. When calibrated with sufficient statistics of high momentum straight tracks, the optical system is able reach a precision of 50 μm.

The sagitta resolution for runs with no magnetic field in the MS can be studied as a function of the muon momentum measured by the ID. The sagitta resolution as a function of the muon momentum was parameterized as

$$\sigma_s(p) = \frac{K_0}{p} \oplus K_1$$

where the first term K_0 is due to multiple scattering in the material of the MS, and the second term K_1 is due to the single tube resolution and chamber-to-chamber alignment. These two terms have been already measured at the MS system test beam [4, 5] and found to be $K_0 = 9$ mm×GeV and $K_1 = 50$ μm. A similar measurement was done with cosmic muons by selecting segment triplets (Inner, Middle

Fig. 26 Distribution of the sagitta (as defined in the text) for straight tracks. (**A**), (**B**): Using alignment corrections derived from the optical system only; (**C**), (**D**): Using track-based alignment corrections. (**A**), (**C**): For a station in a Large barrel sector; (**B**), (**D**): For a station in a Small barrel sector, the optical system corrections of the small sectors have, by design, an accuracy at a level of 1 mm. *In all panels*, the hashed distribution is obtained using the "nominal" geometry. Mean and sigma in the statistical boxes refer to the distributions with alignment corrections, the peak is fitted with a Gaussian in a ±1 sigma interval

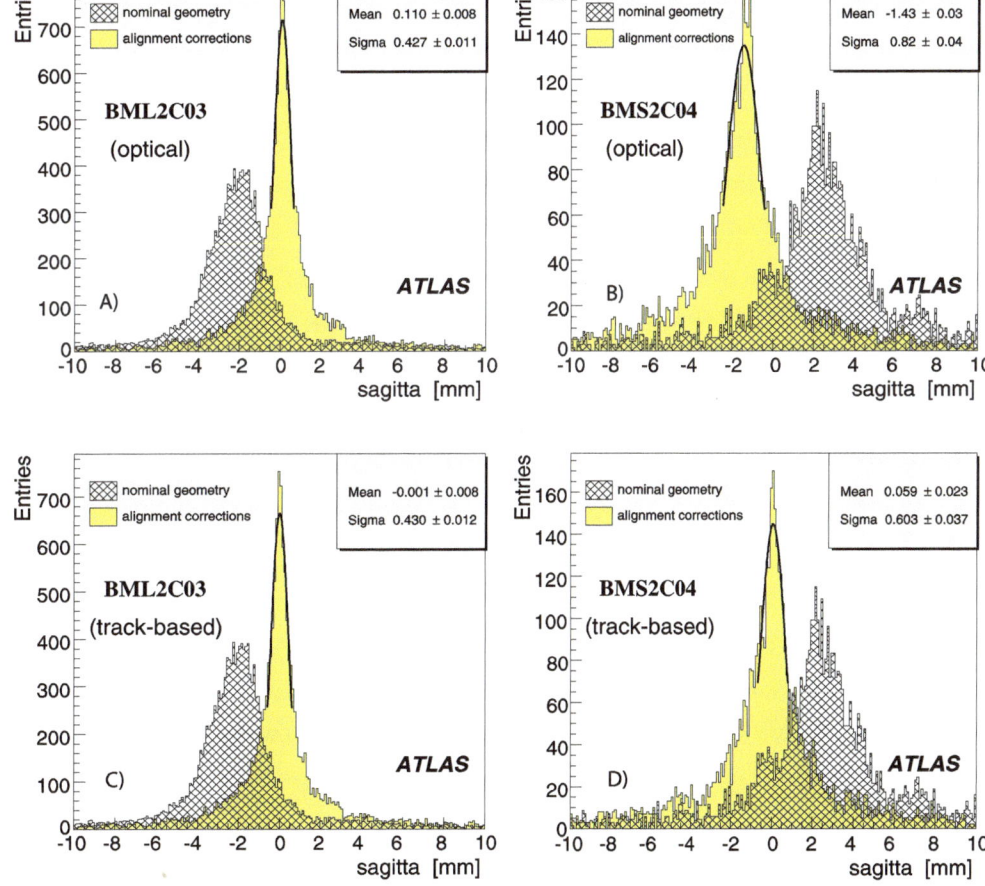

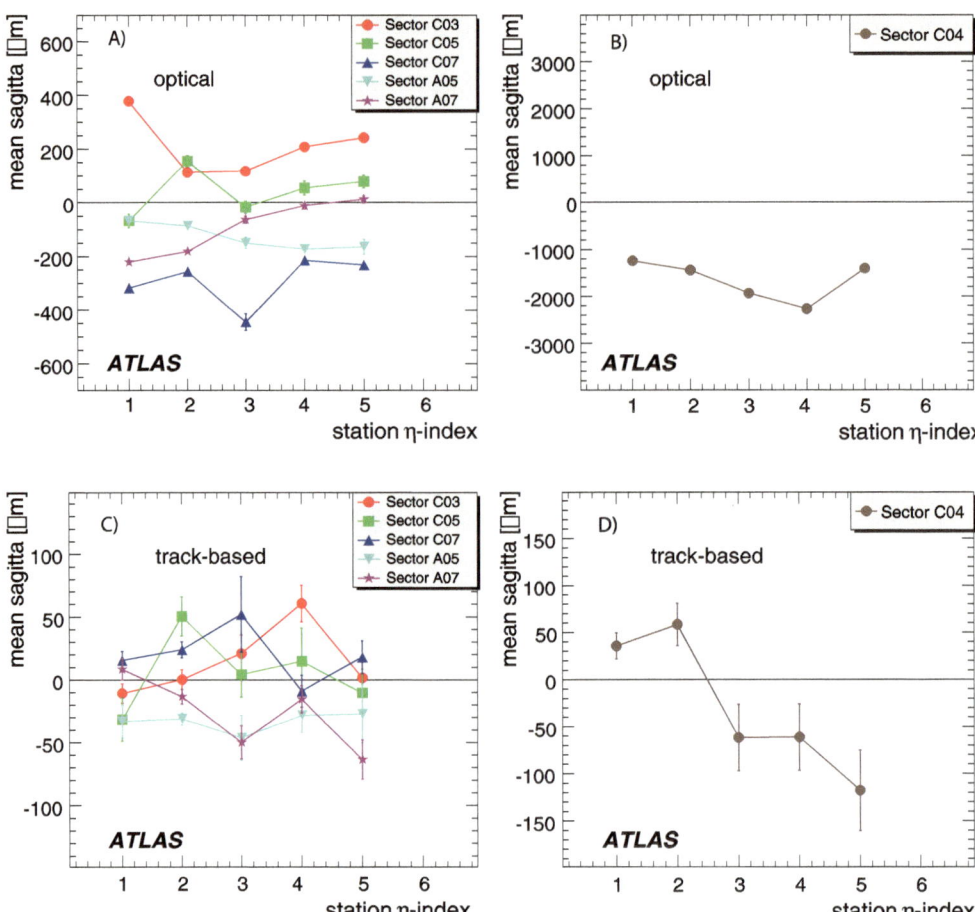

Fig. 27 Mean value of the track sagitta distributions obtained (**A**), (**B**): With the optical alignment system only, (**C**), (**D**): and using the track-based alignment. (**A**), (**C**): For the upper Large barrel sectors. (**B**), (**D**): For a Small barrel sector with $56° < \phi < 79°$

and Outer station) of MS projective towers. The RMS of the sagitta of the Middle station segment with respect to the Outer–Inner straight line extrapolation has been fit in five momentum bins. The result is shown in Fig. 28 for sector 5 (Large sector) with RPC-time corrections applied in the calibration procedure. The fitted value for the two terms is $K_0 = (12.2 \pm 0.7)$ mm×GeV and $K_1 = (107 \pm 21)$ µm. In the MS the multiple scattering term is expected to be worse than the one measured at the test beam setup and larger for

Small sectors due to the presence of the toroid coils between the Inner and Outer chambers. The value of K_1 measured with cosmic muons in sector 5 is only about a factor two worse than that measured at the test beam. Several effects contribute to this, including alignment, chamber deformations, calibration and single tube resolution. Similar studies performed for other sectors show worse results due to the smaller data sample available for alignment and calibration.

These preliminary studies with cosmic rays indicate that the method of track-based alignment is robust and with sufficient muon data from collisions the design alignment precision will be achieved.

8 Pattern recognition and segment reconstruction

The pattern recognition algorithm first groups hits close in space and time for each detector. Each pattern is characterized by a position and a direction and contains all the associated hits. Starting from these patterns, the segments are reconstructed with a straight line fit. The Gt_0-refit is applied at this stage and, if the Gt_0-refit procedure does not converge, the segment parameters are computed with the tube

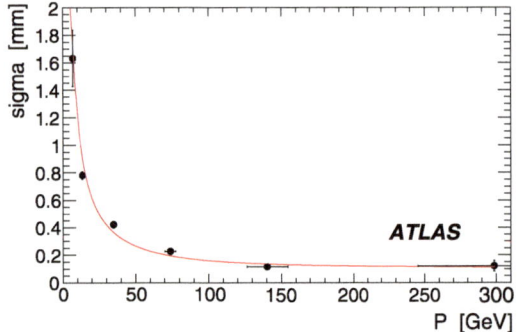

Fig. 28 RMS value of the sagitta distribution in sector 5 as a function of the muon momentum measured by the ID. The fit to the function described in the text is superimposed

t_0 provided by the calibration with tube resolution increased to 2 mm. After this, a drift radius is assigned to each tube with an uncertainty of 2 mm (independent of the drift radius value) in order to keep high track reconstruction efficiency even in the case where no precise alignment constants are available. The minimum number of hits per segment was set to 3 and no cuts were applied on the number of missed hits.

These relaxed requirements tend to increase the number of *fake* segments while keeping a high segment efficiency. Since cosmic ray events are quite clean and have low hit multiplicity this fake rate increase is not considered as a problem. On the other hand, a high reconstruction efficiency allows the use of segments to spot hardware problems in individual chambers or in calibration or decoding software. Most of the fake segments are rejected at the track reconstruction level.

Figure 29(A) shows, on the left, the number of MDT hits per segment for segments associated to a track. In the distribution clear peaks are observed at 6 and 8 hits corresponding to the 6-layer (Middle and Outer) and 8-layer (Inner) chambers.

The efficiency of the segment reconstruction in run 91,060 was determined in the following way. First, cosmic shower events are suppressed by requiring less than 20 segments in the event. Then a pair of segments in two MDT stations (Inner, Medium or Outer) are fitted to a straight line and the line is extrapolated to the third station. In the extrapolation multiple scattering is taken into account assuming a 2 GeV momentum for the cosmic muon. If the extrapolated line crossed the third station, a reconstructed segment is searched for in that station, but it is not required that the hits of this segment be associated to the muon track. The segment efficiency is then computed for each MDT chamber as the fraction of times a segment is found. In order to reduce the effect of the non-instrumented regions a fiducial cut in η was applied for both barrel and end-cap. Chambers that were not operational in the analyzed run were removed from the sample. It was not possible to determine the efficiency for all chambers due to the limited coverage of the trigger for the run used for this analysis (Fall 2008) and flux of cosmic rays. For tracks crossing the overlap region between two adjacent chambers (Small/Large sector overlap) it was not required that two segments be reconstructed, since this may lead to a slight overestimation of the efficiency.

The distribution of the segment efficiency is shown in Fig. 29(B) for 322 chambers in the barrel. The average value is 99.5% and the segment efficiency is uniform over the acceptance as shown in Table 5. In the efficiency for the barrel chambers there is a small loss of about 0.5% due the presence of the support structure of the ATLAS barrel. The Inner chambers have a slightly lower segment efficiency due to the geometry of the trigger and a larger uncertainty in the track extrapolation. Studies on systematic effects in determining the segment efficiency, such as its dependence on the Gt_0-

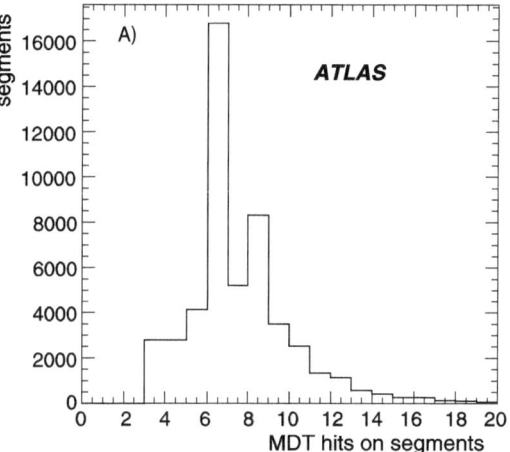

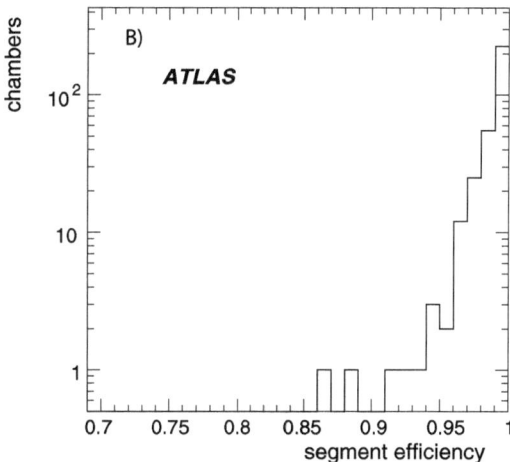

Fig. 29 (**A**): Distribution of the number of MDT hits per segment for segments associated to a track. (**B**): Segment reconstruction efficiency for 322 MDT chambers

Table 5 Average value of the segment reconstruction efficiency in the MDT stations

MDT station	BI	BM	BO	EI	EM	EO
Segment efficiency	0.987	0.992	0.996	0.992	0.998	0.999

refit, on the extrapolation and on the track angle, show that a systematic error of $\sim$0.5% affects the values of efficiency listed in Table 5.

An alternative method to evaluate the segment reconstruction efficiency, almost independent of chamber hardware problems, is described in the following. As in the previous case, this method can be used only with no magnetic field in the MS. All segment pairs in two different MDT stations (Inner, Middle or Outer) with a polar angle difference smaller than 7.5 mrad are considered. The segment pairs are fitted to a straight line and this is extrapolated to the third MDT station. The track is kept if at least three hit tubes are found in the third MDT with a signal charge above the ADC cut and aligned with the track extrapolation within one tube diameter, ± 3 cm. The segment efficiency is then computed as the fraction of selected tracks that have a segment reconstructed with at least 3 hits in the identified drift tubes. Since the normalization already requires the presence of three hits in the tested MDT station, this segment efficiency is almost independent of local hardware problems. A segment reconstruction efficiency higher than 0.99 is found in all MDT stations.

The rate of fake segments was studied with a random trigger. An average rate of 0.06 fake segments per event was found with the relaxed hit association criteria used for cosmic muons. This rate is expected to be strongly reduced to about 2×10^{-3} if the segment reconstruction requirements are made to be more stringent as shown by using an alternative muon tracking algorithm.

9 Track reconstruction

The MOORE and Muonboy programs have been optimized to reconstruct muon tracks originating from the IP. To cope with the different topology of cosmic ray muons they have been slightly modified as explained in Sect. 2. To mimic muons in collision events, the tracks are split at their perigee (point of closest approach to the beam axis), giving, usually, two reconstructed tracks: one in the upper part of the MS and one in the lower part. Events with at least one ID track satisfying the following criteria were selected:

- at least 20 hits in the Transition Radiation Tracker
- the number of hits summed over the SemiConductor Tracker (SCT) and the Pixel detector greater than 4
- the distance of closest approach in the transverse plane $|d_0|$ and along the z axis $|z_0|$ smaller than 1 m
- the value of the muon track $\chi^2/\text{ndf} < 3$
- the value of the reconstructed pseudorapidity $|\eta| < 1$
- reconstructed momentum greater than 5 GeV.

This selection has been applied for all the studies reported in this section with the exception of the momentum resolution results.

9.1 Resolution

The distribution of residuals for MDT hits associated to a track is shown in Fig. 30. The hit residual is defined as the difference between the drift radius measured in a tube and the distance of the track to the tube wire. The distribution refers only to tracks with MDT hits in at least three different muon stations (Inner, Middle and Outer) because these tracks have well constrained parameters and individual hits give a small contribution to the track parameters. The distribution was fitted to a double Gaussian with common mean

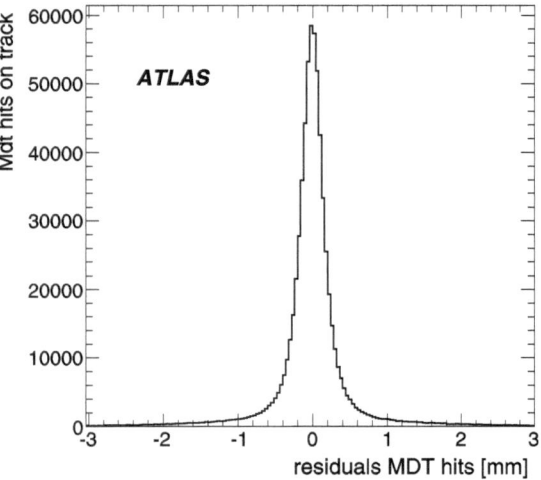

Fig. 30 Distribution of residuals for MDT hits associated to a track. The residuals have been fitted with a double-Gaussian function with common mean. The mean value is 6 μm, the standard deviation of the narrow Gaussian is about 150 μm and the one of the wide Gaussian is about 700 μm

value. The mean of the distribution was 6 μm and the RMS widths was 150 μm for the narrow Gaussian, accounting for 75% of the distribution, and 700 μm for the other. When compared to the distribution of the segment residuals shown in Sect. 6 two additional effects contribute to the broadening of this distribution: the misalignment between stations and multiple scattering in the MS material.

9.2 Efficiency

The track reconstruction efficiency is computed as the fraction of events where a track is reconstructed in the MS top or bottom hemisphere once an ID track was found satisfying the selection criteria described above. In this case also tracks with hits in only two out of three MDT stations (Inner, Middle or Outer) are accepted, even if these tracks have a worse momentum resolution than tracks reconstructed in three stations. About 15% of the selected tracks are in this category. In addition, to compute the track efficiency in the top (bottom) hemisphere, a momentum cut of 5 GeV (9 GeV) on the ID track is applied. The result is shown in Fig. 31 as a function of the pseudorapidity of the ID track, for the top and bottom hemisphere separately.

The value of the efficiency, integrated over the η acceptance, is 94.9% for the top and 93.7% for the bottom hemisphere respectively. If the four central bins are removed in Fig. 31 the efficiency increases to 98.3% and 96.3% respectively. The statistical error on these values is below 0.1%. The lower efficiency in the central detector region, around $|\eta| = 0$, is due to the presence of the main ATLAS service gap while lower efficiency in the Bottom hemisphere is explained by the uninstrumented regions occupied by the support structure of the ATLAS barrel.

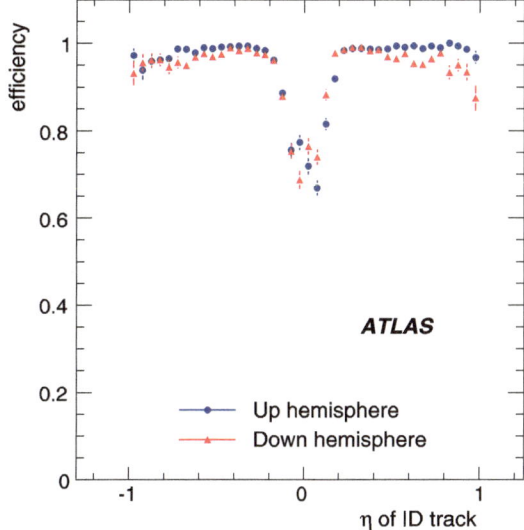

Fig. 31 Track reconstruction efficiency as a function of pseudorapidity. The loss in efficiency in the region near $|\eta| = 0$ is due to the loss of acceptance for detector services. The presence of a track measured in the ID with $|\eta| < 1$ is required

9.3 Momentum measurement

The momentum of cosmic muons was measured in runs with magnetic field. The momentum measurement can be defined at the MS entrance or at the point of closest approach to the IP. In the second case, for tracks crossing the ID, a correction was made for the energy loss in the calorimeters. This correction is based on the average energy loss computed by the track extrapolation algorithm and is on average 3.1 GeV for muons pointing to the interaction region with a distance of closest approach of $|d_0| < 1$ m and $|z_0| < 2$ m.

The distribution of momentum at the MS entrance is shown in Fig. 32—(A) for the top and bottom hemispheres separately. The difference between the two distributions is due to the ID track momentum cut of 5 GeV that translates in a different momentum cut-off in the two MS hemispheres, since cosmic muons are directed downwards. The same distribution extrapolated to the perigee is shown in Fig. 32(B), demonstrating that the correction for the energy loss in the calorimeter removes the offset.

The distribution of the number of MDT hits associated with a track is shown in Fig. 33(A). For this plot tracks measured in three MDT stations have been selected. A clear peak around 20 hits is visible (8 tubes in the Inner stations, 6 in the Middle and Outer stations).

In events with tracks that cross the whole MS, the track is split at the perigee and the two independent momentum measurements, in the top and bottom hemisphere, can be compared. Figure 33(B) shows the distribution of the difference of the two momentum values, top–bottom measured at the MS entrance, for tracks with momenta greater than 15 GeV. In this case the muons cross the calorimeter twice

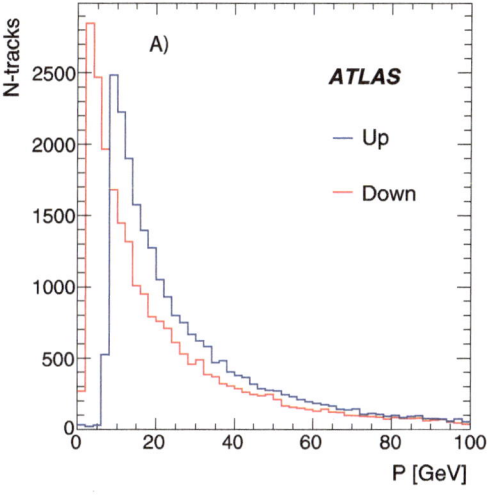

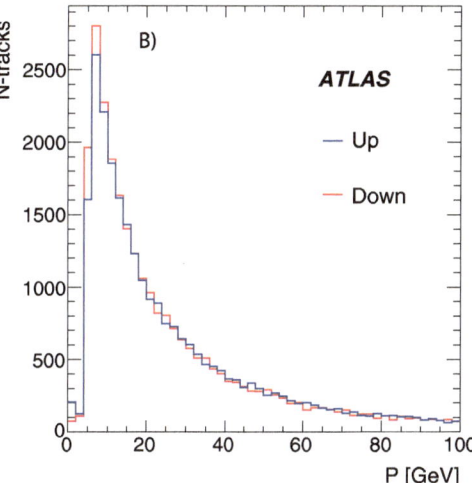

Fig. 32 (**A**): Distribution of momentum of cosmic muons as measured at the MS entrance for the upper and lower hemispheres. The difference between the two distributions is due to the ID track momentum cut of 5 GeV. (**B**): Same distributions with track momentum extrapolated to the IP

and the energy loss is twice the value quoted above, in good agreement with the 6.3 GeV mean value of the distribution.

The MS momentum resolution has been estimated by comparing for each cosmic muon the two independent measurements in the top and bottom hemispheres. In order to increase the available statistics no requirements on the presence of ID tracks were applied in this study. Only events with at least two reconstructed tracks in the MS are considered. Each track is required to have:

- at least 17 MDT hits, of which at least 7 in the Inner and 5 in the Middle and Outer stations of the same ϕ sector
- at least 2 different layers of RPC with a hit in the ϕ projection
- polar angle $65° < \theta < 115°$

– distance of closest approach to the IP $|d_0| < 1$ m and $|z_0| < 2$ m
– polar and azimuthal angles of the MS track pair agree within 10 °.

About 19 K top–bottom track pairs were selected in this way. For each track the value of transverse momentum was evaluated at the IP. The difference between the two values divided by their average

$$\frac{\Delta p_T}{p_T} = 2\, \frac{p_{\text{Tup}} - p_{\text{Tdown}}}{p_{\text{Tup}} + p_{\text{Tdown}}}$$

was measured in eleven bins of p_T. Since the cosmic muon momentum distribution is a steep function (see Fig. 32), the p_T value of each bin was taken as the mean value of the distribution in that bin.

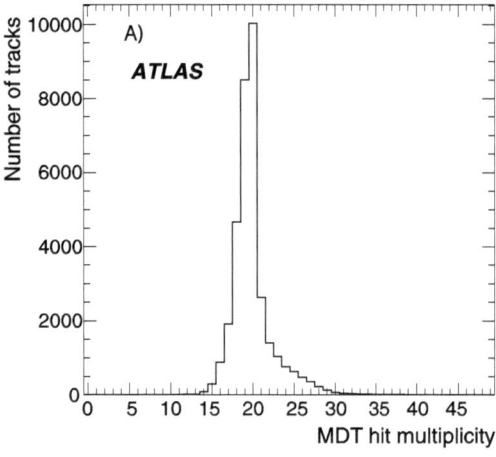

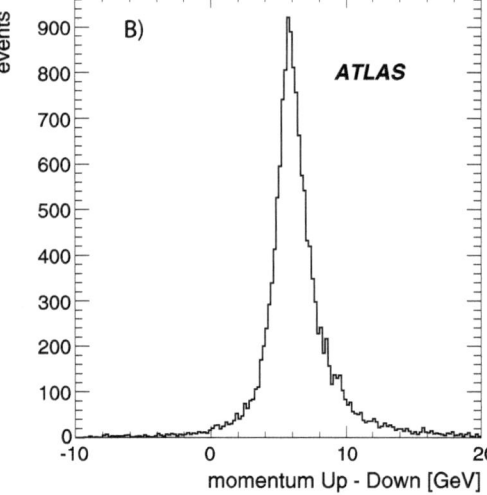

Fig. 33 (**A**): Number of MDT hits on track. (**B**): Momentum difference between momenta measured by the MS in *the top* and *bottom* hemispheres for cosmic muons. The momenta are expressed at the MS entrance and only tracks with momenta bigger than 15 GeV are considered. The mean value of 6.3 GeV is due to the energy loss in the calorimeter material

The distribution of $\Delta p_T/p_T$ was fitted in each bin with a double-Gaussian function with common mean value. The narrow Gaussian was convoluted with a Landau function to account for the distribution of energy loss in the calorimeter. For $p_T < 10$ GeV the normalizations of the two Gaussians were constrained such that 95% of the events are in the narrow Gaussian. Above 10 GeV this constraint was lowered to 70%. The mean value is representative of the difference in the transverse momentum scale between the two MS hemispheres. The RMS of the narrow Gaussian plus the width of the Landau, divided by $\sqrt{2}$, is taken as an estimate of the transverse momentum resolution for each p_T bin. The Landau width is added linearly to the narrow Gaussian RMS since the two quantities are strongly correlated.

The distribution of $\Delta p_T/p_T$ is shown in Fig. 34 for all p_T bins together with the fitted function. For the eleven bins the fit probability is in the range between 45% and 99%, showing that the chosen parametrization is a good representation of the data distribution.

Different fits have been done to study the systematics of the mean and RMS value. (i) The constraint between the two Gaussian areas has been changed by $\pm 10\%$. (ii) A double Gaussian with common mean and asymmetric fit range, with the fit range reduced to two standard deviations on the positive side to avoid the energy loss tail. (iii) A fit with two independent Gaussians with no range constraint. The result is that the estimated resolution is quite independent of the fit assumptions. The variation of the fit resolution ranges between 0.5% at low p_T up to a maximum of 1% in the highest momentum bin.

The fit mean values indicate that the p_T scales in the two MS hemispheres are in agreement within 1%, or better. The relative p_T resolution, $\sigma_{p_T}/p_T = \sigma(\Delta p_T/p_T)/\sqrt{2}$, is shown in Fig. 35, for the two main muon reconstruction algorithms [7, 8], as a function of the transverse momentum. The two results are consistent taking into account the independent statistical uncertainties.

The resolution function can be fitted with the sum in quadrature of three terms, the uncertainty on the energy loss corrections P_0, the multiple scattering term P_1, and the intrinsic resolution P_2.

$$\frac{\sigma_{p_T}}{p_T} = \frac{P_0}{p_T} \oplus P_1 \oplus P_2 \times p_T.$$

The result of the fit is shown in Fig. 35. The values of the parameters are: $P_0 = 0.29 \pm 0.03 \pm 0.01$ GeV, $P_1 = 0.043 \pm 0.002 \pm 0.002$, $P_2 = (4.1 \pm 0.4 \pm 0.6) \times 10^{-4}$ GeV^{-1}. The second uncertainty, due the systematics of the bin-by-bin fit method, was evaluated by changing the fitting assumptions as explained above. The expected values for these parameters were computed in reference [3] on the basis of an analytic calculation of the p_T resolution that takes into account the detailed description of the material in the MS, the

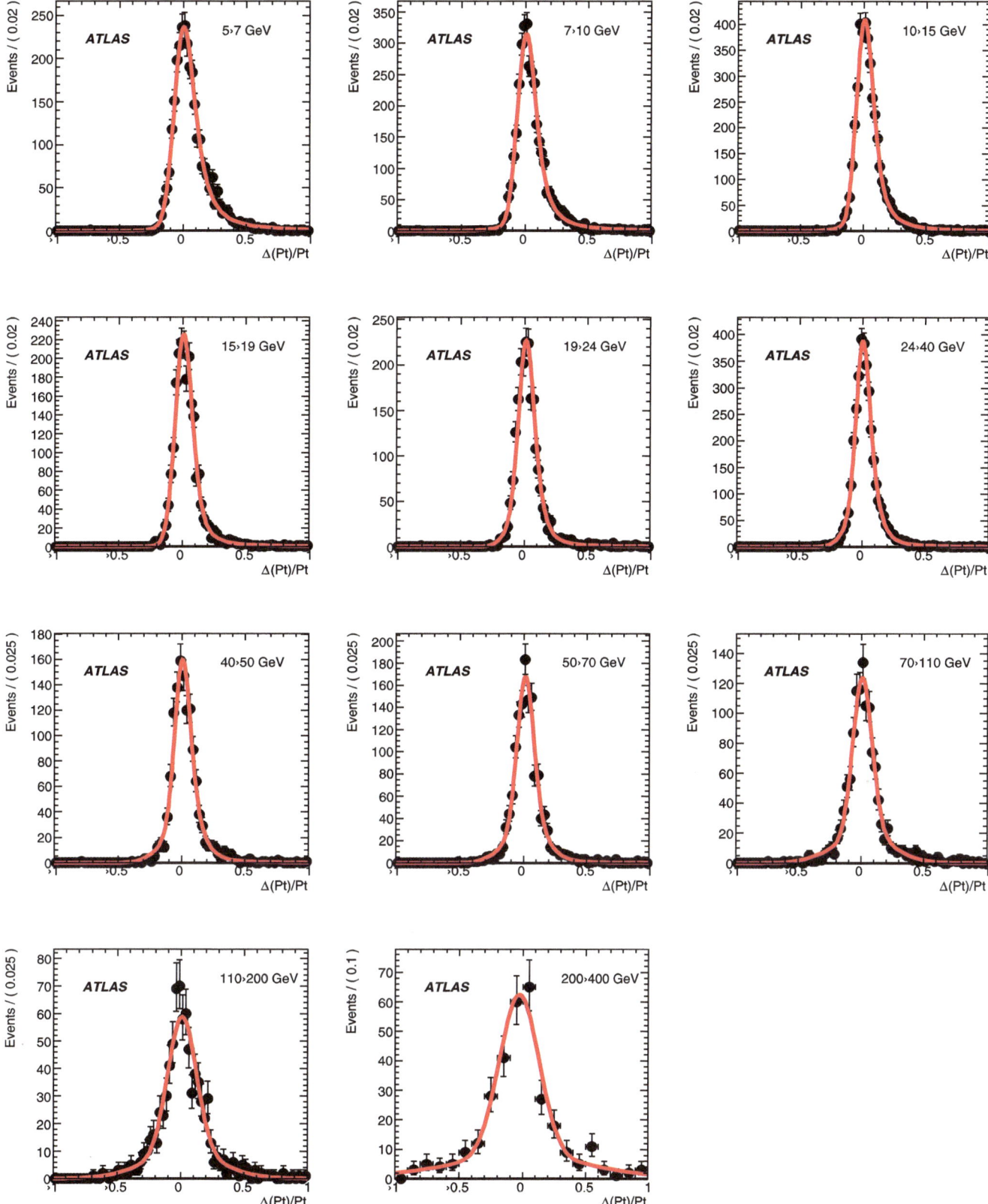

Fig. 34 Distributions of $\Delta p_T/p_T$ in the eleven p_T bins. Fits to the function described in the text are superimposed

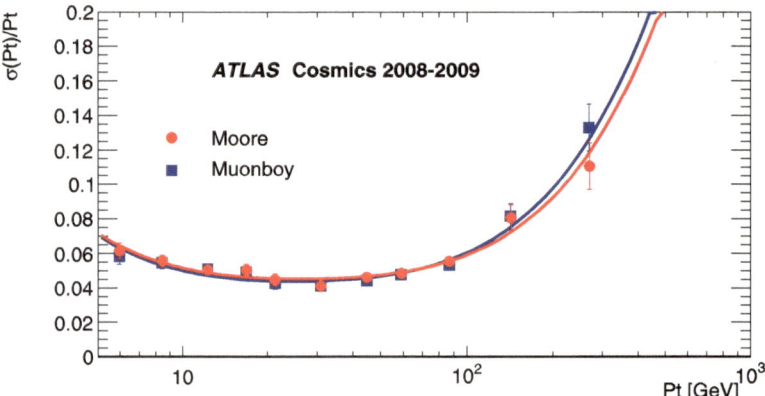

Fig. 35 Transverse momentum resolution evaluated with the top–bottom method explained in the text as a function of p_T, barrel region only ($|\eta| < 1.1$). The fit to the three resolution parameters as described in the text is superimposed

single tube resolution, the alignment accuracy and the magnetic field map. The values obtained for the barrel MS were: $P_0 = 0.35$ GeV, $P_1 = 0.035$ and $P_2 = 1.2 \times 10^{-4}$ GeV^{-1}. The result is in fair agreement with the expected values for the first two terms, while the intrinsic term is worse. The difference has been investigated to trace the effects that contribute to worsen the resolution as determined with cosmic muons.

First, more than 70% of the track pairs considered in the analysis are in the Large sectors 5–13. At high p_T the momentum resolution in the barrel Large sectors is worse than in Small sectors because the field integral is smaller (see Fig. 1). Instead, in the low p_T region dominated by multiple scattering the resolution in Large sectors is better since the magnet coils are in the Small sectors.

Second, the single tube resolution is affected by imperfect calibrations and the additional time jitter is not completely recovered by the Gt_0-refit (see Fig. 12). Part of the tracks in the sample contain segments with a badly converging Gt_0-refit. As a cross-check, all the tracks with bad convergence were removed and the analysis was repeated. The intrinsic term decreased by about 30%.

Third, the alignment in many sectors of the MS is still not at the required level due to the limited statistics of straight tracks in the cosmic ray data sample. Last, several other effects that contribute to resolution have not been removed, such as chamber deformations (due to temperature effects), wire sagging (particularly important in large chambers), single chamber geometrical defects. Each of these effects contribute to worsening the resolution and can be removed with dedicated software tools. At the present stage of commissioning, the momentum resolution is close to the design value for $p_T < 50$ GeV, but is not as good for higher momenta.

10 Summary

The data collected in several months during the 2008–2009 cosmic ray runs have been analyzed to assess the perfor-

mance of the Muon Spectrometer after its installation in the ATLAS experiment. Parts of the detector, the Small Wheels in front of the end-cap toroids, were installed during the runs and the commissioning of the many detectors was proceeding while debugging the data acquisition and the data control systems. The detector coverage during most of the run period was higher than 99%, with the exception of the RPC chambers which were still under commissioning. For this detector subsystem the coverage steadily improved during the commissioning runs reaching more than 95% in Spring 2009. Results on several aspects of the Muon Spectrometer performance have been presented. These include detector coverage, efficiency, resolution and relative timing of trigger and precision tracking chambers, track reconstruction, calibration, alignment and data quality.

Finally, with data collected when the magnetic field was on, a first estimate of the spectrometer momentum resolution was obtained. Efficiency and resolution of single elements have been measured for MDT, RPC and TGC chambers and were found in agreement with results obtained previously with high-momentum muon beams. The trigger chamber timing has been adjusted with enough precision to guarantee that the interaction bunch crossing can be identified with a minimal number of failures. The muon trigger logic, based on fast tracking of pointing muons has been extensively tested in the regions of the detector with good cosmic ray illumination. A slight deterioration of the MDT spatial resolution, compared to test beam results, was observed, which can be understood in terms of an additional time jitter due to the asynchronous timing of cosmic muons and to their non-pointing geometry. These effects were partially removed, modifying the track reconstruction programs with dedicated algorithms. Allowing for an increase of the single hit resolution, to cope with these effects, the track segment efficiency in individual chambers was found to be satisfactory and uniform over the large number of chambers.

The performance of the end-cap and barrel optical alignment systems have been measured using cosmic muon tracks with no magnetic field. The results demonstrate that the end-cap optical system is able to provide the required precision for chamber alignment. The design of the alignment system in the barrel requires additional constraints provided by straight tracks. The method has been tested with good results, but is limited by the statistics of high-momentum muons with the required pointing geometry.

With the geometry corrections provided by the alignment system, tracks in projective geometry were reconstructed in the barrel showing that the reconstruction efficiency is uniform over the entire acceptance and that the sagitta error is in agreement with the detector resolution, the alignment precision and the effect of multiple coulomb scattering.

Finally with magnetic field, tracks crossing the whole spectrometer were used to obtain two independent measurements of the momentum. The momentum resolution was evaluated using the two values in the top and bottom part of the detector and the results were analyzed, fitting the distribution of the difference as function of the momentum. Taking into account the momentum spectrum, the multiple scattering in the spectrometer and the energy loss in traversing the calorimeters, the momentum resolution is in good agreement with results from simulation for transverse momenta smaller than 50 GeV. The statistics of high-momentum pointing tracks limits the accuracy of the individual chamber calibration and the precision of the alignment. At higher momenta, these limitations result in degraded momentum resolution.

During the long period of commissioning with cosmic rays it was possible to optimize the performance of the various hardware and software elements and to reach a level of understanding, such that we can consider the Muon Spectrometer to be ready to efficiently detect muons produced in high-energy proton–proton collisions.

Acknowledgements We are greatly indebted to all CERN's departments and to the LHC project for their immense efforts not only in building the LHC, but also for their direct contributions to the construction and installation of the ATLAS detector and its infrastructure. We acknowledge equally warmly all our technical colleagues in the collaborating Institutions without whom the ATLAS detector could not have been built. Furthermore we are grateful to all the funding agencies which supported generously the construction and the commissioning of the ATLAS detector and also provided the computing infrastructure.

The ATLAS detector design and construction has taken about fifteen years, and our thoughts are with all our colleagues who sadly could not see its final realization.

We acknowledge the support of ANPCyT, Argentina; Yerevan Physics Institute, Armenia; ARC and DEST, Australia; Bundesministerium für Wissenschaft und Forschung, Austria; National Academy of Sciences of Azerbaijan; State Committee on Science & Technologies of the Republic of Belarus; CNPq and FINEP, Brazil; NSERC, NRC, and CFI, Canada; CERN; CONICYT, Chile; NSFC, China; COLCIENCIAS, Colombia; Ministry of Education, Youth and Sports of the Czech Republic, Ministry of Industry and Trade of the Czech Republic, and Committee for Collaboration of the Czech Republic with CERN; Danish Natural Science Research Council and the Lundbeck Foundation; European Commission, through the ARTEMIS Research Training Network; IN2P3-CNRS and Dapnia-CEA, France; Georgian Academy of Sciences; BMBF, HGF, DFG and MPG, Germany; Ministry of Education and Religion, through the EPEAEK program PYTHAGORAS II and GSRT, Greece; ISF, MINERVA, GIF, DIP, and Benoziyo Center, Israel; INFN, Italy; MEXT, Japan; CNRST, Morocco; FOM and NWO, Netherlands; The Research Council of Norway; Ministry of Science and Higher Education, Poland; GRICES and FCT, Portugal; Ministry of Education and Research, Romania; Ministry of Education and Science of the Russian Federation and State Atomic Energy Corporation "Rosatom"; JINR; Ministry of Science, Serbia; Department of International Science and Technology Cooperation, Ministry of Education of the Slovak Republic; Slovenian Research Agency, Ministry of Higher Education, Science and Technology, Slovenia; Ministerio de Educación y Ciencia, Spain; The Swedish Research Council, The Knut and Alice Wallenberg Foundation, Sweden; State Secretariat for Education and Science, Swiss National Science Foundation, and Cantons of Bern and Geneva, Switzerland; National Science Council, Taiwan; TAEK, Turkey; The Science and Technology Facilities Council and The Leverhulme Trust, United Kingdom; DOE and NSF, United States of America.

References

1. The ATLAS Muon Collaboration, *The ATLAS Muon Spectrometer technical design report*. CERN-LHCC/97-22 (31 May 1997). ISBN 92-9083-108-1
2. G. Aad et al. (The ATLAS Collaboration), The ATLAS experiment at the CERN large hadron collider. J. Instrum. **3**, S08003 (2008). 1–437
3. G. Aad et al. (The ATLAS Collaboration), *Expected Performance of the ATLAS Experiment: Detector, Trigger and Physics*. CERN-OPEN 2008-020 (December 2008). ISBN 978-92-9083-321-5
4. C. Adorisio et al., System Test of the ATLAS Muon spectrometer in the H8 beam at the CERN SPS. Nucl. Instrum. Meth. A **593**, 232–254 (2008)
5. C. Adorisio et al., Study of the ATLAS MDT spectrometer using high energy CERN combined test beam data. Nucl. Instrum. Meth. A **598**, 400–415 (2009)
6. The ATLAS Collaboration, *ATLAS computing technical design report*. CERN-LHCC/2005-022 (20 June 2005). ISBN 92-9083-250-9
7. D. Adams et al., Track reconstruction in the ATLAS Muon spectrometer with MOORE. ATL-SOFT-2003-007, 2.10.2003, http://cdsweb.cern.ch/collection/ATLAS
8. S. Hassani et al., A Muon identification and combined reconstruction procedure for the ATLAS detector at the LHC using (Muonboy, STACO, MuTag) reconstruction packages. Nucl. Instrum. Meth. A **572**, 77–79 (2007)
9. The ATLAS Collaboration, *First level trigger technical design report*. CERN-LHCC/98-014 (30 June 1998). ISBN 92-9083-128-6
10. F. Anulli et al., The level-1 Muon barrel trigger of the ATLAS experiment. J. Instrum. **4**, P04010 (2009). 1–35
11. The ATLAS Collaboration, *High-level trigger, data acquisition and controls technical design report*. CERN-LHCC/2003-022 (30 June 2003). ISBN 92-9083-205-3
12. P. Bagnaia et al., *Calibration model for the MDT chambers of the ATLAS Muon spectrometer*. ATL-MUON-PUB-2008-004 (28 March 2008), http://cdsweb.cern.ch/collection/ATLAS
13. M. Verducci, ATLAS database experience with the COOL conditions database project. J. Phys. Conf. Ser. **119**, 042031 (2008)

14. Y. Arai, Development of front end electronics and TDC LSI for the ATLAS MDT. Nucl. Instrum. Meth. A **453**, 365 (2000)

15. Y. Arai et al., ATLAS muon drift tube electronics. J. Instrum. **3**, P09001 (2008). 1–58

16. D.S. Levin et al., Drift time spectrum and gas monitoring in the ATLAS Muon spectrometer precision chambers. Nucl. Instrum. Meth. A **588**, 347–358 (2008)

17. R. Veenhof, Garfield, a drift chamber simulation program. Prepared for International Conference on Programming and Mathematical Methods for Solving Physical Problems, Dubna, Russia, 14–19 June 1993

18. S.F. Biagi, Monte Carlo simulation of electron drift and diffusion in counting gases under the influence of electric and magnetic fields. Nuclear Instrum. Methods Phys. Res., Sect. A, Accel. Spectrom. Detect. Assoc. Equip. **421**(1–2), 234–240 (1999)

19. R.M. Avramidou, E. Gazis, R. Veenhof, Drift properties of the ATLAS MDT chambers. Nucl. Instrum. Meth. A **568**, 672–681 (2006)

20. P. Branchini et al., Global time fit for tracking in an array of drift cells: the drift tubes of the ATLAS experiment. IEEE Trans. Nucl. Sci. **55**, 620 (2008)

21. C. Amelung, The alignment system of the ATLAS Muon spectrometer. Eur. Phys. J. C **33**, 999–1001 (2004)

22. S. Aefsky et al., The optical alignment system of the ATLAS Muon spectrometer endcaps. J. Instrum. **3**, P11005 (2008). pp. 1–59

23. C. Guyot et al., *The alignment system of the barrel part of the ATLAS Muon spectrometer*. ATLAS Note ATL-MUON-PUB-2008-007 (2008), http://cdsweb.cern.ch/collection/ATLAS

24. V. Blobel, Millepede: linear least squares fits with a large number of parameters, http://www.desy.de/~blobel/mptalks.html

The ATLAS Simulation Infrastructure

The ATLAS Collaboration[*,**]

G. Aad[48], B. Abbott[111], J. Abdallah[11], A.A. Abdelalim[49], A. Abdesselam[118], O. Abdinov[10], B. Abi[112], M. Abolins[88], H. Abramowicz[152], H. Abreu[115], B.S. Acharya[163a,163b], D.L. Adams[24], T.N. Addy[56], J. Adelman[174], C. Adorisio[36a,36b], P. Adragna[75], T. Adye[129], S. Aefsky[22], J.A. Aguilar-Saavedra[124b], M. Aharrouche[81], S.P. Ahlen[21], F. Ahles[48], A. Ahmad[147], H. Ahmed[2], M. Ahsan[40], G. Aielli[133a,133b], T. Akdogan[18a], T.P.A. Åkesson[79], G. Akimoto[154], A.V. Akimov[94], A. Aktas[48], M.S. Alam[1], M.A. Alam[76], S. Albrand[55], M. Aleksa[29], I.N. Aleksandrov[65], C. Alexa[25a], G. Alexander[152], G. Alexandre[49], T. Alexopoulos[9], M. Alhroob[20], M. Aliev[15], G. Alimonti[89a], J. Alison[120], M. Aliyev[10], P.P. Allport[73], S.E. Allwood-Spiers[53], J. Almond[82], A. Aloisio[102a,102b], R. Alon[170], A. Alonso[79], M.G. Alviggi[102a,102b], K. Amako[66], C. Amelung[22], A. Amorim[124a], G. Amorós[166], N. Amram[152], C. Anastopoulos[139], T. Andeen[29], C.F. Anders[48], K.J. Anderson[30], A. Andreazza[89a,89b], V. Andrei[58a], X.S. Anduaga[70], A. Angerami[34], F. Anghinolfi[29], N. Anjos[124a], A. Annovi[47], A. Antonaki[8], M. Antonelli[47], S. Antonelli[19a,19b], J. Antos[144b], B. Antunovic[41], F. Anulli[132a], S. Aoun[83], G. Arabidze[8], I. Aracena[143], Y. Arai[66], A.T.H. Arce[14], J.P. Archambault[28], S. Arfaoui[29,a], J.-F. Arguin[14], T. Argyropoulos[9], M. Arik[18a], A.J. Armbruster[87], O. Arnaez[4], C. Arnault[115], A. Artamonov[95], D. Arutinov[20], M. Asai[143], S. Asai[154], R. Asfandiyarov[171], S. Ask[82], B. Åsman[145a,145b], D. Asner[28], L. Asquith[77], K. Assamagan[24], A. Astbury[168], A. Astvatsatourov[52], G. Atoian[174], B. Auerbach[174], K. Augsten[127], M. Aurousseau[4], N. Austin[73], G. Avolio[162], R. Avramidou[9], D. Axen[167], C. Ay[54], G. Azuelos[93,b], Y. Azuma[154], M.A. Baak[29], A.M. Bach[14], H. Bachacou[136], K. Bachas[29], M. Backes[49], E. Badescu[25a], P. Bagnaia[132a,132b], Y. Bai[32a], T. Bain[157], J.T. Baines[129], O.K. Baker[174], M.D. Baker[24], S. Baker[77], F. Baltasar Dos Santos Pedrosa[29], E. Banas[38], P. Banerjee[93], S. Banerjee[168], D. Banfi[89a,89b], A. Bangert[137], V. Bansal[168], S.P. Baranov[94], S. Baranov[65], A. Barashkou[65], T. Barber[27], E.L. Barberio[86], D. Barberis[50a,50b], M. Barbero[20], D.Y. Bardin[65], T. Barillari[99], M. Barisonzi[173], T. Barklow[143], N. Barlow[27], B.M. Barnett[129], R.M. Barnett[14], A. Baroncelli[134a], A.J. Barr[118], F. Barreiro[80], J. Barreiro Guimarães da Costa[57], P. Barrillon[115], R. Bartoldus[143], D. Bartsch[20], R.L. Bates[53], L. Batkova[144a], J.R. Batley[27], A. Battaglia[16], M. Battistin[29], F. Bauer[136], H.S. Bawa[143], M. Bazalova[125], B. Beare[157], T. Beau[78], P.H. Beauchemin[118], R. Beccherle[50a], N. Becerici[18a], P. Bechtle[41], G.A. Beck[75], H.P. Beck[16], M. Beckingham[48], K.H. Becks[173], A.J. Beddall[18c], A. Beddall[18c], V.A. Bednyakov[65], C. Bee[83], M. Begel[24], S. Behar Harpaz[151], P.K. Behera[63], M. Beimforde[99], C. Belanger-Champagne[165], P.J. Bell[49], W.H. Bell[49], G. Bella[152], L. Bellagamba[19a], F. Bellina[29], M. Bellomo[119a], A. Belloni[57], K. Belotskiy[96], O. Beltramello[29], S. Ben Ami[151], O. Benary[152], D. Benchekroun[135a], M. Bendel[81], B.H. Benedict[162], N. Benekos[164], Y. Benhammou[152], G.P. Benincasa[124a], D.P. Benjamin[44], M. Benoit[115], J.R. Bensinger[22], K. Benslama[130], S. Bentvelsen[105], M. Beretta[47], D. Berge[29], E. Bergeaas Kuutmann[41], N. Berger[4], F. Berghaus[168], E. Berglund[49], J. Beringer[14], P. Bernat[115], R. Bernhard[48], C. Bernius[77], T. Berry[76], A. Bertin[19a,19b], M.I. Besana[89a,89b], N. Besson[136], S. Bethke[99], R.M. Bianchi[48], M. Bianco[72a,72b], O. Biebel[98], J. Biesiada[14], M. Biglietti[132a,132b], H. Bilokon[47], M. Bindi[19a,19b], S. Binet[115], A. Bingul[18c], C. Bini[132a,132b], C. Biscarat[179], U. Bitenc[48], K.M. Black[57], R.E. Blair[5], J.-B. Blanchard[115], G. Blanchot[29], C. Blocker[22], A. Blondel[49], W. Blum[81], U. Blumenschein[54], G.J. Bobbink[105], A. Bocci[44], M. Boehler[41], J. Boek[173], N. Boelaert[79], S. Böser[77], J.A. Bogaerts[29], A. Bogouch[90,*], C. Bohm[145a], J. Bohm[125], V. Boisvert[76], T. Bold[162c], V. Boldea[25a], V.G. Bondarenko[96], M. Bondioli[162], M. Boonekamp[136], S. Bordoni[78], C. Borer[16], A. Borisov[128], G. Borissov[71], I. Borjanovic[72a], S. Borroni[132a,132b], K. Bos[105], D. Boscherini[19a], M. Bosman[11], H. Boterenbrood[105], J. Bouchami[93], J. Boudreau[123], E.V. Bouhova-Thacker[71], C. Boulahouache[123], C. Bourdarios[115], A. Boveia[30], J. Boyd[29], I.R. Boyko[65], I. Bozovic-Jelisavcic[12b], J. Bracinik[17], A. Braem[29], P. Branchini[134a], G.W. Brandenburg[57], A. Brandt[7], G. Brandt[41], O. Brandt[54], U. Bratzler[155], B. Brau[84], J.E. Brau[114], H.M. Braun[173], B. Brelier[157], J. Bremer[29], R. Brenner[165], S. Bressler[151], D. Britton[53], F.M. Brochu[27], I. Brock[20], R. Brock[88], E. Brodet[152], C. Bromberg[88], G. Brooijmans[34], W.K. Brooks[31b], G. Brown[82], P.A. Bruckman de Renstrom[38], D. Bruncko[144b], R. Bruneliere[48], S. Brunet[41], A. Bruni[19a], G. Bruni[19a], M. Bruschi[19a], F. Bucci[49], J. Buchanan[118], P. Buchholz[141], A.G. Buckley[45], I.A. Budagov[65], B. Budick[108], V. Büscher[81], L. Bugge[117], O. Bulekov[96], M. Bunse[42], T. Buran[117], H. Burckhart[29], S. Burdin[73], T. Burgess[13],

S. Burke[129], E. Busato[33], P. Bussey[53], C.P. Buszello[165], F. Butin[29], B. Butler[143], J.M. Butler[21], C.M. Buttar[53], J.M. Butterworth[77], T. Byatt[77], J. Caballero[24], S. Cabrera Urbán[166], D. Caforio[19a,19b], O. Cakir[3a], P. Calafiura[14], G. Calderini[78], P. Calfayan[98], R. Calkins[106a], L.P. Caloba[23a], D. Calvet[33], P. Camarri[133a,133b], D. Cameron[117], S. Campana[29], M. Campanelli[77], V. Canale[102a,102b], F. Canelli[30], A. Canepa[158a], J. Cantero[80], L. Capasso[102a,102b], M.D.M. Capeans Garrido[29], I. Caprini[25a], M. Caprini[25a], M. Capua[36a,36b], R. Caputo[147], C. Caramarcu[25a], R. Cardarelli[133a], T. Carli[29], G. Carlino[102a], L. Carminati[89a,89b], B. Caron[2,b], S. Caron[48], G.D. Carrillo Montoya[171], S. Carron Montero[157], A.A. Carter[75], J.R. Carter[27], J. Carvalho[124a], D. Casadei[108], M.P. Casado[11], M. Cascella[122a,122b], A.M. Castaneda Hernandez[171], E. Castaneda-Miranda[171], V. Castillo Gimenez[166], N.F. Castro[124b], G. Cataldi[72a], A. Catinaccio[29], J.R. Catmore[71], A. Cattai[29], G. Cattani[133a,133b], S. Caughron[34], D. Cauz[163a,163c], P. Cavalleri[78], D. Cavalli[89a], M. Cavalli-Sforza[11], V. Cavasinni[122a,122b], F. Ceradini[134a,134b], A.S. Cerqueira[23a], A. Cerri[29], L. Cerrito[75], F. Cerutti[47], S.A. Cetin[18b], A. Chafaq[135a], D. Chakraborty[106a], K. Chan[2], J.D. Chapman[27], J.W. Chapman[87], E. Chareyre[78], D.G. Charlton[17], V. Chavda[82], S. Cheatham[71], S. Chekanov[5], S.V. Chekulaev[158a], G.A. Chelkov[65], H. Chen[24], S. Chen[32c], X. Chen[171], A. Cheplakov[65], V.F. Chepurnov[65], R. Cherkaoui El Moursli[135d], V. Tcherniatine[24], D. Chesneanu[25a], E. Cheu[6], S.L. Cheung[157], L. Chevalier[136], F. Chevallier[136], V. Chiarella[47], G. Chiefari[102a,102b], L. Chikovani[51], J.T. Childers[58a], A. Chilingarov[71], G. Chiodini[72a], V. Chizhov[65], G. Choudalakis[30], S. Chouridou[137], I.A. Christidi[77], A. Christov[48], D. Chromek-Burckhart[29], M.L. Chu[150], J. Chudoba[125], G. Ciapetti[132a,132b], A.K. Ciftci[3a], R. Ciftci[3a], D. Cinca[33], V. Cindro[74], M.D. Ciobotaru[162], C. Ciocca[19a,19b], A. Ciocio[14], M. Cirilli[87], M. Citterio[89a], A. Clark[49], P.J. Clark[45], W. Cleland[123], J.C. Clemens[83], B. Clement[55], C. Clement[145a,145b], Y. Coadou[83], M. Cobal[163a,163c], A. Coccaro[50a,50b], J. Cochran[64], J. Coggeshall[164], E. Cogneras[179], A.P. Colijn[105], C. Collard[115], N.J. Collins[17], C. Collins-Tooth[53], J. Collot[55], G. Colon[84], P. Conde Muiño[124a], E. Coniavitis[165], M. Consonni[104], S. Constantinescu[25a], C. Conta[119a,119b], F. Conventi[102a,d], M. Cooke[34], B.D. Cooper[75], A.M. Cooper-Sarkar[118], N.J. Cooper-Smith[76], K. Copic[34], T. Cornelissen[50a,50b], M. Corradi[19a], F. Corriveau[85,e], A. Corso-Radu[162], A. Cortes-Gonzalez[164], G. Cortiana[99], G. Costa[89a], M.J. Costa[166], D. Costanzo[139], T. Costin[30], D. Côté[41], R. Coura Torres[23a], L. Courneyea[168], G. Cowan[76], C. Cowden[27], B.E. Cox[82], K. Cranmer[108], J. Cranshaw[5], M. Cristinziani[20], G. Crosetti[36a,36b], R. Crupi[72a,72b], S. Crépé-Renaudin[55], C. Cuenca Almenar[174], T. Cuhadar Donszelmann[139], M. Curatolo[47], C.J. Curtis[17], P. Cwetanski[61], Z. Czyczula[174], S. D'Auria[53], M. D'Onofrio[73], A. D'Orazio[99], C. Da Via[82], W. Dabrowski[37], T. Dai[87], C. Dallapiccola[84], S.J. Dallison[129,*], C.H. Daly[138], M. Dam[35], H.O. Danielsson[29], D. Dannheim[99], V. Dao[49], G. Darbo[50a], G.L. Darlea[25b], W. Davey[86], T. Davidek[126], N. Davidson[86], R. Davidson[71], M. Davies[93], A.R. Davison[77], I. Dawson[139], R.K. Daya[39], K. De[7], R. de Asmundis[102a], S. De Castro[19a,19b], P.E. De Castro Faria Salgado[24], S. De Cecco[78], J. de Graat[98], N. De Groot[104], P. de Jong[105], L. De Mora[71], M. De Oliveira Branco[29], D. De Pedis[132a], A. De Salvo[132a], U. De Sanctis[163a,163c], A. De Santo[148], J.B. De Vivie De Regie[115], G. De Zorzi[132a,132b], S. Dean[77], D.V. Dedovich[65], J. Degenhardt[120], M. Dehchar[118], C. Del Papa[163a,163c], J. Del Peso[80], T. Del Prete[122a,122b], A. Dell'Acqua[29], L. Dell'Asta[89a,89b], M. Della Pietra[102a,d], D. della Volpe[102a,102b], M. Delmastro[29], P.A. Delsart[55], C. Deluca[147], S. Demers[174], M. Demichev[65], B. Demirkoz[11], J. Deng[162], W. Deng[24], S.P. Denisov[128], J.E. Derkaoui[135c], F. Derue[78], P. Dervan[73], K. Desch[20], P.O. Deviveiros[157], A. Dewhurst[129], B. DeWilde[147], S. Dhaliwal[157], R. Dhullipudi[24,f], A. Di Ciaccio[133a,133b], L. Di Ciaccio[4], A. Di Domenico[132a,132b], A. Di Girolamo[29], B. Di Girolamo[29], S. Di Luise[134a,134b], A. Di Mattia[88], R. Di Nardo[133a,133b], A. Di Simone[133a,133b], R. Di Sipio[19a,19b], M.A. Diaz[31a], F. Diblen[18c], E.B. Diehl[87], J. Dietrich[48], T.A. Dietzsch[58a], S. Diglio[115], K. Dindar Yagci[39], J. Dingfelder[48], C. Dionisi[132a,132b], P. Dita[25a], S. Dita[25a], F. Dittus[29], F. Djama[83], R. Djilkibaev[108], T. Djobava[51], M.A.B. do Vale[23a], A. Do Valle Wemans[124a], T.K.O. Doan[4], D. Dobos[29], E. Dobson[29], M. Dobson[162], C. Doglioni[118], T. Doherty[53], J. Dolejsi[126], I. Dolenc[74], Z. Dolezal[126], B.A. Dolgoshein[96], T. Dohmae[154], M. Donega[120], J. Donini[55], J. Dopke[173], A. Doria[102a], A. Dos Anjos[171], A. Dotti[122a,122b], M.T. Dova[70], A. Doxiadis[105], A.T. Doyle[53], Z. Drasal[126], M. Dris[9], J. Dubbert[99], E. Duchovni[170], G. Duckeck[98], A. Dudarev[29], F. Dudziak[115], M. Dührssen[29], L. Duflot[115], M.-A. Dufour[85], M. Dunford[30], H. Duran Yildiz[3b], A. Dushkin[22], R. Duxfield[139], M. Dwuznik[37], M. Düren[52], W.L. Ebenstein[44], J. Ebke[98], S. Eckweiler[81], K. Edmonds[81], C.A. Edwards[76], K. Egorov[61], W. Ehrenfeld[41], T. Ehrich[99], T. Eifert[29], G. Eigen[13], K. Einsweiler[14], E. Eisenhandler[75], T. Ekelof[165], M. El Kacimi[4], M. Ellert[165], S. Elles[4], F. Ellinghaus[81], K. Ellis[75], N. Ellis[29], J. Elmsheuser[98], M. Elsing[29], D. Emeliyanov[129], R. Engelmann[147], A. Engl[98], B. Epp[62], A. Eppig[87], J. Erdmann[54], A. Ereditato[16], D. Eriksson[145a], I. Ermoline[88], J. Ernst[1], M. Ernst[24], J. Ernwein[136], D. Errede[164], S. Errede[164], E. Ertel[81], M. Escalier[115], C. Escobar[166], X. Espinal Curull[11], B. Esposito[47], A.I. Etienvre[136], E. Etzion[152], H. Evans[61], L. Fabbri[19a,19b], C. Fabre[29], K. Facius[35], R.M. Fakhrutdinov[128], S. Falciano[132a], Y. Fang[171],

M. Fanti[89a,89b], A. Farbin[7], A. Farilla[134a], J. Farley[147], T. Farooque[157], S.M. Farrington[118], P. Farthouat[29],
P. Fassnacht[29], D. Fassouliotis[8], B. Fatholahzadeh[157], L. Fayard[115], F. Fayette[54], R. Febbraro[33], P. Federic[144a],
O.L. Fedin[121], W. Fedorko[29], L. Feligioni[83], C.U. Felzmann[86], C. Feng[32d], E.J. Feng[30], A.B. Fenyuk[128],
J. Ferencei[144b], J. Ferland[93], B. Fernandes[124a], W. Fernando[109], S. Ferrag[53], J. Ferrando[118], V. Ferrara[41],
A. Ferrari[165], P. Ferrari[105], R. Ferrari[119a], A. Ferrer[166], M.L. Ferrer[47], D. Ferrere[49], C. Ferretti[87], M. Fiascaris[118],
F. Fiedler[81], A. Filipčič[74], A. Filippas[9], F. Filthaut[104], M. Fincke-Keeler[168], M.C.N. Fiolhais[124a], L. Fiorini[11],
A. Firan[39], G. Fischer[41], M.J. Fisher[109], M. Flechl[165], I. Fleck[141], J. Fleckner[81], P. Fleischmann[172],
S. Fleischmann[20], T. Flick[173], L.R. Flores Castillo[171], M.J. Flowerdew[99], T. Fonseca Martin[76], A. Formica[136],
A. Forti[82], D. Fortin[158a], D. Fournier[115], A.J. Fowler[44], K. Fowler[137], H. Fox[71], P. Francavilla[122a,122b],
S. Franchino[119a,119b], D. Francis[29], M. Franklin[57], S. Franz[29], M. Fraternali[119a,119b], S. Fratina[120], J. Freestone[82],
S.T. French[27], R. Froeschl[29], D. Froidevaux[29], J.A. Frost[27], C. Fukunaga[155], E. Fullana Torregrosa[5], J. Fuster[166],
C. Gabaldon[80], O. Gabizon[170], T. Gadfort[24], S. Gadomski[49], G. Gagliardi[50a,50b], P. Gagnon[61], C. Galea[98],
E.J. Gallas[118], M.V. Gallas[29], V. Gallo[16], B.J. Gallop[129], P. Gallus[125], E. Galyaev[40], K.K. Gan[109], Y.S. Gao[143,g],
A. Gaponenko[14], M. Garcia-Sciveres[14], C. García[166], J.E. García Navarro[49], R.W. Gardner[30], N. Garelli[29],
H. Garitaonandia[105], V. Garonne[29], C. Gatti[47], G. Gaudio[119a], V. Gautard[136], P. Gauzzi[132a,132b], I.L. Gavrilenko[94],
C. Gay[167], G. Gaycken[20], E.N. Gazis[9], P. Ge[32d], C.N.P. Gee[129], Ch. Geich-Gimbel[20], K. Gellerstedt[145a,145b],
C. Gemme[50a], M.H. Genest[98], S. Gentile[132a,132b], F. Georgatos[9], S. George[76], A. Gershon[152], H. Ghazlane[135d],
N. Ghodbane[33], B. Giacobbe[19a], S. Giagu[132a,132b], V. Giakoumopoulou[8], V. Giangiobbe[122a,122b], F. Gianotti[29],
B. Gibbard[24], A. Gibson[157], S.M. Gibson[118], L.M. Gilbert[118], M. Gilchriese[14], V. Gilewsky[91], D.M. Gingrich[2,b],
J. Ginzburg[152], N. Giokaris[8], M.P. Giordani[163a,163c], R. Giordano[102a,102b], F.M. Giorgi[15], P. Giovannini[99],
P.F. Giraud[29], P. Girtler[62], D. Giugni[89a], P. Giusti[19a], B.K. Gjelsten[117], L.K. Gladilin[97], C. Glasman[80], A. Glazov[41],
K.W. Glitza[173], G.L. Glonti[65], J. Godfrey[142], J. Godlewski[29], M. Goebel[41], T. Göpfert[43], C. Goeringer[81],
C. Gössling[42], T. Göttfert[99], V. Goggi[119a,119b,,h], S. Goldfarb[87], D. Goldin[39], T. Golling[174], A. Gomes[124a],
L.S. Gomez Fajardo[41], R. Gonçalo[76], L. Gonella[20], C. Gong[32b], S. González de la Hoz[166], M.L. Gonzalez Silva[26],
S. Gonzalez-Sevilla[49], J.J. Goodson[147], L. Goossens[29], H.A. Gordon[24], I. Gorelov[103], G. Gorfine[173], B. Gorini[29],
E. Gorini[72a,72b], A. Gorišek[74], E. Gornicki[38], B. Gosdzik[41], M. Gosselink[105], M.I. Gostkin[65], I. Gough Eschrich[162],
M. Gouighri[135a], D. Goujdami[135a], M.P. Goulette[49], A.G. Goussiou[138], C. Goy[4], I. Grabowska-Bold[162,c],
P. Grafström[29], K.-J. Grahn[146], S. Grancagnolo[15], V. Grassi[147], V. Gratchev[121], N. Grau[34], H.M. Gray[34,i],
J.A. Gray[147], E. Graziani[134a], B. Green[76], T. Greenshaw[73], Z.D. Greenwood[24,f], I.M. Gregor[41], P. Grenier[143],
E. Griesmayer[46], J. Griffiths[138], N. Grigalashvili[65], A.A. Grillo[137], K. Grimm[147], S. Grinstein[11], Y.V. Grishkevich[97],
M. Groh[99], M. Groll[81], E. Gross[170], J. Grosse-Knetter[54], J. Groth-Jensen[79], K. Grybel[141], C. Guicheney[33],
A. Guida[72a,72b], T. Guillemin[4], H. Guler[85,j], J. Gunther[125], B. Guo[157], A. Gupta[30], Y. Gusakov[65], A. Gutierrez[93],
P. Gutierrez[111], N. Guttman[152], O. Gutzwiller[171], C. Guyot[136], C. Gwenlan[118], C.B. Gwilliam[73], A. Haas[143],
S. Haas[29], C. Haber[14], H.K. Hadavand[39], D.R. Hadley[17], P. Haefner[99], R. Härtel[99], Z. Hajduk[38], H. Hakobyan[175],
J. Haller[41,k], K. Hamacher[173], A. Hamilton[49], S. Hamilton[160], L. Han[32b], K. Hanagaki[116], M. Hance[120], C. Handel[81],
P. Hanke[58a], J.R. Hansen[35], J.B. Hansen[35], J.D. Hansen[35], P.H. Hansen[35], T. Hansl-Kozanecka[137], P. Hansson[143],
K. Hara[159], G.A. Hare[137], T. Harenberg[173], R.D. Harrington[21], O.M. Harris[138], K. Harrison[17], J. Hartert[48],
F. Hartjes[105], A. Harvey[56], S. Hasegawa[101], Y. Hasegawa[140], K. Hashemi[22], S. Hassani[136], S. Haug[16],
M. Hauschild[29], R. Hauser[88], M. Havranek[125], C.M. Hawkes[17], R.J. Hawkings[29], T. Hayakawa[67], H.S. Hayward[73],
S.J. Haywood[129], S.J. Head[82], V. Hedberg[79], L. Heelan[28], S. Heim[88], B. Heinemann[14], S. Heisterkamp[35], L. Helary[4],
M. Heller[115], S. Hellman[145a,145b], C. Helsens[11], T. Hemperek[20], R.C.W. Henderson[71], M. Henke[58a], A. Henrichs[54],
A.M. Henriques Correia[29], S. Henrot-Versille[115], C. Hensel[54], T. Henß[173], Y. Hernández Jiménez[166],
A.D. Hershenhorn[151], G. Herten[48], R. Hertenberger[98], L. Hervas[29], N.P. Hessey[105], E. Higón-Rodriguez[166],
J.C. Hill[27], K.H. Hiller[41], S. Hillert[145a,145b], S.J. Hillier[17], I. Hinchliffe[14], E. Hines[120], M. Hirose[116], F. Hirsch[42],
D. Hirschbuehl[173], J. Hobbs[147], N. Hod[152], M.C. Hodgkinson[139], P. Hodgson[139], A. Hoecker[29], M.R. Hoeferkamp[103],
J. Hoffman[39], D. Hoffmann[83], M. Hohlfeld[81], T. Holy[127], J.L. Holzbauer[88], Y. Homma[67], T. Horazdovsky[127],
T. Hori[67], C. Horn[143], S. Horner[48], S. Horvat[99], J.-Y. Hostachy[55], S. Hou[150], A. Hoummada[135a], T. Howe[39],
J. Hrivnac[115], T. Hryn'ova[4], P.J. Hsu[174], S.-C. Hsu[14], G.S. Huang[111], Z. Hubacek[127], F. Hubaut[83], F. Huegging[20],
E.W. Hughes[34], G. Hughes[71], M. Hurwitz[30], U. Husemann[41], N. Huseynov[10], J. Huston[88], J. Huth[57],
G. Iacobucci[102a], G. Iakovidis[9], I. Ibragimov[141], L. Iconomidou-Fayard[115], J. Idarraga[158b], P. Iengo[4],
O. Igonkina[105], Y. Ikegami[66], M. Ikeno[66], Y. Ilchenko[39], D. Iliadis[153], T. Ince[168], P. Ioannou[8], M. Iodice[134a],
A. Irles Quiles[166], A. Ishikawa[67], M. Ishino[66], R. Ishmukhametov[39], T. Isobe[154], V. Issakov[174,*], C. Issever[118],

S. Istin[18a], Y. Itoh[101], A.V. Ivashin[128], W. Iwanski[38], H. Iwasaki[66], J.M. Izen[40], V. Izzo[102a], B. Jackson[120],
J.N. Jackson[73], P. Jackson[143], M.R. Jaekel[29], V. Jain[61], K. Jakobs[48], S. Jakobsen[35], J. Jakubek[127], D.K. Jana[111],
E. Jansen[104], A. Jantsch[99], M. Janus[48], R.C. Jared[171], G. Jarlskog[79], L. Jeanty[57], I. Jen-La Plante[30], P. Jenni[29],
P. Jez[35], S. Jézéquel[4], W. Ji[79], J. Jia[147], Y. Jiang[32b], M. Jimenez Belenguer[29], S. Jin[32a], O. Jinnouchi[156], D. Joffe[39],
M. Johansen[145a,145b], K.E. Johansson[145a], P. Johansson[139], S. Johnert[41], K.A. Johns[6], K. Jon-And[145a,145b],
G. Jones[82], R.W.L. Jones[71], T.J. Jones[73], P.M. Jorge[124a], J. Joseph[14], V. Juranek[125], P. Jussel[62], V.V. Kabachenko[128],
M. Kaci[166], A. Kaczmarska[38], M. Kado[115], H. Kagan[109], M. Kagan[57], S. Kaiser[99], E. Kajomovitz[151], S. Kalinin[173],
L.V. Kalinovskaya[65], A. Kalinowski[130], S. Kama[41], N. Kanaya[154], M. Kaneda[154], V.A. Kantserov[96], J. Kanzaki[66],
B. Kaplan[174], A. Kapliy[30], J. Kaplon[29], D. Kar[43], M. Karagounis[20], M. Karagoz Unel[118], V. Kartvelishvili[71],
A.N. Karyukhin[128], L. Kashif[57], A. Kasmi[39], R.D. Kass[109], A. Kastanas[13], M. Kastoryano[174], M. Kataoka[4],
Y. Kataoka[154], E. Katsoufis[9], J. Katzy[41], V. Kaushik[6], K. Kawagoe[67], T. Kawamoto[154], G. Kawamura[81],
M.S. Kayl[105], F. Kayumov[94], V.A. Kazanin[107], M.Y. Kazarinov[65], J.R. Keates[82], R. Keeler[168], P.T. Keener[120],
R. Kehoe[39], M. Keil[54], G.D. Kekelidze[65], M. Kelly[82], M. Kenyon[53], O. Kepka[125], N. Kerschen[29], B.P. Kerševan[74],
S. Kersten[173], K. Kessoku[154], M. Khakzad[28], F. Khalil-zada[10], H. Khandanyan[164], A. Khanov[112], D. Kharchenko[65],
A. Khodinov[147], A. Khomich[58a], G. Khoriauli[20], N. Khovanskiy[65], V. Khovanskiy[95], E. Khramov[65], J. Khubua[51],
H. Kim[7], M.S. Kim[2], P.C. Kim[143], S.H. Kim[159], O. Kind[15], P. Kind[173], B.T. King[73], J. Kirk[129], G.P. Kirsch[118],
L.E. Kirsch[22], A.E. Kiryunin[99], D. Kisielewska[37], T. Kittelmann[123], H. Kiyamura[67], E. Kladiva[144b], M. Klein[73],
U. Klein[73], K. Kleinknecht[81], M. Klemetti[85], A. Klier[170], A. Klimentov[24], R. Klingenberg[42], E.B. Klinkby[44],
T. Klioutchnikova[29], P.F. Klok[104], S. Klous[105], E.-E. Kluge[58a], T. Kluge[73], P. Kluit[105], M. Klute[54], S. Kluth[99],
N.S. Knecht[157], E. Kneringer[62], B.R. Ko[44], T. Kobayashi[154], M. Kobel[43], B. Koblitz[29], M. Kocian[143], A. Kocnar[113],
P. Kodys[126], K. Köneke[41], A.C. König[104], S. Koenig[81], L. Köpke[81], F. Koetsveld[104], P. Koevesarki[20], T. Koffas[29],
E. Koffeman[105], F. Kohn[54], Z. Kohout[127], T. Kohriki[66], H. Kolanoski[15], V. Kolesnikov[65], I. Koletsou[4], J. Koll[88],
D. Kollar[29], S. Kolos[162,l], S.D. Kolya[82], A.A. Komar[94], J.R. Komaragiri[142], T. Kondo[66], T. Kono[41,k], R. Konoplich[108],
S.P. Konovalov[94], N. Konstantinidis[77], S. Koperny[37], K. Korcyl[38], K. Kordas[153], A. Korn[14], I. Korolkov[11],
E.V. Korolkova[139], V.A. Korotkov[128], O. Kortner[99], P. Kostka[41], V.V. Kostyukhin[20], S. Kotov[99], V.M. Kotov[65],
K.Y. Kotov[107], C. Kourkoumelis[8], A. Koutsman[105], R. Kowalewski[168], H. Kowalski[41], T.Z. Kowalski[37],
W. Kozanecki[136], A.S. Kozhin[128], V. Kral[127], V.A. Kramarenko[97], G. Kramberger[74], M.W. Krasny[78],
A. Krasznahorkay[108], A. Kreisel[152], F. Krejci[127], J. Kretzschmar[73], N. Krieger[54], P. Krieger[157], K. Kroeninger[54],
H. Kroha[99], J. Kroll[120], J. Kroseberg[20], J. Krstic[12a], U. Kruchonak[65], H. Krüger[20], Z.V. Krumshteyn[65],
T. Kubota[154], S. Kuehn[48], A. Kugel[58c], T. Kuhl[173], D. Kuhn[62], V. Kukhtin[65], Y. Kulchitsky[90], S. Kuleshov[31b],
C. Kummer[98], M. Kuna[83], J. Kunkle[120], A. Kupco[125], H. Kurashige[67], M. Kurata[159], L.L. Kurchaninov[158a],
Y.A. Kurochkin[90], V. Kus[125], R. Kwee[15], L. La Rotonda[36a,36b], J. Labbe[4], C. Lacasta[166], F. Lacava[132a,132b],
H. Lacker[15], D. Lacour[78], V.R. Lacuesta[166], E. Ladygin[65], R. Lafaye[4], B. Laforge[78], T. Lagouri[80], S. Lai[48],
M. Lamanna[29], C.L. Lampen[6], W. Lampl[6], E. Lancon[136], U. Landgraf[48], M.P.J. Landon[75], J.L. Lane[82],
A.J. Lankford[162], F. Lanni[24], K. Lantzsch[29], A. Lanza[119a], S. Laplace[4], C. Lapoire[83], J.F. Laporte[136], T. Lari[89a],
A. Larner[118], M. Lassnig[29], P. Laurelli[47], W. Lavrijsen[14], P. Laycock[73], A.B. Lazarev[65], A. Lazzaro[89a,89b],
O. Le Dortz[78], E. Le Guirriec[83], E. Le Menedeu[136], M. Le Vine[24], A. Lebedev[64], C. Lebel[93], T. LeCompte[5],
F. Ledroit-Guillon[55], H. Lee[105], J.S.H. Lee[149], S.C. Lee[150], M. Lefebvre[168], M. Legendre[136], B.C. LeGeyt[120],
F. Legger[98], C. Leggett[14], M. Lehmacher[20], G. Lehmann Miotto[29], X. Lei[6], R. Leitner[126], D. Lellouch[170],
J. Lellouch[78], V. Lendermann[58a], K.J.C. Leney[73], T. Lenz[173], G. Lenzen[173], B. Lenzi[136], K. Leonhardt[43], C. Leroy[93],
J.-R. Lessard[168], C.G. Lester[27], A. Leung Fook Cheong[171], J. Levêque[83], D. Levin[87], L.J. Levinson[170], M. Leyton[14],
H. Li[171], S. Li[41], X. Li[87], Z. Liang[39], Z. Liang[150,m], B. Liberti[133a], P. Lichard[29], M. Lichtnecker[98], K. Lie[164],
W. Liebig[105], J.N. Lilley[17], H. Lim[5], A. Limosani[86], M. Limper[63], S.C. Lin[150], J.T. Linnemann[88], E. Lipeles[120],
L. Lipinsky[125], A. Lipniacka[13], T.M. Liss[164], D. Lissauer[24], A. Lister[49], A.M. Litke[137], C. Liu[28], D. Liu[150,n], H. Liu[87],
J.B. Liu[87], M. Liu[32b], T. Liu[39], Y. Liu[32b], M. Livan[119a,119b], A. Lleres[55], S.L. Lloyd[75], E. Lobodzinska[41], P. Loch[6],
W.S. Lockman[137], S. Lockwitz[174], T. Loddenkoetter[20], F.K. Loebinger[82], A. Loginov[174], C.W. Loh[167], T. Lohse[15],
K. Lohwasser[48], M. Lokajicek[125], R.E. Long[71], L. Lopes[124a], D. Lopez Mateos[34,i], M. Losada[161], P. Loscutoff[14],
X. Lou[40], A. Lounis[115], K.F. Loureiro[109], L. Lovas[144a], J. Love[21], P.A. Love[71], A.J. Lowe[61], F. Lu[32a], H.J. Lubatti[138],
C. Luci[132a,132b], A. Lucotte[55], A. Ludwig[43], D. Ludwig[41], I. Ludwig[48], F. Luehring[61], L. Luisa[163a,163c], D. Lumb[48],
L. Luminari[132a], E. Lund[117], B. Lund-Jensen[146], B. Lundberg[79], J. Lundberg[29], J. Lundquist[35], D. Lynn[24], J. Lys[14],
E. Lytken[79], H. Ma[24], L.L. Ma[171], J.A. Macana Goia[93], G. Maccarrone[47], A. Macchiolo[99], B. Maček[74],
J. Machado Miguens[124a], R. Mackeprang[35], R.J. Madaras[14], W.F. Mader[43], R. Maenner[58c], T. Maeno[24],

P. Mättig[173], S. Mättig[41], P.J. Magalhaes Martins[124a], E. Magradze[51], Y. Mahalalel[152], K. Mahboubi[48],
A. Mahmood[1], C. Maiani[132a,132b], C. Maidantchik[23a], A. Maio[124a], S. Majewski[24], Y. Makida[66], M. Makouski[128],
N. Makovec[115], Pa. Malecki[38], P. Malecki[38], V.P. Maleev[121], F. Malek[55], U. Mallik[63], D. Malon[5], S. Maltezos[9],
V. Malyshev[107], S. Malyukov[65], M. Mambelli[30], R. Mameghani[98], J. Mamuzic[41], L. Mandelli[89a], I. Mandić[74],
R. Mandrysch[15], J. Maneira[124a], P.S. Mangeard[88], I.D. Manjavidze[65], P.M. Manning[137], A. Manousakis-Katsikakis[8],
B. Mansoulie[136], A. Mapelli[29], L. Mapelli[29], L. March[80], J.F. Marchand[4], F. Marchese[133a,133b], G. Marchiori[78],
M. Marcisovsky[125], C.P. Marino[61], F. Marroquim[23a], Z. Marshall[34,i], S. Marti-Garcia[166], A.J. Martin[75],
A.J. Martin[174], B. Martin[29], B. Martin[88], F.F. Martin[120], J.P. Martin[93], T.A. Martin[17], B. Martin dit Latour[49],
M. Martinez[11], V. Martinez Outschoorn[57], A. Martini[47], A.C. Martyniuk[82], F. Marzano[132a], A. Marzin[136],
L. Masetti[20], T. Mashimo[154], R. Mashinistov[96], J. Masik[82], A.L. Maslennikov[107], I. Massa[19a,19b], N. Massol[4],
A. Mastroberardino[36a,36b], T. Masubuchi[154], P. Matricon[115], H. Matsunaga[154], T. Matsushita[67], C. Mattravers[118,o],
S.J. Maxfield[73], A. Mayne[139], R. Mazini[150], M. Mazur[48], M. Mazzanti[89a], J. McDonald[85], S.P. McKee[87],
A. McCarn[164], R.L. McCarthy[147], N.A. McCubbin[129], K.W. McFarlane[56], H. McGlone[53], G. Mchedlidze[51],
S.J. McMahon[129], R.A. McPherson[168,e], A. Meade[84], J. Mechnich[105], M. Mechtel[173], M. Medinnis[41],
R. Meera-Lebbai[111], T.M. Meguro[116], S. Mehlhase[41], A. Mehta[73], K. Meier[58a], B. Meirose[48], C. Melachrinos[30],
B.R. Mellado Garcia[171], L. Mendoza Navas[161], Z. Meng[150,p], S. Menke[99], E. Meoni[11], P. Mermod[118],
L. Merola[102a,102b], C. Meroni[89a], F.S. Merritt[30], A.M. Messina[29], J. Metcalfe[103], A.S. Mete[64], J.-P. Meyer[136],
J. Meyer[172], J. Meyer[54], T.C. Meyer[29], W.T. Meyer[64], J. Miao[32d], S. Michal[29], L. Micu[25a], R.P. Middleton[129],
S. Migas[73], L. Mijović[74], G. Mikenberg[170], M. Mikestikova[125], M. Mikuž[74], D.W. Miller[143], W.J. Mills[167],
C.M. Mills[57], A. Milov[170], D.A. Milstead[145a,145b], D. Milstein[170], A.A. Minaenko[128], M. Miñano[166], I.A. Minashvili[65],
A.I. Mincer[108], B. Mindur[37], M. Mineev[65], Y. Ming[130], L.M. Mir[11], G. Mirabelli[132a], S. Misawa[24], S. Miscetti[47],
A. Misiejuk[76], J. Mitrevski[137], V.A. Mitsou[166], P.S. Miyagawa[82], J.U. Mjörnmark[79], D. Mladenov[22], T. Moa[145a,145b],
S. Moed[57], V. Moeller[27], K. Mönig[41], N. Möser[20], W. Mohr[48], S. Mohrdieck-Möck[99], R. Moles-Valls[166],
J. Molina-Perez[29], J. Monk[77], E. Monnier[83], S. Montesano[89a,89b], F. Monticelli[70], R.W. Moore[2], C. Mora Herrera[49],
A. Moraes[53], A. Morais[124a], J. Morel[54], G. Morello[36a,36b], D. Moreno[161], M. Moreno Llácer[166], P. Morettini[50a],
M. Morii[57], A.K. Morley[86], G. Mornacchi[29], S.V. Morozov[96], J.D. Morris[75], H.G. Moser[99], M. Mosidze[51], J. Moss[109],
R. Mount[143], E. Mountricha[136], S.V. Mouraviev[94], E.J.W. Moyse[84], M. Mudrinic[12b], F. Mueller[58a], J. Mueller[123],
K. Mueller[20], T.A. Müller[98], D. Muenstermann[42], A. Muir[167], Y. Munwes[152], R. Murillo Garcia[162], W.J. Murray[129],
I. Mussche[105], E. Musto[102a,102b], A.G. Myagkov[128], M. Myska[125], J. Nadal[11], K. Nagai[159], K. Nagano[66],
Y. Nagasaka[60], A.M. Nairz[29], K. Nakamura[154], I. Nakano[110], H. Nakatsuka[67], G. Nanava[20], A. Napier[160],
M. Nash[77,q], N.R. Nation[21], T. Nattermann[20], T. Naumann[41], G. Navarro[161], S.K. Nderitu[20], H.A. Neal[87], E. Nebot[80],
P. Nechaeva[94], A. Negri[119a,119b], G. Negri[29], A. Nelson[64], T.K. Nelson[143], S. Nemecek[125], P. Nemethy[108],
A.A. Nepomuceno[23a], M. Nessi[29], M.S. Neubauer[164], A. Neusiedl[81], R.M. Neves[108], P. Nevski[24], F.M. Newcomer[120],
R.B. Nickerson[118], R. Nicolaidou[136], L. Nicolas[139], G. Nicoletti[47], B. Nicquevert[29], F. Niedercorn[115], J. Nielsen[137],
A. Nikiforov[15], K. Nikolaev[65], I. Nikolic-Audit[78], K. Nikolopoulos[8], H. Nilsen[48], P. Nilsson[7], A. Nisati[132a],
T. Nishiyama[67], R. Nisius[99], L. Nodulman[5], M. Nomachi[116], I. Nomidis[153], M. Nordberg[29], B. Nordkvist[145a,145b],
D. Notz[41], J. Novakova[126], M. Nozaki[66], M. Nožička[41], I.M. Nugent[158a], A.-E. Nuncio-Quiroz[20],
G. Nunes Hanninger[20], T. Nunnemann[98], E. Nurse[77], D.C. O'Neil[142], V. O'Shea[53], F.G. Oakham[28,b], H. Oberlack[99],
A. Ochi[67], S. Oda[154], S. Odaka[66], J. Odier[83], H. Ogren[61], A. Oh[82], S.H. Oh[44], C.C. Ohm[145a,145b], T. Ohshima[101],
H. Ohshita[140], T. Ohsugi[59], S. Okada[67], H. Okawa[162], Y. Okumura[101], T. Okuyama[154], A.G. Olchevski[65],
M. Oliveira[124a], D. Oliveira Damazio[24], J. Oliver[57], E. Oliveira Garcia[166], D. Olivito[120], A. Olszewski[38],
J. Olszowska[38], C. Omachi[67,r], A. Onofre[124a], P.U.E. Onyisi[30], C.J. Oram[158a], M.J. Oreglia[30], Y. Oren[152],
D. Orestano[134a,134b], I. Orlov[107], C. Oropeza Barrera[53], R.S. Orr[157], E.O. Ortega[130], B. Osculati[50a,50b],
R. Ospanov[120], C. Osuna[11], J.P. Ottersbach[105], F. Ould-Saada[117], A. Ouraou[136], Q. Ouyang[32a], M. Owen[82],
S. Owen[139], A. Oyarzun[31b], V.E. Ozcan[77], K. Ozone[66], N. Ozturk[7], A. Pacheco Pages[11], C. Padilla Aranda[11],
E. Paganis[139], C. Pahl[63], F. Paige[24], K. Pajchel[117], S. Palestini[29], D. Pallin[33], A. Palma[124a], J.D. Palmer[17],
Y.B. Pan[171], E. Panagiotopoulou[9], B. Panes[31a], N. Panikashvili[87], S. Panitkin[24], D. Pantea[25a], M. Panuskova[125],
V. Paolone[123], Th.D. Papadopoulou[9], S.J. Park[54], W. Park[24,s], M.A. Parker[27], S.I. Parker[14], F. Parodi[50a,50b],
J.A. Parsons[34], U. Parzefall[48], E. Pasqualucci[132a], A. Passeri[134a], F. Pastore[134a,134b], Fr. Pastore[29], G. Pásztor[49,t],
S. Pataraia[99], J.R. Pater[82], S. Patricelli[102a,102b], A. Patwa[24], T. Pauly[29], L.S. Peak[149], M. Pecsy[144a],
M.I. Pedraza Morales[171], S.V. Peleganchuk[107], H. Peng[171], A. Penson[34], J. Penwell[61], M. Perantoni[23a], K. Perez[34,i],
E. Perez Codina[11], M.T. Pérez García-Estañ[166], V. Perez Reale[34], L. Perini[89a,89b], H. Pernegger[29], R. Perrino[72a],

S. Persembe[3a], P. Perus[115], V.D. Peshekhonov[65], B.A. Petersen[29], T.C. Petersen[35], E. Petit[83], C. Petridou[153],
E. Petrolo[132a], F. Petrucci[134a,134b], D. Petschull[41], M. Petteni[142], R. Pezoa[31b], A. Phan[86], A.W. Phillips[27],
G. Piacquadio[29], M. Piccinini[19a,19b], R. Piegaia[26], J.E. Pilcher[30], A.D. Pilkington[82], J. Pina[124a], M. Pinamonti[163a,163c],
J.L. Pinfold[2], B. Pinto[124a], C. Pizio[89a,89b], R. Placakyte[41], M. Plamondon[168], M.-A. Pleier[24], A. Poblaguev[174],
S. Poddar[58a], F. Podlyski[33], P. Poffenberger[168], L. Poggioli[115], M. Pohl[49], F. Polci[55], G. Polesello[119a],
A. Policicchio[138], A. Polini[19a], J. Poll[75], V. Polychronakos[24], D. Pomeroy[22], K. Pommès[29], P. Ponsot[136],
L. Pontecorvo[132a], B.G. Pope[88], G.A. Popeneciu[25a], D.S. Popovic[12a], A. Poppleton[29], J. Popule[125], X. Portell Bueso[48],
R. Porter[162], G.E. Pospelov[99], S. Pospisil[127], M. Potekhin[24], I.N. Potrap[99], C.J. Potter[148], C.T. Potter[85], K.P. Potter[82],
G. Poulard[29], J. Poveda[171], R. Prabhu[20], P. Pralavorio[83], S. Prasad[57], R. Pravahan[7], L. Pribyl[29], D. Price[61],
L.E. Price[5], P.M. Prichard[73], D. Prieur[123], M. Primavera[72a], K. Prokofiev[29], F. Prokoshin[31b], S. Protopopescu[24],
J. Proudfoot[5], X. Prudent[43], H. Przysiezniak[4], S. Psoroulas[20], E. Ptacek[114], C. Puigdengoles[11], J. Purdham[87],
M. Purohit[24,s], P. Puzo[115], Y. Pylypchenko[117], M. Qi[32c], J. Qian[87], W. Qian[129], Z. Qin[41], A. Quadt[54], D.R. Quarrie[14],
W.B. Quayle[171], F. Quinonez[31a], M. Raas[104], V. Radeka[24], V. Radescu[58b], B. Radics[20], T. Rador[18a], F. Ragusa[89a,89b],
G. Rahal[179], A.M. Rahimi[109], S. Rajagopalan[24], M. Rammensee[48], M. Rammes[141], F. Rauscher[98], E. Rauter[99],
M. Raymond[29], A.L. Read[117], D.M. Rebuzzi[119a,119b], A. Redelbach[172], G. Redlinger[24], R. Reece[120], K. Reeves[40],
E. Reinherz-Aronis[152], A. Reinsch[114], I. Reisinger[42], D. Reljic[12a], C. Rembser[29], Z.L. Ren[150], P. Renkel[39],
S. Rescia[24], M. Rescigno[132a], S. Resconi[89a], B. Resende[136], P. Reznicek[126], R. Rezvani[157], A. Richards[77],
R.A. Richards[88], R. Richter[99], E. Richter-Was[38,u], M. Ridel[78], M. Rijpstra[105], M. Rijssenbeek[147],
A. Rimoldi[119a,119b], L. Rinaldi[19a], R.R. Rios[39], I. Riu[11], F. Rizatdinova[112], E. Rizvi[75], D.A. Roa Romero[161],
S.H. Robertson[85,e], A. Robichaud-Veronneau[49], D. Robinson[27], J.E.M. Robinson[77], M. Robinson[114], A. Robson[53],
J.G. Rocha de Lima[106a], C. Roda[122a,122b], D. Roda Dos Santos[29], D. Rodriguez[161], Y. Rodriguez Garcia[15], S. Roe[29],
O. Røhne[117], V. Rojo[1], S. Rolli[160], A. Romaniouk[96], V.M. Romanov[65], G. Romeo[26], D. Romero Maltrana[31a],
L. Roos[78], E. Ros[166], S. Rosati[138], G.A. Rosenbaum[157], L. Rosselet[49], V. Rossetti[11], L.P. Rossi[50a], M. Rotaru[25a],
J. Rothberg[138], D. Rousseau[115], C.R. Royon[136], A. Rozanov[83], Y. Rozen[151], X. Ruan[115], B. Ruckert[98],
N. Ruckstuhl[105], V.I. Rud[97], G. Rudolph[62], F. Rühr[58a], F. Ruggieri[134a], A. Ruiz-Martinez[64], L. Rumyantsev[65],
Z. Rurikova[48], N.A. Rusakovich[65], J.P. Rutherfoord[6], C. Ruwiedel[20], P. Ruzicka[125], Y.F. Ryabov[121], P. Ryan[88],
G. Rybkin[115], S. Rzaeva[10], A.F. Saavedra[149], H.F.-W. Sadrozinski[137], R. Sadykov[65], H. Sakamoto[154],
G. Salamanna[105], A. Salamon[133a], M.S. Saleem[111], D. Salihagic[99], A. Salnikov[143], J. Salt[166],
B.M. Salvachua Ferrando[5], D. Salvatore[36a,36b], F. Salvatore[148], A. Salvucci[47], A. Salzburger[29], D. Sampsonidis[153],
B.H. Samset[117], H. Sandaker[13], H.G. Sander[81], M.P. Sanders[98], M. Sandhoff[173], P. Sandhu[157], R. Sandstroem[105],
S. Sandvoss[173], D.P.C. Sankey[129], B. Sanny[173], A. Sansoni[47], C. Santamarina Rios[85], C. Santoni[33],
R. Santonico[133a,133b], J.G. Saraiva[124a], T. Sarangi[171], E. Sarkisyan-Grinbaum[7], F. Sarri[122a,122b], O. Sasaki[66],
N. Sasao[68], I. Satsounkevitch[90], G. Sauvage[4], P. Savard[157,b], A.Y. Savine[6], V. Savinov[123], L. Sawyer[24,f], D.H. Saxon[53],
L.P. Says[33], C. Sbarra[19a,19b], A. Sbrizzi[19a,19b], D.A. Scannicchio[29], J. Schaarschmidt[43], P. Schacht[99], U. Schäfer[81],
S. Schaetzel[58b], A.C. Schaffer[115], D. Schaile[98], R.D. Schamberger[147], A.G. Schamov[107], V.A. Schegelsky[121],
D. Scheirich[87], M. Schernau[162], M.I. Scherzer[14], C. Schiavi[50a,50b], J. Schieck[99], M. Schioppa[36a,36b], S. Schlenker[29],
E. Schmidt[48], K. Schmieden[20], C. Schmitt[81], M. Schmitz[20], M. Schott[29], D. Schouten[142], J. Schovancova[125],
M. Schram[85], A. Schreiner[63], C. Schroeder[81], N. Schroer[58c], M. Schroers[173], J. Schultes[173], H.-C. Schultz-Coulon[58a],
J.W. Schumacher[43], M. Schumacher[48], B.A. Schumm[137], Ph. Schune[136], C. Schwanenberger[82], A. Schwartzman[143],
Ph. Schwemling[78], R. Schwienhorst[88], R. Schwierz[43], J. Schwindling[136], W.G. Scott[129], J. Searcy[114], E. Sedykh[121],
E. Segura[11], S.C. Seidel[103], A. Seiden[137], F. Seifert[43], J.M. Seixas[23a], G. Sekhniaidze[102a], D.M. Seliverstov[121],
B. Sellden[145a], N. Semprini-Cesari[19a,19b], C. Serfon[98], L. Serin[115], R. Seuster[99], H. Severini[111], M.E. Sevior[86],
A. Sfyrla[164], E. Shabalina[54], M. Shamim[114], L.Y. Shan[32a], J.T. Shank[21], Q.T. Shao[86], M. Shapiro[14], P.B. Shatalov[95],
K. Shaw[139], D. Sherman[29], P. Sherwood[77], A. Shibata[108], M. Shimojima[100], T. Shin[56], A. Shmeleva[94],
M.J. Shochet[30], M.A. Shupe[6], P. Sicho[125], A. Sidoti[15], F. Siegert[77], J. Siegrist[14], Dj. Sijacki[12a], O. Silbert[170],
J. Silva[124a], Y. Silver[152], D. Silverstein[143], S.B. Silverstein[145a], V. Simak[127], Lj. Simic[12a], S. Simion[115], B. Simmons[77],
M. Simonyan[35], P. Sinervo[157], N.B. Sinev[114], V. Sipica[141], G. Siragusa[81], A.N. Sisakyan[65], S.Yu. Sivoklokov[97],
J. Sjoelin[145a,145b], T.B. Sjursen[13], K. Skovpen[107], P. Skubic[111], M. Slater[17], T. Slavicek[127], K. Sliwa[160], J. Sloper[29],
T. Sluka[125], V. Smakhtin[170], S.Yu. Smirnov[96], Y. Smirnov[24], L.N. Smirnova[97], O. Smirnova[79], B.C. Smith[57],
D. Smith[143], K.M. Smith[53], M. Smizanska[71], K. Smolek[127], A.A. Snesarev[94], S.W. Snow[82], J. Snow[111],
J. Snuverink[105], S. Snyder[24], M. Soares[124a], R. Sobie[168,e], J. Sodomka[127], A. Soffer[152], C.A. Solans[166], M. Solar[127],
J. Solc[127], E. Solfaroli Camillocci[132a,132b], A.A. Solodkov[128], O.V. Solovyanov[128], R. Soluk[2], J. Sondericker[24],

V. Sopko[127], B. Sopko[127], M. Sosebee[7], A. Soukharev[107], S. Spagnolo[72a,72b], F. Spanò[34], E. Spencer[137], R. Spighi[19a], G. Spigo[29], F. Spila[132a,132b], R. Spiwoks[29], M. Spousta[126], T. Spreitzer[142], B. Spurlock[7], R.D.St. Denis[53], T. Stahl[141], J. Stahlman[120], R. Stamen[58a], S.N. Stancu[162], E. Stanecka[29], R.W. Stanek[5], C. Stanescu[134a], S. Stapnes[117], E.A. Starchenko[128], J. Stark[55], P. Staroba[125], P. Starovoitov[91], J. Stastny[125], P. Stavina[144a], G. Stavropoulos[14], G. Steele[53], P. Steinbach[43], P. Steinberg[24], I. Stekl[127], B. Stelzer[142], H.J. Stelzer[41], O. Stelzer-Chilton[158a], H. Stenzel[52], K. Stevenson[75], G.A. Stewart[53], M.C. Stockton[29], K. Stoerig[48], G. Stoicea[25a], S. Stonjek[99], P. Strachota[126], A.R. Stradling[7], A. Straessner[43], J. Strandberg[87], S. Strandberg[14], A. Strandlie[117], M. Strauss[111], P. Strizenec[144b], R. Ströhmer[172], D.M. Strom[114], R. Stroynowski[39], J. Strube[129], B. Stugu[13], D.A. Soh[150,v], D. Su[143], Y. Sugaya[116], T. Sugimoto[101], C. Suhr[106a], M. Suk[126], V.V. Sulin[94], S. Sultansoy[3d], T. Sumida[29], X.H. Sun[32d], J.E. Sundermann[48], K. Suruliz[163a,163b], S. Sushkov[11], G. Susinno[36a,36b], M.R. Sutton[139], T. Suzuki[154], Y. Suzuki[66], I. Sykora[144a], T. Sykora[126], T. Szymocha[38], J. Sánchez[166], D. Ta[20], K. Tackmann[29], A. Taffard[162], R. Tafirout[158a], A. Taga[117], Y. Takahashi[101], H. Takai[24], R. Takashima[69], H. Takeda[67], T. Takeshita[140], M. Talby[83], A. Talyshev[107], M.C. Tamsett[76], J. Tanaka[154], R. Tanaka[115], S. Tanaka[131], S. Tanaka[66], S. Tapprogge[81], D. Tardif[157], S. Tarem[151], F. Tarrade[24], G.F. Tartarelli[89a], P. Tas[126], M. Tasevsky[125], E. Tassi[36a,36b], M. Tatarkhanov[14], C. Taylor[77], F.E. Taylor[92], G.N. Taylor[86], R.P. Taylor[168], W. Taylor[158b], P. Teixeira-Dias[76], H. Ten Kate[29], P.K. Teng[150], Y.D. Tennenbaum-Katan[151], S. Terada[66], K. Terashi[154], J. Terron[80], M. Terwort[41k], M. Testa[47], R.J. Teuscher[157,e], M. Thioye[174], S. Thoma[48], J.P. Thomas[17], E.N. Thompson[84], P.D. Thompson[17], P.D. Thompson[157], R.J. Thompson[82], A.S. Thompson[53], E. Thomson[120], R.P. Thun[87], T. Tic[125], V.O. Tikhomirov[94], Y.A. Tikhonov[107], P. Tipton[174], F.J. Tique Aires Viegas[29], S. Tisserant[83], B. Toczek[37], T. Todorov[4], S. Todorova-Nova[160], B. Toggerson[162], J. Tojo[66], S. Tokár[144a], K. Tokushuku[66], K. Tollefson[88], L. Tomasek[125], M. Tomasek[125], M. Tomoto[101], L. Tompkins[14], K. Toms[103], A. Tonoyan[13], C. Topfel[16], N.D. Topilin[65], E. Torrence[114], E. Torró Pastor[166], J. Toth[83,t], F. Touchard[83], D.R. Tovey[139], T. Trefzger[172], L. Tremblet[29], A. Tricoli[29], I.M. Trigger[158a], S. Trincaz-Duvoid[78], T.N. Trinh[78], M.F. Tripiana[70], N. Triplett[64], W. Trischuk[157], A. Trivedi[24,s], B. Trocmé[55], C. Troncon[89a], A. Trzupek[38], C. Tsarouchas[9], J.C.-L. Tseng[118], M. Tsiakiris[105], P.V. Tsiareshka[90], D. Tsionou[139], G. Tsipolitis[9], V. Tsiskaridze[51], E.G. Tskhadadze[51], I.I. Tsukerman[95], V. Tsulaia[123], J.-W. Tsung[20], S. Tsuno[66], D. Tsybychev[147], J.M. Tuggle[30], D. Turecek[127], I. Turk Cakir[3e], E. Turlay[105], P.M. Tuts[34], M.S. Twomey[138], M. Tylmad[145a,145b], M. Tyndel[129], K. Uchida[116], I. Ueda[154], M. Ugland[13], M. Uhlenbrock[20], M. Uhrmacher[54], F. Ukegawa[159], G. Unal[29], A. Undrus[24], G. Unel[162], Y. Unno[66], D. Urbaniec[34], E. Urkovsky[152], P. Urquijo[49,w], P. Urrejola[31a], G. Usai[7], M. Uslenghi[119a,119b], L. Vacavant[83], V. Vacek[127], B. Vachon[85], S. Vahsen[14], P. Valente[132a], S. Valentinetti[19a,19b], S. Valkar[126], E. Valladolid Gallego[166], S. Vallecorsa[151], J.A. Valls Ferrer[166], R. Van Berg[120], H. van der Graaf[105], E. van der Kraaij[105], E. van der Poel[105], D. van der Ster[29], N. van Eldik[84], P. van Gemmeren[5], Z. van Kesteren[105], I. van Vulpen[105], W. Vandelli[29], A. Vaniachine[5], P. Vankov[73], F. Vannucci[78], R. Vari[132a], E.W. Varnes[6], D. Varouchas[14], A. Vartapetian[7], K.E. Varvell[149], L. Vasilyeva[94], V.I. Vassilakopoulos[56], F. Vazeille[33], C. Vellidis[8], F. Veloso[124a], S. Veneziano[132a], A. Ventura[72a,72b], D. Ventura[138], M. Venturi[48], N. Venturi[16], V. Vercesi[119a], M. Verducci[172], W. Verkerke[105], J.C. Vermeulen[105], M.C. Vetterli[142,b], I. Vichou[164], T. Vickey[118], G.H.A. Viehhauser[118], M. Villa[19a,19b], E.G. Villani[129], M. Villaplana Perez[166], E. Vilucchi[47], M.G. Vincter[28], E. Vinek[29], V.B. Vinogradov[65], S. Viret[33], J. Virzi[14], A. Vitale[19a,19b], O. Vitells[170], I. Vivarelli[48], F. Vives Vaque[11], S. Vlachos[9], M. Vlasak[127], N. Vlasov[20], A. Vogel[20], P. Vokac[127], M. Volpi[11], H. von der Schmitt[99], J. von Loeben[99], H. von Radziewski[48], E. von Toerne[20], V. Vorobel[126], V. Vorwerk[11], M. Vos[166], R. Voss[29], T.T. Voss[173], J.H. Vossebeld[73], N. Vranjes[12a], M. Vranjes Milosavljevic[12a], V. Vrba[125], M. Vreeswijk[105], T. Vu Anh[81], D. Vudragovic[12a], R. Vuillermet[29], I. Vukotic[115], P. Wagner[120], J. Walbersloh[42], J. Walder[71], R. Walker[98], W. Walkowiak[141], R. Wall[174], C. Wang[44], H. Wang[171], J. Wang[55], S.M. Wang[150], A. Warburton[85], C.P. Ward[27], M. Warsinsky[48], R. Wastie[118], P.M. Watkins[17], A.T. Watson[17], M.F. Watson[17], G. Watts[138], S. Watts[82], A.T. Waugh[149], B.M. Waugh[77], M.D. Weber[16], M. Weber[129], M.S. Weber[16], P. Weber[58a], A.R. Weidberg[118], J. Weingarten[54], C. Weiser[48], H. Wellenstein[22], P.S. Wells[29], M. Wen[47], T. Wenaus[24], S. Wendler[123], T. Wengler[82], S. Wenig[29], N. Wermes[20], M. Werner[48], P. Werner[29], M. Werth[162], U. Werthenbach[141], M. Wessels[58a], K. Whalen[28], A. White[7], M.J. White[27], S. White[24], S.R. Whitehead[118], D. Whiteson[162], D. Whittington[61], F. Wicek[115], D. Wicke[81], F.J. Wickens[129], W. Wiedenmann[171], M. Wielers[129], P. Wienemann[20], C. Wiglesworth[73], L.A.M. Wiik[48], A. Wildauer[166], M.A. Wildt[41k], H.G. Wilkens[29], E. Williams[34], H.H. Williams[120], S. Willocq[84], J.A. Wilson[17], M.G. Wilson[143], A. Wilson[87], I. Wingerter-Seez[4], F. Winklmeier[29], M. Wittgen[143], M.W. Wolter[38], H. Wolters[124a], B.K. Wosiek[38], J. Wotschack[29], M.J. Woudstra[84], K. Wraight[53], C. Wright[53], D. Wright[143], B. Wrona[73], S.L. Wu[171], X. Wu[49], E. Wulf[34], B.M. Wynne[45], L. Xaplanteris[9], S. Xella[35], S. Xie[48], D. Xu[139], N. Xu[171], M. Yamada[159],

A. Yamamoto[66], K. Yamamoto[64], S. Yamamoto[154], T. Yamamura[154], J. Yamaoka[44], T. Yamazaki[154], Y. Yamazaki[67], Z. Yan[21], H. Yang[87], U.K. Yang[82], Z. Yang[145a,145b], W.-M. Yao[14], Y. Yao[14], Y. Yasu[66], J. Ye[39], S. Ye[24], M. Yilmaz[3c], R. Yoosoofmiya[123], K. Yorita[169], R. Yoshida[5], C. Young[143], S.P. Youssef[21], D. Yu[24], J. Yu[7], L. Yuan[78], A. Yurkewicz[147], R. Zaidan[63], A.M. Zaitsev[128], Z. Zajacova[29], V. Zambrano[47], L. Zanello[132a,132b], A. Zaytsev[107], C. Zeitnitz[173], M. Zeller[174], A. Zemla[38], C. Zendler[20], O. Zenin[128], T. Zenis[144a], Z. Zenonos[122a,122b], S. Zenz[14], D. Zerwas[115], G. Zevi della Porta[57], Z. Zhan[32d], H. Zhang[83], J. Zhang[5], Q. Zhang[5], X. Zhang[32d], L. Zhao[108], T. Zhao[138], Z. Zhao[32b], A. Zhemchugov[65], J. Zhong[150x], B. Zhou[87], N. Zhou[34], Y. Zhou[150], C.G. Zhu[32d], H. Zhu[41], Y. Zhu[171], X. Zhuang[98], V. Zhuravlov[99], R. Zimmermann[20], S. Zimmermann[20], S. Zimmermann[48], M. Ziolkowski[141], L. Živković[34], G. Zobernig[171], A. Zoccoli[19a,19b], M. zur Nedden[15], V. Zutshi[106a]

*CERN, 1211 Geneva 23, Switzerland

[1] University at Albany, 1400 Washington Ave, Albany, NY 12222, USA

[2] University of Alberta, Department of Physics, Centre for Particle Physics, Edmonton, AB T6G 2G7, Canada

[3] Ankara University[a], Faculty of Sciences, Department of Physics, 061000 Tandogan, Ankara; Dumlupinar University[b], Faculty of Arts and Sciences, Department of Physics, Kutahya; Gazi University[c], Faculty of Arts and Sciences, Department of Physics, 06500 Teknikokullar, Ankara; TOBB University of Economics and Technology[d], Faculty of Arts and Sciences, Division of Physics, 06560 Sogutozu, Ankara; Turkish Atomic Energy Authority[e], 06530 Lodumlu, Ankara, Turkey

[4] LAPP, Université de Savoie, CNRS/IN2P3, Annecy-le-Vieux, France

[5] Argonne National Laboratory, High Energy Physics Division, 9700 S. Cass Avenue, Argonne, IL 60439, USA

[6] University of Arizona, Department of Physics, Tucson, AZ 85721, USA

[7] The University of Texas at Arlington, Department of Physics, Box 19059, Arlington, TX 76019, USA

[8] University of Athens, Nuclear & Particle Physics, Department of Physics, Panepistimiopouli, Zografou, 15771 Athens, Greece

[9] National Technical University of Athens, Physics Department, 9-Iroon Polytechniou, 15780 Zografou, Greece

[10] Institute of Physics, Azerbaijan Academy of Sciences, H. Javid Avenue 33, 143 Baku, Azerbaijan

[11] Institut de Física d'Altes Energies, IFAE, Edifici Cn, Universitat Autònoma de Barcelona, 08193 Bellaterra (Barcelona), Spain

[12] University of Belgrade[a], Institute of Physics, P.O. Box 57, 11001 Belgrade; Vinca Institute of Nuclear Sciences[b], Mihajla Petrovica Alasa 12-14, 11001 Belgrade, Serbia

[13] University of Bergen, Department for Physics and Technology, Allegaten 55, 5007 Bergen, Norway

[14] Lawrence Berkeley National Laboratory and University of California, Physics Division, MS50B-6227, 1 Cyclotron Road, Berkeley, CA 94720, USA

[15] Humboldt University, Institute of Physics, Berlin, Newtonstr. 15, 12489 Berlin, Germany

[16] University of Bern, Albert Einstein Center for Fundamental Physics, Laboratory for High Energy Physics, Sidlerstrasse 5, 3012 Bern, Switzerland

[17] University of Birmingham, School of Physics and Astronomy, Edgbaston, Birmingham B15 2TT, UK

[18] Bogazici University[a], Faculty of Sciences, Department of Physics, 80815 Bebek-Istanbul; Dogus University[b], Faculty of Arts and Sciences, Department of Physics, 34722 Kadikoy, Istanbul; [c]Gaziantep University, Faculty of Engineering, Department of Physics Engineering, 27310 Sehitkamil, Gaziantep; Istanbul Technical University[d], Faculty of Arts and Sciences, Department of Physics, 34469 Maslak, Istanbul, Turkey

[19] INFN Sezione di Bologna[a]; Università di Bologna, Dipartimento di Fisica[b], viale C. Berti Pichat, 6/2, 40127 Bologna, Italy

[20] University of Bonn, Physikalisches Institut, Nussallee 12, 53115 Bonn, Germany

[21] Boston University, Department of Physics, 590 Commonwealth Avenue, Boston, MA 02215, USA

[22] Brandeis University, Department of Physics, MS057, 415 South Street, Waltham, MA 02454, USA

[23] Universidade Federal do Rio De Janeiro, COPPE/EE/IF [a], Caixa Postal 68528, Ilha do Fundao, 21945-970 Rio de Janeiro; [b]Universidade de Sao Paulo, Instituto de Fisica, R.do Matao Trav. R.187, Sao Paulo, SP, 05508-900, Brazil

[24] Brookhaven National Laboratory, Physics Department, Bldg. 510A, Upton, NY 11973, USA

[25] National Institute of Physics and Nuclear Engineering[a], Str. Atomistilor 407, P.O. Box MG-6, 077125 Bucharest-Magurele; University Politehnica Bucharest[b], Rectorat AN 001, 313 Splaiul Independentei, sector 6, 060042 Bucuresti; West University[c] in Timisoara, Bd. Vasile Parvan 4, Timisoara, Romania

[26] Universidad de Buenos Aires, FCEyN, Dto. Fisica, Pab I, C. Universitaria, 1428 Buenos Aires, Argentina

[27] University of Cambridge, Cavendish Laboratory, J J Thomson Avenue, Cambridge CB3 0HE, UK

[28] Carleton University, Department of Physics, 1125 Colonel By Drive, Ottawa, ON K1S 5B6, Canada

[29] CERN, 1211 Geneva 23, Switzerland

[30] University of Chicago, Enrico Fermi Institute, 5640 S. Ellis Avenue, Chicago, IL 60637, USA

[31] Pontificia Universidad Católica de Chile, Facultad de Fisica, Departamento de Fisica[a], Avda. Vicuna Mackenna 4860, San Joaquin, Santiago; Universidad Técnica Federico Santa María, Departamento de Física[b], Avda. España 1680, Casilla 110-V, Valparaíso, Chile

[32] Institute of High Energy Physics, Chinese Academy of Sciences[a], P.O. Box 918, 19 Yuquan Road, Shijing Shan District, Beijing 100049; University of Science & Technology of China (USTC), Department of Modern Physics[b], Hefei, Anhui 230026; Nanjing University, Department of Physics[c], 22 Hankou Road, Nanjing, 210093; Shandong University, High Energy Physics Group[d], Jinan, Shandong 250100, China

[33] Laboratoire de Physique Corpusculaire, Clermont Université, Université Blaise Pascal, CNRS/IN2P3, 63177 Aubiere Cedex, France

[34] Columbia University, Nevis Laboratory, 136 So. Broadway, Irvington, NY 10533, USA

[35] University of Copenhagen, Niels Bohr Institute, Blegdamsvej 17, 2100 Kobenhavn 0, Denmark

[36] INFN Gruppo Collegato di Cosenza[a]; Università della Calabria, Dipartimento di Fisica[b], 87036 Arcavacata di Rende, Italy

[37] Faculty of Physics and Applied Computer Science of the AGH-University of Science and Technology, (FPACS, AGH-UST), al. Mickiewicza 30, 30059 Cracow, Poland

[38]The Henryk Niewodniczanski Institute of Nuclear Physics, Polish Academy of Sciences, ul. Radzikowskiego 152, 31342 Krakow, Poland

[39]Southern Methodist University, Physics Department, 106 Fondren Science Building, Dallas, TX 75275-0175, USA

[40]University of Texas at Dallas, 800 West Campbell Road, Richardson, TX 75080-3021, USA

[41]DESY, Notkestr. 85, 22603 Hamburg and Platanenallee 6, 15738 Zeuthen, Germany

[42]TU Dortmund, Experimentelle Physik IV, 44221 Dortmund, Germany

[43]Technical University Dresden, Institut für Kern- und Teilchenphysik, Zellescher Weg 19, 01069 Dresden, Germany

[44]Duke University, Department of Physics, Durham, NC 27708, USA

[45]University of Edinburgh, School of Physics & Astronomy, James Clerk Maxwell Building, The Kings Buildings, Mayfield Road, Edinburgh EH9 3JZ, UK

[46]Fachhochschule Wiener Neustadt; Johannes Gutenbergstrasse 3, 2700 Wiener Neustadt, Austria

[47]INFN Laboratori Nazionali di Frascati, via Enrico Fermi 40, 00044 Frascati, Italy

[48]Albert-Ludwigs-Universität, Fakultät für Mathematik und Physik, Hermann-Herder Str. 3, 79104 Freiburg i.Br., Germany

[49]Université de Genève, Section de Physique, 24 rue Ernest Ansermet, 1211 Geneve 4, Switzerland

[50]INFN Sezione di Genova[a]; Università di Genova, Dipartimento di Fisica[b], via Dodecaneso 33, 16146 Genova, Italy

[51]Institute of Physics of the Georgian Academy of Sciences, 6 Tamarashvili St., 380077 Tbilisi; Tbilisi State University, HEP Institute, University St. 9, 380086 Tbilisi, Georgia

[52]Justus-Liebig-Universität Giessen, II Physikalisches Institut, Heinrich-Buff Ring 16, 35392 Giessen, Germany

[53]University of Glasgow, Department of Physics and Astronomy, Glasgow G12 8QQ, UK

[54]Georg-August-Universität, II. Physikalisches Institut, Friedrich-Hund Platz 1, 37077 Göttingen, Germany

[55]Laboratoire de Physique Subatomique et de Cosmologie, CNRS/IN2P3, Université Joseph Fourier, INPG, 53 avenue des Martyrs, 38026 Grenoble Cedex, France

[56]Hampton University, Department of Physics, Hampton, VA 23668, USA

[57]Harvard University, Laboratory for Particle Physics and Cosmology, 18 Hammond Street, Cambridge, MA 02138, USA

[58]Ruprecht-Karls-Universität Heidelberg: Kirchhoff-Institut für Physik[a], Im Neuenheimer Feld 227, 69120 Heidelberg; Physikalisches Institut[b], Philosophenweg 12, 69120 Heidelberg; ZITI Ruprecht-Karls-University Heidelberg[c], Lehrstuhl für Informatik V, B6, 23-29, 68131 Mannheim, Germany

[59]Hiroshima University, Faculty of Science, 1-3-1 Kagamiyama, Higashihiroshima-shi, Hiroshima 739-8526, Japan

[60]Hiroshima Institute of Technology, Faculty of Applied Information Science, 2-1-1 Miyake Saeki-ku, Hiroshima-shi, Hiroshima 731-5193, Japan

[61]Indiana University, Department of Physics, Swain Hall West 117, Bloomington, IN 47405-7105, USA

[62]Institut für Astro- und Teilchenphysik, Technikerstrasse 25, 6020 Innsbruck, Austria

[63]University of Iowa, 203 Van Allen Hall, Iowa City, IA 52242-1479, USA

[64]Iowa State University, Department of Physics and Astronomy, Ames High Energy Physics Group, Ames, IA 50011-3160, USA

[65]Joint Institute for Nuclear Research, JINR, Dubna 141 980, Moscow Region, Russia

[66]KEK, High Energy Accelerator Research Organization, 1-1 Oho, Tsukuba-shi, Ibaraki-ken 305-0801, Japan

[67]Kobe University, Graduate School of Science, 1-1 Rokkodai-cho, Nada-ku, Kobe 657-8501, Japan

[68]Kyoto University, Faculty of Science, Oiwake-cho, Kitashirakawa, Sakyou-ku, Kyoto-shi, Kyoto 606-8502, Japan

[69]Kyoto University of Education, 1 Fukakusa, Fujimori, fushimi-ku, Kyoto-shi, Kyoto 612-8522, Japan

[70]Universidad Nacional de La Plata, FCE, Departamento de Física, IFLP (CONICET-UNLP), C.C. 67, 1900 La Plata, Argentina

[71]Lancaster University, Physics Department, Lancaster LA1 4YB, UK

[72]INFN Sezione di Lecce[a]; Università del Salento, Dipartimento di Fisica[b] Via Arnesano, 73100 Lecce, Italy

[73]University of Liverpool, Oliver Lodge Laboratory, P.O. Box 147, Oxford Street, Liverpool L69 3BX, UK

[74]Jožef Stefan Institute and University of Ljubljana, Department of Physics, 1000 Ljubljana, Slovenia

[75]Queen Mary University of London, Department of Physics, Mile End Road, London E1 4NS, UK

[76]Royal Holloway, University of London, Department of Physics, Egham Hill, Egham, Surrey TW20 0EX, UK

[77]University College London, Department of Physics and Astronomy, Gower Street, London WC1E 6BT, UK

[78]Laboratoire de Physique Nucléaire et de Hautes Energies, Université Pierre et Marie Curie (Paris 6), Université Denis Diderot (Paris-7), CNRS/IN2P3, Tour 33, 4 place Jussieu, 75252 Paris Cedex 05, France

[79]Lunds universitet, Naturvetenskapliga fakulteten, Fysiska institutionen, Box 118, 221 00 Lund, Sweden

[80]Universidad Autonoma de Madrid, Facultad de Ciencias, Departamento de Fisica Teorica, 28049 Madrid, Spain

[81]Universität Mainz, Institut für Physik, Staudinger Weg 7, 55099 Mainz, Germany

[82]University of Manchester, School of Physics and Astronomy, Manchester M13 9PL, UK

[83]CPPM, Aix-Marseille Université, CNRS/IN2P3, Marseille, France

[84]University of Massachusetts, Department of Physics, 710 North Pleasant Street, Amherst, MA 01003, USA

[85]McGill University, High Energy Physics Group, 3600 University Street, Montreal, Quebec H3A 2T8, Canada

[86]University of Melbourne, School of Physics, Parkville, Victoria 3010, Australia

[87]The University of Michigan, Department of Physics, 2477 Randall Laboratory, 500 East University, Ann Arbor, MI 48109-1120, USA

[88]Michigan State University, Department of Physics and Astronomy, High Energy Physics Group, East Lansing, MI 48824-2320, USA

[89]INFN Sezione di Milano[a]; Università di Milano, Dipartimento di Fisica[b], via Celoria 16, 20133 Milano, Italy

[90]B.I. Stepanov Institute of Physics, National Academy of Sciences of Belarus, Independence Avenue 68, Minsk 220072, Republic of Belarus

[91]National Scientific & Educational Centre for Particle & High Energy Physics, NC PHEP BSU, M. Bogdanovich St. 153, Minsk 220040, Republic of Belarus

[92]Massachusetts Institute of Technology, Department of Physics, Room 24-516, Cambridge, MA 02139, USA

[93]University of Montreal, Group of Particle Physics, C.P. 6128, Succursale Centre-Ville, Montreal, Quebec H3C 3J7, Canada

[94]P.N. Lebedev Institute of Physics, Academy of Sciences, Leninsky pr. 53, 117 924 Moscow, Russia

[95] Institute for Theoretical and Experimental Physics (ITEP), B. Cheremushkinskaya ul. 25, 117 218 Moscow, Russia

[96] Moscow Engineering & Physics Institute (MEPhI), Kashirskoe Shosse 31, 115409 Moscow, Russia

[97] Lomonosov Moscow State University Skobeltsyn Institute of Nuclear Physics (MSU SINP), 1(2), Leninskie gory, GSP-1, Moscow 119991, Russia

[98] Ludwig-Maximilians-Universität München, Fakultät für Physik, Am Coulombwall 1, 85748 Garching, Germany

[99] Max-Planck-Institut für Physik, (Werner-Heisenberg-Institut), Föhringer Ring 6, 80805 München, Germany

[100] Nagasaki Institute of Applied Science, 536 Aba-machi, Nagasaki 851-0193, Japan

[101] Nagoya University, Graduate School of Science, Furo-Cho, Chikusa-ku, Nagoya 464-8602, Japan

[102] INFN Sezione di Napoli[a]; Università di Napoli, Dipartimento di Scienze Fisiche[b], Complesso Universitario di Monte Sant'Angelo, via Cinthia, 80126 Napoli, Italy

[103] University of New Mexico, Department of Physics and Astronomy, MSC07 4220, Albuquerque, NM 87131, USA

[104] Radboud University Nijmegen/NIKHEF, Department of Experimental High Energy Physics, Heyendaalseweg 135, 6525 AJ, Nijmegen, Netherlands

[105] Nikhef National Institute for Subatomic Physics, and University of Amsterdam, Science Park 105, 1098 XG Amsterdam, Netherlands

[106][a] DeKalb, Illinois 60115, USA

[107] Budker Institute of Nuclear Physics (BINP), Novosibirsk 630 090, Russia

[108] New York University, Department of Physics, 4 Washington Place, New York, NY 10003, USA

[109] Ohio State University, 191 West Woodruff Ave, Columbus, OH 43210-1117, USA

[110] Okayama University, Faculty of Science, Tsushimanaka 3-1-1, Okayama 700-8530, Japan

[111] University of Oklahoma, Homer L. Dodge Department of Physics and Astronomy, 440 West Brooks, Room 100, Norman, OK 73019-0225, USA

[112] Oklahoma State University, Department of Physics, 145 Physical Sciences Building, Stillwater, OK 74078-3072, USA

[113] Palacký University, 17.listopadu 50a, 772 07 Olomouc, Czech Republic

[114] University of Oregon, Center for High Energy Physics, Eugene, OR 97403-1274, USA

[115] LAL, Univ. Paris-Sud, IN2P3/CNRS, Orsay, France

[116] Osaka University, Graduate School of Science, Machikaneyama-machi 1-1, Toyonaka, Osaka 560-0043, Japan

[117] University of Oslo, Department of Physics, P.O. Box 1048, Blindern, 0316 Oslo 3, Norway

[118] Oxford University, Department of Physics, Denys Wilkinson Building, Keble Road, Oxford OX1 3RH, UK

[119] INFN Sezione di Pavia[a]; Università di Pavia, Dipartimento di Fisica Nucleare e Teorica[b], Via Bassi 6, 27100 Pavia, Italy

[120] University of Pennsylvania, Department of Physics, High Energy Physics Group, 209 S. 33rd Street, Philadelphia, PA 19104, USA

[121] Petersburg Nuclear Physics Institute, 188 300 Gatchina, Russia

[122] INFN Sezione di Pisa[a]; Università di Pisa, Dipartimento di Fisica E. Fermi[b], Largo B. Pontecorvo 3, 56127 Pisa, Italy

[123] University of Pittsburgh, Department of Physics and Astronomy, 3941 O'Hara Street, Pittsburgh, PA 15260, USA

[124] Laboratorio de Instrumentacao e Fisica Experimental de Particulas – LIP[a], Avenida Elias Garcia 14-1, 1000-149 Lisboa, Portugal; Universidad de Granada, Departamento de Fisica Teorica y del Cosmos and CAFPE[b], 18071 Granada, Spain

[125] Institute of Physics, Academy of Sciences of the Czech Republic, Na Slovance 2, 18221 Praha 8, Czech Republic

[126] Charles University in Prague, Faculty of Mathematics and Physics, Institute of Particle and Nuclear Physics, V Holesovickach 2, 18000 Praha 8, Czech Republic

[127] Czech Technical University in Prague, Zikova 4, 166 35 Praha 6, Czech Republic

[128] State Research Center Institute for High Energy Physics, Pobeda street, 1, Protvino 142281, Moscow Region, Russia

[129] Rutherford Appleton Laboratory, Science and Technology Facilities Council, Harwell Science and Innovation Campus, Didcot OX11 0QX, UK

[130] University of Regina, Physics Department, Canada

[131] Ritsumeikan University, Noji Higashi 1 chome 1-1, Kusatsu, Shiga 525-8577, Japan

[132] INFN Sezione di Roma I[a]; Università La Sapienza, Dipartimento di Fisica[b], Piazzale A. Moro 2, 00185 Roma, Italy

[133] INFN Sezione di Roma Tor Vergata[a]; Università di Roma Tor Vergata, Dipartimento di Fisica[b], via della Ricerca Scientifica, 00133 Roma, Italy

[134] INFN Sezione di Roma Tre[a]; Università Roma Tre, Dipartimento di Fisica[b], via della Vasca Navale 84, 00146 Roma, Italy

[135] Réseau Universitaire de Physique des Hautes Energies (RUPHE): Université Hassan II, Faculté des Sciences Ain Chock[a], B.P. 5366, Casablanca; Centre National de l'Energie des Sciences Techniques Nucleaires (CNESTEN)[b], B.P. 1382 R.P. 10001 Rabat 10001; Université Mohamed Premier[c], LPTPM, Faculté des Sciences, B.P.717. Bd. Mohamed VI, 60000, Oujda; Université Mohammed V, Faculté des Sciences[d], 4 Avenue Ibn Battouta, BP 1014 RP, 10000 Rabat, Morocco

[136] CEA, DSM/IRFU, Centre d'Etudes de Saclay, 91191 Gif-sur-Yvette, France

[137] University of California Santa Cruz, Santa Cruz Institute for Particle Physics (SCIPP), Santa Cruz, CA 95064, USA

[138] University of Washington, Seattle, Department of Physics, Box 351560, Seattle, WA 98195-1560, USA

[139] University of Sheffield, Department of Physics & Astronomy, Hounsfield Road, Sheffield S3 7RH, UK

[140] Shinshu University, Department of Physics, Faculty of Science, 3-1-1 Asahi, Matsumoto-shi, Nagano 390-8621, Japan

[141] Universität Siegen, Fachbereich Physik, 57068 Siegen, Germany

[142] Simon Fraser University, Department of Physics, 8888 University Drive, Burnaby, BC V5A 1S6, Canada

[143] SLAC National Accelerator Laboratory, Stanford, California 94309, USA

[144] Comenius University, Faculty of Mathematics, Physics & Informatics[a], Mlynska dolina F2, 84248 Bratislava; Institute of Experimental Physics of the Slovak Academy of Sciences, Dept. of Subnuclear Physics[b], Watsonova 47, 04353 Kosice, Slovak Republic

[145] Stockholm University, Department of Physics[a]; The Oskar Klein Centre[b], AlbaNova, 106 91 Stockholm, Sweden

[146] Royal Institute of Technology (KTH), Physics Department, 106 91 Stockholm, Sweden

[147] Stony Brook University, Department of Physics and Astronomy, Nicolls Road, Stony Brook, NY 11794-3800, USA

[148] University of Sussex, Department of Physics and Astronomy Pevensey 2 Building, Falmer, Brighton BN1 9QH, UK

[149] University of Sydney, School of Physics, Sydney NSW 2006, Australia

[150] Insitute of Physics, Academia Sinica, Taipei 11529, Taiwan

[151] Technion, Israel Inst. of Technology, Department of Physics, Technion City, Haifa 32000, Israel

[152] Tel Aviv University, Raymond and Beverly Sackler School of Physics and Astronomy, Ramat Aviv, Tel Aviv 69978, Israel

[153] Aristotle University of Thessaloniki, Faculty of Science, Department of Physics, Division of Nuclear & Particle Physics, University Campus, 54124 Thessaloniki, Greece

[154] The University of Tokyo, International Center for Elementary Particle Physics and Department of Physics, 7-3-1 Hongo, Bunkyo-ku, Tokyo 113-0033, Japan

[155] Tokyo Metropolitan University, Graduate School of Science and Technology, 1-1 Minami-Osawa, Hachioji, Tokyo 192-0397, Japan

[156] Tokyo Institute of Technology, 2-12-1-H-34 O-Okayama, Meguro, Tokyo 152-8551, Japan

[157] University of Toronto, Department of Physics, 60 Saint George Street, Toronto, Ontario M5S 1A7, Canada

[158] TRIUMF[(a)], 4004 Wesbrook Mall, Vancouver, BC V6T 2A3; [(b)] York University, Department of Physics and Astronomy, 4700 Keele St., Toronto, Ontario M3J 1P3, Canada

[159] University of Tsukuba, Institute of Pure and Applied Sciences, 1-1-1 Tennoudai, Tsukuba-shi, Ibaraki 305-8571, Japan

[160] Tufts University, Science & Technology Center, 4 Colby Street, Medford, MA 02155, USA

[161] Universidad Antonio Narino, Centro de Investigaciones, Cra 3 Este No. 47A-15, Bogota, Colombia

[162] University of California, Irvine, Department of Physics & Astronomy, CA 92697-4575, USA

[163] INFN Gruppo Collegato di Udine[(a)]; ICTP[(b)], Strada Costiera 11, 34014, Trieste; Università di Udine, Dipartimento di Fisica[(c)], via delle Scienze 208, 33100 Udine, Italy

[164] University of Illinois, Department of Physics, 1110 West Green Street, Urbana, IL 61801, USA

[165] University of Uppsala, Department of Physics and Astronomy, P.O. Box 516, 751 20 Uppsala, Sweden

[166] Instituto de Física Corpuscular (IFIC) Centro Mixto UVEG-CSIC, Apdo. 22085 46071 Valencia, Dept. Física At. Mol. y Nuclear; Univ. of Valencia, and Instituto de Microelectrónica de Barcelona (IMB-CNM-CSIC) 08193 Bellaterra Barcelona, Spain

[167] University of British Columbia, Department of Physics, 6224 Agricultural Road, Vancouver, BC V6T 1Z1, Canada

[168] University of Victoria, Department of Physics and Astronomy, P.O. Box 3055, Victoria, BC V8W 3P6, Canada

[169] Waseda University, WISE, 3-4-1 Okubo, Shinjuku-ku, Tokyo 169-8555, Japan

[170] The Weizmann Institute of Science, Department of Particle Physics, P.O. Box 26, 76100 Rehovot, Israel

[171] University of Wisconsin, Department of Physics, 1150 University Avenue, Madison WI 53706, USA

[172] Julius-Maximilians-University of Würzburg, Physikalisches Institut, Am Hubland, 97074 Würzburg, Germany

[173] Bergische Universität, Fachbereich C, Physik, Postfach 100127, Gauss-Strasse 20, 42097 Wuppertal, Germany

[174] Yale University, Department of Physics, PO Box 208121, New Haven CT, 06520-8121, USA

[175] Yerevan Physics Institute, Alikhanian Brothers Street 2, 375036 Yerevan, Armenia

[176] ATLAS-Canada Tier-1 Data Centre, TRIUMF, 4004 Wesbrook Mall, Vancouver, BC, V6T 2A3, Canada

[177] GridKA Tier-1 FZK, Forschungszentrum Karlsruhe GmbH, Steinbuch Centre for Computing (SCC), Hermann-von-Helmholtz-Platz 1, 76344 Eggenstein-Leopoldshafen, Germany

[178] Port d'Informacio Cientifica (PIC), Universitat Autonoma de Barcelona (UAB), Edifici D, 08193 Bellaterra, Spain

[179] Centre de Calcul CNRS/IN2P3, Domaine scientifique de la Doua, 27 bd du 11 Novembre 1918, 69622 Villeurbanne Cedex, France

[180] INFN-CNAF, Viale Berti Pichat 6/2, 40127 Bologna, Italy

[181] Nordic Data Grid Facility, NORDUnet A/S, Kastruplundgade 22, 1, 2770 Kastrup, Denmark

[182] SARA Reken- en Netwerkdiensten, Science Park 121, 1098 XG Amsterdam, Netherlands

[183] Academia Sinica Grid Computing, Institute of Physics, Academia Sinica, No.128, Sec. 2, Academia Rd., Nankang, Taipei, Taiwan 11529, Taiwan

[184] UK-T1-RAL Tier-1, Rutherford Appleton Laboratory, Science and Technology Facilities Council, Harwell Science and Innovation Campus, Didcot OX11 0QX, UK

[185] RHIC and ATLAS Computing Facility, Physics Department, Building 510, Brookhaven National Laboratory, Upton, NY 11973, USA

[a] Also at CPPM, Marseille, France.

[b] Also at TRIUMF, 4004 Wesbrook Mall, Vancouver, BC V6T 2A3, Canada.

[c] Also at Faculty of Physics and Applied Computer Science of the AGH-University of Science and Technology, (FPACS, AGH-UST), al. Mickiewicza 30, 30059 Cracow, Poland.

[d] Also at Università di Napoli Parthenope, via A. Acton 38, 80133 Napoli, Italy.

[e] Also at Institute of Particle Physics (IPP), Canada.

[f] Louisiana Tech University, 305 Wisteria Street, P.O. Box 3178, Ruston, LA 71272, USA.

[g] At Department of Physics, California State University, Fresno, 2345 E. San Ramon Avenue, Fresno, CA 93740-8031, USA.

[h] Currently at Istituto Universitario di Studi Superiori IUSS, V.le Lungo Ticino Sforza 56, 27100 Pavia, Italy.

[i] Also at California Institute of Technology, Physics Department, Pasadena, CA 91125, USA.

[j] Also at University of Montreal, Canada.

[k] Also at Institut für Experimentalphysik, Universität Hamburg, Luruper Chaussee 149, 22761 Hamburg, Germany.

[l] Also at Petersburg Nuclear Physics Institute, 188 300 Gatchina, Russia.

[m] Also at School of Physics and Engineering, Sun Yat-sen University, China.

[n] Also at School of Physics, Shandong University, Jinan, China.

[o] Also at Rutherford Appleton Laboratory, Science and Technology Facilities Council, Harwell Science and Innovation Campus, Didcot OX11, UK.

[p] Also at School of Physics, Shandong University, Jinan.

[q] Also at Rutherford Appleton Laboratory, Science and Technology Facilities Council, Harwell Science and Innovation Campus, Didcot OX11 0QX, UK.

[r] Now at KEK.

[s] University of South Carolina, Dept. of Physics and Astronomy, 700 S. Main St, Columbia, SC 29208, USA.

[t] Also at KFKI Research Institute for Particle and Nuclear Physics, Budapest, Hungary.

[u] Also at Institute of Physics, Jagiellonian University, Cracow, Poland.

[v] Also at School of Physics and Engineering, Sun Yat-sen University, Taiwan.

[w] Transfer to LHCb 31.01.2010.

[x] Also at Department of Physics, Nanjing University, China.

[*] Deceased.

Abstract The simulation software for the ATLAS Experiment at the Large Hadron Collider is being used for large-scale production of events on the LHC Computing Grid. This simulation requires many components, from the generators that simulate particle collisions, through packages simulating the response of the various detectors and triggers. All of these components come together under the ATLAS simulation infrastructure. In this paper, that infrastructure is discussed, including that supporting the detector description, interfacing the event generation, and combining the GEANT4 simulation of the response of the individual detectors. Also described are the tools allowing the software validation, performance testing, and the validation of the simulated output against known physics processes.

1 Introduction

ATLAS [1], one of the general-purpose detectors at the Large Hadron Collider [2], began operation in 2008. The detector will collect data from proton–proton collisions with center-of-mass energies up to 14 TeV, as well as 5.5 TeV per nucleon pair in heavy ion (Pb–Pb) collisions. During proton–proton collisions at the design luminosity of 10^{34} cm^{-2} s^{-1}, beam bunches will cross every 25 ns (40 MHz) and provide on average 23 collisions per bunch crossing. ATLAS has been designed to record up to 200 bunch crossings per second, keeping only the most interesting interactions for physics analyses, including searches for new physics.

In order to study the detector response for a wide range of physics processes and scenarios, a detailed simulation has been implemented that carries events from the event generation through to output in a format which is identical to that of the true detector. The simulation program is integrated into the ATLAS software framework, Athena [3], and uses the GEANT4 simulation toolkit [4, 5]. The core software and large-scale production infrastructures are discussed further in Sect. 2.

** e-mail: atlas.secretariat@cern.ch

The simulation software chain is generally divided into three steps, though they may be combined into a single job: generation of the event and immediate decays (see Sect. 3), simulation of the detector and physics interactions (see Sect. 5), and digitization of the energy deposited in the sensitive regions of the detector into voltages and currents for comparison to the readout of the ATLAS detector (see Sect. 6). The output of the simulation chain can be presented in either an object-based format or in a format identical to the output of the ATLAS data acquisition system (DAQ). Thus, both the simulated and real data from the detector can then be run through the same ATLAS trigger and reconstruction packages.

The ATLAS detector geometry used for simulation, digitization, and reconstruction is built from databases containing the information describing the physical construction and conditions data. The latter contains all the information needed to emulate a single data-taking run of the real detector (e.g. detector misalignments or temperatures). The same geometry and simulation infrastructure is able to reproduce the test stands and installation configurations of the ATLAS detector. The detector description is discussed in Sect. 4.

Large computing resources are required to accurately model the complex detector geometry and physics descriptions in the standard ATLAS detector simulation. This has led to the development of several varieties of fast simulation. Each is best suited to a particular use-case, and they are described in Sect. 7. Validation of the software, testing of the software performance, and validation of the physics performance and output of each piece of the simulation software chain is discussed in Sect. 8.

This paper reviews the status of the software and geometry used for large-scale production in 2008.

2 ATLAS offline software overview

The ATLAS software framework, Athena [3], uses PYTHON as an object-oriented scripting and interpreter language to configure and load C++ algorithms and objects. Rather than develop an entirely new high-energy physics data processing infrastructure, ATLAS adopted the Gaudi framework [6, 7], originally developed for LHCb and written in C++. Gaudi was created as a flexible framework to support a variety of applications through base classes and basic functionality. As much as possible, the infrastructure relies on the CLHEP common libraries [8], which include utility classes particularly designed for use in high-energy physics software (e.g. vectors and rotations).

Athena releases are divided into major projects by functionality [9], and all of the ATLAS simulation software (including event generation and digitization) resides in a single project. The dependencies of the "simulation" project are the "core" project, which includes the Athena framework, the "conditions" and "detector description" projects, which include all code necessary for the description of the ATLAS detector, and the "event" project, which includes descriptions of persistent objects. The number of lines of code by software language for the simulation project are summarized in Table 1, as calculated using cloc [10] in Athena release 14.4. Lines of code in the upstream Athena projects, excluding external dependencies like Gaudi and CLHEP, are summarized in Table 2.

Table 1 Numbers of files, lines of code, and lines of comments in the ATLAS simulation project, by programming language for major contributors. External dependencies are not included

Language	Files	Comment	Code
C++	930	24,000	120,000
FORTRAN	270	15,000	42,000
C/C++ Header	1,100	13,000	34,000
Python	430	16,000	27,000
HTML	62	130	15,000
Bourne Shell	390	1,000	7,300
C Shell	380	210	3,800
XML	52	1,200	3,400
Sum	3,600	70,000	250,000

Table 2 Numbers of lines of code in each of the projects upstream of the ATLAS simulation project, versus the programming language. Most projects are dominated by C++ and PYTHON code. The most significant exception is the detector project, which contains 70,000 lines of XML and Java code

Project	C/C++ code	C/C++ headers	PYTHON code	Total code
Core	390,000	43,000	240,000	860,000
Event	200,000	110,000	16,000	350,000
Conditions	280,000	90,000	21,000	620,000
Detector	38,000	6,100	8,400	140,000
Sum	910,000	250,000	280,000	2,000,000

All Athena jobs consist of three distinct steps. First, in the initialization step, services and algorithms are loaded on demand using dynamic library loading. Generally, algorithms include methods to be called once per event, whereas services may be accessed many times during a single event. The configuration and initialization is controlled within a common PYTHON infrastructure which allows introspection, particularly useful in debugging and providing help for the users. Also, by using a scripting language for loading and configuring objects, there is no need to recompile C++ code or a script for each job. Small modifications can be made in the scripts (also called "fragments" or "job options"), or even in the midst of the job, without having to stop and recompile the libraries. This scripting method also lightens the load on the user, since there is, under normal circumstances, no need to compile anything prior to running a job. Each algorithm and service can be configured differently for each step of the simulation software chain, allowing maximal sharing of infrastructure among the distinct steps of the chain. Algorithms can be added to a top list of methods to be run during the event loop.

Second, the event loop begins. All algorithms in the top list are run sequentially on each event. An external generator or algorithm controlling GEANT4 may be added to this list, for example. From these main methods, other services and algorithms can be called. A messaging service, called throughout the jobs, controls log file outputs with different levels of verbosity. The user may configure the total logging verbosity or configure the verbosity individually for a single algorithm, particularly useful for debugging.

During the finalization stage of the job, all algorithms are terminated and all objects are deleted. At this point, algorithms may output any statistics (e.g. memory or CPU usage) they track.

These three steps comprise each Athena job, but the infrastructure allows for the insertion of hooks at various places. Each step of the ATLAS simulation chain takes advantage of this infrastructure to provide maximal flexibility for the user. Only requested modules are loaded as plug-ins, keeping each step as light as possible in memory and as fast as possible during the event loop.

For storing data, ATLAS has adopted a scheme for separating transient from persistent objects. Most general C++ types, immediately prior to storage, are converted to a type that requires less space. Although, for example, energy is accumulated in the calorimeter by summing double-precision floating point numbers, at the end of each event and prior to storage, the total energy is converted into a single-precision floating point number (float). Summing with floats was found to alter the total energy because of truncation. For some types, more complicated storage schemes are implemented that rely on properties of the information to be stored (e.g. where it is possible to sacrifice some accuracy). Metadata, general property information for data collected in a file,

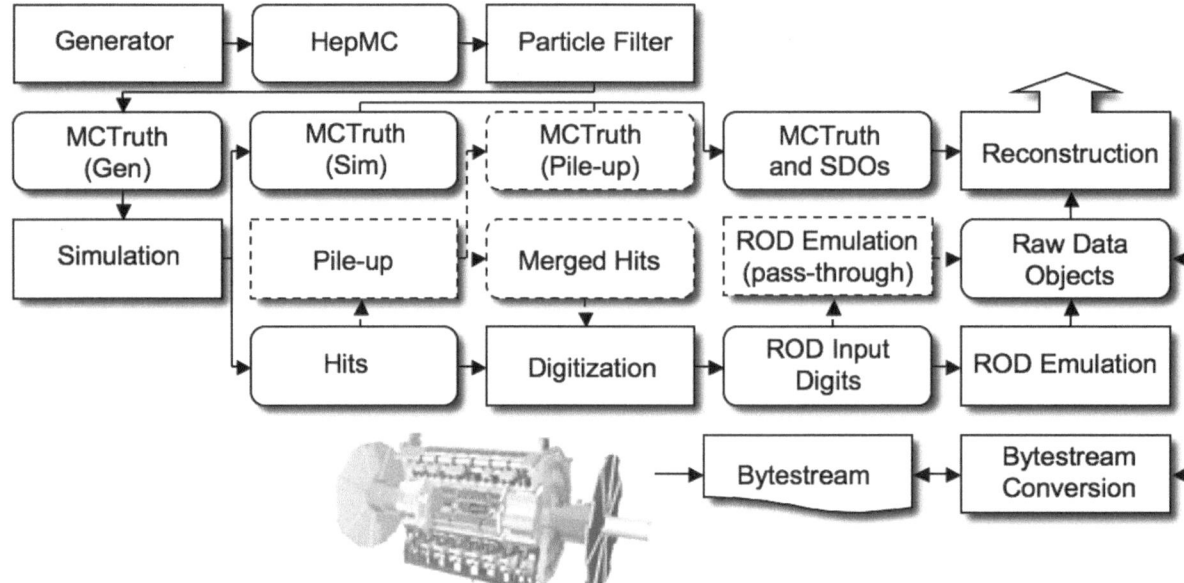

Fig. 1 The flow of the ATLAS simulation software, from event generators (*top left*) through reconstruction (*top right*). Algorithms are placed in *square-cornered boxes* and persistent data objects are placed in *rounded boxes*. The optional pile-up portion of the chain, used only when events are overlaid, is *dashed*. Generators are used to produce data in HepMC format. Monte Carlo truth is saved in addition to energy depositions in the detector (hits). This truth is merged into Simulated Data Objects (SDOs) during the digitization. Also, during the digitization stage, Read Out Driver (ROD) electronics are simulated

are included in the output files for all the stages of the event simulation. The metadata include all configuration information for the job. Athena has also adopted the POOL (Pool Of persistent Objects for LHC) file handling and persistency framework [11–13].

2.1 ATLAS simulation overview

An overview of the ATLAS simulation data flow can be seen in Fig. 1. Algorithms and applications to be run are placed in square-cornered boxes, and persistent data objects are placed in round-cornered boxes. The optional steps required for pile-up or event overlay (see Sect. 6.2) are shown with a dashed outline.

A generator produces events in standard HepMC format [14]. These events can be filtered at generation time so that only events with a certain property (e.g. leptonic decay or missing energy above a certain value) are kept. The generator is responsible for any prompt decays (e.g. Z or W bosons) but stores any "stable" particle expected to propagate through a part of the detector (see Sect. 3). Because it only considers immediate decays, there is no need to consider detector geometry during the generation step, except in controlling what particles are considered stable. During this step, the run number for the simulated data set and event numbers for each event are established. Event numbers are generally ordered in a single job, though events may be omitted because of filtering at each step. Run numbers for simulated data sets derive from the job options used to generate the sample and mimic real run numbers used during data taking.

These generated events are then read into the simulation. A record of all particles produced by the generator is retained in the simulation output file (see Sect. 3.6), but cuts can be applied to select only certain particles to process in the simulation. Each particle is propagated through the full ATLAS detector by GEANT4. The configuration of the detector, including misalignments and distortions, can be set at run time by the user. The energies deposited in the sensitive portions of the detector are recorded as "hits," containing the total energy deposition, position, and time, and are written to a simulation output file, called a hit file.

In both event generation and detector simulation, information called "truth" is recorded for each event. In the generation jobs, the truth is a history of the interactions from the generator, including incoming and outgoing particles. A record is kept for every particle, whether the particle is to be passed through the detector simulation or not. In the simulation jobs, truth tracks and decays for certain particles are stored. This truth contains, for example, the locations of the conversions of photons within the inner detector and the subsequent electron and positron tracks. In the digitization jobs, Simulated Data Objects (SDOs) are created from the truth. These SDOs are maps from the hits in the sensitive regions of the detector to the particles in the simulation truth record that deposited the hits' energy. The truth information is further processed in the reconstruction jobs and can be used during the analysis of simulated data to quantify the success of the reconstruction software.

The digitization takes hit output from simulated events: hard scattering signal, minimum bias, beam halo, beam gas, and cavern background events. Each type of event can be overlaid at a user-specified rate before the detector signal (e.g. voltage or time) is generated. The overlay (called "pile-up") is done during digitization to save the CPU time required by the simulation. At this stage, detector noise is added to the event. The first level trigger, implemented with hardware on the real detector, is also simulated in a "pass" mode. Here no events are discarded but each trigger hypothesis is evaluated. The digitization first constructs "digits," inputs to the read out drivers (RODs) in the detector electronics. The ROD functionality is then emulated, and the output is a Raw Data Object (RDO) file. The output from the ATLAS detector itself is in "byte stream" format, which can be fairly easily converted to and from RDO file format. The two are similar, and in some subdetectors they are almost interchangeable. Truth information is the major exception. It is stripped in the conversion to bytestream.

The simulation software chain, divided in this way, uses resources more effectively than a single-step event simulation and simplifies software validation. Event generation jobs, typically quick and with small output files, can be run for several thousands of events at a time. By storing the output rather than regenerating it each time, it becomes possible to run identical events through different versions of the simulation software or with different detector configurations. The simulation step is particularly slow, and can take several minutes per event (see Sect. 8.2). Simulation jobs are therefore divided into groups of 50 or fewer events; only a few events may be completed in a single heavy ion simulation job. Digitization jobs are generally configured to run ~1000 events. This configuration eases file handling by producing a smaller number of RDO files. Each step is partially configured based on the input files. For example, the detector geometry used for a digitization job is selected based on the input hit file.

The ATLAS high level trigger[1] (HLT) [15] and reconstruction [16] run on these RDO files. The reconstruction is identical for the simulation and the data, with the exception that truth information can be treated and is available only in simulated data. During data taking, the HLT is run on bytestream files, however all hypotheses and additional test hypotheses may be evaluated by translating the RDOs into bytestream format.

---

[1]The ATLAS high level trigger comprises two stages: level 2, and the event filter. Both are software triggers run with the reconstruction, and may be treated as a single unit for the purposes of this discussion.

2.2 Large-scale production system

Because of the significant time consumption of the ATLAS simulation, only minimal jobs can be completed interactively on most computers. It is, therefore, desirable to distribute as much as possible production of the necessary simulated data for ATLAS. Complete software releases are built and distributed to production sites and users every few weeks, providing all Athena software and all external dependencies, including generators and GEANT4. These releases are patched several times with "production caches" before a new clean release is built and distributed. With each release or production cache, a small set of data files are packaged that include database replicas, any necessary external data files, and some sample output files. These sample files can be used to ensure that the locally installed release can be validated by processing events through the entire software chain, from generation through reconstruction.

Large-scale production is then done on the World-wide LHC Computing Grid ("WLCG" or "Grid") [17]. A single task on the Grid (e.g. simulation of 500,000 $t\bar{t}$ events) is separated into many jobs depending on the content and complexity of the task. A job can be completed by a single CPU within the maximum allowed time for a job on the Grid (typically 2–3 days). The output, including log files, of every Grid job is registered with the ATLAS Distributed Data Management system (DDM) [18]. The DDM uses DQ2 [19] for dataset bookkeeping, and allows users to search for datasets on the Grid, analyze them in place, and, if necessary, retrieve them. Separate Grid software controls the distribution of jobs to the various Grid sites. In a typical task, 10 jobs are queued and run as a test sample, and only once they finish successfully are the remainder of the jobs released to the Grid. In the case of a full chain of jobs being run (generation, simulation, and digitization), each subsequent step is automatically held in the queue until the required data are available from the previous step. Frequently, Grid jobs are configured to run two steps (e.g. simulation and digitization together, or digitization and reconstruction together). About one million events per day can be produced using GEANT4 on the Grid.

On the Grid, "job transforms" are run, which may only include well-defined, minor modifications to some standard job configuration after the input events have been specified. A task is given a random number seed, and each job increments the seed in sequence. The modifications to a generation job also may include a configuration file for the selected generator to be run. These configuration files are included with each release and may not be arbitrarily modified by the user during production. The modification to a simulation job may include detector geometry and conditions and specially designed job options fragments that are included with each release. These fragments are typically constructed for a very specific purpose, for example a non-standard vertex smearing, simulation of cavern background, or propagation and late decay of long-lived exotic particles. Many of these modifications can be chained to provide maximal flexibility to the user, but if two fragments are sufficiently complex such chaining becomes impossible. The modifications to a digitization job may include geometry and conditions versions, calorimeter sampling fraction, trigger configuration, and noise control. These modifications are discussed further in the subsequent sections.

3 Event generation overview

Event generation consists of the production of a set of particles which is passed to either full or fast detector simulation. Event generation runs within the Athena framework, but most of the generators themselves are written and maintained by authors external to ATLAS. The ATLAS-specific implementation, therefore, consists mostly of a set of interface packages. These are designed to be as simple as practicable and wherever possible to be factorized from the external packages. This is essential to allow rapid feedback and bug reporting to the authors of the external packages. Most of the well-understood and thoroughly debugged generators are written in FORTRAN. Their interfaces transfer the event information, mostly contained in FORTRAN common blocks, into an object format that can be used by the ATLAS software. This ensures that any downstream algorithms are shielded from details specific to an individual generator. Events can either be stored as POOL files for later use or passed to simulation in the same Athena job.

Details of the framework and comments specific to each generator are listed below. Large-scale production has been run with PYTHIA [20] (including an ATLAS variant, PythiaB [21, 22], used for production of events with B-hadrons), HERWIG [23–25], Sherpa [26], Hijing [27], Alpgen [28], MC@NLO [29], and AcerMC [30]. Tauola [31] and Photos [32] are routinely used to handle tau decays and photon emission. EvtGen [33] is used for B-decays in cases where the physics is sensitive to details of the B hadron decays.[2] ISAJET [34] is used for generating supersymmetric particles in conjunction with HERWIG. The newer C++ generators PYTHIA 8 [35] and HERWIG++ [36] are being tested. Both produce events in the HepMC format, so no translation is needed. They can be passed directly to simulation. As these new generators evolve and undergo extensive testing and validation, they are expected to enter the production shortly and eventually supersede their FORTRAN

predecessors. Some production was also done with Mad-Graph [37] (vector boson scattering), CHARYBDIS [38] (black hole event generation), and CompHep [39, 40] (specific exotic physics models). Discussion of the generation of cavern background, beam halo, and beam gas events follow in Sects. 6.2.1 and 6.2.2. Single particle generators are also used to generate cosmic ray events and single particle events for performance studies and calibration of the detector.

Each generated event contains the particles from a single interaction with a vertex located at the geometric origin. Modifications to account for the beam properties are applied to the event before it is passed to GEANT4 (see Sect. 5.1). Particles with a proper lifetime $c\tau > 10$ mm are considered stable by the event generator. They can propagate far enough to interact with detector material before decaying. Their decays are handled by the simulation. Any particles with $c\tau < 10$ mm are decayed by the event generator, and their interactions with material or curving in the magnetic field of ATLAS are ignored.

3.1 Generator framework

Many external generator packages assume that the parameters for a particular job are set via a main program. This would require recompilation to change parameters. The Athena generator interfaces allow for the passing of all relevant parameters at run time, permitting a fixed software release to be used to produce different physics configurations. During initialization, the relevant parameters are passed via PYTHON fragments. The combination of the fragments, random number seeds, and the software release uniquely identifies the resulting data.[3] The Athena event manager is run for each event, and a run number and an event number are created; then the event generator is asked to produce an event. This event is created in memory in the format specific to the generator itself. The event must then be mapped into a common format so that subsequent algorithms are independent of the generator used.

ATLAS uses the HepMC event record [14], initially developed by the ATLAS collaboration but now supported by WLCG [17]. This is a set of C++ classes which holds the full event as produced by the generator. Stable particles are used as input to simulation; unstable ones can be of use in physics studies and diagnostics. Each event generator produces a very large number of stable particles (e.g. muon, kaons, pions, electrons, photons), a much larger number of unstable particles (e.g. gluons, quarks, B mesons, heavy hyperons), and, possibly, other objects (e.g. "strings" or "color singlet clusters") specific to an individual generator. The

---

[2] PYTHIA remains the default for current inclusive production, but Evt-Gen is likely to be used by default for the long-term production.

[3] Since pseudo-random number generators are chip architecture dependent, jobs are exactly reproducible only when run on the same type of processor (e.g. Intel or AMD).

HepMC record consists of a connected tree, navigation inside of which retains information on the event history including the parents of unstable particles. There is an important caveat here: the event generators are modeling quantum processes, and the event record has the structure of a classical decay chain. It is inevitable that compromises must be made and difficulties can arise from an over-literal interpretation of the tree structure. A very simple example is provided by events containing an e^+e^- pair. The parent of the e^+e^- pair cannot be uniquely specified, as the pair may arise from an intermediate Z boson, photon, or quantum interference. The HepMC event record is also used to contain the particle information from secondaries produced by interactions in the detector. This is discussed below in the section on Monte Carlo truth (see Sect. 3.6). Information about all interacting partons (e.g. momentum fractions x_1 and x_2) is saved, so that parton distribution function reweighting can be done without rerunning the event generation.

The FORTRAN generators usually use the HEPEVT common block [41] to store the information. Unfortunately, the different generators use slightly different structures. A separate translation into HepMC is needed for each one. The C++ generators such as HERWIG++ produce output in the HepMC format. No translation is required and the integrity of the HepMC event record is the responsibility of the generator authors.

3.2 General purpose generators

General purpose generators produce complete events starting from a proton–proton, proton–nucleus or nucleus–nucleus initial state. They are used standalone or with specialized generators that improve the description of certain final states. They have many parameters, some of which are related to fundamental parameters such as the QCD coupling constant and electroweak parameters, and some of which describe the models used to parametrize long distance QCD, soft QCD, and electroweak processes.

3.2.1 PYTHIA and PythiaB

PYTHIA [20] and HERWIG (see below) in their FORTRAN versions have been tested, used, and validated over many years in e^+e^- and hadron colliders. They start with a hard scattering process calculated to lowest order in QCD. They then add addition al QCD and QED radiation in a shower approximation which is most accurate when the radiation is emitted at small angle. The approximation is poorest in those cases with a large number of widely separated emissions of comparable energy. In addition, PYTHIA use a model for hard and soft scattering processes in a single event in order to simulate underlying activity. This model is used

in the simulation of minimum bias events. While other generators may be used for specific final states, PYTHIA and HERWIG are the benchmarks.

ATLAS uses PYTHIA 6.4. There are two models of QCD radiation in PYTHIA. By default, ATLAS uses the showering model introduced in PYTHIA 6.3. This showering model is believed to better match the theoretical description of QCD showers. It produces somewhat more jet activity [42, 43], resulting in "busier" events than the older model which was used, for example, for detailed simulations at the Tevatron (see, for example, [44, 45]). In this model, the multiple scatters which make up the underlying event are interleaved with the parton shower according to the hard scale of the scatter or the emission. At the end of the shower, a phenomenological model is used to combine the quarks and gluons into hadrons. This hadronization model, which has many parameters, has been tuned by comparison with data in e^+e^-, ep, and hadron colliders [46, 47]. The underlying event model was retuned within ATLAS [48] to recover an acceptable description of the Tevatron data [49, 50]. PYTHIA contains a very large number of built-in processes, and new ones can be added by modifying the code. Hard scattering events can also be generated in a separate program in a standard format and fed into PYTHIA for the addition of a parton shower and hadronization. PYTHIA is the default generator in ATLAS: many hundreds of millions of events have been generated using it. Its ease of use, speed, and robustness make it an ideal choice for the default. It is supplemented by other generators, either to obtain some estimate of the uncertainties, or when specialized generators are expected to give a better physical description in certain final states.

PythiaB [21, 22] is an ATLAS-specific modification of PYTHIA aimed at the efficient generation of events related to B-physics. In PYTHIA, most high p_T bottom quarks are produced in the QCD shower of a high p_T light quark or gluon from a hard scattering process. Most showers do not produce such a $b\bar{b}$ pair, so using PYTHIA to generate B-physics events is inefficient. PythiaB reuses those QCD showers that contain a b- or c-quark, hadronizing them several times to increase the probability of producing a b-hadron. Since the probability producing a b- or c-quark in a parton shower is low, this procedure results in more efficient procedure of making b-hadron events without introducing any bias in the distribution of b-hadrons within the event. If a specific decay mode is then required the b-hadron decay can be forced using a modified b-hadron decay table, either in PYTHIA itself or via EvtGen.

3.2.2 HERWIG

ATLAS uses HERWIG 6.5 [23–25], the last release of the FORTRAN HERWIG package which is now superseded by HERWIG++ (see below). It is a flexible generator with a large number of built in processes and has been tuned to agree with the Tevatron data [49, 50]. In particular, most of the generation of supersymmetric processes is done with HERWIG using the ISAWIG package [23–25] with the particle spectra and decay modes generated by ISAJET. ATLAS uses HERWIG with the Jimmy [51] implementation of the underlying event.

3.2.3 Sherpa

Sherpa [26] is a generator written in C++ which implements the CKKW duplicate removal prescription [52] to match fixed-order QCD matrix elements to QCD showers. It uses an interface to PYTHIA's hadronization model and produces complete events. It is expected to give better approximations for final states with large numbers of isolated jets than generators such as PYTHIA and HERWIG based on pure QCD showering. Sherpa generates underlying events using a simple multi-parton interaction model based on that of PYTHIA. For each new process to be generated, Sherpa must be recompiled to incorporate the specific libraries for the process of interest. On the Grid, this implies either recompiling Sherpa at the production site or deploying updated libraries for new production jobs. Instead, Sherpa is run locally to produce event files in Sherpa's native format. These files are then translated into the HepMC format with an additional Athena Grid job. It is also possible to run Sherpa entirely within an Athena job.

3.2.4 Hijing

Hijing [27] is a dedicated generator for the production of heavy ion events at all impact parameters. In a dense nuclear environment, such as appears in central collisions, a particle produced in a primary collision can re-interact several times as it propagates. Hijing models the propagation. It is also the only generator that can be used for proton–nucleus collisions occurring in beam–gas interactions. Hijing uses the PYTHIA hadronization model.

3.2.5 Single particle generators

A single particle event generator is frequently used for calibrating the detector, testing, and evaluating the reconstruction efficiencies. Although unphysical, these generators produce events with a single primary particle, for example a muon, electron, or charged pion, at a specified energy, position, and momentum direction. A range may also be specified for either the energy or direction. No underlying event, proton remnants, or other primary interactions are included when these events are generated.

A specialized single particle generator is used to produce cosmic ray events. Single muons are generated at the earth's surface in a square region (typically 600 m by 600 m) above the ATLAS detector and with the standard cosmic ray p_T spectrum [53, 54]. The upper and lower energy cutoffs of the spectrum are configurable. Those muons pointing to a sphere of configurable size (typically 20 m) centered at the geometric origin are propagated through the bedrock and the ATLAS cavern during simulation.

3.3 Specialized generators

Specialized generators do not produce complete events which can be passed directly to simulation. Rather, they are run in conjunction with one of the general purpose generators to improve the accuracy for specific decays or specific final states. Several of these specialized generators are "Les Houches" type generators. That is, they are run stand-alone using unmodified code from the generator author and produce an ASCII file containing partonic four-vectors in the "Les Houches" format [55, 56]. Athena uses a common interface that reads in these files and prepares them for processing in PYTHIA or HERWIG [55].

3.3.1 ISAJET

The FORTRAN generator ISAJET [34] is not used in large-scale production. However, it is used in conjunction with HERWIG for generation of supersymmetric events. Here, the ISASUGRA component of ISAJET is used to generate consistent sets of masses and decay modes for supersymmetric models. These are then loaded into HERWIG using the ISAWIG translation package, and HERWIG then generates complete final states.

3.3.2 Photos and Tauola

ATLAS uses the dedicated tau decay package Tauola to handle tau decays [31]. General purpose generators are set to treat tau leptons as stable. The events are passed to Tauola for decay. Because Tauola is a FORTRAN package, the events are extracted from the HEPEVT record. The Tauola interface is dependent on the generator that produced the tau, because helicities and helicity correlations are passed in generator-dependent formats. The original generator's results must be replaced, so both the input and output formats of Tauola are in fact generator-dependent. Special attention is paid to the polarization of the tau. In certain cases, for example the decay $W^\pm \rightarrow \tau^\pm \nu_\tau$, the polarization is known for the tau. In others, such as $Z \rightarrow \tau^+ \tau^-$, there is a correlation between the polarization of the taus.

Photos handles electromagnetic radiation [32]. It is used by Tauola, and, therefore, Tauola cannot be used without Photos. Photos is also used to improve the description of electromagnetic radiation in, for example, the decay

$W^{\pm} \to e^{\pm}\nu_e$, where radiation distorts the electron energy distribution. In these cases the final state electromagnetic radiation is switched off in the general purpose generator, usually HERWIG or PYTHIA, to avoid double counting.

3.3.3 EvtGen

EvtGen [33], originally developed by the CLEO collaboration, provides a more complete description of B meson and hadron decays than that provided by defaults in PYTHIA or HERWIG. Recent modifications have been made to handle B_S and b-baryon decays, incorporating measurements from the Tevatron, BaBar, and Belle. In particular, EvtGen incorporates the best measurements of branching ratios and has theoretical models for unmeasured decay modes. It includes angular correlations, which impact the acceptance for certain decay modes of B mesons and baryons. It has been used for ATLAS studies involving the prospects for measurements of exclusive B decays.

3.3.4 Alpgen

Alpgen [28] is a "Les Houches" type generator enabling more sophisticated generation of certain final states. HERWIG or PYTHIA is then used to perform the hadronization and produce final (and initial) state QCD radiation. Alpgen is targeted at final states with several well-separated hadronic jets where the fixed order QCD matrix element is expected to give a better approximation than the shower approximation of PYTHIA or HERWIG. Alpgen is used, for example, to generate final states containing a W or Z and many jets. Alpgen also provides an algorithm to prevent double counting by event rejection. The Athena interface package includes the methods needed to pass events through HERWIG or PYTHIA and veto those events that would contribute to double counting. This process can be very inefficient for final states with large numbers of jets, and generation time can be significant.

3.3.5 MC@NLO

MC@NLO [29], which is also a "Les Houches" type generator, runs standalone to produce ASCII files which are then processed by HERWIG running inside of Athena. MC@NLO uses fundamental (hard scattering) processes evaluated at next to leading order in QCD perturbation theory. It is used, for example, to generate top events as it gives a better representation of the transverse momentum (p_T) distribution of top quarks than PYTHIA or HERWIG. MC@NLO includes one loop corrections, with the consequence that events appear with negative and positive weight which must be taken into account when they are used. Any resulting distribution

will contain entries from both types of event, and, given sufficient statistics, the result will by physical (i.e. positive).[4] MC@NLO has been used for large-scale production of top, W and Z events. Only the parts of MC@NLO needed to read these events and process them via HERWIG are included in Athena releases.

3.3.6 AcerMC

AcerMC [30] is a "Les Houches" type generator aimed primarily at the production of W or Z bosons with several jets, including jets originating from b-quarks. A partonic final state is obtained by running it standalone and making an external ASCII file. Only the parts needed to read these events and process them via PYTHIA are included in Athena releases.

3.4 New C++ generators

3.4.1 PYTHIA 8

PYTHIA 8 [35] is a rewrite of the FORTRAN PYTHIA in C++ with new and expanded physics models. It provides a new user interface, transverse-momentum-ordered showers, and interleaving with multiple interactions. The program is under intensive tests and it will require some further tunings before it can replace the Pythia6 code as a leading generator. It is, however, interfaced to Athena and used for generator studies in ATLAS. It includes support for both "Les Houches" and HepMC event formats.

3.4.2 HERWIG++

HERWIG++ [36] is the C++ based replacement for HERWIG. It contains only important processes from the Standard Model, the universal extra dimensions model, and supersymmetric models (whose details are specified via Supersymmetric Les Houches Accord model files [58, 59]). Additional hard scattering processes can be used via "Les Houches" input from specialized generators, and additional decay models can be added by users.

HERWIG++ will soon be used for generation of some Standard Model processes, notably W and Z production. It will also be used for supersymmetric processes, because it includes full spin correlations and QCD radiation in the supersymmetric decay chains. The current version of HERWIG++ also incorporates an underlying event model based on the extension of Jimmy [51] to include soft scatters [60] and can thus potentially generate minimum bias physics.

---

[4] An alternative tool, POWHEG [57], implements essentially the same physics and produces events with only positive weight. Once it includes all the processes that MC@NLO does and has been validated, it is expected to take the place of MC@NLO.

3.5 Parton distribution functions

Parton distribution functions (PDFs) are used to describe the substructure of the proton and are used by all the event generators as external inputs. ATLAS uses the Les Houches Accord PDF Interface (LHAPDF [61]) library which is a replacement for PDFLIB [62] which provides a large repository of PDFs. CTEQ [63] PDFs are used by default (MC@NLO uses NLO PDFs, and all other generators use LO PDFs). There is a correlation between the PDFs and the tuning of parameters connected to initial state radiation [64, 65]: inconsistent results can be obtained by varying the PDFs in isolation. Therefore, when a new set of PDFs is used, the parameters of the event generator are retuned to produce consistent results [42].

3.6 Monte Carlo truth

The entire connected tree of the HepMC event record is stored as the Monte Carlo truth. Only the stable particles are propagated by the simulation. The various status codes and event history provided by the individual generators are retained within the HepMC event record. Unfortunately, much of this information is specific to a particular generator. Only status codes 1 (stable) and 2 (unstable) have a general meaning: the remaining values are used differently by the individual generators. As remarked in Sect. 3.1, there can be ambiguities resulting from the attempt to represent a quantum process by a classical tree. Some filters have been provided to select HepMC particles that, for example, are stable at the generator level or are non-interacting (e.g. neutrinos).

When the simulation is run, the HepMC tree from the event generator is copied, and some particles resulting from decays within, or interactions simulated by, GEANT4 are added to the copy (see Sect. 5.3). In this way, a complete event including both the generator and simulation information is provided. In order to ensure consistency, a particle decayed by GEANT4 but considered stable by the generator (such as a K_S) has its status code changed when the copy is made. A particle that has status code 2 after simulation will be identified as stable at the generator level, if the decay took place in GEANT4. GEANT4 secondaries are distinguished from those from the generators by an offset applied to their numerical identifier. The resulting Monte Carlo truth record can be large and account for a significant fraction ($\sim$30%) of the disk space used by a simulated event after reconstruction.

3.7 Default parameters, tuning and bug fixing

The generator authors define default parameters. In some cases, however, these parameters are not tuned for use at the Large Hadron Collider and are superseded by parameters obtained by comparisons to data. The criteria for a particle to be considered stable are modified for use in ATLAS, for example. Once high-energy data appear, it is expected that retuning of the parameters will occur. These tunings can be made by varying parameters at run time. Once a new tuning is available, it can be loaded as a PYTHON fragment at run time or hard coding the values into the generator interfaces. In either case, the tuning becomes available as part of the next Athena software release and will be enabled by default. The settings can be overridden if needed or the previous defaults re-established. It is important to note that the parameters are often not independent and a complete set must be used. Arbitrary adjustments of a few parameters may result in inconsistent results. One of the most important sets of tunings is concerned with structure of minimum bias events and spectator processes in a hard scattering event: the underlying event. At present, these tunings are obtained for both PYTHIA and HERWIG [42] by first tuning to the Tevatron data and then extrapolating. The extrapolation from the Tevatron relies on the models used by PYTHIA and HERWIG. This extrapolation has had testing from comparisons of the Tevatron data at 630 and 1800 GeV [46, 66]. A high priority task for the ATLAS simulation as data accumulates is the testing of these tunings and changing of the parameters as needed.

4 ATLAS detector description

The ATLAS detector is described in detail in Ref. [1], but its main features will be summarized here. We discuss the geometry used in the simulation, which as much as possible matches the as-built detector. A cut-away view of the entire detector is shown in Fig. 2. ATLAS comprises several concentric components. The subdetectors are:

- A Beam Conditions Monitor (BCM) and Beam Loss Monitor used for detecting dangerous conditions and triggering an abort in the detector system. The BCM is located 1.84 m from the interaction point (IP) at $|\eta| \sim 4.2$.[5]
- A tracking detector composed of a fine granularity pixel detector with three layers covering $|\eta| < 2.5$, a silicon strip tracker (SCT) with eight layers determining four space points covering $|\eta| < 2.5$, and a transition radiation tracker (TRT) which has 32 space points on a typical track, covering $|\eta| < 2.0$.

---

[5]Pseudorapidity, $\eta \equiv -\ln\tan(\theta/2)$, where θ is the polar angle measured from the beam pipe. The other coordinate variables used are typically r, z and ϕ, where the x-axis points towards the center of the LHC ring, the y-axis points up, the z-axis defines a right-handed coordinate system, $r \equiv \sqrt{x^2 + y^2}$, and ϕ is the azimuthal angle defined such that $\phi = 0$ along the x-axis.

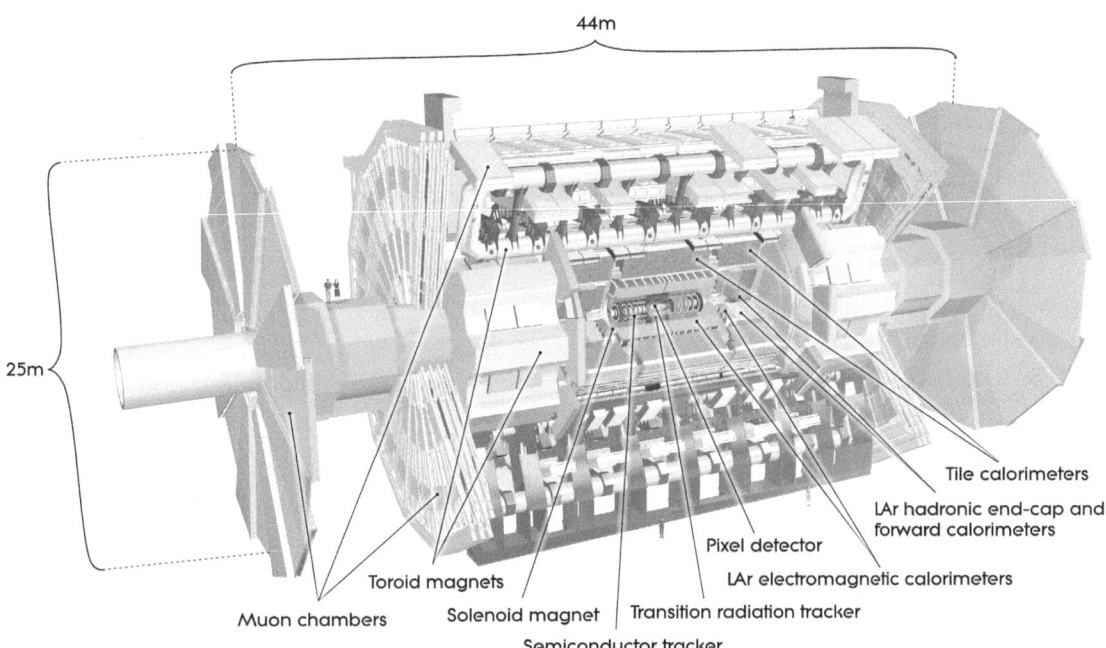

Fig. 2 ATLAS detector view

- Hermetic calorimetry composed of liquid argon (LAr) electromagnetic calorimetry covering $|\eta| < 3.2$, scintillating tile hadronic calorimetry in the barrel ($|\eta| < 1.7$), sampling LAr hadronic calorimetry in the endcap ($1.5 < |\eta| < 3.2$), and LAr electromagnetic and hadronic forward calorimetry covering $3.2 < |\eta| < 4.9$.
- Four different types of muon chambers, two of which are high precision (monitored drift tubes, and cathode strip chambers) and two of which have a rapid response for the muon trigger (thin gap chambers, and resistive plate chambers), covering $|\eta| < 2.7$.
- Luminosity detectors, including a zero-degree calorimeter that sits 140 m from the interaction point, a detector that performs a luminosity measurement using Cherenkov integration (LUCID), and an absolute luminosity detector for ATLAS.

The ATLAS magnetic field is formed by a solenoid, providing a 2.0 T uniform magnetic field in the tracking subdetectors, and a toroidal magnet system, composed of a barrel and two endcap toroid magnets. In the inner detector, the field has small ϕ- and z-asymmetries due to the toroid field and perturbations from the iron nearby. The field in the toroidal system has approximate z- and eight-fold ϕ-symmetry and provides on average 2.5 Tm of bending power in the barrel and 5 Tm in the endcap. During a simulation run, the field map required about 30 MB of memory.

In the standard production simulation (see Sect. 2.2) the luminosity detectors are not included. They can be simulated in dedicated jobs, but keeping particles in such a high

pseudorapidity region increases simulation time by approximately 50% per unit of pseudorapidity ($|\eta|$) per event (see Sect. 5.1).

Several layouts of the complete detector are available, including those that were used for recording cosmic ray events while the detector was being completed. Test stands are also supported with the same infrastructure. All these layouts are described in Sect. 4.4. As much as possible, the details of the detector geometry are preserved in the simulation layout. Some approximations are necessary for describing dead materials, for example bundles of cables and cooling pipes in the service areas of the detector. In these cases, the description only aims to match the general distribution of the material, including inhomogeneities in ϕ.

4.1 Simulated detector geometry

The geometry structure can be viewed in terms of solids, basic shapes without a position in the detector; logical volumes, solids with ad ditional properties (e.g. name or material); and physical volumes, individual placements of logical volumes. Table 3 shows the number of materials, solids, logical volumes, physical volumes, and total volumes created when constructing various pieces of the ATLAS detector. Not all volumes are equivalent, however: in the case of repeating structures, as in the sampling portion of the LAr calorimetry in particular, it is possible to define a single logical volume that is repeated in hundreds of physical volumes (known as volume parameterization). Because of nesting, one can also define dependencies that create many total volumes

Table 3 Numbers of materials, solids, logical volumes, physical volumes, and total volumes required to construct various pieces of the ATLAS detector. "Inner Detector" here includes the beampipe, BCM, pixel tracker, SCT, and TRT

Subsystem	Materials	Solids	Logical vol.	Physical vol.	Total vol.
Beampipe	43	195	152	514	514
BCM	40	131	91	453	453
Pixel	121	7,290	8,133	8,825	16,158
SCT	130	1,297	9,403	44,156	52,414
TRT	68	300	357	4,034	1,756,219
LAr calorimetry	68	674	639	106,519	506,484
Tile calorimetry	8	51,694	35,227	75,745	1,050,977
Inner detector	243	12,501	18,440	56,838	1,824,614
Calorimetry	73	52,366	35,864	182,262	1,557,459
Muon system	22	33,594	9,467	76,945	1,424,768
ATLAS total	327	98,459	63,769	316,043	4,806,839

Table 4 Numbers of physical volumes and memory required to build various pieces of the ATLAS detector in GeoModel. Here "calorimetry" is simply the sum of the liquid argon and tile calorimetry

Subsystem	Phys. volumes	Memory [kB]
Inner detector	56,838	22,268
Calorimetry	182,262	44,116
Muon system	76,945	31,524
ATLAS total	316,043	97,908

from the physical volumes used. In other cases, a single volume can correspond to a piece of shielding or support with a complex shape. One can see in this table the complexity of the ATLAS detector, with hundreds of materials and hundreds of thousands of physical volumes. Such a detailed detector description is crucial for accurately modeling, for example, missing transverse energy, track reconstruction efficiencies, and calorimeter response.

Table 4 shows the number of physical volumes contained in each detector subsystem and the memory required to build each using the GeoModel library (see Sect. 4.3) [67]. As expected, the two are correlated, although differences in volume complexity invalidate a direct correspondence. The entire geometry must be translated into a GEANT4 equivalent, so the total memory required for the geometry of the entire ATLAS detector is almost 300 MB (see Sect. 8).

In creating such a complex, dense geometry, removing volume overlaps and touching surfaces provides a particular challenge. Any overlap of more than 1 picometer and any place in which two volume faces touch can lead to stuck tracks during the simulation, a situation in which a track in GEANT4 may not know in which volume it belongs. These stuck tracks result in a loss of the event, but they can be overcome by introducing small gaps between volumes, at the cost of an extra step for each particle moving through the transition region.

Many layouts are available corresponding to the various revisions of material. The material budget is constantly updated, so that the geometry description is as realistic as possible. During any major updates of detector geometry, the subdetectors are generally required to make all changes backwards-compatible so that all older geometries can be configured and run as normal. This requirement allows for a fair comparisons between software releases with consistent geometries. During any job, the user may choose to enable or disable portions of the detector. Each subdetector is responsible for including any necessary materials and elements for its own construction. In this way, only required elements and materials are used during simulation, and memory loads are reduced when not using the entire ATLAS detector. The switches for disabling portions of the detector generally correspond to the highest level of the tree-structure in the detector geometry (i.e. entire subdetectors, not pieces).

It is possible to apply detector "conditions" modifications to each chosen geometry layout. The detector misalignment can be configured by selecting misaligned layouts either for each subdetector or for the full detector at once. Each ATLAS subdetector sits within a well defined envelope, allowing each to shift and distort without colliding with any other. One set of misalignments is provided specifically for the detector simulation, which includes appropriate levels of misalignment without any overlapping volumes. One may alternately use a misalignment directly derived from the data, but these misalignments may include overlapping volumes, which can cause missed hits. The misalignments are generally applied to more than simply the active volume. That is, if a silicon module in the inner detector is moved, any volumes associated with it (e.g. physically attached electronics) are moved as well.

In digitization and reconstruction jobs, conditions may include detector information beyond misalignments (e.g. dead channels). The infrastructure is in place to record detector conditions in a database and, at run time, allow the user to select conditions from a specific data taking run. Conditions and geometry versions selected by the user can be transferred from the simulation jobs to the digitization and reconstruction jobs so that no additional user interaction is required. These default versions may at any time be overridden by job options.

In order to study the penalties of a poor material description on jet resolution and missing transverse energy bias, a special geometry layout with material distortions was created [68]. Material distortions correspond to additional material added to half of the detector ($y > 0$) to approximate a poor material description.

4.2 Databases and configuration

Two databases are used to construct the detector geometry chosen by the user: one to store basic constants (the ATLAS Geometry database), and one to store various conditions data (e.g. calibrations, dead channel, misalignments) for the specific run chosen (ATLAS Conditions database) [69]. At CERN, large (terabytes) Oracle data bases are used, primarily because they are well supported and straightforward to update. With any stable software release, a small subset of data needed for Athena jobs is replicated from Oracle into SQLite [70–72]—file based databases—and is distributed to the production centers. The large I/O requirements of production jobs can overwhelm a central Oracle server and are better handled by relatively small SQLite files. These files can also be replicated to individual production nodes for local and rapid access. The database replica version to be used can be chosen at run time for each Grid job.

Both the geometry and conditions databases support versioning of the data. The data are organized in a tree consisting of branch and leaf nodes. The nodes in this tree can be "tagged," and one can create a hierarchy of the tags. Such tag hierarchies are uniquely identified by the tag of the root node, which is usually referred to as top level geometry or conditions tag.

A geometry database stores all fundamental constants for detector construction. Volume dimensions, rotations, and positions, as well as element and material properties including density, are all stored as database entries. New detector-specific tags may be created for inclusion in a global ATLAS geometry tag, where different tags generally correspond to different detector geometry revisions. At run time, the user can select a global geometry tag as well as detector-specific geometry tags to create the desired geometry. In addition to constants for detector construction, the geometry database contains links to external data files that may store, for example, magnetic field maps. These files are shipped with software distribution kits to production sites. By using links through the database, it is possible to select a magnetic field map based on the chosen geometry layout. The selection of

field map based upon the name provided in the database, for example, can be overridden with job options.

A separate conditions database stores detector conditions data which are indexed by intervals of validity and tags. The entire detector may be optionally misaligned with a global misalignment tag, and the user may configure the job to use specific misalignment versions for each subdetector. The global misalignment is used frequently to study the performance of the entire ATLAS detector with misalignments of the expected as-built magnitude. The detector-specific misalignments allow studies of the effects of misalignment of a single subdetector assuming ideal alignment of the remainder of ATLAS. The inner detector, for example, has completed an alignment challenge, wherein simulated data was produced with misalignments, and the analysis group was challenged to align the detector as with data. The tile calorimeter, on the other hand, does not use any misalignments in its geometry. A variety of misalignments have been used in the lead-up to data taking in order to speed the process of global detector alignment and improve early physics studies.

During data collection, the alignment constants of the detector are recorded periodically in the central conditions database. The user is able to recreate the misalignment conditions for a specific run by selecting an alignment version, again by subdetector if desired, at run time.

4.3 GeoModel and translations

The ATLAS simulation, digitization, and reconstruction each run in distinct jobs, but they must be able to use the same detector geometry. Therefore, a complete geometry description is maintained that can be used by each step and is not specific to any. By using the geometry databases, it is already possible to read identical detector constants and run conditions.

For these reasons, ATLAS uses GeoModel [67], a library of basic geometrical shapes, to describe and construct the detector. GeoModel contains geometry features similar to those of GEANT4: basic volumes can be constructed, rotated, and shifted in space; subvolumes can be placed inside a volume; boolean volumes can be made by adding or subtracting primitives; volumes can be parameterized and repeated. For the digitization and reconstruction, this detector description is entirely sufficient to place hits, reconstruct tracks and objects, and complete all necessary calculations.

The GeoModel descriptions of most ATLAS subsystems are built using constants in the geometry database. However, a translator has been constructed that parses an XML description of a detector's geometry and builds a transient representation from GeoModel primitives at run time. This generic package can translate any valid XML description of detector geometry into GeoModel format. It has been used for describing the geometry of the muon system's rather complicated dead material.

For the simulation, the geometry is translated entirely from the GeoModel to the GEANT4 format. All volumes and subvolumes are translated, constructed, and properly placed within the "world volume" (the volume allocated for the detector, at the edge of which particles cease to be simulated). All information tied to GeoModel, including position, rotation, and dimensions, are also translated into a GEANT4 equivalent. Once the geometry has been translated, all subsystems rely solely on their GEANT4 description. The GeoModel geometry is currently maintained in memory for the entirety of the job, though it may be released to ease memory pressure in the future. As shown in Table 4, this release can be expected to save 100 MB of memory. Sensitive detectors and particle range cuts (see Sect. 5.5), for example, are tied to the GEANT4 geometry by volume name and can be added at any time after the geometry has been constructed. Each change in detector description is particularly weighty in simulation, because any additional volumes must be built both in GeoModel and in the GEANT4 geometry.

4.4 Alternate layouts

In addition to the standard detector layouts, several commissioning layouts are available to the user for simulation of cosmic ray data taking. During detector assembly, cosmic ray data were taken for several runs using as many subdetectors as were available. Some of these subdetector configurations included calorimeter endcaps shifted out of position while the inner detector was being accessed and were missing large portions of the beam pipe that had not yet been installed. One such commissioning layout is shown in Fig. 3. For studies of cosmic rays and cavern background, it is possible to simulate the ATLAS cavern surrounding the detector as well as the bedrock surrounding the two shafts leading down from the surface.

Several different magnetic field configurations are also available for some of the full detector layouts. Fields with the toroidal magnets on and solenoid off or solenoid on and toroidal magnets off are provided. These magnet configurations have already been used for some cosmic ray data taking runs and may be used for brief periods during high-energy collisions as well. Field maps have also been constructed that reflect the as-built misalignments of the mag-

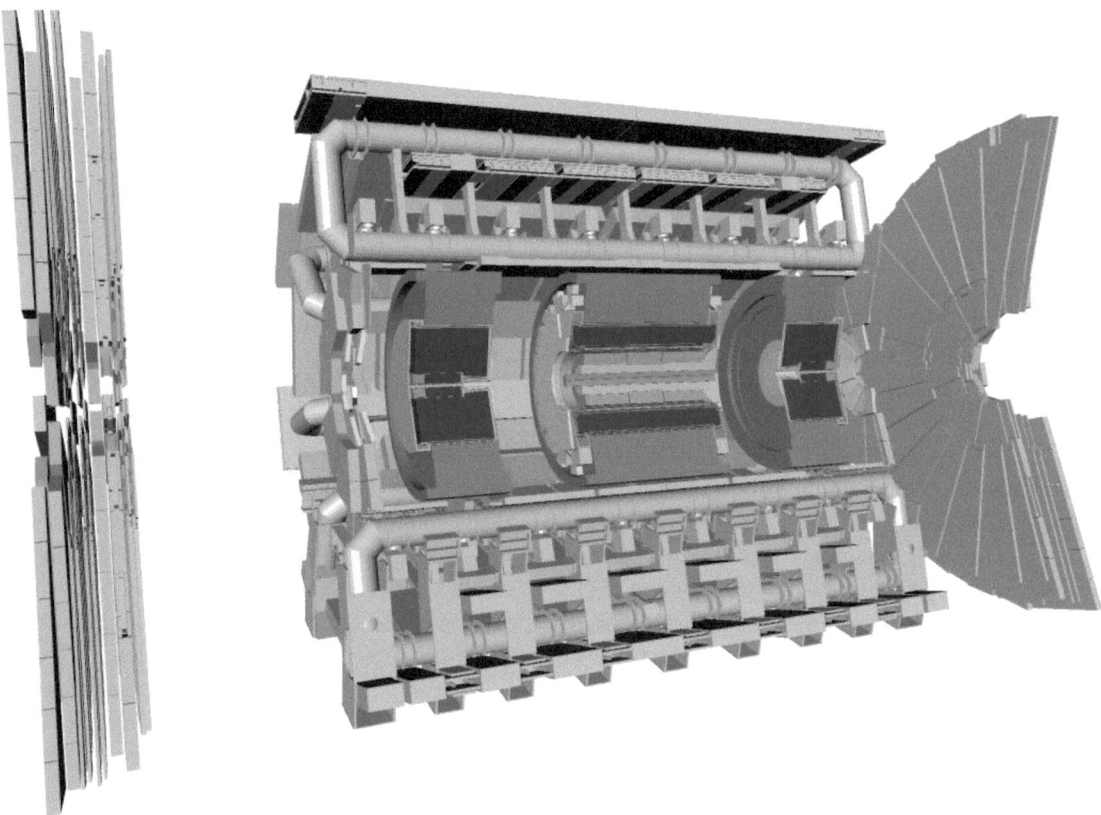

Fig. 3 Commissioning layout of the detector used for cosmic ray data taking during 2008. The endcap toroidal magnets and beampipe are not yet installed. The calorimeter endcaps (*purple*) are shifted by 3.1 m and the muon endcaps (*green*) are shifted to provide access to the inner detector during installation. The barrel toroid magnets are shown in *yellow*, and the inner detector is shown in *blue*

Table 5 Examples of test stands for ATLAS simulated using GEANT4

Subsystem	Incident particle	Energy
Hadronic Endcap calorimeter	$e^{+/-}, \pi^{+/-}, \mu^{+/-}$	6–245 GeV
Electromagnetic Barrel calorimeter	$e^{+/-}$	10–245 GeV
Electromagnetic Endcap calorimeter	$e^{+/-}$	10–200 GeV
Combined Endcap calorimeter	$e^{+/-}, \pi^{+/-}, \mu^{+/-}$	6–200 GeV
Hadronic Barrel calorimeter	$\pi^{+/-}, p$	5–350 GeV
Entire detector endcap wedge	$e^{+/-}, \pi^{+/-}, \mu^{+/-}$	1–350 GeV
Muon Detectors	$\mu^{+/-}$	20–350 GeV
Silicon Pixel Tracker Endcap	Cosmic Rays	0.5–200 GeV
Silicon Strip Tracker Barrel	Cosmic Rays	0.5–200 GeV

net system, for example a vertical shift of a 1.6 mm in the solenoid.

There are also several test stand layouts that were constructed to model test beam and standalone cosmic ray runs. A sample of these various test stands, including subsystems, incident particles, and energies, are listed in Table 5. A combined test beam run was taken with a wedge of the full detector during 2004 [73], and standalone test beams were constructed for the muon detectors [74, 75], tile calorimeter [76], and liquid argon calorimeter subsystems [77, 78]. The combined test beam setup is shown in Fig. 4. Cosmic ray data were also collected with various pieces of the inner detector [79, 80] and with the muon chambers both prior to and after installation.

All test stand and commissioning layouts are available as a part of the same geometry infrastructure and can be selected at run time for simulation. By maintaining all detector configurations as a part of a common infrastructure, it is possible to ensure consistency between, for example, the test beam and full detector simulation. Conclusions drawn from analysis of the test beam simulated data are generally still valid for the full detector simulation. The extensive tuning of the detector simulation and digitization on test beam data can be applied directly to the full detector. As many common elements as possible are kept between the two, including GEANT4 version and physics list (see Sect. 5).

5 Core simulation

The standard simulation of ATLAS relies on the GEANT4 particle simulation toolkit. GEANT4 provides models for physics and infrastructure for particle transportation through a geometry, but several ATLAS-specific pieces are provided as user-code. The detector geometry itself is constructed in the GEANT4 format, and all particle scoring (done in "sensitive detector" classes) are done on the Athena side.

Each subsystem's scoring is optimized and tailored to store only what is necessary for accurately reproducing the performance of that particular subsystem [81–88]. Athena code is necessary to add to the Monte Carlo truth record. Physics models are chosen and parameters optimized for the ATLAS detector. The results shown in this paper used GEANT4 version 8.3 with official patch #2 and two modifications: updates for boundary represented volumes and a patch to the G4Tubs code. The software is continuously evolving, and ATLAS has moved to newer Geant4 versions since the writing of this paper.

The GEANT4 Collaboration and ATLAS Simulation Group have benefitted from 15 years of close collaboration. Frequently, new GEANT4 features have allowed faster or more realistic simulation of the ATLAS detector. Feature requests from the ATLAS collaboration have helped drive the development of GEANT4. The ATLAS simulation has also provided one of the more complicated test-beds for the GEANT4 toolkit, and GEANT4 has been extensively evaluated and validated during large-scale simulation production.

In order to provide PYTHON flexibility to the GEANT4 simulation, an additional layer of infrastructure is necessary. "Standard" GEANT4 simulation typically runs from compiled C++, and in order to modify any of the parameters or the geometry used in the simulation it is necessary to recompile. The Framework for ATLAS Detector Simulation (FADS) [89] wraps several GEANT4 classes in order to allow selection and configuration without recompilation of any libraries. Since a PYTHON interface is used for configuration, all the usual introspection capabilities of PYTHON may be employed. FADS wraps GEANT4 base-classes for volumes, materials, and sensitive detectors for hit processing as well as GEANT4 physics process definitions. These wrappers serve a dual purpose: first, they ease translations between the GEANT4 and Athena standards of geometry, hits, and particle storage. Second, FADS can catalogue the options available to the user, loading only those that will be needed for the desired simulation configuration while still providing all possibilities without any recompilation. Through FADS, a user is free to select a physics list (see Sect. 5.4) for use during the simulation. The user may also modify the physics list by adding particles or processes not included in the GEANT4 toolkit but included in the FADS catalogues, for instance in the simulation of long-lived exotic particles. Similarly, the detector description is configured with PYTHON dictionaries and FADS catalogues before it is built in GEANT4 and may be modified by the user. For example, sensitive detectors may be assigned to any volume in the detector. Range cuts (see Sect. 5.5) may also be added in the PYTHON and FADS layer prior to their being applied to any constructed GEANT4 geometry. Once the PYTHON configuration is complete, FADS objects are translated into their GEANT4 equivalents and loaded. Even after

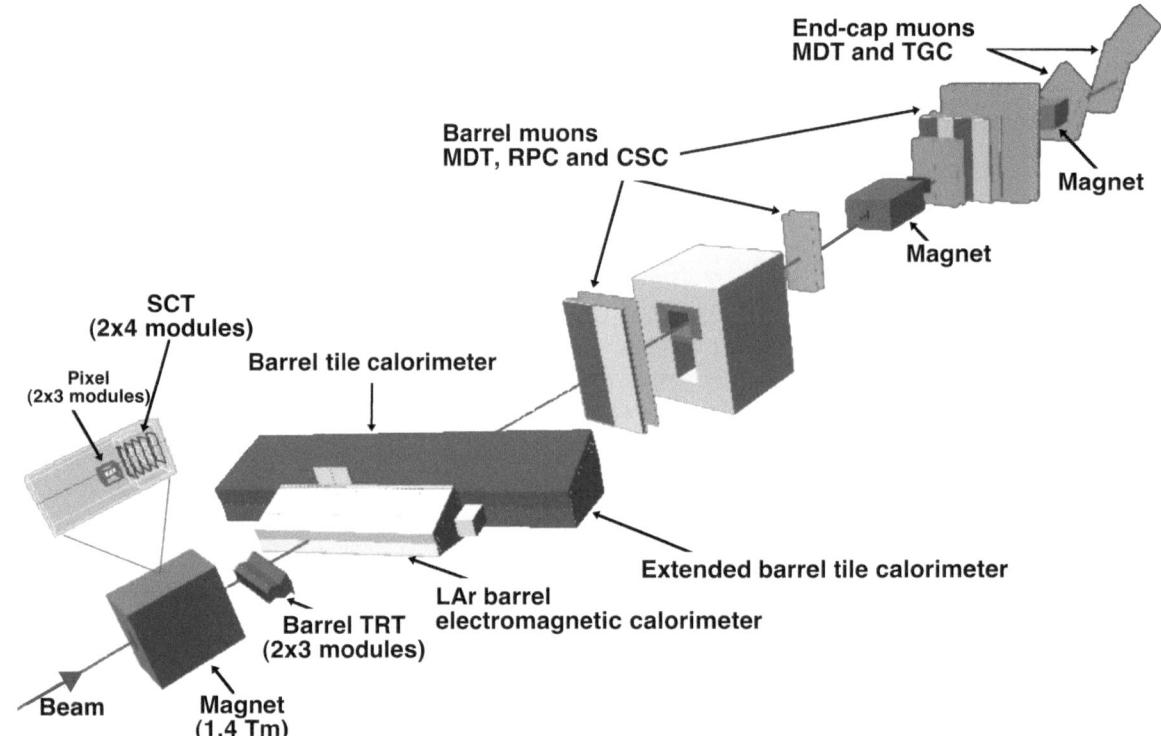

Fig. 4 Combined test beam setup from 2004

this translation, they can be modified through the standard GEANT4 user interface.

In order to fit into the Athena framework, a service for GEANT4 and an algorithm that calls the service during the event loop have been implemented [90]. The service wraps the event loop of GEANT4 and provides a few additional handles for user-configuration in the PYTHON layer. The service also takes care of initialization and finalization of each GEANT4 event. The generated events are translated from HepMC format into the standard GEANT4 event format prior to each event, and at the end of each event an analysis is done to ensure that the simulation finished without errors. Most of the functionality of the standard GEANT4 run manager is included in this service, so that any Athena-specific modifications (e.g. event translation from HepMC to GEANT4 format) to the usual GEANT4 event sequence can be made. The service also provides for interaction with GEANT4 through its standard user interface.

This section describes the possible user inputs, initialization, output, and various parameters of the simulation. Several useful features, including visualization, are also described.

5.1 Simulation input

The ATLAS simulation offers a choice for event generation. Events can be read from a file produced by any of the generators described in Sect. 3; one of the external generators can

be configured and run concurrently; or commands can be provided for a single particle generator. The single particle generator can produce particles by the particle PDG identifier [91] with a configurable position and momentum. Neutral and charged geantinos (pseudo-particles without any interactions) are available for making material depth maps of the detectors and for debugging. The user may also choose to skip a certain number of events at the beginning of an input file, allowing 20 simulation jobs of 50 events each to a 1000 event input file without overlap or repetition.

Several cuts and transformations can then be made to the event. The vertex position is smeared to represent the luminous region within ATLAS.[6] It can be shifted if the user desires, but both the shift and smear are given initial default values that represent ideal collisions within the ATLAS detector. The generated event can be rotated in any direction, though only rotation in ϕ is physical. Primary particles are only passed through the detector simulation if they are within a specified range in η–ϕ. By default, primary particles with $|\eta| \geq 6.0$ are not simulated to save time. For example, typically beam-remnants are not simulated. This cut was chosen to ensure consistent response in the forward calorimeter: a cut at $|\eta| = 6.0$ allows a sufficient number of particles to scatter back into the forward detector from high-

---

[6]During early data taking, the beams will collide head-on. Therefore, no crossing angle is added to the simulated events for the time being.

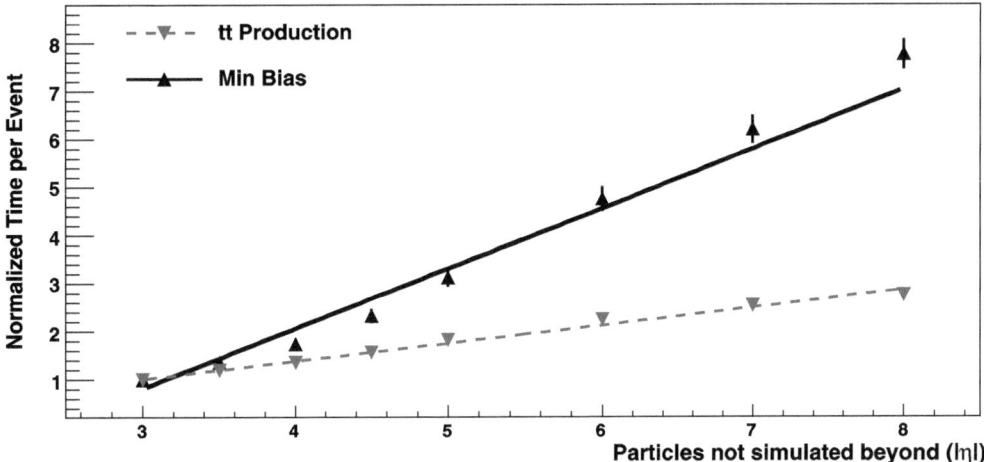

Fig. 5 CPU time per event as a function of the acceptance in $|\eta|$ of primary (generator) particles. If only central response is important, the CPU time required to simulate a single event is significantly different than if forward detectors must be included in the simulation. These two cases approximately represent the extremes of the plot. The time is nor-malized to CPU time for simulation of all primaries inside of $|\eta| < 3.0$ and increases between three-fold and eight-fold for simulation of all primaries in $|\eta| < 8.0$. The average of 200 simulated $t\bar{t}$ and minimum bias events was taken. Linear fits are overlaid

η without requiring an unacceptable amount of CPU time. Generally, an increase in one unit of pseudorapidity corresponds to an increase of 40–120% in CPU time, so that it is not possible to simulate the very forward detectors like LU-CID during a standard simulation job. Figure 5 shows the η-dependence of the CPU time per event. The increase is approximately three-fold in $t\bar{t}$ events and eight-fold in minimum bias events from simulating particles in $|\eta| < 3.0$ to simulating particles in $|\eta| < 8.0$. The difference between the two types of events is primarily because the majority of activity in the minimum bias events is forward, and there is considerable central ($\eta < 3.0$) activity in the $t\bar{t}$ events.

At run time, either through job options or in the production system described in Sect. 2.2, seeds for pseudo-random number generators to be used by GEANT4, Athena, and any particle generators can be set. Different pseudo-random number generators may be configured for each. Since all random number seeds can be controlled, a single job is entirely reproducible. The seeds can also be written to a file or read from a file, providing an additional level of reproducibility.[7]

The user must also select a layout for the detector. As described in Sect. 4, several layouts of the full detector and various test stands are available. Combined test beam simulation requires such a different configuration that an independent but similar core drives the PYTHON configuration and loading of user job options. The distinction is made at run time based on the detector or test stand configuration

selected. The layout of the detector determines what other options are available to the user at run time. During simulation of the entire ATLAS detector, several addition al options are available. For example, a neutron time cut (see Sect. 5.5) may be enabled. The magnetic field may be enabled and the field map may be selected.

The user may optionally select a set of run conditions for the simulation job, through which all options and flags are set and a pre-defined job is run. This option is particularly useful for testing and debugging.

5.2 Simulation initialization

Although the initialization in a standard Athena job occurs in a single step, for an ATLAS simulation job the initialization is broken into three steps to allow additional user intervention. Table 6 summarizes the processes that occur at each one of the three steps. The division of the initialization is such that most modifications to the simulation conditions can be accomplished in job options alone (i.e. without code modification). Normally, the user provides job options and allows the initialization to progress unhindered. Some parts of the job, for example the detector layout, are only loaded after the initialization has begun. In order to modify volumes after the layout is loaded, the user must intervene during the initialization. Only certain commands are effective at each stage of the initialization, since some parts of the GEANT4 simulation have been loaded and created while others have yet to be translated from dictionaries.

Stage one of the initialization occurs as soon as Athena is started. Several external PYTHON modules are loaded that provide basic functionality for any Athena job. The job properties provided by the user are read during this phase and

---

[7]Because of caching in GEANT4, it is not possible to reproduce an individual event or track without starting from the beginning of the job.

Table 6 Initialization sequence for the ATLAS simulation. Dividing the necessary configuration into several distinct steps allows user intervention at critical points

Init. stage	Processes
1	External modules loaded, job properties locked, metadata written, event generation configured, hit file initialized, GEANT4 service created
2	Detector, physics regions, range cuts created, GeoModel geometry translated, truth strategies initialized, magnetic field loaded, physics list selected, user actions initialized
3	Fast simulation models assigned, physics regions constructed, sensitive detectors assigned, GEANT4 run manager and physics models initialized, recording envelopes and visualization initialized

are locked. Once the job properties are locked, any significant modification to the running of a simulation job must be done by directly accessing the affected services and algorithms. This saves propagation of changes in the case of a late modification to a job property. Metadata that will be stored with the hit output file are gathered. External dependencies that require early initialization are loaded, providing a service for GeoModel, a service for database interaction, and a service for frozen showers (see Sect. 7.1). The event generation mode (reading external events, generating events from an external generator, or generating single particles on-the-fly) is determined, and any necessary configuration is included for the generator. A stream is opened for the output hit file, if necessary, and hit containers for each enabled subdetector are added to the new file. Finally, a service is created to interface with and control GEANT4, although at this point GEANT4 is not fully initialized.

Stage two of the initialization begins with the construction of the detector in PYTHON dictionaries. Dictionaries of physics regions, range cuts, and volumes in which to apply step limitation (see Sect. 5.5) are constructed, and all key properties are assigned to the detector facilities or built in dictionaries for later addition to the geometry. Each enabled piece of the ATLAS detector or test setup is then recursively constructed in GeoModel according to the parameters specified in the geometry and conditions databases. After each subdetector has been constructed in GeoModel, it is translated recursively into an equivalent GEANT4 geometry. After this point in the initialization all volumes and regions are available to the user for modification. Next, the Monte Carlo truth strategies (see Sect. 5.3) are added to the simulation. The magnetic field is then loaded. Under normal circumstances, the field is a map loaded from an external data file, the name of which is specified in the geometry database according to the geometry layout selected. The user may optionally choose to load data from one of the magnetic field test configurations, rather than the standard AT-LAS magnetic field, or to create a new, basic magnetic field. The physics list (see Sect. 5.4) to be used for simulation is also set at this point.

User actions are then initialized. GEANT4 allows a user to insert pieces of code in various places throughout the simulation event loop, including after each step, when each track is queued ("stacked"), before and after each event, before and after each track is simulated, and before and after each run.[8] By default, ATLAS includes user actions that monitor simulation time, memory, and the number of tracks generated during each event. A neutrino cut (see Sect. 5.5) is also implemented as a track-stacking action. Whenever a new track is queued, its type is checked. The LAr calorimeters also use end-of-event actions for merging hits to save space prior to storage. All truth storage strategies (see Sect. 5.3) are implemented as stepping actions that store interesting interactions based on the type of process, detector region, and energies of the particles involved. Users may also configure their own actions and add them to the simulation in the same way. Examples have been constructed for integrating interaction lengths or radiation lengths through the detector when making geantino maps, for stopping or killing particles if certain conditions are met, and for turning on additional output only under specific conditions in order to study a bug or issue without having to sift through enormous log files.

Stage three of the initialization completes the job preparation. During this stage, the fast simulation models are built and added to the volumes to which they have been assigned. Any physics regions that will be used are constructed. Sensitive detectors are built and assigned to the regions of the detector that are to be made sensitive (i.e. in which hits will be stored). GEANT4's run manager and physics models are initialized. Recording envelopes are added (see Sect. 5.3), and any visualization that has been enabled by the user is initialized (see Sect. 5.7).

Once the initialization is complete and all the necessary elements have been loaded into memory, the event loop begins.

5.3 Monte Carlo truth information

The GEANT4 simulation adds to the Monte Carlo truth record already defined during generation (see Sect. 3.6).

---

[8]Here, run is used in the GEANT4 sense to refer to a finite set of events within a simulation job. Several runs may comprise a job, and each run may include an arbitrary number of events.

Far too many secondary tracks are produced during detector simulation to store information for every interaction. Only those interactions which are of greatest relevance to physics analyses are saved, according to several saving rules ("strategies"). Most are applicable only to the inner detector. For each interaction that satisfies any of the storage criteria, the incoming particle, step information, vertex, and outgoing particles are included in the truth record. Later in the software chain, individual track segments are recombined so that, for example, a single electron that undergoes several bremsstrahlung events along its path is counted as only one "true" particle.

The strategies include (with all cuts on kinetic energy)[9]:

– In the inner detector, bremsstrahlung vertices are stored if the primary electron or muon has an energy above 500 MeV and the photon produced has an energy above 100 MeV.
– In the inner detector, ionization vertices are stored if the primary particle has an energy above 500 MeV and the electron generated has an energy above 100 MeV.
– In the inner detector, hadronic interaction vertices are stored if the primary particle has an energy above 500 MeV.
– In the inner detector, decay vertices are stored if the decaying particle has an energy above 500 MeV.
– In the inner detector, the conversions of photons above 500 MeV are stored.
– In the calorimeter, muon bremsstrahlung vertices are stored if the primary muon's energy was above 1 GeV and the photon generated is above 500 MeV.

All cuts and regions of applicability are made configurable, so that any energy cut-offs can be modified and a strategy can be assigned to any volume in the simulation. additional rules could be constructed, for example, for tracking shower development within the calorimeter, but many would consume too much CPU time and disk space for use in standard simulation jobs.

Standard simulation jobs also define several volumes that are used to record all particles escaping part of the detector. All tracks above 1 GeV are typically recorded at the end of the inner detector, the end of the calorimeter, and the end of the muon system (and the end of the ATLAS world volume). It is possible for the user to configure the simulation at run time to add additional volumes to the list of these recording volume. In each case, tracks are saved as they exit each volume.

---

[9]For the most recent production, cuts are applied on transverse momentum, $p_T > 100$ MeV, rather than on kinetic energy. The lower cut allows for a study of tracking performance in minimum bias events where it may be possible to reconstruct tracks down to only a few hundred MeV.

5.4 Physics list

Physics lists include all numerical models that describe the particles' interactions in the GEANT4 simulation. Models are generally good for a single type of interaction and over a limited energy range. The GEANT4 collaboration provides several combinations of these models that have been tailored to various scenarios as standard physics lists that ship with each distribution. In order to enhance reproducibility and ensure that validated combinations of models are used, only those physics lists provided by the GEANT4 collaboration are used by the ATLAS simulation. One exception is allowed, namely transition radiation. Transition radiation is crucial for the tracking portion of the inner detector and is added to each physics list.

There are several physics lists that are used by ATLAS:

QGSP_BERT—the physics list used for all simulation production after 2008. The list includes the Quark–Gluon String Precompound model (QGSP) and the Bertini intranuclear cascade model (BERT) [4] as part of the hadronic physics package. The electromagnetic physics package includes step-limiting Multiple Coulomb Scattering (MSC).

QGSP_EMV—the physics list used for simulation production before 2008. This list included the QGSP model, but without the Bertini cascade. The MSC of this list was not allowed to limit the step, so it is labeled an electromagnetic variant (EMV).

QGSP_BERT_HP—the physics list used for neutron fluence studies and comparisons with the FLUKA simulation package [92, 93]. This list includes the QGSP and Bertini models, step-limiting MSC, and additional "high-precision" low-energy neutron physics models.

A step limitation process that controls the maximum allowed step length of a charged particle was added in the inner detector when using the QGSP_EMV physics list. It helped the simulation to better reproduced test beam and cosmic ray data. The step-limiting MSC that is a part of QGSP_BERT was found to agree equally well with data, and therefore the step limitation was removed from simulation with QGSP_BERT.

These physics lists were studied in detail for each subdetector [94]. Table 7 shows the number of steps, number of hits in sensitive detector regions, and number of secondary particles with kinetic energy above 50 MeV and 1 GeV within several regions of the detector and for the whole of ATLAS using both the QGSP_EMV and QGSP_BERT physics lists. Sensitive detectors to record calibration hits (described in Sect. 5.6) are included in the cryostat of the LAr calorimeter. The average was taken of 50 $t\bar{t}$ events, where there were on average 482 primary (generator-level) particles per event. The calorimeter clearly

Table 7 Number of steps, number of hits in sensitive detector (SD) regions, and number of secondary particles with kinetic energy above 50 MeVand 1 GeVwithin several regions of the detector and for the whole of ATLAS, using both the QGSP_EMV and QGSP_BERT physics lists. The average was taken of 50 $t\bar{t}$ events, where the average number of primary tracks per event was 482. Sensitive detectors to record calibration hits are included in the cryostat of the LAr calorimeter

QGSP_EMV	Steps	Hits in SD	Sec. above 50 MeV	Sec. above 1 GeV
Inner detector	1.80×10^6	3.10×10^5	1,570	260
Calorimetry	1.87×10^7	6.87×10^6	39,900	2,040
Muon system	1.90×10^6	1,030	7,820	332
Total ATLAS	2.24×10^7	7.18×10^6	49,300	2,630
QGSP_BERT	Steps	Hits in SD	Sec. above 50 MeV	Sec. above 1 GeV
Inner detector	2.13×10^6	1.98×10^5	1,450	269
Calorimetry	3.93×10^7	1.36×10^7	40,100	2,170
Muon system	2.69×10^6	1,285	8,210	385
Total ATLAS	4.41×10^7	1.38×10^7	49,700	2,820

Table 8 Number of steps for various processes and detector regions during simulation with the QGSP_EMV physics list. The average was taken of 50 $t\bar{t}$ events, where the average number of primary tracks per event was 482. The "other processes" in the inner detector are primarily step limitation processes

Process	Inner detector	Calorimeter	Muon system
Transportation	1.50×10^6	9.33×10^6	2.15×10^5
MSC	4,910	1.09×10^5	5,200
Photoelectric effect	6,060	1.32×10^6	2.03×10^5
Compton scattering	12,800	1.43×10^6	4.26×10^5
Ionization	1.08×10^5	4.97×10^6	8.10×10^5
Bremsstrahlung	6,310	1.28×10^6	1.87×10^5
Conversion	434	82,400	17,300
Annihilation	291	82,800	17,800
Decay	254	2,320	538
Other hadronic interaction	1,710	1.23×10^5	21,600
Other process	1.56×10^5	4,800	831
Total	1.80×10^6	1.87×10^7	1.90×10^6

dominates the total number of steps and hits in sensitive detector for both physics lists. The muon system, though it has a comparable number of hits, consists mostly of shielding and therefore has far fewer hits in sensitive detector regions. The numbers of steps divided into different process types for QGSP_EMV and QGSP_BERT are listed in Tables 8 and 9. In both cases, transportation processes dominate the inner detector simulation, while electromagnetic physics and transportation dominate the calorimeter and the muon system.

Simulation time was also examined for each physics list. Simulation using the QGSP_BERT physics list consumes ~2.5 times more CPU time than does simulation with the QGSP_EMV physics list. However, applying a neutron time cut (see Sect. 5.5) with the QGSP_BERT list reduces simulation time by more than 30%. Simulation with QGSP_BERT_HP requires approximately five times more CPU time than QGSP_EMV. Therefore, the QGSP_BERT_HP physics list cannot be used for standard simulation.

5.5 Simulation optimizations

In order to optimize use of both disk space and CPU time, several other modifications are made to the standard GEANT4 simulation [94, 95].

Comparing the QGSP_BERT physics list to the QGSP_EMV physics list, approximately three times as many neutrons are generated in typical hard scattering events, and they travel approximately three times further. These neutrons cause an increase in the output hit file size of approximately 75% as well as an increase in CPU time per event for hard scattering events. A GEANT4 neutron time cut is, therefore, applied which removes all neutrons

Table 9 Number of steps for various processes and detector regions during simulation with the QGSP_BERT physics list. The average was taken of 50 $t\bar{t}$ events, where the average number of primary tracks per event was 482. The "other processes" in the calorimeter and muon system are primarily neutron killer processes

Process	Inner detector	Calorimeter	Muon system
Transportation	1.76×10^6	1.46×10^7	2.31×10^5
MSC	2.31×10^5	1.48×10^7	5,200
Photoelectric effect	6,760	1.37×10^6	2.32×10^5
Compton scattering	14,800	1.66×10^6	5.03×10^5
Ionization	1.03×10^5	4.81×10^6	9.71×10^5
Bremsstrahlung	6,060	1.22×10^6	1.92×10^5
Conversion	416	86,800	18,100
Annihilation	271	87,000	18,500
Decay	212	1,670	402
Other hadronic interaction	2,190	6.66×10^5	1.23×10^5
Other process	426	25,400	5,720
Total	2.13×10^6	3.93×10^7	2.69×10^6

150 ns after the primary interaction. This cut was found to be sufficient time for the hadronic shower development and did not degrade the energy scale or energy resolution of the calorimeters. The total energy deposition of the shower is changed by ~1%. Output files are the same size when using the QGSP_BERT physics list with this cut enabled as they are when using the QGSP_EMV physics list without a neutron time cut. The simulation time required for QGSP_BERT is reduced by 10–15% when the neutron cut is enabled.

Neutrinos are also removed as soon as they are created in the simulation. No particle is allowed by GEANT4 to step through more than one volume at a time. Therefore, neutrinos may require several thousand steps to exit the entire ATLAS detector. They may therefore consume a noticeable fraction of simulation time, even though their interaction probability is practically null. The removal is done when the particles are stacked.

Range cuts are GEANT4 parameters that control the creation of secondary electrons or photons during bremsstrahlung and ionization processes. If the expected range of the secondary is less than some minimum value, the energy of that secondary particle is deposited at the end of the primary particle's step and no separate secondary is produced. Effectively, this parameter defines an energy scale at which particle propagation may be ignored. By increasing the range cuts throughout the detector one can decrease the CPU time required per event. Particularly near boundaries and thin materials, the detector's sampling fraction may be affected if the range cuts are too large. Range cuts can be specified separately for electrons, positrons, and photons, but in ATLAS the same distance is used for all three. Range cuts are specified as a distance, and for each material the distance is translated into an energy based on the average energy loss of a particle in that material. For the majority of the ATLAS detector, range cuts take a default value of

1 mm. Exceptions are listed in Table 10. Deviations usually occur in sensitive volumes that are very thin, where it is important to correctly calculate the sampling fraction of the detector or model the energy deposition. Reduced range cuts are also applied to very thin volumes that are adjacent to sensitive volumes for the same reason. In the monitored drift tube muon chambers, for example, range cuts are only reduced in the thin aluminum tubes surrounding the sensitive detector (gas)—the gas itself takes the standard 1 mm cuts. In some shielding volumes it may be possible to relax range cuts considerably without degrading physics performance.

GEANT4 uses a set of parameters to control errors and accumulated biases on charged particles transportation through a magnetic field. Because the equation of motion is solved numerically, the user must select the numerical integration method to be used, including the order of integration, and the tolerances on the errors of the step. ATLAS has chosen to use the GEANT4 standard fourth-order Runge–Kutta method with the default error parameters for the majority of the detector. These parameters are generally satisfactory and result in errors and biases that are less than the position resolution of the detector. In the inner detector, however, tracks were found to be shifted sufficiently that detector residuals were affected. Here, stepping parameters were tightened by an order of magnitude. Further optimization of the stepping algorithms of GEANT4 has been undertaken, including the configuration of the choice of stepper and stepping parameters as a function of the initial particle type, energy, and position within the detector. Such a configuration can allow more careful stepping of muons in the calorimetry without degrading the total performance of the simulation significantly. Muons in particular can accumulate a significant bias after passing through all the sampling layers of the calorimeter, making more accurate tracking necessary. As a fourth-order stepper requires four values of the magnetic

Table 10 Range cuts for detectors that do not take the ATLAS default of 1 mm

Subdetector	Range cut value
Silicon pixels and strips in the inner detector	0.05 mm
Gas in the transition radiation tracker	0.05 mm
Electromagnetic Barrel and Endcap calorimeters	0.1 mm
Forward calorimeter (all compartments)	0.03 mm
Aluminum tubing of monitored drift tube muon chambers	0.05 mm

Table 11 Hit collection size, in kB per event, by subdetector. The average was taken of 50 simulated $t\bar{t}$ events. Calorimeter calibration hits are hits in the dead material of the calorimeters stored for studying simulation-based calorimeter calibration schemes

Collection name	Size [kB/event]	Percentage of File
Silicon pixel tracker	82	4%
Silicon strip tracker	356	16%
Transition radiation tracker	921	46%
Electromagnetic Barrel calorimeter	89	4%
Electromagnetic Endcap calorimeter	104	5%
Hadronic Barrel calorimeter	29	1%
Hadronic Endcap calorimeter	22	1%
Forward calorimeter	42	2%
Calorimeter calibration hits	243	12%
Muon system (all collections)	3	<1%
Truth (all collections)	134	7%
Total	1987	100%

field to be calculated, optimization of magnetic field map access will also be key to improving the performance of the simulation's tracking.

5.6 Hit storage format

The output from the simulation is a hit file, containing some metadata describing the configuration of the simulation during the run, all requested truth information, and a collection of hits for each subdetector. The hits are records of energy deposition, with position and time, during the simulation. Each subdetector is responsible for implementing their own sensitive detector for the selection, processing, and recording of these hits. In most subsystems, including the inner detector and muon system, this consists simply of recording all hits that occur in sensitive regions of the detector for subsequent storage. Some additional manipulation is done at the end of each event to compress the output as much as possible; still, the files are typically 2 MB per event for hard scattering events (e.g. $t\bar{t}$ production).

The file size is large, mostly due to the inner detector, for which the majority of hits are independently stored. Merging hits there is difficult, since they tend to be isolated and cannot normally be merged across readout channels. These consume typically 60% of the disk space in a hit file (e.g. 65% of the hit file for $t\bar{t}$ events). In the calorimetry, there are

far too many hits created by electromagnetic and hadronic showers for the individual storage of a four-vector for each (see Sect. 5.4). Instead, hit merging occurs at the end of each event. By optimizing time binning, hits can be compressed to a large extent. About 10% of the hit file is consumed by optional "calibration hits" for the calorimeters, hits in dead material, stored to improve the detector calibration and missing energy calculation and to study simulation-based calorimeter calibration schemes. Under normal circumstances, the muon systems contribute a negligible portion of the hit file. The contributions by subdetector can be found in Table 11 for the average of 50 simulated $t\bar{t}$ events. Here and elsewhere in this paper, file sizes are without compression and are taken from ROOT [96]. In practice, compression reduces the actual disk space required for the files, but file-level metadata adds several hundred kilobytes.

By comparing Tables 7 and 11, one can understand these numbers in terms of hits in the sensitive detector region. Although the muon system is large, the majority of it is shielding. Therefore, it collects far fewer hits than the other subsystems and requires less disk space for the hit records. The calorimetry produces 95% of the hits in sensitive regions during simulation. Because of the compression applied prior to storage, the calorimetry comprises only 25% of the hit file.

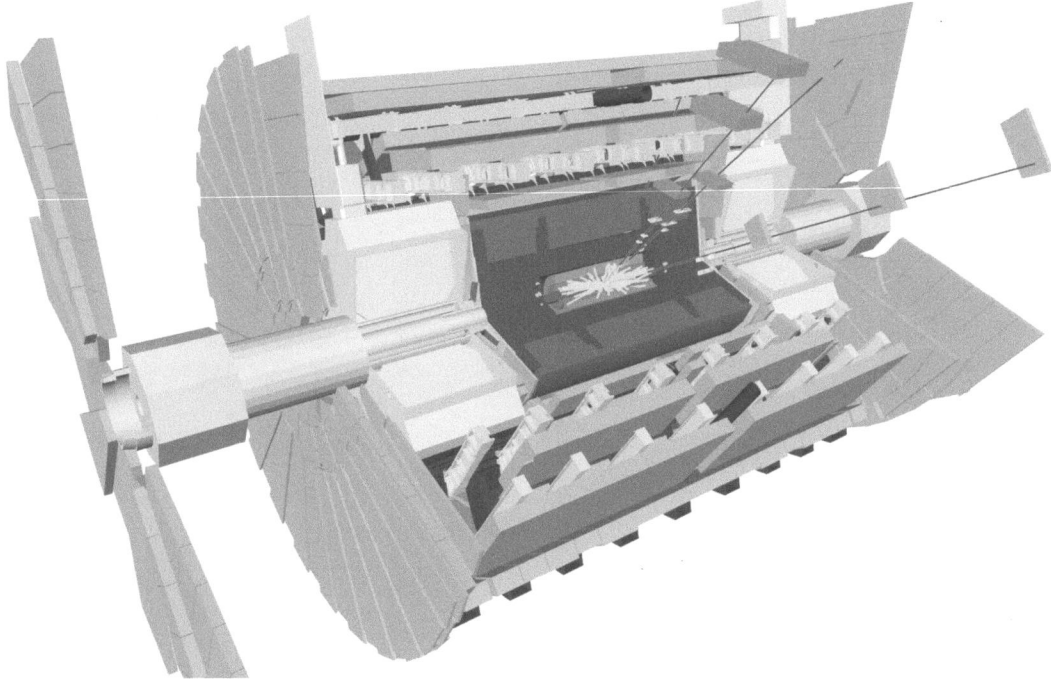

Fig. 6 An event display made with VP1. A Higgs boson decays into four muons (shown in *red*). Inner detector tracks are in *green*, and energy deposited in the calorimeter by the muons is shown in *yellow*

5.7 Visualization

Visualization is used to understand anomalies or features in odd events, occasionally to debug errors due to geometry, and to check for overlaps and touching volumes in the geometry that can be spotted by eye. Although GEANT4 contains viewing software of its own, because the geometry of AT-LAS must be translated from GeoModel into GEANT4 format it is useful to use a viewer that can construct geometry directly from its GeoModel description. A general purpose three-dimensional event display program, VP1[10] [97], has been developed specifically for ATLAS. It is optimized specifically for the visualization of the ATLAS geometry and is arguably the most useful tool for understanding and debugging of detector description across all ATLAS subsystems. Two examples of VP1 event displays are in Figs. 6 and 7. Ray Tracer, the GEANT4 visualization utility, has also been used to visualize portions of the detector containing some exotic shapes.

VP1, as well as other event display programs used in AT-LAS (e.g. Atlantis [98] and Persint [99]), are mostly used for visualizing real and simulated events after they have run through the common reconstruction software. The VP1 viewer can be injected directly into the simulation job in order to visualize events immediately after the simulation step.

---

[10]ATLAS is at Point 1 of the LHC ring. The name VP1 is short for Virtual Point 1.

6 Digitization

The ATLAS digitization software converts the hits produced by the core simulation into detector responses: "digits." Typically, a digit is produced when the voltage or current on a particular readout channel rises above a pre-configured threshold within a particular time window. Some subdetector's digit formats include the signal shape in detail over this time, while others simply record that the threshold has been exceeded within the relevant time window.

The peculiarities of each subdetector's charge collection, including cross-talk, electronic noise and channel-dependent variations in detector response are modelled in subdetector-specific digitization software [79, 82, 100–102]. The various subdetector digitization packages are steered by a top-level PYTHON digitization package which ensures uniform and consistent configuration across the subdetectors. The properties of the digitization algorithms were tuned to reproduce the detector response seen in lab tests, test beam data, and cosmic ray running. Dead channels and noise rates are read from database tables to reproduce conditions seen in a particular run. In some cases, dead channels are removed during the reconstruction step.

The digits of each subdetector are written out as Raw Data Objects (RDOs). For some subdetectors this requires the digits produced to be converted to RDOs by a second algorithm during the digitization process. For others there is no intermediate digit object and RDOs are produced

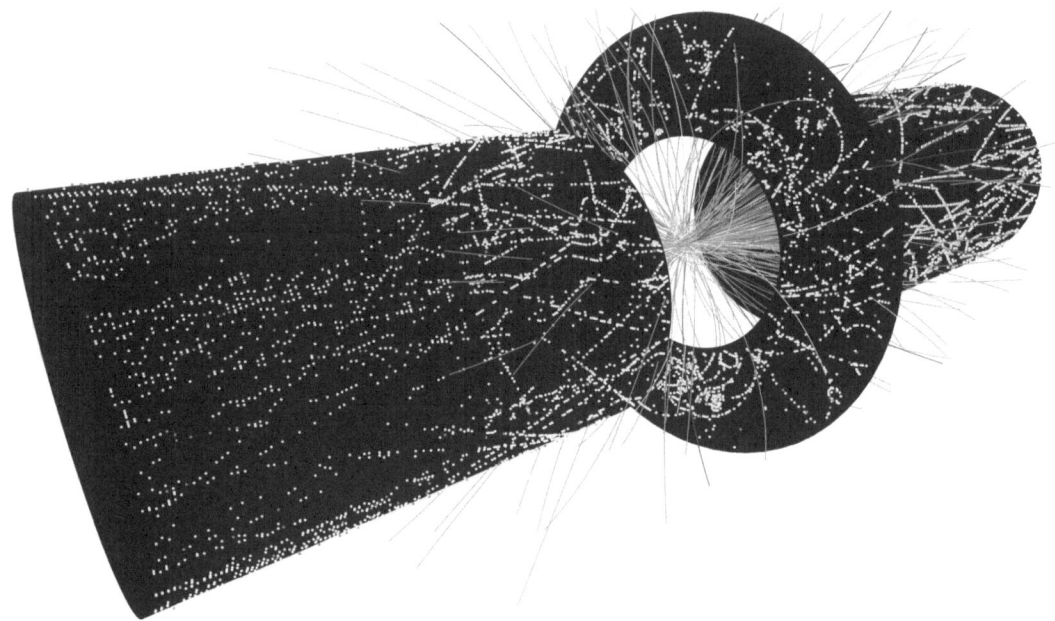

Fig. 7 A Higgs boson decaying into four muons, with only the inner detector tracks and hits in the TRT being displayed by VP1

directly from the hits. In addition to RDOs, the digitization algorithms can also produce Simulated Data Objects (SDOs). These SDOs contain information about all the particles and noise that contributed to the signal produced in the given sensor and the amount of energy contributed to the signal by each. The relationship between RDOs and SDOs depends on the particular subdetector. For example, in the SCT each RDO represents a group of consecutive strips which recorded a hit, whereas one SDO is produced for each strip where energy was deposited by a particle in the Monte Carlo truth tree. No SDOs are created in the calorimeter. SDOs are mainly used for determining tracking efficiency and fake track rates.

Simulating the detector readout in response to a single interesting hard scattering interaction is unrealistic. In reality, for any given bunch crossing there may be multiple proton–proton interactions. In addition to the hard scattering which triggers the detector readout, many inelastic, non-diffractive proton–proton interactions may appear. These interactions must be included in a realistic model of detector response. The effects of beam gas and beam halo interactions, as well as detector response to long-lived particles, must be incorporated. These interactions are treated separately at the event generation and simulation stages. Within a digitization job, hits from the hard scattering are overlaid with those from the requested number of these additional interactions before the detector response is calculated. Because of long signal integration times, most subdetector responses are affected by interactions from neighboring bunch crossings as well. Therefore, additional interactions offset in time are overlaid as

necessary. The overlaying off these various types of events, known collectively as "pile-up," is described in Sect. 6.2.

Before reconstruction can be run, bytestream data from the real detector must be converted into RDO format. As mentioned above, the digitization usually avoids this step by writing out RDOs directly. However, in order to do simulation studies with the High Level Trigger it is necessary to translate the RDO files into bytestream format. There is some loss due to truncation in the first conversion from RDO to bytestream, but the inverse operation is basically lossless. Having the ability to convert output in both directions also allows evaluation of the conversions themselves.

6.1 Digitization configuration

The ATLAS digitization takes as input hit files produced by the ATLAS simulation. For pile-up simulation, there are also input hit files for each type of background interaction to be overlaid. In such cases it is the main hard scattering event which sets the run number and event number. Run and event numbers from overlaid events are ignored.

The digitization steering package exists entirely in the PYTHON layer and configures how the digitization will be performed before the event loop starts. This configuration is highly flexible, but also ensures that sensible default values are given for each configurable property of the job. In the configuration of digitization jobs, the user may specify the number of events to digitize, the number of leading events to skip in the input file, the input hit file(s), and the output file. Digitization and writing out of RDOs may be enabled or disabled by subdetector. In order to ensure consistency, the detector layout version is, by default, read from the hard scattering events' hit file metadata.

Digitization options also include the following:

Detector Noise Simulation: Detector noise simulation can be turned off in the inner detector, calorimeter or muon spectrometer or any combination thereof. This is useful for data overlay jobs where noise is taken from real data events and for studies using a noise-free detector.

Random Number Services: The type of random number engine to be used in all digitization algorithms can be specified (Ranlux64, the default, or Ranecu [103]). Each algorithm has one or more random number streams. Random number seeds can be initialized from a text file or set in job options. The user may alternately specify an offset from the default values of the seeds, to be used in all streams.

Metadata: In the default configuration, metadata from the simulation stage are used to configure the physics list (for setting the sampling fraction of the calorimeters) and the detector layout. The metadata can be overridden.

Pile-up Background Events: The overlay of minimum bias, cavern background, beam gas and beam halo events can all be configured separately. In each case the mean number of events (if any) per bunch crossing to be overlaid and a collection of files containing the events to be overlaid onto the signal events can be specified.

Beam Properties: The LHC beam bunch spacing can be configured, as can the number of bunch crossings to overlay before and after the hard scattering event.

Detector Conditions: Default detector conditions (including, e.g., dead electronics and noisy channels) are associated with each detector layout. Non-default conditions may be specified globally or by subdetector for use in digitization.

After a check to make sure that at least one subdetector has been left switched on, the input and output streams are initialized. GeoModel is initialized using the detector layout and conditions versions read from the hits file metadata or specified by the user. Setting the detector layout version to be different from that used in the simulation is possible, but considered to be an expert action. The magnetic field service is initialized at this point. It is necessary because the magnetic field affects charge propagation from the active regions of the detector to the readout surfaces.

At this point, caches for pile-up events are created and configured with the appropriate collection of hits files as well as the number of events to be overlaid per bunch crossing. These caches are controlled by an overall pile-up manager service. A second pile-up service is created to hold information about the time window within which interactions can affect the response recorded by each subdetector. During the initialization stage, this information can be combined with the bunch spacing to calculate the number of bunch

crossings which should be simulated for each subdetector for each event.

Subsequently, the subdetector digitization algorithms are configured and added to the sequence of algorithms to be run in the job. The collections of RDOs, hits, and truth information which are to be recorded are added to the output stream. Digitization algorithms exist for the following subdetectors:

Inner Detector: BCM, silicon pixel tracker, SCT, and TRT.

Calorimeter: LAr and tile calorimeters. Separate algorithms also exist to simulate the formation of trigger towers in the calorimeters, which serve as inputs to the level one trigger.

Muon Spectrometer: Cathode Strip Chambers, Monitored Drift Tubes, Resistive Plate Chambers and Thin Gap Chambers.

If requested, the level one trigger simulators are added to the algorithm sequence, provided that the digitization of the relevant parts of ATLAS have been turned on. The default mode of simulation production is to run the level one trigger simulation during the reconstruction step rather than as part of the digitization step.

As the digitization algorithm for each subdetector is configured, the names and seeds for the random number streams it requires are added to a list. In the case where seeds are to be read in from a file, the default list of stream names and their seeds are replaced by the file contents. Once all algorithms have been configured, the list is used to configure the random number service. Separate random number streams are used for each subdetector digitization algorithm and give the same result independent of what is used for the other subdetectors.[11]

Much of the job configuration information, along with the detector layout version, is written to the output file as digitization metadata. The run number provided in the simulation metadata is used to establish a validity range for the digitization metadata corresponding to the current run only. At this point the digitization job is fully configured and the event loop begins.

6.2 Pile-up

To simulate pile-up, various types of events are read in, and hits from each are overlaid. The different types considered can be configured at run time, and normally comprise signal, minimum bias, cavern background, beam gas, and beam halo events. The number of events to overlay of each type

---

[11] Here "digitization algorithm" does not include the calorimeter trigger tower simulation algorithms, which require the corresponding calorimeter digitization to be performed. Similarly, the level one trigger simulation requires the simulation and digitization of the expected trigger inputs to give meaningful results.

Table 12 The time window (relative to the current bunch crossing) during which interactions in each subdetector are simulated during pile-up jobs, along with the corresponding numbers of bunch crossing simulated in the case of 25 ns bunch spacing and 75 ns bunch spacing

Subdetector	Simulation window [ns]	No. Bunch crossings (25 ns bunch spacing)	No. Bunch crossings (75 ns bunch spacing)
BCM	$-50, +25$	4	1
Pixel trackers	$-50, +25$	4	1
SCT	$-50, +25$	4	1
TRT	$-50, +50$	5	1
LAr calorimeter	$-801, +126$	38	12
Tile calorimeter	$-200, +200$	17	5
Muon chambers	$-1000, +700$	69	23

per bunch crossing may also be set at run time and is a function of the luminosity to be simulated. The mean number of interactions per bunch crossing (BC), for example 23 at the design luminosity of 10^{34} cm^{-2} s^{-1} with 25 ns bunch spacing, depends linearly on luminosity and bunch spacing. However, this number is Poisson-distributed, with a long tail beyond the most probable value. Thus, a substantial fraction of the bunch crossings will have more than the average number of interactions. In addition, the ATLAS subdetectors are sensitive to hits several bunch crossings before and after the BC that contains the hard scattering event (which triggers the readout). Table 12 shows the simulation window for each detector along with the corresponding number of bunch crossings for 25 ns and 75 ns bunch spacing. All of these detector and electronic effects are taken into account during the pile-up event merging.

6.2.1 Cavern background

Neutrons may propagate through the ATLAS cavern for a few seconds before they are thermalized, thus producing a neutron–photon gas. This gas produces a constant background, called "cavern background," of low-energy electrons and protons from spallation. The cavern background consists mainly of thermalized slow neutrons, long-lived neutral kaons and low-energy photons escaping the calorimeter and the forward shielding elements. Muon detectors are most affected by high cavern-background rates. The radiation levels to be expected in the ATLAS cavern scale with luminosity, and they have been simulated as a function of r and z [104] for the design luminosity of 10^{34} cm^{-2} s^{-1}. Depending on the type of radiation, exact composition of the equipment, and sensitivity of the study, the rates sometimes have to be increased by a safety factor. Cavern background is produced in the following way:

- A standalone dedicated GEANT3/GCALOR-based [105] detector simulation program with improved neutron propagation and a simplified ATLAS detector geometry is run

on proton–proton collisions. The cavern walls are not included in the detector description. The output of this program includes particle fluxes in the envelopes surrounding muon spectrometer chambers. The fluxes are provided as list of particles with all related parameters per proton–proton interaction at the entrance of each envelope.
- The kinematic information of all particles generated by GEANT3/GCALOR is converted to HepMC format, and the flux is modified to be uniform in the time interval of the required bunch spacing (typically [0, 25 ns]).
- The simulation is then carried out using the full detector geometry and GEANT4, and hits are stored.
- The cavern events are mixed, with a safety factor of up to 10, at the digitization level with the minimum bias and signal events.

There are a number of issues with the current simulation of the cavern background. The original primary cavern events were generated in an older version of PYTHIA where the generated particle density is a factor of two lower than in the newer versions of PYTHIA [42, 43]. The statistics for the available cavern events are limited: 40,000 events are available with a safety factor of 1; 10,000 events are available with a safety factor of 2 or 5; and 5,000 events are available with a safety factor of 10. Because of the limited statistics, a number of monitored drift tubes in the muon detector fire more often than expected (i.e. there are spikes in the hit response of the detector). Additionally, neutral particles are tracked through the entire detector during simulation, thus producing additional hits from particles that should have been removed at the edges of the muon chamber envelopes (multiple counting).

In the short term, the problem of limited statistics of the cavern events has been alleviated by taking advantage of the ϕ-symmetry of the muon spectrometer: the cavern events are rotated and re-simulated eight times or more (in multiples of eight). Further improvement in the available cavern statistics can be achieved by repeating the simulation of the cavern events many times with different random number seeds, since the probability of a neutral interaction is very low, of the order 1%.

Table 13 Container size on disk in RDO files. Columns two and three show the average of 50 $t\bar{t}$ events digitized in the absence of pile-up. These events were chosen because they produce large energy deposits throughout the detector. Columns four and five show the average of 50 $t\bar{t}$ events digitized in the presence of pile-up with 10^{33} cm^{-2} luminosity and 25 ns bunch spacing

Category	No Pile-up space on disk [kB/event]	No Pile-up percentage of file	Pile-up space on disk [kB/event]	Pile-up Percentage of file
Inner detector RDOs	187	7.6	322	11.5
Inner detector SDOs	247	10.0	333	11.9
Calorimeter raw channels	995	40.3	1006	35.9
Calorimeter calibration hits	601	24.3	601	21.4
Muon spectrometer RDOs	1	0.04	27	1.0
Muon spectrometer SDOs	1	0.05	59	2.1
Level one trigger	289	11.7	300	10.7
Truth	147	6.0	151	5.4
Headers	>1	0.01	2	0.1
Total	2469	100.00	2801	100.00

6.2.2 Beam halo and beam gas

Beam halo is the background resulting from interactions between the beam and upstream accelerator elements. The flux from upstream (in the tunnel and collimators) is provided by the LHC Machine Division [106, 107]. Beam halo events are generated as discrete particle losses against the upstream collimators. The LHC machine division has estimated the proton loss rate in design conditions as being on the order of 1 MHz. FLUKA simulation of the last 150 m of the beamline indicates that daughter particles from these proton losses will reach the cavern wall (23 m from the interaction point) at a rate of ~400 kHz. This flux is input to the normal GEANT4 simulation to produce hit files.

Beam gas includes the residual hydrogen, oxygen, and carbon gasses in the ATLAS beam pipe. Beam gas interaction events are generated with Hijing (see Sect. 3.2.4) with appropriate time offsets. The interactions are allowed to take place anywhere in the beam pipe of ATLAS, 23 m in either direction from the interaction point.

6.2.3 Pile-up with real data

The pile-up mechanism described above will not work with real data, because it begins at the hit level. One must instead overlay events beginning from detector electronics output (RDOs). One may collect minimum bias, cavern background, beam halo, and beam gas backgrounds from the same "zero bias" trigger used to understand detector electronic noise. Then, one would overlay hits from simulated hard scattering events onto the zero bias trigger data to simulate the pile-up. The zero bias trigger data needed for this type of event overlay can be selected at random from the filled-bunch crossings.[12] The subdetectors should be read out with as little zero-suppression as is possible and with the HLT in "pass-through" mode (i.e. without further filtering). One can use bunch-by-bunch luminosity information to correctly weight the event sample for pile-up studies.

In principle one needs as many zero bias events as generated events, but in practice zero bias events can be reused with independent simulated data sets without introducing any bias. During data taking, zero bias events are sampled at all times, because detector and cavern conditions are likely to vary with time. The zero-bias events could, for example, be collected exactly one orbit after a high-p_T trigger has fired. Appropriately pre-scaled to the output rate needed for simulations (on the order of 1–2 Hz, or about 1% of the recorded events), this means that the rate follows the luminosity and that the bunch structure is guaranteed to be right.

6.3 RDO storage format

The ATLAS detector electronics produce data in bytestream format. The RDO format can be thought of as a POOL-compatible version of the byte stream.[13] The file size on disk is typically around 2.5 MB per event for hard scattering events (e.g. $t\bar{t}$ production) and increases in the presence of pile-up. Table 13 shows an example of the disk consumption by container for 50 $t\bar{t}$ events without pile-up and with pile-up at 10^{33} cm^{-2} s^{-1}. In the absence of pile-up, one of the main consumers of disk space are the calibration hit collections, as described in Sect. 5.6, which are copied directly from the hit file to the RDO. As pile-up luminosity increases however the inner detector containers become increasingly significant.

---

[12]The zero bias trigger is not a minimum bias trigger.

[13]POOL compatibility requires separate transient and persistent object representation.

7 Fast simulations

Because of the complicated detector geometry and detailed physics description used by the ATLAS GEANT4 simulation, it is impossible to achieve the required simulated statistics for many physics studies without faster simulation. To that end, several varieties of fast simulation programs have been developed to complement the GEANT4 simulation. In this section, the standard GEANT4 simulation will be referred to as "full simulation."

Almost 80% of the full simulation time is spent simulating particles traversing the calorimetry, and about 75% of the full simulation time is spent simulating electromagnetic particles. The Fast G4 Simulation aims to speed up this slowest part of the full simulation [108, 109]. The approach taken, therefore, is to remove low energy electromagnetic particles from the calorimeter and replace them with pre-simulated showers stored in memory. Using this approach, CPU time is reduced by a factor of three in hard scattering events (e.g. $t\bar{t}$ production) with little physics penalty. This simulation may eventually become the default simulation for all processes that do not require extremely accurate modeling of calorimeter response or electromagnetic physics.

ATLFAST-I has been developed for physics parameter space scans and studies that require very large statistics but do not require the level of detail contained in the full simulation [110, 111]. Truth objects are smeared by detector resolutions to provide physics objects similar to those of the reconstruction. Object four-vectors are output, without any detailed simulation of efficiencies and fakes. A factor of 1000 speed increase over full simulation is achieved with sufficient detail for many general studies.

ATLFAST-II is a fast simulation meant to provide large statistics to supplement full simulation studies. The aim is to try to simulate events as fast as possible while still being able to run the standard ATLAS reconstruction. ATLFAST-II is made up from two components: the Fast ATLAS Tracking Simulation (Fatras) for the inner detector and muon system simulation [112] and the Fast calorimeter Simulation (FastCaloSim) for the calorimeter simulation. Optionally, any subdetector can be simulated with GEANT4 to provide the higher level of accuracy without the same CPU time consumption as full simulation of the entire detector. An improvement over full simulation time of a factor of 10 is achieved with full GEANT4 inner detector and muon simulation and FastCaloSim, and a factor of 100 is achieved with Fatras and FastCaloSim.

7.1 Fast G4 simulation

The Fast G4 Simulation reduces CPU time consumption without sacrificing accuracy by speeding up the slowest parts of the full simulation. By treating (as described below) electromagnetic showers in the sampling portions of the calorimeters, a reduction in CPU time of a factor of three can be achieved even in hadronic events. Although the calorimetry dominates simulation time for the full simulation, after treatment of the electromagnetic showers the simulation time is evenly distributed throughout all subdetectors. One particular advantage of this fast simulation over the other varieties is that its output file matches identically the format of the output of the full GEANT4 simulation. The data can therefore be run through the identical tests and digitization software following simulation, and the standard ATLAS trigger and reconstruction can be run.

There are three treatments applied to electromagnetic showers. For very high energy (> 10 GeV) electrons and positrons, a tuned shower parameterization is available. For medium energy (10 MeV to 1 GeV) electrons, positrons, and photons, libraries of pre-simulated showers can be applied during the event. For very low energy (< 10 MeV) electrons and positrons, a single hit can be deposited to recreate detector response. Each one of these treatments can be turned on by the user in each compartment of the electromagnetic calori meter and the forward hadronic calorimeter. The energy ranges can be set for each method, compartment, and particle.

For electrons and positrons above a sufficiently high energy, around 10 GeV in the central calorimeters, the sampling calorimeter is sufficiently homogeneous to apply a shower parameterization. Small steps are taken in the direction of the original particle, depositing energy according to several tuned functions as it traverses the detector. The longitudinal profiles of showers are parameterized and normalized with an energy scale to approximate the sampling fraction in each subdetector. The radial profile changes as a function of depth in the shower and is normalized by the longitudinal profile. Energy is deposited in hits in order to mimic the full simulation. Fluctuations are introduced in three separate places, representing the random characteristics of shower length and shape, the sampling resolution of the calorimeter, and the geometric fluctuations in the energy collected.

Particles captured by the fast simulation in the appropriate energy range (typically < 1 GeV) are replaced by a shower from a pre-simulated library, rotated and scaled to match the primary particle. Shower libraries are generated in bins of pseudorapidity and energy for electrons and photons. Only hits in the sensitive detectors are stored in order to save space on disk and in memory. The binning reproduces the fine structure in the calorimeters. The libraries are read into memory the first time they are requested by the simulation, ensuring minimal memory overhead. They consume about 200 MB when all are in use. Showers are randomly selected with linear weighting from the energy bin above or below the primary particle and from the pseudorapidity bin above or below the particle. The shower is then rotated to

Table 14 The default combination of strategies used in the Fast G4 Simulation for each calorimeter compartment

Calorimeter	Parameterization	Shower libraries	Killing
Electromagnetic Barrel	Not used	10–1000 MeV e^+e^- <10 MeV photons	<10 MeV e^+e^-
Electromagnetic Endcap	Not used	10–1000 MeV e^+e^- <10 MeV photons	<10 MeV e^+e^-
Electromagnetic Forward	>9 GeV e^+e^-	10–1000 MeV e^+e^- <10 MeV photons	<10 MeV e^+e^-
Hadronic Forward	>1.5 GeV e^+e^-	Not used	<10 MeV e^+e^-

match the primary particle's original direction, and the energy is scaled to match the primary particle's energy. For example, an electron at pseudorapidity of 2.37 and with an energy of 12 MeV might use a shower from the electron and positron library's 2.4 bin in pseudorapidity and 10 MeV bin in energy. The shower would then be moved and rotated to match the position and direction of the original particle, and its energy would be scaled up by 20%.

Electrons and positrons with energies below about 10 MeV typically deposit only one hit in the sensitive region of the calorimeter. These particles are removed when inside the regular sampling region of any of the calorimeters ("killed"), and a single hit is placed in the calorimeter. The position of the hit is determined by a random exponential number times the radiation length in the detector in order to approximate the particle's range. The energy of the hit is scaled to the response of the detector and smeared by its resolution.

The standard combination of strategies is shown in Table 14. The strategies used in a particular subdetector is optimized for maximum CPU time improvement with minimal complexity. The upper energy bound for shower libraries, 1 GeV, balances memory use with speed. Libraries at higher energies also may not correctly reproduce the tails of electromagnetic shower shape distributions as well as low-energy libraries do. The minimum energy for application of the parameterization model is based purely on CPU time. In most subdetectors, it is faster to allow GEANT4 to produce 1 GeV secondaries and apply shower libraries to those secondaries than it is to apply the shower parameterization. The same argument applies to high energy photons. They pair produce sufficiently quickly that treating them separately only adds complexity to the models. The speed of the parameterization is limited by random number generation and locating hits within the detector geometry.

7.2 ATLFAST-I

ATLFAST-I performs a fast simulation of the ATLAS detector, including object reconstruction, in order to produce high statistics samples of signal and background events. The lowest possible CPU time per event is achieved by replacing detailed detector simulation with parameterizations of the desired detector and reconstruction effects. The high speed of simulation in ATLFAST-I makes it possible to study channels where the statistics involved would otherwise be prohibitive. For example, the background to a $Z \to \tau^+\tau^-$ study from fake taus in di-jet events is expected to require $O(10^9)$ events in 100 fb^{-1} of data. Some searches also require many datasets to be simulated in order to scan across parameter space for the model being tested, such as SUSY.

ATLFAST-I is the least detailed simulation method. There is no realistic detector description, so studies of detector-based quantities, such as calorimeter sampling energies and track hit positions, are not possible. There is no simulation of reconstruction efficiency or misidentification rates, discussed later on, which means the presence of genuine physics objects are overestimated while fake objects are not modeled, with two exceptions. Because jet-flavor tagging efficiencies are applied, fake b-jets and taus are simulated. However, ATLFAST-I provides a useful method of making quick estimates of systematic uncertainties in early data analyses due to the simple process of re-parameterizing the detector and modeling reconstruction effects. The speed of operation enables datasets to be reproduced with different generator configurations, allowing quick estimates of systematic uncertainties arising from generators.

Common to the reconstruction of all objects in ATLFAST-I is that by default no reconstruction efficiencies are applied. These efficiencies can be taken from full simulation and accounted for by the user in the analysis. This applies to electrons, photons and jets as well as to ATLFAST-I tracks. It should be noted that tagging efficiency factors are implicitly taken into account in the tau- and b-tagging procedures. A system to apply a common set of efficiencies and misidentification rates at the analysis stage is in development. The misidentification rates will allow the modeling of fake objects as well.

ATLFAST-I takes input in HepMC format, enabling it to read the output of all ATLAS generators. Generator input is filtered to choose only particles that are useful in the current step. For example, only charged particles are considered in the tracking stage, and all particles are required to be a part of the final state.

The following sections describe steps taken in ATLFAST-I.

7.2.1 Tracks

Charged particle tracks from the generator with $p_T > 500$ MeV and with $|\eta| < 2.5$ are considered as reconstructed ATLFAST-I tracks, and five track parameters[14] are associated to them. These parameters are calculated from the true particle properties by applying parametrized resolution functions which account for the measurement precision, energy loss, and multiple scattering as well as for hadronic interactions in the inner detector material. The resolution functions are taken from fully simulated events. The non-Gaussian tails resulting from hadronic interactions are taken into account by applying a double-Gaussian correlated smearing to the track parameters of hadrons [110, 111]. No vertex smearing is applied. In ATLFAST-I, three types of charged particles are distinguished: hadrons, electrons and muons. Due to the relatively large energy loss from bremsstrahlung, high-p_T electrons are treated separately, and an additional energy loss correction is applied. It should be noted that while these tracks are used for specific studies in B physics, they are not used for lepton identification or b-tagging.

7.2.2 Track-based tau identification

Track-based tau identification is split into two distinct parts, namely reconstruction and identification of tau candidates. The reconstruction part applies a parameterized efficiency to the tracks to calculate the charged component in a tau candidate, while the neutral component is calculated directly from neutral particles in the generated event.

Once a sample of tau candidates has been reconstructed, the identification part is carried out by separating the sample into true and fake taus. True taus are defined as those matched to a hadronic decay in the truth record with $\Delta R \equiv \sqrt{\Delta \eta^2 + \Delta \phi^2} \leq 0.2$, whereas the remainder are considered fakes. Subsequently, a parametrization of the identification efficiency is applied based on the number of tracks.

7.2.3 Calorimetry

Stable charged particles from the event generators are propagated through the magnetic field along a simple helix. The primary vertex is assumed to be at the geometric origin. Using a helix model and assuming a perfectly homogeneous

magnetic field inside the central tracking volume, the impact point on the calorimeter surface is calculated. To calculate this point, no interactions of the particle with the detector material (i.e. no multiple scattering, energy loss, or nuclear interactions) are taken into account. In particular, this implies that no energy loss due to bremsstrahlung for electrons and no pair production from photons in the inner detector media are simulated for the energy depositions in the calorimeters. The effects of these interactions are, however, implicitly taken into account by the application of appropriate resolution functions. For the calculation of track parameters, the four-momenta and the starting point of the particles (e.g. for stable decay products of long-lived particles) are taken from the generator information.

The energies of the electrons, photons, and hadrons are deposited in a calorimeter cell map. The response of the calorimeter is assumed to be unity and uniform over the full detector. No smearing (i.e. no resolution function) is applied. The energy of the particle is entirely deposited in the hit calorimeter cell, assuming a granularity of the calorimeter cell map of $\eta \times \phi = 0.1 \times 0.1$ up to $|\eta| < 3.2$ and $\eta \times \phi = 0.2 \times 0.2$ for $3.2 < |\eta| < 5.0$. Neither lateral nor longitudinal shower development is simulated. Therefore, the longitudinal fine structure of the calorimeters is not taken into account. There is also no separation between the electromagnetic and the hadronic calorimeter compartments.

Based on the map of deposited cell energies, cluster reconstruction is carried out using either SISCone [113] or FastKt algorithms via the FastJet libraries [114]. The default clustering routine is SISCone with a cone size of 0.4. The cluster transverse energy must pass a threshold, typically 5 GeV. The clusters may get re-classified as electrons, photons, taus or jets in one of the following steps. If they are associated to one of these objects, they are removed from the list of clusters.

7.2.4 Electrons and photons

For each true electron or photon, the reconstructed energy is obtained by smearing the true energy according to a resolution calculated by interpolating between resolutions measured in fully simulated events at precise values of η and energy. If, after smearing, the candidate electron or photon has transverse energy exceeding a threshold value, typically 5 GeV, and has $|\eta| < 2.5$, then it is recorded with the η and ϕ directions of the true particle.

Electrons and photons are matched to calorimeter clusters in (η, ϕ) space, with a maximum allowed separation of $\Delta R = 0.15$. If there is a matching cluster then it is removed from the list of clusters to be considered as jet candidates later on.

---

[14] The five parameters are: the azimuthal angle ϕ; longitudinal impact parameter, z_0; transverse impact parameter, $d_0 \equiv \sqrt{x^2 + y^2}$; polar angle in θ; and charge divided by momentum amplitude.

7.2.5 Muons

For each true muon with $p_T > 0.5$ GeV, the reconstructed momentum is calculated from the true muon momentum. A Gaussian resolution function which depends on p_T, η, and ϕ is applied. After smearing, muons with $p_T > 5$ GeV and with $|\eta| < 2.5$ are kept.

7.2.6 Isolation

In order to define isolated electrons, photons, and muons, the following criteria are applied: the difference in the un-smeared energy in a cone of $\Delta R = 0.2$ around the object direction and its smeared energy needs to be below 10 GeV. In addition, there should be no further clusters reconstructed with $\Delta R < 0.4$ around the object direction.

7.2.7 Jets

All clusters that have not been assigned to a true electron or photon are considered jets if their transverse energy exceeds 10 GeV. The jet energy is taken to be the cluster energy, after adding non-isolated muons within $\Delta R = 0.4$, and is smeared according to the jet energy resolution.

These functions do not account for pile-up, although there is a "high luminosity" mode available which adds a pile-up term to the resolution. The pile-up correction is constant with respect to jet transverse energy and is dependent on the size of the jet.

The jet direction is taken to be the cluster direction. Since the response function of the calorimeter is set to one, no jet calibration is needed to correct for the non-compensation of the calorimeter. However, an out-of-cone energy correction is needed. This correction is applied in a separate jet calibration step [110, 111].

7.2.8 Tagging

For each jet found, a label is attached to indicate whether the true jet originated from a light quark, b-quark, c-quark, or tau. This label is based on matching b or c partons or the visible decay products of hadronically-decaying taus at truth-level with $\Delta R < 0.3$ to a reconstructed jet. In the case of hadronically-decaying taus, the ratio between the true visible energy and the jet energy is also required to be larger than unity minus 2σ, where σ is the jet energy resolution as above.

The results of b- and tau-tagging are then simulated by applying identification efficiencies and fake rates to the labels. These efficiencies are determined from full simulation studies and are parameterized as a function of p_T and η.

7.2.9 Missing E_T

The missing transverse energy is calculated from all reconstructed objects: isolated electrons, photons, muons, taus, jets and non-isolated muons, and remaining calorimeter clusters not associated to jets. In addition, cells not associated to clusters are included in the missing E_T calculation. The cell energies are smeared by applying the jet resolution functions.

7.3 ATLFAST-II

ATLFAST-II directly simulates the input to the standard Athena reconstruction algorithms to mimic the full simulation. Unlike ATLFAST-I, which provides only momenta for the reconstructed objects, reconstructed ATLFAST-II output includes all the properties associated with a reconstructed object. In the case of Fatras these include the hits in the inner detector and muon system, and for FastCaloSim these include the energies in the calorimeter cells. Because the standard reconstruction is run, it is possible to work with a combination of full and ATLFAST-II simulated events without modifying any analysis code. Both Fatras and FastCaloSim run together with the event reconstruction. The simulation time is reduced by making use of the simplified detector description used for reconstruction [115]. By default, ATLFAST-II uses full simulation for the inner detector and muon system and FastCaloSim in the calorimetry. ATLFAST-IIF uses FastCaloSim in the calorimetry and Fatras in the inner detector and muon system.

As input, Fatras uses input in HepMC format, performs a smearing of the primary vertex position to represent the luminous region within ATLAS, and records truth information in a way similar to the full simulation. FastCaloSim uses the truth information of all interacting particles at the end of the inner detector volume as input to the calorimeter simulation. In order to simulate pile-up, generated events must be overlaid prior to detector simulation.

7.3.1 Fatras

ATLFAST-II with the fast track simulation engine Fatras (ATLFAST-IIF) reduces simulation time in the inner detector and muon system. Fatras is an ATLAS specific development and establishes a complete simulation within the track reconstruction framework. The reconstruction geometry is a simplified description of the full detector geometry, which keeps the same descriptive accuracy for sensitive detector parts, while approximating all other detector components as simplified layers that carry a high-granularity density map. This detector material description can be sufficient. A factor of 100 reduction in CPU time is obtained with only small physics performance degradation. The propagation of the particles through the tracking detectors is carried out by the extrapolation engine [116] used in the offline track reconstruction applications.

The interactions of the particles with the simplified detector layers are simulated using several methods. Multiple Coulomb scattering is implemented as a Gaussian mixture model to account for tail effects from single large-angle scattering processes; ionization and radiative energy loss are simulated according to the Bethe-Bloch and Bethe-Heitler models; conversion of a photon into an electron and positron is carried out depending on the thickness of the traversed material; hadronic interactions of particles with the detector layers are simulated from a parametric model that has been obtained from GEANT4 simulation results. The decay of unstable particles is enhanced by a dedicated wrapper of the associated GEANT4 modules [112]. The calorimeter simulation of ATLFAST-IIF is typically FastCaloSim, and Fatras provides the input particle collection. Energy deposition for muons in the calorimeter layers is also recorded according to the material description of the reconstruction geometry and is further used for cluster simulation in the FastCaloSim application.

Fatras was first established as a validation tool for the newly deployed inner detector reconstruction sequence. It has already been used for noise studies in the Transition Radiation Tracker and first simulations for a potential future upgrade of the ATLAS inner detector. The validation of Fatras against the full simulation results to be used for first collisions data from LHC is ongoing. An extension of the fast track simulation within the reconstruction geometry has taken place that also allows the use of Fatras in the muon spectrometer. The particles being simulated at the end of the inner detector volume are filtered. Muons are transported through the calorimeter, and their deposited energy is stored as an input to the FastCaloSim module. The trajectories of the muons are then simulated in the muon spectrometer, and the hits within sensitive detector elements are recorded. Standard digitization is applied on top of the simulated hits to account for the detailed calibration that must be included for a comparison to data.

7.3.2 FastCaloSim

Instead of simulating the particle interactions with the detector material, the energy of single particle showers is deposited by FastCaloSim directly using parametrizations of the longitudinal and lateral energy profile. The distribution of active and inactive material in the calorimeter needs to be respected by the parametrization, so a fine binning of the parametrization in the particle energy and pseudorapidity is needed. Furthermore, the energy deposition depends strongly on the origin of the shower in the calorimeter, so all parametrizations are also binned versus the longitudinal depth of the shower center.

The parametrizations are based on a 30 million event sample of fully-simulated (i.e. simulated with GEANT4) single photons and charged pions in an energy range between 200 MeV and 500 GeV, evenly distributed in $|\eta| < 5.0$ and $-\pi < \phi < \pi$. All electron and photon showers are approximated by the photon parametrization and all hadronic showers are approximated by the charged pion parametrization. The simplified reconstruction geometry of the calorimeter is used with details at the level of the readout cells.

The parameterization of the longitudinal energy distribution is constructed from histograms of the total energy in all calorimeter layers, the longitudinal depth of the shower center, and the energy fraction in each layer for the fully-simulated single-particle events. The dominant correlations between fractional energy deposits in each calorimeter layer (i.e. those related to the longitudinal depth of the shower's origin) are accounted for in the parameterization binning. Gaussian correlations between fractional energy deposits in each calorimeter layer (i.e. those describing shower development) are stored in a correlation matrix and are applied to improve the parameterized energy distribution. During fast simulation, the parametrization closest in energy and pseudorapidity to the particle is taken, and then the total shower energy and the shower depth are chosen randomly from the stored histograms and rescaled to match the true particle energy. It was found that after rescaling no interpolation between parametrizations is necessary. Afterwards, the energy fractions in all calorimeter layers are generated randomly, taking into account the correlation matrix. The lateral energy distribution inside each calorimeter layer is simulated using a symmetric average radial shape function. The shape functions are extracted from fits to fully-simulated single-particle events and are constructed for bins of particle type, primary particle energy, position in η, and shower depth in the calorimeter. The asymmetry of shower shapes for particles entering the calorimeter at large incident angles is absorbed in a shape function describing a pseudorapidity-dependent asymmetry term. During simulation, the energy of a calorimeter cell is determined by the integral of the shape function over the cell surface area. Fluctuations derived from the intrinsic resolutions of each calorimeter are applied to the cell energy. The total energy of all cells in one calorimeter layer is normalized to the total energy in the layer making use of the longitudinal shower shape.

The histograms and shape functions needed as input for the parametrizations use about 200 MB of memory. Since no simulation of particle interactions is done, the dominant part of the simulation time is spent on the numerical integration of the lateral shape functions. Overall, the calorimeter simulation time for a single particle is a few microseconds, and a typical (e.g. $t\bar{t}$) event needs a few seconds.

The parameterization of FastCaloSim differs in several important ways from that of the Fast G4 Simulation.

Table 15 Simulation times per event, in kSI2K seconds, for single particles generated with $|\eta| < 3.0$ and with the same transverse momentum. All times are averaged over 500 events. ATLFAST-II uses full simulation for the inner detector and muon system and FastCaloSim in the calorimetry. ATLFAST-IIF uses FastCaloSim in the calorimetry and Fatras in the inner detector and muon system

Sample	Full Sim	Fast G4 Sim	ATLFAST-II	ATLFAST-IIF	ATLFAST-I
5 GeV $\mu^\pm$	0.879	0.899	1.28	0.633	0.011
50 GeV $\mu^\pm$	1.63	1.15	2.71	0.606	0.011
500 GeV $\mu^\pm$	12.0	10.4	11.8	0.615	0.011
1 GeV $e^\pm$	3.62	0.734	0.825	0.513	0.011
5 GeV $e^\pm$	17.8	1.64	1.00	0.542	0.011
50 GeV $e^\pm$	179	4.86	1.25	0.588	0.013
1 GeV $\pi^\pm$	2.40	1.48	0.701	0.515	0.011
5 GeV $\pi^\pm$	10.4	4.27	0.811	0.540	0.011
50 GeV $\pi^\pm$	94.7	30.3	1.04	0.569	0.011

Table 16 Simulation times per event, in kSI2K seconds, for the full simulation, Fast G4 simulation, ATLFAST-II, ATLFAST-IIF, and ATLFAST-I. ATLFAST-II uses full simulation for the inner detector and muon system and FastCaloSim in the calorimetry. ATLFAST-IIF uses FastCaloSim in the calorimetry and Fatras in the inner detector and muon system. All times are averaged over 250 events, except heavy ion times which were averaged over only 50 events. Because the memory required to reconstruct heavy ion events exceeds 3 GB and because FastCaloSim runs during the reconstruction step, the amount of time taken by FastCaloSim could not be measured in that sample. It was estimated as 10% of the full inner detector simulation time, consistent with the other hard scattering events

Sample	Full Sim	Fast G4 Sim	ATLFAST-II	ATLFAST-IIF	ATLFAST-I
Minimum bias	551	246.	31.2	2.13	0.029
$t\bar{t}$	1990	757	101.	7.41	0.097
Jets	2640	832	93.6	7.68	0.084
Photon and jets	2850	639	71.4	5.67	0.063
$W^\pm \to e^\pm \nu_e$	1150	447.	57.0	4.09	0.050
$W^\pm \to \mu^\pm \nu_\mu$	1030	438	55.1	4.13	0.047
Heavy ion	56,000	21,700	~3050	203	5.56

FastCaloSim fills the readout geometry of ATLAS and applies a parameterization from the edge of the inner detector, whereas the Fast G4 Simulation places hits like those of GEANT4 into the full ATLAS detector geometry and is only applied in the sampling portion of the calorimeter (e.g. excluding the cryostats surrounding the calorimetry). As a result, the Fast G4 Simulation output can be run through the standard digitization software, whereas the FastCaloSim output is fed directly into the reconstruction.

7.4 Computing performance

Examples of simulation times in kSI2K seconds [117] for various types of events in the full and fast simulations are provided in Tables 15 and 16.[15] In single central ($|\eta| < 3$)

electron events the simulation time is decreased by a factor of ten or more by the fast G4 simulation, and in hard scattering events the simulation time is decreased by a factor of 2–5. ATLFAST-II without Fatras decreases simulation time by a factor of 20–40, and ATLFAST-IIF decreases simulation time by a factor of 100. FastCaloSim accounts for about 10% of the total simulation time in ATLFAST-II and 60–70% of the total simulation time in ATLFAST-IIF. ATLFAST-I requires a relatively negligible amount of CPU time even for hard scattering events. ATLFAST-I, FastCaloSim, and Fatras run during the reconstruction step (see Sect. 2.1), but for these purposes the time consumed by their methods is included in "simulation time." Figure 8 shows the distribution of simulation times per event for full, Fast G4, and ATLFAST-II simulation of 250 $t\bar{t}$ events. The distributions are similar in shape.

In evaluating these CPU times, it is necessary to keep in mind the ad ditional steps required before analysis of the data can be performed. For both full and fast G4 simulation,

---

[15] Measurements were performed on Sun Fire X2200 M2 units with dual dual-core 2.6 GHz AMD Opteron 2218 processors. All times assume the jobs run on a single core, with the other cores loaded. Normalization was done using the peak specmark int 2000 rating, 1794. For the same system, the peak specmark floating point 2000 rating was 3338. The normalization follows the published results, rather than the

WLCG formula in [118]. Details of cross-platform benchmarking can be found in [119].

Fig. 8 Distributions of CPU time for 250 $t\bar{t}$ events in full, Fast G4, and ATLFAST-II simulations. *Vertical dotted lines* denote the averages of the distributions

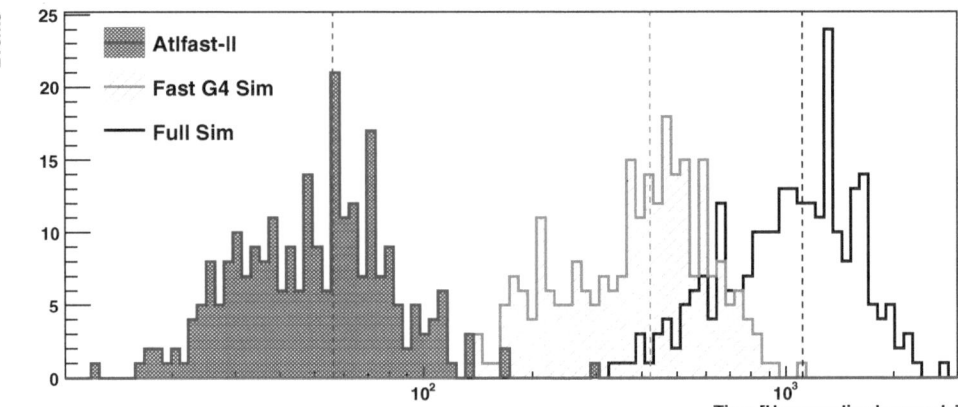

the data must be digitized and reconstructed. For ATLFAST-II, the inner detector and muon system must be digitized[16] and reconstructed, but the calorimeter requires only reconstruction. For ATLFAST-IIF, only the muon system must be digitized before reconstruction is performed. The output of ATLFAST-I is in a format similar to that of the reconstruction and needs no further processing. The CPU time required for these additional steps is discussed in Sect. 8.2.

7.5 Physics performance

The fast simulations have been compared to full simulation in both low-level analyses with single particles entering the calorime ter and high-level analyses of detector observables with jets and active hard scattering events. The Fast G4 Simulation agrees to about 1–2% in jet energy scale after the standard calibration procedure and agrees to within 5% percent in electron identification efficiencies. Due to the simplifications in the calorimeter simulation, FastCaloSim differs at the 5% level from full simulation after reconstruction, especially in properties that are sensitive to the shape of hadronic showers. The jet energy scale differs by 1–2% after recalibration, and electron identification efficiency differs by about 5%. Since all particles are simulated using an average lateral shape function, visible effects like electromagnetic subshowers in charged pion showers are not described. These differences can be reduced by applying additional object-dependent correction functions after reconstruction. Fakes and calorimeter punch-through are not well modeled in ATLFAST-II and ATLFAST-IIF.

Figure 9 shows missing transverse energy along the x-axis for the full and fast simulations in di-jet events with a leading parton p_T between 560 and 1120 GeV, as well as jet p_T resolution as a function of η in $t\bar{t}$ events for jets with $20 < p_T^{\text{True}} < 40$ GeV. ATLFAST-II and the Fast G4

Simulation agree well with full simulation in missing transverse energy spectrum, even in the tails of the distribution. ATLFAST-I does not sufficiently populate the tails of the missing transverse energy distribution, and ATLFAST-IIF has too wide a distribution. ATLFAST-I, ATLFAST-IIF, and ATLFAST-II show 10–20% deviations from full simulation in jet transverse momentum resolution. These fast simulations mostly deal with the response of individual hadrons, which is not always sufficient to model the threshold effects common in jet and missing transverse energy measurements. Fast G4 simulation is consistent with full simulation through the entire range in pseudorapidity.

Figure 10 shows reconstructed muon p_T resolution as a function of muon p_T in $Z \to \mu^+\mu^-$ events. Muons reconstructed using the muon spectrometer alone and those reconstructed using both the muon spectrometer ("standalone") and inner detector ("combined") are shown. Only one type of muon is provided by ATLFAST-I, so it is only included in the combined reconstruction plot. The difference between full simulation and ATLFAST-I is due to an older parameterization. In the cases of ATLFAST-II and the Fast G4 simulation, muon spectrometer simulation is done by GEANT4 and should, therefore, be identical to full simulation. The fast simulations show generally good agreement over the entire range of p_T. ATLFAST-IIF has standalone muon resolution that is 10% better than full simulation in some bins of p_T, but since the muon system simulation of ATLFAST-IIF is still under development, the agreement is expected to improve. It is generally left to the physics groups to evaluate the fast simulations with their analyses and determine which is acceptable.

8 Validation

Validation of the ATLAS simulation chain is done in two distinct phases. First, the software performance must be assessed. Then, the physics performance must be tested and

---

[16]The inner detector and muon system together require about 2/3 of the total digitization time.

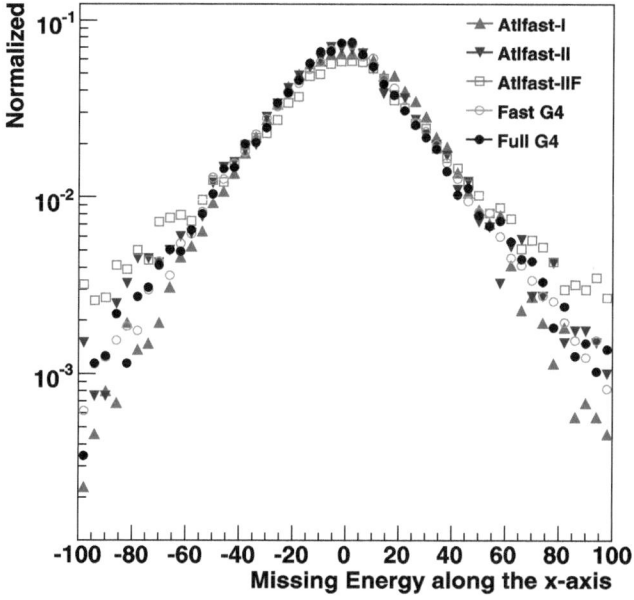

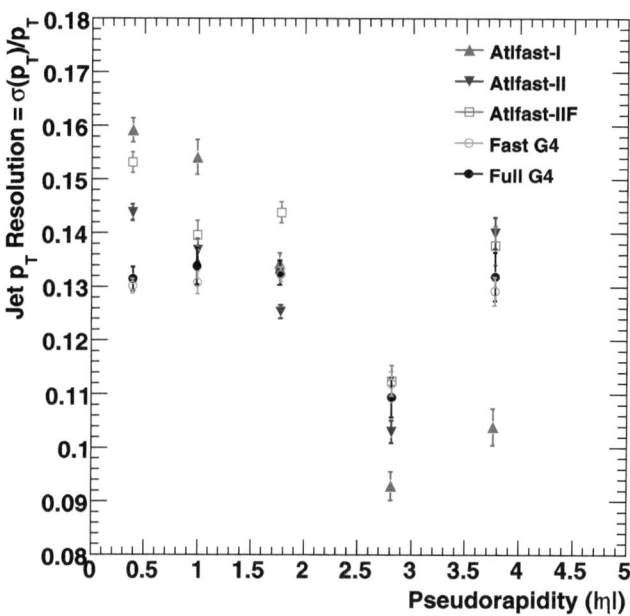

Fig. 9 *Left*, fast simulations (*color*) and full simulation (*black*) comparison of missing transverse energy along the x-axis in di-jet events with a leading parton p_T between 560 and 1120 GeV. *Right*, a compar- ison of jet p_T resolution as a function of pseudo rapidity in $t\bar{t}$ events for jets with $20 < p_T^{\text{True}} < 40$ GeV

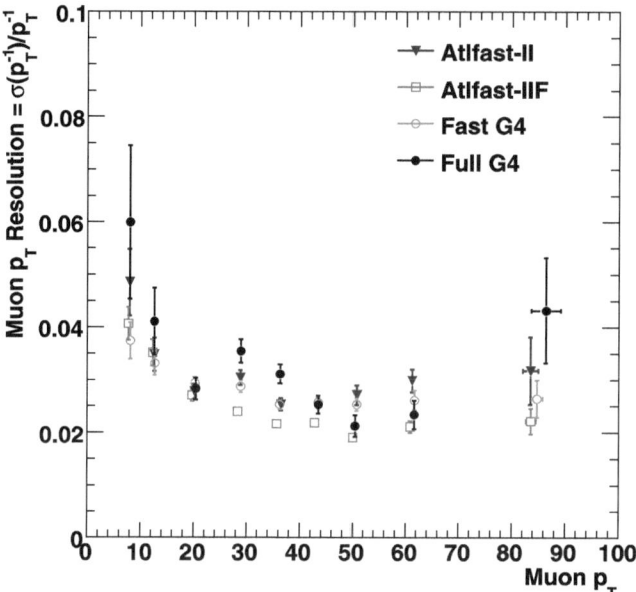

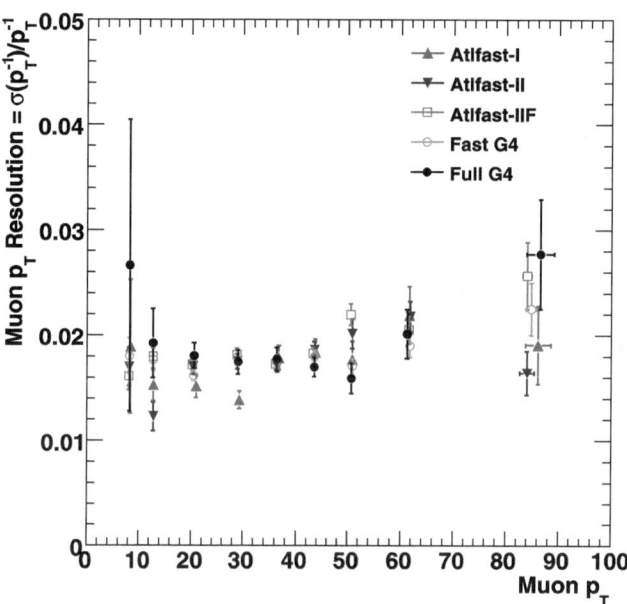

Fig. 10 Fast simulations (*color*) and full simulation (*black*) comparison of reconstructed muon p_T resolution as a function of muon p_T for central ($|\eta| < 1.2$) muons in $Z \rightarrow \mu^+\mu^-$ events for muons recon- structed using only the muon spectrometer (*left*) and using both the inner detector and spectrometer (*right*). ATLFAST-I only provides one type of muon, which is included in the right plot

compared to available data. The first step includes testing robustness, testing software performance, and testing basic functionality. The second step includes comparison to test beam, cosmic data, and physics results obtained from previous simulation productions. In this section the infrastructure for each stage of validation is described.

ATLAS has a fresh software build every night for co- ordinating software development. Each nightly build is run through a rigorous test cycle, and as a release deadline approaches the test results are increasingly scrutinized to evaluate stability and performance. Thanks to the evaluation prior to release, generally only rare bugs appear in production for the first time. The automatic testing infrastructure also allows evaluation of many different versions of the Athena software. Separate bug-fixing and development branches are employed, for example, and significant interface changes or low-level code migrations take place in separate branches until they are sufficiently stable to be merged into the main branch. Each version of the software comes in several flavors for different system architectures, operating systems, compilers, and so on. The simulation production in 2008 used 32-bit builds with gcc 3.4.6 [120] on CERN's Scientific Linux 4 [121]. External dependencies include CLHEP 1.9.3.1 and WLCG 54G.

A web portal has been constructed, using the Savannah bug tracking software [122], for monitoring problems with the various facets of the simulation software. Bugs can be reported and tracked as they are diagnosed and solutions are found, and new features can be requested.

8.1 Automated testing

The software performance of the simulation is monitored in three types of automated tests: ATLAS Testing Nightly (ATN) tests, Run Time Tester (RTT) tests, and Full Chain Tests (FCT). ATN tests are run every night on every software build and are basic functionality tests. RTT tests are run on a subset of builds and include 50 simulation tests to ensure functionality and, in some cases, consistent results. FCT tests are run on only a few builds each night and test the entire software chain in a production-like environment. Releases are required to pass a minimal number of milestones before being declared ready for production. For details about the ATN and RTT, refer to [9].

FCT tests are run daily on a small set of jobs. The aim of the FCT is to verify the readiness of a production cache release candidate for Grid production. The FCT runs jobs that test the functionalities of generation, the different flavors of fast and full simulation, digitization, bytestream conversion, and reconstruction of Standard Model processes, black hole production, and heavy ion collisions. In the case of Standard Model processes, a full chain[17] of jobs are run per release, with hard scattering events that stress the software. If successful, the output from each day's run is saved for use in the next step of the test the following day (e.g. Monday's generation provides input for Tuesday's simulation, which provides input for Wednesday's digitization). The typical number of events processed (50) is limited by the CPU requirements for the full simulation.

As part of the FCT, 1000 events that were simulated with an old validated release are reconstructed. This long test allows better evaluation of the reconstruction's stability. Moreover, the relatively large sample is used to make a preliminary check on the quality of the reconstruction for final state objects (jets, electrons, muons, etc.). All the other tests only check for the success or failure of the job, the number of events in the output file, and unknown error messages in the log file. If any of these checks fail, the release candidate is rejected, and an additional iteration of bug fixing is undertaken. Only once a release succeeds in all FCT tests is it distributed to the Grid.

8.2 Computing performance benchmarking

Event generation jobs are typically fast enough that not a large effort is made to test their software performance. In general, the jobs take a tens of milliseconds per event. Generation with PYTHIA or HERWIG requires about 450 MB of memory, and generation with Hijing requires about 170 MB of memory. The files produced in generation jobs are tens of kB per event (e.g. 40 kB/event for $t\bar{t}$ events).

CPU time, memory consumption, and output file size for the simulation are tested in each stable release using a variety of physics processes. Single muons, electrons, and charged pions are used, as well as di-jets in bins of leading parton p_T, a supersymmetric benchmark point,[18] minimum bias, Higgs boson decaying to four leptons, $Z \rightarrow e^+e^-$, $Z \rightarrow \mu^+\mu^-$, and $Z \rightarrow \tau^+\tau^-$ events. The same input events are used every time to ensure fair comparison of simulation results independent of changes to the event generation.

Simulation of the full detector requires typically ~750 MB of memory (i.e. VSIZE) and includes loading of almost 400 libraries into memory. Memory consumed during simulation is also broken down into its three key components by taking snapshots of memory use during initialization: GeoModel, the ATLAS-side detector geometry, typically about ~100 MB; G4Atlas, the purely GEANT4 component of the memory, typically about ~300 MB; and load modules, the remaining algorithms and services loaded during the job, typically about ~300 MB. Significant changes in any of these can indicate the proper source of a change in memory. The memory requirement is independent of the number of events in the job and varies by only a few percent for different physics processes. Although up to 2 GB of memory may be reserved for a Grid job, by keeping the memory requirements of a typical simulation job under 1 GB, more machines can be used. The memory required to build each piece of the detector can be found in Table 4. Of major concern is any increase in memory ("leaks") during the event loop once all libraries have been loaded and setup is complete. Some increase due to caching is expected

---

[17]A single chain of jobs runs all steps from event generation through reconstruction, sequentially, using the output of one step as the input of the next.

[18]ATLAS mSUGRA benchmark point SU3: $m_0 = 100$ GeV, $m_{1/2} = 300$ GeV, $A_0 = -300$, $\tan\beta = 6$, and $\mu > 0$.

Table 17 Generation, simulation, digitization, and reconstruction times per event, in kSI2K seconds. Generation times are averaged over 5000 events, except generation of heavy ion events, which were averaged over 250 events. Simulation, digitization, and reconstruction times are averaged over 250 events, except simulation of heavy ion events, which were averaged over 50 events. The heavy ion event sim- ulation time is for events with a random impact parameter. Central collisions require on average 3.4 times longer to simulate. Reconstruction times should be taken as indicative of the order of magnitude, rather than as a precise measurement. Based on a previous release, heavy ion collision reconstruction is estimated to take ~ 10 times longer than $t\bar{t}$ event reconstruction

Sample	Generation	Simulation	Digitization	Reconstruction
Minimum bias	0.0267	551	19.6	8.06
$t\bar{t}$ Production	0.226	1990	29.1	47.4
Jets	0.0457	2640	29.2	78.4
Photon and jets	0.0431	2850	25.3	44.7
$W^{\pm} \to e^{\pm}\nu_e$	0.0788	1150	23.5	8.07
$W^{\pm} \to \mu^{\pm}\nu_\mu$	0.0768	1030	23.1	13.6
Heavy ion	2.08	56,000	267	–

Table 18 Digitization computing resources for 50 $t\bar{t}$ events as they scale with luminosity. CPU times are normalized to the time required by a no pile-up job. Cavern background was overlaid during these jobs with a safety factor of one. Beam gas and beam halo were ignored

Resource	No Pile-up	10^{33} cm^{-2} s^{-1}	3.5×10^{33} cm^{-2} s^{-1}	10^{34} cm^{-2} s^{-1}
CPU time factor	1.0	2.3	5.8	160
Memory leak [kB/event]	10	270	800	2,100
Virtual memory [MB]	770	1,000	1,300	2,000
Allocated memory [MB/event]	12	21	40	985

during the processing of the first few events. However, if the memory required by the application continues to grow beyond the system limits, memory corruption and memory pressure can result in serious problems. The memory required by the ATLAS simulation has been found to increase by less than 0.25 MB per event under normal circumstances. The increases are not steady, but come in large ($\sim$10 MB) and sporadic jumps. The source of these increases is not fully understood, but a 50 event simulation job still consumes well under 1 GB of memory.

The CPU time consumed by generation, full simulation, digitization, and reconstruction for various types of events is shown in Tables 15 and 17 and is typically several minutes per hard scattering event. All times are normalized to kSI2K seconds [117]. For the purposes of testing, logfile output was suppressed and no output files (e.g. hit or RDO files) were created. CPU time is also measured as a function of other simulation input parameters prior to significant changes, for example using different physics lists. For these runs, output files were disabled; in simulating $t\bar{t}$ events the time per event is increased by $\sim$0.5% when file writing is enabled. The hard scattering events shown in Table 17 were generated with a 14 TeV center of mass energy; for 10 TeV center of mass energy the simulation time is reduced by 17% for $t\bar{t}$ events. The distribution of CPU time for simulation of 250 $t\bar{t}$ events is shown in Fig. 8.

For the samples in Tables 16 and 17, event generation of W production, minimum bias interactions, di-jet events, and photon and jet events was done using PYTHIA. $t\bar{t}$ production was done using MC@NLO for the hard scattering and Herwig for hadronization and showering. In this case, the generation time includes only the time required by Herwig. Heavy ion event production was done using Hijing.

Digitization jobs are generally fast, but memory consumption can be a serious concern during jobs with many overlaid events. Table 18 shows how resource consumption during digitization of 50 $t\bar{t}$ events scales with pile-up luminosity.[19] The memory required for digitizing with a luminosity of 10^{34} cm^{-2} s^{-1} is sufficiently large that the memory limit of the testing machine was reached, and, therefore, swapping resulted in a significant increase in CPU time. The allocated memory per event is provided for some benchmark of the change in memory over the course of a single event.

8.3 Physics validation

Once a new release is distributed to the Grid sites, a set of several physics samples is produced. Typically, a "valida-

---

[19] The ATLAS software is under a continuous process of improvement, with improving performance in terms of calculation speed and memory profile, and problems such as memory leaks being identified and eliminated.

tion sample" includes 10,000 events for each process, a total of 110,000 single particle events and 250,000 hard scattering events. This standard validation sample includes single muons, pions, and electrons, Standard Model processes ($t\bar{t}$ production, vector boson production, B-physics), and exotic processes (e.g. supersymmetric events and black hole production). The composition of the validation sample has been chosen to test all aspects of the event reconstruction.

The running of the validation sample on the Grid usually exposes rare software problems in the release. It is unlikely that software bugs that appear with a frequency much lower than 1/1000 events are caught by the automatic validation procedure (1000 events is the size of the "long" jobs of the FCT, described in Sect. 8.1). This first round of production provides a feedback mechanism for the developers, who produce bug fixes before the next production cycle.

The last step before using a release for production is physics validation. A dedicated group of experts, including representatives from every detector performance (e.g. tracking, b-tagging, and jet reconstruction groups) and physics group (e.g. Standard Model, supersymmetry, and exotics search groups) in ATLAS, runs physics analyses on the validation samples. Their task is to verify the quality of the single object reconstruction (e.g. jets, electrons, and muons) and the results of more complex physics analyses (e.g. mass reconstruction in $Z \rightarrow \mu^+\mu^-$, $Z \rightarrow e^+e^-$, and $t\bar{t}$ events). The relatively large validation samples may expose minor problems that could not be found with lower statistics, for example a shift of a few percent in the reconstructed energy. In order to properly validate each version of the software, the results from each release are typically compared to those of previous validated releases. The software must, therefore, maintain backwards-compatibility in order to allow fair comparisons. Shifts in file format are carefully coordinated, and maintenance of the old format is continued for as long as necessary to ensure result consistency. The physics validation procedure is also used for checking major changes in the fast and full simulation (detector description, change in the simulation parameters, etc.).

The GEANT4 simulation has also been validated in a physics sense with all available detector data. Combined test beam studies have proven invaluable in understanding the performance of each of the subsystems, and the standalone test beam analyses have provided crucial input towards the optimization of the simulation and choice of parameters [123–126]. In 2008, a significant sample of cosmic ray data was collected with multiple subdetectors. The data have provided an important test of the simulation [127].

Although the detector simulation relies heavily on GEANT4, a significant effort was put into comparing tile calorimeter test stand response with the FLUKA simulation package [128]. For this comparison, the test stand geometry was translated into the FLUKA geometry format, and the output from the FLUKA simulation was translated back into a format comparable to that of GEANT4. It was eventually concluded that little would be gained by attempting a transition to FLUKA that could not already by gained by modifications to parameters and a different choice of physics models within GEANT4. FLUKA has also been used to study neutron flux and radiation levels throughout the detector [104], but many of these studies are being updated in GEANT4 with the high-precision neutron physics list (see Sect. 5.4).

Extensive efforts are underway to compare simulated data to real data and validate the output from each detector. Thanks to the multiple detector descriptions, several analyses have already been prepared and tested to find discrepancies between the detector description of the simulation and that of the as-built detector. For example, subdetectors can be "weighed" in the simulation to ensure that the amount of material is within a few percent of the constructed detector. Although the agreement with first high-energy collision data is not expected to be perfect, a great deal of experience has been gained. The effects of modifications to GEANT4 parameters have also been studied in some detail, so that differences between data and simulation might be remedied rapidly.

Digitization algorithms have been tuned against laboratory test results, test beam data, and, where possible, cosmic ray data taken during the detector commissioning. The studies continue with the data.

9 Summary and conclusions

We have presented the status of the ATLAS simulation project, including all steps from event generation to digitization. A robust and flexible framework is required to cope with the demands of complex detector descriptions and physics models. The software project has been prepared for data since late 2008 and is ready for data.

A variety of event generators are available to provide the user with a complete set of tools for testing new physics models. The simulation is highly configurable to ensure maximal flexibility in the face of the uncertain challenges approaching. The detector description itself, conditions of the detector, and many parameters used in the simulation can be modified at run time. The digitization is also made configurable to cope with uncertainty in machine performance, detector conditions, and cavern conditions. Three varieties of fast simulation have been made available to ease the difficulties caused by the time consumption of the full detector simulation. They each complement the full simulation.

Generation, simulation, and digitization tasks are running continually on the Grid. The validation program has produced a high quality simulation sample for the ATLAS experiment data.

Acknowledgements We are greatly indebted to all CERN's departments and to the LHC project for their immense efforts not only in building the LHC, but also for their direct contributions to the construction and installation of the ATLAS detector and its infrastructure. We acknowledge equally warmly all our technical colleagues in the collaborating Institutions without whom the ATLAS detector could not have been built. Furthermore we are grateful to all the funding agencies which supported generously the construction and the commissioning of the ATLAS detector and also provided the computing infrastructure.

The ATLAS detector design and construction has taken about fifteen years, and our thoughts are with all our colleagues who sadly could not see its final realisation.

We acknowledge the support of ANPCyT, Argentina; Yerevan Physics Institute, Armenia; ARC and DEST, Australia; Bundesministerium für Wissenschaft und Forschung, Austria; National Academy of Sciences of Azerbaijan; State Committee on Science & Technologies of the Republic of Belarus; CNPq and FINEP, Brazil; NSERC, NRC, and CFI, Canada; CERN; CONICYT, Chile; NSFC, China; COLCIENCIAS, Colombia; Ministry of Education, Youth and Sports of the Czech Republic, Ministry of Industry and Trade of the Czech Republic, and Committee for Collaboration of the Czech Republic with CERN; Danish Natural Science Research Council and the Lundbeck Foundation; European Commission, through the ARTEMIS Research Training Network; IN2P3-CNRS and Dapnia-CEA, France; Georgian Academy of Sciences; BMBF, DFG, HGF and MPG, Germany; Ministry of Education and Religion, through the EPEAEK program PYTHAGORAS II and GSRT, Greece; ISF, MINERVA, GIF, DIP, and Benoziyo Center, Israel; INFN, Italy; MEXT, Japan; CNRST, Morocco; FOM and NWO, Netherlands; The Research Council of Norway; Ministry of Science and Higher Education, Poland; GRICES and FCT, Portugal; Ministry of Education and Research, Romania; Ministry of Education and Science of the Russian Federation, Russian Federal Agency of Science and Innovations, and Russian Federal Agency of Atomic Energy; JINR; Ministry of Science, Serbia; Department of International Science and Technology Cooperation, Ministry of Education of the Slovak Republic; Slovenian Research Agency, Ministry of Higher Education, Science and Technology, Slovenia; Ministerio de Educación y Ciencia, Spain; The Swedish Research Council, The Knut and Alice Wallenberg Foundation, Sweden; State Secretariat for Education and Science, Swiss National Science Foundation, and Cantons of Bern and Geneva, Switzerland; National Science Council, Taiwan; TAEK, Turkey; The Science and Technology Facilities Council and The Leverhulme Trust, UK; DOE and NSF, USA; Gordon and Betty Moore Foundation, USA.

The authors would like to thank the GEANT4 developers for their work and support.

References

1. G. Aad et al., The ATLAS experiment at the CERN Large Hadron Collider. J. Instrum. **3**, S08003 (2008)
2. L. Evans, The Large Hadron Collider. New J. Phys. **9**, 335 (2007)
3. ATLAS Collaboration, ATLAS computing technical design report. ATLAS-TDR-017, CERN-LHCC-2005-022 (2005). See also http://atlas-computing.web.cern.ch/atlas-computing/packages/athenaCore/athenaCore.php
4. S. Agostinelli et al., Geant4—a simulation toolkit. Nucl. Instrum. Methods Phys. Res. A **506**, 250–303 (2003)
5. J. Allison et al., Geant4 developments and applications. IEEE Trans. Nucl. Sci. **53**, 270–278 (2006)
6. M. Cattaneo et al., Status of the GAUDI event-processing framework, in *Computing in High Energy and Nuclear Physics 2001 Conference (CHEP2001)*, IHEP, Beijing, China, September 3–7, 2001, ed. by H.S. Chen, (e-article, 2001)
7. G. Barrand et al., GAUDI—a software architecture and framework for building LHCb data processing applications, in *Computing in High Energy and Nuclear Physics 2000 Conference (CHEP2000)*, ed. by M. Mazzucato, Pavia, Italy, February 7–11, 2000 (Padua, INFN, 2000)
8. L. Lonnblad, CLHEP: a project for designing a C++ class library for high-energy physics. Comput. Phys. Commun. **84**, 307–316 (1994)
9. E. Obreshkov et al., Organization and management of ATLAS offline software releases. Nucl. Instrum. Methods A **584**, 244–251 (2008)
10. CLOC—Count Lines of Code, http://cloc.sourceforge.net, 2006
11. D. Dullmann et al., POOL development status and plans, in *Computing in High Energy and Nuclear Physics 2004 (CHEP2004)*, ed. by A. Aimar, J. Harvey, N. Knoors, Interlaken, Switzerland, September 27–October 1, 2004 (CERN, Geneva, 2004)
12. R. Chytracek et al., POOL development status and production experience. IEEE Trans. Nucl. Sci. **52**, 2827–2831 (2005)
13. D. Duellmann, The LCG POOL project, general overview and project structure, in *Computing in High Energy and Nuclear Physics 2003 (CHEP2003)*, La Jolla, CA, March 24–28, 2003. (CERN, Geneva, 2003). physics/0306129
14. M. Dobbs, J.B. Hansen, The HepMC C++ Monte Carlo event record for high energy physics. Comput. Phys. Commun. **134**, 41–46 (2001)
15. M. Grothe et al., Architecture of the ATLAS high level trigger event selection software. Nucl. Instrum. Methods Phys. Res. A **518**, 537–541 (2003). In Proceedings of the 2003 Conference for Computing in High Energy and Nuclear Physics, La Jolla, CA, USA, 24–28 March 2003
16. V. Boisvert et al., Final report of the ATLAS reconstruction task force. ATL-SOFT-2003-010 (2003)
17. I. Bird et al. (eds.), LHC computing Grid. Technical design report, CERN-LHCC-2005-024 (2005)
18. P. Nevski, Large scale data movement on the GRID, (2006)
19. M. Branco et al., Managing ATLAS data on a petabyte-scale with DQ2. J. Phys. Conf. Ser. **119**, 062017 (2007). In Proceedings of Computing in High Energy and Nuclear Physics 2007 Conference (CHEP2007), Victoria, BC, Canada, September 2–7, 2007, ed. R. Sobie, R. Tafirout, J. Thomson
20. T. Sjöstrand, S. Mrenna, P. Skands, PYTHIA 6.4 physics and manual. J. High Energy Phys. **05**, 026 (2006). hep-ph/0603175
21. M. Smizanska, S.P. Baranov, J. Hrivnac, E. Kneringer, Overview of Monte Carlo simulations for ATLAS B-physics in the period 1996–1999. ATL-PHYS-2000-025 (2000)
22. C. Anastopoulos et al., Physics analysis tools for beauty physics in ATLAS. J. Phys. Conf. Ser. **119**, 032003 (2007). In Proceedings of Computing in High Energy and Nuclear Physics 2007 Conference (CHEP2007), Victoria, BC, Canada, September 2–7, 2007, ed. R. Sobie, R. Tafirout, J. Thomson
23. G. Marchesini, B.R. Webber, G. Abbiendi, I.G. Knowles, M.H. Seymour, L. Stanco, HERWIG: a Monte Carlo event generator for simulating hadron emission reactions with interfering gluons. Version 5.1, April 1991. Comput. Phys. Commun. **67**, 465 (1992)
24. G. Corcella et al., HERWIG 6: an event generator for hadron emission reactions with interfering gluons (including supersymmetric processes). J. High Energy Phys. **01**, 010 (2001). hep-ph/0011363
25. G. Corcella et al., HERWIG 6.5 release note. CERN-TH/2002-270 (2005). hep-ph/0210213v2
26. T. Gleisberg et al., SHERPA 1.alpha, a proof-of-concept version. J. High Energy Phys. **02**, 056 (2004). hep-ph/0311263

27. M. Gyulassy, X.N. Wang, HIJING 1.0: a Monte Carlo program for parton and particle production in high-energy hadronic and nuclear collisions. Comput. Phys. Commun. **83**, 307 (1994). nucl-th/9502021

28. M.L. Mangano et al., ALPGEN, a generator for hard multiparton processes in hadronic collisions. J. High Energy Phys. **07**, 001 (2003). hep-ph/0206293

29. S. Frixione, P. Nason, B.R. Webber, Matching NLO QCD and parton showers in heavy flavour production. J. High Energy Phys. **08**, 007 (2003). hep-ph/0305252

30. B.P. Kersevan, E. Richter-Was, The Monte Carlo event generator AcerMC version 2.0 with interfaces to PYTHIA 6.2 and HERWIG 6.5 (2004). hep-ph/0405247

31. S. Jadach, J.H. Kuhn, Z. Was, TAUOLA: a Library of Monte Carlo programs to simulate decays of polarized tau leptons. Comput. Phys. Commun. **64**, 275–299 (1990)

32. E. Barberio, B. van Eijk, Z. Was, PHOTOS: a universal Monte Carlo for QED radiative corrections in decays. Comput. Phys. Commun. **66**, 115–128 (1991)

33. D.J. Lange, The EvtGen particle decay simulation package. Nucl. Instrum. Methods A **462**, 152–155 (2001)

34. F.E. Paige, S.D. Protopopescu, H. Baer, X. Tata, ISAJET 7.69: a Monte Carlo event generator for p p, anti-p p, and e+ e- reactions (2003). hep-ph/0312045

35. T. Sjöstrand, S. Mrenna, P. Skands, A brief introduction to PYTHIA 8.1. Comput. Phys. Commun. **178**, 852–867 (2008). 0710.3820 [hep-ph]

36. M. Bahr et al., Herwig++ physics and manual. Eur. Phys. J. C **58**, 639–707 (2008). 0803.0883 [hep-ph]

37. T. Stelzer, W.F. Long, Automatic generation of tree level helicity amplitudes. Comput. Phys. Commun. **81**, 357–371 (1994). hep-ph/9401258

38. C.M. Harris, P. Richardson, B.R. Webber, CHARYBDIS: a black hole event generator. J. High Energy Phys. **08**, 033 (2003). hep-ph/0307305

39. E. Boos et al. (CompHep Collaboration), CompHEP 4.4: Automatic computations from Lagrangians to events. Nucl. Instrum. Methods A **534**, 250 (2004). hep-ph/0403113

40. A. Pukhov et al., CompHEP—a package for evaluation of Feynman diagrams and integration over multi-particle phase space. User's manual for version 3.3. INP MSU report 98-41/542 (1999). hep-ph/9908288. See also http://comphep.sinp.msu.ru

41. T. Sjöstrand et al., in *Z Physics at LEP 1*, vol. 3 (Geneva, 1989), p. 143

42. A. Moraes, C. Buttar, I. Dawson, Prediction for minimum bias and the underlying event at LHC energies. Eur. Phys. J. C **50**, 435–466 (2007)

43. A. Moraes, C. Buttar, I. Dawson, Comparison of predictions for minimum bias event generators and consequences for ATLAS radiation background. ATL-PHYS-2003-020 (2002)

44. V.M. Abazov et al. (DZero Collaboration), Search for large extra dimensions in the monojet + MET channel with the DZero detector. Phys. Rev. Lett. **90**, 251802 (2003). hep-ex/0302014

45. T. Aaltonen et al. (CDF Collaboration), Measurement of the inclusive jet cross section at the Fermilab Tevatron p anti-p collider using a cone-based jet algorithm. Phys. Rev. D **78**, 052006 (2008)

46. R. Field, R.C. Group (CDF Collaboration), PYTHIA Tune A, HERWIG, and JIMMY in Run 2 at CDF, CDF-ANAL-CDF-PUBLIC-7822 (2005). hep-ph/0510198

47. E. Norrbin, T. Sjostrand, Production and hadronization of heavy quarks. Eur. Phys. J. C **17**, 137–161 (2000). hep-ph/0005110

48. A. Moraes, in *Proceedings of the First International Workshop on Multiple Partonic Interactions at the LHC (MPI@LHC08)*, Perugia, Italy, October 27–31, 2008, ed. by P. Bartalini, L. Fano (DESY-PROC, 2008), to appear

49. A.A. Affolder et al. (CDF Collaboration), Charged jet evolution and the underlying event in $p\bar{p}$ collisions at 1.8 TeV. Phys. Rev. D **65**, 092002 (2002)

50. D.E. Acosta et al. (CDF Collaboration), The underlying event in hard interactions at the Tevatron $\bar{p}p$ collider. Phys. Rev. D **70**, 072002 (2004) hep-ex/0404004

51. J.M. Butterworth, J.R. Forshaw, M.H. Seymour, Z. Phys. C **72**, 637–646 (1996)

52. S. Catani, F. Krauss, R. Kuhn, B.R. Webber, QCD matrix elements + parton showers. J. High Energy Phys. **11**, 63 (2001). hep-ph/0109231

53. M. Boonekamp et al. Cosmic ray, beam-halo and beam-gas rate studies for ATLAS commissioning. ATL-GEN-2004-001 (2004)

54. A. Dar, Atmospheric neutrinos, astrophysical neutrons, and proton-decay experiments. Phys. Rev. Lett. **51**, 227 (1983)

55. E. Boos et al., Generaic user process interface for event generators, in *Proceedings of Les Houches 2001: Physics at TeV Colliders*, Les Houches, France, May 21–June 1, 2001, ed. by W. Giele et al. (2001). hep-ph/0109068v1, The QCD/SM Working Group Report

56. J. Alwall et al., A standard format for Les Houches event files. Comput. Phys. Commun. **176**, 300–304 (2007). hep-ph/0609017

57. S. Frixione, P. Nason, C. Oleari, Matching NLO QCD computations with parton shower simulations: the POWHEG method. J. High Energy Phys. **11**, 070 (2007). 0709.2092 [hep-ph]

58. P. Skands et al., SUSY Les Houches accord: interfacing SUSY spectrum calculators, decay packages, and event generators. J. High Energy Phys. **0407**, 36 (2004). hep-ph/0311123

59. B.C. Allanach et al., SUSY Les Houches Accord 2. Comput. Phys. Commun. **180**, 8–25 (2009)

60. I. Borozan, M.H. Seymour, An eikonal model for multiparticle production in hadron hadron interactions. J. High Energy Phys. **09**, 015 (2002). hep-ph/0207283

61. D. Bourilkov, R.C. Group, M.R. Whalley, LHAPDF: PDF use from the Tevatron to the LHC, 2006. hep-ph/0605240

62. H. Plothow-Besch, PDFLIB: a library of all available parton density functions of the nucleon, the pion and the photon and the corresponding alpha-s calculations. Comput. Phys. Commun. **75**, 396–416 (1993)

63. W.K. Tung et al., Global QCD Analysis and Collider Phenomenology–CTEQ, in *Proceedings of the DIS2007 Workshop*, Munich, Germany, April 2007. 0707.0275 [hep-ph]

64. S. Chekanov et al. (ZEUS Collaboration), An NLO QCD analysis of inclusive cross-section and jet-production data from the ZEUS experiment. Eur. Phys. J. C **42**, 1–16 (2005). hep-ph/0503274

65. Albrow M. et al. (TeV4LHC QCD Working Group), Tevatron-for-LHC report of the QCD working group. FERMILAB-Conf-06-359 (2006). hep-ph/0610012

66. P. Skands, Color connections, multijets, in *Proceedings of the First International Workshop on Multiple Partonic Interactions at the LHC (MPI@LHC08)*, Perugia, Italy, October 27–31, 2008, ed. by P. Bartalini and L. Fano, (DESY-PROC, 2008), to appear

67. J. Boudreau, V. Tsulaia, The GeoModel toolkit for detector description, in *Computing in High Energy and Nuclear Physics 2004 (CHEP2004)*, ed. by A. Aimar, J. Harvey, N. Knoors, Interlaken, Switzerland, September 27–October 1, 2004 (CERN, Geneva, 2004)

68. G. Aad et al. (ATLAS Collaboration), Expected performance of the ATLAS experiment, detector, trigger, and physics. CERN-OPEN-2008-020 (2008)

69. F. Viegas, R. Hawkings, G. Dimitrov, Relational databases for conditions data and event selection in ATLAS. J. Phys. Conf. Ser. **119**, 001001 (2007). In Proceedings of Computing in High Energy and Nuclear Physics 2007 Conference (CHEP2007), Victoria, BC, Canada, September 2–7, 2007, ed. R. Sobie, R. Tafirout, J. Thomson

70. SQLite, http://www.sqlite.org, 2008
71. R.F. van der Lans, *The SQL Guide to SQLite*, 1st edn. http://lulu.com, 2009
72. C. Newman, *SQLite (Developer's Library)*, 1st edn. (Sams, 2004)
73. B. Di Girolamo et al., Beamline instrumentation in the 2004 combined ATLAS testbeam. ATL-TECH-PUB-2005-001 (2005)
74. C. Adorisio et al., Nucl. Instrum. Methods Phys. Res. A **593**, 232–254 (2008)
75. C. Adorisio et al., System test of the ATLAS muon spectrometer in the H8 beam at the CERN SPS. Nucl. Instrum. Methods A **593**, 232–254 (2008)
76. M. Hurwitz, Performance of ATLAS tile calorimeter production modules in calibration testbeams. Nucl. Instrum. Methods Phys. Res. A **572**, 80–81 (2007)
77. P. Strizenec, A. Minaenko (ATLAS LAr Endcap group), J. Phys. Conf. Ser. **160**, 012078 (2009)
78. H. Bartko, Performance of the combined ATLAS liquid argon end-cap calorimeters in beam tests at the CERN SPS. Diploma Thesis ATL-LARG-2004-007, 2004
79. ATLAS Pixel Collaboration, Pixel offline analysis for endcap A cosmic data. ATL-INDET-PUB-2008-003 (2008)
80. B. Demirkoz, Construction and performance of the ATLAS SCT barrels and cosmic tests. CERN-THESIS-2008-001 (2008)
81. Z. Broklova, Z. Dolezal, Simulations of ATLAS silicon strip detector modules in ATHENA framework. CERN-THESIS-2005-007 (2004)
82. T.H. Kittelmann, Slepton spin determination and simulation of the transition radiation tracker at the ATLAS experiment. PhD in physics, Niels Bohr Institute, University of Copenhagen, 2007
83. D. Salihagic, Comparison of beam test results of the combined ATLAS liquid argon endcap calorimeters with Geant3 and Geant4 simulations, in *11th International Conference on Calorimetry in High Energy Physics*, ed. by C. Cecci, P. Cenci, P. Lubrano, M. Pepe-Altarelli (World Scientific, Singapore, 2004), p. 314
84. A. Dotti, A. Lupi, C. Roda, Results from ATLAS tile calorimeter: a comparison between data and Geant4 simulation. Nucl. Phys. B, Proc. Suppl. **150**, 106–109 (2006)
85. N.C. Benekos, D. Rebuzzi, GEANT4 simulation of the ATLAS muon spectrometer, in *9th ICATPP Conference on Astroparticle, Particle, Space Physics, Detectors and Medical Physics Applications*, ed. by M. Barone, E. Borchi, A. Gaddi, C. Leroy, L. Price, Villa Erba, Como, Italy, 17–21 October, 2005 (World Scientific, Singapore, 2005)
86. K. Kordas, G. Parrour, S. Simion, GEANT4 for the ATLAS electromagnetic calorimeter, in *9th Conference on Calorimetry in High Energy Physics (CALOR 2000)*, ed. by B. Aubert, J. Colas, P. Nédélec, L. Poggioli, Annecy, France, 9–14 October 2000 (Lab. Annecy Phys. Part., Annecy-le-Vieux, 2000)
87. A. Kiryunin, D. Salihagic, P. Strizenec (ATLAS/LAr-HEC Collaboration), GEANT4 simulations of the ATLAS hadronic end cap calorimeter, in *9th Conference on Calorimetry in High Energy Physics (CALOR 2000)*, ed. by B. Aubert, J. Colas, P. Nédélec, L. Poggioli, Annecy, France, 9–14 October 2000 (Lab. Annecy Phys. Part., Annecy-le-Vieux, 2000)
88. P. Loch, R. Mazini, Comparison of experimental electron signals with GEANT3 and GEANT4 simulations for the ATLAS forward calorimeter prototype, in *9th Conference on Calorimetry in High Energy Physics (CALOR 2000)*, ed. by B. Aubert, J. Colas, P. Nédélec, L. Poggioli, Annecy, France, 9–14 October 2000 (Lab. Annecy Phys. Part., Annecy-le-Vieux, 2000)
89. A. Dell'Acqua et al., Development of the ATLAS simulation framework, in *Computing in High Energy and Nuclear Physics 2001 Conference (CHEP2001)*, IHEP, Beijing, China, September 3–7, 2001, ed. by H.S. Chen, (e-article, 2001)
90. A. Dell'Acqua et al., ATLAS simulation readiness for first data at LHC. J. Phys. Conf. Ser. **119**, 001001 (2007). In Proceedings of Computing in High Energy and Nuclear Physics 2007 Conference (CHEP2007), Victoria, BC, Canada, September 2–7, 2007, ed. R. Sobie, R. Tafirout, J. Thomson
91. C. Amsler et al.(Particle Data Group), Phys. Lett. B **667**, 1 (2008)
92. G. Battistoni et al., The FLUKA code: description and benchmarking. AIP Conf. Proc. **896**, 31–49 (2007). In Proceedings of the Hadronic Shower Simulation Workshop 2006, Fermilab 6–8 September 2006
93. A. Fassò, A. Ferrari, J. Ranft, P.R. Sala, FLUKA: a multi-particle transport code. CERN 2005-10, 2005, Also INFN/TC_05/11, SLAC-R-773
94. A. Rimoldi et al., First report of the simulation optimization group. ATL-SOFT-PUB-2008-002 (2008)
95. A. Rimoldi et al., Final report of the simulation optimization task force. ATL-SOFT-PUB-2008-004 (2008)
96. R. Brun, F. Rademakers, ROOT—an object oriented data analysis framework. Nucl. Instrum. Methods Phys. Res. A **389**, 81–86 (1997). In Proceedings AIHENP '96 Workshop, Lausanne, Sep. 1996. See also http://root.cern.ch
97. T. Kittelmann (ATLAS Collaboration), The virtual point 1 event display for the ATLAS experiment, in *Proceedings of Computing in High Energy and Nuclear Physics 2009 Conference (CHEP2009)*, Prague, Czech Republic, March 22–28, 2009
98. G. Taylor, Visualizing the ATLAS inner detector with Atlantis. Nucl. Instrum. Methods Phys. Res. A **549**, 183–187 (2005)
99. M. Virchaux, D. Pomarède, The PERSINT manual. ATL-SOFT-2001-003 (2001)
100. S. Gadomski, Model of the SCT detectors and electronics for the ATLAS simulation using Geant4. ATL-SOFT-2001-005 (2001)
101. W. Lampl et al., Digitization of LAr calorimeter for CSC simulations. ATL-LARG-PUB-2007-011 (2007)
102. D. Rebuzzi et al., Geant4 muon digitization in the ATHENA framework. ATL-SOFT-PUB-2007-001 (2007)
103. F. James, Comput. Phys. Commun. **60**, 329–344 (1990)
104. S. Baranov et al., ATLAS radiation background task force summary. ATL-GEN-2005-001 (2005)
105. C. Zeitnitz, T.A. Gabriel, The GEANT-CALOR interface and benchmark calculations for ZEUS calorimeters. Nucl. Instrum. Methods A **349**, 106–111 (1994)
106. I. Azhgirey, I. Baishev, K.M. Potter, V. Talanov, Cascade simulations for the machine induced background study in the IR1 of the LHC. LHC Project Note 324 (2003)
107. I. Azhgirey, I. Baishev, K.M. Potter, V. Talanov, Machine induced background in the low luminosity insertions of the LHC. LHC Project Report 567 (2002)
108. E. Barberio et al., The Geant4-based ATLAS fast electromagnetic shower simulation. ATL-SOFT-CONF-2007-002 (2007)
109. E. Barberio et al., The fast shower simulation in the ATLAS calorimeter. J. Phys. Conf. Ser. **119**, 001001 (2007). In Proceedings of Computing in High Energy and Nuclear Physics 2007 Conference (CHEP2007), Victoria, BC, Canada, September 2–7, 2007, ed. R. Sobie, R. Tafirout, J. Thomson
110. E. Richter-Was, D. Froidevaux, L. Poggioli, ATLFAST 2.0 a fast simulation package for ATLAS. CERN-ATL-PHYS-98-131 (1998)
111. S. Dean, P. Sherwood, Athena-Atlfast. http://www.hep.ucl.ac.uk/atlas/atlfast/, 2008
112. K. Edmonds et al., TheFast ATLAS track simulation (FATRAS). ATL-SOFT-PUB-2008-01 (2008)
113. G. Soyez, The SISCone and anti-kt jet algorithms (2008). 0807.0021 [hep-ph]
114. M. Cacciari, G.P. Salam, Dispelling the N**3 myth for the k(t) jet-finder. Phys. Lett. B **641**, 57–61 (2006). hep-ph/0512210

115. A. Salzburger, S. Todorova, M. Wolter, The ATLAS tracking geometry description. ATL-SOFT-PUB-2007-004 (2007)
116. A. Salzburger, The ATLAS track extrapolation package. ATL-SOFT-PUB-2007-05 (2007)
117. J.L. Henning, S.P.E.C. CPU2000, Measuring CPU performance in the new millenium. IEEE Comput. 28–35 (2000)
118. C.P.U. Normalization Criterion, http://www.egee.cesga.es/EGEE-SA1-SWE/accounting/guides/NormalizationCriterion, 2009
119. A. De Salvo, F. Brasolin, Benchmarking the ATLAS software through the kit validation engine, in *Proceedings of Computing in High Energy and Nuclear Physics 2009 Conference (CHEP2009)*, Prague, Czech Republic, March 22–28, 2009
120. GCC, the GNU Compiler Collection. http://gcc.gnu.org, 2008
121. Scientific Linux CERN, 4 (SLC4). http://linux.web.cern.ch/linux/scientific4, 2008
122. Y. Perrin, F. Orellana, M. Roy, D. Feichtinger, The LCG Savannah software development portal. CERN Yellow Report CERN 2005-002 (2005), pp. 609–612
123. T. Carli, H. Hakobyan, A. Henriques-Correia, M. Simonyan, Measurement of pion and proton longitudinal shower profiles up to 20 nuclear interaction lengths with the ATLAS time calorimeter. ATL-TILECAL-PUB-2007-008 (2007)
124. P. Adragna et al., Testbeam studies of production modules of the ATLAS tile calorimeter. ATL-TILECAL-PUB-2009-002 (2009)
125. P. Speckmayer, Comparison of data with Monte Carlo simulations at the ATLAS barrel combined testbeam 2004. J. Phys. Conf. Ser. **160** (2009). In Proceedings of the XIII International Conference on Calorimetry in High Energy Physics (CALOR 08), Pavia, Italy, May 26–30, 2008, ed. M. Fraternali, G. Gaudio, M. Livan
126. A.E. Kiryunin et al., GEANT4 physics evaluation with testbeam data of the ATLAS hadronic end-cap calorimeter, J. Phys., Conf. Ser. **160** (2009). In Proceedings of the XIII International Conference on Calorimetry in High Energy Physics (CALOR 08), Pavia, Italy, May 26–30, 2008, ed. M. Fraternali, G. Gaudio, M. Livan
127. M. Cooke et al., In situ commissioning of the ATLAS electromagnetic calorimeter with cosmic muons. ATL-LARG-PUB-2007-013 (2007)
128. M. Campanella, A. Ferrari, P.R. Sala, S. Vanini, First calorimeter simulation with the FLUGG prototype. ATL-SOFT-99-004 (1999)

Readiness of the ATLAS Tile Calorimeter for LHC collisions

The ATLAS Collaboration[*,**]

G. Aad[48], B. Abbott[111], J. Abdallah[11], A.A. Abdelalim[49], A. Abdesselam[118], O. Abdinov[10], B. Abi[112], M. Abolins[88], H. Abramowicz[153], H. Abreu[115], B.S. Acharya[164a,164b], D.L. Adams[24], T.N. Addy[56], J. Adelman[175], C. Adorisio[36a,36b], P. Adragna[75], T. Adye[129], S. Aefsky[22], J.A. Aguilar-Saavedra[124b,a], M. Aharrouche[81], S.P. Ahlen[21], F. Ahles[48], A. Ahmad[148], M. Ahsan[40], G. Aielli[133a,133b], T. Akdogan[18a], T.P.A. Åkesson[79], G. Akimoto[155], A.V. Akimov[94], A. Aktas[48], M.S. Alam[1], M.A. Alam[76], S. Albrand[55], M. Aleksa[29], I.N. Aleksandrov[65], C. Alexa[25a], G. Alexander[153], G. Alexandre[49], T. Alexopoulos[9], M. Alhroob[20], M. Aliev[15], G. Alimonti[89a], J. Alison[120], M. Aliyev[10], P.P. Allport[73], S.E. Allwood-Spiers[53], J. Almond[82], A. Aloisio[102a,102b], R. Alon[171], A. Alonso[79], M.G. Alviggi[102a,102b], K. Amako[66], C. Amelung[22], A. Amorim[124a,b], G. Amorós[167], N. Amram[153], C. Anastopoulos[139], T. Andeen[29], C.F. Anders[48], K.J. Anderson[30], A. Andreazza[89a,89b], V. Andrei[58a], X.S. Anduaga[70], A. Angerami[34], F. Anghinolfi[29], N. Anjos[124a], A. Annovi[47], A. Antonaki[8], M. Antonelli[47], S. Antonelli[19a,19b], J. Antos[144b], B. Antunovic[41], F. Anulli[132a], S. Aoun[83], G. Arabidze[8], I. Aracena[143], Y. Arai[66], A.T.H. Arce[44], J.P. Archambault[28], S. Arfaoui[29,c], J.-F. Arguin[14], T. Argyropoulos[9], M. Arik[18a], A.J. Armbruster[87], O. Arnaez[4], C. Arnault[115], A. Artamonov[95], D. Arutinov[20], M. Asai[143], S. Asai[155], R. Asfandiyarov[172], S. Ask[82], B. Åsman[146a,146b], D. Asner[28], L. Asquith[77], K. Assamagan[24], A. Astvatsatourov[52], G. Atoian[175], B. Auerbach[175], K. Augsten[127], M. Aurousseau[4], N. Austin[73], G. Avolio[163], R. Avramidou[9], C. Ay[54], G. Azuelos[93,d], Y. Azuma[155], M.A. Baak[29], A.M. Bach[14], H. Bachacou[136], K. Bachas[29], M. Backes[49], E. Badescu[25a], P. Bagnaia[132a,132b], Y. Bai[32a], T. Bain[158], J.T. Baines[129], O.K. Baker[175], M.D. Baker[24], S. Baker[77], F. Baltasar Dos Santos Pedrosa[29], E. Banas[38], P. Banerjee[93], S. Banerjee[169], D. Banfi[89a,89b], A. Bangert[137], V. Bansal[169], S.P. Baranov[94], A. Barashkou[65], T. Barber[27], E.L. Barberio[86], D. Barberis[50a,50b], M. Barbero[20], D.Y. Bardin[65], T. Barillari[99], M. Barisonzi[174], T. Barklow[143], N. Barlow[27], B.M. Barnett[129], R.M. Barnett[14], A. Baroncelli[134a], A.J. Barr[118], F. Barreiro[80], J. Barreiro Guimarães da Costa[57], P. Barrillon[115], R. Bartoldus[143], D. Bartsch[20], R.L. Bates[53], L. Batkova[144a], J.R. Batley[27], A. Battaglia[16], M. Battistin[29], F. Bauer[136], H.S. Bawa[143], M. Bazalova[125], B. Beare[158], T. Beau[78], P.H. Beauchemin[118], R. Beccherle[50a], P. Bechtle[41], G.A. Beck[75], H.P. Beck[16], M. Beckingham[48], K.H. Becks[174], A.J. Beddall[18c], A. Beddall[18c], V.A. Bednyakov[65], C. Bee[83], M. Begel[24], S. Behar Harpaz[152], P.K. Behera[63], M. Beimforde[99], C. Belanger-Champagne[166], P.J. Bell[49], W.H. Bell[49], G. Bella[153], L. Bellagamba[19a], F. Bellina[29], M. Bellomo[119a], A. Belloni[57], K. Belotskiy[96], O. Beltramello[29], S. Ben Ami[152], O. Benary[153], D. Benchekroun[135a], M. Bendel[81], B.H. Benedict[163], N. Benekos[165], Y. Benhammou[153], D.P. Benjamin[44], M. Benoit[115], J.R. Bensinger[22], K. Benslama[130], S. Bentvelsen[105], M. Beretta[47], D. Berge[29], E. Bergeaas Kuutmann[41], N. Berger[4], F. Berghaus[169], E. Berglund[49], J. Beringer[14], P. Bernat[115], R. Bernhard[48], C. Bernius[77], T. Berry[76], A. Bertin[19a,19b], M.I. Besana[89a,89b], N. Besson[136], S. Bethke[99], R.M. Bianchi[48], M. Bianco[72a,72b], O. Biebel[98], J. Biesiada[14], M. Biglietti[132a,132b], H. Bilokon[47], M. Bindi[19a,19b], A. Bingul[18c], C. Bini[132a,132b], C. Biscarat[180], U. Bitenc[48], K.M. Black[57], R.E. Blair[5], J.-B. Blanchard[115], G. Blanchot[29], C. Blocker[22], A. Blondel[49], W. Blum[81], U. Blumenschein[54], G.J. Bobbink[105], A. Bocci[44], M. Boehler[41], J. Boek[174], N. Boelaert[79], S. Böser[77], J.A. Bogaerts[29], A. Bogouch[90,*], C. Bohm[146a], J. Bohm[125], V. Boisvert[76], T. Bold[163,e], V. Boldea[25a], V.G. Bondarenko[96], M. Bondioli[163], M. Boonekamp[136], S. Bordoni[78], C. Borer[16], A. Borisov[128], G. Borissov[71], I. Borjanovic[12a], S. Borroni[132a,132b], K. Bos[105], D. Boscherini[19a], M. Bosman[11], H. Boterenbrood[105], J. Bouchami[93], J. Boudreau[123], E.V. Bouhova-Thacker[71], C. Boulahouache[123], C. Bourdarios[115], A. Boveia[30], J. Boyd[29], I.R. Boyko[65], I. Bozovic-Jelisavcic[12b], J. Bracinik[17], A. Braem[29], P. Branchini[134a], A. Brandt[7], G. Brandt[41], O. Brandt[54], U. Bratzler[156], B. Brau[84], J.E. Brau[114], H.M. Braun[174], B. Brelier[158], J. Bremer[29], R. Brenner[166], S. Bressler[152], D. Britton[53], F.M. Brochu[27], I. Brock[20], R. Brock[88], E. Brodet[153], C. Bromberg[88], G. Brooijmans[34], W.K. Brooks[31b], G. Brown[82], P.A. Bruckman de Renstrom[38], D. Bruncko[144b], R. Bruneliere[48], S. Brunet[41], A. Bruni[19a], G. Bruni[19a], M. Bruschi[19a], F. Bucci[49], J. Buchanan[118], P. Buchholz[141], A.G. Buckley[45], I.A. Budagov[65], B. Budick[108], V. Büscher[81], L. Bugge[117], O. Bulekov[96], M. Bunse[42], T. Buran[117], H. Burckhart[29], S. Burdin[73], T. Burgess[13], S. Burke[129], E. Busato[33], P. Bussey[53], C.P. Buszello[166], F. Butin[29], B. Butler[143], J.M. Butler[21], C.M. Buttar[53], J.M. Butterworth[77], T. Byatt[77], J. Caballero[24], S. Cabrera

Urbán[167], D. Caforio[19a,19b], O. Cakir[3a], P. Calafiura[14], G. Calderini[78], P. Calfayan[98], R. Calkins[106], L.P. Caloba[23a], D. Calvet[33], P. Camarri[133a,133b], D. Cameron[117], S. Campana[29], M. Campanelli[77], V. Canale[102a,102b], F. Canelli[30], A. Canepa[159a], J. Cantero[80], L. Capasso[102a,102b], M.D.M. Capeans Garrido[29], I. Caprini[25a], M. Caprini[25a], M. Capua[36a,36b], R. Caputo[148], C. Caramarcu[25a], R. Cardarelli[133a], T. Carli[29], G. Carlino[102a], L. Carminati[89a,89b], B. Caron[2,f], S. Caron[48], G.D. Carrillo Montoya[172], S. Carron Montero[158], A.A. Carter[75], J.R. Carter[27], J. Carvalho[124a,g], D. Casadei[108], M.P. Casado[11], M. Cascella[122a,122b], A.M. Castaneda Hernandez[172], E. Castaneda-Miranda[172], V. Castillo Gimenez[167], N.F. Castro[124b,a], G. Cataldi[72a], A. Catinaccio[29], J.R. Catmore[71], A. Cattai[29], G. Cattani[133a,133b], S. Caughron[34], P. Cavalleri[78], D. Cavalli[89a], M. Cavalli-Sforza[11], V. Cavasinni[122a,122b], F. Ceradini[134a,134b], A.S. Cerqueira[23a], A. Cerri[29], L. Cerrito[75], F. Cerutti[47], S.A. Cetin[18b], A. Chafaq[135a], D. Chakraborty[106], K. Chan[2], J.D. Chapman[27], J.W. Chapman[87], E. Chareyre[78], D.G. Charlton[17], V. Chavda[82], S. Cheatham[71], S. Chekanov[5], S.V. Chekulaev[159a], G.A. Chelkov[65], H. Chen[24], S. Chen[32c], X. Chen[172], A. Cheplakov[65], V.F. Chepurnov[65], R. Cherkaoui El Moursli[135d], V. Tcherniatine[24], D. Chesneanu[25a], E. Cheu[6], S.L. Cheung[158], L. Chevalier[136], F. Chevallier[136], G. Chiefari[102a,102b], L. Chikovani[51], J.T. Childers[58a], A. Chilingarov[71], G. Chiodini[72a], V. Chizhov[65], G. Choudalakis[30], S. Chouridou[137], I.A. Christidi[77], A. Christov[48], D. Chromek-Burckhart[29], M.L. Chu[151], J. Chudoba[125], G. Ciapetti[132a,132b], A.K. Ciftci[3a], R. Ciftci[3a], D. Cinca[33], V. Cindro[74], M.D. Ciobotaru[163], C. Ciocca[19a,19b], A. Ciocio[14], M. Cirilli[87,h], A. Clark[49], P.J. Clark[45], W. Cleland[123], J.C. Clemens[83], B. Clement[55], C. Clement[146a,146b], Y. Coadou[83], M. Cobal[164a,164c], A. Coccaro[50a,50b], J. Cochran[64], J. Coggeshall[165], E. Cogneras[180], A.P. Colijn[105], C. Collard[115], N.J. Collins[17], C. Collins-Tooth[53], J. Collot[55], G. Colon[84], P. Conde Muiño[124a], E. Coniavitis[166], M.C. Conidi[11], M. Consonni[104], S. Constantinescu[25a], C. Conta[119a,119b], F. Conventi[102a,i], M. Cooke[34], B.D. Cooper[75], A.M. Cooper-Sarkar[118], N.J. Cooper-Smith[76], K. Copic[34], T. Cornelissen[50a,50b], M. Corradi[19a], F. Corriveau[85,j], A. Corso-Radu[163], A. Cortes-Gonzalez[165], G. Cortiana[99], G. Costa[89a], M.J. Costa[167], D. Costanzo[139], T. Costin[30], D. Côté[29], R. Coura Torres[23a], L. Courneyea[169], G. Cowan[76], C. Cowden[27], B.E. Cox[82], K. Cranmer[108], J. Cranshaw[5], M. Cristinziani[20], G. Crosetti[36a,36b], R. Crupi[72a,72b], S. Crépe-Renaudin[55], C. Cuenca Almenar[175], T. Cuhadar Donszelmann[139], M. Curatolo[47], C.J. Curtis[17], P. Cwetanski[61], Z. Czyczula[175], S. D'Auria[53], M. D'Onofrio[73], A. D'Orazio[99], C. Da Via[82], W. Dabrowski[37], T. Dai[87], C. Dallapiccola[84], S.J. Dallison[129,*], C.H. Daly[138], M. Dam[35], H.O. Danielsson[29], D. Dannheim[99], V. Dao[49], G. Darbo[50a], G.L. Darlea[25b], W. Davey[86], T. Davidek[126], N. Davidson[86], R. Davidson[71], M. Davies[93], A.R. Davison[77], I. Dawson[139], R.K. Daya[39], K. De[7], R. de Asmundis[102a], S. De Castro[19a,19b], P.E. De Castro Faria Salgado[24], S. De Cecco[78], J. de Graat[98], N. De Groot[104], P. de Jong[105], L. De Mora[71], M. De Oliveira Branco[29], D. De Pedis[132a], A. De Salvo[132a], U. De Sanctis[164a,164c], A. De Santo[149], J.B. De Vivie De Regie[115], S. Dean[77], D.V. Dedovich[65], J. Degenhardt[120], M. Dehchar[118], C. Del Papa[164a,164c], J. Del Peso[80], T. Del Prete[122a,122b], A. Dell'Acqua[29], L. Dell'Asta[89a,89b], M. Della Pietra[102a,k], D. della Volpe[102a,102b], M. Delmastro[29], P.A. Delsart[55], C. Deluca[148], S. Demers[175], M. Demichev[65], B. Demirkoz[11], J. Deng[163], W. Deng[24], S.P. Denisov[128], J.E. Derkaoui[135c], F. Derue[78], P. Dervan[73], K. Desch[20], P.O. Deviveiros[158], A. Dewhurst[129], B. DeWilde[148], S. Dhaliwal[158], R. Dhullipudi[24,l], A. Di Ciaccio[133a,133b], L. Di Ciaccio[4], A. Di Girolamo[29], B. Di Girolamo[29], S. Di Luise[134a,134b], A. Di Mattia[88], R. Di Nardo[133a,133b], A. Di Simone[133a,133b], R. Di Sipio[19a,19b], M.A. Diaz[31a], F. Diblen[18c], E.B. Diehl[87], J. Dietrich[48], T.A. Dietzsch[58a], S. Diglio[115], K. Dindar Yagci[39], J. Dingfelder[48], C. Dionisi[132a,132b], P. Dita[25a], S. Dita[25a], F. Dittus[29], F. Djama[83], R. Djilkibaev[108], T. Djobava[51], M.A.B. do Vale[23a], A. Do Valle Wemans[124a], T.K.O. Doan[4], D. Dobos[29], E. Dobson[29], M. Dobson[163], C. Doglioni[118], T. Doherty[53], J. Dolejsi[126], I. Dolenc[74], Z. Dolezal[126], B.A. Dolgoshein[96], T. Dohmae[155], M. Donega[120], J. Donini[55], J. Dopke[174], A. Doria[102a], A. Dos Anjos[172], A. Dotti[122a,122b], M.T. Dova[70], A. Doxiadis[105], A.T. Doyle[53], Z. Drasal[126], M. Dris[9], J. Dubbert[99], E. Duchovni[171], G. Duckeck[98], A. Dudarev[29], F. Dudziak[115], M. Dührssen[29], L. Duflot[115], M.-A. Dufour[85], M. Dunford[30], H. Duran Yildiz[3b], R. Duxfield[139], M. Dwuznik[37], M. Düren[52], W.L. Ebenstein[44], J. Ebke[98], S. Eckweiler[81], K. Edmonds[81], C.A. Edwards[76], K. Egorov[61], W. Ehrenfeld[41], T. Ehrich[99], T. Eifert[29], G. Eigen[13], K. Einsweiler[14], E. Eisenhandler[75], T. Ekelof[166], M. El Kacimi[4], M. Ellert[166], S. Elles[4], F. Ellinghaus[81], K. Ellis[75], N. Ellis[29], J. Elmsheuser[98], M. Elsing[29], D. Emeliyanov[129], R. Engelmann[148], A. Engl[98], B. Epp[62], A. Eppig[87], J. Erdmann[54], A. Ereditato[16], D. Eriksson[146a], I. Ermoline[88], J. Ernst[1], M. Ernst[24], J. Ernwein[136], D. Errede[165], S. Errede[165], E. Ertel[81], M. Escalier[115], C. Escobar[167], X. Espinal Curull[11], B. Esposito[47], A.I. Etienvre[136], E. Etzion[153], H. Evans[61], L. Fabbri[19a,19b], C. Fabre[29], K. Facius[35], R.M. Fakhrutdinov[128], S. Falciano[132a], Y. Fang[172], M. Fanti[89a,89b], A. Farbin[7], A. Farilla[134a], J. Farley[148], T. Farooque[158], S.M. Farrington[118], P. Farthouat[29], P. Fassnacht[29], D. Fassouliotis[8], B. Fatholahzadeh[158], L. Fayard[115], F. Fayette[54], R. Febbraro[33], P. Federic[144a], O.L. Fedin[121], W. Fedorko[29], L. Feligioni[83], C.U. Felzmann[86], C. Feng[32d], E.J. Feng[30],

A.B. Fenyuk[128], J. Ferencei[144b], J. Ferland[93], B. Fernandes[124a,m], W. Fernando[109], S. Ferrag[53], J. Ferrando[118], V. Ferrara[41], A. Ferrari[166], P. Ferrari[105], R. Ferrari[119a], A. Ferrer[167], M.L. Ferrer[47], D. Ferrere[49], C. Ferretti[87], M. Fiascaris[118], F. Fiedler[81], A. Filipčič[74], A. Filippas[9], F. Filthaut[104], M. Fincke-Keeler[169], M.C.N. Fiolhais[124a,g], L. Fiorini[11], A. Firan[39], G. Fischer[41], M.J. Fisher[109], M. Flechl[48], I. Fleck[141], J. Fleckner[81], P. Fleischmann[173], S. Fleischmann[20], T. Flick[174], L.R. Flores Castillo[172], M.J. Flowerdew[99], T.Fonseca Martin[76], J. Fopma[118], A. Formica[136], A. Forti[82], D. Fortin[159a], D. Fournier[115], A.J. Fowler[44], K. Fowler[137], H. Fox[71], P. Francavilla[122a,122b], S. Franchino[119a,119b], D. Francis[29], M. Franklin[57], S. Franz[29], M. Fraternali[119a,119b], S. Fratina[120], J. Freestone[82], S.T. French[27], R. Froeschl[29], D. Froidevaux[29], J.A. Frost[27], C. Fukunaga[156], E. Fullana Torregrosa[5], J. Fuster[167], C. Gabaldon[80], O. Gabizon[171], T. Gadfort[24], S. Gadomski[49], G. Gagliardi[50a,50b], P. Gagnon[61], C. Galea[98], E.J. Gallas[118], V. Gallo[16], B.J. Gallop[129], P. Gallus[125], E. Galyaev[40], K.K. Gan[109], Y.S. Gao[143,n], A. Gaponenko[14], M. Garcia-Sciveres[14], C. García[167], J.E. García Navarro[49], R.W. Gardner[30], N. Garelli[29], H. Garitaonandia[105], V. Garonne[29], C. Gatti[47], G. Gaudio[119a], V. Gautard[136], P. Gauzzi[132a,132b], I.L. Gavrilenko[94], C. Gay[168], G. Gaycken[20], E.N. Gazis[9], P. Ge[32d], C.N.P. Gee[129], Ch. Geich-Gimbel[20], K. Gellerstedt[146a,146b], C. Gemme[50a], M.H. Genest[98], S. Gentile[132a,132b], F. Georgatos[9], S. George[76], A. Gershon[153], H. Ghazlane[135d], N. Ghodbane[33], B. Giacobbe[19a], S. Giagu[132a,132b], V. Giakoumopoulou[8], V. Giangiobbe[122a,122b], F. Gianotti[29], B. Gibbard[24], A. Gibson[158], S.M. Gibson[118], L.M. Gilbert[118], V. Gilewsky[91], D.M. Gingrich[2,o], J. Ginzburg[153], N. Giokaris[8], M.P. Giordani[164a,164c], R. Giordano[102a,102b], F.M. Giorgi[15], P. Giovannini[99], P.F. Giraud[136], P. Girtler[62], D. Giugni[89a], P. Giusti[19a], B.K. Gjelsten[117], L.K. Gladilin[97], C. Glasman[80], A. Glazov[41], K.W. Glitza[174], G.L. Glonti[65], J. Godfrey[142], J. Godlewski[29], M. Goebel[41], T. Göpfert[43], C. Goeringer[81], C. Gössling[42], T. Göttfert[99], V. Goggi[119a,119b,p], S. Goldfarb[87], D. Goldin[39], T. Golling[175], A. Gomes[124a,q], L.S. Gomez Fajardo[41], R. Gonçalo[76], L. Gonella[20], C. Gong[32b], S. González de la Hoz[167], M.L. Gonzalez Silva[26], S. Gonzalez-Sevilla[49], J.J. Goodson[148], L. Goossens[29], H.A. Gordon[24], I. Gorelov[103], G. Gorfine[174], B. Gorini[29], E. Gorini[72a,72b], A. Gorišek[74], E. Gornicki[38], B. Gosdzik[41], M. Gosselink[105], M.I. Gostkin[65], I. Gough Eschrich[163], M. Gouighri[135a], D. Goujdami[135a], M.P. Goulette[49], A.G. Goussiou[138], C. Goy[4], I. Grabowska-Bold[163,r], P. Grafström[29], K.-J. Grahn[147], S. Grancagnolo[15], V. Grassi[148], V. Gratchev[121], N. Grau[34], H.M. Gray[34,s], J.A. Gray[148], E. Graziani[134a], B. Green[76], T. Greenshaw[73], Z.D. Greenwood[24,t], I.M. Gregor[41], P. Grenier[143], E. Griesmayer[46], J. Griffiths[138], N. Grigalashvili[65], A.A. Grillo[137], K. Grimm[148], S. Grinstein[11], Y.V. Grishkevich[97], M. Groh[99], M. Groll[81], E. Gross[171], J. Grosse-Knetter[54], J. Groth-Jensen[79], K. Grybel[141], C. Guicheney[33], A. Guida[72a,72b], T. Guillemin[4], H. Guler[85,u], J. Gunther[125], B. Guo[158], L. Gurriana[124a], Y. Gusakov[65], A. Gutierrez[93], P. Gutierrez[111], N. Guttman[153], O. Gutzwiller[172], C. Guyot[136], C. Gwenlan[118], C.B. Gwilliam[73], A. Haas[143], S. Haas[29], C. Haber[14], H.K. Hadavand[39], D.R. Hadley[17], P. Haefner[99], S. Haider[29], Z. Hajduk[38], H. Hakobyan[176], J. Haller[41,v], K. Hamacher[174], A. Hamilton[49], S. Hamilton[161], L. Han[32b], K. Hanagaki[116], M. Hance[120], C. Handel[81], P. Hanke[58a], J.R. Hansen[35], J.B. Hansen[35], J.D. Hansen[35], P.H. Hansen[35], T. Hansl-Kozanecka[137], P. Hansson[143], K. Hara[160], G.A. Hare[137], T. Harenberg[174], R.D. Harrington[21], O.M. Harris[138], K. Harrison[17], J. Hartert[48], F. Hartjes[105], A. Harvey[56], S. Hasegawa[101], Y. Hasegawa[140], S. Hassani[136], S. Haug[16], M. Hauschild[29], R. Hauser[88], M. Havranek[125], C.M. Hawkes[17], R.J. Hawkings[29], T. Hayakawa[67], H.S. Hayward[73], S.J. Haywood[129], S.J. Head[82], V. Hedberg[79], L. Heelan[28], S. Heim[88], B. Heinemann[14], S. Heisterkamp[35], L. Helary[4], M. Heller[115], S. Hellman[146a,146b], C. Helsens[11], T. Hemperek[20], R.C.W. Henderson[71], M. Henke[58a], A. Henrichs[54], A.M. Henriques Correia[29], S. Henrot-Versille[115], C. Hensel[54], T. Henß[174], Y. Hernández Jiménez[167], A.D. Hershenhorn[152], G. Herten[48], R. Hertenberger[98], L. Hervas[29], N.P. Hessey[105], E. Higón-Rodriguez[167], J.C. Hill[27], K.H. Hiller[41], S. Hillert[146a,146b], S.J. Hillier[17], I. Hinchliffe[14], E. Hines[120], M. Hirose[116], F. Hirsch[42], D. Hirschbuehl[174], J. Hobbs[148], N. Hod[153], M.C. Hodgkinson[139], P. Hodgson[139], A. Hoecker[29], M.R. Hoeferkamp[103], J. Hoffman[39], D. Hoffmann[83], M. Hohlfeld[81], D. Hollander[30], T. Holy[127], J.L. Holzbauer[88], Y. Homma[67], T. Horazdovsky[127], T. Hori[67], C. Horn[143], S. Horner[48], S. Horvat[99], J.-Y. Hostachy[55], S. Hou[151], A. Hoummada[135a], T. Howe[39], J. Hrivnac[115], T. Hryn'ova[4], P.J. Hsu[175], S.-C. Hsu[14], G.S. Huang[111], Z. Hubacek[127], F. Hubaut[83], F. Huegging[20], T.B. Huffman[118], E.W. Hughes[34], G. Hughes[71], M. Hurwitz[30], U. Husemann[41], N. Huseynov[10], J. Huston[88], J. Huth[57], G. Iacobucci[102a], G. Iakovidis[9], I. Ibragimov[141], L. Iconomidou-Fayard[115], J. Idarraga[159b], P. Iengo[4], O. Igonkina[105], Y. Ikegami[66], M. Ikeno[66], Y. Ilchenko[39], D. Iliadis[154], T. Ince[20], P. Ioannou[8], M. Iodice[134a], A. Irles Quiles[167], A. Ishikawa[67], M. Ishino[66], R. Ishmukhametov[39], T. Isobe[155], C. Issever[118], S. Istin[18a], Y. Itoh[101], A.V. Ivashin[128], W. Iwanski[38], H. Iwasaki[66], J.M. Izen[40], V. Izzo[102a], B. Jackson[120], J.N. Jackson[73], P. Jackson[143], M.R. Jaekel[29], V. Jain[61], K. Jakobs[48], S. Jakobsen[35], J. Jakubek[127], D.K. Jana[111], E. Jankowski[158], E. Jansen[77], A. Jantsch[99], M. Janus[48], G. Jarlskog[79], L. Jeanty[57], I. Jen-La Plante[30], P. Jenni[29], P. Jež[35], S. Jézéquel[4], W. Ji[79],

J. Jia[148], Y. Jiang[32b], M. Jimenez Belenguer[29], S. Jin[32a], O. Jinnouchi[157], D. Joffe[39], M. Johansen[146a,146b],
K.E. Johansson[146a], P. Johansson[139], S. Johnert[41], K.A. Johns[6], K. Jon-And[146a,146b], G. Jones[82], R.W.L. Jones[71],
T.J. Jones[73], P.M. Jorge[124a,b], J. Joseph[14], V. Juranek[125], P. Jussel[62], V.V. Kabachenko[128], M. Kaci[167],
A. Kaczmarska[38], M. Kado[115], H. Kagan[109], M. Kagan[57], S. Kaiser[99], E. Kajomovitz[152], S. Kalinin[174],
L.V. Kalinovskaya[65], S. Kama[41], N. Kanaya[155], M. Kaneda[155], V.A. Kantserov[96], J. Kanzaki[66], B. Kaplan[175],
A. Kapliy[30], J. Kaplon[29], D. Kar[43], M. Karagounis[20], M. Karagoz Unel[118], M. Karnevskiy[41], V. Kartvelishvili[71],
A.N. Karyukhin[128], L. Kashif[57], A. Kasmi[39], R.D. Kass[109], A. Kastanas[13], M. Kastoryano[175], M. Kataoka[4],
Y. Kataoka[155], E. Katsoufis[9], J. Katzy[41], V. Kaushik[6], K. Kawagoe[67], T. Kawamoto[155], G. Kawamura[81],
M.S. Kayl[105], F. Kayumov[94], V.A. Kazanin[107], M.Y. Kazarinov[65], J.R. Keates[82], R. Keeler[169], P.T. Keener[120],
R. Kehoe[39], M. Keil[54], G.D. Kekelidze[65], M. Kelly[82], M. Kenyon[53], O. Kepka[125], N. Kerschen[29], B.P. Kerševan[74],
S. Kersten[174], K. Kessoku[155], M. Khakzad[28], F. Khalil-zada[10], H. Khandanyan[165], A. Khanov[112], D. Kharchenko[65],
A. Khodinov[148], A. Khomich[58a], G. Khoriauli[20], N. Khovanskiy[65], V. Khovanskiy[95], E. Khramov[65], J. Khubua[51],
H. Kim[7], M.S. Kim[2], P.C. Kim[143], S.H. Kim[160], O. Kind[15], P. Kind[174], B.T. King[73], J. Kirk[129], G.P. Kirsch[118],
L.E. Kirsch[22], A.E. Kiryunin[99], D. Kisielewska[37], T. Kittelmann[123], H. Kiyamura[67], E. Kladiva[144b], M. Klein[73],
U. Klein[73], K. Kleinknecht[81], M. Klemetti[85], A. Klier[171], A. Klimentov[24], R. Klingenberg[42], E.B. Klinkby[44],
T. Klioutchnikova[29], P.F. Klok[104], S. Klous[105], E.-E. Kluge[58a], T. Kluge[73], P. Kluit[105], M. Klute[54], S. Kluth[99],
N.S. Knecht[158], E. Kneringer[62], B.R. Ko[44], T. Kobayashi[155], M. Kobel[43], B. Koblitz[29], M. Kocian[143], A. Kocnar[113],
P. Kodys[126], K. Köneke[41], A.C. König[104], S. Koenig[81], L. Köpke[81], F. Koetsveld[104], P. Koevesarki[20], T. Koffas[29],
E. Koffeman[105], F. Kohn[54], Z. Kohout[127], T. Kohriki[66], H. Kolanoski[15], V. Kolesnikov[65], I. Koletsou[4], J. Koll[88],
D. Kollar[29], S. Kolos[163,w], S.D. Kolya[82], A.A. Komar[94], J.R. Komaragiri[142], T. Kondo[66], T. Kono[41,x],
R. Konoplich[108], S.P. Konovalov[94], N. Konstantinidis[77], S. Koperny[37], K. Korcyl[38], K. Kordas[154], A. Korn[14],
I. Korolkov[11], E.V. Korolkova[139], V.A. Korotkov[128], O. Kortner[99], P. Kostka[41], V.V. Kostyukhin[20], S. Kotov[99],
V.M. Kotov[65], K.Y. Kotov[107], C. Kourkoumelis[8], A. Koutsman[105], R. Kowalewski[169], H. Kowalski[41],
T.Z. Kowalski[37], W. Kozanecki[136], A.S. Kozhin[128], V. Kral[127], V.A. Kramarenko[97], G. Kramberger[74],
M.W. Krasny[78], A. Krasznahorkay[108], J. Kraus[88], A. Kreisel[153], F. Krejci[127], J. Kretzschmar[73], N. Krieger[54],
P. Krieger[158], K. Kroeninger[54], H. Kroha[99], J. Kroll[120], J. Kroseberg[20], J. Krstic[12a], U. Kruchonak[65], H. Krüger[20],
Z.V. Krumshteyn[65], T. Kubota[155], S. Kuehn[48], A. Kugel[58c], T. Kuhl[174], D. Kuhn[62], V. Kukhtin[65], Y. Kulchitsky[90],
S. Kuleshov[31b], C. Kummer[98], M. Kuna[83], J. Kunkle[120], A. Kupco[125], H. Kurashige[67], M. Kurata[160],
Y.A. Kurochkin[90], V. Kus[125], R. Kwee[15], A. La Rosa[29], L. La Rotonda[36a,36b], J. Labbe[4], C. Lacasta[167],
F. Lacava[132a,132b], H. Lacker[15], D. Lacour[78], V.R. Lacuesta[167], E. Ladygin[65], R. Lafaye[4], B. Laforge[78], T. Lagouri[80],
S. Lai[48], M. Lamanna[29], C.L. Lampen[6], W. Lampl[6], E. Lancon[136], U. Landgraf[48], M.P.J. Landon[75], J.L. Lane[82],
A.J. Lankford[163], F. Lanni[24], K. Lantzsch[29], A. Lanza[119a], S. Laplace[4], C. Lapoire[83], J.F. Laporte[136], T. Lari[89a],
A. Larner[118], M. Lassnig[29], P. Laurelli[47], W. Lavrijsen[14], P. Laycock[73], A.B. Lazarev[65], A. Lazzaro[89a,89b],
O. Le Dortz[78], E. Le Guirriec[83], E. Le Menedeu[136], A. Lebedev[64], C. Lebel[93], T. LeCompte[5], F. Ledroit-Guillon[55],
H. Lee[105], J.S.H. Lee[150], S.C. Lee[151], M. Lefebvre[169], M. Legendre[136], B.C. LeGeyt[120], F. Legger[98], C. Leggett[14],
M. Lehmacher[20], G. Lehmann Miotto[29], X. Lei[6], R. Leitner[126], D. Lellouch[171], J. Lellouch[78], V. Lendermann[58a],
K.J.C. Leney[73], T. Lenz[174], G. Lenzen[174], B. Lenzi[136], K. Leonhardt[43], C. Leroy[93], J.-R. Lessard[169], C.G. Lester[27],
A. Leung Fook Cheong[172], J. Levêque[83], D. Levin[87], L.J. Levinson[171], M. Leyton[15], H. Li[172], X. Li[87], Z. Liang[39],
Z. Liang[151,y], B. Liberti[133a], P. Lichard[29], M. Lichtnecker[98], K. Lie[165], W. Liebig[105], J.N. Lilley[17], A. Limosani[86],
M. Limper[63], S.C. Lin[151], J.T. Linnemann[88], E. Lipeles[120], L. Lipinsky[125], A. Lipniacka[13], T.M. Liss[165],
D. Lissauer[24], A. Lister[49], A.M. Litke[137], C. Liu[28], D. Liu[151,z], H. Liu[87], J.B. Liu[87], M. Liu[32b], T. Liu[39], Y. Liu[32b],
M. Livan[119a,119b], A. Lleres[55], S.L. Lloyd[75], E. Lobodzinska[41], P. Loch[6], W.S. Lockman[137], S. Lockwitz[175],
T. Loddenkoetter[20], F.K. Loebinger[82], A. Loginov[175], C.W. Loh[168], T. Lohse[15], K. Lohwasser[48], M. Lokajicek[125],
R.E. Long[71], L. Lopes[124a,b], D. Lopez Mateos[34,aa], M. Losada[162], P. Loscutoff[14], X. Lou[40], A. Lounis[115],
K.F. Loureiro[109], L. Lovas[144a], J. Love[21], P.A. Love[71], A.J. Lowe[61], F. Lu[32a], H.J. Lubatti[138], C. Luci[132a,132b],
A. Lucotte[55], A. Ludwig[43], D. Ludwig[41], I. Ludwig[48], F. Luehring[61], D. Lumb[48], L. Luminari[132a], E. Lund[117],
B. Lund-Jensen[147], B. Lundberg[79], J. Lundberg[29], J. Lundquist[35], D. Lynn[24], J. Lys[14], E. Lytken[79], H. Ma[24],
L.L. Ma[172], J.A. Macana Goia[93], G. Maccarrone[47], A. Macchiolo[99], B. Maček[74], J. Machado Miguens[124a,b],
R. Mackeprang[35], R.J. Madaras[14], W.F. Mader[43], R. Maenner[58c], T. Maeno[24], P. Mättig[174], S. Mättig[41],
P.J. Magalhaes Martins[124a,g], E. Magradze[51], Y. Mahalalel[153], K. Mahboubi[48], A. Mahmood[1], C. Maiani[132a,132b],
C. Maidantchik[23a], A. Maio[124a,q], S. Majewski[24], Y. Makida[66], M. Makouski[128], N. Makovec[115], Pa. Malecki[38],
P. Malecki[38], V.P. Maleev[121], F. Malek[55], U. Mallik[63], D. Malon[5], S. Maltezos[9], V. Malyshev[107], S. Malyukov[65],

M. Mambelli[30], R. Mameghani[98], J. Mamuzic[41], L. Mandelli[89a], I. Mandić[74], R. Mandrysch[15], J. Maneira[124a], P.S. Mangeard[88], L. Manhaes de Andrade Filho[23a], I.D. Manjavidze[65], P.M. Manning[137], A. Manousakis-Katsikakis[8], B. Mansoulie[136], A. Mapelli[29], L. Mapelli[29], L. March[80], J.F. Marchand[4], F. Marchese[133a,133b], G. Marchiori[78], M. Marcisovsky[125], C.P. Marino[61], F. Marroquim[23a], Z. Marshall[34,aa], S. Marti-Garcia[167], A.J. Martin[75], A.J. Martin[175], B. Martin[29], B. Martin[88], F.F. Martin[120], J.P. Martin[93], T.A. Martin[17], B. Martin dit Latour[49], M. Martinez[11], V. Martinez Outschoorn[57], A.C. Martyniuk[82], F. Marzano[132a], A. Marzin[136], L. Masetti[20], T. Mashimo[155], R. Mashinistov[96], J. Masik[82], A.L. Maslennikov[107], I. Massa[19a,19b], N. Massol[4], A. Mastroberardino[36a,36b], T. Masubuchi[155], P. Matricon[115], H. Matsunaga[155], T. Matsushita[67], C. Mattravers[118,ab], S.J. Maxfield[73], A. Mayne[139], R. Mazini[151], M. Mazur[48], J. Mc Donald[85], S.P. Mc Kee[87], A. McCarn[165], R.L. McCarthy[148], N.A. McCubbin[129], K.W. McFarlane[56], H. McGlone[53], G. Mchedlidze[51], S.J. McMahon[129], R.A. McPherson[169,j], A. Meade[84], J. Mechnich[105], M. Mechtel[174], M. Medinnis[41], R. Meera-Lebbai[111], T.M. Meguro[116], S. Mehlhase[41], A. Mehta[73], K. Meier[58a], B. Meirose[48], C. Melachrinos[30], B.R. Mellado Garcia[172], L. Mendoza Navas[162], Z. Meng[151,ac], S. Menke[99], E. Meoni[11], P. Mermod[118], L. Merola[102a,102b], C. Meroni[89a], F.S. Merritt[30], A.M. Messina[29], J. Metcalfe[103], A.S. Mete[64], J.-P. Meyer[136], J. Meyer[173], J. Meyer[54], T.C. Meyer[29], W.T. Meyer[64], J. Miao[32d], S. Michal[29], L. Micu[25a], R.P. Middleton[129], S. Migas[73], L. Mijović[74], G. Mikenberg[171], M. Mikestikova[125], M. Mikuž[74], D.W. Miller[143], M. Miller[30], W.J. Mills[168], C.M. Mills[57], A. Milov[171], D.A. Milstead[146a,146b], D. Milstein[171], A.A. Minaenko[128], M. Miñano[167], I.A. Minashvili[65], A.I. Mincer[108], B. Mindur[37], M. Mineev[65], Y. Ming[130], L.M. Mir[11], G. Mirabelli[132a], S. Misawa[24], A. Misiejuk[76], J. Mitrevski[137], V.A. Mitsou[167], P.S. Miyagawa[82], J.U. Mjörnmark[79], T. Moa[146a,146b], S. Moed[57], V. Moeller[27], K. Mönig[41], N. Möser[20], W. Mohr[48], S. Mohrdieck-Möck[99], R. Moles-Valls[167], J. Molina-Perez[29], J. Monk[77], E. Monnier[83], S. Montesano[89a,89b], F. Monticelli[70], R.W. Moore[2], C. Mora Herrera[49], A. Moraes[53], A. Morais[124a,b], J. Morel[54], G. Morello[36a,36b], D. Moreno[162], M. Moreno Llácer[167], P. Morettini[50a], M. Morii[57], A.K. Morley[86], G. Mornacchi[29], S.V. Morozov[96], J.D. Morris[75], H.G. Moser[99], M. Mosidze[51], J. Moss[109], R. Mount[143], E. Mountricha[136], S.V. Mouraviev[94], E.J.W. Moyse[84], M. Mudrinic[12b], F. Mueller[58a], J. Mueller[123], K. Mueller[20], T.A. Müller[98], D. Muenstermann[42], A. Muir[168], Y. Munwes[153], R. Murillo Garcia[163], W.J. Murray[129], I. Mussche[105], E. Musto[102a,102b], A.G. Myagkov[128], M. Myska[125], J. Nadal[11], K. Nagai[160], K. Nagano[66], Y. Nagasaka[60], A.M. Nairz[29], K. Nakamura[155], I. Nakano[110], H. Nakatsuka[67], G. Nanava[20], A. Napier[161], M. Nash[77,ad], N.R. Nation[21], T. Nattermann[20], T. Naumann[41], G. Navarro[162], S.K. Nderitu[20], H.A. Neal[87], E. Nebot[80], P. Nechaeva[94], A. Negri[119a,119b], G. Negri[29], A. Nelson[64], T.K. Nelson[143], S. Nemecek[125], P. Nemethy[108], A.A. Nepomuceno[23a], M. Nessi[29], M.S. Neubauer[165], A. Neusiedl[81], R.M. Neves[108], P. Nevski[24], F.M. Newcomer[120], R.B. Nickerson[118], R. Nicolaidou[136], L. Nicolas[139], G. Nicoletti[47], B. Nicquevert[29], F. Niedercorn[115], J. Nielsen[137], A. Nikiforov[15], K. Nikolaev[65], I. Nikolic-Audit[78], K. Nikolopoulos[8], H. Nilsen[48], P. Nilsson[7], A. Nisati[132a], T. Nishiyama[67], R. Nisius[99], L. Nodulman[5], M. Nomachi[116], I. Nomidis[154], M. Nordberg[29], B. Nordkvist[146a,146b], D. Notz[41], J. Novakova[126], M. Nozaki[66], M. Nožička[41], I.M. Nugent[159a], A.-E. Nuncio-Quiroz[20], G. Nunes Hanninger[20], T. Nunnemann[98], E. Nurse[77], D.C. O'Neil[142], V. O'Shea[53], F.G. Oakham[28,f], H. Oberlack[99], A. Ochi[67], S. Oda[155], S. Odaka[66], J. Odier[83], H. Ogren[61], A. Oh[82], S.H. Oh[44], C.C. Ohm[146a,146b], T. Ohshima[101], H. Ohshita[140], T. Ohsugi[59], S. Okada[67], H. Okawa[163], Y. Okumura[101], T. Okuyama[155], A.G. Olchevski[65], M. Oliveira[124a,g], D. Oliveira Damazio[24], E. Oliveira Garcia[167], D. Olivito[120], A. Olszewski[38], J. Olszowska[38], C. Omachi[67,ae], A. Onofre[124a,af], P.U.E. Onyisi[30], C.J. Oram[159a], M.J. Oreglia[30], Y. Oren[153], D. Orestano[134a,134b], I. Orlov[107], C. Oropeza Barrera[53], R.S. Orr[158], E.O. Ortega[130], B. Osculati[50a,50b], R. Ospanov[120], C. Osuna[11], J.P Ottersbach[105], F. Ould-Saada[117], A. Ouraou[136], Q. Ouyang[32a], M. Owen[82], S. Owen[139], A. Oyarzun[31b], V.E. Ozcan[77], K. Ozone[66], N. Ozturk[7], A. Pacheco Pages[11], C. Padilla Aranda[11], E. Paganis[139], C. Pahl[63], F. Paige[24], K. Pajchel[117], S. Palestini[29], D. Pallin[33], A. Palma[124a,b], J.D. Palmer[17], Y.B. Pan[172], E. Panagiotopoulou[9], B. Panes[31a], N. Panikashvili[87], S. Panitkin[24], D. Pantea[25a], M. Panuskova[125], V. Paolone[123], Th.D. Papadopoulou[9], S.J. Park[54], W. Park[24,ag], M.A. Parker[27], F. Parodi[50a,50b], J.A. Parsons[34], U. Parzefall[48], E. Pasqualucci[132a], A. Passeri[134a], F. Pastore[134a,134b], Fr. Pastore[29], G. Pásztor[49,ah], S. Pataraia[99], J.R. Pater[82], S. Patricelli[102a,102b], T. Pauly[29], L.S. Peak[150], M. Pecsy[144a], M.I. Pedraza Morales[172], S.V. Peleganchuk[107], H. Peng[172], A. Penson[34], J. Penwell[61], M. Perantoni[23a], K. Perez[34,aa], E. Perez Codina[11], M.T. Pérez García-Estañ[167], V. Perez Reale[34], L. Perini[89a,89b], H. Pernegger[29], R. Perrino[72a], S. Persembe[3a], P. Perus[115], V.D. Peshekhonov[65], B.A. Petersen[29], T.C. Petersen[35], E. Petit[83], C. Petridou[154], E. Petrolo[132a], F. Petrucci[134a,134b], D. Petschull[41], M. Petteni[142], R. Pezoa[31b], A. Phan[86], A.W. Phillips[27], G. Piacquadio[29], M. Piccinini[19a,19b], R. Piegaia[26], J.E. Pilcher[30], A.D. Pilkington[82], J. Pina[124a,q], M. Pinamonti[164a,164c], J.L. Pinfold[2],

B. Pinto[124a,b], C. Pizio[89a,89b], R. Placakyte[41], M. Plamondon[169], M.-A. Pleier[24], A. Poblaguev[175], S. Poddar[58a], F. Podlyski[33], L. Poggioli[115], M. Pohl[49], F. Polci[55], G. Polesello[119a], A. Policicchio[138], A. Polini[19a], J. Poll[75], V. Polychronakos[24], D. Pomeroy[22], K. Pommès[29], P. Ponsot[136], L. Pontecorvo[132a], B.G. Pope[88], G.A. Popeneciu[25a], D.S. Popovic[12a], A. Poppleton[29], J. Popule[125], X. Portell Bueso[48], R. Porter[163], G.E. Pospelov[99], S. Pospisil[127], M. Potekhin[24], I.N. Potrap[99], C.J. Potter[149], C.T. Potter[85], K.P. Potter[82], G. Poulard[29], J. Poveda[172], R. Prabhu[20], P. Pralavorio[83], S. Prasad[57], R. Pravahan[7], L. Pribyl[29], D. Price[61], L.E. Price[5], P.M. Prichard[73], D. Prieur[123], M. Primavera[72a], K. Prokofiev[29], F. Prokoshin[31b], S. Protopopescu[24], J. Proudfoot[5], X. Prudent[43], H. Przysiezniak[4], S. Psoroulas[20], E. Ptacek[114], J. Purdham[87], M. Purohit[24,ai], P. Puzo[115], Y. Pylypchenko[117], M. Qi[32c], J. Qian[87], W. Qian[129], Z. Qin[41], A. Quadt[54], D.R. Quarrie[14], W.B. Quayle[172], F. Quinonez[31a], M. Raas[104], V. Radeka[24], V. Radescu[58b], B. Radics[20], T. Rador[18a], F. Ragusa[89a,89b], G. Rahal[180], A.M. Rahimi[109], S. Rajagopalan[24], M. Rammensee[48], M. Rammes[141], F. Rauscher[98], E. Rauter[99], M. Raymond[29], A.L. Read[117], D.M. Rebuzzi[119a,119b], A. Redelbach[173], G. Redlinger[24], R. Reece[120], K. Reeves[40], E. Reinherz-Aronis[153], A. Reinsch[114], I. Reisinger[42], D. Reljic[12a], C. Rembser[29], Z.L. Ren[151], P. Renkel[39], S. Rescia[24], M. Rescigno[132a], S. Resconi[89a], B. Resende[136], P. Reznicek[126], R. Rezvani[158], N. Ribeiro[124a], A. Richards[77], R. Richter[99], E. Richter-Was[38,aj], M. Ridel[78], M. Rijpstra[105], M. Rijssenbeek[148], A. Rimoldi[119a,119b], L. Rinaldi[19a], R.R. Rios[39], I. Riu[11], F. Rizatdinova[112], E. Rizvi[75], D.A. Roa Romero[162], S.H. Robertson[85,j], A. Robichaud-Veronneau[49], D. Robinson[27], J.E.M. Robinson[77], M. Robinson[114], A. Robson[53], J.G. Rocha de Lima[106], C. Roda[122a,122b], D. Roda Dos Santos[29], D. Rodriguez[162], Y. Rodriguez Garcia[15], S. Roe[29], O. Røhne[117], V. Rojo[1], S. Rolli[161], A. Romaniouk[96], V.M. Romanov[65], G. Romeo[26], D. Romero Maltrana[31a], L. Roos[78], E. Ros[167], S. Rosati[138], G.A. Rosenbaum[158], L. Rosselet[49], V. Rossetti[11], L.P. Rossi[50a], M. Rotaru[25a], J. Rothberg[138], D. Rousseau[115], C.R. Royon[136], A. Rozanov[83], Y. Rozen[152], X. Ruan[115], B. Ruckert[98], N. Ruckstuhl[105], V.I. Rud[97], G. Rudolph[62], F. Rühr[58a], F. Ruggieri[134a], A. Ruiz-Martinez[64], L. Rumyantsev[65], Z. Rurikova[48], N.A. Rusakovich[65], J.P. Rutherfoord[6], C. Ruwiedel[20], P. Ruzicka[125], Y.F. Ryabov[121], P. Ryan[88], G. Rybkin[115], S. Rzaeva[10], A.F. Saavedra[150], H.F.-W. Sadrozinski[137], R. Sadykov[65], F. Safai Tehrani[132a,132b], H. Sakamoto[155], G. Salamanna[105], A. Salamon[133a], M.S. Saleem[111], D. Salihagic[99], A. Salnikov[143], J. Salt[167], B.M. Salvachua Ferrando[5], D. Salvatore[36a,36b], F. Salvatore[149], A. Salvucci[47], A. Salzburger[29], D. Sampsonidis[154], B.H. Samset[117], H. Sandaker[13], H.G. Sander[81], M.P. Sanders[98], M. Sandhoff[174], P. Sandhu[158], R. Sandstroem[105], S. Sandvoss[174], D.P.C. Sankey[129], B. Sanny[174], A. Sansoni[47], C. Santamarina Rios[85], C. Santoni[33], R. Santonico[133a,133b], J.G. Saraiva[124a,q], T. Sarangi[172], E. Sarkisyan-Grinbaum[7], F. Sarri[122a,122b], O. Sasaki[66], N. Sasao[68], I. Satsounkevitch[90], G. Sauvage[4], P. Savard[158,f], A.Y. Savine[6], V. Savinov[123], L. Sawyer[24,ak], D.H. Saxon[53], L.P. Says[33], C. Sbarra[19a,19b], A. Sbrizzi[19a,19b], D.A. Scannicchio[29], J. Schaarschmidt[43], P. Schacht[99], U. Schäfer[81], S. Schaetzel[58b], A.C. Schaffer[115], D. Schaile[98], R.D. Schamberger[148], A.G. Schamov[107], V. Scharf[58a], V.A. Schegelsky[121], D. Scheirich[87], M. Schernau[163], M.I. Scherzer[14], C. Schiavi[50a,50b], J. Schieck[99], M. Schioppa[36a,36b], S. Schlenker[29], E. Schmidt[48], K. Schmieden[20], C. Schmitt[81], M. Schmitz[20], A. Schönig[58b], M. Schott[29], D. Schouten[142], J. Schovancova[125], M. Schram[85], A. Schreiner[63], C. Schroeder[81], N. Schroer[58c], M. Schroers[174], J. Schultes[174], H.-C. Schultz-Coulon[58a], J.W. Schumacher[43], M. Schumacher[48], B.A. Schumm[137], Ph. Schune[136], C. Schwanenberger[82], A. Schwartzman[143], Ph. Schwemling[78], R. Schwienhorst[88], R. Schwierz[43], J. Schwindling[136], W.G. Scott[129], J. Searcy[114], E. Sedykh[121], E. Segura[11], S.C. Seidel[103], A. Seiden[137], F. Seifert[43], J.M. Seixas[23a], G. Sekhniaidze[102a], D.M. Seliverstov[121], B. Sellden[146a], N. Semprini-Cesari[19a,19b], C. Serfon[98], L. Serin[115], R. Seuster[99], H. Severini[111], M.E. Sevior[86], A. Sfyrla[165], E. Shabalina[54], M. Shamim[114], L.Y. Shan[32a], J.T. Shank[21], Q.T. Shao[86], M. Shapiro[14], P.B. Shatalov[95], K. Shaw[139], D. Sherman[29], P. Sherwood[77], A. Shibata[108], M. Shimojima[100], T. Shin[56], A. Shmeleva[94], M.J. Shochet[30], M.A. Shupe[6], P. Sicho[125], A. Sidoti[15], F. Siegert[77], J. Siegrist[14], Dj. Sijacki[12a], O. Silbert[171], J. Silva[124a,al], Y. Silver[153], D. Silverstein[143], S.B. Silverstein[146a], V. Simak[127], Lj. Simic[12a], S. Simion[115], B. Simmons[77], M. Simonyan[35], P. Sinervo[158], N.B. Sinev[114], V. Sipica[141], G. Siragusa[81], A.N. Sisakyan[65], S.Yu. Sivoklokov[97], J. Sjoelin[146a,146b], T.B. Sjursen[13], K. Skovpen[107], P. Skubic[111], M. Slater[17], T. Slavicek[127], K. Sliwa[161], J. Sloper[29], V. Smakhtin[171], S.Yu. Smirnov[96], Y. Smirnov[24], L.N. Smirnova[97], O. Smirnova[79], B.C. Smith[57], D. Smith[143], K.M. Smith[53], M. Smizanska[71], K. Smolek[127], A.A. Snesarev[94], S.W. Snow[82], J. Snow[111], J. Snuverink[105], S. Snyder[24], M. Soares[124a], R. Sobie[169,j], J. Sodomka[127], A. Soffer[153], C.A. Solans[167], M. Solar[127], J. Solc[127], E. Solfaroli Camillocci[132a,132b], A.A. Solodkov[128], O.V. Solovyanov[128], J. Sondericker[24], V. Sopko[127], B. Sopko[127], M. Sosebee[7], A. Soukharev[107], S. Spagnolo[72a,72b], F. Spanò[34], R. Spighi[19a], G. Spigo[29], F. Spila[132a,132b], R. Spiwoks[29], M. Spousta[126], T. Spreitzer[142], B. Spurlock[7], R.D. St. Denis[53], T. Stahl[141], J. Stahlman[120], R. Stamen[58a], S.N. Stancu[163], E. Stanecka[29], R.W. Stanek[5], C. Stanescu[134a], S. Stapnes[117], E.A. Starchenko[128], J. Stark[55], P. Staroba[125], P. Starovoitov[91], J. Stastny[125],

P. Stavina[144a], G. Steele[53], P. Steinbach[43], P. Steinberg[24], I. Stekl[127], B. Stelzer[142], H.J. Stelzer[41],
O. Stelzer-Chilton[159a], H. Stenzel[52], K. Stevenson[75], G.A. Stewart[53], M.C. Stockton[29], K. Stoerig[48], G. Stoicea[25a],
S. Stonjek[99], P. Strachota[126], A.R. Stradling[7], A. Straessner[43], J. Strandberg[87], S. Strandberg[14], A. Strandlie[117],
M. Strauss[111], P. Strizenec[144b], R. Ströhmer[173], D.M. Strom[114], R. Stroynowski[39], J. Strube[129], B. Stugu[13],
P. Sturm[174], D.A. Soh[151,am], D. Su[143], Y. Sugaya[116], T. Sugimoto[101], C. Suhr[106], M. Suk[126], V.V. Sulin[94],
S. Sultansoy[3d], T. Sumida[29], X.H. Sun[32d], J.E. Sundermann[48], K. Suruliz[164a,164b], S. Sushkov[11], G. Susinno[36a,36b],
M.R. Sutton[139], T. Suzuki[155], Y. Suzuki[66], I. Sykora[144a], T. Sykora[126], T. Szymocha[38], J. Sánchez[167], D. Ta[20],
K. Tackmann[29], A. Taffard[163], R. Tafirout[159a], A. Taga[117], Y. Takahashi[101], H. Takai[24], R. Takashima[69],
H. Takeda[67], T. Takeshita[140], M. Talby[83], A. Talyshev[107], M.C. Tamsett[76], J. Tanaka[155], R. Tanaka[115], S. Tanaka[131],
S. Tanaka[66], S. Tapprogge[81], D. Tardif[158], S. Tarem[152], F. Tarrade[24], G.F. Tartarelli[89a], P. Tas[126], M. Tasevsky[125],
E. Tassi[36a,36b], M. Tatarkhanov[14], C. Taylor[77], F.E. Taylor[92], G.N. Taylor[86], R.P. Taylor[169], W. Taylor[159b],
P. Teixeira-Dias[76], H. Ten Kate[29], P.K. Teng[151], Y.D. Tennenbaum-Katan[152], S. Terada[66], K. Terashi[155], J. Terron[80],
M. Terwort[41,v], M. Testa[47], R.J. Teuscher[158,j], J. Therhaag[20], M. Thioye[175], S. Thoma[48], J.P. Thomas[17],
E.N. Thompson[84], P.D. Thompson[17], P.D. Thompson[158], R.J. Thompson[82], A.S. Thompson[53], E. Thomson[120],
R.P. Thun[87], T. Tic[125], V.O. Tikhomirov[94], Y.A. Tikhonov[107], P. Tipton[175], F.J. Tique Aires Viegas[29], S. Tisserant[83],
B. Toczek[37], T. Todorov[4], S. Todorova-Nova[161], B. Toggerson[163], J. Tojo[66], S. Tokár[144a], K. Tokushuku[66],
K. Tollefson[88], L. Tomasek[125], M. Tomasek[125], M. Tomoto[101], L. Tompkins[14], K. Toms[103], A. Tonoyan[13], C. Topfel[16],
N.D. Topilin[65], I. Torchiani[29], E. Torrence[114], E. Torró Pastor[167], J. Toth[83,ah], F. Touchard[83], D.R. Tovey[139],
T. Trefzger[173], L. Tremblet[29], A. Tricoli[29], I.M. Trigger[159a], S. Trincaz-Duvoid[78], T.N. Trinh[78], M.F. Tripiana[70],
N. Triplett[64], W. Trischuk[158], A. Trivedi[24,an], B. Trocmé[55], C. Troncon[89a], A. Trzupek[38], C. Tsarouchas[9],
J.C.-L. Tseng[118], M. Tsiakiris[105], P.V. Tsiareshka[90], D. Tsionou[139], G. Tsipolitis[9], V. Tsiskaridze[51],
E.G. Tskhadadze[51], I.I. Tsukerman[95], V. Tsulaia[123], J.-W. Tsung[20], S. Tsuno[66], D. Tsybychev[148], J.M. Tuggle[30],
C.D. Tunnell[30], D. Turecek[127], I. Turk Cakir[3e], E. Turlay[105], P.M. Tuts[34], M.S. Twomey[138], M. Tylmad[146a,146b],
M. Tyndel[129], K. Uchida[116], I. Ueda[155], R. Ueno[28], M. Ugland[13], M. Uhlenbrock[20], M. Uhrmacher[54], F. Ukegawa[160],
G. Unal[29], A. Undrus[24], G. Unel[163], Y. Unno[66], D. Urbaniec[34], E. Urkovsky[153], P. Urquijo[49,ao], P. Urrejola[31a],
G. Usai[7], M. Uslenghi[119a,119b], L. Vacavant[83], V. Vacek[127], B. Vachon[85], S. Vahsen[14], P. Valente[132a],
S. Valentinetti[19a,19b], A. Valero[167], S. Valkar[126], E. Valladolid Gallego[167], S. Vallecorsa[152], J.A. Valls Ferrer[167],
R. Van Berg[120], H. van der Graaf[105], E. van der Kraaij[105], E. van der Poel[105], D. van der Ster[29], N. van Eldik[84],
P. van Gemmeren[5], Z. van Kesteren[105], I. van Vulpen[105], W. Vandelli[29], A. Vaniachine[5], P. Vankov[73], F. Vannucci[78],
R. Vari[132a], E.W. Varnes[6], D. Varouchas[14], A. Vartapetian[7], K.E. Varvell[150], L. Vasilyeva[94], V.I. Vassilakopoulos[56],
F. Vazeille[33], C. Vellidis[8], F. Veloso[124a], S. Veneziano[132a], A. Ventura[72a,72b], D. Ventura[138], M. Venturi[48], N. Venturi[16],
V. Vercesi[119a], M. Verducci[173], W. Verkerke[105], J.C. Vermeulen[105], M.C. Vetterli[142,f], I. Vichou[165], T. Vickey[145b,ap],
G.H.A. Viehhauser[118], M. Villa[19a,19b], E.G. Villani[129], M. Villaplana Perez[167], E. Vilucchi[47], M.G. Vincter[28],
E. Vinek[29], V.B. Vinogradov[65], S. Viret[33], J. Virzi[14], A. Vitale[19a,19b], O. Vitells[171], I. Vivarelli[48], F. Vives Vaque[11],
S. Vlachos[9], M. Vlasak[127], N. Vlasov[20], A. Vogel[20], P. Vokac[127], M. Volpi[11], H. von der Schmitt[99], J. von Loeben[99],
H. von Radziewski[48], E. von Toerne[20], V. Vorobel[126], V. Vorwerk[11], M. Vos[167], R. Voss[29], T.T. Voss[174],
J.H. Vossebeld[73], N. Vranjes[12a], M. Vranjes Milosavljevic[12a], V. Vrba[125], M. Vreeswijk[105], T. Vu Anh[81],
D. Vudragovic[12a], R. Vuillermet[29], I. Vukotic[115], P. Wagner[120], J. Walbersloh[42], J. Walder[71], R. Walker[98],
W. Walkowiak[141], R. Wall[175], C. Wang[44], H. Wang[172], J. Wang[55], S.M. Wang[151], A. Warburton[85], C.P. Ward[27],
M. Warsinsky[48], R. Wastie[118], P.M. Watkins[17], A.T. Watson[17], M.F. Watson[17], G. Watts[138], S. Watts[82],
A.T. Waugh[150], B.M. Waugh[77], M.D. Weber[16], M. Weber[129], M.S. Weber[16], P. Weber[58a], A.R. Weidberg[118],
J. Weingarten[54], C. Weiser[48], H. Wellenstein[22], P.S. Wells[29], T. Wenaus[24], S. Wendler[123], Z. Weng[151,aq],
T. Wengler[82], S. Wenig[29], N. Wermes[20], M. Werner[48], P. Werner[29], M. Werth[163], U. Werthenbach[141], M. Wessels[58a],
K. Whalen[28], A. White[7], M.J. White[27], S. White[24], S.R. Whitehead[118], D. Whiteson[163], D. Whittington[61],
F. Wicek[115], D. Wicke[81], F.J. Wickens[129], W. Wiedenmann[172], M. Wielers[129], P. Wienemann[20], C. Wiglesworth[73],
L.A.M. Wiik[48], A. Wildauer[167], M.A. Wildt[41,v], H.G. Wilkens[29], E. Williams[34], H.H. Williams[120], S. Willocq[84],
J.A. Wilson[17], M.G. Wilson[143], A. Wilson[87], I. Wingerter-Seez[4], F. Winklmeier[29], M. Wittgen[143], M.W. Wolter[38],
H. Wolters[124a,g], B.K. Wosiek[38], J. Wotschack[29], M.J. Woudstra[84], K. Wraight[53], C. Wright[53], D. Wright[143],
B. Wrona[73], S.L. Wu[172], X. Wu[49], E. Wulf[34], B.M. Wynne[45], L. Xaplanteris[9], S. Xella[35], S. Xie[48], D. Xu[139], N. Xu[172],
M. Yamada[160], A. Yamamoto[66], K. Yamamoto[64], S. Yamamoto[155], T. Yamamura[155], J. Yamaoka[44], T. Yamazaki[155],
Y. Yamazaki[67], Z. Yan[21], H. Yang[87], U.K. Yang[82], Z. Yang[146a,146b], W.-M. Yao[14], Y. Yao[14], Y. Yasu[66], J. Ye[39], S. Ye[24],
M. Yilmaz[3c], R. Yoosoofmiya[123], K. Yorita[170], R. Yoshida[5], C. Young[143], S.P. Youssef[21], D. Yu[24], J. Yu[7], L. Yuan[78],

A. Yurkewicz[148], R. Zaidan[63], A.M. Zaitsev[128], Z. Zajacova[29], V. Zambrano[47], L. Zanello[132a,132b], A. Zaytsev[107], C. Zeitnitz[174], M. Zeller[175], A. Zemla[38], C. Zendler[20], O. Zenin[128], T. Zenis[144a], Z. Zenonos[122a,122b], S. Zenz[14], D. Zerwas[115], G. Zevi della Porta[57], Z. Zhan[32d], H. Zhang[83], J. Zhang[5], Q. Zhang[5], X. Zhang[32d], L. Zhao[108], T. Zhao[138], Z. Zhao[32b], A. Zhemchugov[65], J. Zhong[151,ar], B. Zhou[87], N. Zhou[34], Y. Zhou[151], C.G. Zhu[32d], H. Zhu[41], Y. Zhu[172], X. Zhuang[98], V. Zhuravlov[99], R. Zimmermann[20], S. Zimmermann[20], S. Zimmermann[48], M. Ziolkowski[141], L. Živković[34], G. Zobernig[172], A. Zoccoli[19a,19b], M. zur Nedden[15], V. Zutshi[106]

[*]CERN, 1211 Geneva 23, Switzerland

[1]University at Albany, 1400 Washington Ave, Albany, NY 12222, United States of America

[2]University of Alberta, Department of Physics, Centre for Particle Physics, Edmonton, AB T6G 2G7, Canada

[3]Ankara University[(a)], Faculty of Sciences, Department of Physics, TR 061000 Tandogan, Ankara; Dumlupinar University[(b)], Faculty of Arts and Sciences, Department of Physics, Kutahya; Gazi University[(c)], Faculty of Arts and Sciences, Department of Physics, 06500, Teknikokullar, Ankara; TOBB University of Economics and Technology[(d)], Faculty of Arts and Sciences, Division of Physics, 06560, Sogutozu, Ankara; Turkish Atomic Energy Authority[(e)], 06530, Lodumlu, Ankara, Turkey

[4]LAPP, Université de Savoie, CNRS/IN2P3, Annecy-le-Vieux, France

[5]Argonne National Laboratory, High Energy Physics Division, 9700 S. Cass Avenue, Argonne IL 60439, United States of America

[6]University of Arizona, Department of Physics, Tucson, AZ 85721, United States of America

[7]The University of Texas at Arlington, Department of Physics, Box 19059, Arlington, TX 76019, United States of America

[8]University of Athens, Nuclear & Particle Physics, Department of Physics, Panepistimiopouli, Zografou, GR 15771 Athens, Greece

[9]National Technical University of Athens, Physics Department, 9-Iroon Polytechniou, GR 15780 Zografou, Greece

[10]Institute of Physics, Azerbaijan Academy of Sciences, H. Javid Avenue 33, AZ 143 Baku, Azerbaijan

[11]Institut de Física d'Altes Energies, IFAE, Edifici Cn, Universitat Autònoma de Barcelona, ES-08193 Bellaterra (Barcelona), Spain

[12]University of Belgrade[(a)], Institute of Physics, P.O. Box 57, 11001 Belgrade; Vinca Institute of Nuclear Sciences[(b)] M. Petrovica Alasa 12-14, 11000 Belgrade, Serbia, Serbia

[13]University of Bergen, Department for Physics and Technology, Allegaten 55, NO-5007 Bergen, Norway

[14]Lawrence Berkeley National Laboratory and University of California, Physics Division, MS50B-6227, 1 Cyclotron Road, Berkeley, CA 94720, United States of America

[15]Humboldt University, Institute of Physics, Berlin, Newtonstr. 15, D-12489 Berlin, Germany

[16]University of Bern, instein Center for Fundamental Physics, ry for High Energy Physics, Sidlerstrasse 5, CH-3012 Bern, Switzerland

[17]University of Birmingham, School of Physics and Astronomy, Edgbaston, Birmingham B15 2TT, United Kingdom

[18]Bogazici University[(a)], Faculty of Sciences, Department of Physics, TR-80815 Bebek-Istanbul; Dogus University[(b)], Faculty of Arts and Sciences, Department of Physics, 34722, Kadikoy, Istanbul; [(c)]Gaziantep University, Faculty of Engineering, Department of Physics Engineering, 27310, Sehitkamil, Gaziantep, Turkey; Istanbul Technical University[(d)], Faculty of Arts and Sciences, Department of Physics, 34469, Maslak, Istanbul, Turkey

[19]INFN Sezione di Bologna[(a)]; Università di Bologna, Dipartimento di Fisica[(b)], viale C. Berti Pichat, 6/2, IT-40127 Bologna, Italy

[20]University of Bonn, Physikalisches Institut, Nussallee 12, D-53115 Bonn, Germany

[21]Boston University, Department of Physics, 590 Commonwealth Avenue, Boston, MA 02215, United States of America

[22]Brandeis University, Department of Physics, MS057, 415 South Street, Waltham, MA 02454, United States of America

[23]Universidade Federal do Rio De Janeiro, COPPE/EE/IF [(a)], Caixa Postal 68528, Ilha do Fundao, BR-21945-970 Rio de Janeiro; [(b)]Universidade de Sao Paulo, Instituto de Fisica, R.do Matao Trav. R.187, Sao Paulo-SP, 05508-900, Brazil

[24]Brookhaven National Laboratory, Physics Department, Bldg. 510A, Upton, NY 11973, United States of America

[25]National Institute of Physics and Nuclear Engineering[(a)], Bucharest-Magurele, Str. Atomistilor 407, P.O. Box MG-6, R-077125, Romania; University Politehnica Bucharest[(b)], Rectorat-AN 001, 313 Splaiul Independentei, sector 6, 060042 Bucuresti; West University[(c)] in Timisoara, Bd. Vasile Parvan 4, Timisoara, Romania

[26]Universidad de Buenos Aires, FCEyN, Dto. Fisica, Pab I-C. Universitaria, 1428 Buenos Aires, Argentina

[27]University of Cambridge, Cavendish Laboratory, J J Thomson Avenue, Cambridge CB3 0HE, United Kingdom

[28]Carleton University, Department of Physics, 1125 Colonel By Drive, Ottawa ON K1S 5B6, Canada

[29]CERN, CH-1211 Geneva 23, Switzerland

[30]University of Chicago, Enrico Fermi Institute, 5640 S. Ellis Avenue, Chicago, IL 60637, United States of America

[31]Pontificia Universidad Católica de Chile, Facultad de Fisica, Departamento de Fisica[(a)], Avda. Vicuna Mackenna 4860, San Joaquin, Santiago; Universidad Técnica Federico Santa María, Departamento de Física[(b)], Avda. Espāna 1680, Casilla 110-V, Valparaíso, Chile

[32]Institute of High Energy Physics, Chinese Academy of Sciences[(a)], P.O. Box 918, 19 Yuquan Road, Shijing Shan District, CN-Beijing 100049; University of Science & Technology of China (USTC), Department of Modern Physics[(b)], Hefei, CN-Anhui 230026; Nanjing University, Department of Physics[(c)], 22 Hankou Road, Nanjing, 210093; Shandong University, High Energy Physics Group[(d)], Jinan, CN-Shandong 250100, China

[33]Laboratoire de Physique Corpusculaire, Clermont Université, Université Blaise Pascal, CNRS/IN2P3, FR-63177 Aubiere Cedex, France

[34]Columbia University, Nevis Laboratory, 136 So. Broadway, Irvington, NY 10533, United States of America

[35]University of Copenhagen, Niels Bohr Institute, Blegdamsvej 17, DK-2100 Kobenhavn 0, Denmark

[36]INFN Gruppo Collegato di Cosenza[(a)]; Università della Calabria, Dipartimento di Fisica[(b)], IT-87036 Arcavacata di Rende, Italy

[37]Faculty of Physics and Applied Computer Science of the AGH-University of Science and Technology, (FPACS, AGH-UST), al. Mickiewicza 30, PL-30059 Cracow, Poland

[38]The Henryk Niewodniczanski Institute of Nuclear Physics, Polish Academy of Sciences, ul. Radzikowskiego 152, PL-31342 Krakow, Poland

[39]Southern Methodist University, Physics Department, 106 Fondren Science Building, Dallas, TX 75275-0175, United States of America

[40]University of Texas at Dallas, 800 West Campbell Road, Richardson, TX 75080-3021, United States of America

[41]DESY, Notkestr. 85, D-22603 Hamburg and Platanenallee 6, D-15738 Zeuthen, Germany

[42] TU Dortmund, Experimentelle Physik IV, DE-44221 Dortmund, Germany

[43] Technical University Dresden, Institut für Kern- und Teilchenphysik, Zellescher Weg 19, D-01069 Dresden, Germany

[44] Duke University, Department of Physics, Durham, NC 27708, United States of America

[45] University of Edinburgh, School of Physics & Astronomy, James Clerk Maxwell Building, The Kings Buildings, Mayfield Road, Edinburgh EH9 3JZ, United Kingdom

[46] Fachhochschule Wiener Neustadt; Johannes Gutenbergstrasse 3 AT-2700 Wiener Neustadt, Austria

[47] INFN Laboratori Nazionali di Frascati, via Enrico Fermi 40, IT-00044 Frascati, Italy

[48] Albert-Ludwigs-Universität, Fakultät für Mathematik und Physik, Hermann-Herder Str. 3, D-79104 Freiburg i.Br., Germany

[49] Université de Genève, Section de Physique, 24 rue Ernest Ansermet, CH-1211 Geneve 4, Switzerland

[50] INFN Sezione di Genova[a]; Università di Genova, Dipartimento di Fisica[b], via Dodecaneso 33, IT-16146 Genova, Italy

[51] Institute of Physics of the Georgian Academy of Sciences, 6 Tamarashvili St., GE-380077 Tbilisi; Tbilisi State University, HEP Institute, University St. 9, GE-380086 Tbilisi, Georgia

[52] Justus-Liebig-Universität Giessen, II Physikalisches Institut, Heinrich-Buff Ring 16, D-35392 Giessen, Germany

[53] University of Glasgow, Department of Physics and Astronomy, Glasgow G12 8QQ, United Kingdom

[54] Georg-August-Universität, II. Physikalisches Institut, Friedrich-Hund Platz 1, D-37077 Göttingen, Germany

[55] Laboratoire de Physique Subatomique et de Cosmologie, CNRS/IN2P3, Université Joseph Fourier, INPG, 53 avenue des Martyrs, FR-38026 Grenoble Cedex, France

[56] Hampton University, Department of Physics, Hampton, VA 23668, United States of America

[57] Harvard University, Laboratory for Particle Physics and Cosmology, 18 Hammond Street, Cambridge, MA 02138, United States of America

[58] Ruprecht-Karls-Universität Heidelberg: Kirchhoff-Institut für Physik[a], Im Neuenheimer Feld 227, D-69120 Heidelberg; Physikalisches Institut[b], Philosophenweg 12, D-69120 Heidelberg; ZITI Ruprecht-Karls-University Heidelberg[c], Lehrstuhl für Informatik V, B6, 23-29, DE-68131 Mannheim, Germany

[59] Hiroshima University, Faculty of Science, 1-3-1 Kagamiyama, Higashihiroshima-shi, JP-Hiroshima 739-8526, Japan

[60] Hiroshima Institute of Technology, Faculty of Applied Information Science, 2-1-1 Miyake Saeki-ku, Hiroshima-shi, JP-Hiroshima 731-5193, Japan

[61] Indiana University, Department of Physics, Swain Hall West 117, Bloomington, IN 47405-7105, United States of America

[62] Institut für Astro- und Teilchenphysik, Technikerstrasse 25, A-6020 Innsbruck, Austria

[63] University of Iowa, 203 Van Allen Hall, Iowa City, IA 52242-1479, United States of America

[64] Iowa State University, Department of Physics and Astronomy, Ames High Energy Physics Group, Ames, IA 50011-3160, United States of America

[65] Joint Institute for Nuclear Research, JINR Dubna, RU-141 980 Moscow Region, Russia

[66] KEK, High Energy Accelerator Research Organization, 1-1 Oho, Tsukuba-shi, Ibaraki-ken 305-0801, Japan

[67] Kobe University, Graduate School of Science, 1-1 Rokkodai-cho, Nada-ku, JP Kobe 657-8501, Japan

[68] Kyoto University, Faculty of Science, Oiwake-cho, Kitashirakawa, Sakyou-ku, Kyoto-shi, JP-Kyoto 606-8502, Japan

[69] Kyoto University of Education, 1 Fukakusa, Fujimori, fushimi-ku, Kyoto-shi, JP-Kyoto 612-8522, Japan

[70] Universidad Nacional de La Plata, FCE, Departamento de Física, IFLP (CONICET-UNLP), C.C. 67, 1900 La Plata, Argentina

[71] Lancaster University, Physics Department, Lancaster LA1 4YB, United Kingdom

[72] INFN Sezione di Lecce[a]; Università del Salento, Dipartimento di Fisica[b] Via Arnesano IT-73100 Lecce, Italy

[73] University of Liverpool, Oliver Lodge Laboratory, P.O. Box 147, Oxford Street, Liverpool L69 3BX, United Kingdom

[74] Jožef Stefan Institute and University of Ljubljana, Department of Physics, SI-1000 Ljubljana, Slovenia

[75] Queen Mary University of London, Department of Physics, Mile End Road, London E1 4NS, United Kingdom

[76] Royal Holloway, University of London, Department of Physics, Egham Hill, Egham, Surrey TW20 0EX, United Kingdom

[77] University College London, Department of Physics and Astronomy, Gower Street, London WC1E 6BT, United Kingdom

[78] Laboratoire de Physique Nucléaire et de Hautes Energies, Université Pierre et Marie Curie (Paris 6), Université Denis Diderot (Paris-7), CNRS/IN2P3, Tour 33, 4 place Jussieu, FR-75252 Paris Cedex 05, France

[79] Lunds universitet, Naturvetenskapliga fakulteten, Fysiska institutionen, Box 118, SE-221 00 Lund, Sweden

[80] Universidad Autonoma de Madrid, Facultad de Ciencias, Departamento de Fisica Teorica, ES-28049 Madrid, Spain

[81] Universität Mainz, Institut für Physik, Staudinger Weg 7, DE-55099 Mainz, Germany

[82] University of Manchester, School of Physics and Astronomy, Manchester M13 9PL, United Kingdom

[83] CPPM, Aix-Marseille Université, CNRS/IN2P3, Marseille, France

[84] University of Massachusetts, Department of Physics, 710 North Pleasant Street, Amherst, MA 01003, United States of America

[85] McGill University, High Energy Physics Group, 3600 University Street, Montreal, Quebec H3A 2T8, Canada

[86] University of Melbourne, School of Physics, AU-Parkville, Victoria 3010, Australia

[87] The University of Michigan, Department of Physics, 2477 Randall Laboratory, 500 East University, Ann Arbor, MI 48109-1120, United States of America

[88] Michigan State University, Department of Physics and Astronomy, High Energy Physics Group, East Lansing, MI 48824-2320, United States of America

[89] INFN Sezione di Milano[a]; Università di Milano, Dipartimento di Fisica[b], via Celoria 16, IT-20133 Milano, Italy

[90] B.I. Stepanov Institute of Physics, National Academy of Sciences of Belarus, Independence Avenue 68, Minsk 220072, Republic of Belarus

[91] National Scientific & Educational Centre for Particle & High Energy Physics, NC PHEP BSU, M. Bogdanovich St. 153, Minsk 220040, Republic of Belarus

[92] Massachusetts Institute of Technology, Department of Physics, Room 24-516, Cambridge, MA 02139, United States of America

[93] University of Montreal, Group of Particle Physics, C.P. 6128, Succursale Centre-Ville, Montreal, Quebec, H3C 3J7, Canada

[94] P.N. Lebedev Institute of Physics, Academy of Sciences, Leninsky pr. 53, RU-117 924 Moscow, Russia

[95] Institute for Theoretical and Experimental Physics (ITEP), B. Cheremushkinskaya ul. 25, RU 117 218 Moscow, Russia

[96] Moscow Engineering & Physics Institute (MEPhI), Kashirskoe Shosse 31, RU-115409 Moscow, Russia

[97] Lomonosov Moscow State University Skobeltsyn Institute of Nuclear Physics (MSU SINP), 1(2), Leninskie gory, GSP-1, Moscow 119991 Russian Federation, Russia

[98] Ludwig-Maximilians-Universität München, Fakultät für Physik, Am Coulombwall 1, DE-85748 Garching, Germany

[99] Max-Planck-Institut für Physik, (Werner-Heisenberg-Institut), Föhringer Ring 6, 80805 München, Germany

[100] Nagasaki Institute of Applied Science, 536 Aba-machi, JP Nagasaki 851-0193, Japan

[101] Nagoya University, Graduate School of Science, Furo-Cho, Chikusa-ku, Nagoya, 464-8602, Japan

[102] INFN Sezione di Napoli[(a)]; Università di Napoli, Dipartimento di Scienze Fisiche[(b)], Complesso Universitario di Monte Sant'Angelo, via Cinthia, IT-80126 Napoli, Italy

[103] University of New Mexico, Department of Physics and Astronomy, MSC07 4220, Albuquerque, NM 87131 USA, United States of America

[104] Radboud University Nijmegen/NIKHEF, Department of Experimental High Energy Physics, Heyendaalseweg 135, NL-6525 AJ, Nijmegen, Netherlands

[105] Nikhef National Institute for Subatomic Physics, and University of Amsterdam, Science Park 105, 1098 XG Amsterdam, Netherlands

[106] Department of Physics, Northern Illinois University, LaTourette Hall ad, DeKalb, IL 60115, United States of America

[107] Budker Institute of Nuclear Physics (BINP), RU-Novosibirsk 630 090, Russia

[108] New York University, Department of Physics, 4 Washington Place, New York, NY 10003, USA

[109] Ohio State University, 191 West Woodruff Ave, Columbus, OH 43210-1117, United States of America

[110] Okayama University, Faculty of Science, Tsushimanaka 3-1-1, Okayama 700-8530, Japan

[111] University of Oklahoma, Homer L. Dodge Department of Physics and Astronomy, 440 West Brooks, Room 100, Norman, OK 73019-0225, United States of America

[112] Oklahoma State University, Department of Physics, 145 Physical Sciences Building, Stillwater, OK 74078-3072, United States of America

[113] Palacký University, 17.listopadu 50a, 772 07 Olomouc, Czech Republic

[114] University of Oregon, Center for High Energy Physics, Eugene, OR 97403-1274, United States of America

[115] LAL, Univ. Paris-Sud, IN2P3/CNRS, Orsay, France

[116] Osaka University, Graduate School of Science, Machikaneyama-machi 1-1, Toyonaka, Osaka 560-0043, Japan

[117] University of Oslo, Department of Physics, P.O. Box 1048, Blindern, NO-0316 Oslo 3, Norway

[118] Oxford University, Department of Physics, Denys Wilkinson Building, Keble Road, Oxford OX1 3RH, United Kingdom

[119] INFN Sezione di Pavia[(a)]; Università di Pavia, Dipartimento di Fisica Nucleare e Teorica[(b)], Via Bassi 6, IT-27100 Pavia, Italy

[120] University of Pennsylvania, Department of Physics, High Energy Physics Group, 209 S. 33rd Street, Philadelphia, PA 19104, United States of America

[121] Petersburg Nuclear Physics Institute, RU-188 300 Gatchina, Russia

[122] INFN Sezione di Pisa[(a)]; Università di Pisa, Dipartimento di Fisica E. Fermi[(b)], Largo B. Pontecorvo 3, IT-56127 Pisa, Italy

[123] University of Pittsburgh, Department of Physics and Astronomy, 3941 O'Hara Street, Pittsburgh, PA 15260, United States of America

[124] Laboratorio de Instrumentacao e Fisica Experimental de Particulas-LIP[(a)], Avenida Elias Garcia 14-1, PT-1000-149 Lisboa, Portugal; Universidad de Granada, Departamento de Fisica Teorica y del Cosmos and CAFPE[(b)], E-18071 Granada, Spain

[125] Institute of Physics, Academy of Sciences of the Czech Republic, Na Slovance 2, CZ-18221 Praha 8, Czech Republic

[126] Charles University in Prague, Faculty of Mathematics and Physics, Institute of Particle and Nuclear Physics, V Holesovickach 2, CZ-18000 Praha 8, Czech Republic

[127] Czech Technical University in Prague, Zikova 4, CZ-166 35 Praha 6, Czech Republic

[128] State Research Center Institute for High Energy Physics, Moscow Region, 142281, Protvino, Pobeda street, 1, Russia

[129] Rutherford Appleton Laboratory, Science and Technology Facilities Council, Harwell Science and Innovation Campus, Didcot OX11 0QX, United Kingdom

[130] University of Regina, Physics Department, Canada

[131] Ritsumeikan University, Noji Higashi 1 chome 1-1, JP-Kusatsu, Shiga 525-8577, Japan

[132] INFN Sezione di Roma I[(a)]; Università La Sapienza, Dipartimento di Fisica[(b)], Piazzale A. Moro 2, IT- 00185 Roma, Italy

[133] INFN Sezione di Roma Tor Vergata[(a)]; Università di Roma Tor Vergata, Dipartimento di Fisica[(b)], via della Ricerca Scientifica, IT-00133 Roma, Italy

[134] INFN Sezione di Roma Tre[(a)]; Università Roma Tre, Dipartimento di Fisica[(b)], via della Vasca Navale 84, IT-00146 Roma, Italy

[135] Réseau Universitaire de Physique des Hautes Energies (RUPHE): Université Hassan II, Faculté des Sciences Ain Chock[(a)], B.P. 5366, MA-Casablanca; Centre National de l'Energie des Sciences Techniques Nucleaires (CNESTEN)[(b)], B.P. 1382 R.P. 10001 Rabat 10001; Université Mohamed Premier[(c)], LPTPM, Faculté des Sciences, B.P.717. Bd. Mohamed VI, 60000, Oujda; Université Mohammed V, Faculté des Sciences[(d)]4 Avenue Ibn Battouta, BP 1014 RP, 10000 Rabat, Morocco

[136] CEA, DSM/IRFU, Centre d'Etudes de Saclay, FR-91191 Gif-sur-Yvette, France

[137] University of California Santa Cruz, Santa Cruz Institute for Particle Physics (SCIPP), Santa Cruz, CA 95064, United States of America

[138] University of Washington, Seattle, Department of Physics, Box 351560, Seattle, WA 98195-1560, United States of America

[139] University of Sheffield, Department of Physics & Astronomy, Hounsfield Road, Sheffield S3 7RH, United Kingdom

[140] Shinshu University, Department of Physics, Faculty of Science, 3-1-1 Asahi, Matsumoto-shi, JP-Nagano 390-8621, Japan

[141] Universität Siegen, Fachbereich Physik, D 57068 Siegen, Germany

[142] Simon Fraser University, Department of Physics, 8888 University Drive, CA-Burnaby, BC V5A 1S6, Canada

[143] SLAC National Accelerator Laboratory, Stanford, California 94309, United States of America

[144] Comenius University, Faculty of Mathematics, Physics & Informatics[(a)], Mlynska dolina F2, SK-84248 Bratislava; Institute of Experimental Physics of the Slovak Academy of Sciences, Dept. of Subnuclear Physics[(b)], Watsonova 47, SK-04353 Kosice, Slovak Republic

[145] [(a)]University of Johannesburg, Department of Physics, PO Box 524, Auckland Park, Johannesburg 2006; [(b)]School of Physics, University of the Witwatersrand, Private Bag 3, Wits 2050, Johannesburg, South Africa, South Africa

[146] Stockholm University: Department of Physics[(a)]; The Oskar Klein Centre[(b)], AlbaNova, SE-106 91 Stockholm, Sweden

[147]Royal Institute of Technology (KTH), Physics Department, SE-106 91 Stockholm, Sweden

[148]Stony Brook University, Department of Physics and Astronomy, Nicolls Road, Stony Brook, NY 11794-3800, United States of America

[149]University of Sussex, Department of Physics and Astronomy 2 Building, Falmer, Brighton BN1 9QH, United Kingdom

[150]University of Sydney, School of Physics, AU-Sydney NSW 2006, Australia

[151]Insitute of Physics, Academia Sinica, TW-Taipei 11529, Taiwan

[152]Technion, Israel Inst. of Technology, Department of Physics, Technion City, IL-Haifa 32000, Israel

[153]Tel Aviv University, Raymond and Beverly Sackler School of Physics and Astronomy, Ramat Aviv, IL-Tel Aviv 69978, Israel

[154]Aristotle University of Thessaloniki, Faculty of Science, Department of Physics, Division of Nuclear & Particle Physics, University Campus, GR-54124, Thessaloniki, Greece

[155]The University of Tokyo, International Center for Elementary Particle Physics and Department of Physics, 7-3-1 Hongo, Bunkyo-ku, JP-Tokyo 113-0033, Japan

[156]Tokyo Metropolitan University, Graduate School of Science and Technology, 1-1 Minami-Osawa, Hachioji, Tokyo 192-0397, Japan

[157]Tokyo Institute of Technology, 2-12-1-H-34 O-Okayama, Meguro, Tokyo 152-8551, Japan

[158]University of Toronto, Department of Physics, 60 Saint George Street, Toronto M5S 1A7, Ontario, Canada

[159]TRIUMF[a], 4004 Wesbrook Mall, Vancouver, B.C. V6T 2A3; [b]York University, Department of Physics and Astronomy, 4700 Keele St., Toronto, Ontario, M3J 1P3, Canada

[160]University of Tsukuba, Institute of Pure and Applied Sciences, 1-1-1 Tennoudai, Tsukuba-shi, JP-Ibaraki 305-8571, Japan

[161]Tufts University, Science & Technology Center, 4 Colby Street, Medford, MA 02155, United States of America

[162]Universidad Antonio Narino, Centro de Investigaciones, Cra 3 Este No. 47A-15, Bogota, Colombia

[163]University of California, Irvine, Department of Physics & Astronomy, CA 92697-4575, United States of America

[164]INFN Gruppo Collegato di Udine[a]; ICTP[b], Strada Costiera 11, IT-34014, Trieste; Università di Udine, Dipartimento di Fisica[c], via delle Scienze 208, IT-33100 Udine, Italy

[165]University of Illinois, Department of Physics, 1110 West Green Street, Urbana, Illinois 61801, United States of America

[166]University of Uppsala, Department of Physics and Astronomy, P.O. Box 516, SE-751 20 Uppsala, Sweden

[167]Instituto de Física Corpuscular (IFIC) Centro Mixto UVEG-CSIC, Apdo. 22085 ES-46071 Valencia, Dept. Física At. Mol. y Nuclear; Dept. Ing. Electrónica; Univ. of Valencia, and Inst. de Microelectrónica de Barcelona (IMB-CNM-CSIC) 08193 Bellaterra, Spain

[168]University of British Columbia, Department of Physics, 6224 Agricultural Road, CA-Vancouver, B.C. V6T 1Z1, Canada

[169]University of Victoria, Department of Physics and Astronomy, P.O. Box 3055, Victoria B.C., V8W 3P6, Canada

[170]Waseda University, WISE, 3-4-1 Okubo, Shinjuku-ku, Tokyo, 169-8555, Japan

[171]The Weizmann Institute of Science, Department of Particle Physics, P.O. Box 26, IL-76100 Rehovot, Israel

[172]University of Wisconsin, Department of Physics, 1150 University Avenue, WI 53706 Madison, Wisconsin, United States of America

[173]Julius-Maximilians-University of Würzburg, Physikalisches Institut, Am Hubland, 97074 Würzburg, Germany

[174]Bergische Universität, Fachbereich C, Physik, Postfach 100127, Gauss-Strasse 20, D-42097 Wuppertal, Germany

[175]Yale University, Department of Physics, PO Box 208121, New Haven CT, 06520-8121, United States of America

[176]Yerevan Physics Institute, Alikhanian Brothers Street 2, AM-375036 Yerevan, Armenia

[177]ATLAS-Canada Tier-1 Data Centre, TRIUMF, 4004 Wesbrook Mall, Vancouver, BC, V6T 2A3, Canada

[178]GridKA Tier-1 FZK, Forschungszentrum Karlsruhe GmbH, Steinbuch Centre for Computing (SCC), Hermann-von-Helmholtz-Platz 1, 76344 Eggenstein-Leopoldshafen, Germany

[179]Port d'Informacio Cientifica (PIC), Universitat Autonoma de Barcelona (UAB), Edifici D, E-08193 Bellaterra, Spain

[180]Centre de Calcul CNRS/IN2P3, Domaine scientifique de la Doua, 27 bd du 11 Novembre 1918, 69622 Villeurbanne Cedex, France

[181]INFN-CNAF, Viale Berti Pichat 6/2, 40127 Bologna, Italy

[182]Nordic Data Grid Facility, NORDUnet A/S, Kastruplundgade 22, 1, DK-2770 Kastrup, Denmark

[183]SARA Reken- en Netwerkdiensten, Science Park 121, 1098 XG Amsterdam, Netherlands

[184]Academia Sinica Grid Computing, Institute of Physics, Academia Sinica, No. 128, Sec. 2, Academia Rd., Nankang, Taipei, Taiwan 11529, Taiwan

[185]UK-T1-RAL Tier-1, Rutherford Appleton Laboratory, Science and Technology Facilities Council, Harwell Science and Innovation Campus, Didcot OX11 0QX, United Kingdom

[186]RHIC and ATLAS Computing Facility, Physics Department, Building 510, Brookhaven National Laboratory, Upton, New York 11973, United States of America

[a]Also at LIP, Portugal.

[b]Also at Faculdade de Ciencias, Universidade de Lisboa, Portugal.

[c]Also at CPPM, Marseille, France.

[d]Also at TRIUMF, Vancouver, Canada.

[e]Also at FPACS, AGH-UST, Cracow, Poland.

[f]Also at TRIUMF, Vancouver, Canada.

[g]Also at Department of Physics, University of Coimbra, Portugal.

[h]Now at CERN.

[i]Also at Università di Napoli Parthenope, Napoli, Italy.

[j]Also at Institute of Particle Physics (IPP), Canada.

[k]Also at Università di Napoli Parthenope, via A. Acton 38, IT-80133 Napoli, Italy.

[l]Louisiana Tech University, 305 Wisteria Street, P.O. Box 3178, Ruston, LA 71272, United States of America.

[m]Also at Universidade de Lisboa, Portugal.

[n]At California State University, Fresno, USA.

[o]Also at TRIUMF, 4004 Wesbrook Mall, Vancouver, B.C. V6T 2A3, Canada.

[p]Currently at Istituto Universitario di Studi Superiori IUSS, Pavia, Italy.

[q] Also at Faculdade de Ciencias, Universidade de Lisboa, Portugal and at Centro de Fisica Nuclear da Universidade de Lisboa, Portugal.
[r] Also at FPACS, AGH-UST, Cracow, Poland.
[s] Also at California Institute of Technology, Pasadena, USA.
[t] Louisiana Tech University, Ruston, USA.
[u] Also at University of Montreal, Montreal, Canada.
[v] Also at Institut für Experimentalphysik, Universität Hamburg, Hamburg, Germany.
[w] Also at Petersburg Nuclear Physics Institute, Gatchina, Russia.
[x] Also at Institut für Experimentalphysik, Universität Hamburg, Luruper Chaussee 149, 22761 Hamburg, Germany.
[y] Also at School of Physics and Engineering, Sun Yat-sen University, China.
[z] Also at School of Physics, Shandong University, Jinan, China.
[aa] Also at California Institute of Technology, Pasadena, USA.
[ab] Also at Rutherford Appleton Laboratory, Didcot, UK.
[ac] Also at school of physics, Shandong University, Jinan.
[ad] Also at Rutherford Appleton Laboratory, Didcot, UK.
[ae] Now at KEK.
[af] Also at Departamento de Fisica, Universidade de Minho, Portugal.
[ag] University of South Carolina, Columbia, USA.
[ah] Also at KFKI Research Institute for Particle and Nuclear Physics, Budapest, Hungary.
[ai] University of South Carolina, Dept. of Physics and Astronomy, 700 S. Main St, Columbia, SC 29208, United States of America.
[aj] Also at Institute of Physics, Jagiellonian University, Cracow, Poland.
[ak] Louisiana Tech University, Ruston, USA.
[al] Also at Centro de Fisica Nuclear da Universidade de Lisboa, Portugal.
[am] Also at School of Physics and Engineering, Sun Yat-sen University, Taiwan.
[an] University of South Carolina, Columbia, USA.
[ao] Transfer to LHCb 31.01.2010.
[ap] Also at Department of Physics, Oxford University, Oxford, United Kingdom.
[aq] Also at Sun Yat-sen University, Guangzhou, PR China.
[ar] Also at Nanjing University, China.
[*] Deceased

Abstract The Tile hadronic calorimeter of the ATLAS detector has undergone extensive testing in the experimental hall since its installation in late 2005. The readout, control and calibration systems have been fully operational since 2007 and the detector has successfully collected data from the LHC single beams in 2008 and first collisions in 2009. This paper gives an overview of the Tile Calorimeter performance as measured using random triggers, calibration data, data from cosmic ray muons and single beam data. The detector operation status, noise characteristics and performance of the calibration systems are presented, as well as the validation of the timing and energy calibration carried out with minimum ionising cosmic ray muons data. The calibration systems' precision is well below the design value of 1%. The determination of the global energy scale was performed with an uncertainty of 4%.

1 Introduction

The ATLAS Tile Calorimeter (TileCal) [1] is the barrel hadronic calorimeter of the ATLAS experiment [2] at the

CERN Large Hadron Collider [3]. Calorimeters have a primary role in a general-purpose hadron collider detector. The ATLAS calorimeter system provides accurate energy and position measurements of electrons, photons, isolated hadrons, taus and jets. It also contributes in particle identification and in muon momentum reconstruction. In the barrel part of ATLAS, together with the electromagnetic barrel calorimeter, TileCal focuses on precise measurements of hadrons, jets, taus and the missing transverse energy (E_T^{miss}). The performance requirements are driven by the ATLAS physics programme:

- The energy resolution for jets of $\sigma/E = 50\%/\sqrt{E(\mathrm{GeV})} \oplus 3\%$ guarantees good sensitivity for measurements of physics processes at the TeV scale, e.g. quark compositeness and heavy bosons decaying to jets. While one cannot separate the individual calorimeter performance issues, studies have shown that a random 10% non-uniformity on the TileCal cells energy response would add no more than 1% to the jet energy resolution constant term [4].

- For precision measurements such as the top quark mass, it will be desirable to reach a systematic uncertainty on the jet energy scale of 1%. Since about a third of the jet transverse energy is deposited in TileCal [5], its energy scale uncertainty should ultimately be below a 3% requirement.

** e-mail: atlas.secretariat@cern.ch

- The response linearity within 2% up to about 4 TeV is crucial for observing new physics phenomena (e.g. quark compositeness).
- A good measurement of E_T^{miss} is important for many physics signatures, in particular for SUSY particle searches and new physics. In addition to sufficient total calorimeter thickness and a large coverage in pseudorapidity, this very sensitive measurement requires also a small fraction of dead detector regions which create fake E_T^{miss}. The requirement depends on the signal to background ratio of the search.

The Tile Calorimeter has been installed in the experimental hall since 2005 and since then has undergone through several phases of commissioning and integration in the AT-LAS detector system. The main goal of this paper is to present the outcome of this commissioning phase, at the start of the LHC collisions data-taking. The paper is organised as follows: Section 2 gives a brief description of the Tile Calorimeter and discusses the overall detector status and the data-taking conditions after the commissioning was carried out. Section 3 presents the method for the channel signal reconstruction, the overall quality of the detector in coverage, noise characteristics and conditions stability. Section 4 shows the details on the three calibration systems used to set and maintain the cell energy scale and set the timing offsets, as well as results on the precision and stability of each system. The related energy scale uncertainties and the inter-calibration issues are also discussed. The last section (Sect. 5) is devoted to the validation of the performance using data from cosmic muons produced in cosmic ray showers in the atmosphere, referred to in short form throughout this paper as "cosmic muons" or "cosmic ray muons". Results are presented on energy and time reconstruction, uniformity across the calorimeter and comparison with Monte Carlo simulations. A subsection is devoted to the intercalibration of the scintillators that are located in the gap between barrel and extended barrels.

2 Detector and data taking setup

2.1 Overview of the Tile Calorimeter

TileCal is a large hadronic sampling calorimeter using plastic scintillator as the active material and low-carbon steel (iron) as the absorber. Spanning the pseudorapidity[1] region $-1.7 < \eta < 1.7$, the calorimeter is sub-divided into the barrel, also called long barrel (LB), in the central region

$(-1.0 < \eta < 1.0)$ and the two extended barrels (EB) that flank it on both sides $(0.8 < |\eta| < 1.7)$, as shown in Fig. 1. Both the barrel and extended barrel cylinders are segmented into 64 wedges (modules) in ϕ, corresponding to a $\Delta\phi$ granularity of $\sim$0.1 radians. Radially, each module is further segmented into three layers which are approximately 1.5, 4.1 and 1.8 λ (nuclear interaction length for protons) thick for the barrel and 1.5, 2.6 and 3.3 for the extended barrel. The $\Delta\eta$ segmentation for each module is 0.1 in the first two radial layers and 0.2 in the third layer (Fig. 2). The ϕ, η and radial segmentation define the three dimensional TileCal cells. Each cell volume is made of dozens of iron plates and scintillating tiles. Wavelength shifting fibres coupled to the tiles on either ϕ edge of the cells, as shown in Fig. 3, collect the produced light and are read out via square light guides by two different photomultiplier tubes (PMTs), each linked to one readout channel. Light attenuation in the scintillating tiles themselves would cause a response non-uniformity of up to 40% in the case of a single readout, for particles entering at different impact positions across ϕ. The double readout improves the response uniformity to within a few percent, in addition to providing redundancy.

In addition to the standard cells, the Intermediate Tile Calorimeter (ITC) covers the region $0.8 < \eta < 1.0$ (labelled D4 and C10 in Fig. 2). To accommodate services and readout electronics for other ATLAS detector systems, several of the ITC cells have a special construction: per side, three D4 cells have reduced thickness and eight C10 cells are plain scintillator plates. Located on the remaining, inner radius surface of the extended barrel modules, the gap scintillators cover the region of $1.0 < \eta < 1.2$ (labelled E1 and E2 in the figure), while the crack scintillators are located on the front of the Liquid Argon endcap and cover the region $1.2 < \eta < 1.6$ (labelled E3 and E4).

In the present (initial) configuration, eight pairs of crack scintillators have been removed to permit routing of signal cables from the 16 Minimum Bias Trigger Scintillators (MBTS), in each side. Located on the front face of the Liquid Argon end-cap cryostat, the MBTS span an η range of $2.12 < |\eta| < 3.85$ and are readout by the TileCal EB electronics. They are used mainly for triggering on collisions in the very early stage of LHC operation and for rate measurements of halo muons, beam-gas and minimum bias events during the low-luminosity running.

The Tile Calorimeter readout architecture divides the detector in four partitions, a definition that is widely used in this paper. The barrel is divided in two partitions (LBA and LBC) by the plane perpendicular to the beam line and crossing the interaction point, and each of the two extended barrels is a separate partition (EBA and EBC).

The TileCal readout electronics is contained in "drawers" which slide into the structural girders at the outer radius of the calorimeter. Barrel modules are read out by two drawers

---

[1] The pseudorapidity η is defined as $\eta = -\ln(\tan\frac{\theta}{2})$, where θ is the polar angle measured from the beam axis. The azimuthal angle ϕ is measured around the beam axis, with positive (negative) values corresponding to the top (bottom) part of the detector.

Fig. 1 A cut-away drawing of the ATLAS inner detector and calorimeters. The Tile Calorimeter consists of one barrel and two extended barrel sections and surrounds the Liquid Argon barrel electromagnetic and endcap hadronic calorimeters. In the innermost radii of ATLAS, the inner detector (*shown in grey*) is used for precision tracking of charged particles

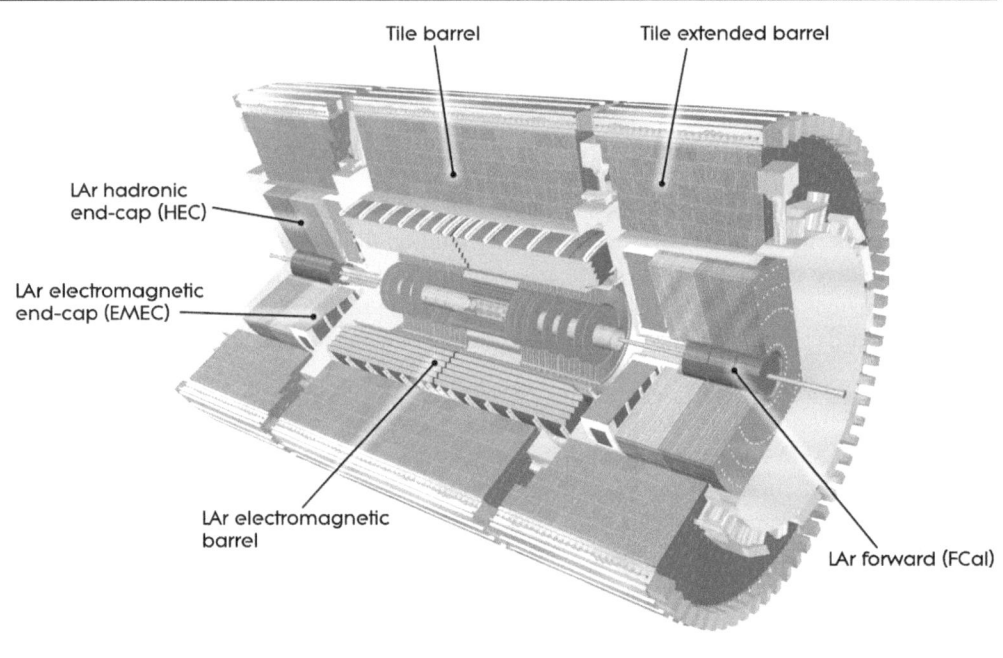

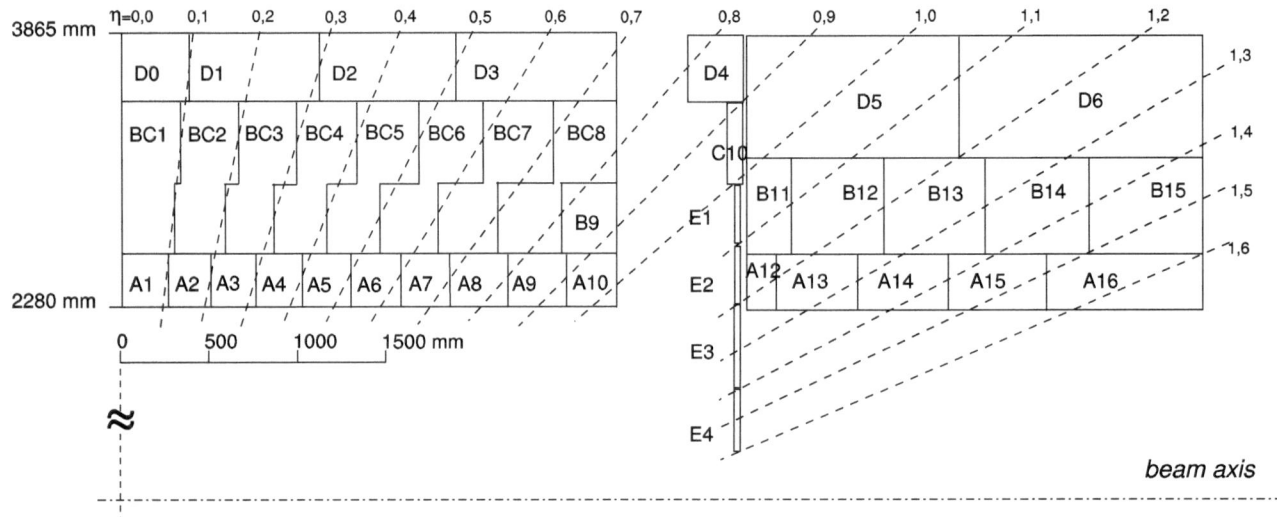

Fig. 2 Segmentation in depth and η of the Tile Calorimeter modules in the barrel (*left*) and extended barrel (*right*). The bottom of the picture corresponds to the inner radius of the cylinder. The Tile Calorimeter is symmetric with respect to the interaction point. The cells between two consecutive dashed lines form the first level trigger calorimeter tower

(one inserted from each face) and extended barrel modules are read out by one drawer each. Each drawer typically contains 45 (32) readout channels in the barrel (extended barrel) and a summary of the channels, cells and trigger outputs in TileCal is shown in Table 1.[2]

The front-end electronics as well as the drawers' Low Voltage Power Supplies (LVPS) are located on the calorimeter itself and are designed to operate under the conditions of magnetic fields and radiation. One drawer with its LVPS reads out a region of $\Delta\eta \times \Delta\phi = 0.8 \times 0.1$ in the barrel and 0.7×0.1 in the extended barrel.

In the electronics readout, the signals from the PMT are first shaped using a passive shaping circuit. The shaped pulse is amplified in separate high (HG) and low (LG) gain branches, with a nominal gain ratio of 64:1. The shaper, the charge injection calibration system (CIS), and the gain splitting are all located on a small printed circuit board known as the 3-in-1 card [6]. The HG and LG signals are sampled with the LHC bunch-crossing frequency of 40 MHz using a 10-bit ADC in the Tile Data Management Unit (DMU) chip

---

[2]The 16 reduced thickness extended barrel C10 cells are readout by only one PMT. Two extended barrel D4 cells are merged with the corresponding D5 cells and have a common readout.

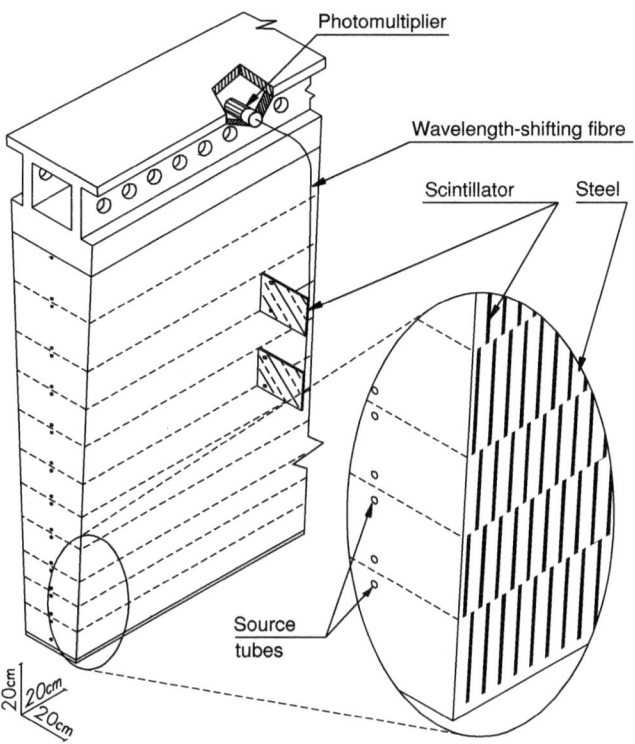

Fig. 3 Schematic showing the mechanical assembly and the optical readout of the Tile Calorimeter, corresponding to a ϕ wedge. The various components of the optical readout, namely the tiles, the fibres and the photomultipliers, are shown. The trapezoidal scintillating tiles are oriented radially and normal to the beam line and are read out by fibres coupled to their non-parallel sides

Table 1 Number of channels, cells and trigger outputs of the Tile Calorimeter. The gap and crack and MBTS channels are readout in the extended barrel drawers

	Channels	Cells	Trigger Outputs
Long barrel	5760	2880	1152
Extended barrel	3564	1790	768
Gap and crack	480	480	128
MBTS	32	32	32
Total	9836	5182	2080

which is located on the digitiser board [7]. This chip contains a pipeline memory that stores the sampled data for up to 6.4 µs. The pipeline memory can be adjusted in coarse timing steps of 25 ns. The digitisation timing of the ADCs can be adjusted in multiples of ~0.1 ns so that the central sample is as close to the PMT pulse peak as possible and to make sure the full extension of the pulse is sampled. However, this adjustment is possible only for groups of six channels, so a residual offset remains, that must be dealt with at the signal reconstruction level (see Sect. 3.2). Due to bandwidth requirements, only seven samples from one gain are read out from the front-end electronics. A gain

switch is used to determine if the high or low gain is sent. The digitised samples are sent via optical fibres to the back-end electronics which are located outside the experimental hall. From the digitised samples, the back-end electronics determine the time and energy of the channel's signal as described in Sect. 3.2.

In addition to the digital readout of the PMT signal, a millisecond-timescale integrator circuit is also located on the 3-in-1 card. The Tile integrator is designed to measure the PMT current during ^{137}Cs calibrations (see Sect. 4) and also to measure the current from minimum bias proton-proton interactions at the LHC. The integration period is approximately 14 ms and a 12-bit ADC is used for the readout.

Adder boards are distributed along the drawer. Each adder board receives the analogue signals from up to six 3-in-1 cards corresponding to cells of the same η. The trigger signal corresponding to a "tower" (see Fig. 2) of cells with $\Delta\eta \times \Delta\phi = 0.1 \times 0.1$ is formed by an analogue sum of the input signals and, together with the signals from the other calorimeters, are sent via long cables to the Level-1 (L1) calorimeter trigger system to identify jets, taus, total calorimeter energy and E_T^{miss} signatures. The signal from all four gap and crack scintillators is also summed by the adder board and passed to the L1 calorimeter trigger. A second output of the adder boards (so-called muon output), that can be used at a later stage to reduce the muon background rates, contains only the signal from cells of the outermost calorimeter layer. Presently a fraction of the muon outputs is used for carrying the MBTS signals to the L1 trigger system.

2.2 Detector and data taking overview

The detector performance and stability results exposed in this paper are based on calibration systems' data and random triggered events which cover extended periods from mid-2008 up to the end of 2009 excluding the maintenance period between December 2008 and May 2009. The results from cosmic muons and single beam are from the autumn 2008 data-taking period, with the exception of the single beam data for timing studies, for which the winter 2009 and spring 2010 data is also used.

The Tile Calorimeter at the end of 2008 data-taking period was fully operational with approximately 1.5% dead cells. The majority of the dead cells were due to three drawers that were non-operational because of power supply problems or data corruption, amounting to 60 cells or 1.2%. The remaining dead cells were randomly distributed throughout TileCal. During the 2009 data-taking period there were 48 unusable cells, fewer than 1%. The number of dead L1 trigger towers is less than 0.5% and they are uniformly distributed throughout the detector. For details on how non-operational cells are defined and the breakdown of their problems for the 2009 data-taking, see Sect. 3.1.

The cosmic data used for performance validation was collected mainly between September and October 2008 using the full ATLAS detector, including the inner detector and muon systems, with around one million events used for the present paper. The cosmic trigger configuration during this run period consisted of L1 triggers from the muon spectrometer[3] (both the Resistive Plate Chamber (RPC) and the Thin Gap Chambers (TGC)), the L1 calorimeter trigger and the MBTS. For much of the cosmic ray analysis discussed in Sect. 5, the data sample was selected by requiring a L1 trigger and at least one track reconstructed in the inner detector, from the Pixel, SemiConductor Tracker (SCT) and Transition Radiation Tracker (TRT).[4] The majority of the events came from the L1 muon spectrometer triggers. During this running period, the ATLAS magnets were run in four different configurations; no magnetic field, solenoid magnet on only, toroid magnet on only and both solenoid and toroid magnets on. The results exposed here were obtained with the full ATLAS fields on.

From the single beam data used in this paper the "splash" events and "scraping" events are used for time and energy studies. The former term is used for events occurring when the LHC beam hits the closed tertiary collimators positioned 140 m up-stream of the detector and are characterised by millions of high-energy particles arriving simultaneously in the ATLAS detector. The latter occur when the open collimators are scraping the LHC beam, allowing a moderate number of particles to the detector.

3 Detector performance and signal reconstruction

3.1 Detector and data quality status overview

The TileCal detector operated at the end of 2009 with 99.1% of cells functional for the digital readout and 99.7% of trigger towers functional for the L1. The numbers and fractions of non-operational cells, channels and trigger towers in the four calorimeter partitions are shown in Table 2.

The problematic channels belong to two categories: channels with fatal problems and channels with data quality problems. The so-called fatal problems are channels deemed unusable and are masked for the offline reconstruction and at the High Level Trigger (HLT). These channels include:

1. 44 cells (88 channels) due to two drawers with non-functional LVPS.
2. 10 channels with no response due to failures of one or more components in the readout chain, such as 3-in-1 cards, PMTs or ADCs.

---

[3]See Ref. [2], Fig. 1.4, for details on the layout.

[4]See Ref. [2], Fig. 1.1, for details on the layout.

Table 2 Summary of the number of masked channels and cells in TileCal as of November 9th, 2009. The number of dead trigger towers quoted is towers that are non-operational due to problems in TileCal's front-end electronics, not counting those related to LVPS (18 towers)

Partition	Masked Channels	Masked Cells	Dead Trigger Towers
Barrel A-side	59 (2.05%)	23 (1.60%)	2 (0.3%)
Barrel C-side	58 (2.01%)	25 (1.74%)	0 (0.0%)
Ext. barrel A-side	6 (0.29%)	0 (0.00%)	2 (0.5%)
Ext. barrel C-side	1 (0.05%)	0 (0.00%)	1 (0.3%)
Total	124 (1.26%)	48 (0.93%)	5 (0.3%)

3. 24 channels with digital data errors (17 channels with a high occurrence rate of corrupted data and 7 with gain switching problems).
4. 2 channels with high noise

The position in (η, ϕ) as of November 2009 of the unusable masked cells as described above, are shown in Fig. 4 and are summarised in Table 2. One can notice the majority of the masked cells concentrated in two non-functional front-end drawers.

Channels with data quality problems are flagged as such for the reconstruction, but they are not masked. These channels include:

1. Channels with occasional data-corruption problems, mainly due to front-end electronics malfunction or bad configuration. These are excluded from the reconstruction by checking a quality fragment in the data record on

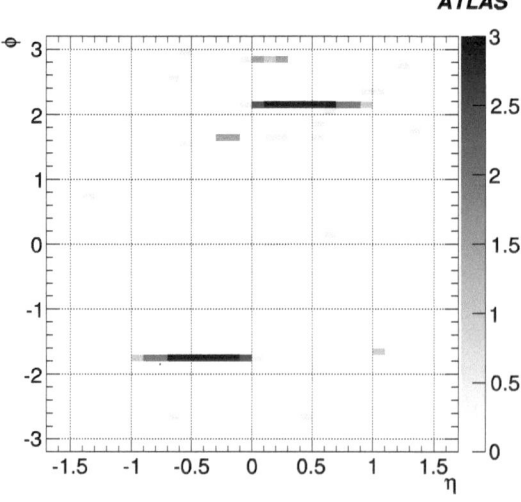

Fig. 4 Position in η and ϕ of the masked cells representing the status on November 9th, 2009. The colours corresponding to numbers 1, 2, 3 show the number of layers masked for this (η, ϕ) region. The non-integer numbers indicate that one readout channel of the cell is masked

an event by event basis. A fraction of the channels can be recovered by resetting the front-end between LHC fills.

2. Channels which cannot be calibrated with one of the calibration systems (see Sect. 4). These are flagged as poorly calibrated channels.

3. Noisy channels, which are treated by describing appropriately in the database their higher-than-average noise level to take into account while reconstructing their energy.

4. Channels where the response varies significantly over time. These are also flagged for the offline use as poor quality channels but their response can be corrected over time if the source of variation is understood. Typical cases include channels with varying response due to changes over time of the high voltage applied to the photomultipliers.

The parameters that directly affect the measured response of a channel are the temperature in the drawer and the applied high voltage because the PMT gain depends on them. The PMT gain G is proportional to V^7, where V is the applied high voltage (HV), and decreases with temperature by 0.2% per °C. The operating conditions of the detector have

been constantly monitored online and recorded by the Detector Control System (DCS). The operating values of voltages, currents, temperatures at the LVPS and at the front-end have been very stable. Figure 5 gives a measure of the long term evolution of the high voltage applied on the PMTs for two periods of 3 and 6 months separated by the maintenance period. The HV values, which are typically close to ~670 V, have shown on average a difference of 0.17 V with respect to the value set during intercalibration with an RMS of 0.37 V during the considered period. This average stability within 0.4 V for the whole calorimeter represents a 0.4% reproducibility in the gain of the PMTs due to this factor alone. Figure 6 shows the stability of the temperature measured by a probe installed in one PMT block for the same period as for the HV measurements. The average over all the calorimeter PMT probes is 24.1°C with an RMS of 0.2°C for a period of 9 months interleaved by the maintenance period.

3.2 Energy and time reconstruction

The channel signal properties—pulse amplitude, time and pedestal—for all TileCal channels are reconstructed with

Fig. 5 Stability of the PMT high voltage with respect to its set value, averaging over all PMTs for two periods of 3 and 6 months (*left*) separated by the maintenance period. The distribution of the differences of the measured and the set HV values for all PMTs over the period considered is also shown (*right*)

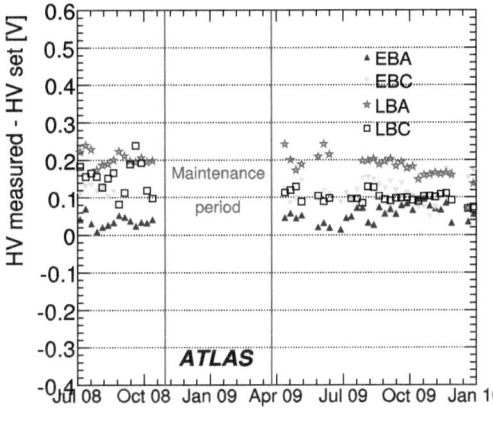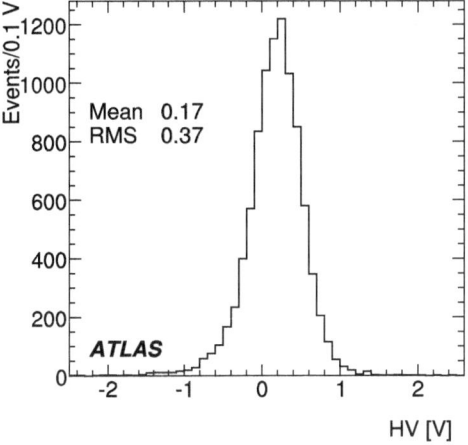

Fig. 6 Stability of the temperature, as measured at one PMT in each drawer, averaging over all drawers and presented for two periods of 3 and 6 months separated by the maintenance period (*left*). The distribution of the values for individual drawers over the whole period is also shown (*right*)

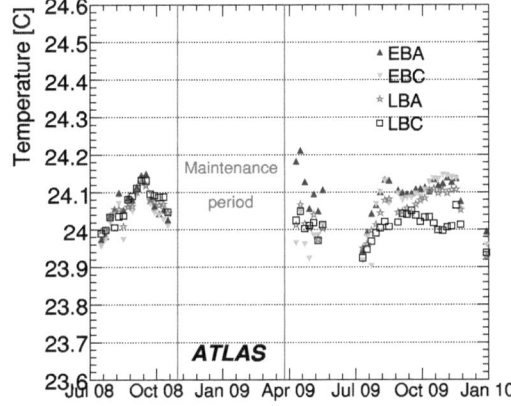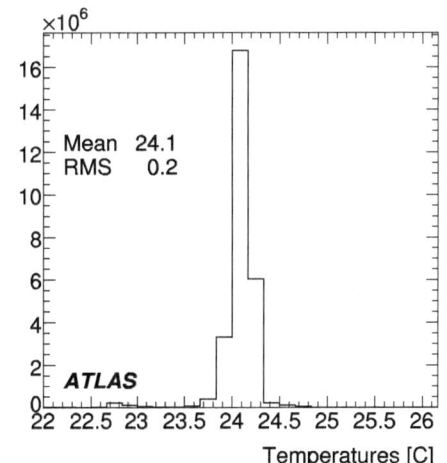

the Optimal Filtering (OF) method [8], which makes use of weighted linear combinations of the digitised signal samples (spaced by 25 ns). Due to the simplicity of its mathematical formulation, OF is implemented in the Digital Signal Processors (DSPs) of the ReadOut Driver boards (RODs) [9] and therefore provides energy and time information to the HLT of ATLAS during the online data-taking. At present, since the data-taking rate allows it, the seven digitised samples are also available offline for all the events together with the results of the OF reconstruction from the RODs. The procedure to compute the energy (given by the amplitude A) and time (τ) are given by the equations:

$$A = \sum_{i=1}^{n=7} a_i S_i, \qquad \tau = \frac{1}{A} \sum_{i=1}^{n=7} b_i S_i, \qquad (1)$$

where S_i is the sample taken at time t_i ($i = 1, \ldots, n$). The coefficients of these combinations, a_i and b_i, known as the OF weights, are obtained from knowledge of the pulse shape and noise autocorrelation matrix, and are chosen in such a way that the impact of the noise to the calorimeter resolution is minimised. Figure 7 shows the pulse shape extracted from data taken at the testbeam, selecting a channel with a given value of deposited energy for each gain. This pulse shape is the reference used in the estimation of the OF weights.

The reconstructed channel energy used by the HLT and offline is:

$$E_{\text{channel}} = A \cdot C_{\text{ADC} \to \text{pC}} \cdot C_{\text{pC} \to \text{GeV}} \cdot C_{\text{Cs}} \cdot C_{\text{Laser}}. \qquad (2)$$

The signal amplitude A, described in more detail above, represents the measured energy in ADC counts as in (1). The factor $C_{\text{ADC} \to \text{pC}}$ is the conversion factor of ADC counts to charge and it is determined for each channel using a well defined injected charge with the CIS (Charge Injection System) calibration system. The factor $C_{\text{pC} \to \text{GeV}}$ is the conver-

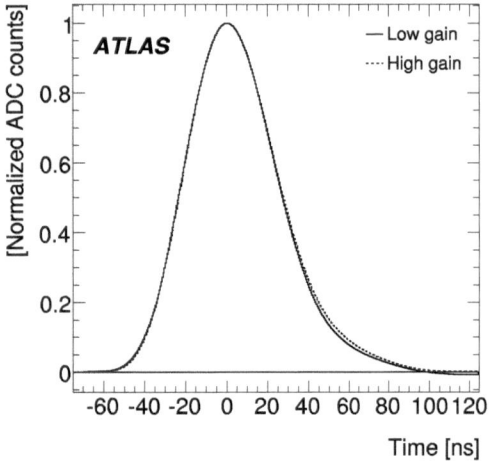

Fig. 7 Pulse shape for high and low gain from testbeam data, used as reference for the OF weights calculation

sion factor of charge to energy in GeV and it has been defined at testbeam for a subset of modules via the response to electron beams of known momentum in the first radial layer. This factor is globally applied to all cells after being adjusted for a dependence on the radial layer (see Sect. 4.4). The factor C_{Cs} corrects for residual non-uniformities after the gain equalisation of all channels has been performed by the Cs radioactive source system. The factor C_{Laser}, not currently implemented, corrects for non-linearities of the PMT response measured by the Laser calibration system. The derived time dependence of the last two factors will be applied to preserve the energy scale of TileCal. The details of the calibration procedures are discussed in Sect. 4.

The channel time, τ in (1), is the time difference between the peak of the reconstructed pulse and the peak of the reference pulse. The OF weights used in the reconstruction were calculated based on this reference pulse shifted by a time phase that depends on each channel's timing offsets measured with the calibration systems (and single-beam data), the time-of-flight from the interaction point to that cell and the hardware time adjustments mentioned in Sect. 2.1. Thus the reconstructed time τ should be compatible with zero for energy depositions coming from the interaction point. If the time residual is not well known, for small deviations ($|\tau| < 15$ ns) the uncertainty of the reconstructed amplitude depends on τ through a well-defined parabolic function, that can be used for an energy correction at the level of the HLT or offline reconstruction.

The OF results rely on having, for each channel, a fixed and known time phase between the pulse peak and the 40 MHz LHC clock signal. This is not the case during the commissioning phase of the detector, where signals caused by cosmic rays are completely asynchronous with respect to the LHC clock. Nevertheless OF can still be applied in this case and an accurate reconstruction may be obtained by applying the proper weights for each event according to the time position of the signal. The estimation of the signal time is achieved through an iterative procedure provided by a set of OF weights calculated at different phases from -75 ns to $+75$ ns in steps of 1 ns. Figure 8 presents the relative difference between the reconstructed offline energy and the energy calculated in the DSPs for cosmic muon data and shows the effect of the limited numerical precision of the DSPs. The results in the following sections are based on channel energies reconstructed offline with the iterative procedure to define the phase.

The Fit method is another signal reconstruction algorithm. It is based on a three parameter fit to the known pulse shape function $g(t)$, as expressed by:

$$S_i = A g(t_i - \tau) + ped. \qquad (3)$$

The meaning of the variables S_i and t_i and the parameters A and τ is the same as for the OF method, while ped is a

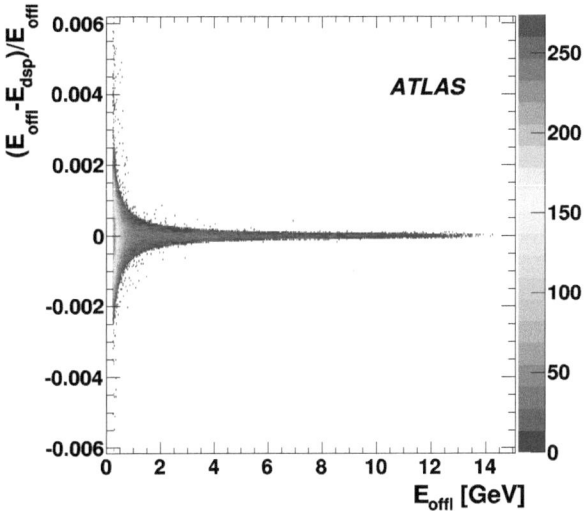

Fig. 8 Difference between the reconstructed offline energy, E_{offl}, and the energy given by the DSP E_{DSP} relative to E_{offl} and as a function of E_{offl} (in GeV), extracted from cosmic muon runs

free parameter that defines the baseline of the pulse. The Fit method is mathematically equivalent to OF in the absence of pile-up and noise, but it is not suitable for fast online signal processing in DSPs. Results from the Fit and OF methods were compared with testbeam data and were found to be equivalent [10]. Since the autumn of 2008 data-taking, the Fit method is used only for CIS calibration data, where the pulse is a superposition of charge-proportional and charge-independent components [10].

The cell energy is the sum, and the cell time the average, of the respective measurements by the two corresponding readout channels. In cases of single readout cells, or if one of the channels is masked out, the cell energy is twice the energy measured in the single available channel. The measurement of the cell's energy is thus robust to failures in a single readout channel.

3.3 Noise performance

The noise in TileCal was measured in dedicated bi-gain standalone runs with empty events (often called pedestal runs) and in random triggered events within ATLAS physics runs (often called random triggers). The noise of each channel was derived from the seven digitised samples using the same method that was used for signal reconstruction in cosmic and single beam events, i.e. using the OF with iterations.[5] In Fig. 9 the evolution during the running periods of 2008 and 2009 of the average noise, in ADC counts, is

---

[5]Note that the level of noise depends on the OF method used. The non-iterative OF method results in lower noise than the OF with iterations by ∼14%. Note also that the non-iterative OF will be applied for the data-taking during the collision phase, since the timing will be fixed by the LHC 40 MHz clock frequency.

shown for all channels and for an individual channel. The channel noise is estimated as the RMS of the single digitised samples averaged over the events in dedicated TileCal pedestal runs. The overall stability is better than 1%.

The cell noise in MeV as a function of η is shown in Fig. 10 averaged over all modules in ϕ for cells in a given η position. The cell noise is estimated as the RMS of the cell's energy distribution using the iterative OF signal reconstruction in random triggered events during a physics run with LHC single beam in 2008. Different colours are used to indicate cells in different longitudinal layers. The noise values vary between 30 and 60 MeV. The channels with higher noise are principally at the proximity of the LVPS which are located at the outer boundaries of the TileCal barrel and extended barrel modules.

The cell noise probability distribution is an important component in the ATLAS calorimeter's energy clustering algorithm. It is determined from the cell energy in empty events recorded through the standard ATLAS data acquisition chain within physics runs and it is characterised by the σ of a fitted single Gaussian to the energy (E) distribution. The ratio E/σ is used to judge if a cell has a noise-like or a signal-like energy deposition. Figure 11 shows the ratio E/σ for all TileCal cells (squares). One can observe the existence of non-Gaussian tails that could lead to fake signal cells if a criterion of $E/\sigma > 4$ is used. However, since a double Gaussian distribution provides a good description of the data, the two Gaussian σ's and the relative amplitudes are used to construct a probability density function on the basis of which a new "effective σ" (σ_{eff}) for every cell is defined at the significance level of 68.3%. The improvement is shown in Fig. 11 where the triangles represent the ratio E/σ_{eff} for

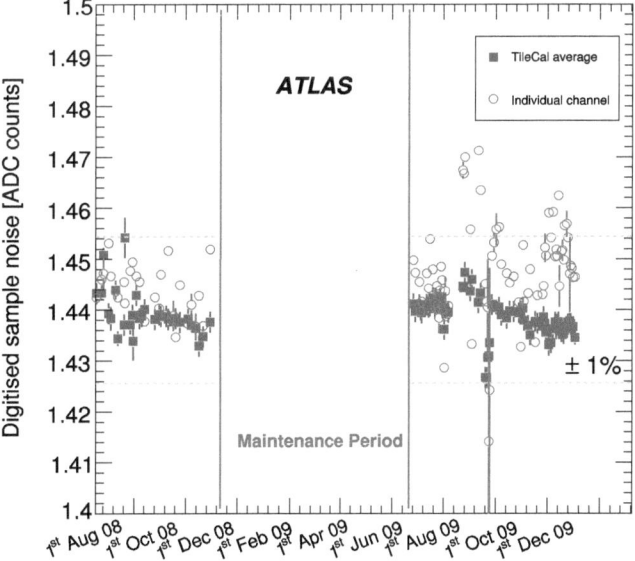

Fig. 9 Stability of average noise (RMS of the single digitised samples averaged over events and channels), in ADC counts, for all channels and for an individual channel

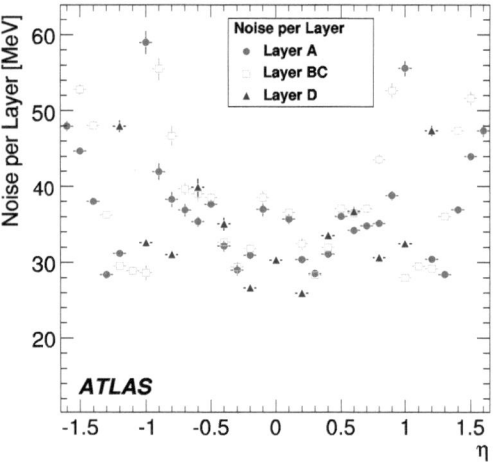

Fig. 10 Average cell noise in random triggered events as a function of the cell η and radial layer. The noise is represented by the RMS of the cell's energy distribution and the error bar shows its spread over all cells in the same pseudorapidity bin

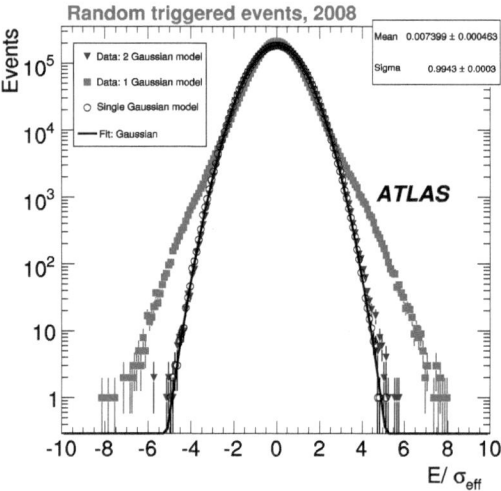

Fig. 11 Significance level of the cell energy as compared to noise (Energy/Gaussian σ) using the single and the double Gaussian descriptions of noise in random triggered events

all the calorimeter cells. One can observe that there are no tails when compared to a Gaussian fit (line) or to a toy Monte Carlo noise generator, that randomly attributes to cells energies from a single Gaussian model (circles). Thus the ratio E/σ_{eff} can be safely used to distinguish signal from noise in a TileCal cell.

4 Calibration

This section describes the calibration procedures and data sets used in TileCal to establish the reference detector response. Furthermore, the calibration results obtained in the years 2008 and 2009, during the commissioning of the Tile

Calorimeter in the ATLAS cavern, and the cross-checks related to the current understanding of its calibration are also discussed. The main objectives of the calibration procedures in TileCal are to:

- Establish the global electro-magnetic (EM) scale and the uncertainty associated with it. The EM scale calibration factor converts the calorimeter signals, measured as electric charge in pC, to the energy deposited by electrons, which would produce these signals.
- Minimise, measure and correct the cell-to-cell variations at the EM scale.
- Measure and correct the non-linearity of the calorimeter response.
- Measure the average time offset between the signal detection and the collision time for every readout channel.
- Monitor the stability of these quantities in time.

The Tile Calorimeter calibrations systems treat different sections of the readout chain as illustrated in Fig. 12. They provide:

- Calibration of the initial part of the signal readout path (including the optics elements and the PMTs) with movable radioactive ^{137}Cs γ-sources [11], hereafter to be called simply Cs.
- Monitoring of the gains of the photomultipliers by illuminating all of them with a laser system [4, 12].
- Calibration of the front-end electronic gains with a charge injection system (CIS) [6].

In order to detect non-uniformities or degradation in the detector elements (optical and otherwise), the calibration systems are specified to meet a precision of 1% on the measurement of the response of a cell.

The number of channels that cannot be calibrated by each individual calibration system is well below 1%. This is additional to the number of channels that are unusable due to LVPS problems or other issues not related to the given calibration system. In the following sections the performance distributions appear sometimes with fewer channels due to the fact that not all could be available for all the calibration periods.

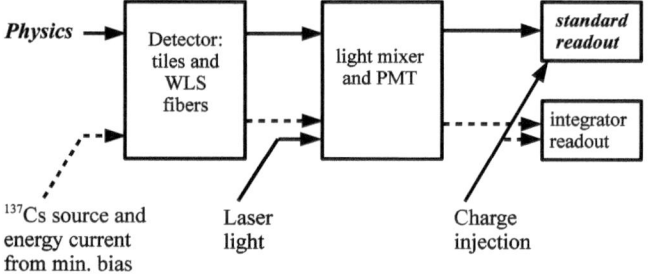

Fig. 12 Flow diagram of the readout signal paths of the different TileCal calibration tools. The paths are partially overlapping, allowing for cross-checks and an easier identification of component failures

The current calibration protocol includes a number of dedicated calibration runs performed with a frequency derived from experience gained during the detector commissioning. The CIS constants are very stable in time and are only updated twice per year. For monitoring and identification of bad channels, CIS runs are performed between physics runs twice per week. For monitoring, laser runs are also performed twice per week. The resulting laser constants will be used only for monitoring purposes until the stability of this calibration system is fully understood. The Cs scans are performed outside beam periods, with a periodicity of weeks or months, depending on the machine schedule since a full scan takes 6 to 8 hours. Starting from 2010, every Cs run is expected to result in new constants that adjust the global EM energy scale which will be updated accordingly. Laser runs accompany Cs runs in order to disentangle between changes related to the optical system and PMTs. Since the laser runs are more frequent than the Cs scans, the former provide information on the PMT gain changes between two Cs scans.

A dedicated monitoring system based on slow integrators [6] records signals in the Tile readout channels over thousands of bunch crossings during the physics runs and is also a part of the Tile calibration framework. As this measurement requires experience with collisions it is still being commissioned.

4.1 Charge injection system and gain calibration in the readout electronics

The circuitry for the Charge Injection system is a permanent part of each front-end electronics channel [6] and it is used to measure the pC/ADC conversion factor for the digital readout of the laser calibration and physics data and to determine the conversion factor for the slow integrator readout, measured in ohms.

To reconstruct the amplitude for each injected charge, a three-parameter fit is performed as described at the end of Sect. 3, with the amplitude being one of the parameters of the fit [10]. To determine the values of the gains for each channel, dedicated CIS calibration runs are taken frequently, in which a scan is performed over the full range of charges for both gains. The typical channel-to-channel variation of these constants is measured to be approximately 1.5%, as shown in Fig. 13. This spread indicates the level of corrections for which the CIS constants are applied.

The stability in time of the average high gain and low gain readout calibration constants from August 2008 to October 2009 is shown in Fig. 14 for 99.4% of the total number of ADCs. The time stability of a typical channel is also shown for each gain. Over this period, the RMS variation for the high and low gain detector-wide averages and for the single channels shown, is less than 0.1%. The superimposed bands of ±0.7% represent the systematic uncertainty for the individual channel calibration constants, mainly due to the uncertainty on the injected charge.

The distributions of high gain and low gain readout calibration constants for individual ADC channels were compared for the sample of channels calibrated during the Tile-Cal standalone testbeam period of 2002 to 2003 and for the full detector in the cavern in 2009. No significant change in the calibration constants was observed, thus limiting the contribution from the CIS calibration to the systematic uncertainty on transferring the EM scale from testbeam to ATLAS to below 0.1%.

To determine the values of the gains for each channel for the current integrator readout, dedicated calibration runs are periodically taken, in which a scan is performed over the full range of currents for all six integrator gains. The channel gain is extracted as a slope from a 2-parameter fit performed on the measured channel response in voltage to each applied current. The typical channel-to-channel variation of these integrator gain constants is measured to be approximately 0.9%, as shown in Fig. 15 (left) for the gain used during calorimeter calibration with the Cs radioactive source. The 12-bit ADCs used to digitise the PMT currents were pro-

Fig. 13 Channel-to-channel variation of the high gain (*left*) and of the low gain (*right*) readout calibration constants as measured by the CIS, prior to any correction. The measured HG/LG gain ratio of 62.9 corresponds to the nominal of 64 (see Sect. 2.1) within tolerances of individual electronics components

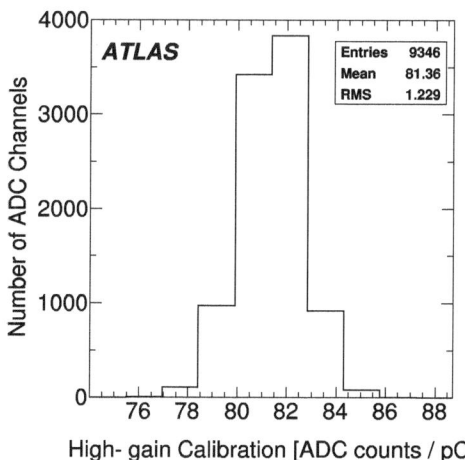
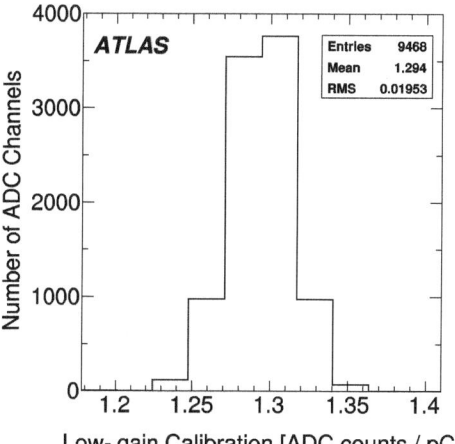

Fig. 14 Stability in time of the average high gain (*left*) and low gain (*right*) readout calibration constants from August 2008 to October 2009

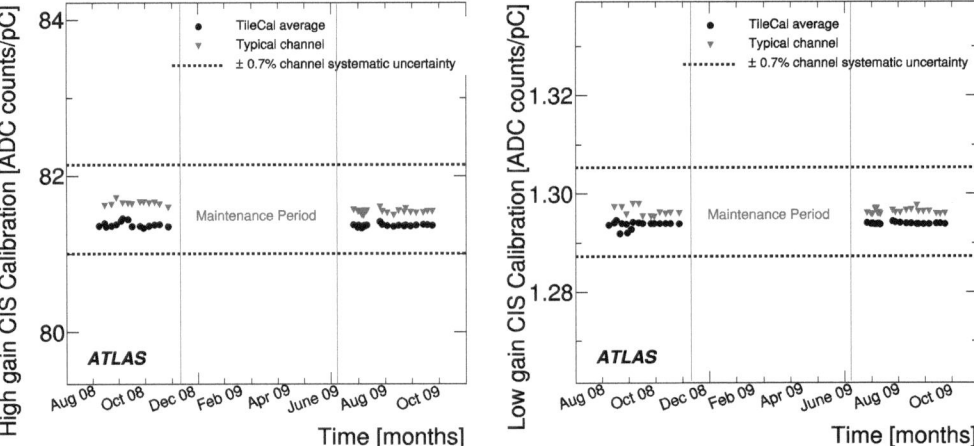

duced in two unequal batches with about 2% difference in amplifier gains, which can be clearly seen in the distribution of the integrator gains in Fig. 15 (left).

The relative variation of the integrator gains used by the Cs calibration system is shown in the right part of Fig. 15. The measurements in 95.9% of the integrators performed at different dates are compared to the reference measurements of January 2008. The error bars represent the dispersion of the individual channel measurements relative to their reference values in the first run. The stability of individual channels is better than 0.05% while the stability of the average integrator gain is better than 0.01% over the considered period of time of 26 months.

The variation of the integrator gains for individual channels used in the Cs calibration system readout from 2001 to 2009 was studied on the sample of channels calibrated in both instances. No significant change in the calibration constants was observed over eight years. The contribution from the integrator gain calibration to the systematic uncertainty on setting the EM scale of TileCal in ATLAS as compared to the testbeam was found to be below 0.2%.

4.2 Calibration with laser system

The Tile Calorimeter is equipped with a custom-made laser calibration system [12] dedicated to the monitoring and calibration of the Tile photomultiplier properties, including the gain and linearity of each PMT. The frequency doubled infrared laser providing a 532 nm green light beam is located in the ATLAS USA15 electronics room, 100 m from the detector. The laser emits short pulses, which reasonably resemble those from the physics signals, with a nominal energy of a few mJ. This power is sufficient to simultaneously saturate all Tile readout channels, and thus to probe their linearity over the full readout dynamic range. A dedicated set of optical elements insures proper attenuation, partial de-coherence and propagation of the original light beam to every photomultiplier used in the Tile Calorimeter readout. This calibration system was commissioned until September 2009 and since then it is operating in a stable configuration. By varying the voltages applied to the photomultipliers it was shown that the system sensitivity to the relative gain variations is of 0.3% on data sets recorded over few hours. The long term stability of the laser calibration system is under study.

Fig. 15 Distribution of the integrator gain used by the Cesium calibration system is shown *on the left*. Relative stability over twenty-two months of the same integrator gain is shown *on the right*

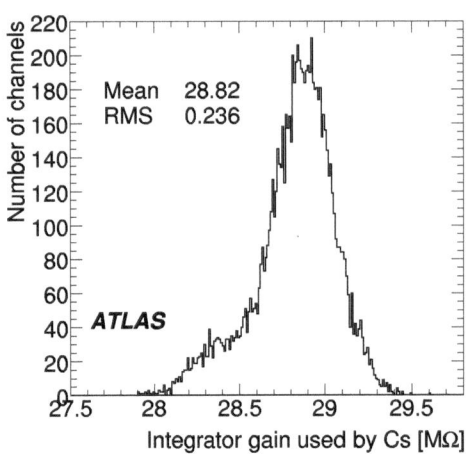

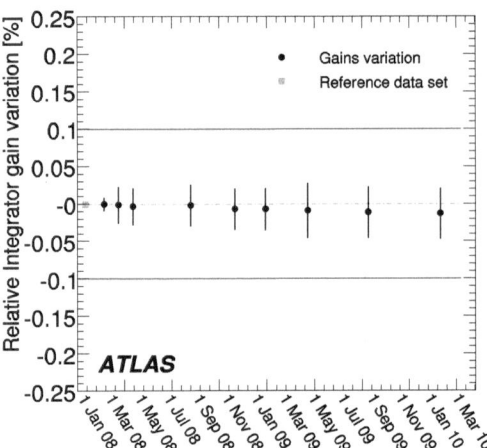

The time stability of the PMT gains was evaluated using dedicated laser runs and averaging over 98.8% of the Tile-Cal channels. An estimation of the relative gain variation in time was based on the analysis of the shape of the distribution of the PMT responses to the signal induced by the laser system at many instances. The average gain variation as a function of time over 40 days is shown on Fig. 16. This variation is found to be within 1.0% over the considered period of time. The displayed error bars of 0.5% account for both the statistical uncertainty and the systematic effects and are entirely dominated by the latter. The systematic uncertainty comes from the limited reproducibility of the light intensity on the photomultipliers downstream of the full optical chain through which the laser beam propagates to the detector. The design goal of the laser system is to monitor the relative gain stability with 0.5% accuracy for time periods of months to years. The results mentioned above set the precision with which the PMT response stability can be monitored by the laser system between two Cs scans that are typically one month apart and monitor the combined response of the optics elements and PMTs.

Once the global variation of the laser signal is accounted for, the gain stability per individual channel can be studied. A typical channel to channel variation for HG and LG is shown in Fig. 17, where the relative gain variations for two laser calibration runs separated by 50 days are presented. The shaded sidebands represent channels with relative variation above 1%. The observed RMS of 0.3% (0.2%) in the HG (LG) is a convolution of residual fluctuations of the laser system and variations of the PMT response. Therefore, this RMS can be considered as an upper limit on possible stochastic variations in photomultiplier gains.

Once the intrinsic stability of the laser calibration system is understood, this system will be used to calibrate the gain and linearity[6] of each PMT.

4.3 Calibration based on ^{137}Cs radioactive γ-source

The Tile Calorimeter includes the capability of moving through each scintillator tile a Cs radioactive γ-source along the Z-direction of the ATLAS detector. Capsules containing the Cs sources with activities of about 330 MBq emitting 0.662 MeV γ-rays are hydraulically driven through a system of 10 km of steel tubes that traverses every scintillating tile in every module [13]. Three sources of similar intensity are deployed in the three cylinders of the Tile Calorimeter. When a capsule traverses a given cell, the integrator circuit located on the 3-in-1 cards (Sect. 2.1), reads out the current signal in the PMTs. The total area under the integrator current vs capsule position curve corresponding to the source path length in a cell, is calculated and normalised to the cell size. This estimator of the cell response to Cs is used throughout this section.

Source scans provide the means to diagnose optical instrumentation defects [14] and to measure the response of each individual cell. The precision of the Cs based calibration was evaluated from the reproducibility of multiple measurements under the same conditions and was found to be about 0.3% for a typical cell [11]. The precision is 0.5% for the cells on the edge of the TileCal cylinders and a few percent for the narrow cells C10 and D4 in the gap region (see Fig. 3). As discussed in Sect. 5.4, cosmic ray muons are used to cross-check the calibration factors for the cells of this type.

4.3.1 Intercalibration and EM scale factor via the Cs system

The Cs calibration has proceeded in two distinct phases.

– Photomultiplier gain equalisation to a chosen level of Cs response was performed for every individual channel on the 11% of production TileCal modules that were tested with particle beams during 2001–2004 [10]. The next step was to measure the numerical value for the fixed EM scale with electron beams. With the electrons entering the calorimeter modules at an incidence angle of 20°, the average cell response normalised to beam energy was measured to be (1.050 ± 0.003) pC/GeV, defining the TileCal EM scale factor. This factor was determined for the cells of the first layer and propagated via the gain equalisation

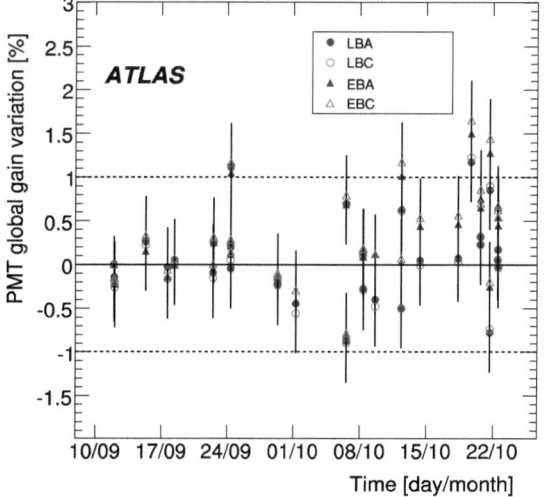

Fig. 16 Average PMT gain variation measured by the laser calibration system as a function of time over forty days in 2009

_____

[6]All the photomultipliers used in TileCal were characterised on their arrival from Hamamatsu at dedicated test benches with LED light sources. No PMT was found with non-linearity worse than 3% up to 800 pC of collected charge.

Fig. 17 Channel-to-channel variation of the relative gain of the photomultipliers for two Laser calibration runs taken in HG (LG) mode, shown *on the left* (*right*)

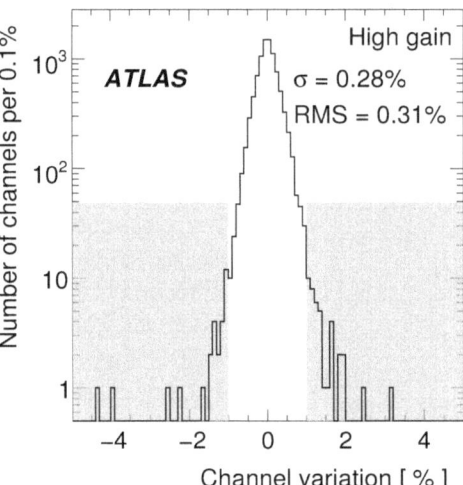 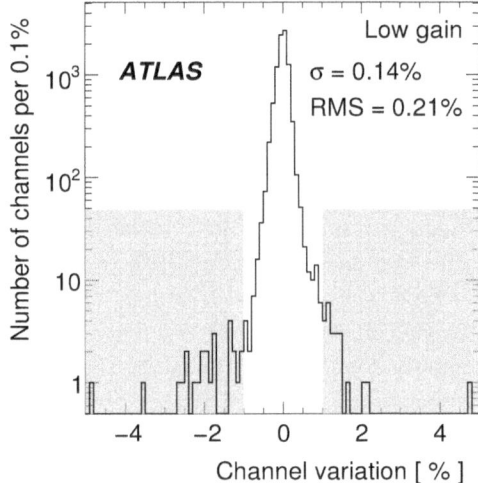

to all the other cells. The RMS spread of $(2.4 \pm 0.1)\%$ was found to be due to local variations in individual tile responses and tile-fibre optical couplings. The above two steps effectively resulted in setting the cell EM scale in the subset of TileCal modules exposed to electron beams.

– The second phase in the calibration was to reproduce the above PMT gain equalisation on the full set of the Tile Calorimeter modules in the ATLAS environment and to transfer via the Cs response the EM scale factor as defined in the testbeam. This took place in the second half of 2008. In some cases the PMT gain is intentionally higher by 20% (D0, D1, D2, D3, D4 and C10 cells) in order to improve on signal to noise ratio for the detection of muons (see also Sect. 5.1). For the central barrel cells of the third radial layer this improvement will facilitate their possible usage in the L1 muon trigger. The EM scale for these cells is recovered by applying appropriate corrections to the cell energy reconstruction.

To set the EM scale as defined at the testbeam, the target response to Cs for 2008 and 2009 was defined as the response measured at the testbeam scaled by the ratio of the activities of the testbeam source to the sources used in the cavern. These ratios were measured by intercalibrating the sources using two TileCal modules that are kept outside the experimental hall. The source activity decay time between the testbeam and the ATLAS scans was taken into account. By adjusting PMT gains in order to have equal response to Cs between the testbeam and the ATLAS setup, the numerical factor that converts charge to GeV is preserved. It is evident that the comparison of the source activities is of utmost importance in order to preserve the absolute energy scale as set with electrons.

Five ^{137}Cs radioactive sources of different ages and activities were used over the last years. Three sources are currently used in the ATLAS cavern and two different sources were used for checks on instrumentation quality and for the

calibration at the testbeam. In spring of 2009, one long barrel and one extended barrel module were scanned sixty times under the same conditions with all five radioactive sources. With the reproducibility of a single measurement better than 0.1%, a full set of ratios of the source activities was evaluated with the precision of 0.05%. The results for these ratios after averaging over all data sets available are shown in the last column of Table 3. It should be noted that the third column of the table gives an initial estimation of the activities as measured by the manufacturer with a ±15% uncertainty. We plan to exchange the sources between the Tile Calorimeter cylinders in the cavern for future checks on reproducibility of the responses and also to monitor the ratios of the source activity in time.

4.3.2 Effect of magnetic field

Comparing the EM scale response between the testbeam and full detector, the magnetic field configuration has to be considered. During the testbeam no magnetic field was present while during data-taking in ATLAS, TileCal operates in the presence of magnetic field. The calorimeter iron, mainly the

Table 3 Activity of five ^{137}Cs radioactive sources as of April 2009, and ratios with respect to the reference source RP3713 of the measured activities averaged over all data sets collected in the spring of 2009. Source RP3713 was used in calibrations during the test beam period. Source RP3712, kept in Building 175, is used for ageing tests

Source	Location in 2009	Activity in April 2009 (±15%)	Measured activity, normalised to RP3713
RP3713	Storage	264 MBq	–
RP4091	LB	372 MBq	1.1860 ± 0.0005
RP4090	EBA	363 MBq	1.1590 ± 0.0005
RP4089	EBC	377 MBq	1.2180 ± 0.0007
RP3712	Bld. 175	319 MBq	1.2200 ± 0.0005

Fig. 18 Ratio of the TileCal cell response to the radioactive Cs source in full ATLAS magnetic field to the TileCal cell response to the Cs source without the field, shown as function of η *(left)*. Ratio of the TileCal D3 cell response to radioactive Cs source in full ATLAS magnetic field over its response to the Cs source without the field, shown as function of ϕ *(right)*. The *vertical lines* indicate the position of the ATLAS toroid coils

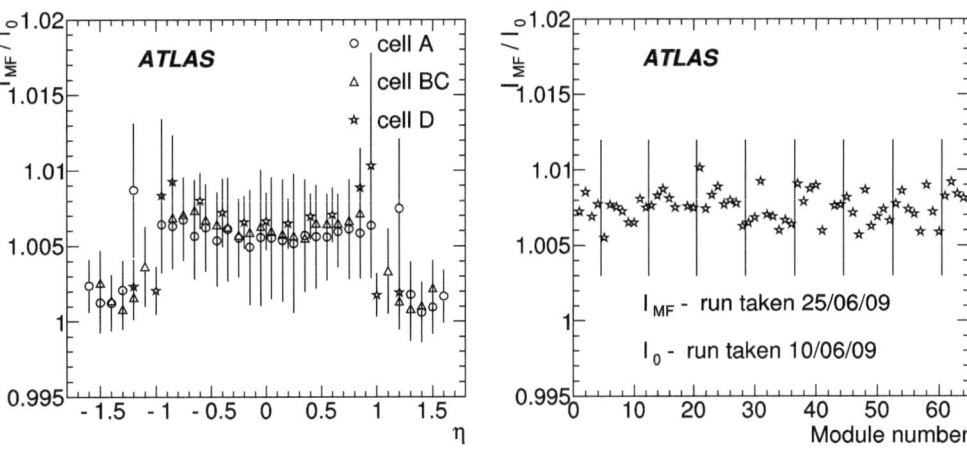

girder volume at the outer radius, serves as the flux return of the solenoid field. The general behaviour of iron-scintillator calorimeters in magnetic field is known from other experiments [15–17]. A small increase in the scintillator light yield, which also varies modestly over a broad range of the applied field is expected.

The impact of the full ATLAS magnetic field on the Tile Calorimeter response was studied using the Cs calibration system. The ratio of the TileCal cell response to a radioactive Cs source in the full ATLAS magnetic field to its response to the Cs source without the field is given in Fig. 18 (left) as a function of η for two consecutive Cs runs. The cells in individual radial layers are shown with different symbols. The error bars represent the RMS of the above ratio over the sample of the sixty four identical cells in the full ϕ range.

As expected, the effect of magnetic field is stronger in the barrel partitions, where the flux of the solenoid field return is the most intense, and where the increase in calorimeter response is on average $\sim0.6\%$. A small increase of $\sim0.2\%$ is observed for the extended barrel. This increase was not fully reproducible in every instance of the magnetic field turn-on in 2008, which contributes 0.5% to the systematic uncertainty of propagating the EM scale from the testbeam to the ATLAS running configuration. The ratio of the D3 cell[7] response to radioactive Cs source with and without the full AT-LAS magnetic field is shown in the right part of Fig. 18 as function of ϕ. The vertical lines illustrate the positions of the Toroid coils. No clear structure in ϕ is observed, indicating that in the final ATLAS configuration the full magnetic field does not significantly affect the Tile Calorimeter response uniformity in ϕ. Starting from 2010, Cs calibrations will be exclusively based on the data taken with the full magnetic field.

---

[7] A cell through which the Toroid field return is the strongest.

4.3.3 Monitoring with Cs in ATLAS

Once the EM scale was established and reproduced in AT-LAS, periodic scans are performed to monitor the stability of the detector response to the radioactive source in time. This is the final step that insures the monitoring of the known EM scale in time.

The Tile Calorimeter response to the Cs source as a function of time is shown in Fig. 19. The first scan was taken approximately two weeks after the original PMT gain equalisation in July 2008. Around 55 calibration runs with the radioactive source are considered for the time period from August 2008 to February 2010. The maintenance period of

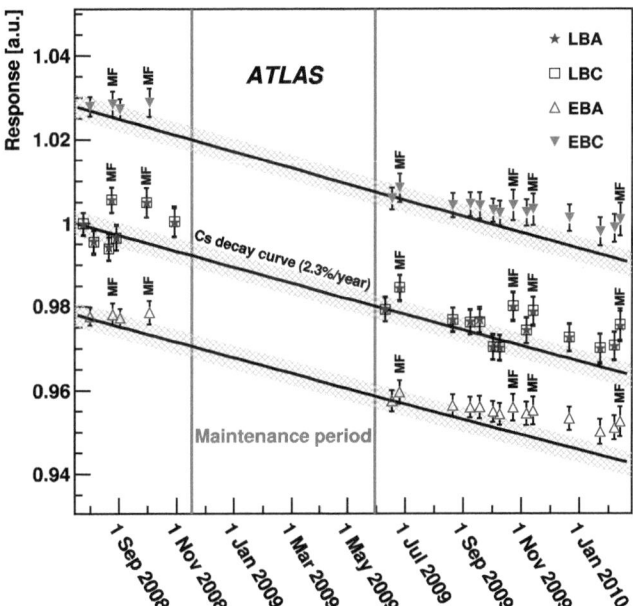

Fig. 19 TileCal response to radioactive Cs sources in all four calorimeter partitions not corrected for the difference in the source activities as a function of time, averaging over all channels in a partition. The error bars represent the RMS spread in the responses of the sample of channels used. The "MF" symbol stands for the Cs calibration data taken with magnetic field on

six months is indicated by the vertical lines on Fig. 19 and is excluded from the studies. The very first points after the maintenance period correspond to the second gain equalisation to the same target value, corrected for the expected decrease in the source activity in time, as indicated on the figure. The average response to the radioactive sources in the four calorimeter partitions is shown by the points of different colours. Since three sources with about 3% difference in their activity are used in the barrel and two extended barrel cylinders, the data points follow three distinct paths in time. The error bars, which are always below 0.4%, represent the RMS spread in responses over the full set of channels in a given partition. The number of cells with unreliable Cs calibration or with unstable HV level is below 0.2% of the total and they are excluded from the present study. The shaded bands along the lines indicate the level of reproducibility of the Cs measurements. The "MF" label indicates that the corresponding Cs calibration run was taken with both the ATLAS toroid and solenoid fields on. The response increase due to magnetic field is larger in the barrel partitions. Details on the magnetic field effects were already discussed in Sect. 4.3.2.

The relative deviation of the measured Cs response from the expected values due to the decrease in the source activity is shown in Fig. 20 (left) for the same set of the calibration runs reported above. Similarly, the maintenance period is excluded and the "MF" marks are used when the magnetic field was present during the calibration. The overall TileCal

response to the radioactive sources follows the expected Cs decay within 1% when no magnetic field is applied. Within this 1%, there is a visible deviation from the expected decay line with increasing average response over time. A study of the Cs calibration procedure has been unable to attribute this increase to any subtle systematic effect, therefore it is attributed to an increase in the detector response and it is under investigation. A conservative time dependent systematic uncertainty on the calibration of the EM scale of about 0.1% per month is adopted to account for this effect. It is estimated from the Cs data with no magnetic field within two periods of 3 and 7 months in 2008, 2009 and 2010. The ratio of RMS/mean of the TileCal response to radioactive Cs sources in all four calorimeter partitions is shown as function of time in Fig. 20 (right). The spread in the measured Cs responses stays within 0.4% over seventeen months indicating that the cell-to-cell intercalibration does not significantly change over this period of time. A small effect of the magnetic field on the Cs response spread is also clearly seen.

4.4 Calorimeter intercalibration

In this section the understanding of the cell and layer intercalibration acquired from the testbeam and from calibration and single beam data is exposed. The intercalibration as validated by cosmic muons is exposed later in Sect. 5.3.1.

The intercalibration with Cs sources in the ATLAS cavern reports channel response non-uniformities at the level of

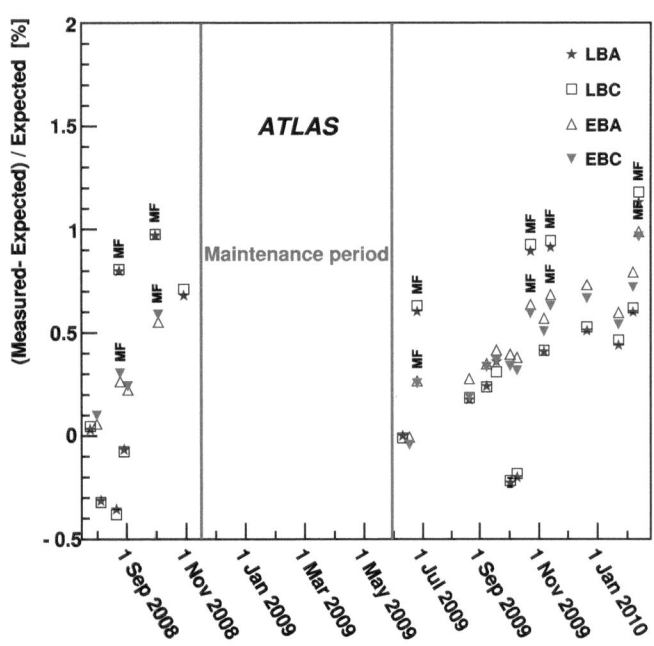
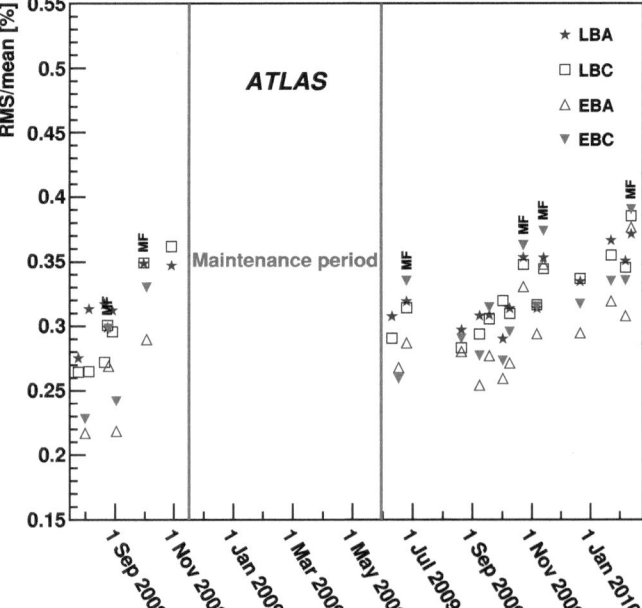

Fig. 20 Relative deviations of the TileCal response to Cs sources from the expected value for all four calorimeter partitions, shown as a function of time (*left*). The ratio of RMS/mean of the TileCal response to radioactive Cs source in all four calorimeter partitions is shown as

function of time (*right*). The "MF" symbol stands for the Cs calibration data taken in the magnetic field. The response is averaged over the channels of each partition

Fig. 21 The average energy measured in the single beam events recorded in September 2008. *Left*: average energy measured in individual radial layers after the radial layer corrections were applied (A is the inner radius layer). *Right*: the average energy measured in individual partitions, demonstrating good intercalibration between them.

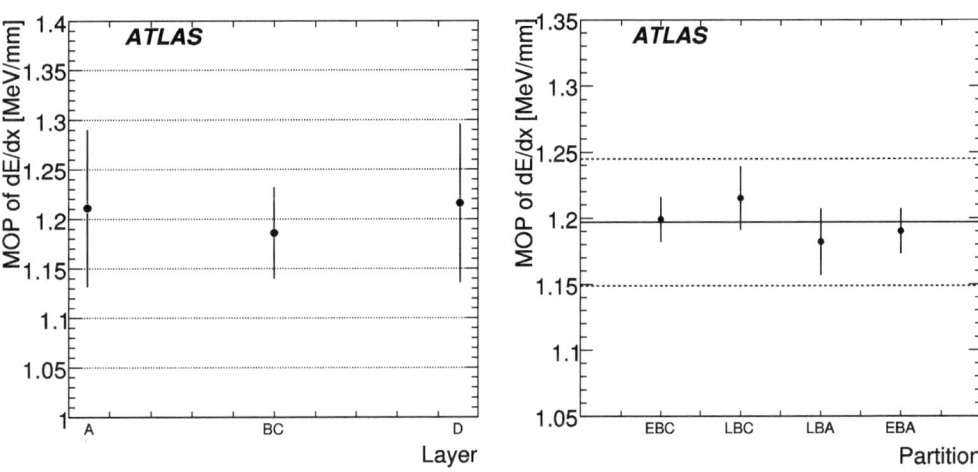

$\sim$0.4% in each Cs-scan, which is compatible with the precision of this calibration system. Since the Cs system uses a different readout path than what is used for the physics signal induced by particles (digital readout), other calibration uncertainties also have to be considered. The charge injection system reports negligible non-uniformity after the channel-to-channel corrections. The integrators contribute at the level of 0.05%, which is negligible. Altogether, the non-uniformities are given mostly by the Cs system. The Cs scans of the whole Tile Calorimeter revealed the same level of uniformity among individual optical elements in a cell as was measured during the optics instrumentation period.

In the testbeam, a difference in the response to the Cs source and to particles was observed, increasing for layers at larger radius [10]. This is due to the increasing size of the scintillator tiles for the external layers, and the resulting few percent layer miscalibration is accounted for by applying radial depth weights in the energy scale calibration. The details of this procedure are described in Ref. [18]. Figure 21 (left) shows the dE/dx for muons crossing the calorimeter parallel to the beam axis along its whole length from scraping events[8] in 2008. The dE/dx response for the muons from single beam events was estimated as the peak of the fit to the convolution of a Landau function with a Gaussian (most probable value, referred throughout the paper as MOP). Within a large statistical uncertainty, the response vs radial layer is flat. Given the fact that if the radial depth weights had not been applied the ratio of responses between layers A and D would be 1.088, this observation gives confidence in their use. Figure 21 (right) shows the mean response of the four TileCal partitions to muons. Data are from 2008 single beam runs. The precision is limited by the systematic uncertainty of $\sim$4%, while the statistical uncertainty is $\sim$2%.

4.5 Uncertainty on the propagation of the EM scale from testbeam

The EM scale of TileCal in ATLAS is set by adjusting the PMT HV to reproduce the calorimeter response to the Cs radioactive source to the level it had during the tests with electron beams, where the EM scale was determined and measured.[9] After correcting for the expected decrease in the Cs source intensity, the HV levels currently set in ATLAS are expected to reproduce those used at the testbeam. Any difference in the detector parameters from that observed at the testbeam, if not fully understood or disproved and if it affects the EM scale setting, should be considered as the systematic uncertainty on the EM scale determination.

The following sources of systematic uncertainties on the EM scale, as discussed in the previous sections, are only related to the transfer of the EM calibration factor from the testbeam to ATLAS because they originate from differences between the two setups:

– 0.1% from the calibration of the digital readout (HG, LG) by CIS.
– 0.2% from the calibration of the Cs readout gains.
– 0.5% from the non-reproducibility of the calorimeter response after the magnetic field is turned off, as reported by Cs measurements in 2008 (see Sect. 4.3.2).
– 0.3% from the uncertainty to the radial depth weights, briefly mentioned in Sect. 4.4.

The first two uncertainties were evaluated by comparing calibration results on a fixed sample of channels which were calibrated during the testbeam and then re-calibrated recently in ATLAS. The two latter are related to observations with limited understanding of the underlying phenomena.

---

[8]Events produced by the proton beam hitting the edge of the collimators located at about 140 m upstream ATLAS.

[9]The modules that were calibrated with the beams were carefully chosen to give a representative sample of the full TileCal module population. Thus no significant uncertainty on the EM scale is expected to result from data obtained with the electron beams.

When these uncertainties are combined in quadrature with the statistical uncertainty on the EM scale derived at the testbeams, the result is a systematic uncertainty of $\pm 0.7\%$.

˙ In addition to the above, there is a systematic uncertainty from the observed increase of the calorimeter response to the Cs source with respect to the expected value by about 0.1% per month as observed during 10 months of frequent monitoring in 2008 and 2009–2010. This is a time dependent uncertainty increasing since the initial EM scale setting in ATLAS in June 2008.

– Presently (early 2010) we assign an uncertainty of -1.5% due to the increasing response of roughly 0.1%/month as measured by the Cs system during 2008 to 2010.
– During the data-taking period from which cosmic muon results are presented in this paper (September to October 2008), the same uncertainty was -0.8%.

After setting the EM scale in ATLAS, the high voltage values applied on the PMTs were compared between the testbeam periods (2002 to 2004) and June 2008. While the TileCal response has been calibrated reliably with the Cs system to match the response measured during beam calibrations and hence to transfer EM scale to the ATLAS cavern, the PMT high voltages for the LB partition in June 2008 had to be lowered on average by (6.5 ± 0.2) V compared to those used during testbeam calibrations. This was due to the fact that the Cs system measured an increased response in June 2008 for the beam calibrated modules with respect to their response in testbeams. If this response increase had not been a detector effect but an artifact of the Cs calibration system, a corresponding bias of -5.3% (the true energy being higher than the measured one) would have to be considered as an uncertainty for the cosmic data taken in autumn 2008. This would be added to the uncertainty from the observed increase of roughly 0.1% per month since June 2008, as mentioned above.

The energy response from muons is a handle to assess this uncertainty or bias. A full description on the energy scale analysis with cosmic and testbeam is given in Sect. 5.3. The comparison between the testbeam and ATLAS EM scale is performed via the double ratio of dE/dx Data/MC ratios of cosmic over testbeam muons for LB modules. In other words, the agreement of data to the MC energy scale between testbeam and ATLAS is compared. Table 6 presents the values and the uncertainties of the above mentioned double ratio per layer. Among the calibration related uncertainties, the contributions from the non-reproducibility of the response increase due to magnetic field and from the unexplained response increase measured by the Cs during 2008 are comprised. The reported ratios show an agreement of the EM scale set in 2008 and the expected scale as it was transported from the testbeam within the uncertainty range.

However, the possible calibration bias mentioned in the previous paragraph, that would be represented by a double ratio of 0.95, can be excluded only at a $\lesssim 2\sigma$ level.

If the uncertainty coming from the reduced high voltage settings with respect to the testbeam is not taken into account, the overall estimate of the EM scale systematic uncertainty from the calibrations is $(-1.7\%, +0.7\%)$ in early 2010.[10]

4.6 Timing calibration

To allow for optimal reconstruction of the energy deposited in the calorimeter by the OF signal reconstruction method (see Sect. 3), the time difference between the digitising sampling clock and the peak of the PMT pulses must be minimised and measured with a precision of 1 ns. To achieve this, the clock phases in the DMUs in the front-end hardware (see Sect. 2.1) are adjusted in multiples of 0.1 ns. Ideally all PMT signals would be sampled at the peak but several factors limit the ability to do this. First, the clock phase is defined per digitiser board which corresponds to six readout channels. Second, only one clock phase can be defined for both gains and there is a 2.3 ns difference between the HG and LG pulse peaks. Therefore in the front-end hardware, the accuracy of phase synchronisation for individual channels is limited to be within 3 ns. Any residual time differences between the clock phase and the pulse peak are measured for each channel and accounted for in the OF signal reconstruction algorithm.

The time phase and the residual offsets for all channels can be measured using the laser calibration system, cosmic-ray events, beam splash and collision events. What is exposed in this section is the procedure to only pre-set the timing in order to synchronise the detector with the trigger signals and with the other detectors prior to the final detailed adjustments, to be carried out with collisions data.

Prior to beam, the laser was the primary source used to measure the channel timing. Since the laser light is asynchronous with respect to the clock, a single reference channel in each partition was selected and all other channels' timing was defined with respect to that reference [19]. The timing precision for channels in the same module is 0.6 ns for 99% of the Tile Calorimeter readout channels. In addition, the mean time difference between the HG and the LG was measured to be (2.3 ± 0.4) ns. One limitation in the laser system for timing calibration is understanding the propagation time in the laser fibres from the laser source to the PMTs. For this reason, the inter-partition timing and global timing with respect to the rest of ATLAS were coarsely set using cosmic-ray data and more accurately using 2008 beam data.

---

[10] This uncertainty is $(-1.1\%, +0.7\%)$ for October 2008, the period in which the cosmic muon data of this paper were collected.

Fig. 22 Timing of TileCal signals recorded with single beam data in September 2008 (**a** and **b**), November 2009 (**c**) and February 2010 (**d**). The time is averaged over the full range of the azimuthal angle ϕ for all cells with the same Z-coordinate (along beam axis), shown separately for the three radial layers. Corrections for the muon time-of-flight along the z axis are applied in the (**b**), (**c**) and (**d**) figures, but not on the top left (**a**)

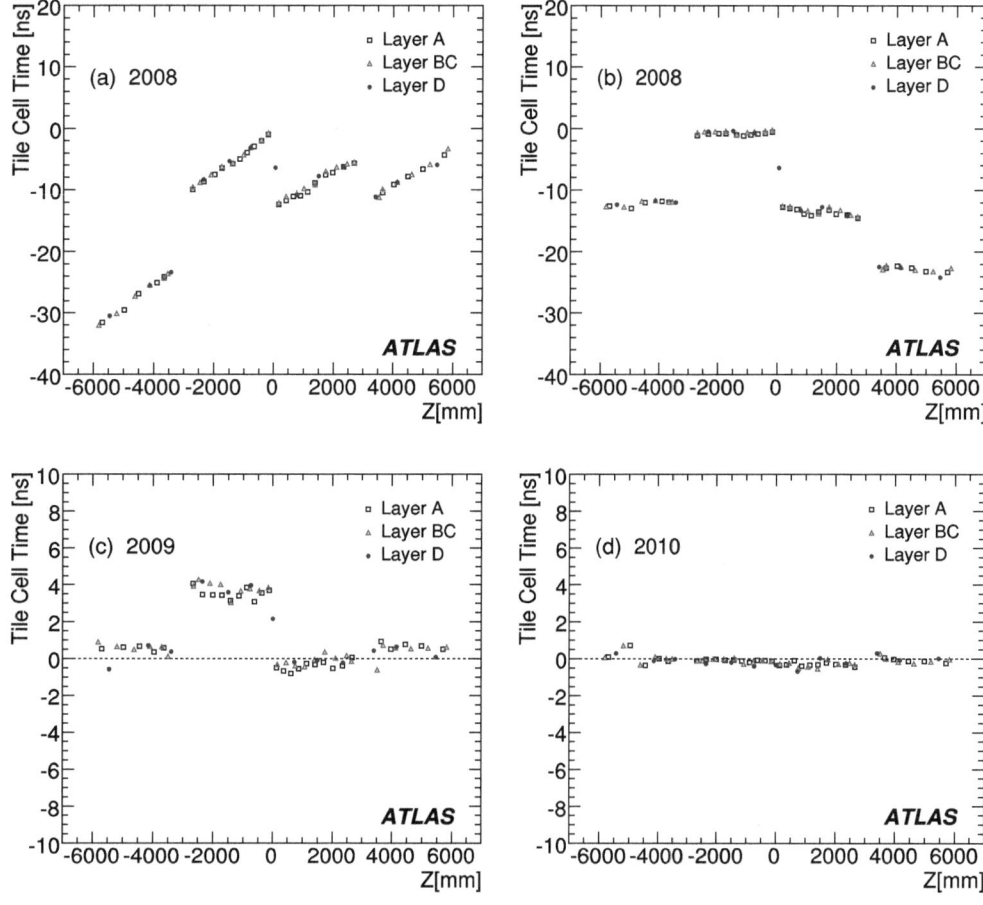

The timing calibration based on laser data was validated using beam splash events. These events contain millions of high-energy particles arriving simultaneously in the ATLAS detector. Since the total deposited energy is large, it is only possible to study the timing response in the LG. Using these events, the time intercalibration of individual channels in the same module was confirmed to be 0.6 ns.

Figure 22 shows the cell time measured in beam splash events, averaged over the full range of the azimuthal angle ϕ for all cells with the same z-coordinate of ATLAS (along the beam axis). The visible discontinuities at $Z = 0$ and $Z = \pm 3000$ mm for the 2008 data are due to the uncorrected time differences between the four TileCal partitions. These were calculated using the 2008 data and adjusted for the 2009 running period. After the muon time-of-flight corrections (b), the timing shows an almost flat distribution within 2 ns in each partition, confirming a good intercalibration between modules with the laser system. The residual slopes, present in all modules, were corrected for by comparing the 2008 single beam data to the laser data and optimising the effective speed of light in the calibration system optical fibres. Consequently, in 2009, the TOF-corrected timing distribution (c) is even more uniform. In preparation for the 2010 run, the 2009 single beam results were used to pro-

vide the offsets for all cells and, as is shown in Fig. 22(d) for the 2010 single beam results, all remaining disuniformities were corrected for. The spread of the TileCal cell timing distribution at the start of the 7 TeV collisions is of 0.5 ns.[11]

5 Performance with cosmic ray muons

The calorimeter response to muons is an important issue since isolated muons will provide a signature of interesting physics events in the LHC collisions phase. For example, semileptonic $t\bar{t}$ decays, the so-called "gold-plated" Higgs decay channel $H \rightarrow Z^0 + Z^0$ and some SUSY processes involve high-p_T muons in their final states, while low-p_T muons originate from B-meson decays [20]. In addition, since the interaction of muons with matter is well understood, the prediction of this response is reliable, and its investigation with data can provide information on the detector performance and intercalibration.

The TileCal energy response performance was studied using cosmic muon data collected in 2008, with the goal of

---

[11] This value takes into account 97% of the TileCal channels. The timing for the remaining outliers was adjusted offline.

verifying the calibration in terms of EM scale and its uniformity over the whole calorimeter. After an initial comparison of the muon energy signal and the corresponding noise in the same set of cells (in Sect. 5.1), the methods and results of the studies of muon response versus path length are described. These studies were based on the extrapolation into TileCal of cosmic muon tracks reconstructed by the Inner Detector, which is described in Sect. 5.2.2. The performance of the energy response to testbeam muons was also checked at low energy, for comparison.

Muon response results and comparison to Monte Carlo simulations are presented in Sect. 5.3. This Section focuses on several key issues: the response uniformity versus radial layer, η and ϕ, the propagation of the EM scale from testbeam to the full detector configuration in the ATLAS cavern, and a discussion on the systematic uncertainties, such as the ones arising from possible biases of the muon response estimation with the muon momentum and path length. A separate Sect. 5.4 is devoted to calibration of special TileCal cells (ITC, gap and crack scintillators).

The measurement of the time-of-flight of particles in TileCal can be used either for background removal (cosmic and non-collision events) or physics analyses [21]. A good synchronisation of the TileCal cells is important for that, and its validation with cosmic ray muons is described in Sect. 5.5.

5.1 Muon response compared to noise

The TileCal readout system is designed so that even small signals induced by muons are well separated from the noise. This feature has been demonstrated with testbeam data [10]. Nevertheless the performance has to be confirmed with data taken with the full ATLAS detector, since the environment is more noisy and changes to the powering system have been made.

This exercise was performed on a large statistics run, with the data sample described in Sect. 2.2: events from various first level triggers were required to have at least one reconstructed Inner Detector track. However, these tracks were not used in any further event or cell selection, for this study. Instead, a different method was used, based on track reconstruction using only TileCal data. This algorithm, named TileMuonFitter, was developed for the data analysis and monitoring of TileCal in the cosmic muon commissioning phase [22, 23]. It uses no external tracking information and uses the set of TileCal cells with energy above a 250 MeV threshold to fit a straight line from the top to the bottom cells (it therefore also ignores the track curvature inside the solenoid magnetic field). In order to reproduce as closely as possible the signal from muons originating in physics collisions, a loose projectivity requirement was imposed. Tracks were selected according to the coordinates of their intersection with the horizontal plane (within ±400 mm) and to their

angle with respect to the vertical, corresponding to a pseudorapidity range of $0.3 < |\eta| < 0.4$.

The signal is either the total energy in TileCal summed up over cells selected by the TileMuonFitter algorithm, or the response in the last radial compartment for the D-cells selected by that algorithm. The noise is evaluated from random triggers using the same cells as for signal. The results are shown in Fig. 23 for tracks entering barrel modules within the pseudorapidity range $0.3 < |\eta| < 0.4$. Top and bottom module responses are considered as two independent entries, so the signal corresponds to that of one module. The signal and noise distributions are well separated for both the total calorimeter response and the last radial layer signal.

In order to estimate the signal-to-noise ratio, the energy distribution is fit to the convolution of a Landau function with a Gaussian. Considering the peak of that convolution fit as the signal, and the RMS of the random trigger distribution as the noise, the signal-to-noise ratio is then $S/N = 29$ for the total response and $S/N = 16$ for D cells. Since muons are the smallest energy signals that TileCal will measure, these values show a good performance of the calorimeter. The obtained values are lower than for testbeam,[12] but the difference is consistent with a higher noise level in the ATLAS cavern and with a higher number of cells being summed.

5.2 Methods for muon response studies

A brief overview of the analysis methods applied to investigated data samples is provided in this Section. First, we briefly describe the dedicated testbeam (TB) studies with low-energy muons (Sect. 5.2.1). The algorithms and event selection used in the cosmic data analysis are then reported in Sect. 5.2.2.

5.2.1 Analysis of low energy testbeam muons

The TB setup, operating conditions and results are summarised in Ref. [10]. Since most of the previous muon TB results were obtained with 180 GeV beams and this energy is too high for the comparison with cosmic ray data, a dedicated study was performed with low-energy muons selected from a pion beam at a nominal energy of 20 GeV. These muons originate from pion decay, the distribution of their momenta is calculated to range from 11.5 GeV/c to 20 GeV/c, peaking at around 17 GeV/c. Data was collected from ten runs with pion beams impinging on one barrel module at different projective incidences, from $-0.65 \leq \eta \leq -0.15$ and $0.15 \leq \eta \leq 0.45$.

---

[12] In testbeam [10], muon beams at a nominal energy of 180 GeV were used for this study. Taking into account the 20 GeV to 180 GeV response ratio, the testbeam S/N ratios at 20 GeV for the tower and the D cells should amount to 42 and 17 respectively.

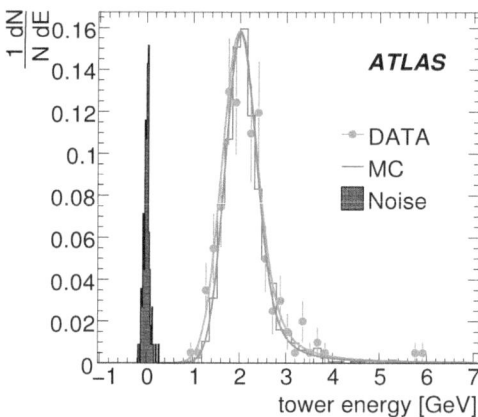
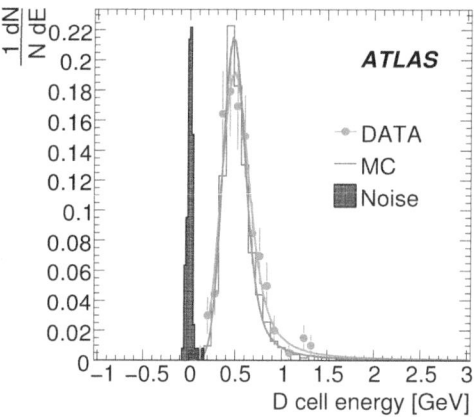
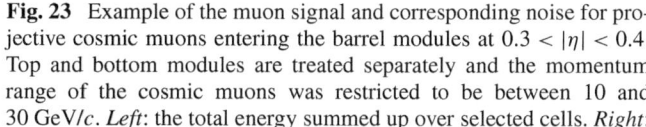

tower energy [GeV]

Fig. 23 Example of the muon signal and corresponding noise for projective cosmic muons entering the barrel modules at $0.3 < |\eta| < 0.4$. Top and bottom modules are treated separately and the momentum range of the cosmic muons was restricted to be between 10 and 30 GeV/c. *Left*: the total energy summed up over selected cells. *Right*: the similar distribution of last radial compartments that can be eventually used to assist in muon identification. The signal (*data points with error bars*) comes from the cosmic muon data sample (see text), the corresponding noise (*filled histogram*) is obtained with the random trigger sample

Two sets of cuts were applied to select muons from the nominal pion beam:

- Single particle events were selected by requiring a MIP-like response in the beam scintillators upstream of the calorimeter modules. Particles with large angle with respect to the beam axis and/or halo particles were removed by applying suitable cuts in the upstream beam chambers.
- Contrary to muons, pions produce hadronic showers that leave signal also in towers surrounding the one hit by the beam. This feature is exploited in the muon/pion selection—events with signal above noise ($E \gtrsim 3\sigma_{\text{noise}}$) in neighbouring towers were considered pions and were removed from further analysis. Moreover, an upper limit on the response in the impact cell in the first calorimeter radial layer was imposed, in order to avoid pion showers with large electromagnetic shower fraction, whose typical lateral (in $\eta \times \phi$) size is smaller than that of a cell.

As the projective beams hit the centre of the given calorimeter tower, the muon response was summed up only from cells in the impact tower. The selection criteria mentioned above guarantee a muon to impinge on the selected tower, therefore no further cut to reject noise events was needed.

The muon track length in the given cell was considered as the radial size of that cell divided by the cosine of the beam incident angle. This approach is fully adequate for projective muons entering the calorimeter at a cell's centre in both η and ϕ direction, see also Fig. 2.

The Monte Carlo simulation of the TB setup takes into account the detailed detector and beam geometry as well as the momentum distribution of the incident muons.

5.2.2 Analysis of the cosmic ray muons with tracks reconstructed by the Inner Detector

The performance of the calorimeter was analysed by taking advantage of the information provided by the central tracking. This is an important handle for the study of the calorimeter cell response which is sensitive to the muon path length and momentum.

Track extrapolation and event selection Events were triggered at the first level trigger by RPC and TGC. The tracking information is obtained from the Inner Detector reconstruction, without further contribution from the Muon Spectrometer. Selected events are required to have one reconstructed track in the SCT volume. Events with reconstructed multiple tracks are rejected. Tracks in the TRT do not have η information and are not used in the study. The quality of the tracks is enhanced by requiring at least eight hits in the silicon detectors (Pixel and SCT). The tracking requirements introduce some cut-off in the distributions of transverse and longitudinal impact parameters. These are $|d_0| \leq 380$ mm and $|z_0| \leq 800$ mm, respectively.[13]

The tracks are extrapolated through the volume of the calorimeters using the tool described in Ref. [24], which uses propagation of the track parameters and covariances that take into account material and magnetic field. Extrapolation is performed in both directions, along the muon momentum and opposite to it. This allows to study the response of modules in the top and bottom part of the detector. Since

---

[13]The transverse impact parameter is defined as the distance to the beam axis of the point of the closest approach of the track to the coordinate origin. The longitudinal impact parameter is the z-coordinate (along the beam axis) of the same point.

the track parameters are measured in the centre the method could be sensitive to systematic differences top/bottom.

Figure 24 demonstrates the correct TileCal cell geometry description. It shows the response of cells in the second layer as a function of the ϕ-coordinate measured at the inner-radius impact point in the given cell. The cells' response average is computed over tracks along the η directions in the central barrel region. The responses corresponding to cells of individual modules (width of $\Delta\phi \approx 0.1$) are shown with symbols of different colours/styles. The match with the nominal position of the cell edges, displayed by vertical lines, is evident. The total response summed over all modules is superimposed as well and it is reasonably uniform across ϕ.

The alignment between Tracker and Calorimeter was investigated using tracks with a limited transverse impact parameter ($|d_0| < 100$ mm). The alignment between tracks and nominal cell edges in the second layer of TileCal is within the selected bin size ($\sim$5 mm). This precision is fully adequate for the correct identification of the cells under study and computation of quantities relevant to the analysis.

One of the key parameters of the track is the path length through a given cell. The track extrapolation provides crossing points of the muon track in each radial layer. Additional linear interpolations are performed using the detailed cell geometry to define the entry and exit points for every cell. The track path length is then evaluated as the distance between the entry and exit points for every cell crossed by the muon. In the analysis we consider, for each event, only cells with path length $L > 20$ cm.

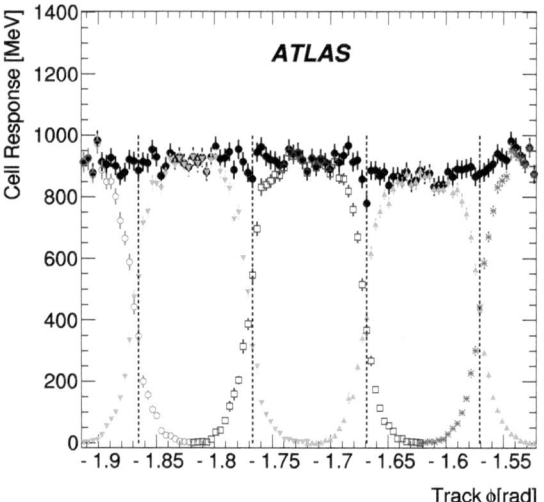

Fig. 24 (Color online) Mean response of cells in the second layer as a function of track ϕ-coordinate for the bottom central region of the calorimeter. Tracks with $10 < p < 30$ GeV/c were selected. The average response over all central region cells in the given module is shown by symbols of different colours/styles, whereas the total response summed over all modules is shown with *black full circles*. *Vertical lines* denote nominal edges of the modules

An upper limit of 30 GeV/c on the muon momentum is used in the analysis in order to restrict the muon radiative energy losses which show considerable fluctuations and can have an impact on data/MC comparisons. In a small fraction of events the cell response is compatible with the pedestal level although the cells should be hit by a muon. The muon actually hits a neighbouring module. This is consistent with the expected deviation from the muon trajectory due to multiple scattering. In order to limit this effect we restrict the analysis to muons with momenta as measured in the Inner Detector larger than 10 GeV/c and apply a fiducial volume cut requiring the track to be well within the module (that has a half width of $\Delta\phi = 0.049$):

$$|\phi_{\text{track}} - \phi_{\text{cell}}| < 0.045. \tag{4}$$

In order to remove residual noise contribution, a cell energy cut of 60 MeV is applied.

Muon tracks close to the vertical direction are badly measured in the Tile Calorimeter due to the strong sampling fraction variation caused by the vertical orientation of the scintillating tiles. To ensure more stable results, tracks are required to enter in the cells with a minimal angle with respect to $\eta = 0$ direction. Given the crossing points at the inner and outer cell radial edges we require

$$|z_{\text{inner}} - z_{\text{outer}}| \geq 6 \text{ cm}. \tag{5}$$

This cut has an appreciable effect only on very central cells, within the vertical coverage of the ID.

Approximately 100 k data events satisfied the above mentioned selection criteria and were further analysed. The corresponding statistics available in the MC sample was about twice higher.

Performance checks   The track path length is the main handle to study the muon response. Figure 25 shows the response of cells in the second layer as a function of the path length x. It includes cosmic events crossing the BC cells over the entire barrel and over all accepted angles. A clear edge at the path length of 840 mm is visible in the figure. This represents the radial depth ΔR of the BC layer cells. Since most cosmic rays are vertical, a large fraction of the muons crossing the central region have a reconstructed path length equal or slightly larger than the layer radius. This is very evident for all cells with a z-coordinate within the vertical coverage of the SCT detector $|z_0| < 1$ m. A linear fit to the corresponding distribution of mean values shows that the muon response scales approximately linearly with the path length, as expected. Figure 25 suggests that the ratio of the cell response with the track path length, i.e. the slope of dE/dx, is one of the quantities that can be used to study the cell/layers intercalibration. This will be discussed in more detail in Sect. 5.3.

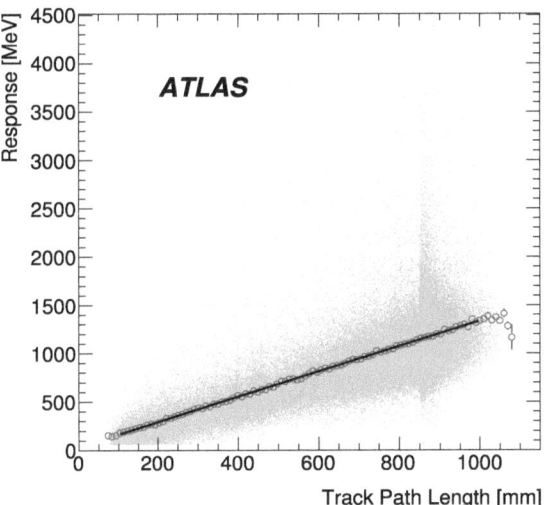

Fig. 25 Mean response of the barrel module BC cells as a function of track path length for tracks with $10 < p < 30$ GeV/c. A linear fit to the corresponding distribution of mean values is superimposed. The excess of events at around the track path length of 840 mm (radial size of the barrel module BC cells) is a purely statistical effect, since most of the cosmic ray muons enter the calorimeter at small zenith angle

5.3 Performance of energy response

In this subsection, the results of the calorimeter energy response studies carried out with cosmic muons are reported. The main aim is to cross-check the energy scale set with testbeam and the calibration systems, both in terms of the EM scale and of its uniformity across the detector cells. The uniformity of the response per cell and as a function of pseudorapidity and azimuthal angle is addressed in Sect. 5.3.1, while the layer intercalibration and EM scale issues are discussed in Sects. 5.3.2 and 5.3.3 respectively.

The energy response of TileCal to cosmic muons is probed by estimating the muon energy loss per unit length of detector material, which is obtained by dividing the energy measured by the path length crossed in a given cell (calculated with the method described in Sect. 5.2.2). For simplicity, we call this quantity dE/dx, even if this is not rigorous, since it is measured in a non-continuous way, and the TileCal cells are made of two different materials, with a direction-dependent sampling fraction.

Our estimator for the muon response is the truncated mean of dE/dx, defined as the mean in which 1% of the events in the high-energy tails of the distribution are removed (the number is rounded to the lowest integer). The statistics of the data sample is limited and rare processes like bremsstrahlung or energetic δ-rays can cause large fluctuations of the full mean. The truncated mean is chosen since it is less sensitive to high-energy tails in the cells' response distribution, that are caused by the muon's radiative energy loss. For testbeam, the truncated mean estimator has an additional advantage over the full mean, since it removes resid-

ual pion signal contamination. The truncated mean also removes muon events with very large energy deposits (high-energy radiation and/or muon nuclear interactions), therefore the muon/pion selection criterion (see Sect. 5.2.1) does not introduce any bias.

The truncated mean of the energy distribution does not scale linearly with the path length, so there is a small residual dependence of the dE/dx on the path length. This is evaluated as a systematic uncertainty and, furthermore, it largely cancels when the ratio of Data/MC is considered.

The dependency of the cell response to the muon momentum was investigated. As can be seen in Fig. 26 (left), the response increases with the momentum as expected, by about 20% between $p = 10$ GeV/c and $p = 100$ GeV/c and there is very good agreement between data and MC from 6 GeV/c to $\sim$100 GeV/c. Figure 26 (right) shows that the MC simulations predict a steeper dependence on the muon momentum for the full mean, and some disagreement even for the truncated mean at the higher energies, which could imply some imprecision in the modelling of the muon radiative energy losses.

The real energy loss by muons is typically 10% lower than the corresponding signal on EM scale and the ratio, known as e/μ, slightly scales with energy [10, 25]. However, in this paper, the validation of the EM scale is carried out by comparing data and Monte Carlo, and response to cosmic and testbeam muons, so this correction factor is not necessary.

5.3.1 Uniformity of the cell response

The studies addressed here measure the response uniformity per cell in a layer, as a function of pseudorapidity η and azimuthal angle ϕ (i.e. per module). Since our estimator is the 1% truncated mean, we require a minimum of 100 events in each set—η or ϕ bin, or cell. For the η and ϕ uniformity analyses, the data is not divided in cells—all cells corresponding to that bin are accumulated and the truncation is applied to the single dE/dx distribution for that bin. This approach allows the usage of the largest possible number of cells per bin while minimising biases from fluctuations in the tails. These results comprise all partitions, but exclude the ITC cells (see Sect. 5.4). In addition, we exclude from this study two cells from the D layer with an unusually high dE/dx.

Muons traverse cells in any direction and at any angle, so the local variations in the optics system (light yield of individual tiles, tile-to-fibre couplings, etc.) are supposed to be averaged out.

Uniformity per cell The uniformity of the cell response is shown in Fig. 27 for each radial layer and the RMS values are summarised in Table 4. The selection criteria, especially

Fig. 26 (*Left*) Muon response dE/dx as a function of momentum as measured in the Inner Detector, estimated with the truncated mean for both data and Monte Carlo. (*Right*) Ratio of Data over Monte Carlo for the muon response dE/dx as a function of momentum, shown for the truncated and full mean. For both distributions the response is averaged over the D5 cells in the bottom of the extended barrel (A side)

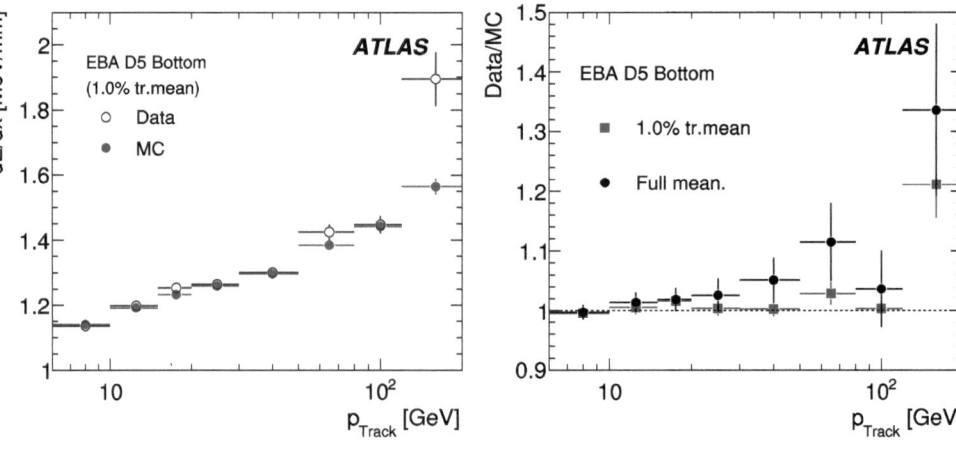

Fig. 27 Distribution of the truncated mean dE/dx per cell, shown separately for each radial layer A, BC and D, for data and Monte Carlo. The momentum range of the cosmic muons was restricted to be between 10 and 30 GeV/c

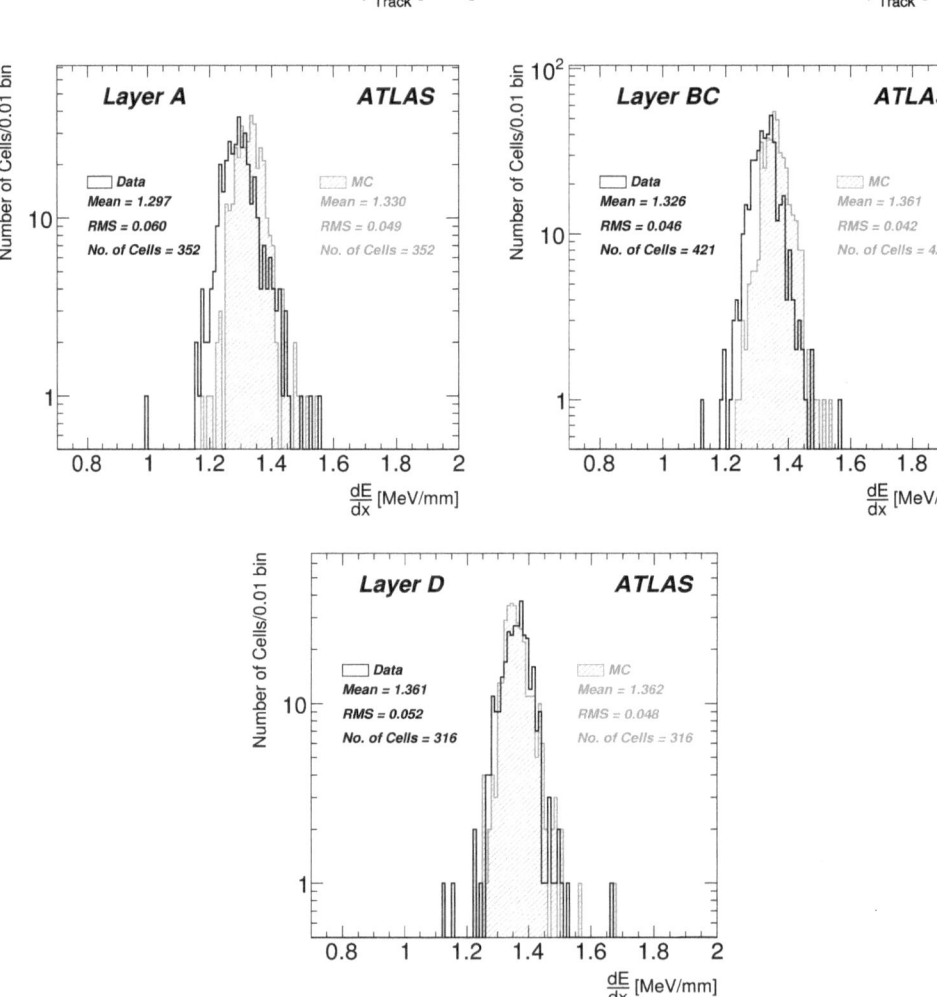

the requirement of 100 events per cell, limit the number of measured cells to the values shown in the figure and table, but still a quite representative fraction of 23% of the total cells is considered. The statistical population for the simulated and real data used for this study is identical.

The observed spread is the combination of different factors: statistical fluctuations, systematic errors due to the inherent limitations of measuring the cell response with the

dE/dx of cosmic muons, and the spread in the cell EM scale inter-calibration.

The Monte Carlo simulation has no variation in the quality of the optical components of the calorimeter or in the channel signal shape. Such variations are present in the data but it is difficult to disentangle between the spread due to them or to the statistical fluctuations from an underlying systematic due to the measurement method. Since the MC

Fig. 28 Momentum of the selected cosmic muon tracks as a function of pseudorapidity η, for both data and Monte Carlo. No momentum selection is applied in the left side distribution, while on the right, only tracks with momenta within 10 GeV/c and 30 GeV/c are shown. The vertical error bars in the upper part show the RMS of the momentum distribution in each η bin; in the lower part the error bars represent the uncertainty on the data/MC value shown

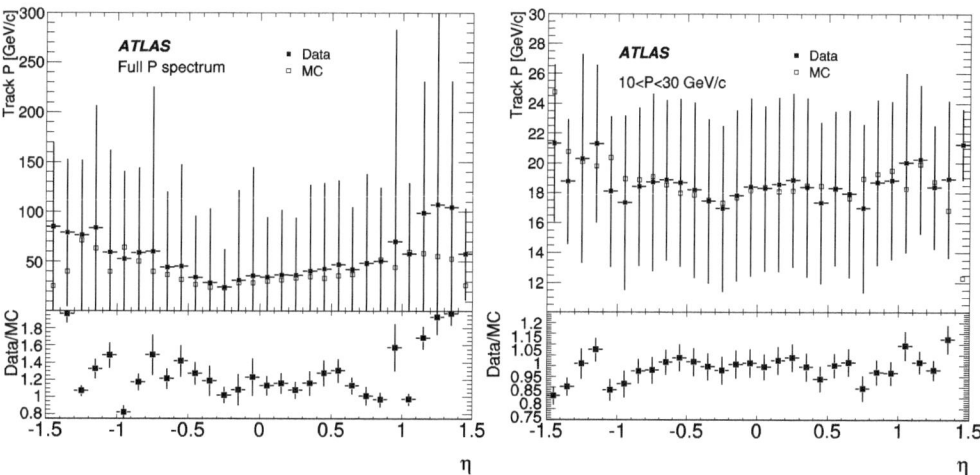

Table 4 The uniformity at the cell level for individual radial compartments. The listed values represent the RMS of the respective distribution of the truncated mean dE/dx for that layer, shown for data and Monte Carlo. The number of cells considered, and the fraction of the total that they represent, are also shown

Layer	Number of cells	Fraction of cells	RMS (MeV/mm) Data	RMS (MeV/mm) MC
A	352	18%	0.060	0.049
BC	421	22%	0.046	0.043
D	316	38%	0.052	0.048

shows an RMS in every layer compatible with that of data, it indicates that cells are well intercalibrated within layers.

From the mean of the dE/dx distributions per layer it is observed that there is a response discrepancy of 5.0% between layer A and layer D (2.3% between layer A and BC) for the cosmic muon data, an issue which is further discussed in Sect. 5.3.2.

The variations as a function of pseudorapidity and azimuthal angle, presented in the following paragraphs, were studied separately in each layer, since they appear to be smaller than the dominating inter-layer differences just shown. Another reason is that the cosmic muons are in general non-projective, so most muon tracks cross the calorimeter in each radial layer at different values of η and/or ϕ. Dealing with the total response as a function of η, ϕ would require projective muons only, thus significantly limiting the available statistics. The results are presented here relative to the average dE/dx.

Uniformity per pseudorapidity When investigating the uniformity as a function of pseudorapidity, the signal distribution includes all cells with the same azimuthal angle. A possible residual dependency of the muon momentum on the pseudorapidity of the detector cells (that could be due to the access shafts) was investigated. Figure 28 (left) shows

that the observed muon momentum distribution is harder than what expected by the Monte Carlo simulation, especially at high values of pseudorapidity. However, in the low momentum region that was selected for the analysis (between 10 and 30 GeV/c, see Sect. 5.2.2), the agreement is much better and the variations of momentum with η ($\sim$10%) are quite tolerable for this study.

The tracks identified in the ID are required to point to the cell centre, as specified in (4), as well as the other selection procedures of Sect. 5.2.2. The results are shown in Fig. 29 separately for each radial compartment. It can be seen that, for all layers, the values for the long barrel (central region, $|\eta| < 1$) are scattered within a $\pm$2% band around the average. At high η, in the extended barrel, the statistical uncertainties are larger due to worse coverage than in the central regions. Still these values are for the most part distributed within a $\pm$3% band.

Uniformity vs. module The uniformity over modules has also been investigated. The response in every module was integrated over all cells in the given radial layer. Studies combine all partitions, barrel and extended barrels.

The results are shown in Fig. 30. Again the same cut on momentum $10 < p < 30$ GeV/c as measured in the Inner Detector was applied. This condition plays two roles—apart from the reason mentioned in Sect. 5.2.2 it also ensures a similar initial momentum distribution in different ϕ-regions.

Both experimental data and MC exhibit an essentially flat response as a function of azimuthal angle ϕ. A residual pattern observed with data matches the MC: this small increase of dE/dx in horizontal ($\phi \to 0$, $\phi \to \pm\pi$) modules is likely due to a difference in muon momentum in events passing the selection criteria. Nevertheless, the data show a good uniformity over ϕ and, except a few cases in the horizontal region, most modules are well within a $\pm$3% band. In particular the average response in top ($\phi \approx \pi/2$) and bottom ($\phi \approx -\pi/2$) modules appears to be within 1%.

Fig. 29 Uniformity of the cell response to cosmic muons, expressed in terms of normalised truncated mean of dE/dx, as a function of pseudorapidity η for each radial layer. The response is integrated over all cells in each pseudorapidity bin in the given radial layer. The results for data are compared to MC simulations and both are normalised to their averages for each layer. Data are shown with *closed circles*, *open circles* indicate MC predictions. Statistical uncertainties only. *Horizontal lines* limiting a $\pm 3\%$ band are shown

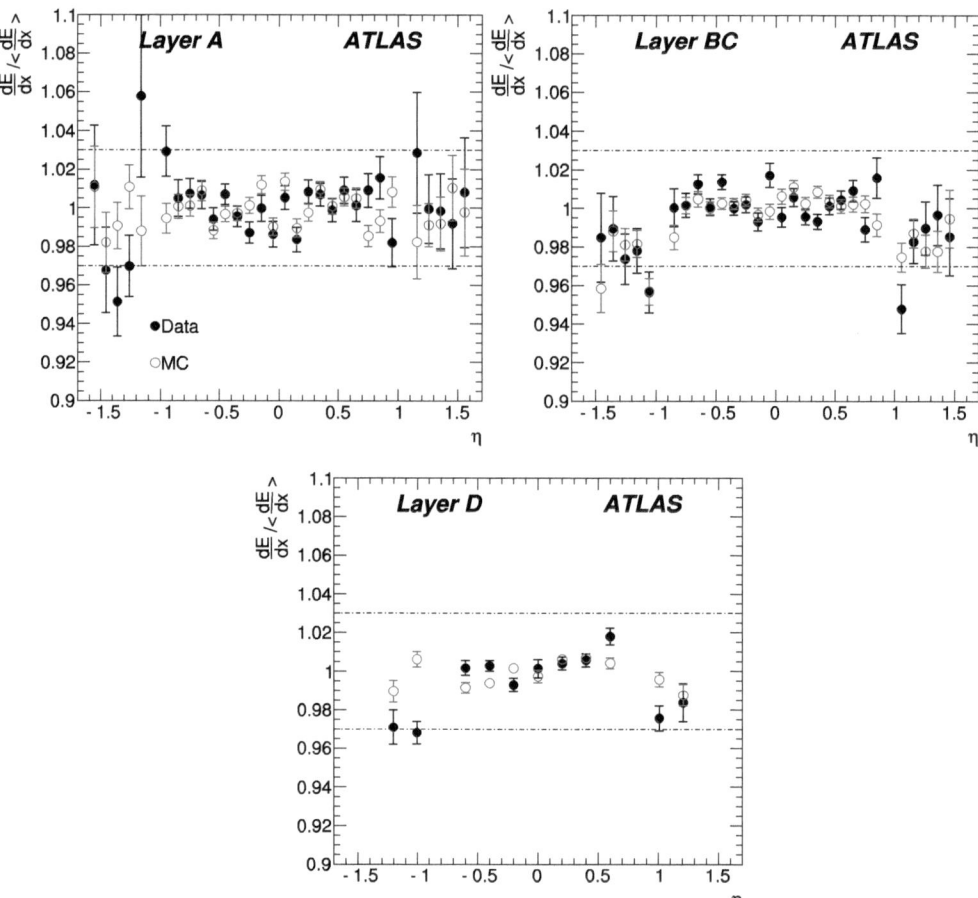

5.3.2 Muon response and layer intercalibration

The results discussed in Sect. 5.3.1 show that the cells are reasonably intercalibrated within a given layer, while there are differences observed between individual layers. In order to better quantify these differences, the layer response was calculated as the truncated mean of a single dE/dx distribution for all cells in a given layer. This approach reduces the statistical uncertainties, with respect to taking the truncated mean in each cell or η, ϕ bin. In addition, only events in the bottom of the detector are used, to avoid a bias from the muon momentum cut—in this way, the muon momentum for all the events is measured in the Inner Detector just prior to their incidence in TileCal.

In the cosmic muon analysis, various sources of systematic uncertainties associated with the truncated mean of dE/dx have been carefully studied. For every contribution, the associated parameter was varied in the given range and the systematic uncertainty contribution was evaluated as half of the difference between the maximum and minimum resulting truncated mean, unless explicitly stated otherwise.

The following contributions were identified:

– As already shown in Fig. 26 (right), data and MC exhibit a slightly different behaviour in function of the muon mo-

mentum. Because of this, the variation of the data/MC ratio over the analysis range (10–30 GeV/c) is considered as the systematic uncertainty due to the response dependence on the muon momentum.

– As the muon momentum is measured in the Inner Detector located in the centre of ATLAS, the response in the top and bottom part of TileCal can be different. Although the difference is well below 1% (see also Sect. 5.3.1), we consider its half as the contribution to the systematic uncertainty.

– Another contribution is associated with the residual dependence of the truncated mean on the path length. The truncated mean dE/dx was evaluated for several path length bins, and the above mentioned difference was calculated.

– The truncation itself represents another source of systematic uncertainty, that is associated with uncertainties in the description of the energy response shape. The uncertainty was estimated by comparing the resulting truncated mean of dE/dx for several values of truncation between 0% and 2.5%. This contribution does not fully cancel for the Data/MC ratio due to the difference that is observed in the tails of the dE/dx distribution between data and MC.

Fig. 30 Uniformity of the cell response to cosmic muons, expressed in terms of normalised truncated mean of dE/dx, as a function of azimuthal angle (module) for each radial layer. The response is integrated over all cells in each module in the given radial layer. All partitions are combined. The results for data are compared to MC simulations and both are normalised to their averages for each layer. Data are shown with *closed circles*, *open circles* indicate MC predictions. Statistical uncertainties only. The gap around $\phi = 0$ corresponds to horizontal modules that are poorly populated by cosmic ray muons passing through the Inner Detector. *Horizontal lines* limiting a $\pm 3\%$ band are shown

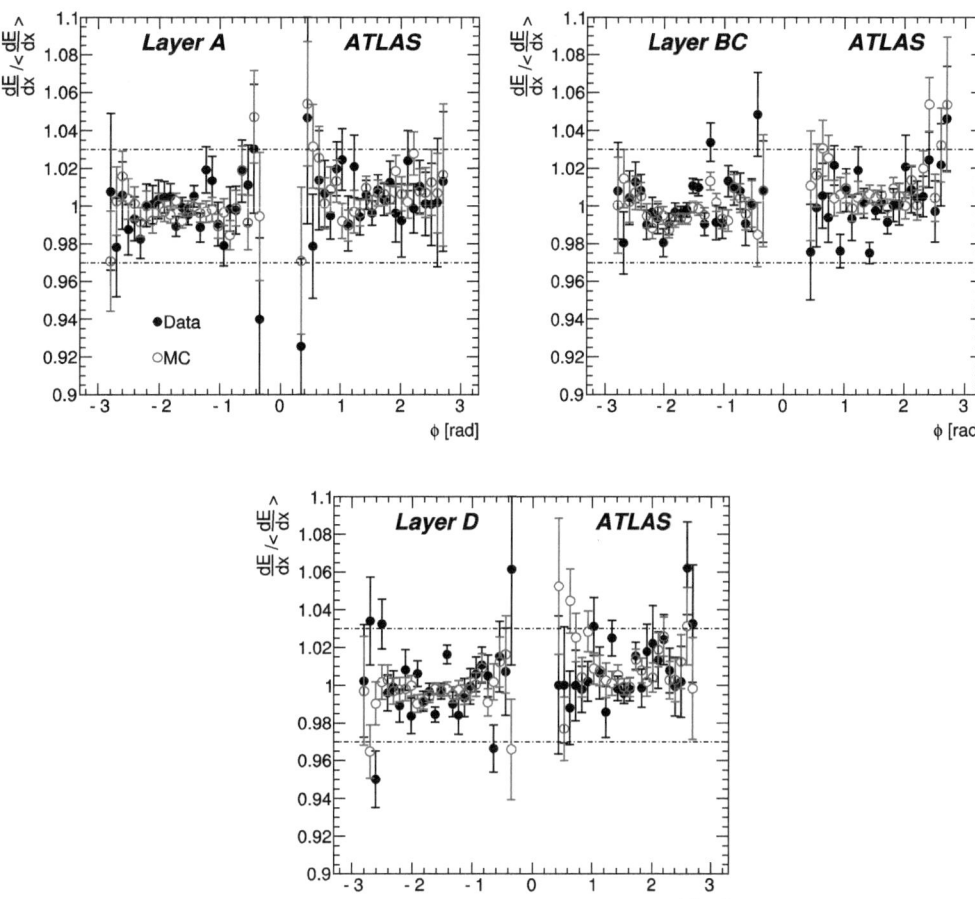

- The impact of the noise cut was studied as well, varying it between 30 MeV and 90 MeV (approximately 1σ and 3σ, where σ is the average noise RMS). The associated systematics appears to be very small.
- The measured response was also compared for various triggers, whose efficiencies depend on the muon momentum and also event topology. The data triggered by TGC and RPC indicate a good match within uncertainties, therefore the associated systematics is considered to be negligible with respect to other contributions mentioned above.
- The EM scale was transferred from testbeam to the AT-LAS cavern by means of the Cs source calibration procedure. Since the scale was set when the magnetic field was switched off and data were collected with magnetic field on, the appropriate correction has to be applied. Moreover, the Cs data show a response increase with time (see Sect. 4.3). Most of the cosmic data were acquired in September/October 2008, therefore we used the last Cs measurement with magnetic field on before the cosmic data taking to correct for both effects mentioned. The combined effect of these two corrections (magnetic field and response increase) amounts to -1% for the barrel and -0.6% for the extended barrel between June and Septem-

ber/October 2008 as shown in Fig. 20. Since the origin of the Cs response variation in time is not yet fully understood, the maximum and minimum of the Cs response in 2008 is considered as input for the corresponding asymmetric systematic uncertainty.

The uncertainties were evaluated separately for the LB and EB partitions and per individual radial layer for data, Monte Carlo, and the data/MC ratio (some contributions cancel in the ratio). The correlations among the radial layers are not taken into account and only the square roots of the diagonal terms of the error matrix are considered, and listed in Table 5.

The results on the longitudinal layer intercalibration are presented in Table 6 and displayed in Fig. 31, the error bars representing the total uncertainty based on the quadratic sum of the statistical and systematic uncertainties.

The differences in the cosmic muon response among individual layers are present even after correcting for the residual dependencies on the path length, momentum, impact angle, impact point, by considering the ratio of data over Monte Carlo. The resulting values are strongly correlated, therefore the maximum difference of 4% between the individual

Table 5 The individual contributions to the systematic uncertainty of the truncated mean dE/dx in cosmic muon Data and Monte Carlo. The listed values correspond to the diagonal terms of the error matrix. Analyses were performed with the ID-track method. The uncertainties on the global EM scale factor are discussed in Sect. 4.5

Systematic Uncertainties [MeV/mm] for Data and MC

Uncertainty source		Long Barrel			Extended Barrel		
		A	BC	D	A	B	D
Path	Data	±0.016	±0.030	±0.019	±0.046	±0.030	±0.017
	MC	±0.006	±0.008	±0.013	±0.014	±0.015	±0.022
	Data/MC ratio	±0.008	±0.016	±0.009	±0.033	±0.021	±0.019
Momentum	Data	±0.024	±0.034	±0.033	±0.037	±0.043	±0.044
	MC	±0.032	±0.042	±0.035	±0.020	±0.042	±0.044
	Data/MC ratio	±0.008	±0.007	±0.004	±0.024	±0.005	±0.009
Noise	Data	±0.007	±0.002	±0.002	±0.009	±0.004	±0.003
	MC	±0.004	±0.002	±0.003	±0.003	±0.002	±0.002
	Data/MC ratio	±0.002	±0.000	±0.001	±0.005	±0.001	±0.000
Truncation	Data	±0.013	±0.014	±0.013	±0.013	±0.013	±0.013
	MC	±0.014	±0.014	±0.014	±0.014	±0.014	±0.014
	Data/MC ratio	±0.004	±0.005	±0.005	±0.003	±0.001	±0.001
Top/Bottom	Data	±0.007	±0.006	±0.012	±0.008	±0.009	±0.008
	MC	±0.015	±0.014	±0.014	±0.016	±0.037	±0.006
	Data/MC ratio	±0.006	±0.014	±0.002	±0.006	±0.021	±0.010
Global EM scale factor	Data	+0.005 / −0.013	+0.005 / −0.013	+0.005 / −0.014	+0.000 / −0.008	+0.000 / −0.008	+0.000 / −0.008
	MC	–	–	–	–	–	–
	Data/MC ratio	+0.004 / −0.010	+0.004 / −0.010	+0.004 / −0.010	+0.000 / −0.006	+0.000 / −0.006	+0.000 / −0.006
Total	Data	+0.033 / −0.035	+0.047 / −0.049	+0.042 / −0.044	+0.062 / −0.063	+0.055 / −0.056	+0.050 / −0.051
	MC	±0.039	±0.047	±0.042	±0.033	±0.060	±0.052
	Data/MC ratio	+0.015 / −0.017	+0.023 / −0.025	+0.012 / −0.015	+0.042 / −0.042	+0.030 / −0.031	+0.023 / −0.024

Table 6 The truncated mean of dE/dx (MeV/mm, see text), measured with cosmic ray muons in barrel (LB) and extended barrel (EB), and projective testbeam muons. Results are shown for both data and Monte Carlo as well as for each radial layer. For cosmic ray muons, only modules in the bottom part are used. Total uncertainties are quoted. For cosmic data the statistical component is negligible. The systematic uncertainty corresponds to the diagonal terms of the error matrix

Radial layer		A	BC	D
Cosmic muons, LB	Data	$1.28^{+0.03}_{-0.04}$	1.32 ± 0.05	1.35 ± 0.04
	MC	1.32 ± 0.04	1.35 ± 0.05	1.34 ± 0.04
	Data/MC	$0.97^{+0.01}_{-0.02}$	0.98 ± 0.02	1.01 ± 0.01
Cosmic muons, EB	Data	1.27 ± 0.06	1.29 ± 0.06	1.32 ± 0.05
	MC	1.31 ± 0.03	1.32 ± 0.06	1.34 ± 0.05
	Data/MC	0.97 ± 0.04	0.98 ± 0.03	0.99 ± 0.02
Testbeam, LB	Data	1.25 ± 0.03	1.39 ± 0.04	1.39 ± 0.03
	MC	1.30 ± 0.02	1.37 ± 0.03	1.36 ± 0.02
	Data/MC	0.96 ± 0.02	1.02 ± 0.04	1.02 ± 0.02
Double ratio $\frac{(\text{Data/MC})_{\text{Cosmic muons, LB}}}{(\text{Data/MC})_{\text{TB, LB}}}$		1.01 ± 0.03	0.96 ± 0.04	0.98 ± 0.03

measurements with the cosmic muon data indicates the layer response discrepancy.

5.3.3 Validation of the EM scale propagation from testbeam

The ratio data/MC mentioned above also depends on the absolute EM scale of the MC simulated energy loss in the calorimeter. Due to the uncertainties in this quantity, the double ratio of data/MC, cosmic muon/TB, is adopted for comparison of the muon response and hence the EM scale between cosmic and TB data in the long barrel. For testbeam data and Monte Carlo, the truncated mean of the dE/dx distribution was obtained for each run, and then averaged over all runs. These are the values already presented in Table 6 and Fig. 31. The evaluation of the systematic uncertainties is briefly described below.

We consider the spread of the dE/dx values over the different incidence angles as the main uncertainty of the measurement, an approach that effectively combines the statistical and part of systematic uncertainties. On top of them, we consider the following subdominant contributions:

- The bias due to the truncation in the dE/dx distribution was estimated in the same way as for cosmic data (mentioned above).
- The uncertainty in the global EM scale due to the non-calibrated integrators (see Sects. 4.3 and 4.4) at that time. This uncertainty applies only to data, not to Monte Carlo.

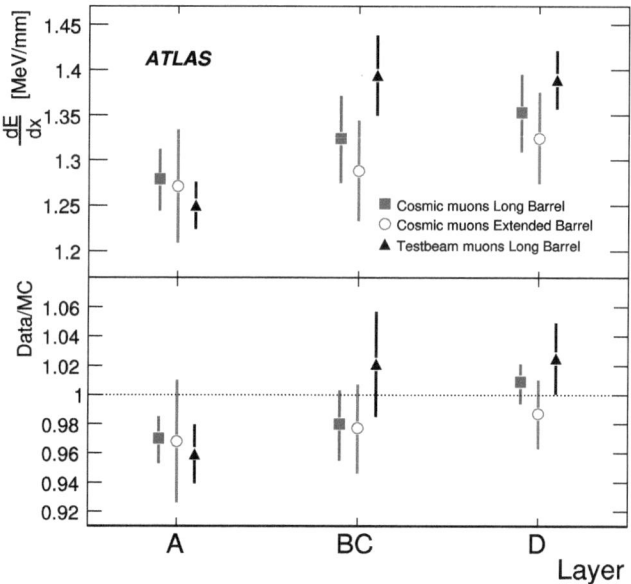

Fig. 31 The truncated mean of the dE/dx for cosmic and testbeam muons shown per radial compartment and, at the bottom, compared to Monte Carlo. For the cosmic muon data, the results were obtained for modules at the bottom part of the calorimeter. The error bars shown combine in quadrature both the statistical and the systematic uncertainties, considering only the diagonal terms of the error matrix

The individual uncertainties were evaluated for each radial layer and the resulting total uncertainties, shown in Table 6, were obtained by summing the individual contributions in quadrature.

The double ratio of data/MC, cosmic muons/TB, is presented in the last row of Table 6. The uncertainty contributions are computed by propagating in quadrature the TB uncertainties just described and the cosmic muon uncertainties mentioned in the previous section, that only take into account the error matrix diagonal terms. The EM scale measured with cosmic muons, relative to that determined at testbeam in the long barrel, amounts to 1.01, 0.96 and 0.98 for the A, BC and D layers respectively. Since the uncertainties per layer are at most 4%, these values are consistent with 1.0, showing that, within the precision limits of the analysis, the propagation of the EM scale from testbeam to ATLAS was performed successfully.

It should be noted that the LHC collisions will provide extra tools to check the EM scale calibration. Isolated muons and single hadrons developing their shower only in TileCal will provide two data samples for which a direct comparison to the testbeam scale will be possible.

5.4 ITC and gap/crack scintillator calibration

Understanding the response of the intermediate Tile Calorimeter (ITC) and the gap and crack scintillators (see Sect. 2.1 and Fig. 2) to cosmic ray muons is essential for their calibration. The gap and crack scintillators can not be calibrated using the Cs calibration source and therefore have arbitrary calibration factors applied to them. This study with cosmic muons gives the first clues for their in-situ performance.

These detectors are calibrated in two steps. The first step is the intercalibration in ϕ among the cells of the same detector type and to determine the calibration factors for each cell. The second step is the absolute calibration and to determine a scale factor defined relative to the MC for each detector type. Since the absolute energy scale in the scintillators is not known, the simulation is used as a reference in this case.

The event selection follows the same procedures as indicated in Sect. 5.2.2, with the exception that only events with a single muon track with a momentum above 5 GeV are considered and that, for the ITC cells, the entry and exit points of the track in the cell must be separated by at least 4 cm in the z direction. These requirements accept 8% of RPC triggered events, 80% of TGC triggered events and 7% of L1 calorimeter triggered events. Problematic cells and scintillators[14] are excluded in this analysis.

The geometrical path length is defined as a straight line between the two surfaces of the cell or scintillator. The muon

---

[14]Cells or scintillators that, even though matched with extrapolated tracks, appear too noisy or show very small signal.

energy loss per unit path length is used to evaluate the response. It is referred to as dE/dx for the ITC cells (C10, D4), which have the same elementary structure as ordinary TileCal cells (as in Sect. 5.3). For the gap and crack scintillators (E1–E4), the muon energy loss estimator is the signal (expressed in units of charge) normalised to the muon path length through the scintillator and, for distinction, it is referred to as $\Delta E/\Delta L$. Figure 32 is an example of the dE/dx or $\Delta E/\Delta L$ distribution for the cells in one module for cosmic ray data and MC. The cells generally show good signal-to-noise separation except for crack scintillators (E3, E4). The signals in the crack scintillators are found to be too small for good separation from noise distributions and the HV of the PMT has been accordingly increased. The noise distribution in the gap scintillators (E1, E2) in the data is mainly due to grooves and holes in these scintillators that accommodate the ^{137}Cs source pipes.

For each cell (scintillator), the dE/dx ($\Delta E/\Delta L$) distributions were fitted with the convolution of a Landau function with a Gaussian. The average and the RMS of the peak positions (MOP) of the fitted functions are summarised in Fig. 33 and shown with the results from the MC. For comparison, results for the extended barrel cells D5 and B11 are also shown with ITC cells in the figure. Cells

with insufficient statistics or with poor fits are excluded and 30%–50% of ITC cells and ~25% of gap scintillators remain.

The average values indicate that the response for the ITC cells is consistent with the cell response of ordinary TileCal cells, which are well calibrated with the standard Tile Calorimeter calibration procedure. The response of the ITC cells is also consistent with MC to within ~5%. In the gap scintillators (E1, E2), where the scale is arbitrary, the observed differences of roughly 20% imply an additional scale factor to adjust data relative to MC.

The uniformity of the response was also determined with these data. The RMS values are ~10% in ITC cells (C10 and D4), while in gap scintillators (E1, E2) the RMS values amount to 15%–20%.

Based on this study, no changes were made to the ITC cells since their response is consistent with the response of the ordinary Tile cells. For the gap scintillators, correction factors for ϕ intercalibration and global scale factors were measured relative to MC. As a result of this analysis, the HV values for the crack scintillators (E3, E4) have been increased to improve the separation between signal and noise. The expected improvement has been verified.

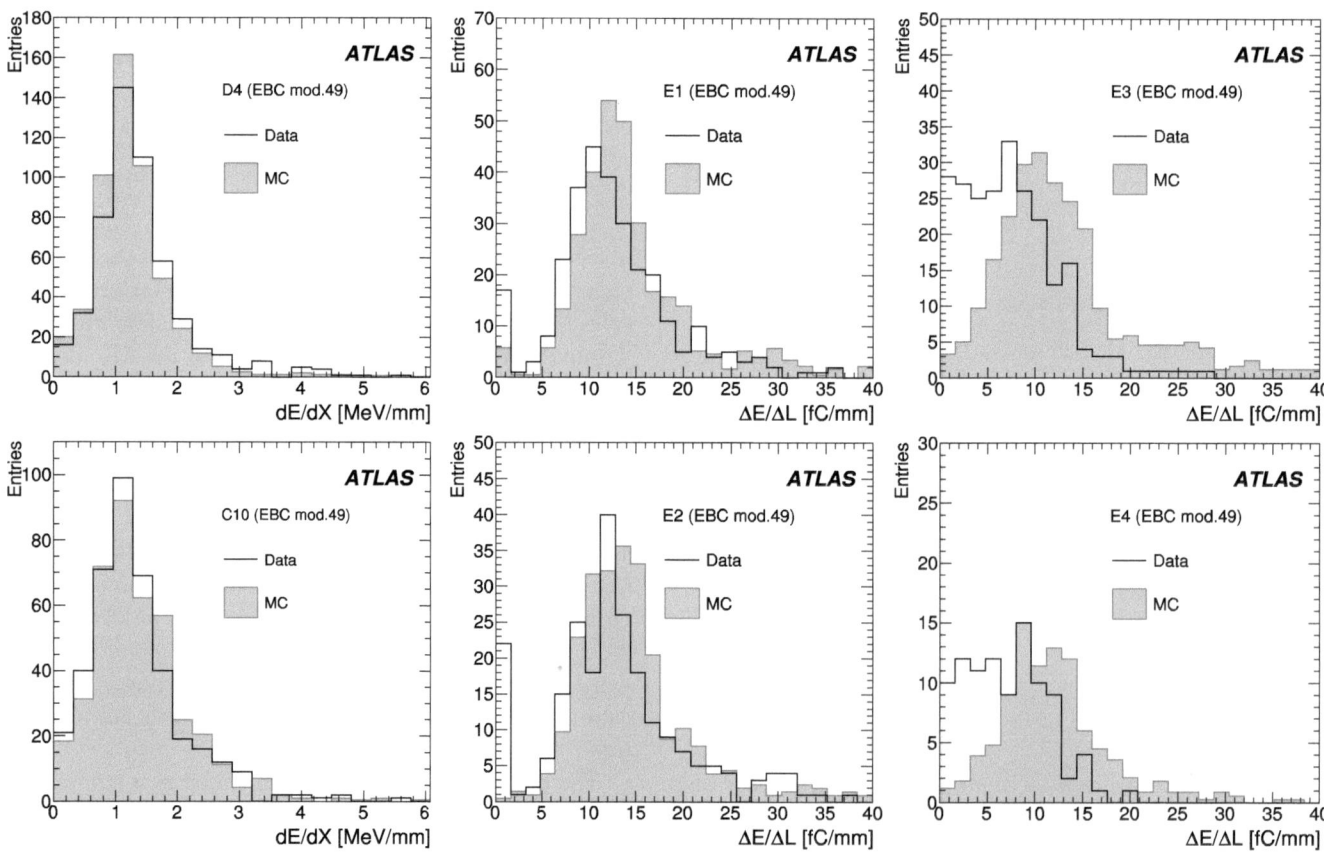

Fig. 32 Responses of ITC cells (D4 and C10), gap scintillator cells (E1 and E2) and crack scintillator cells (E3 and E4) to cosmic ray muons in EBC module 49. They are shown in terms of dE/dx for the ITC cells and $\Delta E/\Delta L$ for the gap and crack scintillators

Fig. 33 Responses of gap and crack scintillators (*left*) and ITC cells (*right*) to cosmic muons. Shown are the average values of the peak positions (MOP) of the fitted functions on the $\Delta E/\Delta L$ and dE/dx distributions respectively. The *vertical bars* indicate the RMS values

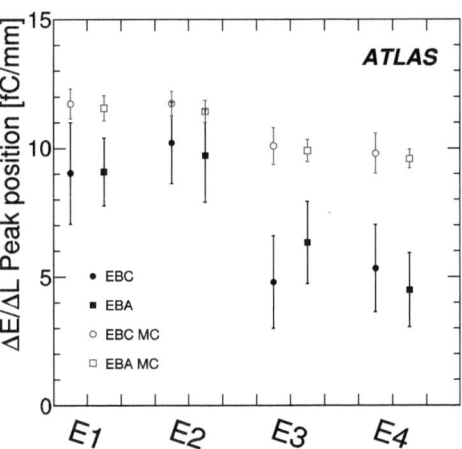 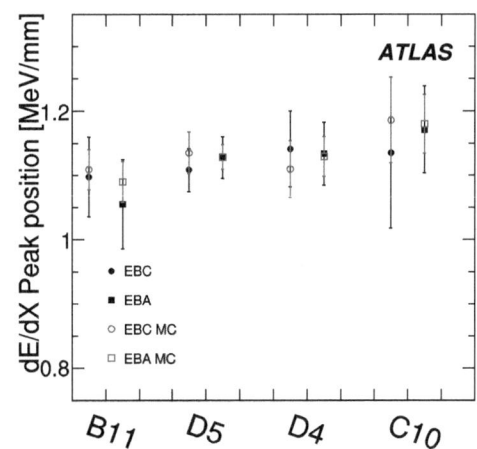

5.5 Performance of time response

Before the start of the LHC in September 2008, cosmic muons provided the only way to verify the accuracy of the time calibration of TileCal at the cell level. In addition to the online monitoring of detector synchronisation, that used distributions of average event time in function of position, detailed analyses of the data, described in this section, were able to measure the timing corrections for a large fraction of the TileCal channels. These analyses, based on the measurement of the muon time-of-flight between the top and bottom cells, have been validated using the data from the 2008 LHC single beam.

5.5.1 Extraction of time corrections

Two methods have been developed to extract the time corrections using the cosmic data [26, 27]. They are based on the comparison of the time determined in the top and bottom modules with the time-of-flight of the cosmic muon through the detector.

The iterative method [26] was successfully applied during the 2007 data takings. The very top barrel module (LBA16) was taken as a reference and the time offsets of the other modules (taken as single values for a whole module) were measured relative to this one. Since not all modules can be directly calibrated with respect to the reference one, an iterative procedure has been adopted, determining first the time of modules in the bottom sector opposite to the reference. In subsequent steps, the time of other modules in the top was determined relatively to those in the bottom already measured in the first step, and so on until all modules were analysed. The results of this method showed at an early stage that the laser-based inter-module time offsets had an accuracy of about ±2 ns. The systematic uncertainty due to the method itself was studied by adding known offsets to the input data, and determined to be 0.5 ns. In principle this

method could also be used at the cell level, but for this a different method was used.

The global matrix method [27] obtains the timing offsets also from comparison of data from top and bottom of the detector, but does that in an integrated way, by solving a system of equations that relates the time offsets of each cell to the measured time differences between those cells. If m and n are, respectively, the numbers of selected cells in the top and bottom part of the detector, and k is the number of valid pairs (see selection criteria in next paragraph) between them, the problem can be posed in matrix form as:

$$Mt = \Delta T \tag{6}$$

in which t is the $(m + n)$-size vector of unknown offsets, ΔT is the k-size vector of measured time differences (averaged over all events, and corrected for time-of-flight). M is a $(m + n) \times k$ matrix, and each line (of k) contains 1 for the element of the top part and -1 for the each element of the bottom part corresponding to the pair identified by that line. In order to properly weigh the results for different pairs, each element in M and ΔT are divided by the standard deviation of the pair time difference measurement. Since $k > (m + n)$, this system of equations is overdetermined, so the (approximate) solution is the least-squares minimum of $\|Mt - \Delta T\|$.

This method was applied to 0.5 M events from the RPC trigger sample of a long run taken in 2008. The event selection required to have at least one energy deposit above 250 MeV both on the top and bottom cells. For each event, cells were selected by requiring an energy between 200 MeV and 20 GeV, and a time difference between both PMTs of less than 6 ns. A final selection required that at least 5 events contribute to a cell pair average, and that the RMS of the measurements is smaller than 5 ns. The efficiency of these selections is of 40%, 75% and 82% for, respectively, the A, BC and D cells. To avoid memory limitations due to the large number of pairs (more than 30 k), the offset extraction was carried out separately for four sets of pairs. To

Fig. 34 (*Left*) Average of the time corrections per module as measured with the global matrix method with cosmic muons, for all cells. (*Right*) Difference of those values with respect to the results from the 2008 single beam data, removing the cells from the first layer. Different symbols correspond to modules in different partitions, as indicated

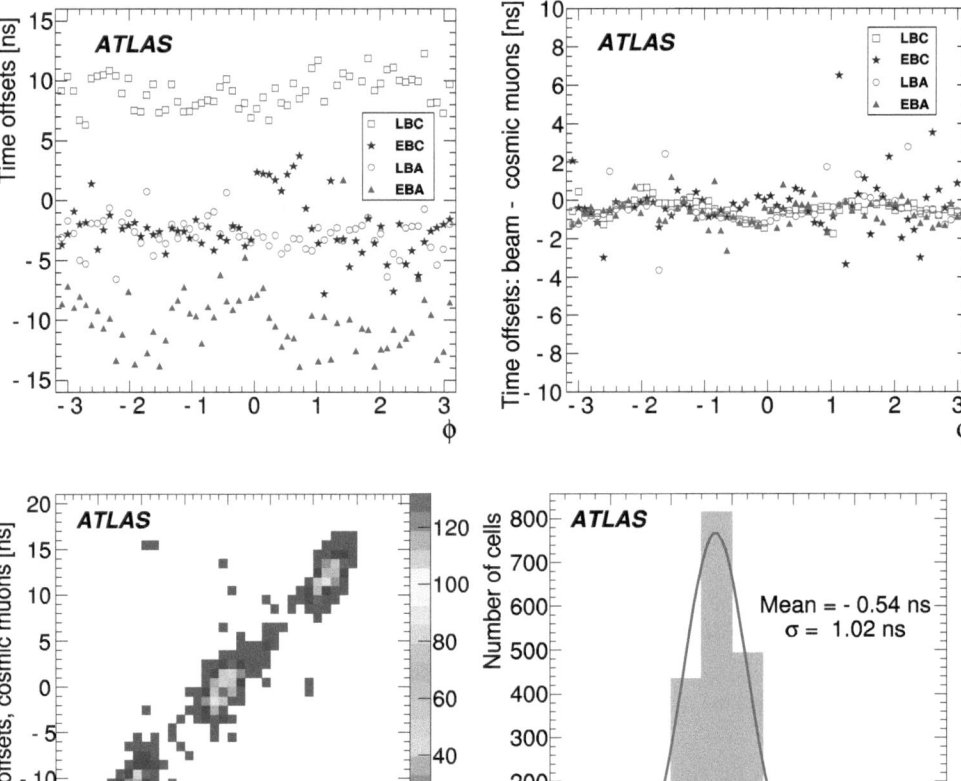

Fig. 35 Correlation (*left*) and difference (*right*) between the time corrections from cosmic muons and the 2008 single beam results. The cells from the first radial layer were removed

ensure consistency, these sets have a partial overlap, and the results are integrated at the end. The results were compared with those obtained with the 2008 single beam data (see Sect. 4.6), which were taken very close in time (less than 1 month) to the cosmic muon run analysed.

5.5.2 Results and comparison with 2008 LHC single beam

The average for each module of the cell offsets measured with the global matrix method is shown in Fig. 34 (left) and the comparison with the single beam data is shown in Figs. 34 (right) and 35.

The results clearly show differences of 10 ns between each partition (Fig. 34 left), but an otherwise good uniformity, of 2 ns, for all the cells in the second and third radial layers within each partition (Fig. 34 right). The results for the first layer are more scattered (this is reflected in the module average distributions, in particular for the EBA partition), in disagreement with the single beam measurements (see also Sect. 4.6). Due to the small size of the cells, the energy deposition with cosmic muons in this layer is small (peaking at roughly half of the value for the second layer), and consequently the signal-to-noise ratio is worse. Since

the single beam energy deposition is significantly larger, those results are more reliable, and so only the cosmic muon results from the second and third layers are considered valid.

It was expected to have differences between partitions, since the laser calibration had not been performed at this level.[15] The difference of 5–8 ns for the first 8 modules of EBC (Fig. 34 left, between 0 and 0.8 in ϕ) was unexpected, but confirmed with single beam data, and traced to an incorrect measurement of laser fibre lengths. So the inter-partition and inter-module results confirmed and validated the results from single beam, which were subsequently used to set the calibration time offsets, as described in Sect. 4.6. Within each partition, the agreement with the single beam data for the second and third layers, both at the level of module averages and single cells, is about 1 ns. Since this is smaller than the spread of the average offsets, these results provide a measurement of the accuracy of the laser-based time calibration, of about 2 ns.

---

[15]This is because the laser calibration data was taken in Tile standalone configuration, which has different delays than the global ATLAS online configuration.

6 Conclusions

The Tile hadronic calorimeter of the ATLAS detector underwent extensive testing during its commissioning and cosmic muon data-taking periods. The calorimeter has 99.1% (December 2009) of its cells operational and conditions that can affect the PMT gains have been monitored to be very stable over one year, such that no corrections are needed. The noise, being within the expectations and requirements, has a non-Gaussian component which has been taken into account in the reconstruction of clusters and physics objects. The noise magnitude has been stable over time within 1%.

The electromagnetic energy scale has been transferred from 11% of modules calibrated at testbeam to the full Tile Calorimeter in the ATLAS cavern setting by means of the TileCal calibration systems. The precision of all calibration systems is remarkable and has proven to follow the systems' design requirements. Regular calibration data-taking has demonstrated the stability of individual systems at levels well below 1%.

The single beam data proved to be very useful in complementing the calibration systems for the synchronisation of the calorimeter cells. The timing intercalibration capability is at the level of 1 ns within a TileCal module and 2 ns within a partition. Cosmic muons provided an independent cross-check of the time calibration settings, having verified a large fraction of the second and third layer cells with 2 ns precision.

The analysis of the cosmic muon data has been a very useful validation procedure to assess the performance with particles at the full calorimeter scale and to compare with Monte Carlo expectations. The separation between signal and noise is very good, with an S/N ratio of ~29 for the sum of the three radial layers. The cell response uniformity, as measured with the muon track dE/dx, is at the level of 4.6%, 3.5% and 3.8% within, respectively, the A, BC and D layers. The energy response shows a maximum difference among the radial layers of 4%.

The estimator of the EM scale relative to the testbeam calibration period as determined by the cosmic muons analysis is consistent with 1, with an uncertainty of 4%. A possible bias of -5% in the EM scale calibration due to lower HV settings as compared to the testbeam cannot therefore be totally excluded. However, the measurements with cosmic ray muons are compatible with a successful propagation of the EM scale factor from testbeam to the full ATLAS configuration.

Acknowledgements We are greatly indebted to all CERN's departments and to the LHC project for their immense efforts not only in building the LHC, but also for their direct contributions to the construction and installation of the ATLAS detector and its infrastructure. We acknowledge equally warmly all our technical colleagues in the collaborating Institutions without whom the ATLAS detector could not have been built. Furthermore we are grateful to all the funding agencies which supported generously the construction and the commissioning of the ATLAS detector and also provided the computing infrastructure.

The ATLAS detector design and construction has taken about fifteen years, and our thoughts are with all our colleagues who sadly could not see its final realisation.

We acknowledge the support of ANPCyT, Argentina; Yerevan Physics Institute, Armenia; ARC and DEST, Australia; Bundesministerium für Wissenschaft und Forschung, Austria; National Academy of Sciences of Azerbaijan; State Committee on Science & Technologies of the Republic of Belarus; CNPq and FINEP, Brazil; NSERC, NRC, and CFI, Canada; CERN; CONICYT, Chile; NSFC, China; COLCIENCIAS, Colombia; Ministry of Education, Youth and Sports of the Czech Republic, Ministry of Industry and Trade of the Czech Republic, and Committee for Collaboration of the Czech Republic with CERN; Danish Natural Science Research Council and the Lundbeck Foundation; European Commission, through the ARTEMIS Research Training Network; IN2P3-CNRS and CEA-DSM/IRFU, France; Georgian Academy of Sciences; BMBF, DFG, HGF and MPG, Germany; Ministry of Education and Religion, through the EPEAEK program PYTHAGORAS II and GSRT, Greece; ISF, MINERVA, GIF, DIP, and Benoziyo Center, Israel; INFN, Italy; MEXT, Japan; CNRST, Morocco; FOM and NWO, Netherlands; The Research Council of Norway; Ministry of Science and Higher Education, Poland; FCT co-financed by QREN/COMPETE of European Union ERDF fund, Portugal; Ministry of Education and Research, Romania; Ministry of Education and Science of the Russian Federation and State Atomic Energy Corporation ROSATOM; JINR; Ministry of Science, Serbia; Department of International Science and Technology Cooperation, Ministry of Education of the Slovak Republic; Slovenian Research Agency, Ministry of Higher Education, Science and Technology, Slovenia; Ministerio de Educación y Ciencia, Spain; The Swedish Research Council, The Knut and Alice Wallenberg Foundation, Sweden; State Secretariat for Education and Science, Swiss National Science Foundation, and Cantons of Bern and Geneva, Switzerland; National Science Council, Taiwan; TAEK, Turkey; The Science and Technology Facilities Council and The Leverhulme Trust, United Kingdom; DOE and NSF, United States of America.

References

1. F. Ariztizabal et al. (TileCal Collaboration) Construction and performance of an iron-scintillator hadron calorimeter with longitudinal tile configuration. Nucl. Instrum. Methods A **349**, 384–397 (1994). http://cdsweb.cern.ch/record/262630
2. G. Aad et al. (ATLAS Collaboration), The ATLAS experiment at the CERN Large Hadron Collider. J. Instrum. **3**, S08003 (2008). http://cdsweb.cern.ch/record/1129811
3. L. Evans et al., LHC Machine. J. Instrum. **3**, S08001 (2008). http://cdsweb.cern.ch/record/1129806
4. ATLAS/Tile Calorimeter Collaboration, *Tile Calorimeter Technical Design Report*, CERN/LHCC 96-42, 1996. http://cdsweb.cern.ch/record/331062
5. P. Mermod et al., Effects of ATLAS Tile calorimeter failures on jets and missing transverse energy measurement, ATLAS Note ATL-TILECAL-PUB-2008-011-1, 2008. http://cdsweb.cern.ch/record/1120460
6. K. Anderson et al., Design of the front-end analog electronics for the ATLAS tile calorimeter. Nucl. Instrum. Methods A **551**, 469–476 (2005)

7. S. Berglund et al., The ATLAS Tile Calorimeter digitizer. J. Instrum. **3**, P01004 (2008). http://cdsweb.cern.ch/record/1071920

8. A. Valero, on behalf of the ATLAS Tile Calorimeter System, The ATLAS TileCal read-out drivers signal reconstruction, in *IEEE Nucl. Sci. Symp. Conference Record*, 2009. http://cdsweb.cern.ch/record/1223960

9. J. Poveda et al., Atlas TileCal read-out driver system production and initial performance results. IEEE Trans. Nucl. Sci. **54**, 2629–2636 (2007)

10. P. Adragna et al. (TileCal Collaboration), Testbeam studies of production modules of the ATLAS Tile Calorimeter. Nucl. Instrum. Methods A **606**, 362–394 (2009). http://cdsweb.cern.ch/record/1161354

11. E. Starchenko et al., Cesium monitoring system for ATLAS Tile Hadron Calorimeter. Nucl. Instrum. Methods A **494**, 281–284 (2002). http://cdsweb.cern.ch/record/685349

12. S. Viret, for the LPC ATLAS group, LASER monitoring system for the ATLAS Tile Calorimeter. Nucl. Instrum. Methods A **617**, 120–122 (2010). Proceedings of the 11th Pisa Meeting on Advanced Detectors

13. N. Shalanda et al., Radioactive source control and electronics for the ATLAS tile calorimeter cesium calibration system. Nucl. Instrum. Methods A **508**, 276–286 (2003)

14. J. Abdallah et al. (TileCal Collaboration), The optical Instrumentation of the ATLAS Tile Calorimeter, ATLAS Note ATL-TILECAL-PUB-2008-005, 2007. http://cdsweb.cern.ch/record/1073936

15. S. Bertolucci et al., Influence of magnetic fields on the response of acrylic scintillators. Nucl. Instrum. Methods A **254**, 561–562 (1987)

16. J. Cumalat et al., Effects of magnetic fields on the light yield of scintilators. Nucl. Instrum. Methods A **293**, 606–614 (1990)

17. J.-M. Chapuis, M. Nessi, The measurements of magnetic field effects on scintillating tiles, ATLAS Note ATL-TILECAL-94-040, 1994. http://cdsweb.cern.ch/record/683495

18. K. Anderson et al., Calibration of ATLAS Tile Calorimeter at electromagnetic scale, ATLAS Note ATL-TILECAL-PUB-2009-001, 2009. http://cdsweb.cern.ch/record/1139228

19. C. Clement, B. Nordkvist, O. Solovyanov, I. Vivarelli, Time calibration of the ATLAS Hadronic Tile Calorimeter using the laser system, ATLAS Note ATL-TILECAL-PUB-2009-003, 2009. http://cdsweb.cern.ch/record/1143376

20. G. Aad et al. (ATLAS Collaboration), Expected performance of the ATLAS experiment—detector, trigger and physics, CERN Report CERN-OPEN-2008-020, 2009. http://cdsweb.cern.ch/record/1125884

21. R. Leitner, V. Shmakova, P. Tas, Time resolution of the ATLAS Tile calorimeter and its performance for a measurement of heavy stable particles, ATLAS Note ATL-TILECAL-PUB-2007-002, 2007. http://cdsweb.cern.ch/record/1024672

22. L. de Andrade Filho, J. de Seixas, Combining Hough transform and optimal filtering for efficient cosmic ray detection with a hadronic calorimeter, in *XII International Workshop on Advanced Computing Analysis Techniques in Physics Research* (Science, 2008)

23. J. Illingworth, A survey of the Hough transform. Comput. Vis. Graph. Image Process. **44**, 87–116 (1988)

24. A. Salzburger, The ATLAS Track Extrapolation Package, ATLAS Note ATL-SOFT-PUB-2007-005, 2007. http://cdsweb.cern.ch/record/1038100

25. Z. Ajaltouni et al., Response of the ATLAS Tile calorimeter prototype to muons. Nucl. Instrum. Methods A **388**, 64–78 (1997)

26. L. Fiorini, I. Korolkov, F. Vives, Tile Calibration of TileCal Modules with Cosmic Muons, ATLAS Note ATL-TILECAL-PUB-2008-010, 2008. http://cdsweb.cern.ch/record/1109974

27. J. Saraiva, on behalf of the ATLAS Tile Calorimeter System, Commissioning of the ATLAS Tile calorimeter with Single Beam and First collisions, in *Proceedings of the 12th Topical Seminar on Innovative Particle and Radiation Detectors, Siena (2010), in Nuclear Physics B (Proceedings Supplement)*, 2010, to appear. http://cdsweb.cern.ch/record/1281689

Studies of the performance of the ATLAS detector using cosmic-ray muons

The ATLAS Collaboration*

CERN, 1211 Geneva 23, Switzerland

Abstract Muons from cosmic-ray interactions in the atmosphere provide a high-statistics source of particles that can be used to study the performance and calibration of the ATLAS detector. Cosmic-ray muons can penetrate to the cavern and deposit energy in all detector subsystems. Such events have played an important role in the commissioning of the detector since the start of the installation phase in 2005 and were particularly important for understanding the detector performance in the time prior to the arrival of the first LHC beams. Global cosmic-ray runs were undertaken in both 2008 and 2009 and these data have been used through to the early phases of collision data-taking as a tool for calibration, alignment and detector monitoring. These large datasets have also been used for detector performance studies, including investigations that rely on the combined performance of different subsystems. This paper presents the results of performance studies related to combined tracking, lepton identification and the reconstruction of jets and missing transverse energy. Results are compared to expectations based on a cosmic-ray event generator and a full simulation of the detector response.

1 Introduction

The ATLAS detector [1] was constructed to provide excellent physics performance in the difficult environment of the Large Hadron Collider (LHC) at CERN [2], which will collide protons at center-of-mass energies up to 14 TeV, with unprecedented luminosity. It is designed to be sensitive to any experimental signature that might be associated with physics at this new high-energy frontier. This includes precision measurements of high p_T leptons and jets, as well as large transverse-energy imbalances attributable to the production of massive weakly interacting particles. Such particles are predicted in numerous theories of physics beyond the Standard Model, for example those invoking weak-scale supersymmetry or the existence of large extra dimensions.

Prior to the start of data-taking, understanding of the expected performance of individual subsystems relied on beam test results and on detailed GEANT4 [3, 4] simulations [5], including the modeling of inactive material both in the detector components and in the detector services and support structure. While extensive beam testing provided a great deal of information about the performance of the individual detector subsystems, a detailed understanding of the full detector could only be achieved after the system was in place and physics signals could be used for performance studies and for validation or tuning of the simulation.

In both 2008 and 2009 the ATLAS detector collected large samples of cosmic-ray events. These extended periods of operation allowed for the training of shift crews, the exercising of the trigger and data acquisition systems as well as of other infrastructure such as the data-handling system, reconstruction software, and tools for hardware and data-quality monitoring. The large data samples accumulated have also been used for a number of commissioning studies. Because cosmic-ray muons interact with the detector mainly as minimum-ionizing particles (MIPs), most traverse all of the subdetectors along their flight path. So, in addition to subdetector-specific cosmic-ray studies, these cosmic-ray data samples provide the first opportunity to study the combined performance of different detector components. Subsystem-specific cosmic-ray commissioning results have been documented in a series of separate publications [6–9]. This paper presents the results of studies relevant to combined tracking performance, lepton identification and calorimeter performance for the reconstruction of jets and missing transverse energy. Where simulation results are available, results are compared to expectations based on a dedicated cosmic-ray event generator, implemented in the detector simulation.

* e-mail: atlas.secretariat@cern.ch

2 The ATLAS detector

The ATLAS detector is described in detail elsewhere [1] and illustrated in Fig. 1. ATLAS uses a right-handed coordinate system with its origin at the nominal interaction point (IP). The beam direction defines the z-axis, the positive x-axis points from the IP towards the center of the LHC ring and the positive y-axis points upwards. Cylindrical coordinates (r, ϕ) are used in the transverse plane and the pseudorapidity η is defined in terms of the polar angle θ as $\eta = -\ln \tan(\theta/2)$.

The ATLAS detector is made up of a barrel region and two endcaps, with each region consisting of several detector subsystems. Closest to the interaction point is the Inner Detector (ID), which performs charged particle tracking out to $|\eta|$ of 2.5. It consists of two silicon detectors—the Pixel Detector and the SemiConductor Tracker (SCT)—and the Transition Radiation Tracker (TRT), all immersed in a 2T axial magnetic field provided by a superconducting solenoid magnet. The TRT is based on individual drift tubes with radiators, which provide for electron identification. The ID is surrounded by barrel and endcap liquid argon electromagnetic (EM) calorimeters which provide coverage out to $|\eta|$ of 3.1. These, in turn, are surrounded by hadronic calorimeters. In the barrel region, the Tile Calorimeter is composed of steel and scintillating tiles, with a central barrel and two extended-barrel regions providing coverage out to $|\eta|$ of 1.7. In the endcap region the Hadronic Endcap Calorimeter (HEC) is based on liquid argon and covers the

region $1.5 < |\eta| < 3.2$. The calorimetric coverage is extended into the region $3.2 < |\eta| < 4.9$ by a liquid argon Forward Calorimeter (FCal) which occupies the same cryostat as the endcap EM calorimeter and the HEC. Beyond the calorimeter system is the Muon Spectrometer (MS), which relies on a set of massive superconducting air-core toroid magnets to produce a toroidal magnetic field in the barrel and endcap regions. In both regions, planes of interleaved muon detectors provide tracking coverage out to $|\eta|$ of 2.7 and triggering to $|\eta|$ of 2.4. The tracking studies presented in this paper are restricted to the barrel region of the detector, where precision measurements of the (r, z) hit coordinates are provided by the Monitored Drift Tube (MDT) system. The remaining ϕ coordinate is measured by the Resistive Plate Chambers, or RPCs, which are primarily used for triggering.

ATLAS employs a three-level trigger system, with the Level-1 (L1) trigger relying primarily on information from the Muon and Calorimeter systems. For cosmic-ray running there was additionally a TRT-based trigger at L1 [10]. There is also a trigger based on signals from scintillators mounted in the endcap region, which are intended for triggering of collision events during the initial low-luminosity data-taking. This, however, plays no significant role in the triggering of cosmic-ray events. For the MS, the triggering in the barrel region of the detector is based on hits in the RPCs; in the endcap region, the Thin Gap Chambers (TGCs) are used. The L1 Calorimeter trigger (L1Calo) is based on analog sums provided directly from the calorime-

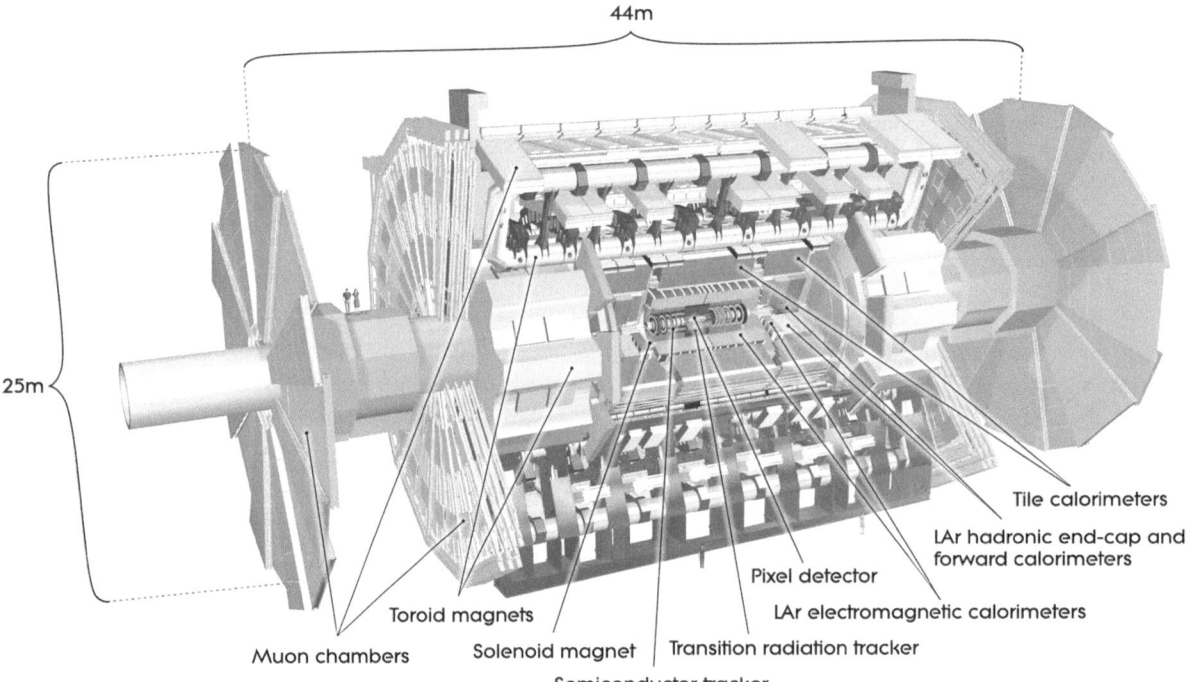

Fig. 1 The ATLAS detector and subsystems

ter front-end readout, from collections of calorimeter cells forming roughly projective trigger towers. In each case, the L1 trigger identifies a region of interest (ROI) and information from this ROI is transmitted to L2. In normal operation, events accepted by the L2 trigger are sent to the Event Filter which performs the L3 triggering, based on full event reconstruction with algorithms similar to those used offline. The L2 and L3 trigger systems are jointly referred to as the High Level Trigger, or HLT. For the cosmic-ray data taking, events were triggered only at L1. Information from the HLT was used only to split the data into different samples.

2.1 Tracking in ATLAS

The two tracking systems, the ID and the MS, provide precision measurements of charged particle tracks. Reconstructed tracks are characterized by a set of parameters (d_0, z_0, ϕ_0, θ_0, q/p) defined at the perigee, the point of closest approach of the track to the z-axis. The parameters d_0 and z_0 are the transverse and longitudinal coordinates of the perigee, ϕ_0 and θ_0 are the azimuthal and polar angles of the track at this point, and q/p is the inverse momentum signed by the track charge. Analyses typically employ track quality cuts on the number of hits in a given tracking subsystem. The track reconstruction algorithms account for the possibility of energy loss and multiple scattering both in the material of the tracking detector itself, and in the material located between the tracking system and the particle production point. For the combined tracking of muons, which reconstructs the particle trajectories through both the ID and the MS, this requires an accurate modeling of the energy losses in the calorimeter. This will be discussed in Sect. 4.2.

3 Cosmic-ray events in ATLAS

Cosmic rays in ATLAS come mostly from above, and arrive mainly via two large access shafts used for the detector installation, as illustrated in Fig. 2.

In proton-proton collisions, the actual beam-spot position varies from the nominal IP by distances that are of order mm in the transverse plane and cm along the beam direction. Tracks produced in proton-proton collisions at the IP are said to be projective, that is, emanating from (or near, in the case of particles arising from secondary vertices) the IP. Cosmic-ray muons passing through the volume of the detector do not normally mimic such a trajectory. However, in a large sample of events, some do pass close to the center of the detector. By placing requirements on track impact parameters with respect to the nominal IP, it is possible to select a sample of approximately projective muons from those passing through the barrel region of the detector. Such cosmic-ray muons are referred to below

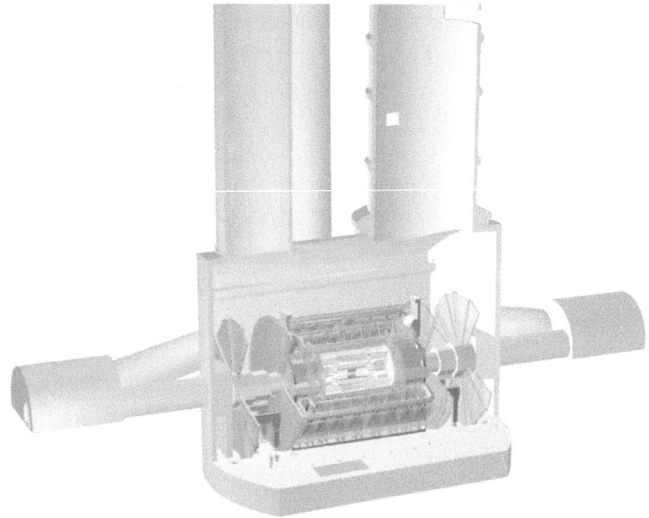

Fig. 2 The ATLAS detector in the experimental cavern. Above the cavern are the two access shafts used for the detector installation

as pseudo-projective. Due to the typical downward trajectory of the incoming cosmic-ray muons this cannot be done for those passing through the endcap region. For that reason, for those analyses presented here that rely on tracking, there is a requirement that the muons pass through the Inner Detector, which occupies a volume extending to about 1.15 m in radius and ±2.7 m in z. The rate of such cosmic-ray muons is of order several Hz. Most analyses further restrict the acceptance to the barrel region of the ID, which has a smaller extent, in z, of ±71.2 cm. Some analyses additionally place requirements on the presence of hits in the SCT or Pixel detectors, further restricting the volume around the nominal IP through which the cosmic-ray muons are required to pass. Track-based event selection criteria are not applied in the case of the jet and missing transverse energy studies presented in Sect. 5, which focus on the identification of fake missing transverse energy due to cosmic-ray events or to cosmic-ray interactions that overlap with triggered events. While calorimeter cells are approximately projective towards the IP,[1] energy deposits in the calorimeter can come from muons that pass through the calorimeter at any angle, including, for example, the highly non-projective up-down trajectory typical of cosmic muons passing through the endcap. While muons usually traverse the detector as MIPs, leaving only small energy deposits along their paths, in rare events they leave a larger fraction of their energy in the detector, particularly in the case of energy losses via bremsstrahlung. These can be particularly important in the case of high-energy muons, which can lose a significant amount of energy between the two tracking detectors. Such

---

[1]This is not the case for the FCal, which covers $3.2 < |\eta| < 4.9$, but that is not relevant to the analyses presented here.

events have been previously exploited for pulse shape studies of the LAr calorimeter and as a source of photons used to validate the photon-identification capabilities of the ATLAS EM calorimeter [7, 11].

The reconstruction of cosmic-ray events is also complicated by the fact that they occur at random times with respect to the 40 MHz readout clock, which is synchronized to the LHC clock during normal operation. For each subsystem, reconstruction of these events therefore first requires some measure of the event time with respect to the readout clock. An added complication, particularly for tracking, is that in the upper half of the detector, cosmic-ray muons travel from the outside in, rather than from the inside out, as would be the case for collisions. These differences can be addressed in the event reconstruction and data analysis. The modifications required for reconstruction of these events in the different detector components are discussed in the subsystem-specific cosmic-ray commissioning papers [6–9].

3.1 Data samples

ATLAS recorded data from global cosmic-ray runs during two extended periods, one in the fall of 2008 and another in the summer and fall of 2009. The analyses presented in this paper are each based on particular subsets of the available data.

For studies involving only the calorimeter, events triggered by L1Calo are used. Studies relying on tracking require that both the MS and ID were operational, and that the associated toroidal and solenoid fields both were at nominal strength. All L1-triggered events taken under those conditions were checked for the presence of a track in the ID. Events with at least one such track were streamed by the HLT to what is referred to here as the Pseudo-projective Cosmic-ray Muon (PCM) dataset, which forms the basic event sample for all of the studies presented in Sect. 4. These events are mainly triggered at L1 by the RPCs. Hundreds of millions of cosmic-ray events were recorded during the 2008 and 2009 cosmic-ray runs. However, the requirement of a track in the Inner Detector reduces the available statistics dramatically, as does the requirement of nominal magnetic field strengths for the MS and ID, which is necessary for studies of the nominal tracking performance.

3.2 Cosmic-ray event simulation

Cosmic-ray events in ATLAS are simulated using a dedicated event generator and the standard GEANT4 detector simulation, with the modeling of the readout electronics adapted to account for the difference in timing. The simulation includes the cavern overburden, the layout of the access shafts and an approximation of the material of the surface buildings. The event generator is based on flux calculations in reference [12] and uses a standard cosmic-muon

momentum spectrum [13]. Single muons are generated near ground level, above the cavern in a 600 m × 600 m region centered above the detector, with angles up to 70° from vertical. Muons pointing to the cavern volume are propagated through up to 100 m of rock overburden, using GEANT4. Measurements of the cosmic-ray flux at different positions in the cavern were used to validate the predictions of this simulation [14]. Once a muon has been propagated to the cavern, additional filters are applied; only events with at least one hit in a given volume of the detector are retained, depending on the desired event sample. Note that only single-muon cosmic-ray events are simulated. No attempt is made to model events in which cosmic-ray interactions produce an air shower that can deposit large amounts of energy in the detector. However, the rate of such events (in data) has been shown to be sufficiently low that they do not produce significant discrepancies in, for example, the agreement between data and Monte Carlo (MC) for the distribution of the summed transverse energy in cosmic-ray events [15].

4 Lepton identification and reconstruction studies using cosmic-ray events

Cosmic-ray muons are an important tool for the commissioning of the muon spectrometer, which is the largest ATLAS subsystem, occupying over 95% of the total detector volume. As the rate of production of high-p_T muons in collision events is rather low, the cosmic-ray data will continue to be relevant to the MS commissioning for some time to come. ATLAS continues to record data from cosmic-ray interactions when LHC beams are not present.

While the cosmic rays are primarily a source of muons, analysis of these data also allows for checks of the algorithms used to identify other leptons. The cosmic-ray muons serve as a source of electrons, mainly δ-electrons but with smaller contributions from the conversion of muon bremsstrahlung photons and muon decays in flight. The identification of a sample of electrons allowed for an examination of the performance of the electron identification algorithms, prior to first collisions. Similarly, although no τ-leptons are expected in the cosmic-ray data sample, the tools designed for τ-identification have been exercised using these data and checked against the simulation.

The analyses discussed in this section rely on the PCM dataset described earlier, which contains cosmic-ray muon events with tracks reconstructed in the ID. Most analyses also require the presence of hits in the Pixel Detector. These differ slightly for different analyses, as will be described below.

4.1 Combined muon tracking performance

This section describes studies of the performance of the combined tracking for muons, using cosmic-ray data recorded in 2009. The investigation uses the PCM dataset in order to have tracks that resemble, as much as possible, tracks from collision data. Selected events are required to have a topology consistent with that expected for the passage of a cosmic-ray muon through the detector, which is illustrated by a typical event in Fig. 3. The requirements are:

– exactly 1 track reconstructed in the ID
– 1 or 2 tracks reconstructed in the MS
– exactly 1 combined track crossing both subdetectors

A special ID pattern recognition algorithm was used to reconstruct cosmic-ray muons as single tracks. Because of the topology of these events, the analysis is restricted to the barrel region of the detector. Good quality ID and MS tracks are ensured using requirements on the number of hits in the different subsystems. Events are required to have been triggered by the RPC chambers, since these also provide measurements along the ϕ coordinate (ϕ hits), which is not measured by the MDTs. Following the procedure used in the ID commissioning with cosmic-ray muons [6], a requirement is also placed on the timing from the TRT, to ensure that the event was triggered in a good ID time window.

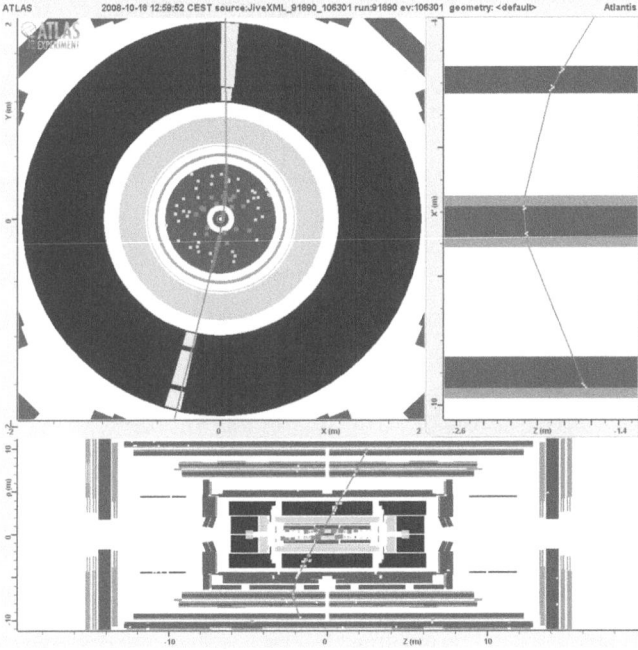

Fig. 3 Event display of a cosmic-ray muon crossing the entire ATLAS detector, close to the nominal IP, leaving hits in all tracking subsystems and significant energy deposits in the calorimeter. The *upper left* view shows the projection into the $r\phi$ plane. The *lower plot* shows the projection in the rz plane. The *upper right projection* is a longitudinal slice through the central part of the Muon Spectrometer at the ϕ value of the MDT planes in which the muon hits were recorded

The track parameter resolutions for Combined Muon (CM) tracks have been investigated in the same manner as used for similar studies of the ID [6] and MS [9] performance, by comparing the two reconstructed tracks left by a single cosmic-ray muon passing through the upper and then the lower half of the detector. In the case of the ID and combined tracks, this involves separately fitting the hits in these two regions, to form what are referred to below as "split tracks" from the track created by the passage of a single muon.

Prior to a study of combined tracking, it is necessary to establish that the relative alignment of the two tracking systems is adequate. Checks were performed by comparing the track parameters for standalone tracks reconstructed by the two separate tracking systems, in the upper and lower halves of ATLAS. Tracks in the MS were reconstructed using a least-squares method that directly incorporates the effects of the material that sits between the MS detector planes and the point at which the track parameters are defined [16]. ID tracking was also performed by standard tracking algorithms [17, 18].

The alignment check relies on the study of three different classes of tracks: split ID tracks, MS standalone tracks, and split CM tracks. In what follows these will be referred to simply as ID, MS and CM tracks, respectively. Different quality cuts are placed on the three track types. For ID and CM tracks $|d_0|$ and $|z_0|$ are required to be less than 400 mm and 500 mm respectively. For MS tracks, for which these parameters must be extrapolated from the MS back to the perigee, the requirements are $|d_0| < 1000$ mm and $|z_0| < 2000$ mm. ID and CM tracks are required to have at least 1, 6 and 20 hits in the Pixel, SCT and TRT detectors, respectively. MS and CM tracks are required to have hits in all three MS layers, with more than four RPC hits, at least two of which are ϕ hits, and a χ^2 per degree of freedom less than 3. All tracks are required to have momentum larger than 5 GeV.

Figure 4 shows the correlation between the ϕ_0 and θ_0 parameters determined from MS and ID tracks in the bottom half of ATLAS. Very good consistency is evident and similar results are obtained in the other hemisphere. The level of agreement between the two systems is better quantified by distributions of the difference between the track parameters obtained from the two systems. These are shown in Fig. 5 for d_0, z_0, ϕ_0 and θ_0, separately for tracks in the upper and lower halves of the detector. The somewhat narrower distributions obtained from the upper half of the detector are attributed to the higher average momentum of the cosmic-ray muons in this part of the detector, since those in the bottom have lost energy passing through the lower half of the calorimeter before reaching the MS. Small biases are observed for the d_0 and ϕ_0 parameters. These are consistent with a slight translational misalignment between the MS and ID that is of order

Fig. 4 Correlations between the track parameters ϕ_0 and θ_0 obtained from standalone ID and MS tracks, in the bottom half of ATLAS

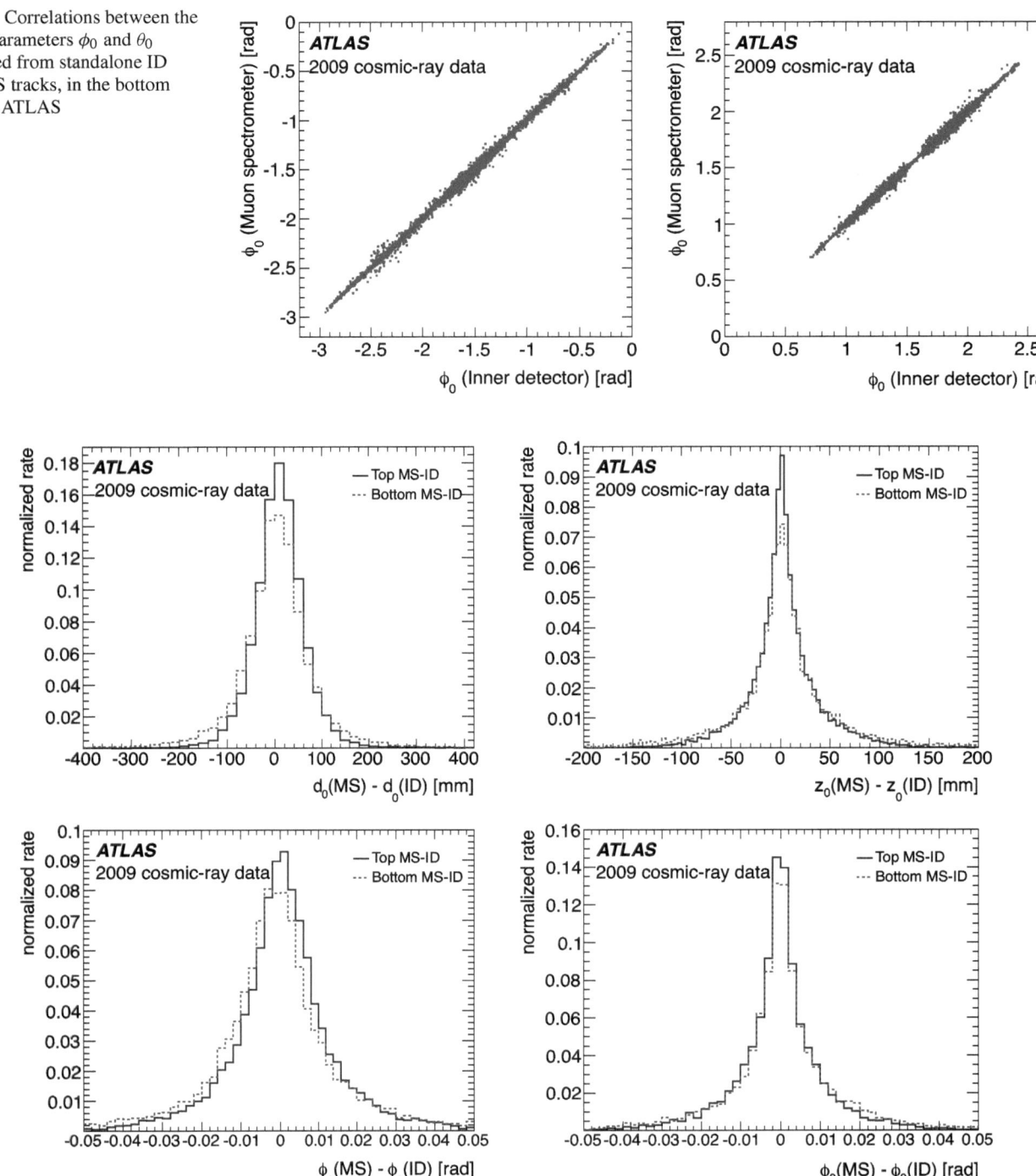

Fig. 5 Difference distributions of the track parameters, d_0, z_0, ϕ_0 and θ_0 obtained from standalone ID and MS tracks, for the top and bottom halves of the detector

1 mm. However, the combined tracking study presented below was performed without any relative ID-MS alignment corrections.

The track parameter resolutions for combined tracking have been investigated in the manner discussed above, using CM tracks passing through the barrel part of the detector, which are split into separate tracks in the upper and lower halves. The two resulting tracks are then fitted using the same combined track fit procedure. For studies of the angular and impact parameter resolution, the track quality cuts are tightened somewhat, with the requirements of at least two pixel hits, $|d_0| < 100$ mm and $|z_0| < 400$ mm. An estimate of the resolution on each track parameter, λ, is obtained from the corresponding distribution of the difference

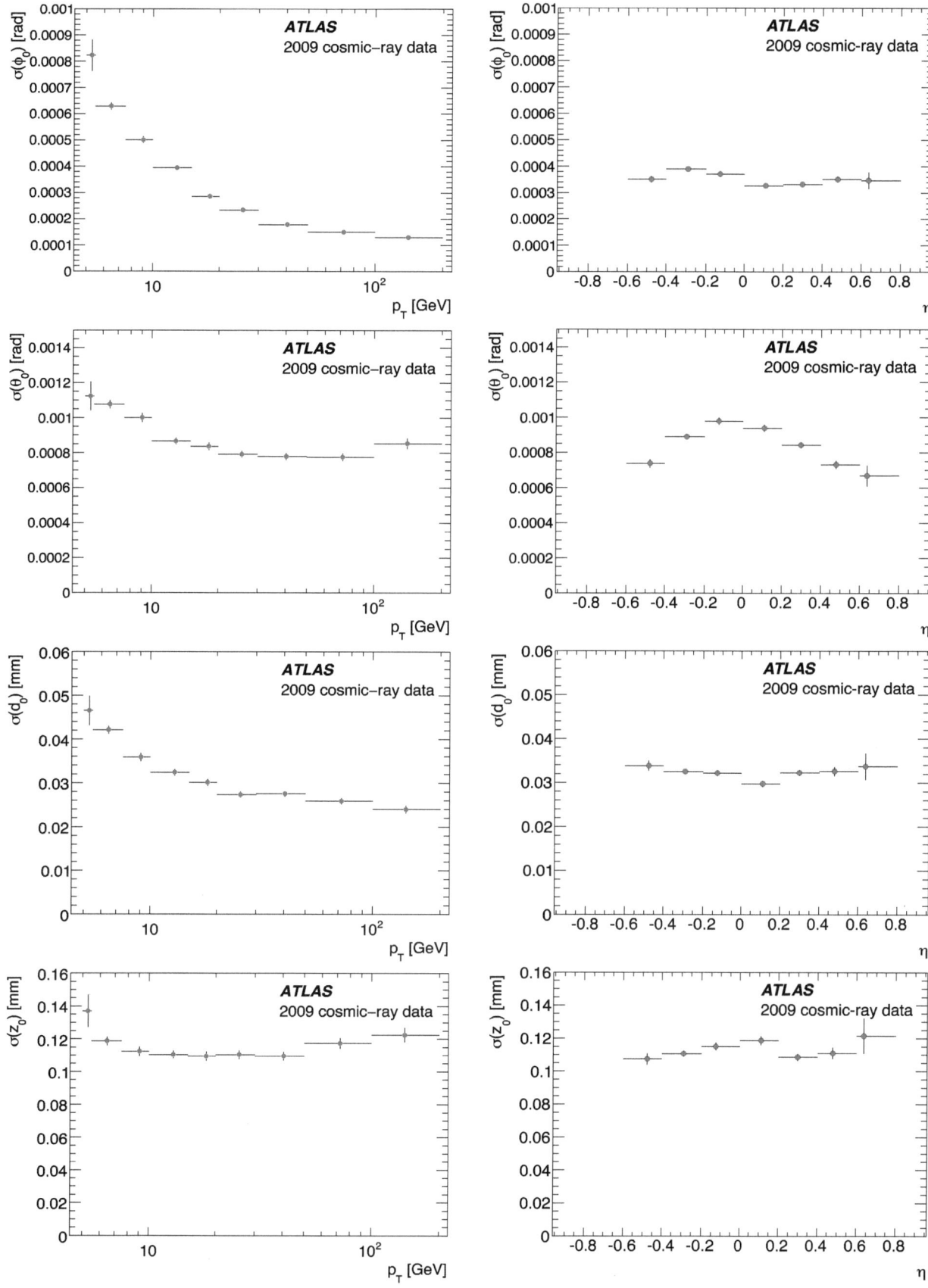

Fig. 6 Resolution on track parameters ϕ_0, θ_0, d_0 and z_0, obtained from split tracks, as a function of p_T (*left column*) and η (*right column*)

in the track parameters obtained from the two split tracks, $\Delta\lambda = \lambda_{up} - \lambda_{low}$. Each such distribution has an expectation value of 0 and a variance equal to two times the square of the parameter resolution: $\text{var}(\Delta\lambda) = 2\sigma^2(\lambda)$. For each parameter, the mean and resolution of this difference distribution have been studied in bins of p_T and η. Since the cosmic-ray muon momentum distribution is a steeply falling function, the p_T value for each bin is taken as the mean of the p_T distribution in that bin. For the resolutions, the results are shown in Fig. 6. The absence of data points in the range of $-0.8 < \eta < -0.6$ is due to a requirement that there be at least 50 muons per bin. The means are roughly independent of p_T and η and show no significant bias, with the exception of the z_0 distribution. That shows a small bias that varies with η, but with a magnitude that is less than about 60 μm over the η-range investigated. This is negligible relative to the MS-ID bias already discussed. The means and resolutions obtained from tracks with $p_T > 30$ GeV are shown in Table 1.

A similar study of the track momentum reconstructed in the upper and lower halves of the detector shows that the mean of the momentum-difference distribution ($p_{up} - p_{low}$) is consistent with zero and flat as a function of p_T and η. For studies of the p_T resolution, slightly looser cuts are employed in order to increase the statistics, particularly in the high-momentum region. For tracks having momenta above 50 GeV the requirement of a pixel hit is removed and the cuts on $|d_0|$ and $|z_0|$ are loosened to 1000 mm. Figure 7 shows the relative p_T resolution for ID, MS and CM tracks as a function of p_T. For each pair of upper/lower tracks, the value of the transverse momentum was evaluated at the perigee. The difference between the values obtained from the upper and lower parts of the detector, divided by their average

$$\frac{\Delta p_T}{p_T} = 2\frac{p_{Tup} - p_{Tdown}}{p_{Tup} + p_{Tdown}}$$

was measured and plotted in eleven bins of p_T. As above, the plotted p_T value is the mean of the p_T distribution in that bin. The results of this procedure have been fitted to

parametrizations appropriate to each particular track class. For the ID the fit function was:

$$\frac{\sigma_{p_T}}{p_T} = P_1 \oplus P_2 \times p_T,$$

where P_1 is related to the multiple scattering term and P_2 to the ID intrinsic resolution. For the MS tracks, the same function is used but with an additional term (coefficient P_0) related to uncertainties on the energy loss corrections associated with the extrapolation of the MS track parameters to the perigee:

$$\frac{\sigma_{p_T}}{p_T} = \frac{P_0}{p_T} \oplus P_1 \oplus P_2 \times p_T.$$

For the combined resolution a more complex function is needed:

$$\frac{\sigma_{p_T}}{p_T} = P_1 \oplus \frac{P_0 \times p_T}{\sqrt{1 + (P_3 \times p_T)^2}} \oplus P_2 \times p_T,$$

where P_1 is related to the multiple scattering term, P_2 to the intrinsic resolution at very high momentum and the P_3 term describes the intermediate region where ID and MS resolutions are comparable.

Table 2 compares the fitted sizes of the multiple scattering and intrinsic resolution terms for the ID, MS and CM tracks. For the CM tracks the multiple scattering term is determined mainly by the ID contribution while the intrinsic high-energy resolution comes mainly from the MS measurement.

Extrapolation of the fit result yields an ID momentum resolution of about 1.6% at low momenta and of about 50% at 1 TeV. The MS standalone results are improved over those previously obtained [9]: the resolution extrapolated to 1 TeV is about 20%. As expected the ID and MS systems dominate the resolution at low and high p_T, respectively. However, at intermediate momenta from about 50 to 150 GeV both systems are required for the best resolution. The $\pm 1\sigma$ region returned by the fit to the resolution for the CM tracks is shown as the shaded region in Fig. 7.

4.2 Muon energy loss in the ATLAS calorimeters

Muons traverse more than 100 radiation lengths between the two tracking systems. Interactions with the calorime-

Table 1 Overview of the track parameter bias and resolution for CM tracks obtained with the track-splitting method for 2009 cosmic-ray data, for tracks with $p_T > 30$ GeV

Parameter	Mean	Resolution
ϕ_0 (mrad)	-0.053 ± 0.005	0.164 ± 0.004
θ_0 (mrad)	0.27 ± 0.03	0.80 ± 0.02
d_0 (μm)	-0.9 ± 0.7	26.8 ± 0.8
z_0 (μm)	2.0 ± 3.7	116.6 ± 2.9

Table 2 Fitted values of the multiple scattering and intrinsic momentum resolution terms (as described in the text) for ID, MS and CM tracks

Fitted Resolution	P_1	P_2
Inner Detector	$1.6 \pm 0.1\%$	$(53 \pm 2) \times 10^{-5}$ GeV^{-1}
Muon Spectrometer	$3.8 \pm 0.1\%$	$(20 \pm 3) \times 10^{-5}$ GeV^{-1}
Combined Muon	$1.6 \pm 0.1\%$	$(23 \pm 3) \times 10^{-5}$ GeV^{-1}

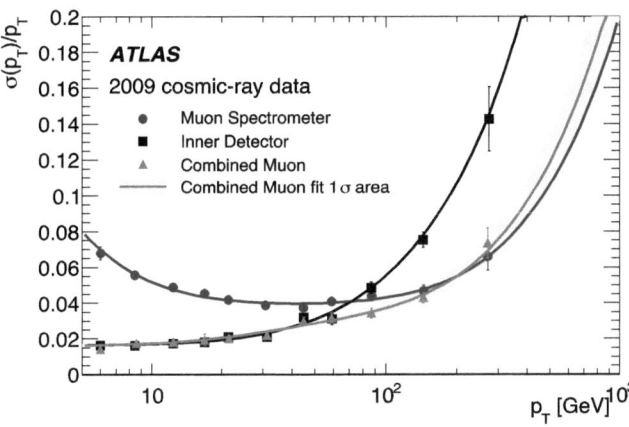

Fig. 7 Resolution on relative p_T as a function of p_T for ID and MS standalone tracks, and for CM tracks. The shaded region shows the $\pm 1\sigma$ region of the fit to the resolution curve for the CM tracks

ter material result in energy losses. These losses are typically around 3 GeV, mainly due to ionization, but are subject to fluctuations, especially for high momentum muons which can deposit a large fraction of their energy via bremsstrahlung. Muon reconstruction in collision events depends on a correct accounting for these losses, as does determination of the missing transverse energy in the event. A parametrization of these losses is normally used for extrapolating the track parameters measured by the MS to the perigee where they are defined. However, since 80% of the material between the trackers is instrumented by the calorimeters, studies of the associated energy deposits in the calorimeter should allow improvements to the resolution in the case of large losses.

This possibility has been investigated using cosmic-ray muons traversing the barrel part of ATLAS. The analysis is based on the PCM sample from a single 2009 cosmic-ray run, consisting of about one million events. Strict criteria were applied to ensure pseudo-projective trajectories that are well measured in the relevant tracking subsystems: the SCT and the TRT in the Inner Detector, and the MDT and RPC systems in the Muon Spectrometer. The analysis was restricted to tracks crossing the bottom part of the Tile Calorimeter, in the region $|\eta| < 0.65$. A track-based algorithm [19–21] was used to collect the muon energy deposits in the calorimeters. The trajectory of the particle was followed using the ATLAS extrapolator [22], which, using the ATLAS tracking geometry [23], takes into account the magnetic field, as well as material effects, to define the position at which the muon crossed each calorimeter layer. The cells within a predefined 'core' region around these points were used for the measurement of the energy loss. This region was optimized according to the granularity and the geometry of each calorimeter layer. Only cells with $|E| > 2\sigma_{\text{noise}}$ were considered. Here σ_{noise} is the electronics noise for the channel and $|E|$ is used instead of E to avoid biases. As

a check that this procedure properly reconstructs the muon energy deposits, the total transverse energies reconstructed in calorimeter cells within cones of $\Delta R = \sqrt{(\Delta \eta)^2 + (\Delta \phi)^2}$ of 0.2, 0.3 and 0.4, around the particle trajectory, were determined. From these, the sum of the transverse energy inside the core region (E_T^{core}) was subtracted. In collision events these quantities can be used to define the muon isolation, while in this analysis they indicate how much energy is deposited outside the core. The distributions of these quantities, shown in the top plot of Fig. 8, are reasonably centered around zero with widths that increase with the cone size, as expected due to the inclusion of a larger number of cells. Small energy losses outside the E_T^{core} region shift the distributions to slightly positive values, due to either uncertainties in the extrapolation process or to radiative losses.

As a measure of the energy deposited by the muon, E_T^{core} is used with no additional correction. Monte Carlo simulations of single muons in the barrel region show that this method provides a nearly unbiased energy determination, with 2% scale uncertainty and 11% resolution for the energy deposited by 100 GeV muons. To allow comparison of these losses with the difference between the momenta reconstructed in the two tracking systems, a parametrization of the losses in the dead (uninstrumented) material, E_{dead}, is added to the calorimeter measurement. The tracking geometry provides this information in combination with the extrapolator. The energy measured in the calorimeter, corrected for the dead material, is compared with the momentum difference between Inner Detector and Muon Spectrometer tracks in the middle plot of Fig. 8. The mean values of the momentum-difference and energy-sum distributions are 3.043 GeV and 3.044 GeV respectively. The typical momentum of the selected tracks is 16 (13) GeV in the Inner Detector (Muon Spectrometer), measured (see Fig. 7) with a resolution of about 2% (4%), while the energy collected in the calorimeters, E_{calo}, is on average 2.4 GeV, with a precision of about 10–20%. The RMS values of the two distributions are 1.081 GeV and 0.850 GeV respectively. In simulation the two distributions have means of 3.10 GeV and 3.12 GeV compared to a true energy loss distribution with a mean of 3.11 GeV and an RMS of 0.750 GeV. The resolutions were 0.950 GeV and 0.820 GeV, respectively, roughly consistent with the measured values. The bottom plot in Fig. 8 shows the distribution of $(P_{\text{ID}} - P_{\text{MS}}) - (E_{\text{calo}} + E_{\text{dead}})$, which has a mean of -0.012 GeV and an RMS of 1.4 GeV. This distribution is dominated by contributions from rather low-momentum tracks. Restricting to the momentum region of 10–25 GeV retains about 40% of the statistics and yields a distribution with mean and RMS of -0.004 GeV and 1.0 GeV, respectively.

Although the tracking systems are relatively more precise than the calorimeters, in both data and Monte Carlo simulation, the RMS of the energy-sum distribution from the

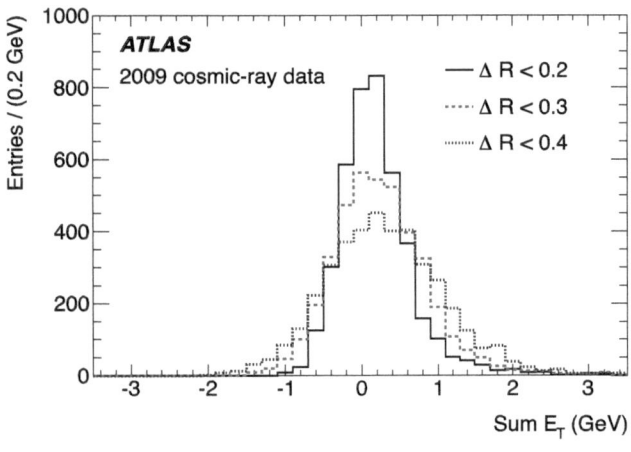

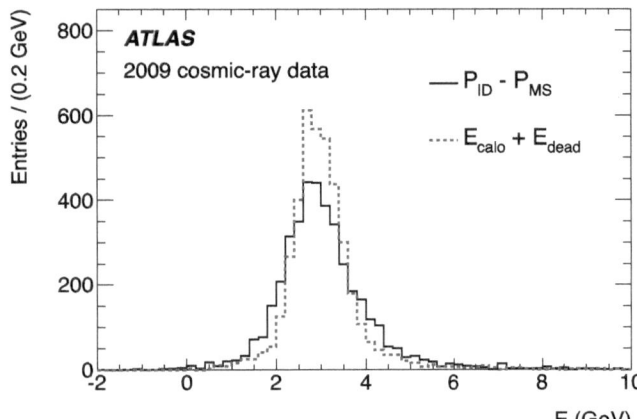

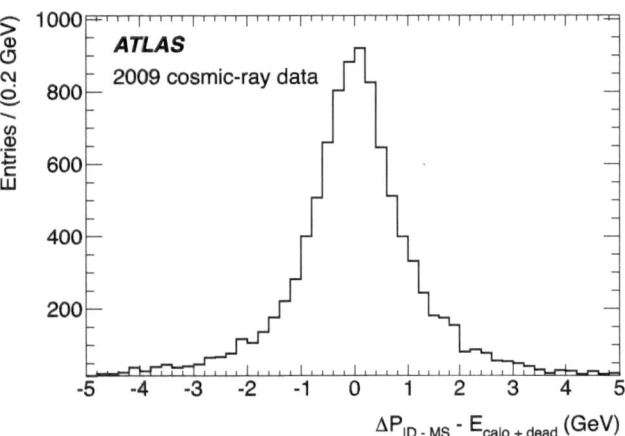

Fig. 8 The *upper plot* shows the sum of the transverse energy around muon tracks, outside the core region, for cones of $\Delta R = 0.2$, 0.3 and 0.4. The *middle plot* compares the momentum difference between Inner Detector and Muon Spectrometer tracks ($P_{ID} - P_{MS}$) with the sum of the energy loss measured in the calorimeters, E_{calo}, and the parametrized energy loss in the inert material, E_{dead}. The *lower plot* shows the distribution of $(P_{ID} - P_{MS}) - (E_{calo} + E_{dead})$

calorimeter is smaller than that of the momentum-difference distribution from the tracking systems. Use of the calorimeter information may therefore allow future improvements to the combined tracking for muons.

4.3 Identification of electrons

The identification of electrons is performed by algorithms relying on information from both the EM calorimeter and the ID. Two methods are used, one seeded by tracks and the other by EM calorimeter clusters. The cluster-based algorithm is the standard identification tool, with clusters seeded using a sliding-window algorithm [24]. This algorithm, used only for the identification of electromagnetic (e/γ) objects in the EM calorimeter, uses a fixed grid of calorimeter cells in $\eta \times \phi$, centered on a seed cell having a signal-to-noise ratio exceeding a set threshold. For a cluster to form an electron candidate, there must normally be an ID track nearby in η and ϕ. However, in cosmic-ray events many tracks have only barrel TRT ($r - \phi$) hit information, in which case the association is done only in ϕ. The threshold for the reconstruction of an e/γ object with the standard selection is about 3 GeV. To improve the identification of electrons with lower p_T, a track-seeded algorithm is employed. This first searches for tracks in the ID with $p_T > 2$ GeV and hits in both the SCT and Pixel Detectors. These tracks are extrapolated to the second layer of the EM calorimeter and a 3×7 ($\eta \times \phi$) cell cluster is formed about this point; the cell size in this layer varies with η but is 0.025×0.025 in $\eta \times \phi$ over the acceptance for this analysis. In both algorithms, the track momentum and the energy of the associated calorimeter cluster are required to satisfy $E/p < 10$. This section describes the use of these standard techniques for the selection of a sample of δ-electrons, which are used to investigate the calorimeter response to electrons with energies in the 5 GeV range. Section 4.4 will describe an alternative low-p_T selection which can identify electrons down to p_T of about 500 MeV, using a more sophisticated clustering algorithm for determination of the energy of the associated electromagnetic calorimeter cluster.

Electron identification relies in part on the particle identification abilities of the TRT. Transition radiation (TR) is produced by a charged particle crossing the boundary between two materials having different dielectric constants. The probability of producing TR photons depends on the Lorentz factor ($\gamma = E/m$) of the particle. The effect commences at γ factors around 1000 which makes it particularly useful for electron identification, since this value is reached for electrons with energies above about 500 MeV. For muons, these large γ factors occur only for energies above about 100 GeV. The TR photons are detected by absorption in the chamber gas which is a xenon mixture characterized by a short absorption length for photons in the relevant energy range. The absorption leads to high electronic pulses; pulses due to energy deposits from particles which do not produce transition radiation are normally much lower. A distinction between the two classes of particles can therefore be made by comparing the pulse heights against high and low thresholds, and looking at the fraction of high-threshold hits for a given track. This fraction is referred to below as the TR ratio.

The production of electrons in cosmic-ray events is expected to be dominated by knock-on or δ-electrons produced by ionization caused by cosmic-ray muons. The energy distribution of such electrons is typically rather soft, but has a tail extending out into the GeV region, where the standard electron identification tools can be employed. The experimental signature of such an event consists of a muon track traversing the muon chambers at the top and bottom of the detector, having corresponding MIP-like energy deposits in the calorimeters, accompanied by a second lower-momentum track in the ID associated with a cluster in the EM calorimeter, as illustrated by the event displayed in Fig. 9. In the upper view, the incoming and outgoing muon tracks, are seen to leave hits in three muon layers on the top of ATLAS and in two layers below, as well as in the Inner Detector. In the lower, expanded view of the ID region the muon track and the electron candidate track are shown with the associated hits in the silicon detectors as well and those in the TRT, which are illustrated by either light or dark markers, depending on whether they are low- or high-threshold. The candidate electron track clearly displays a larger number of high-threshold TRT hits, as expected for an electron, as well as an association to a cluster of energy in calorimeter (at the bottom). Other low-energy deposits in the calorimeter have been suppressed.

The search was performed using data from the PCM sample obtained from cosmic-ray running in the fall of 2008. Based on the expected topology, events were selected if they satisfied the following requirements:

- 2 or more ID tracks.
- 1 electron in the bottom of the detector (since the muons come from above).
- 1 or more muon tracks: if there is more than one there must be at least one track in the top and bottom halves of the detector, consistent with coming from a muon of the same charge.

The events so selected are referred to below as the signal sample, or the ionization sample.

There is one important background for which this selection can lead to the identification of fake electron candidates. A highly energetic muon can emit a bremsstrahlung photon that does not convert within the ID. This photon will produce a cluster in the EM Calorimeter that can be incorrectly associated with the muon track if the track and cluster are nearby, creating a fake electron candidate. The signature for this process is one incoming and one outgoing muon track in the MS, one track in the ID and a cluster in the lower part of the EM Calorimeter. This signature can be clearly distinguished from the true electron production processes by the

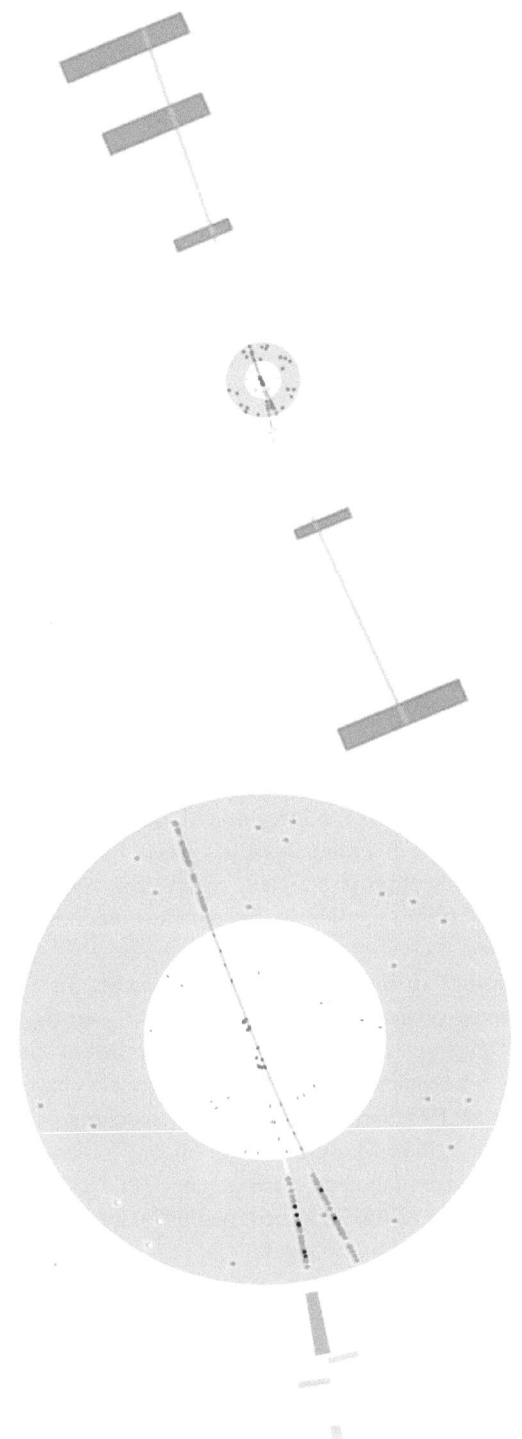

Fig. 9 Event display of a typical δ-electron candidate event. The *upper figure* shows a view that includes the three layers of muon detectors on either side, while the *lower plot* shows a close-up view of the Inner Detector. The *shaded region* represents the volume of the TRT, while the *inner region* is occupied by the SCT and Pixel detectors. The two ID tracks, and associated hits, are clearly visible. High- and low-threshold TRT hits are displayed with the *dark* and *light markers*, respectively. The calorimeter cluster associated with the electron candidate is also shown. Other low-energy deposits in the calorimeter are suppressed

number of tracks in the ID (except for muon decays in flight which are expected to contribute only a very small fraction of the electrons of interest in this analysis). Nevertheless, for muon bremsstrahlung events, an additional track may be present (due for example to an overlapping cosmic ray event or a correlated cosmic-ray muon due to an air shower event) leading to an event with the same signature as the signal process. This background source should produce equal numbers of electron and positron candidates in contrast to true δ-electrons events, where only negatively charged electrons are produced. To study this background (in which electron candidates are actually muons), a sample of events depleted in δ-electrons and enriched muon bremsstrahlung events, was selected using the requirements:

- exactly 1 ID track
- 1 electron in the bottom of the detector
- 1 or more muon tracks

In the analysis of the signal and background samples, slightly modified versions of standard algorithms were used to identify electrons. The standard selection [1] defines three classes of candidates: loose, medium and tight, according to increasingly stringent cuts on the typical properties of electron tracks and their associated EM showers, particularly quantities related to the longitudinal and transverse shower development. For the analysis discussed here, a "modified medium" selection is adopted, which is a combination of selection criteria applied in the standard medium and tight selections, with slight modifications to allow for the different topology of the cosmic-ray muon events. In particular, since most of the muons do not pass through the SCT or Pixel Detector, requirements on the number of hits in the silicon detectors are replaced with quality cuts based on the number of TRT barrel hits and the ϕ matching of the electron track to the EM cluster. A cut on $|z_0|$ is made to ensure that tracks are in the barrel part of the TRT.

In addition to this modified medium selection, a tight selection is defined by two additional requirements:

- $0.8 < E/p < 2.5$
- TR ratio > 0.08

Note that both of these cuts are actually slightly η-dependent, following the standard tight selection. The values quoted above are those applied over most of the acceptance. After application of the modified medium selection, there are 81 events in the signal sample and 1147 in the background sample. Since the background candidates arise dominantly from the case where the EM cluster is associated to the cosmic-ray muon, this sample can be used to model the properties of the corresponding background events in the signal sample, in which the requirement of an additional ID track greatly reduces the number of events. Because E/p and the TR ratio are correlated, these quantities are shown plotted against

one another in the upper plots of Fig. 10, separately for the signal and background samples. The open and solid markers together show the distribution of candidates passing the modified medium selection. The solid markers show the candidates that also survive the tight selection. In each plot, the dotted lines show the cuts applied (as quoted above) on each quantity, for the majority of the candidates. These define the signal region which is enclosed by the overlaid solid lines. The open markers in the signal region and solid markers in the background region arise due to the slight η-dependence of the cuts. There are 34 events from the signal sample passing all cuts, compared to 13 from the background sample. Of the 34 events in the signal region, 4 are positively charged.

The sample of 34 candidates was investigated further in order to confirm the identification of these as electrons and to determine the number of δ-electrons by estimating the background in the signal sample. This was done by performing a three-parameter, binned maximum-likelihood fit to the two-dimensional TR ratio vs. E/p distribution for the background sample and then fitting the resulting background shape to the ionization sample in the regions outside the signal acceptance. The results of this procedure are displayed in the lower plots of Fig. 10. Note that the fit uses finer binning than is used for these projections. The plot on the left shows the distribution of the TR ratio for the 81 candidates passing the modified medium cuts (points with error bars) while the dashed histogram shows the 47 events in the background region and the solid curve shows the projection of the two-dimensional binned maximum likelihood fit, which provides a good description of the distribution from candidates in the background region (dashed histogram). The right-hand plot shows the distribution of E/p for all candidates remaining after the additional application of the tight-selection cut on the TR ratio. The solid curve again shows the projection of the two-dimensional background fit leading to an estimate for the background contribution in the signal region (indicated by the dotted vertical lines) of (8.3 ± 3.0) events. This is consistent with the hypothesis that the dominant background is muon bremsstrahlung, which should produce equal numbers of positive and negative candidates, and the observation of 4 positively charged candidates in the signal sample.

As a final check on the candidate events, several distributions related to shower profiles were compared to expectations based on a Monte Carlo simulation of projective electrons (produced at the nominal IP) with transverse energy of 5 GeV, in the region $|\eta| < 0.8$ which is appropriate for comparison with the cosmic-ray electron sample obtained with this selection. These comparisons are shown in Fig. 11. The upper left plot shows the lateral containment, in the ϕ direction, of energy in the cells of the second layer of the EM calorimeter, as defined by the ratio $E_{3\times3}/E_{3\times7}$ where $E_{i\times j}$ represents the energy deposited in a collection of cells

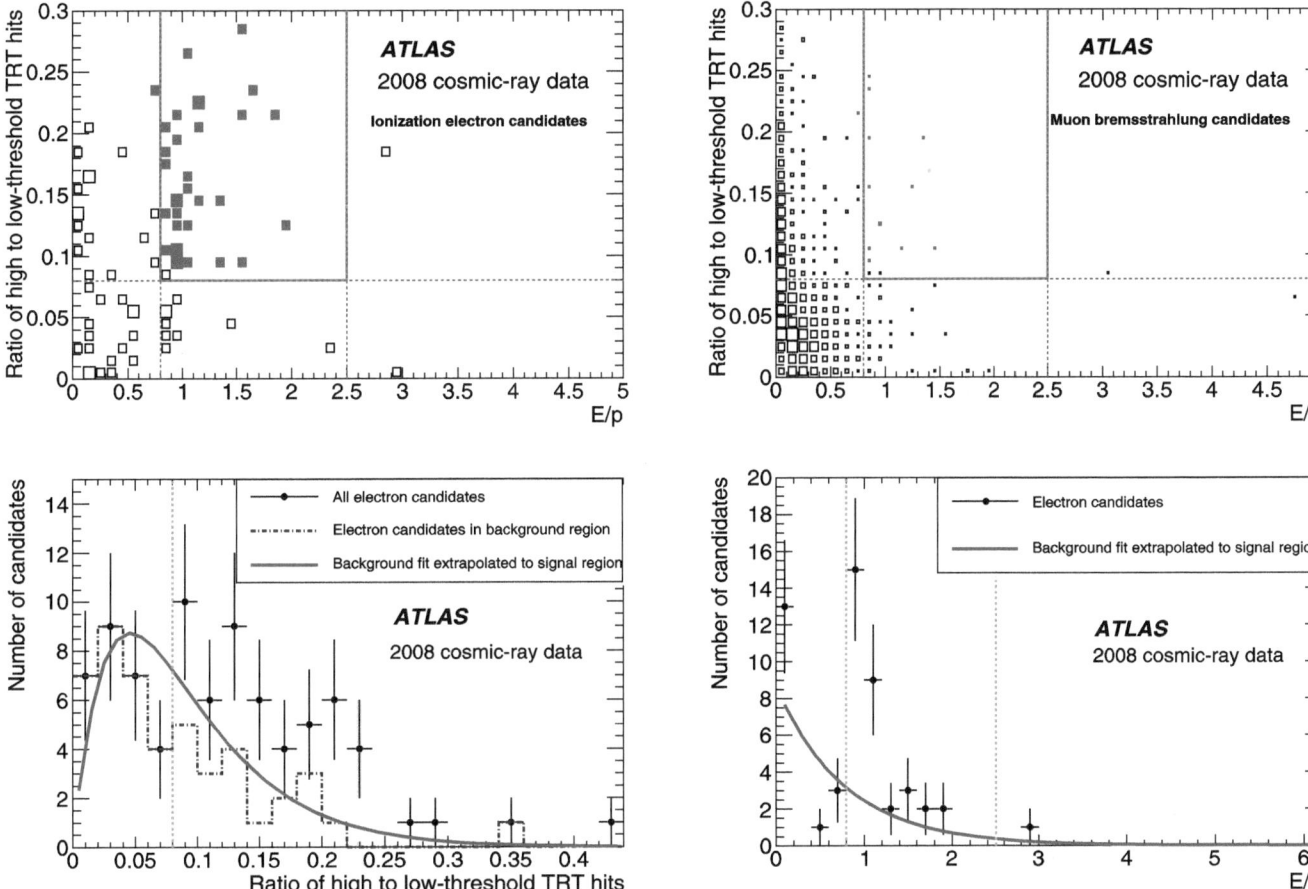

Fig. 10 The *upper plots* show the two-dimensional distributions of the TR ratio vs. E/p for the ionization sample (*left*) and the background sample (*right*). The *open* and *solid markers* together show the distribution of electron candidates passing the modified medium cuts. The *solid markers* indicate the candidates which also survive the tight selection. The *dotted lines* show the cuts applied to most of the events having $\eta \approx 0$ and low transverse energy: $0.8 < E/p < 2.5$ and TR ratio > 0.08. The *solid lines* indicate the corresponding signal region. Two outliers at high TR ratio (1 in signal, 1 in background region), and two outliers at high E/p are not shown. The *lower plots* show projections of the fit result for the ionization sample. The *left plot* shows the distribution of the TR ratio for all 81 electron and positron candidates after the modified medium cuts (*points with error bars*). The *dashed histogram* shows the 47 events in the background region and the *curve* shows the projection of the two-dimensional binned maximum likelihood fit. The *dotted vertical line* indicates the lower selection cut applied to the bulk of the events. The *right plot* shows the distribution of E/p for all modified medium electron candidates after the additional application of the tight-selection cut on the TR ratio. The *curve* shows the projection of the two-dimensional background fit from which the number of background events under the signal region is estimated. The *dotted vertical lines* represent the upper and lower selection cuts on E/p, applied to the bulk of the data

of size $i \times j$ in $\eta \times \phi$. A large mean value is observed for both data and Monte Carlo, as expected since electrons tend to have a small lateral shower width. The upper right plot shows the lateral extent of the shower in η, in the first layer of the EM calorimeter, as measured by the sum of the cell-cluster η separations, weighted by the cell energy. This also shows good agreement between data and Monte Carlo. The other quantities plotted are related to the longitudinal shower shape: the lower left plot shows the fraction of the total cluster energy deposited in the first layer of the EM calorimeter while the lower right plot shows the fraction of energy in the second layer. In both cases the average value should be about 40% for electrons, as these tend to start showering early in the calorimeter. There is reasonable agreement between

data and Monte Carlo, but both show some small discrepancies. These arise from the fact that several of the data events have much larger energies than were used for the Monte Carlo sample, which consists entirely of electrons with a transverse energy of 5 GeV. The deviations are consistent with what would be expected from the bremsstrahlung background in the sample. Those events can be of higher energy than the electron events, affecting the energy distributions of the showers, particularly the longitudinal energy profiles. Distributions of the fractions of energy deposited in the presampler and in layer 3 of the EM calorimeter show a similar level of agreement with the distributions from the projective-electron Monte Carlo sample.

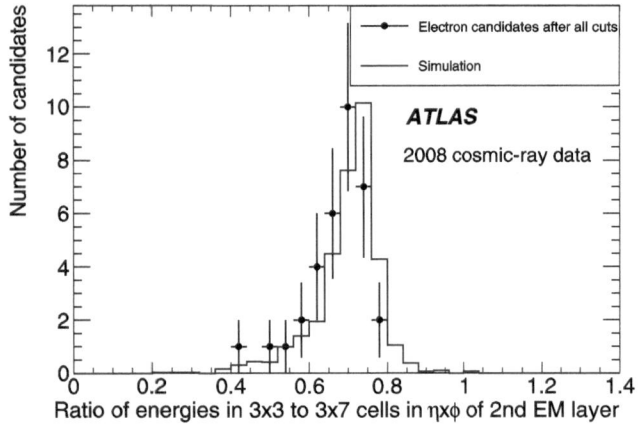

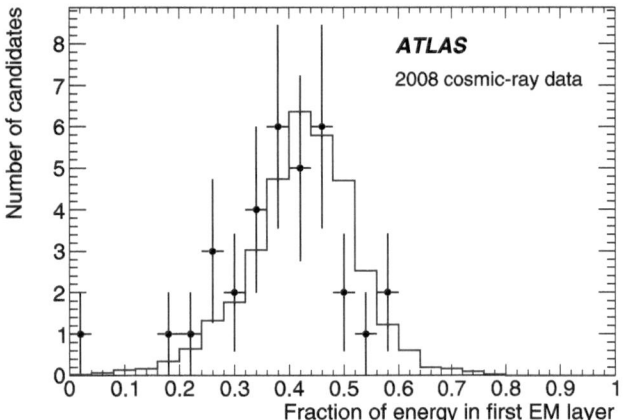

Fig. 11 Comparison of shower profiles for all 34 e$^\pm$ candidates to those from simulated projective electrons with a transverse energy of 5 GeV and $|\eta| < 0.8$. The *data points* indicate the electrons from the cosmic-ray data, while the *histograms* indicate distributions obtained from the simulated electrons. The *upper left plot* shows the ratio of energies in 3×3 over 3×7 cells in $\eta \times \phi$ in the second layer of the EM calorimeter. The *upper right plot* shows the energy-weighted shower width in η, in the first layer of the EM calorimeter. The *lower left (right)* plot shows the distribution of the fraction of energy in the first (second) layer of the EM calorimeter. The Monte Carlo distributions are normalized to the number of data events

4.4 Identification of low momentum electrons

The majority of the electrons in the cosmic-ray data are expected to be of low energy, of the order of a few hundred MeV. The probability of producing an electron with sufficiently high momentum to produce a standard e/γ cluster in the EM calorimeter is rather small, as reflected in the relatively low statistics available using the selection described in the previous section.

In addition to the sliding-window cluster used for the standard electron identification, ATLAS employs a topological clustering algorithm [25] which groups adjacent cells with energies above certain thresholds into clusters which are thus composed of varying number of cells, providing for better noise suppression. Each topological cluster is seeded by a cell having a signal-to-noise ratio ($|E|/\sigma_{noise}$) above a threshold t_{seed}, and is then expanded by iteratively adding neighboring cells having $|E|/\sigma_{noise} > t_{neighbor}$. Following the iterative step, the cluster is completed by adding all

direct neighbor cells along the perimeter having signal-to-noise above $|E|/\sigma_{noise} > t_{cell}$. Several types of topological clusters (differing in t_{seed}, $t_{neighbor}$ and t_{cell}) are used by ATLAS, for the reconstruction of calorimeter energy deposits from hadrons, electrons and photons, over the full range of η.

A selection based on the matching of an ID track to an EM topological cluster was applied to the cosmic-ray data. This analysis, run on data from both the 2008 and 2009 cosmic-ray data-taking periods, is similar to the one described in the previous section, also focusing on events in the barrel part of the detector. The topological signature of the electron events is the same as described in Sect. 4.3 and the data sample is separated into signal and background samples in a similar way, based on the number of tracks; electrons are again searched for in events with at least 2 ID tracks, while events with only one reconstructed track are used as a background sample. Candidate tracks must match an EM cluster from the topological clustering algorithm with $t_{seed} = 4$, $t_{neighbor} = 3$ and $t_{cell} = 0$. This allows the reconstruction of electromagnetic clusters with energies down to about 500 MeV.

Electron candidate tracks are required to be in the barrel region of the TRT and to have at least 25 TRT hits to ensure good quality tracks. There is no requirement of silicon hits. The TR ratio is required to exceed 0.1. Further suppression of backgrounds is achieved using various moments of the calorimeter cluster designed to select the compact clusters typical of electromagnetic objects. For example, Fig. 12 shows data and Monte Carlo distributions for the topological cluster moment λ_{center}, defined as the distance from the calorimeter front face to the shower center, along the shower axis. The two plots show distributions for signal and background events accepted by the low-p_T electron selection, before (left) and after (right) application of the cluster-moment-based selection criteria. The left-hand plot shows the distribution obtained with the signal selection applied to the cosmic-ray data and Monte Carlo along with the expected distribution for true electrons from the Monte Carlo. The MC distribution has been normalized to the data. The cut of $\lambda_{center} < 220$ mm is indicated by the dotted vertical line. Muons which traverse the calorimeter as MIPs leave their energy uniformly distributed in depth, producing a peak in the distribution at the point which corresponds to half the depth of the EM calorimeter. The right-hand plot shows the selected region after all cuts, for the signal events, the events from the background sample, and for those events from the Monte Carlo which are matched to real ("Monte Carlo truth") electrons. Good agreement is observed between data and Monte Carlo.

As in the electron analysis described in the previous section, signal and background regions are defined in the plane of the TR ratio vs. E/p. A fit is performed to the data in the background region of the background sample and then used to estimate the background in the signal sample. Selected events from both samples are shown in Fig. 13 for data and Monte Carlo. The Monte Carlo plots also include the distributions of electron candidates that are matched to Monte Carlo truth electrons, corresponding to 97% of the candidates selected from that sample. The upper plots show the E/p distributions for the selected events. The final selection cut of $E/p > 0.5$ is illustrated by the dashed line. This lower E/p cut, relative to the analysis described in Sect. 4.3, is needed as the lower p_T electrons suffer relatively more energy loss in the detector material before reaching the calorimeter. The lower plots show the momentum distributions of the electron candidates passing the full selection, and show acceptance down to ~500 MeV.

In general, ATLAS does not attempt to identify electrons down to such low energies. This commissioning analysis is intended to illustrate the flexibility that exists for the identification of electrons. While the topological clustering technique discussed here is not part of the standard electron identification algorithm for most of the detector acceptance, it is the default technique in the forward region ($2.5 < |\eta| < 4.9$). This region is beyond the tracking acceptance, so in that case no matching is done to tracks. Instead, electrons are identified by topological clusters having properties (e.g. cluster moments) that are typical of electromagnetic energy deposits.

4.5 Commissioning of the τ reconstruction and identification algorithms

As discussed earlier, the cosmic-ray data have also been used to examine the tools used for the identification of τ leptons. A leptonically decaying τ, where the visible final state is either an electron or muon, is difficult to distinguish from a primary electron or muon. The τ identification algorithm therefore focuses on hadronically decaying τ leptons, for which the dominant final states consists of either one or three charged hadrons and some number of neutrals. Reconstruction of these final states typically involves several subdetectors: one expects ID tracks associated with the charged hadrons and energy deposits in the calorimeter, from both charged and neutral hadrons. The neutrals are dominantly pions which decay to two photons and leave their energy in the EM calorimeter. Hadronically-decaying τ leptons are often referred to as τ-jets.

The identification of τ leptons is primarily concerned with distinguishing these from a large background due to QCD jets. The identification algorithm relies upon features such as the track multiplicity, which should be low for τ leptons, and the transverse profile of the energy deposits in the detector, which is typically narrower for τ-jets than for those from QCD. A τ will almost always have a final state with either one or three tracks, though some allowance is made for imperfect track reconstruction in the ID. Finally, the τ final state will often result in a prominent deposit in the electromagnetic calorimeter, associated with photons produced by the decays of neutral pions.

The identification of τ leptons is performed by an algorithm that can be seeded either by a track from the ID or by an energetic jet in the calorimeter. Track-based τ candidates are seeded by one good quality track having $p_T > 6$ GeV and can incorporate up to seven additional tracks with $p_T > 1$ GeV within $\Delta R < 0.2$ of the seed track. Once the full set of tracks for a τ candidate is established, an associated calorimeter cluster is searched for within $\Delta R < 0.2$ of the p_T-weighted track barycenter. The existence of an associated cluster is not required. Calorimeter-based candidates are seeded by jets reconstructed from calibrated topological clusters [25] with $\Delta R < 0.4$ and $E_T > 10$ GeV. Once a seed jet is established the algorithm searches for associated ID tracks having $p_T > 1$ GeV, within a cone of radius

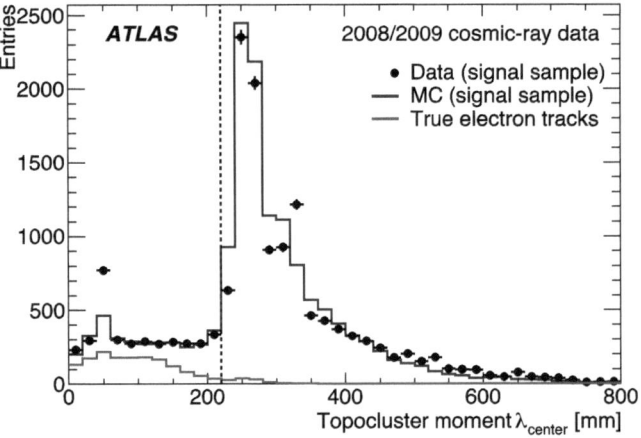

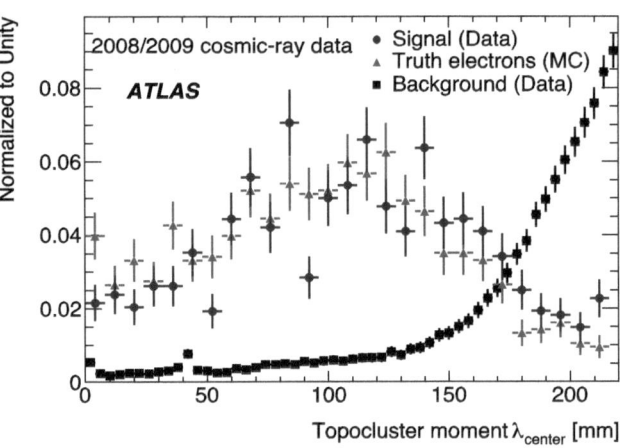

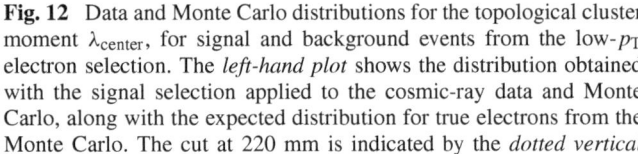

Fig. 12 Data and Monte Carlo distributions for the topological cluster moment λ_{center}, for signal and background events from the low-p_T electron selection. The *left-hand plot* shows the distribution obtained with the signal selection applied to the cosmic-ray data and Monte Carlo, along with the expected distribution for true electrons from the Monte Carlo. The cut at 220 mm is indicated by the *dotted vertical line*. For this plot, none of the cluster shape cuts have been applied. The *right hand plot* shows the selected region after all cuts, for the signal events, the events from the background sample, and for the truth electron distribution from Monte Carlo. The distributions are normalized to unity

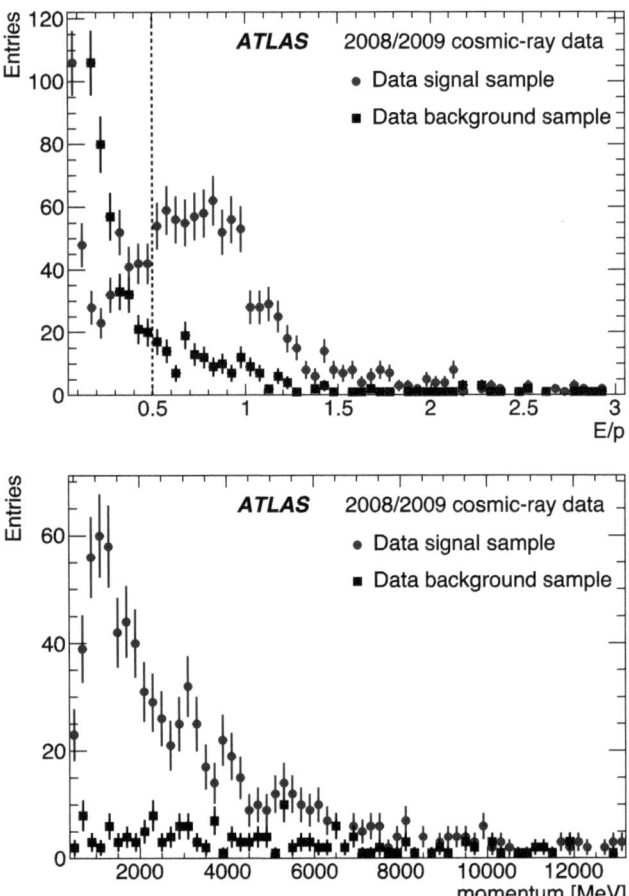

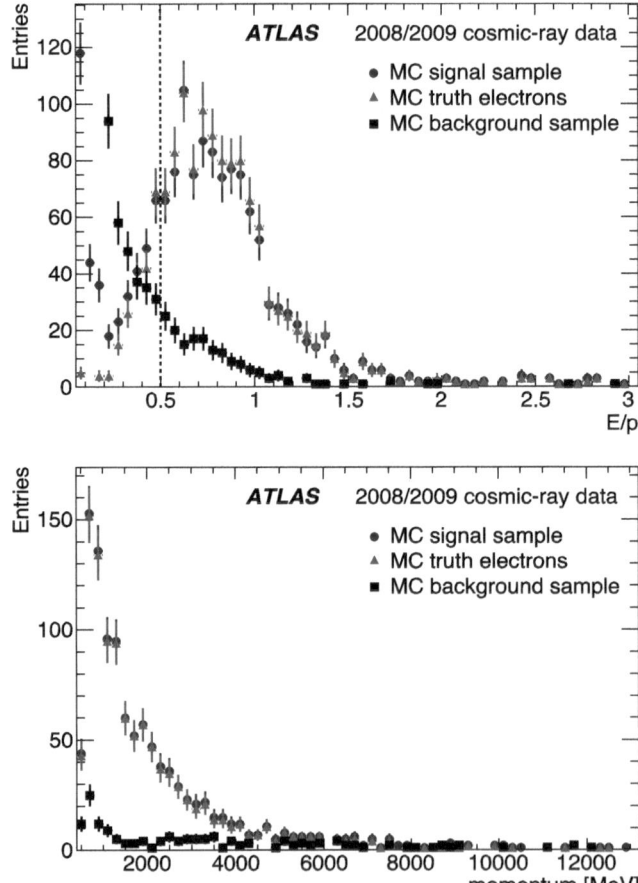

Fig. 13 Results of the selection of low p_T electrons from the cosmic-ray data samples. The *upper plots* show the E/p distributions for selected events in data (*left*) and Monte Carlo (*right*), for both the signal (ionization) and background (muon-bremsstrahlung) samples, and for the signal candidates matched to true electrons in the case of the Monte Carlo. The *lower plots* show the corresponding momentum distributions, for events passing the E/p cut, illustrated by the *dashed lines in the upper plots*

$\Delta R < 0.3$. The existence of such accompanying tracks is not required.

Since no τ leptons are expected in the cosmic-ray data sample, the focus of the study described here was simply to exercise the algorithms designed to identify them, and to investigate how well the quantities used for the selection are modeled in the simulation. Since τ leptons produced in proton-proton collisions originate from the interaction point, these algorithms normally impose tight requirements on the d_0 and z_0 parameters of the τ tracks. However, since application of too tight a selection on these quantities (here with respect to the nominal IP) severely limits the available statistics, in this study acceptance cuts of $|d_0| \leq 40$ mm and $|z_0| \leq 200$ mm were used. These define a region which is well within the sensitive volume of the barrel part of the Pixel Detector, which extends to $r = 123$ mm and $z = \pm 400$ mm.

The analysis described here uses the cosmic-ray data from the fall 2008 run. The PCM dataset was used as the starting point for each study. Additional requirements were placed on the presence of pixel hits, differently for the two seeding methods. The track-based selection required that the seed track have at least one pixel hit. For studies of the calorimeter-seeded algorithm, while there was no explicit requirement on the association of a track to the seed jet, there was a requirement that there be at least one ID track in the event with at least one pixel hit. This track would normally be from the muon responsible for the calorimeter cluster around which the seed jet is formed. However, in cosmic-ray events these tracks are often not associated with the cluster. The pixel hit requirement is thus intended to ensure that the shower shapes (which are used by the identification algorithm) are approximately as expected for particles originating from the IP.

The τ-identification algorithm is designed to reconstruct τ leptons over a wide spectrum of energies. However, the relative performance of the two seeding methods varies as a function of energy with the track-seeding having better performance at lower energies while for higher energies, the calorimeter-seeding is superior. Because of this, the type of cosmic-ray event producing fake τ candidates differs for the two seed types. Most fake track-seeded candidates come from minimum-ionizing muons with low momentum, which produce an ID track that fakes a one-prong candidate. The dominant source of calorimeter-seeded fakes is cosmic-ray muons that undergo hard bremsstrahlung in the calorimeter. When considering real τ leptons reconstructed from collision data, ideally one would like to have candidates seeded simultaneously by the track and cluster-based algorithms. In cosmic-ray data, however, since the origin of fake τ leptons differs for each algorithm, very few candidates fulfil the criteria for both. For this reason, track-seeded and calorimeter-seeded τ candidates have been examined separately.

Results are presented here to illustrate the agreement between data and cosmic-ray Monte Carlo for the properties of the two types of τ candidates, in particular for those quantities used in the identification algorithms. In what follows it should be understood that "τ candidate" refers to a fake candidate that passes the selection described above, in which nominal selection criteria have been loosened to ensure sufficient statistics to allow for a meaningful comparison of the data and the cosmic-ray Monte Carlo simulation. For track-seeded candidates, Fig. 14 shows the E_T distribution and the distribution of the invariant mass of the charged and neutral constituents. These quantities are both obtained via an energy-flow algorithm [24] in which the energies of calorimeter clusters associated with tracks have been replaced by the corresponding track momenta as measured in the ID. Good agreement is seen between the cosmic-ray data and the simulation.

Figure 15 shows data versus Monte Carlo comparisons for some of the quantities used for the identification of calorimeter-seeded τ candidates. The upper left plot shows the E_T distribution. The upper right plot shows the isolation fraction, which is a measure of the collimation of the τ-jet, defined as a ratio in which the denominator is the energy deposited within a cone (around the τ direction) of $\Delta R < 0.4$ and the numerator is the energy deposited in the region $0.1 < \Delta R < 0.2$. The lower left plot shows the centrality fraction, defined as the ratio of the energy within a cone of $\Delta R < 0.1$ to that within a cone of $\Delta R < 0.4$. The lower right plot shows the distribution of the hadronic radius, which is the energy-weighted width of the cluster, calculated from the energy and positions of the constituent calorimeter cells, relative to the cluster center. All distributions show good agreement between the data and the simulation. In the upper right plot of Fig. 15 there are entries at negative values that are attributable to the noise. This is also the cause of the entries at values greater than 1 in the plot of the centrality fraction. The agreement between data and simulation in these regions illustrates that the modeling of the electronic noise in the simulation is reasonable.

5 Jet and missing transverse energy studies using cosmic-ray events

Numerous theories of physics beyond the Standard Model predict the existence of massive weakly interacting particles that escape detection and thus leave a large energy imbalance in the detector. For this reason, detailed understanding of the detector performance for missing transverse energy (E_T^{miss}) is extremely important. The most important input to the calculation of the E_T^{miss} comes from the calorimeter, which provides coverage in the region of $|\eta| < 4.9$. Cosmic-ray energy deposits in the calorimeter typically lead to an

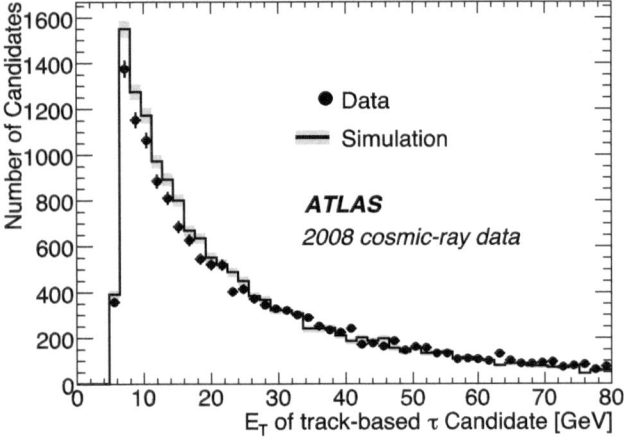
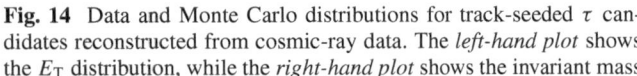
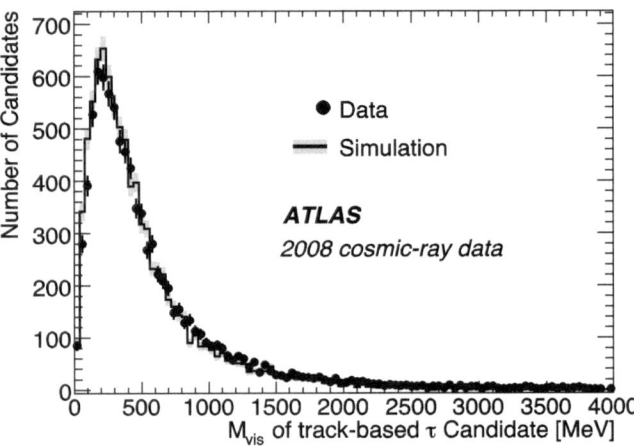

Fig. 14 Data and Monte Carlo distributions for track-seeded τ candidates reconstructed from cosmic-ray data. The *left-hand plot* shows the E_T distribution, while the *right-hand plot* shows the invariant mass of the charged and neutral constituents. In each case, an energy-flow algorithm was employed, as described in the text

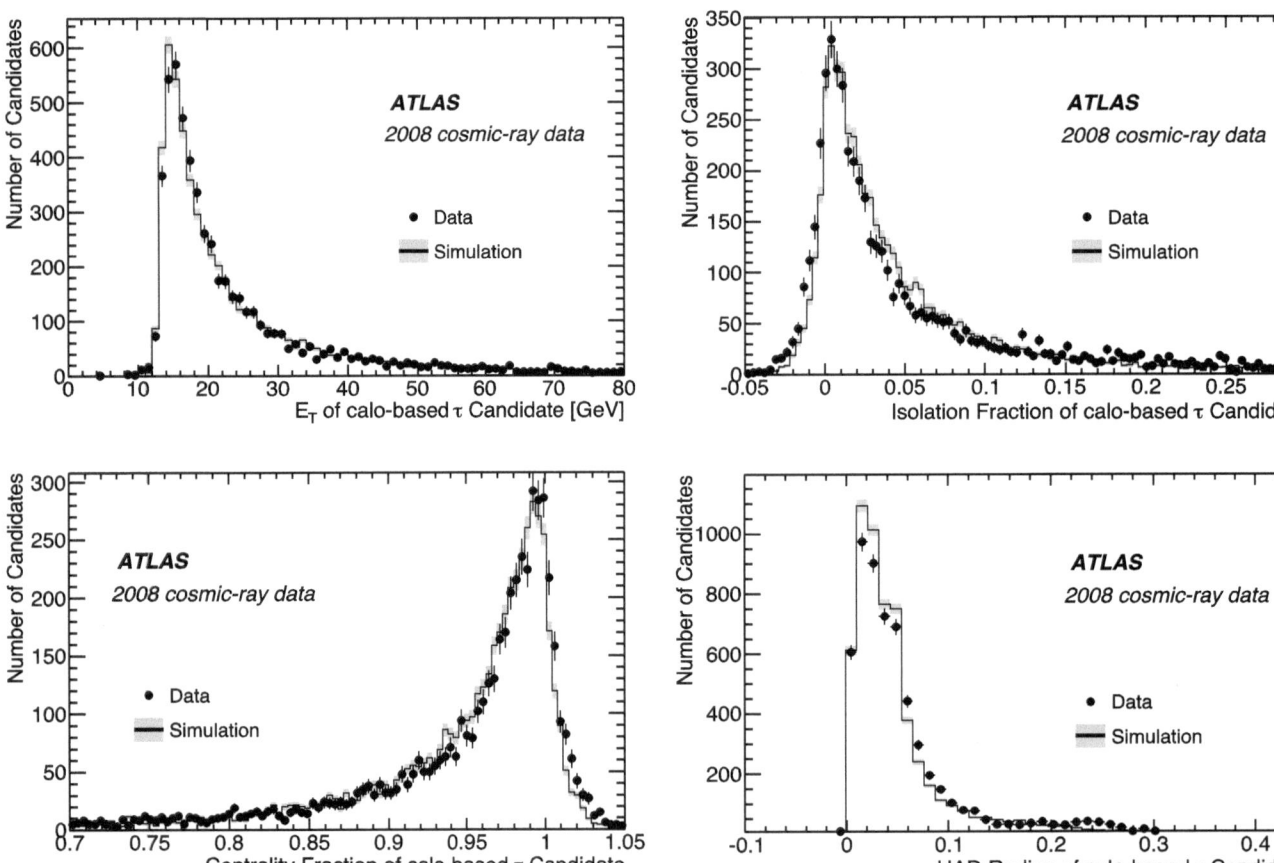

Fig. 15 Data and Monte Carlo distributions for calorimeter-seeded τ candidates reconstructed from cosmic-ray data. The *upper left plot* shows the E_T distribution of all candidates, while the other three plots show distributions of quantities used by the τ-identification algorithm. The *upper right plot* shows the isolation fraction, defined as the ratio in which the denominator is the energy deposited within a cone (around the τ direction) of $\Delta R < 0.4$ and the numerator is the energy deposited in the region $0.1 < \Delta R < 0.2$. The *lower left plot* shows the centrality fraction, defined as the ratio of the energy within a cone of $\Delta R < 0.1$ to that within a cone of $\Delta R < 0.4$. The *lower right plot* shows the distribution of the hadronic radius, the energy-weighted width of the cluster, calculated from the energy and positions of the constituent calorimeter cells, relative to the cluster center

imbalance in the transverse energy in the event. This effect can be large in the case of high-energy cosmic rays that lose a large amount of energy via bremsstrahlung. The energy deposits from cosmic-ray muons (or cosmic-induced air-shower events) can be reconstructed as jets, creating back-grounds to jet selections in many analyses. The properties of jets and E_T^{miss} reconstructed from cosmic-ray data are presented below, along with a discussion of techniques that have been developed to suppress such contributions in the analysis of collision data.

5.1 Missing transverse energy in randomly-triggered events

As is the case when running with proton-proton collisions, during cosmic-ray data-taking randomly triggered events are also recorded. The large sample of such events collected during the global cosmic-ray running allows investigations of the detector performance for the measurement of missing transverse energy. No energy imbalance is expected in these events. However, global quantities such as E_T^{miss} and $\sum E_T$ (defined below) result from the sum of energy deposits in ~200k calorimeter channels, each with its own electronic noise. A proper determination of these quantities relies on a good understanding of the cell-level noise in all calorime-ter channels, and, in particular, a proper treatment of a few very noisy cells and cells having non-nominal high-voltage. There are currently two standard methods for reconstruct-ing missing transverse energy in ATLAS. The first is a cell-level method that takes as input all calorimeter cells with $|E| > 2\sigma_{noise}$. The second method takes as input calibrated topological clusters built with $t_{seed} = 4$, $t_{neighbor} = 2$ and $t_{cell} = 0$. The reconstructed quantities are:

$$E_x^{miss} = -\sum E \sin\theta \cos\phi,$$

$$E_y^{miss} = -\sum E \sin\theta \sin\phi,$$

$$E_T^{miss} = \left((E_x^{miss})^2 + (E_y^{miss})^2\right)^{1/2},$$

$$\sum E_T = \sum E \sin\theta,$$

where in each case the sum is over all cells included in the cluster.

Figure 16 shows the results of both calculations applied to the random triggers recorded during a 2008 cosmic-ray run, illustrating the superior noise suppression of the method using the topological clustering. Tails in the distribution (be-yond 8 GeV for topological-cluster-based, and 16 GeV for cell-based definition), contributing less than 0.1% of events, are due to coherent noise in a specific region of the LAr pre-sampler. This problem was repaired prior to first collisions, as can be seen in [26]. The time stability of the E_T^{miss} cal-culation was also investigated and found to be very good.

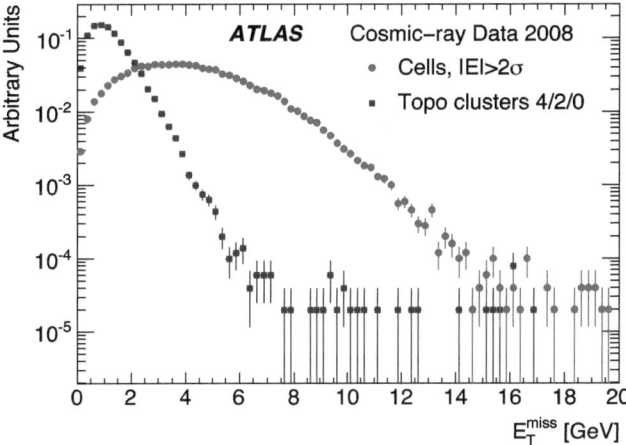

Fig. 16 Distribution of E_T^{miss} from analysis of random triggers recorded during the 2008 global cosmic-ray running, for the two meth-ods described in the text

For the topological-cluster-based method, which provides the best resolution, the mean and width of the distributions of the x and y components of E_T^{miss} were stable to within about 100 MeV over the 45 days of data-taking.

5.2 Jets and missing transverse energy in cosmic-ray events

The reconstruction of jets and E_T^{miss} in cosmic-ray events has been studied using the L1Calo-triggered data taken in the 2008 and 2009 cosmic-ray runs. For jet reconstruction an anti-k_t algorithm [27] is employed, with calibrated topo-logical clusters as input. Figure 17 shows the distributions of missing transverse energy and summed transverse en-ergy from cosmic-ray events having a reconstructed jet with $p_T > 20$ GeV. The 2008 and 2009 data samples are shown separately to demonstrate the consistency of the two sam-ples. The distributions from the 2008 data and the cosmic-ray Monte Carlo are normalized to that of the 2009 data in the region of $100 < E_T^{miss} < 300$ GeV. This is in order to avoid any threshold effects, since the trigger was not sim-ulated in the cosmic-ray Monte Carlo sample. In each case there is agreement with the shape expected from the Monte Carlo, which requires an understanding of the electronic noise in each calorimeter channel. The upper left plot in Fig. 18 shows the corresponding p_T distribution of the jets reconstructed in this sample.

Suppression of these fake jet candidates can be per-formed using a selection based on three quantities:

$$R_J = \sum_{i=1}^{N} \sqrt{(\eta_i - \eta_{jet})^2 + (\phi_i - \phi_{jet})^2} \cdot E_i \Big/ \sum_{i=1}^{N} E_i,$$

$$R_{LC} = \left(\sum_{i=1}^{2} E_i^{Had} + \sum_{i=1}^{32} E_i^{EM}\right) \Big/ \sum_{i=1}^{N} E_i,$$

$$f_{EM} = E_{EM}/(E_{EM} + E_{Had}).$$

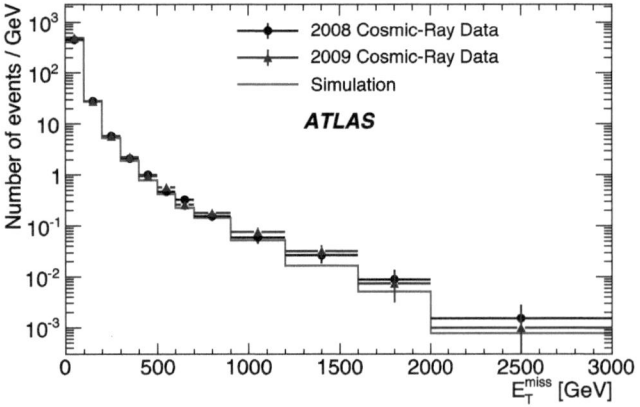 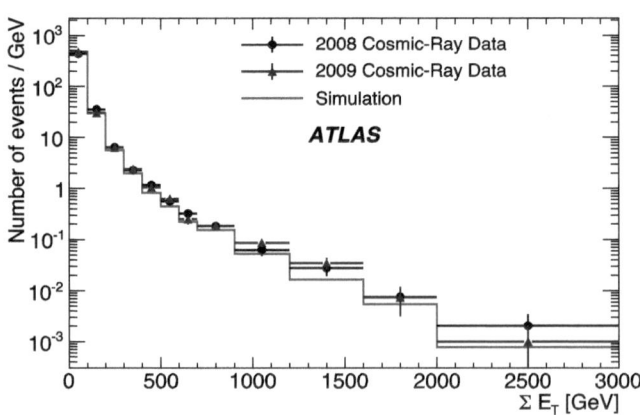

Fig. 17 Distributions of $E_{\mathrm{T}}^{\mathrm{miss}}$ (*left*) and $\sum E_{\mathrm{T}}$ (*right*) from analysis of the 2008 and 2009 L1Calo-triggered cosmic-ray data and from the cosmic-ray Monte Carlo sample. The 2008 and Monte Carlo distributions are normalized to the 2009 data distribution in the region $100 < E_{\mathrm{T}}^{\mathrm{miss}} < 300\,\mathrm{GeV}$

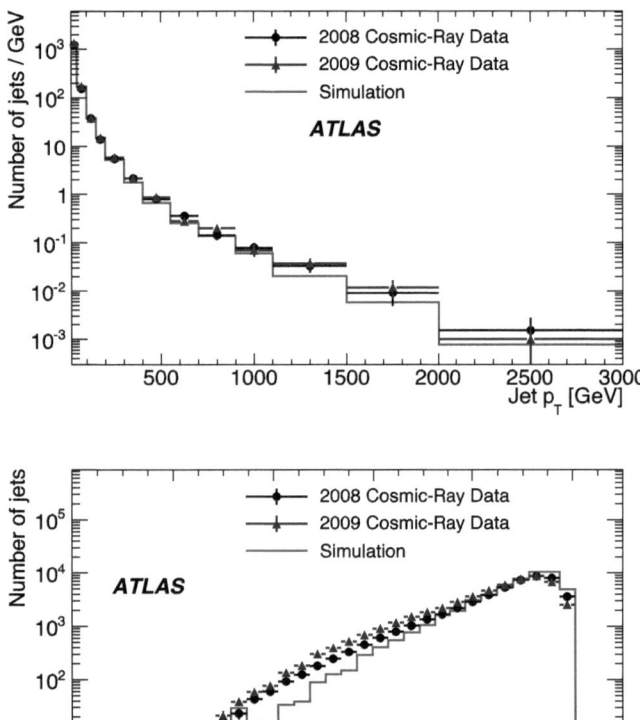 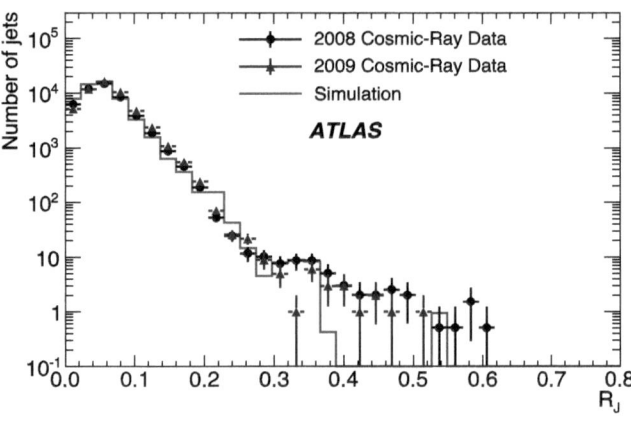

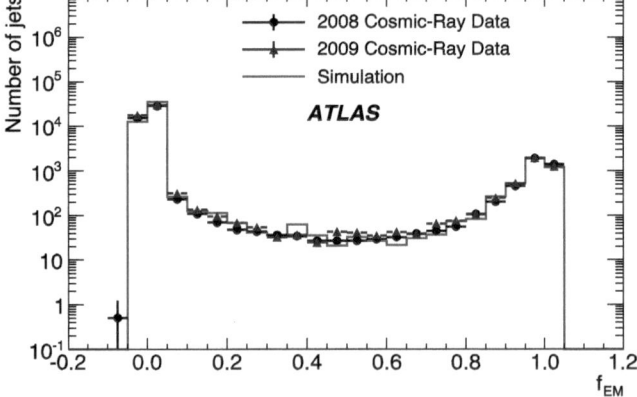

Fig. 18 Properties of fake jets reconstructed from the 2008 and 2009 L1Calo-triggered cosmic-ray data samples: The *upper left plot* shows the jet p_{T} in the acceptance region above 20 GeV, while the other three plots show distributions in quantities used to suppress these contributions in collision data, as described in the text. The normalizations are the same as used in Fig. 17

Here R_{J} represents the energy-weighted lateral extent of the jet, in $\eta \times \phi$ space. R_{LC} represents the fraction of the jet energy contained in the "leading cells", defined as the two most energetic cells in the hadronic calorimeter and the 32 most energetic cells in the EM calorimeter, where the sum in the denominator is over all N calorimeter cells associated with the jet candidate. Finally, f_{EM} represents the electromagnetic fraction of the jet, defined as the fraction of the jet energy that is deposited in the EM calorimeter. The distributions of these three quantities for the selected jets are also shown in Fig. 18. Again there is good agreement between the 2008 and 2009 cosmic-ray data as well as reasonable agreement with the cosmic-ray Monte Carlo. The normalization of the distributions in Fig. 18 is the same as used in Fig. 17.

When operating ATLAS for proton-proton collisions, contributions from cosmic-ray events can either trigger readout of the detector, or overlap with a triggered collision

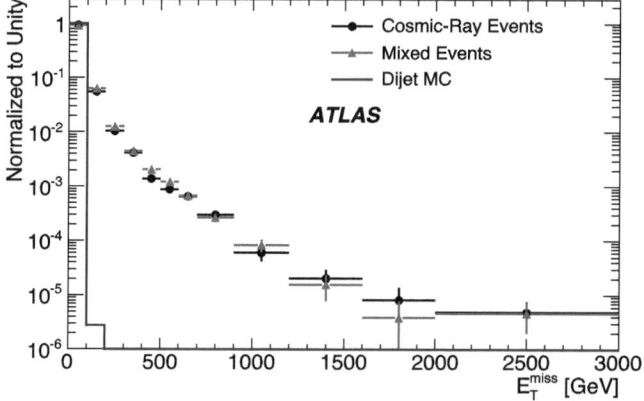
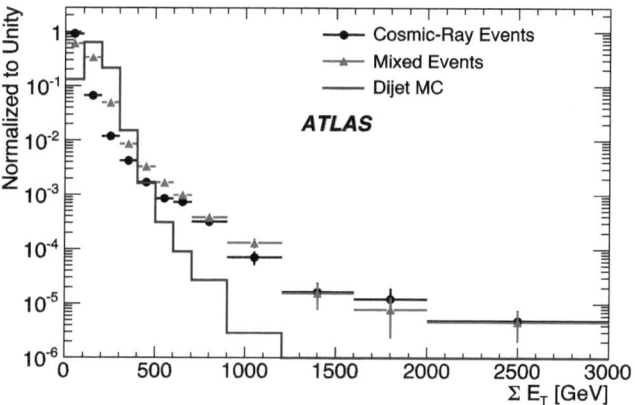

Fig. 19 The same distributions as presented in Fig. 17, obtained from the cosmic-ray data and from the mixed data sample described in the text. The plots are normalized to allow comparison of the shapes of the two distributions. Also shown are the corresponding distributions from dijet Monte Carlo events

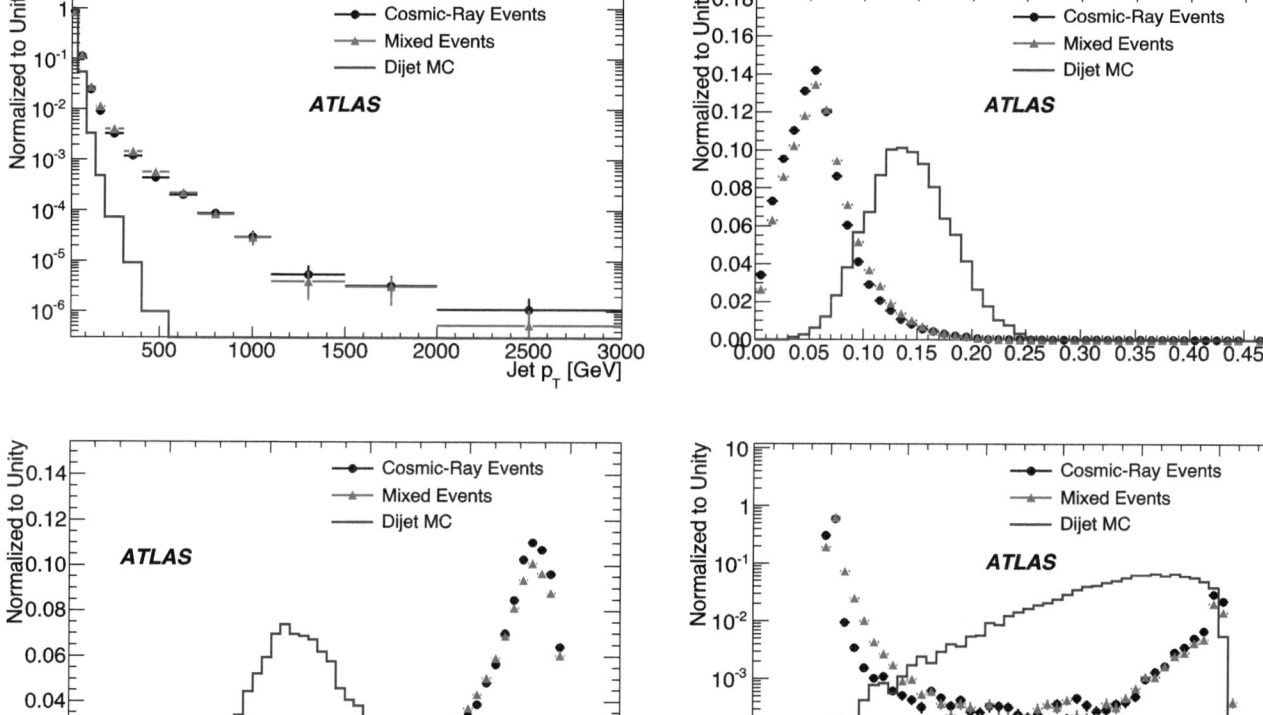

Fig. 20 The same distributions as presented in Fig. 18, obtained from the cosmic-ray data and from the mixed data sample. The plots are normalized to allow comparison of the shapes. Also shown are the corresponding distributions from dijet Monte Carlo events

event. Since cosmic-ray energy deposits in the latter category may be more difficult to identify, this scenario has been studied using a special data sample in which cosmic-ray events from the 2008 data were overlaid with Monte Carlo minimum-bias events. The overlay is done only with single minimum-bias events, so cannot account for events with pileup. However, in terms of faking a missing E_T signal, one might expect that the relative contribution, of a single cosmic-ray event, to a collision event would be highest in the case of overlap with a single collision. The effect of this additional energy on the E_T^{miss} and $\sum E_T$ distributions is illustrated in Fig. 19 which compares the distributions obtained from the mixed sample to those obtained from cosmic-ray data alone. The corresponding distributions obtained from a dijet Monte Carlo sample are also shown. In each case the distributions are obtained from all events having a jet with $p_T > 20$ GeV and $|\eta| < 2.5$. They are shown normalized to unity to allow better comparison of the shapes. The effect of the additional energy from the minimum-bias event is apparent in the $\sum E_T$ distribution, at low values.

The mixed data sample was used to investigate the robustness of the jet-discrimination variables in the case where a cosmic-ray event is overlaid with a minimum-bias event. The distributions shown in Fig. 20 are for the same quantities shown in Fig. 18, now normalized to unity. Each plot shows the distribution obtained from cosmic-ray data, from the mixed sample and from a sample of Monte Carlo dijet events. For the three variables introduced earlier, comparison of the distributions obtained from the two samples shows these variables to be robust against the presence of the additional energy due to the minimum-bias event. This was not the case for other discriminating variables (e.g., the number of clusters or tracks included in jets) that were also investigated. Rejection of fake jets from cosmic-ray events can be performed using a log-likelihood ratio (LLR) based on input probability distribution functions (pdfs) from the mixed sample and the Monte Carlo dijet sample. As investigations of the three discriminating variables showed a high degree of correlation between R_J and R_{LC}, a 2-dimensional pdf for these two variables was employed along with a one-dimensional pdf for f_{EM}.

Figure 21 illustrates the effects of different applications of "cleaning cuts" based on these pdfs. The upper plot shows the cumulative effect of successive applications of the two LLR cuts on the p_T distribution from the dijet sample and on the fake jet p_T distribution from cosmic-ray events. For the chosen cuts, the effect of each cut on the dijet sample is at the 2% level in each of the p_T bins. The middle plot compares the effect of the same cuts on the mixed and cosmic-ray data samples. The lower plot shows the rejection factor for events with jets produced by cosmic-ray interactions plotted against the efficiency for the selection of Monte Carlo dijet events, in the acceptance region previously defined, for three different scenarios:

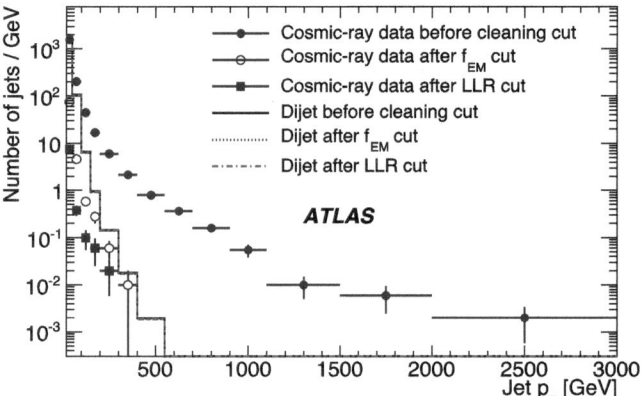

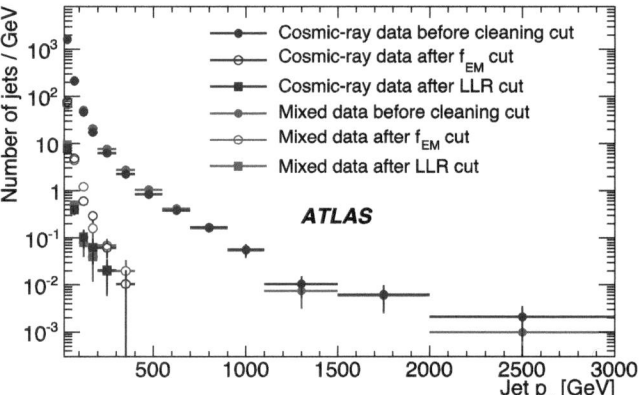

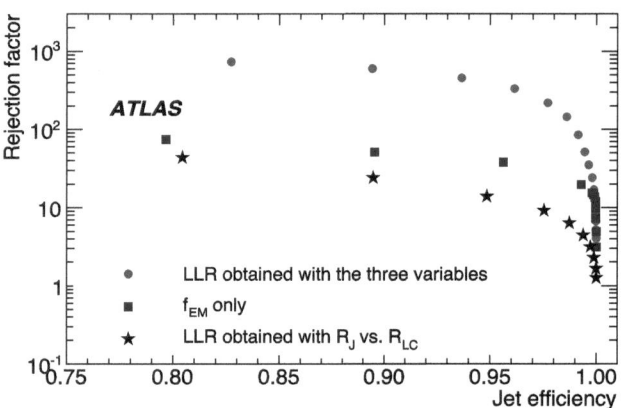

Fig. 21 Performance of the cleaning cuts for the suppression of fake jets from cosmic-ray events. The *upper plot* shows the effect of the cleaning cuts on the p_T distribution of accepted jets, for different cleaning cuts, for the cosmic-ray and dijet Monte Carlo samples. The *middle plot* shows similar distributions, this time comparing the mixed events to those obtained using only cosmic-ray data. The *lower plot* shows the achievable cosmic-ray rejection vs. the efficiency for dijet events

– application of an LLR cut based on f_{EM} only
– application of a LLR cut based only on R_J and R_{LC}
– application of the full, three-variable LLR.

The rejection factor is obtained from an analysis of the mixed sample while the efficiency is derived from application of the selection to the dijet Monte Carlo sample. An overall rejection factor of about 400 can be obtained with 95% efficiency for jets from the dijet Monte Carlo sample. For cosmic-ray events without overlaid minimum-bias energy, the situation is somewhat better, with a rejection factor (again for 95% efficiency for jets in dijet events) of around 550.

6 Summary

Cosmic-ray interactions provide a source of physics signals in the ATLAS detector that have allowed for investigations of the detector alignment, calibration and performance prior to the arrival of first LHC beams. Such events have been used to exercise the detector readout and associated data-handling infrastructure, and the accumulated datasets have been exploited for both standalone and combined performance studies of the detector subsystems. Cosmic-ray data will continue to be relevant to the commissioning of the muon spectrometer until a sufficient number of high-p_T muons have been accumulated from proton-proton collisions. In this paper, results relevant to lepton identification and reconstruction as well as the measurement of missing transverse energy were presented, along with studies related to the rejection of background from cosmic-ray events in collision data. These results, along with those presented in the publications describing results from subsystem-specific cosmic-ray commissioning, demonstrate that ATLAS was prepared for the first collisions from the LHC. Measured distributions obtained from analysis of the cosmic-ray data agree well with the predictions of the detector simulation and a dedicated cosmic-muon event generator, demonstrating that the modeling of the detector response was also in good shape prior to first collisions.

Acknowledgements We are greatly indebted to all CERN's departments and to the LHC project for their immense efforts not only in building the LHC, but also for their direct contributions to the construction and installation of the ATLAS detector and its infrastructure. We acknowledge equally warmly all our technical colleagues in the collaborating Institutions without whom the ATLAS detector could not have been built. Furthermore we are grateful to all the funding agencies which supported generously the construction and the commissioning of the ATLAS detector and also provided the computing infrastructure.

The ATLAS detector design and construction has taken about fifteen years, and our thoughts are with all our colleagues who sadly could not see its final realisation.

We acknowledge the support of ANPCyT, Argentina; YerPhI, Armenia; ARC, Australia; BMWF, Austria; ANAS, Azerbaijan; SSTC, Belarus; CNPq and FAPESP, Brazil; NSERC, NRC and CFI, Canada; CERN; CONICYT, Chile; CAS, MOST and NSFC, China; COLCIENCIAS, Colombia; MEYS (MSMT), MPO and CCRC, Czech Republic; DNRF, DNSRC and Lundbeck Foundation, Denmark; ARTEMIS, European Union; IN2P3-CNRS, CEA-DSM/IRFU, France; GNAS, Georgia; BMBF, DFG, HGF, MPG and AvH Foundation, Germany; GSRT, Greece; ISF, MINERVA, GIF, DIP and Benoziyo Center, Israel; INFN, Italy; MEXT and JSPS, Japan; CNRST, Morocco; FOM and NWO, Netherlands; RCN, Norway; MNiSW, Poland; GRICES and FCT, Portugal; MERYS (MECTS), Romania; MES of Russia and ROSATOM, Russian Federation; JINR; MSTD, Serbia; MSSR, Slovakia; ARRS and MVZT, Slovenia; DST/NRF, South Africa; MICINN, Spain; SRC and Wallenberg Foundation, Sweden; SER, SNSF and Cantons of Bern and Geneva, Switzerland; NSC, Taiwan; TAEK, Turkey; STFC, the Royal Society and Leverhulme Trust, United Kingdom; DOE and NSF, United States of America.

References

1. G. Aad et al., The ATLAS experiment at the CERN large hadron collider. J. Instrum. **3**, S08003 (2008)
2. L. Evans, P. Bryant, LHC machine. J. Instrum. **3**, S08001 (2008)
3. S. Agostinelli et al., GEANT4: A simulation toolkit. Nucl. Instrum. Methods A **506**, 250–303 (2003)
4. J. Allison et al., Geant4 developments and applications. IEEE Trans. Nucl. Sci. **53**, 270 (2006)
5. G. Aad et al., The ATLAS simulation infrastructure. Eur. Phys. J. C **70**, 823 (2010)
6. G. Aad et al., The ATLAS inner detector commissioning and calibration. Eur. Phys. J. C **70**, 787 (2010)
7. G. Aad et al., Readiness of the ATLAS liquid argon calorimeter for LHC collision. Eur. Phys. J. C **70**, 723 (2010)
8. G. Aad et al., Readiness of the ATLAS tile calorimeter for LHC collisions. Eur. Phys. J. C **70**, 1193 (2010)
9. G. Aad et al., Commissioning of the ATLAS muon spectrometer with cosmic ray muons. Eur. Phys. J. C **70**, 875 (2010)
10. S. Fratina et al., The ATLAS transition radiation detector (TRT) fast-OR trigger, ATLAS-INDET-PUB-2009-002, CERN, Geneva, 2009
11. G. Aad et al., Drift time measurement in the ATLAS liquid argon electromagnetic calorimeter using cosmic muons. Eur. Phys. J. C **70**, 755 (2010)
12. A. Dar, Atmospheric neutrinos, astrophysical neutrinos and proton decay experiments. Phys. Rev. Lett. **51**, 227 (1983)
13. C. Amsler et al., Review of particle physics. Phys. Lett. B **667**, 1 (2008)
14. M. Boonekamp et al., Cosmic ray, beam halo and beam gas rate studies for ATLAS commissioning, ATL-GEN-2004-001, CERN, Geneva, 2004
15. H. Okawa, Commissioning of the ATLAS calorimeters at the large hadron collider and prospects towards new physics search, Ph.D. thesis, University of Tokyo, Japan, 2010
16. D. Adams et al., Track reconstruction in the ATLAS muon spectrometer with MOORE, ATL-SOFT-2003-007, CERN, Geneva, Oct 2003
17. T. Cornelissen et al., Concepts, design and implementation of the ATLAS new tracking (NEWT), ATL-SOFT-PUB-2007-007, CERN, Geneva, 2007
18. T. Cornelissen et al., The new ATLAS track reconstruction (NEWT). J. Phys. Conf. Ser. **119**, 032013 (2008)
19. K. Bachas, S. Hassani, Track based software package for measurement of the energy deposited by muons in the calorimeters of the ATLAS detector. J. Phys. Conf. Ser. **119**, 042002 (2008)
20. B. Lenzi, R. Nicolaidou, S. Hassani, TrackInCaloTools: A package for measuring muon energy loss and calorimetric isolation in ATLAS. J. Phys. Conf. Ser. **119**, 032049 (2010)

21. B. Lenzi, Search for the Higgs boson decaying to four leptons in the ATLAS detector at LHC and studies of muon isolation and energy loss, Ph.D. thesis, Université Paris Sud Orsay, France, 2010

22. A. Salzburger, The ATLAS track extrapolation package, ATL-SOFT-PUB-2007-005, CERN, Geneva, Jun 2007

23. A. Salzburger, S. Todorova, M. Wolter, The ATLAS tracking geometry description, ATL-SOFT-PUB-2007-004, CERN, Geneva, Jun 2007

24. G. Aad et al., *Expected Performance of the ATLAS Experiment—Detector, Trigger and Physics*, CERN-OPEN 2008-020, Dec. 2008, ISBN 978-92-9083-321-5, arXiv:0901.0512 [hep-ex] (2009)

25. T. Barillari et al., Local hadronic calibration, ATL-LARG-PUB-2009-001-2 CERN, Geneva, 2010

26. G. Aad et al., Performance of the ATLAS detector using first collision data. J. High Energy Phys. **1009**, 056 (2010)

27. M. Cacciari, G.P. Salam, G. Soyez, The anti-k_t jet clustering algorithm. J. High Energy Phys. **04**, 063 (2008)

The ATLAS Collaboration

G. Aad[48], B. Abbott[111], J. Abdallah[11], A.A. Abdelalim[49], A. Abdesselam[118], O. Abdinov[10], B. Abi[112], M. Abolins[88], H. Abramowicz[153], H. Abreu[115], B.S. Acharya[164a,164b], D.L. Adams[24], T.N. Addy[56], J. Adelman[175], S. Adomeit[98], P. Adragna[75], T. Adye[129], S. Aefsky[22], J.A. Aguilar-Saavedra[124b,a], M. Aharrouche[81], S.P. Ahlen[21], F. Ahles[48], A. Ahmad[148], M. Ahsan[40], G. Aielli[133a,133b], T. Akdogan[18a], T.P.A. Åkesson[79], G. Akimoto[155], A.V. Akimov[94], A. Aktas[48], M.S. Alam[1], M.A. Alam[76], S. Albrand[55], M. Aleksa[29], I.N. Aleksandrov[65], C. Alexa[25a], G. Alexander[153], G. Alexandre[49], T. Alexopoulos[9], M. Alhroob[20], M. Aliev[15], G. Alimonti[89a], J. Alison[120], M. Aliyev[10], P.P. Allport[73], S.E. Allwood-Spiers[53], J. Almond[82], A. Aloisio[102a,102b], R. Alon[171], A. Alonso[79], M.G. Alviggi[102a,102b], K. Amako[66], C. Amelung[22], A. Amorim[124a,b], G. Amorós[167], N. Amram[153], C. Anastopoulos[139], T. Andeen[29], C.F. Anders[48], K.J. Anderson[30], A. Andreazza[89a,89b], V. Andrei[58a], X.S. Anduaga[70], A. Angerami[34], F. Anghinolfi[29], N. Anjos[124a], A. Annovi[47], A. Antonaki[8], M. Antonelli[47], S. Antonelli[19a,19b], J. Antos[144b], B. Antunovic[41], F. Anulli[132a], S. Aoun[83], G. Arabidze[8], I. Aracena[143], Y. Arai[66], A.T.H. Arce[44], J.P. Archambault[28], S. Arfaoui[29,c], J.-F. Arguin[14], T. Argyropoulos[9], M. Arik[18a], A.J. Armbruster[87], O. Arnaez[4], C. Arnault[115], A. Artamonov[95], D. Arutinov[20], M. Asai[143], S. Asai[155], J. Silva[124a,d], R. Asfandiyarov[172], S. Ask[82], B. Åsman[146a,146b], D. Asner[28], L. Asquith[77], K. Assamagan[24], A. Astvatsatourov[52], G. Atoian[175], B. Auerbach[175], K. Augsten[127], M. Aurousseau[4], N. Austin[73], G. Avolio[163], R. Avramidou[9], C. Ay[54], G. Azuelos[93,e], Y. Azuma[155], M.A. Baak[29], A.M. Bach[14], H. Bachacou[136], K. Bachas[29], M. Backes[49], E. Badescu[25a], P. Bagnaia[132a,132b], Y. Bai[32a], T. Bain[158], J.T. Baines[129], O.K. Baker[175], M.D. Baker[24], S. Baker[77], F. Baltasar Dos Santos Pedrosa[29], E. Banas[38], P. Banerjee[93], Sw. Banerjee[169], D. Banfi[89a,89b], A. Bangert[137], V. Bansal[169], S.P. Baranov[94], A. Barashkou[65], T. Barber[27], E.L. Barberio[86], D. Barberis[50a,50b], M. Barbero[20], D.Y. Bardin[65], T. Barillari[99], M. Barisonzi[174], T. Barklow[143], N. Barlow[27], B.M. Barnett[129], R.M. Barnett[14], A. Baroncelli[134a], A.J. Barr[118], F. Barreiro[80], J. Barreiro Guimarães da Costa[57], P. Barrillon[115], R. Bartoldus[143], D. Bartsch[20], R.L. Bates[53], L. Batkova[144a], J.R. Batley[27], A. Battaglia[16], M. Battistin[29], F. Bauer[136], H.S. Bawa[143], B. Beare[158], T. Beau[78], P.H. Beauchemin[118], R. Beccherle[50a], P. Bechtle[41], G.A. Beck[75], H.P. Beck[16], M. Beckingham[48], K.H. Becks[174], A.J. Beddall[18c], A. Beddall[18c], V.A. Bednyakov[65], C. Bee[83], M. Begel[24], S. Behar Harpaz[152], P.K. Behera[63], M. Beimforde[99], C. Belanger-Champagne[166], P.J. Bell[49], W.H. Bell[49], G. Bella[153], L. Bellagamba[19a], F. Bellina[29], M. Bellomo[119a], A. Belloni[57], K. Belotskiy[96], O. Beltramello[29], S. Ben Ami[152], O. Benary[153], D. Benchekroun[135a], M. Bendel[81], B.H. Benedict[163], N. Benekos[165], Y. Benhammou[153], D.P. Benjamin[44], M. Benoit[115], J.R. Bensinger[22], K. Benslama[130], S. Bentvelsen[105], M. Beretta[47], D. Berge[29], E. Bergeaas Kuutmann[41], N. Berger[4], F. Berghaus[169], E. Berglund[49], J. Beringer[14], P. Bernat[115], R. Bernhard[48], C. Bernius[77], T. Berry[76], A. Bertin[19a,19b], M.I. Besana[89a,89b], N. Besson[136], S. Bethke[99], R.M. Bianchi[48], M. Bianco[72a,72b], O. Biebel[98], J. Biesiada[14], M. Biglietti[132a,132b], H. Bilokon[47], M. Bindi[19a,19b], A. Bingul[18c], C. Bini[132a,132b], C. Biscarat[180], U. Bitenc[48], K.M. Black[57], R.E. Blair[5], J.-B. Blanchard[115], G. Blanchot[29], C. Blocker[22], A. Blondel[49], W. Blum[81], U. Blumenschein[54], G.J. Bobbink[105], A. Bocci[44], M. Boehler[41], J. Boek[174], N. Boelaert[79], S. Böser[77], J.A. Bogaerts[29], A. Bogouch[90,*], C. Bohm[146a], V. Boisvert[76], T. Bold[163,f], V. Boldea[25a], M. Bondioli[163], M. Boonekamp[136], S. Bordoni[78], C. Borer[16], A. Borisov[128], G. Borissov[71], I. Borjanovic[12a], S. Borroni[132a,132b], K. Bos[105], D. Boscherini[19a], M. Bosman[11], H. Boterenbrood[105], J. Bouchami[93], J. Boudreau[123], E.V. Bouhova-Thacker[71], C. Boulahouache[123], C. Bourdarios[115], A. Boveia[30], J. Boyd[29],

I.R. Boyko[65], I. Bozovic-Jelisavcic[12b], J. Bracinik[17], A. Braem[29], P. Branchini[134a], A. Brandt[7], G. Brandt[41], O. Brandt[54], U. Bratzler[156], B. Brau[84], J.E. Brau[114], H.M. Braun[174], B. Brelier[158], J. Bremer[29], R. Brenner[166], S. Bressler[152], D. Britton[53], F.M. Brochu[27], I. Brock[20], R. Brock[88], E. Brodet[153], G. Brooijmans[34], W.K. Brooks[31b], G. Brown[82], P.A. Bruckman de Renstrom[38], D. Bruncko[144b], R. Bruneliere[48], S. Brunet[41], A. Bruni[19a], G. Bruni[19a], M. Bruschi[19a], F. Bucci[49], J. Buchanan[118], P. Buchholz[141], A.G. Buckley[45], I.A. Budagov[65], B. Budick[108], V. Büscher[81], L. Bugge[117], O. Bulekov[96], M. Bunse[42], T. Buran[117], H. Burckhart[29], S. Burdin[73], T. Burgess[13], S. Burke[129], E. Busato[33], P. Bussey[53], C.P. Buszello[166], F. Butin[29], B. Butler[143], J.M. Butler[21], C.M. Buttar[53], J.M. Butterworth[77], T. Byatt[77], J. Caballero[24], S. Cabrera Urbán[167], D. Caforio[19a,19b], O. Cakir[3a], P. Calafiura[14], G. Calderini[78], P. Calfayan[98], R. Calkins[106], L.P. Caloba[23a], D. Calvet[33], P. Camarri[133a,133b], D. Cameron[117], S. Campana[29], M. Campanelli[77], V. Canale[102a,102b], F. Canelli[30], A. Canepa[159a], J. Cantero[80], L. Capasso[102a,102b], M.D.M. Capeans Garrido[29], I. Caprini[25a], M. Caprini[25a], M. Capua[36a,36b], R. Caputo[148], C. Caramarcu[25a], R. Cardarelli[133a], T. Carli[29], G. Carlino[102a], L. Carminati[89a,89b], B. Caron[2,g], S. Caron[48], G.D. Carrillo Montoya[172], S. Carron Montero[158], A.A. Carter[75], J.R. Carter[27], J. Carvalho[124a,h], D. Casadei[108], M.P. Casado[11], M. Cascella[122a,122b], A.M. Castaneda Hernandez[172], E. Castaneda-Miranda[172], V. Castillo Gimenez[167], N.F. Castro[124b,a], G. Cataldi[72a], A. Catinaccio[29], J.R. Catmore[71], A. Cattai[29], G. Cattani[133a,133b], S. Caughron[34], P. Cavalleri[78], D. Cavalli[89a], M. Cavalli-Sforza[11], V. Cavasinni[122a,122b], F. Ceradini[134a,134b], A.S. Cerqueira[23a], A. Cerri[29], L. Cerrito[75], F. Cerutti[47], S.A. Cetin[18b], A. Chafaq[135a], D. Chakraborty[106], K. Chan[2], J.D. Chapman[27], J.W. Chapman[87], E. Chareyre[78], D.G. Charlton[17], V. Chavda[82], S. Cheatham[71], S. Chekanov[5], S.V. Chekulaev[159a], G.A. Chelkov[65], H. Chen[24], S. Chen[32c], X. Chen[172], A. Cheplakov[65], V.F. Chepurnov[65], R. Cherkaoui El Moursli[135d], V. Tcherniatine[24], D. Chesneanu[25a], E. Cheu[6], S.L. Cheung[158], L. Chevalier[136], F. Chevallier[136], G. Chiefari[102a,102b], L. Chikovani[51], J.T. Childers[58a], A. Chilingarov[71], G. Chiodini[72a], M.V. Chizhov[65], G. Choudalakis[30], S. Chouridou[137], I.A. Christidi[77], A. Christov[48], D. Chromek-Burckhart[29], M.L. Chu[151], J. Chudoba[125], G. Ciapetti[132a,132b], A.K. Ciftci[3a], R. Ciftci[3a], D. Cinca[33], V. Cindro[74], M.D. Ciobotaru[163], C. Ciocca[19a,19b], A. Ciocio[14], M. Cirilli[87,i], A. Clark[49], P.J. Clark[45], W. Cleland[123], J.C. Clemens[83], B. Clement[55], C. Clement[146a,146b], Y. Coadou[83], M. Cobal[164a,164c], A. Coccaro[50a,50b], J. Cochran[64], J. Coggeshall[165], E. Cogneras[180], A.P. Colijn[105], C. Collard[115], N.J. Collins[17], C. Collins-Tooth[53], J. Collot[55], G. Colon[84], P. Conde Muiño[124a], E. Coniavitis[166], M.C. Conidi[11], M. Consonni[104], S. Constantinescu[25a], C. Conta[119a,119b], F. Conventi[102a,j], M. Cooke[34], B.D. Cooper[75], A.M. Cooper-Sarkar[118], N.J. Cooper-Smith[76], K. Copic[34], T. Cornelissen[50a,50b], M. Corradi[19a], F. Corriveau[85,k], A. Corso-Radu[163], A. Cortes-Gonzalez[165], G. Cortiana[99], G. Costa[89a], M.J. Costa[167], D. Costanzo[139], T. Costin[30], D. Côté[29], R. Coura Torres[23a], L. Courneyea[169], G. Cowan[76], C. Cowden[27], B.E. Cox[82], K. Cranmer[108], J. Cranshaw[5], M. Cristinziani[20], G. Crosetti[36a,36b], R. Crupi[72a,72b], S. Crépé-Renaudin[55], C. Cuenca Almenar[175], T. Cuhadar Donszelmann[139], M. Curatolo[47], C.J. Curtis[17], P. Cwetanski[61], Z. Czyczula[175], S. D'Auria[53], M. D'Onofrio[73], A. D'Orazio[99], C. Da Via[82], W. Dabrowski[37], T. Dai[87], C. Dallapiccola[84], S.J. Dallison[129,*], C.H. Daly[138], M. Dam[35], H.O. Danielsson[29], D. Dannheim[99], V. Dao[49], G. Darbo[50a], G.L. Darlea[25b], W. Davey[86], T. Davidek[126], N. Davidson[86], R. Davidson[71], M. Davies[93], A.R. Davison[77], I. Dawson[139], R.K. Daya[39], K. De[7], R. de Asmundis[102a], S. De Castro[19a,19b], P.E. De Castro Faria Salgado[24], S. De Cecco[78], J. de Graat[98], N. De Groot[104], P. de Jong[105], L. De Mora[71], M. De Oliveira Branco[29], D. De Pedis[132a], A. De Salvo[132a], U. De Sanctis[164a,164c], A. De Santo[149], J.B. De Vivie De Regie[115], S. Dean[77], D.V. Dedovich[65], J. Degenhardt[120], M. Dehchar[118], C. Del Papa[164a,164c], J. Del Peso[80], T. Del Prete[122a,122b], A. Dell'Acqua[29], L. Dell'Asta[89a,89b], M. Della Pietra[102a,l], D. della Volpe[102a,102b], M. Delmastro[29], P.A. Delsart[55], C. Deluca[148], S. Demers[175], M. Demichev[65], B. Demirkoz[11], J. Deng[163], W. Deng[24], S.P. Denisov[128], J.E. Derkaoui[135c], F. Derue[78], P. Dervan[73], K. Desch[20], P.O. Deviveiros[158], A. Dewhurst[129], B. DeWilde[148], S. Dhaliwal[158], R. Dhullipudi[24,m], A. Di Ciaccio[133a,133b], L. Di Ciaccio[4], A. Di Girolamo[29], B. Di Girolamo[29], S. Di Luise[134a,134b], A. Di Mattia[88], R. Di Nardo[133a,133b], A. Di Simone[133a,133b], R. Di Sipio[19a,19b], M.A. Diaz[31a], F. Diblen[18c], E.B. Diehl[87], J. Dietrich[48], T.A. Dietzsch[58a], S. Diglio[115], K. Dindar Yagci[39], J. Dingfelder[48], C. Dionisi[132a,132b], P. Dita[25a], S. Dita[25a], F. Dittus[29], F. Djama[83], R. Djilkibaev[108], T. Djobava[51], M.A.B. do Vale[23a], T.K.O. Doan[4], D. Dobos[29], E. Dobson[29], M. Dobson[163], C. Doglioni[118], T. Doherty[53], J. Dolejsi[126], I. Dolenc[74], Z. Dolezal[126], B.A. Dolgoshein[96], T. Dohmae[155], M. Donega[120], J. Donini[55], J. Dopke[174], A. Doria[102a], A. Dotti[122a,122b], M.T. Dova[70], A.D. Doxiadis[105], A.T. Doyle[53], Z. Drasal[126], M. Dris[9], J. Dubbert[99], S. Dube[14], E. Duchovni[171], G. Duckeck[98], A. Dudarev[29], F. Dudziak[115], M. Dührssen[29], L. Duflot[115], M.-A. Dufour[85], M. Dunford[30], H. Duran Yildiz[3b], R. Duxfield[139], M. Dwuznik[37], M. Düren[52], J. Ebke[98], S. Eckweiler[81], K. Edmonds[81], C.A. Edwards[76], K. Egorov[61], W. Ehrenfeld[41], T. Ehrich[99], T. Eifert[29], G. Eigen[13], K. Einsweiler[14], E. Eisenhandler[75], T. Ekelof[166], M. El Kacimi[4], M. Ellert[166], S. Elles[4], F. Ellinghaus[81], K. Ellis[75], N. Ellis[29], J. Elmsheuser[98], M. Elsing[29], D. Emeliyanov[129], R. Engelmann[148], A. Engl[98], B. Epp[62], A. Eppig[87], J. Erdmann[54], A. Ereditato[16], D. Eriksson[146a], J. Ernst[1], M. Ernst[24], J. Ernwein[136], D. Errede[165], S. Errede[165], E. Ertel[81], M. Escalier[115], C. Escobar[167], X. Espinal Curull[11], B. Esposito[47], A.I. Etienvre[136], E. Etzion[153], H. Evans[61], L. Fabbri[19a,19b], C. Fabre[29], K. Facius[35], R.M. Fakhrutdinov[128], S. Falciano[132a], Y. Fang[172], M. Fanti[89a,89b], A. Farbin[7], A. Farilla[134a], J. Farley[148], T. Farooque[158], S.M. Farrington[118], P. Farthouat[29], P. Fassnacht[29], D. Fassouliotis[8], B. Fatholahzadeh[158], L. Fayard[115], R. Febbraro[33], P. Federic[144a],

O.L. Fedin[121], W. Fedorko[29], L. Feligioni[83], C.U. Felzmann[86], C. Feng[32d], E.J. Feng[30], A.B. Fenyuk[128], J. Ferencei[144b], J. Ferland[93], B. Fernandes[124a,n], W. Fernando[109], S. Ferrag[53], J. Ferrando[118], V. Ferrara[41], A. Ferrari[166], P. Ferrari[105], R. Ferrari[119a], A. Ferrer[167], M.L. Ferrer[47], D. Ferrere[49], C. Ferretti[87], M. Fiascaris[118], F. Fiedler[81], A. Filipčič[74], A. Filippas[9], F. Filthaut[104], M. Fincke-Keeler[169], M.C.N. Fiolhais[124a,h], L. Fiorini[11], A. Firan[39], G. Fischer[41], M.J. Fisher[109], M. Flechl[48], I. Fleck[141], J. Fleckner[81], P. Fleischmann[173], S. Fleischmann[20], T. Flick[174], L.R. Flores Castillo[172], M.J. Flowerdew[99], T. Fonseca Martin[76], J. Fopma[118], A. Formica[136], A. Forti[82], D. Fortin[159a], D. Fournier[115], A.J. Fowler[44], K. Fowler[137], H. Fox[71], P. Francavilla[122a,122b], S. Franchino[119a,119b], D. Francis[29], M. Franklin[57], S. Franz[29], M. Fraternali[119a,119b], S. Fratina[120], J. Freestone[82], S.T. French[27], R. Froeschl[29], D. Froidevaux[29], J.A. Frost[27], C. Fukunaga[156], E. Fullana Torregrosa[5], J. Fuster[167], C. Gabaldon[80], O. Gabizon[171], T. Gadfort[24], S. Gadomski[49], G. Gagliardi[50a,50b], P. Gagnon[61], C. Galea[98], E.J. Gallas[118], V. Gallo[16], B.J. Gallop[129], P. Gallus[125,o], E. Galyaev[40], K.K. Gan[109], Y.S. Gao[143,p], A. Gaponenko[14], M. Garcia-Sciveres[14], C. García[167], J.E. García Navarro[49], R.W. Gardner[30], N. Garelli[29], H. Garitaonandia[105], V. Garonne[29], C. Gatti[47], G. Gaudio[119a], P. Gauzzi[132a,132b], I.L. Gavrilenko[94], C. Gay[168], G. Gaycken[20], E.N. Gazis[9], P. Ge[32d], C.N.P. Gee[129], Ch. Geich-Gimbel[20], K. Gellerstedt[146a,146b], C. Gemme[50a], M.H. Genest[98], S. Gentile[132a,132b], F. Georgatos[9], S. George[76], A. Gershon[153], H. Ghazlane[135d], N. Ghodbane[33], B. Giacobbe[19a], S. Giagu[132a,132b], V. Giakoumopoulou[8], V. Giangiobbe[122a,122b], F. Gianotti[29], B. Gibbard[24], A. Gibson[158], S.M. Gibson[118], L.M. Gilbert[118], M. Gilchriese[14], V. Gilewsky[91], D.M. Gingrich[2,q], J. Ginzburg[153], N. Giokaris[8], M.P. Giordani[164c], R. Giordano[102a,102b], F.M. Giorgi[15], P. Giovannini[99], P.F. Giraud[136], D. Giugni[89a], P. Giusti[19a], B.K. Gjelsten[117], L.K. Gladilin[97], C. Glasman[80], A. Glazov[41], K.W. Glitza[174], G.L. Glonti[65], J. Godfrey[142], J. Godlewski[29], M. Goebel[41], T. Göpfert[43], C. Goeringer[81], C. Gössling[42], T. Göttfert[99], S. Goldfarb[87], D. Goldin[39], T. Golling[175], A. Gomes[124a,r], L.S. Gomez Fajardo[41], R. Gonçalo[76], L. Gonella[20], C. Gong[32b], S. González de la Hoz[167], M.L. Gonzalez Silva[26], S. Gonzalez-Sevilla[49], J.J. Goodson[148], L. Goossens[29], H.A. Gordon[24], I. Gorelov[103], G. Gorfine[174], B. Gorini[29], E. Gorini[72a,72b], A. Gorišek[74], E. Gornicki[38], B. Gosdzik[41], M. Gosselink[105], M.I. Gostkin[65], I. Gough Eschrich[163], M. Gouighri[135a], D. Goujdami[135a], M.P. Goulette[49], A.G. Goussiou[138], C. Goy[4], I. Grabowska-Bold[163,s], P. Grafström[29], K.-J. Grahn[147], S. Grancagnolo[15], V. Grassi[148], V. Gratchev[121], N. Grau[34], H.M. Gray[34,t], J.A. Gray[148], E. Graziani[134a], B. Green[76], T. Greenshaw[73], Z.D. Greenwood[24,u], I.M. Gregor[41], P. Grenier[143], E. Griesmayer[46], J. Griffiths[138], N. Grigalashvili[65], A.A. Grillo[137], K. Grimm[148], S. Grinstein[11], Y.V. Grishkevich[97], M. Groh[99], M. Groll[81], E. Gross[171], J. Grosse-Knetter[54], J. Groth-Jensen[79], K. Grybel[141], C. Guicheney[33], A. Guida[72a,72b], T. Guillemin[4], H. Guler[85,v], J. Gunther[125], B. Guo[158], Y. Gusakov[65], A. Gutierrez[93], P. Gutierrez[111], N. Guttman[153], O. Gutzwiller[172], C. Guyot[136], C. Gwenlan[118], C.B. Gwilliam[73], A. Haas[143], S. Haas[29], C. Haber[14], H.K. Hadavand[39], D.R. Hadley[17], P. Haefner[99], S. Haider[29], Z. Hajduk[38], H. Hakobyan[176], J. Haller[41,w], K. Hamacher[174], A. Hamilton[49], S. Hamilton[161], L. Han[32b], K. Hanagaki[116], M. Hance[120], C. Handel[81], P. Hanke[58a], J.R. Hansen[35], J.B. Hansen[35], J.D. Hansen[35], P.H. Hansen[35], P. Hansson[143], K. Hara[160], G.A. Hare[137], T. Harenberg[174], R.D. Harrington[21], O.M. Harris[138], K. Harrison[17], J. Hartert[48], F. Hartjes[105], A. Harvey[56], S. Hasegawa[101], Y. Hasegawa[140], S. Hassani[136], S. Haug[16], M. Hauschild[29], R. Hauser[88], M. Havranek[125,o], C.M. Hawkes[17], R.J. Hawkings[29], T. Hayakawa[67], H.S. Hayward[73], S.J. Haywood[129], S.J. Head[17], V. Hedberg[79], L. Heelan[28], S. Heim[88], B. Heinemann[14], S. Heisterkamp[35], L. Helary[4], M. Heller[115], S. Hellman[146a,146b], C. Helsens[11], T. Hemperek[20], R.C.W. Henderson[71], M. Henke[58a], A. Henrichs[54], A.M. Henriques Correia[29], S. Henrot-Versille[115], C. Hensel[54], T. Henß[174], Y. Hernández Jiménez[167], A.D. Hershenhorn[152], G. Herten[48], R. Hertenberger[98], L. Hervas[29], N.P. Hessey[105], E. Higón-Rodriguez[167], J.C. Hill[27], K.H. Hiller[41], S. Hillert[146a,146b], S.J. Hillier[17], I. Hinchliffe[14], E. Hines[120], M. Hirose[116], F. Hirsch[42], D. Hirschbuehl[174], J. Hobbs[148], N. Hod[153], M.C. Hodgkinson[139], P. Hodgson[139], A. Hoecker[29], M.R. Hoeferkamp[103], J. Hoffman[39], D. Hoffmann[83], M. Hohlfeld[81], T. Holy[127], J.L. Holzbauer[88], Y. Homma[67], T. Horazdovsky[127], C. Horn[143], S. Horner[48], J.-Y. Hostachy[55], S. Hou[151], A. Hoummada[135a], T. Howe[39], J. Hrivnac[115], T. Hryn'ova[4], P.J. Hsu[175], S.-C. Hsu[14], G.S. Huang[111], Z. Hubacek[127], F. Hubaut[83], F. Huegging[20], T.B. Huffman[118], E.W. Hughes[34], G. Hughes[71], M. Huhtinen[29], M. Hurwitz[30], U. Husemann[41], N. Huseynov[10], J. Huston[88], J. Huth[57], G. Iacobucci[102a], G. Iakovidis[9], I. Ibragimov[141], L. Iconomidou-Fayard[115], J. Idarraga[159b], P. Iengo[4], O. Igonkina[105], Y. Ikegami[66], M. Ikeno[66], Y. Ilchenko[39], D. Iliadis[154], T. Ince[20], P. Ioannou[8], M. Iodice[134a], A. Irles Quiles[167], A. Ishikawa[67], M. Ishino[66], R. Ishmukhametov[39], T. Isobe[155], C. Issever[118], S. Istin[18a], Y. Itoh[101], A.V. Ivashin[128], W. Iwanski[38], H. Iwasaki[66], J.M. Izen[40], V. Izzo[102a], B. Jackson[120], J.N. Jackson[73], P. Jackson[143], M.R. Jaekel[29], V. Jain[61], K. Jakobs[48], S. Jakobsen[35], J. Jakubek[127], D.K. Jana[111], E. Jankowski[158], E. Jansen[77], A. Jantsch[99], M. Janus[48], G. Jarlskog[79], L. Jeanty[57], I. Jen-La Plante[30], P. Jenni[29], P. Jež[35], S. Jézéquel[4], W. Ji[79], J. Jia[148], Y. Jiang[32b], M.Jimenez Belenguer[29], S. Jin[32a], O. Jinnouchi[157], D. Joffe[39], M. Johansen[146a,146b], K.E. Johansson[146a], P. Johansson[139], S. Johnert[41], K.A. Johns[6], K. Jon-And[146a,146b], G. Jones[82], R.W.L. Jones[71], T.J. Jones[73], P.M. Jorge[124a,b], J. Joseph[14], V. Juranek[125], P. Jussel[62], V.V. Kabachenko[128], M. Kaci[167], A. Kaczmarska[38], M. Kado[115], H. Kagan[109], M. Kagan[57], S. Kaiser[99], E. Kajomovitz[152], S. Kalinin[174], L.V. Kalinovskaya[65], S. Kama[41], N. Kanaya[155], M. Kaneda[155], V.A. Kantserov[96], J. Kanzaki[66], B. Kaplan[175], A. Kapliy[30], J. Kaplon[29], D. Kar[43],

M. Karagounis[20], M. Karagoz[118], M. Karnevskiy[41], V. Kartvelishvili[71], A.N. Karyukhin[128], L. Kashif[57], A. Kasmi[39], R.D. Kass[109], A. Kastanas[13], M. Kataoka[4], Y. Kataoka[155], E. Katsoufis[9], J. Katzy[41], V. Kaushik[6], K. Kawagoe[67], T. Kawamoto[155], G. Kawamura[81], M.S. Kayl[105], V.A. Kazanin[107], M.Y. Kazarinov[65], J.R. Keates[82], R. Keeler[169], R. Kehoe[39], M. Keil[54], G.D. Kekelidze[65], M. Kelly[82], M. Kenyon[53], O. Kepka[125], N. Kerschen[29], B.P. Kerševan[74], S. Kersten[174], K. Kessoku[155], M. Khakzad[28], F. Khalil-zada[10], H. Khandanyan[165], A. Khanov[112], D. Kharchenko[65], A. Khodinov[148], A. Khomich[58a], G. Khoriauli[20], N. Khovanskiy[65], V. Khovanskiy[95], E. Khramov[65], J. Khubua[51], H. Kim[7], M.S. Kim[2], P.C. Kim[143], S.H. Kim[160], O. Kind[15], B.T. King[73], M. King[67], J. Kirk[129], G.P. Kirsch[118], L.E. Kirsch[22], A.E. Kiryunin[99], D. Kisielewska[37], T. Kittelmann[123], E. Kladiva[144b], M. Klein[73], U. Klein[73], K. Kleinknecht[81], M. Klemetti[85], A. Klier[171], A. Klimentov[24], R. Klingenberg[42], E.B. Klinkby[44], T. Klioutchnikova[29], P.F. Klok[104], S. Klous[105], E.-E. Kluge[58a], T. Kluge[73], P. Kluit[105], S. Kluth[99], N.S. Knecht[158], E. Kneringer[62], B.R. Ko[44], T. Kobayashi[155], M. Kobel[43], B. Koblitz[29], M. Kocian[143], A. Kocnar[113], P. Kodys[126], K. Köneke[41], A.C. König[104], S. Koenig[81], L. Köpke[81], F. Koetsveld[104], P. Koevesarki[20], T. Koffas[29], E. Koffeman[105], F. Kohn[54], Z. Kohout[127], T. Kohriki[66], T. Koi[143], H. Kolanoski[15], V. Kolesnikov[65], I. Koletsou[4], J. Koll[88], D. Kollar[29], S.D. Kolya[82], A.A. Komar[94], J.R. Komaragiri[142], T. Kondo[66], T. Kono[41,x], R. Konoplich[108], N. Konstantinidis[77], S. Koperny[37], K. Korcyl[38], K. Kordas[154], A. Korn[14], I. Korolkov[11], E.V. Korolkova[139], V.A. Korotkov[128], O. Kortner[99], S. Kortner[99], P. Kostka[41], V.V. Kostyukhin[20], S. Kotov[99], V.M. Kotov[65], C. Kourkoumelis[8], A. Koutsman[105], R. Kowalewski[169], T.Z. Kowalski[37], W. Kozanecki[136], A.S. Kozhin[128], V. Kral[127], V.A. Kramarenko[97], G. Kramberger[74], M.W. Krasny[78], A. Krasznahorkay[108], J. Kraus[88], J.K. Kraus[20], A. Kreisel[153], F. Krejci[127], J. Kretzschmar[73], N. Krieger[54], P. Krieger[158], K. Kroeninger[54], H. Kroha[99], J. Kroll[120], J. Kroseberg[20], J. Krstic[12a], U. Kruchonak[65], H. Krüger[20], Z.V. Krumshteyn[65], A. Kruth[20], T. Kubota[155], S. Kuehn[48], A. Kugel[58c], T. Kuhl[174], D. Kuhn[62], V. Kukhtin[65], Y. Kulchitsky[90], S. Kuleshov[31b], C. Kummer[98], M. Kuna[83], J. Kunkle[120], A. Kupco[125], H. Kurashige[67], M. Kurata[160], Y.A. Kurochkin[90], V. Kus[125], M. Kuze[157], R. Kwee[15], A. La Rosa[29], L. La Rotonda[36a,36b], J. Labbe[4], C. Lacasta[167], F. Lacava[132a,132b], H. Lacker[15], D. Lacour[78], V.R. Lacuesta[167], E. Ladygin[65], R. Lafaye[4], B. Laforge[78], T. Lagouri[80], S. Lai[48], M. Lamanna[29], C.L. Lampen[6], W. Lampl[6], E. Lancon[136], U. Landgraf[48], M.P.J. Landon[75], J.L. Lane[82], A.J. Lankford[163], F. Lanni[24], K. Lantzsch[29], A. Lanza[119a], S. Laplace[4], C. Lapoire[83], J.F. Laporte[136], T. Lari[89a], A. Larner[118], M. Lassnig[29], P. Laurelli[47], W. Lavrijsen[14], P. Laycock[73], A.B. Lazarev[65], A. Lazzaro[89a,89b], O. Le Dortz[78], E. Le Guirriec[83], E. Le Menedeu[136], A. Lebedev[64], C. Lebel[93], T. LeCompte[5], F. Ledroit-Guillon[55], H. Lee[105], J.S.H. Lee[150], S.C. Lee[151], M. Lefebvre[169], M. Legendre[136], B.C. LeGeyt[120], F. Legger[98], C. Leggett[14], M. Lehmacher[20], G. Lehmann Miotto[29], X. Lei[6], R. Leitner[126], D. Lellouch[171], J. Lellouch[78], V. Lendermann[58a], K.J.C. Leney[73], T. Lenz[174], G. Lenzen[174], B. Lenzi[136], K. Leonhardt[43], C. Leroy[93], J.-R. Lessard[169], C.G. Lester[27], A. Leung Fook Cheong[172], J. Levêque[83], D. Levin[87], L.J. Levinson[171], M. Leyton[15], H. Li[172], X. Li[87], Z. Liang[39], Z. Liang[151,y], B. Liberti[133a], P. Lichard[29], M. Lichtnecker[98], K. Lie[165], W. Liebig[105], J.N. Lilley[17], A. Limosani[86], M. Limper[63], S.C. Lin[151], J.T. Linnemann[88], E. Lipeles[120], L. Lipinsky[125], A. Lipniacka[13], T.M. Liss[165], D. Lissauer[24], A. Lister[49], A.M. Litke[137], C. Liu[28], D. Liu[151,z], H. Liu[87], J.B. Liu[87], M. Liu[32b], Y. Liu[32b], M. Livan[119a,119b], A. Lleres[55], S.L. Lloyd[75], E. Lobodzinska[41], P. Loch[6], W.S. Lockman[137], S. Lockwitz[175], T. Loddenkoetter[20], F.K. Loebinger[82], A. Loginov[175], C.W. Loh[168], T. Lohse[15], K. Lohwasser[48], M. Lokajicek[125], R.E. Long[71], L. Lopes[124a,b], D. Lopez Mateos[34,aa], M. Losada[162], P. Loscutoff[14], X. Lou[40], A. Lounis[115], K.F. Loureiro[109], L. Lovas[144a], J. Love[21], P.A. Love[71], A.J. Lowe[61], F. Lu[32a], H.J. Lubatti[138], C. Luci[132a,132b], A. Lucotte[55], A. Ludwig[43], D. Ludwig[41], I. Ludwig[48], F. Luehring[61], D. Lumb[48], L. Luminari[132a], E. Lund[117], B. Lund-Jensen[147], B. Lundberg[79], J. Lundberg[29], J. Lundquist[35], D. Lynn[24], J. Lys[14], E. Lytken[79], H. Ma[24], L.L. Ma[172], J.A. Macana Goia[93], G. Maccarrone[47], A. Macchiolo[99], B. Maček[74], J. Machado Miguens[124a,b], R. Mackeprang[35], R.J. Madaras[14], W.F. Mader[43], R. Maenner[58c], T. Maeno[24], P. Mättig[174], S. Mättig[41], P.J. Magalhaes Martins[124a,h], E. Magradze[51], Y. Mahalalel[153], K. Mahboubi[48], A. Mahmood[1], C. Maiani[132a,132b], C. Maidantchik[23a], A. Maio[124a,r], S. Majewski[24], Y. Makida[66], M. Makouski[128], N. Makovec[115], P. Mal[6], Pa. Malecki[38], P. Malecki[38], V.P. Maleev[121], F. Malek[55], U. Mallik[63], D. Malon[5], S. Maltezos[9], V. Malyshev[107], S. Malyukov[65], R. Mameghani[98], J. Mamuzic[41], L. Mandelli[89a], I. Mandić[74], R. Mandrysch[15], J. Maneira[124a], P.S. Mangeard[88], I.D. Manjavidze[65], A. Mann[54], P.M. Manning[137], A. Manousakis-Katsikakis[8], B. Mansoulie[136], A. Mapelli[29], L. Mapelli[29], L. March[80], J.F. Marchand[29], F. Marchese[133a,133b], G. Marchiori[78], M. Marcisovsky[125,o], C.P. Marino[61], F. Marroquim[23a], Z. Marshall[34,aa], S. Marti-Garcia[167], A.J. Martin[75], B. Martin[29], B. Martin[88], F.F. Martin[120], J.P. Martin[93], T.A. Martin[17], B. Martin dit Latour[49], M. Martinez[11], V. Martinez Outschoorn[57], A.C. Martyniuk[82], F. Marzano[132a], A. Marzin[136], L. Masetti[81], T. Mashimo[155], R. Mashinistov[96], J. Masik[82], A.L. Maslennikov[107], I. Massa[19a,19b], N. Massol[4], A. Mastroberardino[36a,36b], T. Masubuchi[155], P. Matricon[115], H. Matsunaga[155], T. Matsushita[67], C. Mattravers[118,ab], S.J. Maxfield[73], A. Mayne[139], R. Mazini[151], M. Mazur[48], S.P. Mc Kee[87], A. McCarn[165], R.L. McCarthy[148], N.A. McCubbin[129], K.W. McFarlane[56], H. McGlone[53], G. Mchedlidze[51], S.J. McMahon[129], R.A. McPherson[169,k], A. Meade[84], J. Mechnich[105], M. Mechtel[174], M. Medinnis[41], R. Meera-Lebbai[111], T. Meguro[116], S. Mehlhase[41], A. Mehta[73], K. Meier[58a], B. Meirose[48], C. Melachrinos[30], B.R. Mellado Garcia[172], L. Mendoza Navas[162], Z. Meng[151,ac], S. Menke[99], E. Meoni[11], P. Mermod[118], L. Merola[102a,102b], C. Meroni[89a], F.S. Merritt[30],

A.M. Messina[29], J. Metcalfe[103], A.S. Mete[64], J.-P. Meyer[136], J. Meyer[173], J. Meyer[54], T.C. Meyer[29], W.T. Meyer[64], J. Miao[32d], S. Michal[29], L. Micu[25a], R.P. Middleton[129], S. Migas[73], L. Mijović[74], G. Mikenberg[171], M. Mikestikova[125], M. Mikuž[74], D.W. Miller[143], W.J. Mills[168], C. Mills[57], A. Milov[171], D.A. Milstead[146a,146b], D. Milstein[171], A.A. Minaenko[128], M. Miñano[167], I.A. Minashvili[65], A.I. Mincer[108], B. Mindur[37], M. Mineev[65], Y. Ming[130], L.M. Mir[11], G. Mirabelli[132a], S. Misawa[24], A. Misiejuk[76], J. Mitrevski[137], V.A. Mitsou[167], S. Mitsui[160], P.S. Miyagawa[82], K. Miyazaki[67], J.U. Mjörnmark[79], T. Moa[146a,146b], V. Moeller[27], K. Mönig[41], N. Möser[20], W. Mohr[48], S. Mohrdieck-Möck[99], R. Moles-Valls[167], J. Molina-Perez[29], J. Monk[77], E. Monnier[83], S. Montesano[89a,89b], F. Monticelli[70], R.W. Moore[2], C.Mora Herrera[49], A. Moraes[53], A. Morais[124a,b], J. Morel[54], G. Morello[36a,36b], D. Moreno[81], M.Moreno Llácer[167], P. Morettini[50a], M. Morii[57], A.K. Morley[86], G. Mornacchi[29], J.D. Morris[75], H.G. Moser[99], M. Mosidze[51], J. Moss[109], R. Mount[143], E. Mountricha[136], S.V. Mouraviev[94], E.J.W. Moyse[84], M. Mudrinic[12b], F. Mueller[58a], J. Mueller[123], K. Mueller[20], T.A. Müller[98], D. Muenstermann[42], A. Muir[168], Y. Munwes[153], W.J. Murray[129], I. Mussche[105], E. Musto[102a,102b], A.G. Myagkov[128], M. Myska[125,o], J. Nadal[11], K. Nagai[160], K. Nagano[66], Y. Nagasaka[60], A.M. Nairz[29], K. Nakamura[155], I. Nakano[110], G. Nanava[20], A. Napier[161], M. Nash[77,ad], N.R. Nation[21], T. Nattermann[20], T. Naumann[41], G. Navarro[162], S.K. Nderitu[20], H.A. Neal[87], E. Nebot[80], P. Nechaeva[94], A. Negri[119a,119b], G. Negri[29], A. Nelson[64], S. Nelson[143], T.K. Nelson[143], S. Nemecek[125], P. Nemethy[108], A.A. Nepomuceno[23a], M. Nessi[29], M.S. Neubauer[165], A. Neusiedl[81], R.M. Neves[108], P. Nevski[24], R.B. Nickerson[118], R. Nicolaidou[136], L. Nicolas[139], G. Nicoletti[47], B. Nicquevert[29], F. Niedercorn[115], J. Nielsen[137], A. Nikiforov[15], K. Nikolaev[65], I. Nikolic-Audit[78], K. Nikolopoulos[8], H. Nilsen[48], P. Nilsson[7], A. Nisati[132a], T. Nishiyama[67], R. Nisius[99], L. Nodulman[5], M. Nomachi[116], I. Nomidis[154], M. Nordberg[29], B. Nordkvist[146a,146b], D. Notz[41], J. Novakova[126], M. Nozaki[66], M. Nožička[41], I.M. Nugent[159a], A.-E. Nuncio-Quiroz[20], G. Nunes Hanninger[20], T. Nunnemann[98], E. Nurse[77], D.C. O'Neil[142], V. O'Shea[53], F.G. Oakham[28,g], H. Oberlack[99], A. Ochi[67], S. Oda[155], S. Odaka[66], J. Odier[83], H. Ogren[61], A. Oh[82], S.H. Oh[44], C.C. Ohm[146a,146b], T. Ohshima[101], T. Ohsugi[59], S. Okada[67], H. Okawa[163], Y. Okumura[101], T. Okuyama[155], A.G. Olchevski[65], M. Oliveira[124a,h], D. Oliveira Damazio[24], E. Oliver Garcia[167], D. Olivito[120], A. Olszewski[38], J. Olszowska[38], C. Omachi[67,ae], A. Onofre[124a,af], P.U.E. Onyisi[30], C.J. Oram[159a], M.J. Oreglia[30], Y. Oren[153], D. Orestano[134a,134b], I. Orlov[107], C. Oropeza Barrera[53], R.S. Orr[158], E.O. Ortega[130], B. Osculati[50a,50b], R. Ospanov[120], C. Osuna[11], G. Otero y Garzon[26], J.P Ottersbach[105], F. Ould-Saada[117], A. Ouraou[136], Q. Ouyang[32a], M. Owen[82], S. Owen[139], A. Oyarzun[31b], V.E. Ozcan[77], N. Ozturk[7], A. Pacheco Pages[11], C. Padilla Aranda[11], E. Paganis[139], F. Paige[24], K. Pajchel[117], S. Palestini[29], D. Pallin[33], A. Palma[124a,b], J.D. Palmer[17], Y.B. Pan[172], E. Panagiotopoulou[9], B. Panes[31a], N. Panikashvili[87], S. Panitkin[24], D. Pantea[25a], M. Panuskova[125], V. Paolone[123], Th.D. Papadopoulou[9], S.J. Park[54], W. Park[24,ag], M.A. Parker[27], F. Parodi[50a,50b], J.A. Parsons[34], U. Parzefall[48], E. Pasqualucci[132a], A. Passeri[134a], F. Pastore[134a,134b], Fr. Pastore[29], G. Pásztor[49,ah], S. Pataraia[99], N. Patel[150], J.R. Pater[82], S. Patricelli[102a,102b], T. Pauly[29], M. Pecsy[144a], M.I. Pedraza Morales[172], S.V. Peleganchuk[107], H. Peng[172], A. Penson[34], J. Penwell[61], M. Perantoni[23a], K. Perez[34,aa], E. Perez Codina[11], M.T. Pérez García-Estañ[167], V. Perez Reale[34], L. Perini[89a,89b], H. Pernegger[29], R. Perrino[72a], S. Persembe[3a], P. Perus[115], V.D. Peshekhonov[65], B.A. Petersen[29], T.C. Petersen[35], E. Petit[83], C. Petridou[154], E. Petrolo[132a], F. Petrucci[134a,134b], D. Petschull[41], M. Petteni[142], R. Pezoa[31b], B. Pfeifer[48], A. Phan[86], A.W. Phillips[27], G. Piacquadio[29], E. Piccaro[75], M. Piccinini[19a,19b], R. Piegaia[26], J.E. Pilcher[30], A.D. Pilkington[82], J. Pina[124a,r], M. Pinamonti[164a,164c], J.L. Pinfold[2], B. Pinto[124a,b], C. Pizio[89a,89b], R. Placakyte[41], M. Plamondon[169], M.-A. Pleier[24], A. Poblaguev[175], S. Poddar[58a], F. Podlyski[33], L. Poggioli[115], M. Pohl[49], F. Polci[55], G. Polesello[119a], A. Policicchio[138], A. Polini[19a], J. Poll[75], V. Polychronakos[24], D. Pomeroy[22], K. Pommès[29], L. Pontecorvo[132a], B.G. Pope[88], G.A. Popeneciu[25a], D.S. Popovic[12a], A. Poppleton[29], X. Portell Bueso[48], R. Porter[163], G.E. Pospelov[99], S. Pospisil[127], M. Potekhin[24], I.N. Potrap[99], C.J. Potter[149], C.T. Potter[85], K.P. Potter[82], G. Poulard[29], J. Poveda[172], R. Prabhu[20], P. Pralavorio[83], S. Prasad[57], R. Pravahan[7], L. Pribyl[29], D. Price[61], L.E. Price[5], P.M. Prichard[73], D. Prieur[123], M. Primavera[72a], K. Prokofiev[29], F. Prokoshin[31b], S. Protopopescu[24], J. Proudfoot[5], X. Prudent[43], H. Przysieznik[4], S. Psoroulas[20], E. Ptacek[114], J. Purdham[87], M. Purohit[24,ai], P. Puzo[115], Y. Pylypchenko[117], J. Qian[87], W. Qian[129], Z. Qin[41], A. Quadt[54], D.R. Quarrie[14], W.B. Quayle[172], F. Quinonez[31a], M. Raas[104], V. Radeka[24], V. Radescu[58b], B. Radics[20], T. Rador[18a], F. Ragusa[89a,89b], G. Rahal[180], A.M. Rahimi[109], S. Rajagopalan[24], M. Rammensee[48], M. Rammes[141], F. Rauscher[98], E. Rauter[99], M. Raymond[29], A.L. Read[117], D.M. Rebuzzi[119a,119b], A. Redelbach[173], G. Redlinger[24], R. Reece[120], K. Reeves[40], E. Reinherz-Aronis[153], A. Reinsch[114], I. Reisinger[42], D. Reljic[12a], C. Rembser[29], Z.L. Ren[151], P. Renkel[39], S. Rescia[24], M. Rescigno[132a], S. Resconi[89a], B. Resende[136], P. Reznicek[126], R. Rezvani[158], A. Richards[77], R. Richter[99], E. Richter-Was[38,aj], M. Ridel[78], M. Rijpstra[105], M. Rijssenbeek[148], A. Rimoldi[119a,119b], L. Rinaldi[19a], R.R. Rios[39], I. Riu[11], F. Rizatdinova[112], E. Rizvi[75], D.A. Roa Romero[162], S.H. Robertson[85,k], A. Robichaud-Veronneau[49], D. Robinson[27], J.E.M. Robinson[77], M. Robinson[114], A. Robson[53], J.G. Rocha de Lima[106], C. Roda[122a,122b], D. Roda Dos Santos[29], D. Rodriguez[162], Y. Rodriguez Garcia[15], S. Roe[29], O. Røhne[117], V. Rojo[1], S. Rolli[161], A. Romaniouk[96], V.M. Romanov[65], G. Romeo[26], D. Romero Maltrana[31a], L. Roos[78], E. Ros[167], S. Rosati[138], G.A. Rosenbaum[158], L. Rosselet[49], V. Rossetti[11], L.P. Rossi[50a], M. Rotaru[25a], J. Rothberg[138], D. Rousseau[115], C.R. Royon[136], A. Rozanov[83], Y. Rozen[152], X. Ruan[115], B. Ruckert[98], N. Ruckstuhl[105], V.I. Rud[97],

G. Rudolph[62], F. Rühr[58a], F. Ruggieri[134a], A. Ruiz-Martinez[64], L. Rumyantsev[65], Z. Rurikova[48], N.A. Rusakovich[65], J.P. Rutherfoord[6], C. Ruwiedel[20], P. Ruzicka[125,o], Y.F. Ryabov[121], P. Ryan[88], G. Rybkin[115], S. Rzaeva[10], A.F. Saavedra[150], H.F-W. Sadrozinski[137], R. Sadykov[65], F. Safai Tehrani[132a,132b], H. Sakamoto[155], G. Salamanna[105], A. Salamon[133a], M. Saleem[111], D. Salihagic[99], A. Salnikov[143], J. Salt[167], B.M. Salvachua Ferrando[5], D. Salvatore[36a,36b], F. Salvatore[149], A. Salvucci[47], A. Salzburger[29], D. Sampsonidis[154], B.H. Samset[117], H. Sandaker[13], H.G. Sander[81], M.P. Sanders[98], M. Sandhoff[174], P. Sandhu[158], R. Sandstroem[105], S. Sandvoss[174], D.P.C. Sankey[129], A. Sansoni[47], C. Santamarina Rios[85], C. Santoni[33], R. Santonico[133a,133b], J.G. Saraiva[124a,r], T. Sarangi[172], E. Sarkisyan-Grinbaum[7], F. Sarri[122a,122b], O. Sasaki[66], N. Sasao[68], I. Satsounkevitch[90], G. Sauvage[4], P. Savard[158,g], A.Y. Savine[6], V. Savinov[123], L. Sawyer[24,ak], D.H. Saxon[53], L.P. Says[33], C. Sbarra[19a,19b], A. Sbrizzi[19a,19b], D.A. Scannicchio[29], J. Schaarschmidt[43], P. Schacht[99], U. Schäfer[81], S. Schaetzel[58b], A.C. Schaffer[115], D. Schaile[98], R.D. Schamberger[148], A.G. Schamov[107], V. Scharf[58a], V.A. Schegelsky[121], D. Scheirich[87], M. Schernau[163], M.I. Scherzer[14], C. Schiavi[50a,50b], J. Schieck[99], M. Schioppa[36a,36b], S. Schlenker[29], E. Schmidt[48], K. Schmieden[20], C. Schmitt[81], M. Schmitz[20], A. Schöning[58b], M. Schott[29], D. Schouten[142], J. Schovancova[125], M. Schram[85], A. Schreiner[63], C. Schroeder[81], N. Schroer[58c], M. Schroers[174], J. Schultes[174], H.-C. Schultz-Coulon[58a], J.W. Schumacher[43], M. Schumacher[48], B.A. Schumm[137], Ph. Schune[136], C. Schwanenberger[82], A. Schwartzman[143], Ph. Schwemling[78], R. Schwienhorst[88], R. Schwierz[43], J. Schwindling[136], W.G. Scott[129], J. Searcy[114], E. Sedykh[121], E. Segura[11], S.C. Seidel[103], A. Seiden[137], F. Seifert[43], J.M. Seixas[23a], G. Sekhniaidze[102a], D.M. Seliverstov[121], B. Sellden[146a], N. Semprini-Cesari[19a,19b], C. Serfon[98], L. Serin[115], R. Seuster[99], H. Severini[111], M.E. Sevior[86], A. Sfyrla[29], E. Shabalina[54], M. Shamim[114], L.Y. Shan[32a], J.T. Shank[21], Q.T. Shao[86], M. Shapiro[14], P.B. Shatalov[95], K. Shaw[139], D. Sherman[29], P. Sherwood[77], A. Shibata[108], M. Shimojima[100], T. Shin[56], A. Shmeleva[94], M.J. Shochet[30], M.A. Shupe[6], P. Sicho[125], A. Sidoti[15], F. Siegert[77], J. Siegrist[14], Dj. Sijacki[12a], O. Silbert[171], Y. Silver[153], D. Silverstein[143], S.B. Silverstein[146a], V. Simak[127], Lj. Simic[12a], S. Simion[115], B. Simmons[77], M. Simonyan[35], P. Sinervo[158], N.B. Sinev[114], V. Sipica[141], G. Siragusa[81], A.N. Sisakyan[65], S.Yu. Sivoklokov[97], J. Sjölin[146a,146b], T.B. Sjursen[13], K. Skovpen[107], P. Skubic[111], M. Slater[17], T. Slavicek[127], K. Sliwa[161], J. Sloper[29], V. Smakhtin[171], S.Yu. Smirnov[96], Y. Smirnov[24], L.N. Smirnova[97], O. Smirnova[79], B.C. Smith[57], D. Smith[143], K.M. Smith[53], M. Smizanska[71], K. Smolek[127], A.A. Snesarev[94], S.W. Snow[82], J. Snow[111], J. Snuverink[105], S. Snyder[24], M. Soares[124a], R. Sobie[169,k], J. Sodomka[127], A. Soffer[153], C.A. Solans[167], M. Solar[127], J. Solc[127], E. Solfaroli Camillocci[132a,132b], A.A. Solodkov[128], O.V. Solovyanov[128], J. Sondericker[24], V. Sopko[127], B. Sopko[127], M. Sosebee[7], A. Soukharev[107], S. Spagnolo[72a,72b], F. Spanò[34], R. Spighi[19a], G. Spigo[29], F. Spila[132a,132b], R. Spiwoks[29], M. Spousta[126], B. Spurlock[7], R.D. St. Denis[53], T. Stahl[141], J. Stahlman[120], R. Stamen[58a], E. Stanecka[29], R.W. Stanek[5], C. Stanescu[134a], S. Stapnes[117], E.A. Starchenko[128], J. Stark[55], P. Staroba[125], P. Starovoitov[91], P. Stavina[144a], G. Steele[53], P. Steinbach[43], P. Steinberg[24], I. Stekl[127], B. Stelzer[142], H.J. Stelzer[41], O. Stelzer-Chilton[159a], H. Stenzel[52], K. Stevenson[75], G.A. Stewart[53], M.C. Stockton[29], K. Stoerig[48], G. Stoicea[25a], S. Stonjek[99], P. Strachota[126], A.R. Stradling[7], A. Straessner[43], J. Strandberg[87], S. Strandberg[14], A. Strandlie[117], M. Strang[109], M. Strauss[111], P. Strizenec[144b], R. Ströhmer[173], D.M. Strom[114], R. Stroynowski[39], J. Strube[129], B. Stugu[13], P. Sturm[174], D.A. Soh[151,al], D. Su[143], Y. Sugaya[116], T. Sugimoto[101], C. Suhr[106], K. Suita[67], M. Suk[126], V.V. Sulin[94], S. Sultansoy[3d], T. Sumida[29], X. Sun[32d], J.E. Sundermann[48], K. Suruliz[164a,164b], S. Sushkov[11], G. Susinno[36a,36b], M.R. Sutton[139], Y. Suzuki[66], I. Sykora[144a], T. Sykora[126], T. Szymocha[38], J. Sánchez[167], D. Ta[20], K. Tackmann[29], A. Taffard[163], R. Tafirout[159a], A. Taga[117], Y. Takahashi[101], H. Takai[24], R. Takashima[69], H. Takeda[67], T. Takeshita[140], M. Talby[83], A. Talyshev[107], M.C. Tamsett[76], J. Tanaka[155], R. Tanaka[115], S. Tanaka[131], S. Tanaka[66], K. Tani[67], S. Tapprogge[81], D. Tardif[158], S. Tarem[152], F. Tarrade[24], G.F. Tartarelli[89a], P. Tas[126], M. Tasevsky[125], E. Tassi[36a,36b], M. Tatarkhanov[14], C. Taylor[77], F.E. Taylor[92], G.N. Taylor[86], W. Taylor[159b], M.Teixeira Dias Castanheira[75], P. Teixeira-Dias[76], H. Ten Kate[29], P.K. Teng[151], Y.D. Tennenbaum-Katan[152], S. Terada[66], K. Terashi[155], J. Terron[80], M. Terwort[41,w], M. Testa[47], R.J. Teuscher[158,k], J. Therhaag[20], M. Thioye[175], S. Thoma[48], J.P. Thomas[17], E.N. Thompson[84], P.D. Thompson[17], P.D. Thompson[158], R.J. Thompson[82], A.S. Thompson[53], E. Thomson[120], R.P. Thun[87], T. Tic[125], V.O. Tikhomirov[94], Y.A. Tikhonov[107], P. Tipton[175], F.J. Tique Aires Viegas[29], S. Tisserant[83], B. Toczek[37], T. Todorov[4], S. Todorova-Nova[161], B. Toggerson[163], J. Tojo[66], S. Tokár[144a], K. Tokunaga[67], K. Tokushuku[66], K. Tollefson[88], M. Tomoto[101], L. Tompkins[14], K. Toms[103], A. Tonoyan[13], C. Topfel[16], N.D. Topilin[65], I. Torchiani[29], E. Torrence[114], E. Torró Pastor[167], J. Toth[83,ah], F. Touchard[83], D.R. Tovey[139], T. Trefzger[173], L. Tremblet[29], A. Tricoli[29], I.M. Trigger[159a], S. Trincaz-Duvoid[78], T.N. Trinh[78], M.F. Tripiana[70], N. Triplett[64], W. Trischuk[158], A. Trivedi[24,am], B. Trocmé[55], C. Troncon[89a], A. Trzupek[38], C. Tsarouchas[9], J.C-L. Tseng[118], M. Tsiakiris[105], P.V. Tsiareshka[90], D. Tsionou[139], G. Tsipolitis[9], V. Tsiskaridze[51], E.G. Tskhadadze[51], I.I. Tsukerman[95], V. Tsulaia[123], J.-W. Tsung[20], S. Tsuno[66], D. Tsybychev[148], J.M. Tuggle[30], D. Turecek[127], I. Turk Cakir[3e], E. Turlay[105], P.M. Tuts[34], M.S. Twomey[138], M. Tylmad[146a,146b], M. Tyndel[129], K. Uchida[116], I. Ueda[155], R. Ueno[28], M. Ugland[13], M. Uhlenbrock[20], M. Uhrmacher[54], F. Ukegawa[160], G. Unal[29], A. Undrus[24], G. Unel[163], Y. Unno[66], D. Urbaniec[34], E. Urkovsky[153], P. Urquijo[49,an], P. Urrejola[31a], G. Usai[7], M. Uslenghi[119a,119b], L. Vacavant[83], V. Vacek[127], B. Vachon[85], S. Vahsen[14], P. Valente[132a], S. Valentinetti[19a,19b], S. Valkar[126], E. Valladolid Gallego[167], S. Vallecorsa[152], J.A. Valls Ferrer[167], H. van der Graaf[105], E. van der Kraaij[105], E. van der Poel[105], D. van der Ster[29],

N. van Eldik[84], P. van Gemmeren[5], Z. van Kesteren[105], I. van Vulpen[105], W. Vandelli[29], A. Vaniachine[5], P. Vankov[73], F. Vannucci[78], R. Vari[132a], E.W. Varnes[6], D. Varouchas[14], A. Vartapetian[7], K.E. Varvell[150], V.I. Vassilakopoulos[56], F. Vazeille[33], C. Vellidis[8], F. Veloso[124a], S. Veneziano[132a], A. Ventura[72a,72b], D. Ventura[138], M. Venturi[48], N. Venturi[16], V. Vercesi[119a], M. Verducci[138], W. Verkerke[105], J.C. Vermeulen[105], M.C. Vetterli[142,g], I. Vichou[165], T. Vickey[118], G.H.A. Viehhauser[118], M. Villa[19a,19b], E.G. Villani[129], M. Villaplana Perez[167], E. Vilucchi[47], M.G. Vincter[28], E. Vinek[29], V.B. Vinogradov[65], S. Viret[33], J. Virzi[14], A. Vitale[19a,19b], O. Vitells[171], I. Vivarelli[48], F. Vives Vaque[11], S. Vlachos[9], M. Vlasak[127], N. Vlasov[20], A. Vogel[20], P. Vokac[127], M. Volpi[11], H. von der Schmitt[99], J. von Loeben[99], H. von Radziewski[48], E. von Toerne[20], V. Vorobel[126], V. Vorwerk[11], M. Vos[167], R. Voss[29], T.T. Voss[174], J.H. Vossebeld[73], N. Vranjes[12a], M. Vranjes Milosavljevic[12a], V. Vrba[125], M. Vreeswijk[105], T. Vu Anh[81], D. Vudragovic[12a], R. Vuillermet[29], I. Vukotic[115], P. Wagner[120], J. Walbersloh[42], J. Walder[71], R. Walker[98], W. Walkowiak[141], R. Wall[175], C. Wang[44], H. Wang[172], J. Wang[55], S.M. Wang[151], A. Warburton[85], C.P. Ward[27], M. Warsinsky[48], R. Wastie[118], P.M. Watkins[17], A.T. Watson[17], M.F. Watson[17], G. Watts[138], S. Watts[82], A.T. Waugh[150], B.M. Waugh[77], M.D. Weber[16], M. Weber[129], M.S. Weber[16], P. Weber[54], A.R. Weidberg[118], J. Weingarten[54], C. Weiser[48], H. Wellenstein[22], P.S. Wells[29], T. Wenaus[24], S. Wendler[123], Z. Weng[151,ao], T. Wengler[82], S. Wenig[29], N. Wermes[20], M. Werner[48], P. Werner[29], M. Werth[163], U. Werthenbach[141], M. Wessels[58a], K. Whalen[28], A. White[7], M.J. White[27], S. White[24], S.R. Whitehead[118], D. Whiteson[163], D. Whittington[61], F. Wicek[115], D. Wicke[81], F.J. Wickens[129], W. Wiedenmann[172], M. Wielers[129], P. Wienemann[20], C. Wiglesworth[73], L.A.M. Wiik[48], A. Wildauer[167], M.A. Wildt[41,w], H.G. Wilkens[29], E. Williams[34], H.H. Williams[120], S. Willocq[84], J.A. Wilson[17], M.G. Wilson[143], A. Wilson[87], I. Wingerter-Seez[4], F. Winklmeier[29], M. Wittgen[143], M.W. Wolter[38], H. Wolters[124a,h], B.K. Wosiek[38], J. Wotschack[29], M.J. Woudstra[84], K. Wraight[53], C. Wright[53], D. Wright[143], B. Wrona[73], S.L. Wu[172], X. Wu[49], E. Wulf[34], B.M. Wynne[45], L. Xaplanteris[9], S. Xella[35], S. Xie[48], D. Xu[139], M. Yamada[160], A. Yamamoto[66], K. Yamamoto[64], S. Yamamoto[155], T. Yamamura[155], J. Yamaoka[44], T. Yamazaki[155], Y. Yamazaki[67], Z. Yan[21], H. Yang[87], U.K. Yang[82], Z. Yang[146a,146b], W.-M. Yao[14], Y. Yao[14], Y. Yasu[66], J. Ye[39], S. Ye[24], M. Yilmaz[3c], R. Yoosoofmiya[123], K. Yorita[170], R. Yoshida[5], C. Young[143], S.P. Youssef[21], D. Yu[24], J. Yu[7], L. Yuan[78], A. Yurkewicz[148], R. Zaidan[63], A.M. Zaitsev[128], Z. Zajacova[29], V. Zambrano[47], L. Zanello[132a,132b], A. Zaytsev[107], C. Zeitnitz[174], M. Zeller[175], A. Zemla[38], C. Zendler[20], O. Zenin[128], T. Ženiš[144a], Z. Zenonos[122a,122b], S. Zenz[14], D. Zerwas[115], G. Zevi della Porta[57], Z. Zhan[32d], H. Zhang[83], J. Zhang[5], Q. Zhang[5], X. Zhang[32d], L. Zhao[108], T. Zhao[138], Z. Zhao[32b], A. Zhemchugov[65], J. Zhong[151,ap], B. Zhou[87], N. Zhou[34], Y. Zhou[151], C.G. Zhu[32d], H. Zhu[41], Y. Zhu[172], X. Zhuang[98], V. Zhuravlov[99], R. Zimmermann[20], S. Zimmermann[20], S. Zimmermann[48], M. Ziolkowski[141], L. Živković[34], G. Zobernig[172], A. Zoccoli[19a,19b], M. zur Nedden[15], V. Zutshi[106]

[1] University at Albany, 1400 Washington Ave, Albany, NY 12222, United States of America
[2] University of Alberta, Department of Physics, Centre for Particle Physics, Edmonton, AB T6G 2G7, Canada
[3] Ankara University[(a)], Faculty of Sciences, Department of Physics, TR 061000 Tandogan, Ankara; Dumlupinar University[(b)], Faculty of Arts and Sciences, Department of Physics, Kutahya; Gazi University[(c)], Faculty of Arts and Sciences, Department of Physics, 06500, Teknikokullar, Ankara; TOBB University of Economics and Technology[(d)], Faculty of Arts and Sciences, Division of Physics, 06560, Sogutozu, Ankara; Turkish Atomic Energy Authority[(e)], 06530, Lodumlu, Ankara, Turkey
[4] LAPP, Université de Savoie, CNRS/IN2P3, Annecy-le-Vieux, France
[5] Argonne National Laboratory, High Energy Physics Division, 9700 S. Cass Avenue, Argonne, IL 60439, United States of America
[6] University of Arizona, Department of Physics, Tucson, AZ 85721, United States of America
[7] The University of Texas at Arlington, Department of Physics, Box 19059, Arlington, TX 76019, United States of America
[8] University of Athens, Nuclear & Particle Physics, Department of Physics, Panepistimiopouli, Zografou, GR 15771 Athens, Greece
[9] National Technical University of Athens, Physics Department, 9-Iroon Polytechniou, GR 15780 Zografou, Greece
[10] Institute of Physics, Azerbaijan Academy of Sciences, H. Javid Avenue 33, AZ 143 Baku, Azerbaijan
[11] Institut de Física d'Altes Energies, IFAE, Edifici Cn, Universitat Autònoma de Barcelona, ES - 08193 Bellaterra (Barcelona), Spain
[12] University of Belgrade[(a)], Institute of Physics, P.O. Box 57, 11001 Belgrade; Vinca Institute of Nuclear Sciences[(b)], M. Petrovica Alasa 12-14, 11000 Belgrade, Serbia
[13] University of Bergen, Department for Physics and Technology, Allegaten 55, NO - 5007 Bergen, Norway
[14] Lawrence Berkeley National Laboratory and University of California, Physics Division, MS50B-6227, 1 Cyclotron Road, Berkeley, CA 94720, United States of America
[15] Humboldt University, Institute of Physics, Berlin, Newtonstr. 15, D-12489 Berlin, Germany

[16]University of Bern, Albert Einstein Center for Fundamental Physics, Laboratory for High Energy Physics, Sidlerstrasse 5, CH - 3012 Bern, Switzerland

[17]University of Birmingham, School of Physics and Astronomy, Edgbaston, Birmingham B15 2TT, United Kingdom

[18]Bogazici University[a], Faculty of Sciences, Department of Physics, TR - 80815 Bebek-Istanbul; Dogus University[b], Faculty of Arts and Sciences, Department of Physics, 34722, Kadikoy, Istanbul; [c]Gaziantep University, Faculty of Engineering, Department of Physics Engineering, 27310, Sehitkamil, Gaziantep; Istanbul Technical University[d], Faculty of Arts and Sciences, Department of Physics, 34469, Maslak, Istanbul, Turkey

[19]INFN Sezione di Bologna[a]; Università di Bologna, Dipartimento di Fisica[b], viale C. Berti Pichat, 6/2, IT - 40127 Bologna, Italy

[20]University of Bonn, Physikalisches Institut, Nussallee 12, D - 53115 Bonn, Germany

[21]Boston University, Department of Physics, 590 Commonwealth Avenue, Boston, MA 02215, United States of America

[22]Brandeis University, Department of Physics, MS057, 415 South Street, Waltham, MA 02454, United States of America

[23]Universidade Federal do Rio De Janeiro, COPPE/EE/IF[a], Caixa Postal 68528, Ilha do Fundao, BR - 21945-970 Rio de Janeiro; [b]Universidade de Sao Paulo, Instituto de Fisica, R.do Matao Trav. R.187, Sao Paulo - SP, 05508 - 900, Brazil

[24]Brookhaven National Laboratory, Physics Department, Bldg. 510A, Upton, NY 11973, United States of America

[25]National Institute of Physics and Nuclear Engineering[a], Str. Atomistilor 407, P.O. Box MG-6, R-077125, Bucharest-Magurele; University Politehnica Bucharest[b], Rectorat - AN 001, 313 Splaiul Independentei, sector 6, 060042 Bucuresti; West University in Timisoara[c], Bd. Vasile Parvan 4, Timisoara, Romania

[26]Universidad de Buenos Aires, FCEyN, Dto. Fisica, Pab I - C. Universitaria, 1428 Buenos Aires, Argentina

[27]University of Cambridge, Cavendish Laboratory, J J Thomson Avenue, Cambridge CB3 0HE, United Kingdom

[28]Carleton University, Department of Physics, 1125 Colonel By Drive, Ottawa ON K1S 5B6, Canada

[29]CERN, CH - 1211 Geneva 23, Switzerland

[30]University of Chicago, Enrico Fermi Institute, 5640 S. Ellis Avenue, Chicago, IL 60637, United States of America

[31]Pontificia Universidad Católica de Chile, Facultad de Fisica, Departamento de Fisica[a], Avda. Vicuna Mackenna 4860, San Joaquin, Santiago; Universidad Técnica Federico Santa María, Departamento de Física[b], Avda. Espãna 1680, Casilla 110-V, Valparaíso, Chile

[32]Institute of High Energy Physics, Chinese Academy of Sciences[a], P.O. Box 918, 19 Yuquan Road, Shijing Shan District, CN - Beijing 100049; University of Science & Technology of China (USTC), Department of Modern Physics[b], Hefei, CN - Anhui 230026; Nanjing University, Department of Physics[c], Nanjing, CN - Jiangsu 210093; Shandong University, High Energy Physics Group[d], Jinan, CN - Shandong 250100, China

[33]Laboratoire de Physique Corpusculaire, Clermont Université, Université Blaise Pascal, CNRS/IN2P3, FR - 63177 Aubiere Cedex, France

[34]Columbia University, Nevis Laboratory, 136 So. Broadway, Irvington, NY 10533, United States of America

[35]University of Copenhagen, Niels Bohr Institute, Blegdamsvej 17, DK - 2100 Kobenhavn 0, Denmark

[36]INFN Gruppo Collegato di Cosenza[a]; Università della Calabria, Dipartimento di Fisica[b], IT-87036 Arcavacata di Rende, Italy

[37]Faculty of Physics and Applied Computer Science of the AGH-University of Science and Technology, (FPACS, AGH-UST), al. Mickiewicza 30, PL-30059 Cracow, Poland

[38]The Henryk Niewodniczanski Institute of Nuclear Physics, Polish Academy of Sciences, ul. Radzikowskiego 152, PL - 31342 Krakow, Poland

[39]Southern Methodist University, Physics Department, 106 Fondren Science Building, Dallas, TX 75275-0175, United States of America

[40]University of Texas at Dallas, 800 West Campbell Road, Richardson, TX 75080-3021, United States of America

[41]DESY, Notkestr. 85, D-22603 Hamburg and Platanenallee 6, D-15738 Zeuthen, Germany

[42]TU Dortmund, Experimentelle Physik IV, DE - 44221 Dortmund, Germany

[43]Technical University Dresden, Institut für Kern- und Teilchenphysik, Zellescher Weg 19, D-01069 Dresden, Germany

[44]Duke University, Department of Physics, Durham, NC 27708, United States of America

[45]University of Edinburgh, School of Physics & Astronomy, James Clerk Maxwell Building, The Kings Buildings, Mayfield Road, Edinburgh EH9 3JZ, United Kingdom

[46]Fachhochschule Wiener Neustadt; Johannes Gutenbergstrasse 3, AT - 2700 Wiener Neustadt, Austria

[47]INFN Laboratori Nazionali di Frascati, via Enrico Fermi 40, IT-00044 Frascati, Italy

[48]Albert-Ludwigs-Universität, Fakultät für Mathematik und Physik, Hermann-Herder Str. 3, D - 79104 Freiburg i.Br., Germany

[49] Université de Genève, Section de Physique, 24 rue Ernest Ansermet, CH - 1211 Geneve 4, Switzerland

[50] INFN Sezione di Genova[a]; Università di Genova, Dipartimento di Fisica[b], via Dodecaneso 33, IT - 16146 Genova, Italy

[51] Institute of Physics of the Georgian Academy of Sciences, 6 Tamarashvili St., GE - 380077 Tbilisi; Tbilisi State University, HEP Institute, University St. 9, GE - 380086 Tbilisi, Georgia

[52] Justus-Liebig-Universität Giessen, II Physikalisches Institut, Heinrich-Buff Ring 16, D-35392 Giessen, Germany

[53] University of Glasgow, Department of Physics and Astronomy, Glasgow G12 8QQ, United Kingdom

[54] Georg-August-Universität, II. Physikalisches Institut, Friedrich-Hund Platz 1, D-37077 Göttingen, Germany

[55] Laboratoire de Physique Subatomique et de Cosmologie, CNRS/IN2P3, Université Joseph Fourier, INPG, 53 avenue des Martyrs, FR - 38026 Grenoble Cedex, France

[56] Hampton University, Department of Physics, Hampton, VA 23668, United States of America

[57] Harvard University, Laboratory for Particle Physics and Cosmology, 18 Hammond Street, Cambridge, MA 02138, United States of America

[58] Ruprecht-Karls-Universität Heidelberg, Kirchhoff-Institut für Physik[a], Im Neuenheimer Feld 227, D-69120 Heidelberg; Physikalisches Institut[b], Philosophenweg 12, D-69120 Heidelberg; ZITI Ruprecht-Karls-University Heidelberg[c], Lehrstuhl für Informatik V, B6, 23-29, DE - 68131 Mannheim, Germany

[59] Hiroshima University, Faculty of Science, 1-3-1 Kagamiyama, Higashihiroshima-shi, JP - 739-8526 Hiroshima, Japan

[60] Hiroshima Institute of Technology, Faculty of Applied Information Science, 2-1-1 Miyake Saeki-ku, Hiroshima-shi, JP - 731-5193 Hiroshima, Japan

[61] Indiana University, Department of Physics, Swain Hall West 117, Bloomington, IN 47405-7105, United States of America

[62] Institut für Astro- und Teilchenphysik, Technikerstrasse 25, A - 6020 Innsbruck, Austria

[63] University of Iowa, 203 Van Allen Hall, Iowa City, IA 52242-1479, United States of America

[64] Iowa State University, Department of Physics and Astronomy, Ames High Energy Physics Group, Ames, IA 50011-3160, United States of America

[65] Joint Institute for Nuclear Research, JINR, Dubna, RU - 141 980 Moscow Region, Russia

[66] KEK, High Energy Accelerator Research Organization, 1-1 Oho, Tsukuba-shi, Ibaraki-ken 305-0801, Japan

[67] Kobe University, Graduate School of Science, 1-1 Rokkodai-cho, Nada-ku, JP 657-8501 Kobe, Japan

[68] Kyoto University, Faculty of Science, Oiwake-cho, Kitashirakawa, Sakyou-ku, Kyoto-shi, JP - 606-8502 Kyoto, Japan

[69] Kyoto University of Education, 1 Fukakusa, Fujimori, fushimi-ku, Kyoto-shi, JP - 612-8522 Kyoto, Japan

[70] Universidad Nacional de La Plata, FCE, Departamento de Física, IFLP (CONICET-UNLP), C.C. 67, 1900 La Plata, Argentina

[71] Lancaster University, Physics Department, Lancaster LA1 4YB, United Kingdom

[72] INFN Sezione di Lecce[a]; Università del Salento, Dipartimento di Fisica[b], Via Arnesano, IT - 73100 Lecce, Italy

[73] University of Liverpool, Oliver Lodge Laboratory, P.O. Box 147, Oxford Street, Liverpool L69 3BX, United Kingdom

[74] Jožef Stefan Institute and University of Ljubljana, Department of Physics, SI-1000 Ljubljana, Slovenia

[75] Queen Mary University of London, Department of Physics, Mile End Road, London E1 4NS, United Kingdom

[76] Royal Holloway, University of London, Department of Physics, Egham Hill, Egham, Surrey TW20 0EX, United Kingdom

[77] University College London, Department of Physics and Astronomy, Gower Street, London WC1E 6BT, United Kingdom

[78] Laboratoire de Physique Nucléaire et de Hautes Energies, Université Pierre et Marie Curie (Paris 6), Université Denis Diderot (Paris-7), CNRS/IN2P3, Tour 33, 4 place Jussieu, FR - 75252 Paris Cedex 05, France

[79] Fysiska institutionen, Lunds universitet, Box 118, SE - 221 00 Lund, Sweden

[80] Universidad Autonoma de Madrid, Facultad de Ciencias, Departamento de Fisica Teorica, ES - 28049 Madrid, Spain

[81] Universität Mainz, Institut für Physik, Staudinger Weg 7, DE - 55099 Mainz, Germany

[82] University of Manchester, School of Physics and Astronomy, Manchester M13 9PL, United Kingdom

[83] CPPM, Aix-Marseille Université, CNRS/IN2P3, Marseille, France

[84] University of Massachusetts, Department of Physics, 710 North Pleasant Street, Amherst, MA 01003, United States of America

[85] McGill University, High Energy Physics Group, 3600 University Street, Montreal, Quebec H3A 2T8, Canada

[86] University of Melbourne, School of Physics, AU - Parkville, Victoria 3010, Australia

[87] The University of Michigan, Department of Physics, 2477 Randall Laboratory, 500 East University, Ann Arbor, MI 48109-1120, United States of America

[88] Michigan State University, Department of Physics and Astronomy, High Energy Physics Group, East Lansing, MI 48824-2320, United States of America

[89] INFN Sezione di Milano[a]; Università di Milano, Dipartimento di Fisica[b], via Celoria 16, IT - 20133 Milano, Italy

[90] B.I. Stepanov Institute of Physics, National Academy of Sciences of Belarus, Independence Avenue 68, Minsk 220072, Republic of Belarus

[91] National Scientific & Educational Centre for Particle & High Energy Physics, NC PHEP BSU, M. Bogdanovich St. 153, Minsk 220040, Republic of Belarus

[92] Massachusetts Institute of Technology, Department of Physics, Room 24-516, Cambridge, MA 02139, United States of America

[93] University of Montreal, Group of Particle Physics, C.P. 6128, Succursale Centre-Ville, Montreal, Quebec, H3C 3J7, Canada

[94] P.N. Lebedev Institute of Physics, Academy of Sciences, Leninsky pr. 53, RU - 117 924 Moscow, Russia

[95] Institute for Theoretical and Experimental Physics (ITEP), B. Cheremushkinskaya ul. 25, RU 117 218 Moscow, Russia

[96] Moscow Engineering & Physics Institute (MEPhI), Kashirskoe Shosse 31, RU - 115409 Moscow, Russia

[97] Lomonosov Moscow State University Skobeltsyn Institute of Nuclear Physics (MSU SINP), 1(2), Leninskie gory, GSP-1, Moscow 119991, Russian Federation, Russia

[98] Ludwig-Maximilians-Universität München, Fakultät für Physik, Am Coulombwall 1, DE - 85748 Garching, Germany

[99] Max-Planck-Institut für Physik, (Werner-Heisenberg-Institut), Föhringer Ring 6, 80805 München, Germany

[100] Nagasaki Institute of Applied Science, 536 Aba-machi, JP 851-0193 Nagasaki, Japan

[101] Nagoya University, Graduate School of Science, Furo-Cho, Chikusa-ku, Nagoya, 464-8602, Japan

[102] INFN Sezione di Napoli[a]; Università di Napoli, Dipartimento di Scienze Fisiche[b], Complesso Universitario di Monte Sant'Angelo, via Cinthia, IT - 80126 Napoli, Italy

[103] University of New Mexico, Department of Physics and Astronomy, MSC07 4220, Albuquerque, NM 87131, United States of America

[104] Radboud University Nijmegen/NIKHEF, Department of Experimental High Energy Physics, Heyendaalseweg 135, NL-6525 AJ, Nijmegen, Netherlands

[105] Nikhef National Institute for Subatomic Physics and University of Amsterdam, Science Park 105, 1098 XG Amsterdam, Netherlands

[106] Department of Physics, Northern Illinois University, LaTourette Hall Normal Road, DeKalb, IL 60115, United States of America

[107] Budker Institute of Nuclear Physics (BINP), RU - Novosibirsk 630 090, Russia

[108] New York University, Department of Physics, 4 Washington Place, New York NY 10003, United States of America

[109] Ohio State University, 191 West Woodruff Ave, Columbus, OH 43210-1117, United States of America

[110] Okayama University, Faculty of Science, Tsushimanaka 3-1-1, Okayama 700-8530, Japan

[111] University of Oklahoma, Homer L. Dodge Department of Physics and Astronomy, 440 West Brooks, Room 100, Norman, OK 73019-0225, United States of America

[112] Oklahoma State University, Department of Physics, 145 Physical Sciences Building, Stillwater, OK 74078-3072, United States of America

[113] Palacký University, 17.listopadu 50a, 772 07 Olomouc, Czech Republic

[114] University of Oregon, Center for High Energy Physics, Eugene, OR 97403-1274, United States of America

[115] LAL, Univ. Paris-Sud, IN2P3/CNRS, Orsay, France

[116] Osaka University, Graduate School of Science, Machikaneyama-machi 1-1, Toyonaka, Osaka 560-0043, Japan

[117] University of Oslo, Department of Physics, P.O. Box 1048, Blindern, NO - 0316 Oslo 3, Norway

[118] Oxford University, Department of Physics, Denys Wilkinson Building, Keble Road, Oxford OX1 3RH, United Kingdom

[119] INFN Sezione di Pavia[a]; Università di Pavia, Dipartimento di Fisica Nucleare e Teorica[b], Via Bassi 6, IT-27100 Pavia, Italy

[120] University of Pennsylvania, Department of Physics, High Energy Physics Group, 209 S. 33rd Street, Philadelphia, PA 19104, United States of America

[121] Petersburg Nuclear Physics Institute, RU - 188 300 Gatchina, Russia

[122] INFN Sezione di Pisa[a]; Università di Pisa, Dipartimento di Fisica E. Fermi[b], Largo B. Pontecorvo 3, IT - 56127 Pisa, Italy

[123] University of Pittsburgh, Department of Physics and Astronomy, 3941 O'Hara Street, Pittsburgh, PA 15260, United States of America

[124]Laboratorio de Instrumentacao e Fisica Experimental de Particulas - LIP[a], Avenida Elias Garcia 14-1, PT - 1000-149 Lisboa, Portugal; Universidad de Granada, Departamento de Fisica Teorica y del Cosmos and CAFPE[b], E-18071 Granada, Spain

[125]Institute of Physics, Academy of Sciences of the Czech Republic, Na Slovance 2, CZ - 18221 Praha 8, Czech Republic

[126]Charles University in Prague, Faculty of Mathematics and Physics, Institute of Particle and Nuclear Physics, V Holesovickach 2, CZ - 18000 Praha 8, Czech Republic

[127]Czech Technical University in Prague, Zikova 4, CZ - 166 35 Praha 6, Czech Republic

[128]State Research Center Institute for High Energy Physics, Moscow Region, 142281, Protvino, Pobeda street, 1, Russia

[129]Rutherford Appleton Laboratory, Science and Technology Facilities Council, Harwell Science and Innovation Campus, Didcot OX11 0QX, United Kingdom

[130]University of Regina, Physics Department, Regina, Canada

[131]Ritsumeikan University, Noji Higashi 1 chome 1-1, JP - Kusatsu, Shiga 525-8577, Japan

[132]INFN Sezione di Roma I[a]; Università La Sapienza, Dipartimento di Fisica[b], Piazzale A. Moro 2, IT- 00185 Roma, Italy

[133]INFN Sezione di Roma Tor Vergata[a]; Università di Roma Tor Vergata, Dipartimento di Fisica[b], via della Ricerca Scientifica, IT-00133 Roma, Italy

[134]INFN Sezione di Roma Tre[a]; Università Roma Tre, Dipartimento di Fisica[b], via della Vasca Navale 84, IT-00146 Roma, Italy

[135]Réseau Universitaire de Physique des Hautes Energies (RUPHE): Université Hassan II, Faculté des Sciences Ain Chock[a], B.P. 5366, MA - Casablanca; Centre National de l'Energie des Sciences Techniques Nucleaires (CNESTEN)[b], B.P. 1382 R.P. 10001 Rabat 10001; Université Mohamed Premier[c], LPTPM, Faculté des Sciences, B.P.717. Bd. Mohamed VI, 60000, Oujda; Université Mohammed V, Faculté des Sciences[d], 4 Avenue Ibn Battouta, BP 1014 RP, 10000 Rabat, Morocco

[136]CEA, DSM/IRFU, Centre d'Etudes de Saclay, FR - 91191 Gif-sur-Yvette, France

[137]University of California Santa Cruz, Santa Cruz Institute for Particle Physics (SCIPP), Santa Cruz, CA 95064, United States of America

[138]University of Washington, Seattle, Department of Physics, Box 351560, Seattle, WA 98195-1560, United States of America

[139]University of Sheffield, Department of Physics & Astronomy, Hounsfield Road, Sheffield S3 7RH, United Kingdom

[140]Shinshu University, Department of Physics, Faculty of Science, 3-1-1 Asahi, Matsumoto-shi, JP - Nagano 390-8621, Japan

[141]Universität Siegen, Fachbereich Physik, D 57068 Siegen, Germany

[142]Simon Fraser University, Department of Physics, 8888 University Drive, CA - Burnaby, BC V5A 1S6, Canada

[143]SLAC National Accelerator Laboratory, Stanford, California 94309, United States of America

[144]Comenius University, Faculty of Mathematics, Physics & Informatics[a], Mlynska dolina F2, SK - 84248 Bratislava; Institute of Experimental Physics of the Slovak Academy of Sciences, Dept. of Subnuclear Physics[b], Watsonova 47, SK - 04353 Kosice, Slovak Republic

[145][a]University of Johannesburg, Department of Physics, PO Box 524, Auckland Park, Johannesburg 2006; [b]School of Physics, University of the Witwatersrand, Private Bag 3, Wits 2050, Johannesburg, South Africa

[146]Stockholm University, Department of Physics[a]; The Oskar Klein Centre[b], AlbaNova, SE - 106 91 Stockholm, Sweden

[147]Royal Institute of Technology (KTH), Physics Department, SE - 106 91 Stockholm, Sweden

[148]Stony Brook University, Department of Physics and Astronomy, Nicolls Road, Stony Brook, NY 11794-3800, United States of America

[149]University of Sussex, Department of Physics and Astronomy Pevensey, 2 Building, Falmer, Brighton BN1 9QH, United Kingdom

[150]University of Sydney, School of Physics, AU - 2006 Sydney NSW, Australia

[151]Insitute of Physics, Academia Sinica, TW - 11529 Taipei, Taiwan

[152]Technion, Israel Inst. of Technology, Department of Physics, Technion City, IL - Haifa 32000, Israel

[153]Tel Aviv University, Raymond and Beverly Sackler School of Physics and Astronomy, Ramat Aviv, IL - Tel Aviv 69978, Israel

[154]Aristotle University of Thessaloniki, Faculty of Science, Department of Physics, Division of Nuclear & Particle Physics, University Campus, GR - 54124, Thessaloniki, Greece

[155]The University of Tokyo, International Center for Elementary Particle Physics and Department of Physics, 7-3-1 Hongo, Bunkyo-ku, JP - Tokyo 113-0033, Japan

[156] Tokyo Metropolitan University, Graduate School of Science and Technology, 1-1 Minami-Osawa, Hachioji, Tokyo 192-0397, Japan

[157] Tokyo Institute of Technology, 2-12-1-H-34 O-Okayama, Meguro, Tokyo 152-8551, Japan

[158] University of Toronto, Department of Physics, 60 Saint George Street, Toronto M5S 1A7, Ontario, Canada

[159] TRIUMF[a], 4004 Wesbrook Mall, Vancouver, B.C. V6T 2A3; [b] York University, Department of Physics and Astronomy, 4700 Keele St., Toronto, Ontario, M3J 1P3, Canada

[160] University of Tsukuba, Institute of Pure and Applied Sciences, 1-1-1 Tennoudai, Tsukuba-shi, JP - Ibaraki 305-8571, Japan

[161] Tufts University, Science & Technology Center, 4 Colby Street, Medford, MA 02155, United States of America

[162] Universidad Antonio Narino, Centro de Investigaciones, Cra 3 Este No.47A-15, Bogota, Colombia

[163] University of California, Department of Physics & Astronomy, Irvine, CA 92697-4575, United States of America

[164] INFN Gruppo Collegato di Udine[a], IT-33100 Udine; ICTP[b], Strada Costiera 11, IT-34014, Trieste; Università di Udine, Dipartimento di Fisica[c], via delle Scienze 208, IT - 33100 Udine, Italy

[165] University of Illinois, Department of Physics, 1110 West Green Street, Urbana, Illinois 61801, United States of America

[166] University of Uppsala, Department of Physics and Astronomy, P.O. Box 516, SE -751 20 Uppsala, Sweden

[167] Instituto de Física Corpuscular (IFIC) Centro Mixto UVEG-CSIC, Apdo. 22085 ES-46071 Valencia, Dept. Física At. Mol. y Nuclear; Dept. Ing. Electrónica; Univ. of Valencia, and Inst. de Microelectrónica de Barcelona (IMB-CNM-CSIC), 08193 Bellaterra, Spain

[168] University of British Columbia, Department of Physics, 6224 Agricultural Road, CA - Vancouver, B.C. V6T 1Z1, Canada

[169] University of Victoria, Department of Physics and Astronomy, P.O. Box 3055, Victoria B.C., V8W 3P6, Canada

[170] Waseda University, WISE, 3-4-1 Okubo, Shinjuku-ku, Tokyo, 169-8555, Japan

[171] The Weizmann Institute of Science, Department of Particle Physics, P.O. Box 26, IL - 76100 Rehovot, Israel

[172] University of Wisconsin, Department of Physics, 1150 University Avenue, WI 53706 Madison, Wisconsin, United States of America

[173] Julius-Maximilians-University of Würzburg, Physikalisches Institute, Am Hubland, 97074 Würzburg, Germany

[174] Bergische Universität, Fachbereich C, Physik, Postfach 100127, Gauss-Strasse 20, D- 42097 Wuppertal, Germany

[175] Yale University, Department of Physics, PO Box 208121, New Haven CT, 06520-8121, United States of America

[176] Yerevan Physics Institute, Alikhanian Brothers Street 2, AM - 375036 Yerevan, Armenia

[177] ATLAS-Canada Tier-1 Data Centre, TRIUMF, 4004 Wesbrook Mall, Vancouver, BC, V6T 2A3, Canada

[178] GridKA Tier-1 FZK, Forschungszentrum Karlsruhe GmbH, Steinbuch Centre for Computing (SCC), Hermann-von-Helmholtz-Platz 1, 76344 Eggenstein-Leopoldshafen, Germany

[179] Port d'Informacio Cientifica (PIC), Universitat Autonoma de Barcelona (UAB), Edifici D, E-08193 Bellaterra, Spain

[180] Centre de Calcul CNRS/IN2P3, Domaine scientifique de la Doua, 27 bd du 11 Novembre 1918, 69622 Villeurbanne Cedex, France

[181] INFN-CNAF, Viale Berti Pichat 6/2, 40127 Bologna, Italy

[182] Nordic Data Grid Facility, NORDUnet A/S, Kastruplundgade 22, 1, DK-2770 Kastrup, Denmark

[183] SARA Reken- en Netwerkdiensten, Science Park 121, 1098 XG Amsterdam, Netherlands

[184] Academia Sinica Grid Computing, Institute of Physics, Academia Sinica, No.128, Sec. 2, Academia Rd., Nankang, Taipei, Taiwan 11529, Taiwan

[185] UK-T1-RAL Tier-1, Rutherford Appleton Laboratory, Science and Technology Facilities Council, Harwell Science and Innovation Campus, Didcot OX11 0QX, United Kingdom

[186] RHIC and ATLAS Computing Facility, Physics Department, Building 510, Brookhaven National Laboratory, Upton, New York 11973, United States of America

[a] Also at LIP, Portugal

[b] Also at Faculdade de Ciencias, Universidade de Lisboa, Portugal

[c] Also at CPPM, Marseille, France

[d] Also at Centro de Fisica Nuclear da Universidade de Lisboa, Portugal

[e] Also at TRIUMF, Vancouver, Canada

[f] Also at FPACS, AGH-UST, Cracow, Poland

[g] Also at TRIUMF, Vancouver, Canada

[h] Also at Department of Physics, University of Coimbra, Portugal

[i] Now at CERN

[j] Also at Università di Napoli Parthenope, Napoli, Italy

[k] Also at Institute of Particle Physics (IPP), Canada

[l] Also at Università di Napoli Parthenope, via A. Acton 38, IT - 80133 Napoli, Italy

[m] Louisiana Tech University, 305 Wisteria Street, P.O. Box 3178, Ruston, LA 71272, United States of America

[n] Also at Universidade de Lisboa, Portugal

[o] also Czech Technical University in Prague, Faculty of Nuclear Science and Physical Engineering

[p] At California State University, Fresno, USA

[q] Also at TRIUMF, 4004 Wesbrook Mall, Vancouver, B.C. V6T 2A3, Canada

[r] Also at Faculdade de Ciencias, Universidade de Lisboa, Portugal and at Centro de Fisica Nuclear da Universidade de Lisboa, Portugal

[s] Also at FPACS, AGH-UST, Cracow, Poland

[t] Also at California Institute of Technology, Pasadena, USA

[u] Louisiana Tech University, Ruston, USA

[v] Also at University of Montreal, Montreal, Canada

[w] Also at Institut für Experimentalphysik, Universität Hamburg, Hamburg, Germany

[x] Also at Institut für Experimentalphysik, Universität Hamburg, Luruper Chaussee 149, 22761 Hamburg, Germany

[y] Also at School of Physics and Engineering, Sun Yat-sen University, China

[z] Also at School of Physics, Shandong University, Jinan, China

[aa] Also at California Institute of Technology, Pasadena, USA

[ab] Also at Rutherford Appleton Laboratory, Didcot, UK

[ac] Also at school of physics, Shandong University, Jinan

[ad] Also at Rutherford Appleton Laboratory, Didcot, UK

[ae] Now at KEK

[af] Also at Departamento de Fisica, Universidade de Minho, Portugal

[ag] University of South Carolina, Columbia, USA

[ah] Also at KFKI Research Institute for Particle and Nuclear Physics, Budapest, Hungary

[ai] University of South Carolina, Dept. of Physics and Astronomy, 700 S. Main St, Columbia, SC 29208, United States of America

[aj] Also at Institute of Physics, Jagiellonian University, Cracow, Poland

[ak] Louisiana Tech University, Ruston, USA

[al] Also at School of Physics and Engineering, Sun Yat-sen University, Taiwan

[am] University of South Carolina, Columbia, USA

[an] Transfer to LHCb 31.01.2010

[ao] Also at school of physics and engineering, Sun Yat-sen University

[ap] Also at Nanjing University, China

[*] Deceased